BIOGEOGRAPHY
Introduction to Space, Time, and Life

Second Edition

GLEN M. MacDONALD

WILEY Blackwell

Library of Congress Cataloging-in-Publication Data
Names: MacDonald, Glen Michael, author. | John Wiley & Sons, publisher.
Title: Biogeography : introduction to space, time, and life / Glen
  MacDonald.
Description: Second edition. | Hoboken, New Jersey : Wiley-Blackwell,
  [2025] | Includes index.
Identifiers: LCCN 2024031608 (print) | LCCN 2024031609 (ebook) | ISBN
  9781119904588 (paperback) | ISBN 9781119904601 (adobe pdf) | ISBN
  9781119904595 (epub)
Subjects: LCSH: Biogeography.
Classification: LCC QH84 .M24 2025 (print) | LCC QH84 (ebook) | DDC
  578.09–dc23/eng20241001
LC record available at https://lccn.loc.gov/2024031608
LC ebook record available at https://lccn.loc.gov/2024031609

Cover design: Wiley
Cover image: © Glen M. MacDonald

Set in 10/12pt TimesTenLTStd by Straive, Pondicherry, India

Printed and bound by CPI Group (UK) Ltd, Croydon, CR0 4YY

C9781119904588_270325

CONTENTS

PREFACE TO THE 2ND EDITION   vii

ABOUT THE COMPANION WEBSITE   ix

CHAPTER 1   *AN INTRODUCTION*   1

PART I   SPACE AND LIFE

CHAPTER 2   *SOME BASICS*   9

    Biology and the Hierarchies of Life   9
        Taxonomic Hierarchy   9
        Ecological Hierarchy   14
        Trophic Hierarchy   15
    Physical Geography and the Functioning of the Earth   20
        Global Climate   20
        Microclimate   27
        World Soils   27
        The Physical Environment of Lakes   30
        The Physical Environment of Oceans   32

CHAPTER 3   *THE PHYSICAL ENVIRONMENT AND THE DISTRIBUTION OF LIFE*   37

    Light   38
    Temperature   41
        Plants   41
        Animals   46
    Moisture   51
        Plants   51
        Animals   54
    Other Physical Factors   55
    Interacting Physical Controls on Geographic Distributions   57
    Environmental Gradients and Species Niches   58

CHAPTER 4   *BIOLOGICAL INTERACTIONS AND THE DISTRIBUTION OF LIFE*   66

    Predation   66
    Competition   71
    Symbiosis: Mutualism, Commensalism, Parasitism, and Mimicry   75
    Combined Physical and Biological Controls on Distribution   78
    Interactions, Gradients, and Niches   79

CHAPTER 5   *DISTURBANCE*   85

    Fire   92
    Wind   100
    Flooding   103

Other Physical Disturbances   109
Pathogens   112
Marine Disturbances   114

CHAPTER 6   *COMMUNITIES, FORMATIONS, AND BIOMES*   121

Communities   121
Plant Physiognomy, Vegetation Structure, and Formations   127
Ecological Equivalents, Life Zones, and the Biomes   128
    Tropical Rainforest   133
    Tropical Seasonal Forest   140
    Tropical Savanna   142
    Desert   146
    The Mediterranean Biome   151
    Temperate Grassland   154
    Temperate Forests   158
    Temperate Rainforest   162
    Coniferous Boreal (Taiga), Subalpine, and Montane Forests   165
    Tundra   168

PART II TIME AND LIFE

CHAPTER 7   *CHANGING CONTINENTS AND CLIMATES*   179

Life and the Geologic Time Scale   179
Shifting Continents   183
Quaternary Climatic Change   192
Future Changes in Continents and Climate   203

CHAPTER 8   *DISPERSAL, COLONIZATION, AND INVASION*   215

Dispersal   216
Colonization, Seasonal Migrations, and Irruptions   222
Diffusion Versus Jump Dispersal   228
Barriers, Corridors, Filters, Stepping Stones, and Sweepstakes   233
Recent Introductions and Invasions by Nonnative Species   237

CHAPTER 9   *EVOLUTION, SPECIATION, AND EXTINCTION*   251

Evolution and Speciation   251
    Some Basic Genetics   252
    Historical Development of Evolutionary Theory   256
    Isolation and Speciation   260
    The Temporal Pattern of Evolution   264
    Direction in Evolution   265
    Perfection in Evolution   266
    Increasing Global Species Diversity   267
Geography and Evolution: Founder Effects, Bottlenecks, Vicariance Events, Adaptive Radiation, and Evolutionary Convergence   268
Extinction   274
The Relationship Between Extinction, Evolution, and Diversity   282

CHAPTER 10   *REALMS, REGIONS, KINGDOMS, AND PROVINCES: THE BIOGEOGRAPHIC SUBDIVISIONS OF THE EARTH*   289

Defining Biogeographic Realms, Regions, Kingdoms, and Provinces   289
Determining the Boundaries Between Regions   294
Factors Behind the Biogeographic Regions   295
    Evolution of the Mammals   296
    Evolution of the Flowering Plants   300

The Biogeographic Regions   303
    Nearctic and Palearctic Regions—The Holarctic   303
    Neotropical Region   305
    Ethiopian (African/Paleotropical) Region   306
    Oriental Region   307
    Australian Region   308
    Marine Biogeographic Regions   308

CHAPTER 11   *BIOGEOGRAPHY AND HUMAN EVOLUTION*   312

The Primate Linkage   312
Early Primates   315
The Hominids: *Australopithecus*   317
The Hominids: Early *Homo*   324
The Hominids: *Homo Sapiens* and Recent Cousins   327
The Geographic Expansion of Modern Humans   331

CHAPTER 12   *HUMANS AS A FORCE IN EVOLUTION AND EXTINCTION*   340

Humans as an Evolutionary Force   340
    Animal and Plant Domestication   341
    Questions of the Origin and Spread of Agriculture   346
Modern Humans as a Force of Extinction   352
    Prehistoric Extinctions   353
    Historic Extinctions   358

PART III   THEORY AND PRACTICE

CHAPTER 13   *DESCRIPTION AND INTERPRETATION OF BIOGEOGRAPHIC
    DISTRIBUTIONS*   371

Mapping Biogeographic Distributions   371
Biogeography of Range Size and Range Shape   376
Common Biogeographic Distributional Patterns   380
    Endemic and Cosmopolitan Distributions Revisited   381
    Continuous Zonal Biogeographic Distributions   381
    Disjunct Distributions   382
    Dispersal Disjunctions   382
    Climatic Disjunctions   383
    Geologic Disjunctions   384
    Evolutionary Disjunctions   385
    Biogeographic Relicts   385
Biogeographic Distributions and the Reconstruction of Evolutionary History   386
    Centers of Origin and the Dispersalist Model   387
    Cladistic Biogeography   388
    Panbiogeography and Vicariance Models   390
Beyond Strict Panbiogeography/Vicariance Biogeography: DNA and the Phylogeographic Revolution   392

CHAPTER 14   *THE GEOGRAPHY OF BIOLOGICAL DIVERSITY*   399

What is Biological Diversity?   399
How Many Different Species are There on Earth?   402
Latitudinal and Altitudinal Diversity Gradients   404
Controls on Geographic Gradients of Species Diversity   407
    Historical Theories   408
    Equilibrium Theories   410
    A Synthesis   418

Island Biogeography   420

    Geographic Patterns of Island Biodiversity   420

    The Equilibrium Theory of Island Biogeography: Historical Roots   424

    The Theory of Island Biogeography Today   429

CHAPTER 15   *BIOGEOGRAPHY AND THE CONSERVATION CHALLENGES OF THE ANTHROPOCENE*   443

The Value of Conservation   445

Endangered and Threatened Species   448

Biogeography and Endangered Species   452

Biogeography and Conservation Planning   457

Geographic Strategies for Species Conservation and Biodiversity Conservation   462

    Geographic Tools for Species Conservation and Biodiversity Conservation   467

    Habitat Restoration and Conservation   471

    The Anthropocene and Climate Change Challenges Ahead   475

Reasons for Hope   480

Final Reflections   483

GLOSSARY OF KEY WORDS AND TERMS   493

INDEX   501

PREFACE TO THE 2ND EDITION

The general goals of this 2nd Edition remain similar to the 1st edition. These are to provide an introduction to biogeography that reflects the fascination and importance of the field, but is understandable and engaging to university students from a wide variety of educational backgrounds and interests. The overall organization of the book has also remained similar to the 1st edition. In Part I we review some relevant concepts from the life and physical sciences. We then embark on an exploration of the current geographic distribution of life on the planet and its relationship to the environment. In Part II we explore the trajectory of the planet and life over geologic time. We consider the responses of species to changing environments through processes of dispersal, evolution and extinction. We include our own species in this consideration and examine the evolution of humans and our impact on the climate and the biosphere. In Part III we examine some of the important practices that biogeographers employ to collect data on plant and animal distributions, and some theories that have arisen in biogeography to provide general explanations for the patterns that are revealed. An important focus of Part III is studying the geographic patterns seen in global biodiversity. The final chapter of Part III tackles how biogeographic practices and theory can be applied to assist in addressing the pressing conservation challenges the world faces today. Biogeography is a field of study that is by its very nature global in scope. Accordingly, the research and examples presented here are drawn from throughout the world. Although the general goals and organization of the book remain the same, the world, and the discipline of biography, in important ways that must also be reflected in the new edition.

The 1st edition of *Biogeography: Space, Time and Life* was published two decades ago. Over that time the science of biogeography has advanced significantly. Techniques and data-sets related to field surveying, geographic positioning systems, remote sensing, geographical information sciences, molecular genetics analysis, species distribution modelling and so on have seen remarkable growth. Citizen scientists are using mobile phone apps to greatly increase our knowledge of the geographical distributions of species and the environments they inhabit. Studies of the geologic history of the Earth, past environments, and the evolution and extinction of species have added incredible amounts of new information and refined our perspectives of life over time. The growth of biogeography and the new information on the spatial and temporal patterns of life on Earth has allowed biogeographers to refine previous theory, develop new theory, and in some cases discard previous ideas. It is hoped that this new edition, with many new examples, figures, tables and sources, captures and conveys this exciting growth. As a discipline, biogeography has also become more diverse in the past twenty years. Scholars from throughout the world are contributing important new information and insights. The ideas and examples presented in this text represent the work of thousands of scientists and scholars. These are cited at the end of each chapter. Hundreds are new to this edition. I encourage the readers to dive into these primary sources and learn and be inspired by the works I have drawn upon and been inspired by.

Over the past twenty years the impact of humans on the environment and the species we share the planet with has intensified. Significant forest and woodland clearances have occurred in many countries across the globe. It has been estimated that over 77,000 species of animals and plants are threatened by extinction, and it is widely acknowledged that this may be just the tip of the iceberg. Emissions of greenhouse gasses by human activities have increased and the global surface temperature has risen by about 0.2°C per-decade. The acceleration of human impacts on the Earth's environment have prompted some scientists to declare that we have entered a new geologic Epoch – the Anthropocene. The challenges posed to the sustainability of our planet are considerable. The current state of many of these challenges is provided in this new edition. Biogeography is playing an important role in tackling the sustainability challenges of the Anthropocene. The need for the insights and conservation solutions provided by biogeography is pressing. It is hoped that this book will help inspire you to take up these challenges, as a concerned citizen or a scientist yourself. You can make a very real and positive difference to the fate of life on our planet.

In closing, I wish to thank all of those who have made this second edition possible. The editorial and production staff at John Wiley & Sons who believed in the project, and provided support and patience during its long gestation are owed a debt of gratitude. Many colleagues at UCLA and elsewhere have kept me intellectually stimulated and generously shared their knowledge with me. I continue to learn so much from my undergraduate and graduate students. I thank them for sharing their questions, knowledge and energy with me. In particular, I thank my graduate students Joan Chimezie, Elly Fard, Jessie George, Jiwoo Han, Scott Lydon, Lisa Martinez and Ben Nauman for keeping me inspired during the heaviest research and writing for this edition. Finally, I owe a huge debt of gratitude to another biogeographer who also hails from the University of Toronto, my wife Joanne. Her understanding of the work, her support and her patience have been invaluable.

ABOUT THE COMPANION WEBSITE

This book is accompanied by a companion website:

www.wiley.com/go/macdonald/biogeography2e

The website includes:

- PowerPoints of the figures
- Gallery of images and videos

CHAPTER 1

AN INTRODUCTION

There cannot be many people who have never marveled at an unusual flower or animal, or contemplated the deeper history of the human race and how they, themselves, originated. Even those with the most casual interest in the environment know that today the earth abounds with an incredible diversity of interesting plants and animals, but many are facing grave threats due to habitat destruction, overhunting, and climate change. You may have wondered: How did this wonderful diversity of life, including our own species, arise; how did we get to this present state of environmental crisis; and what can we do about it? These are big questions, and in picking up this book you might be wondering what the science of biogeography can tell us about all of this.

If this is your first exposure to the science of biogeography, you probably have three basic questions: What exactly is biogeography? Why should I bother studying it? How will this book help me understand biogeography? Let's start with a simple answer to the first question. **Biogeography** is the study of the past and present geographic distributions of plants and animals and other organisms.[1] Of course, there is much more to it than that. We will explore this question further and expand on this definition presently, but first let us consider why you might want to study biogeography as an aid in understanding both the diversity of life and how we might conserve that diversity.

More than most sciences, biogeography helps us to understand and appreciate the living environment that we experience every single day. Biogeography helps us answer questions such as how the great diversity of life that we experience today arose, where the modern human species came from, and what we can do to preserve the natural environment in the face of increasing human population growth and environmental change. How does answering such questions directly affect you? When we think of nature we tend to think about distant national parks and wildlands. These are important, but let's think about closer to home. You cannot set foot outside your door without seeing plants that are growing around you. Many of the plants you see are native to the area in which you live, but many others are exotic species that have been recently introduced by humans. You cannot escape hearing the calls of wild birds, some of which are native and some of which have been introduced. Even if you do not know the scientific names of the plants and animals near your home, you are familiar with the way they look and sound. At some point you must have wondered how this diversity of life around you arose and specifically where all of these different plants and animals that live near you originated. These plants and animals fill your neighborhood with beauty and interest. They can also provide more tangible benefits such as the trees that provide shade or shelter from winds, or the small fish that keep populations of mosquitos in check. These tangible benefits that arise from the diverse plants and animals around you are called **ecosystem services**. Biogeography helps you understand the plant and animal life that you encounter everyday where you live and that provide important ecosystem services to you on a daily basis.

If you have ventured far from your home, perhaps visiting another state, province, or country, you have undoubtedly noticed differences in the vegetation and animal life you encountered. For example, during the winter millions of people travel from the northeastern United States and Canada to enjoy the sunshine and palm-lined beaches of southern Florida. The green vegetation of Florida contrasts greatly with the cold and leafless winter forests of the Northeast. Many animals

---

[1] Throughout the text, key terms and concepts will appear in bold lettering. Short definitions for these terms and concepts can be found at the end of the book.

found in Florida, such as alligators, are not found in the northeastern United States or in Canada. Why are alligators and most other plants and animals found in southern Florida not found in the Northeast? The obvious answer might be that these plants and animals require the warm and humid environment of Florida to survive. However, many plant and animal species from Florida are also absent from the other warm tropical and subtropical areas of the earth. Travel to South America, West Africa, Southeast Asia, or northeastern Australia and you will see palm trees and many interesting animals, including relatives of alligators such as crocodiles, but not a single native alligator (Fig. 1.1). Surprisingly, however, you will find native alligators in southern China. Why are certain plant and animal species limited to relatively small areas of the earth? Why are other types of organisms widely distributed? Why can't you grow many of the plants found in Florida in a garden in New York?

An important role of biogeography is to record such geographic differences in vegetation and wildlife and seek to explain them. Biogeography can tell us why North American alligators *(Alligator mississipiensis)* are found only in warm, moist areas such as Florida and adjacent states. Biogeography can also help us to understand why alligators are also found in China *(Alligator sinensis),* and not in tropical regions of Africa and Australia. In studying biogeography, you will come to understand that a combination of geographical, environmental, and historical factors led to the great diversity and current geographic distribution of plant and animal life. In studying biogeography, you can develop a much greater awareness, understanding, and appreciation of the wonderful diversity of plant and animal life found on a global basis as well as the diversity of life that you encounter every single day.

The study of biogeography also helps us to appreciate the development and history of our own species. We do not stand as far apart from the natural world as we might think. Our development, both biologically and culturally, is influenced by geography, the earth's physical environment, and interactions with other organisms. The evolution and spread of humans around the world make up one of the great stories of biogeography. At present, humans play an increasingly greater role in changing the world's environments and have a significant impact on the lives of plants and animals with which they share the earth. We have helped some species spread throughout the world while driving others from much of their former habitat. Some plants and animals owe their existence to human activity. Sadly, many more species owe their extinction to us. Since our existence depends on our relationships with the other species of the earth, our future will ultimately be determined by how we treat them. Biogeography helps us understand where living organisms, including humans, have come from and where we and earth's other inhabitants are potentially headed.

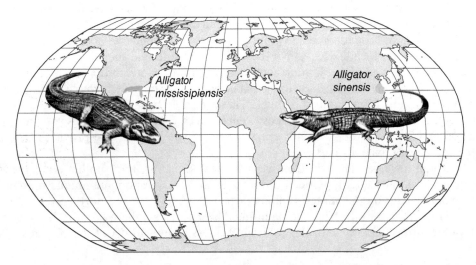

FIGURE 1.1   The world distribution of alligator species *(Alligator mississipiensis* and *Alligator sinensis).* Understanding why alligators are found only in North America and China is a classic biogeographical question. Alligators are found in warm subtropical regions because they cannot tolerate prolonged exposure to cold and freezing temperatures. In the geologic past, Europe, Asia, and North America were joined together in one large continent that experienced relatively warm climates. Therefore, the geographic range of the ancestors of modern alligators was much larger than it is today. Over time, as continents shifted, climate changed, and alligators faced competition from crocodiles and caimans, the geographic range of alligators became fragmented and small.

The ability of humans to alter the environment and affect the survival of plants and animals bestows a great deal of responsibility upon us. We are responsible for making sure our actions cause as little damage as possible to the natural environment. In this effort we will always face tradeoffs between the needs of people for food, resources, living space, and recreational areas, and the preservation and conservation of nature. In some cases, we can alter our activities to lessen the damage we cause. In other instances, we might try to restore damaged areas to a semblance of their natural state. In these enterprises, biogeography can provide important guidance. By examining the long-term histories of plant and animal communities, biogeographers provide information on how humans have altered the environment. In studying the natural distributions of plants and animals, biogeographers can help preserve nature. Biogeography provides important guidance on how we are affecting the environment and what actions we can take to conserve resources and preserve natural environments.

The contrasting state of health of the North American alligator and its relative, the Chinese alligator, underscores the conservation challenges facing biogeographers. According to the Wildlife Conservation Society, there are only 120 Chinese alligators remaining in the wild. This makes it the most endangered crocodilian (crocodiles and their relatives) in the world. Why are the North American alligator populations relatively healthy today and their Chinese relatives so endangered? What might be done to alleviate the risk to the Chinese alligator? Biogeography has much to contribute to such questions.

Let us now turn to further defining biogeography. In studying the present and past distributions of life, biogeographers have two basic tasks: description and explanation. Describing where a plant or animal species occurs today is the easiest part, and yet this task remains incomplete for vast numbers of plants, animals, and other organisms. So far, less than 3 million living species of plants, animals, and other organisms have been identified. It is thought that there may be a total of over 8 million eukaryotic species (organisms such as mammals, birds, insects, and plants that have cells contained in membranes) of which many remain to be discovered and scientifically described. When prokaryotic organisms such as bacteria are included, the total may jump to 2 billion or more. If our knowledge of present-day species remains incomplete, it is even more difficult to reconstruct the past distributions of organisms because we must rely on fossil records that are often fragmentary and difficult to interpret. Explaining exactly how present and past distributions are controlled by the complex geographic, environmental, and historical factors that affect all organisms presents the greatest challenge. In order to understand how the physical and biological environment controls the distribution of plants and animals today, biogeographers must be familiar with concepts and techniques in physiology, anatomy, genetics, ecology, geomorphology, pedology (the study of soils), climatology, limnology (the study of lakes), and oceanography. In order to study how plants and animals were distributed in the past, biogeographers must know some basic geology, paleontology, and evolutionary biology. In studying our own biogeographic history and the relationship between humans and the other species of the earth, biogeographers also intersect in their interests with disciplines such as anthropology and archaeology. Since biogeography requires knowledge from a wide variety of other disciplines, it is called a synthetic science. The interests of the biogeographer blend and overlap with the interests of the ecologist, the evolutionary biologist, and the paleontologist to such an extent as to make the identification of firm boundaries between biogeography and these disciplines difficult. Indeed, it might be argued that from a life sciences perspective, all of these disciplines are subfields of biogeography. Given the synthetic nature of the field, it is not surprising that biogeographers can be found working in many different university departments, including geography, environmental science, sustainability, biology, geology, paleontology, and anthropology. Biogeographers can also be found working in park services, forestry services, environmental services, conservation groups, and private consulting firms.

Clearly, biogeography encompasses a huge area of the natural sciences. One way in which biogeographers tackle the complexity of their subject is to break it down into different subdisciplines. **Phytogeography** studies the present and past distributions of plants. **Zoogeography** examines the present and past distributions of animals. The biogeographic study of the modern relationships between organisms and the environment is called **ecological biogeography.** Some biogeographers concentrate on studying past distributions and the evolution of life. This line of inquiry is called **phylogeography** or **historical biogeography.** The field of **analytical biogeography** is concerned with developing general rules that explain how geography affects the evolution and distribution of plants and animals and how past distributions and evolutionary history are reflected in modern distributions. The application of the lessons learned from ecological, historical, and analytical biogeography, including the protection or restoration of the natural environment, is called **conservation biogeography.** In recent years much attention has been placed on the expanded concept of **ecological sustainability,** which is the capacity of the biosphere to meet the present and future needs of humans and other species. Biogeography has an important role to play in achieving this important goal for our planet.

So, what are the aims of this textbook, and how can it help you understand this fascinating and diverse field? The book is written to provide you with a solid introduction to the field of biogeography. It does not assume extensive background knowledge of biology, geography, or geology. There are, however, fundamental terms and concepts from these three fields that all biogeographers must know, and so they are outlined in this book. In Chapter 2 you will be introduced to some important basics of biology and physical geography including climatology, oceanography, and limnology. Later, in Chapter 7, we will explore some important concepts from geology regarding the history of the earth. Chapter 9 will provide some further primer material on genetics. Throughout the text, you will also be introduced to widely applied concepts from ecology and evolutionary biology. These concepts will be explained in detail, and their importance to biogeography will be highlighted.

In recognition of the ecological, historical, and analytical facets of biogeography, the book is divided into four major sections. The first section, called Space and Life, is concerned with ecological biogeography. In these chapters we will examine how present physical and biological conditions affect organisms and their distributions. Chapter 3 describes how physical conditions affect organisms and control their geographic distributions. In Chapter 4 we will look at the interactions that occur between different organisms and discuss how these biological factors can control geographic distributions. In Chapter 5 we will examine how disturbances such as fires or floods impact ecosystems and influence the geographic distribution of organisms. In Chapter 6 we will in particular explore the classification of life into global biomes that reflect the strong control of climate on vegetation.

The second section of the book concerns historical biogeography and is entitled Time and Life. In Chapter 7 we will discover how the physical geography of the earth has changed over the past 500 million years. We will also explore how the earth's climate has undergone natural periodic changes. We will pay special attention to the shifts between glacial and nonglacial conditions that have occurred over the past 2 million years. We will conclude the chapter with a consideration of how human activity, particularly that related to increases in greenhouse gases, is now changing the climate of the planet. Chapter 8 will delve into how organisms change their distributions in response to changing geographic and environmental conditions. In Chapter 9 we will explore how the processes and patterns of evolution and extinction relate to changing environmental conditions and how geography affects these processes. Chapter 10 discusses how past and present processes play out in determining the major biogeographic regions of the earth. Chapters 11 and 12 examine the fascinating story of human evolution and some of the impacts of our species on the biosphere.

In the third section, entitled Theory and Practice, we put the concepts we have learned together to understand global distributions of plants and animals and how biogeography can be applied to their conservation and the sustainability of the earth. In Chapter 13 we will examine different general geographic patterns observed in the spatial distributions of species. Chapter 14 explores how overall biodiversity is distributed across the planet and the roles the environment, geography, and past history have on biodiversity.

Some people have argued that humans have so altered the planet that we have entered a new geological epoch referred to as the **Anthropocene.** Chapter 15 provides an exploration of conservation and ecological sustainability concepts and potential solutions to some of today's challenges provided by biogeography. The final chapter will provide an outline of some of the current scientific tools and approaches that biogeographers use in conservation work.

One of the greatest attractions of biogeography comes from the wonderful variety of life and the fantastic adaptations that allow organisms to live in environments ranging from the coldest depths of the ocean to the hot sands of the Sahara Desert. Biogeography is a science concerned with life, and it should be lively! In this book you will be introduced to many examples of fantastic and wonderful plants and animals. These organisms, and the many interesting environments and communities in which they live, will be used to illustrate the concepts and theories of biogeography. We will also examine many real-world examples of how biogeography has impacted people and how human activity has affected other organisms. These examples should make your studies more enjoyable and allow you to translate your knowledge of biogeography to the real world. In the end, when you travel across the world, walk down your own street, or look at yourself in the mirror, you will realize that you and every single plant, animal, and person you encounter is a result of, and contributor to, the biogeographical story of life.

KEY WORDS AND TERMS

Analytical biogeography
Anthropocene
Biogeography
Conservation biogeography

Ecological biogeography
Ecological sustainability
Ecosystem services
Historical biogeography

Phylogeography
Phytogeography
Zoogeography

REFERENCES AND FURTHER READING

Ding, Y., Wang, X., He, L., Shao, M., Xie, W., John, B. T., and Scott, T. M. (2001). Study on the current population and habitat of the wild Chinese alligator (Alligator sinensis). *Chinese Biodiversity* **9**(2), 102–108.

Grunewald, K., and Bastian, O. (Eds.) (2015). *Ecosystem Services— Concept, Methods and Case Studies*. Springer, Berlin.

Larsen, B. B., Miller, E. C., Rhodes, M. K., and Wiens, J. J. (2017). Inordinate fondness multiplied and redistributed: the number of species on Earth and the new pie of life. *Quarterly Review of Biology* **92**, 229–265.

Manolis, C. S., and Stevenson, C. (Eds.) (2010). *Crocodiles: Status Survey and Conservation Action Plan*, 3rd edition. Crocodile Specialist Group, Darwin, Australia.

Mora, C., Tittensor, D. P., Adl, S., Simpson, A. G., and Worm, B. (2011). How many species are there on Earth and in the ocean? *PLoS Biology* **9**, e1001127.

Rands, M. R., Adams, W. M., Bennun, L., Butchart, S. H., Clements, A., Coomes, D., Entwistle, A., Hodge, I., Kapos, V., Scharlemann, J. P., and Sutherland, W. J. (2010). Biodiversity conservation: challenges beyond 2010. *Science* **329**, 1298–1303.

Thorbjarnarson, J., Wang, X., Ming, S., He, L., Ding, Y., Wu, Y., and McMurry, S. T. (2002). Wild populations of the Chinese alligator approach extinction. *Biological Conservation* **103**, 93–102.

PART 1

SPACE AND LIFE

CHAPTER 2

SOME BASICS

Biogeography is a very wide ranging science, and accordingly biogeographers are a highly diverse group of scientists in terms of interests and perspectives. This multidisciplinary scope is one of the most satisfying aspects of biogeography—and one of the most challenging. Anthropology, biology, climatology, geography, and geology all contribute concepts, specialized vocabularies, and ways of classifying information that are important to biogeography. In order to understand and appreciate the science of biogeography, we must therefore have some knowledge of its foundations in these other sciences. Before we get down to the business of biogeography, we will review some key concepts from some of these associated disciplines. In this chapter, we will concentrate on aspects of biology and physical geography that are fundamental to biogeography. Later in this book we will discuss some concepts from geology and physical anthropology that are also important to biogeography. Much of the material that we cover here will be familiar to anyone who has taken introductory courses in biology and physical geography. However, even for those with such a background, a little review won't hurt.

BIOLOGY AND THE HIERARCHIES OF LIFE

Biology can be defined as the science of life and all of its phenomena. Life on earth includes millions of different types of organisms, ranging from viruses to whales. The study of any of these organisms could include examination of myriad phenomena, from biochemical reactions within plant cells to the social behavior of pods of whales. It is clear that biologists face a daunting challenge! The science of biology tackles this immense task by breaking it down into smaller components that can be studied individually. One way to categorize and organize the constituent units of a large entity is to develop a hierarchy. A **hierarchy** is a system of organization in which components are ordered by rank. Most corporations are good examples of hierarchies. A large group of office staff is under the direction of one office manager, who, along with several other managers, is under the direction of the vice president of operations. The vice president, along with two or three other vice presidents, reports to the president of the firm. The president occupies the upper-most level of the corporate hierarchy, and the office staff provides its base. Every biogeographer should be familiar with three important biological hierarchies: the taxonomic hierarchy, the ecological hierarchy, and the trophic hierarchy. We will examine each of these.

Taxonomic Hierarchy

Taxonomy is the subdiscipline of biology concerned with the classification and naming of organisms. Taxonomy is also known as **systematics** when the main goal is to determine the evolutionary relationship between groups of organisms. In this case, taxonomists are also called systematisists. The evolutionary histories of organisms that are reconstructed by systematisists are referred to as phylogenies. Taxonomists use observable traits, referred to as characters or character states, to group similar individual organisms together and to separate groups of different organisms. The groups that taxonomists develop are called **taxa** (the singular form is **taxon**). The characteristics that taxonomists use to classify organisms usually start with physical differences, such as color variations in flowers or in the plumage of different birds. Taxonomists may also consider differences that are apparent through chemical analysis of the tissue or fluids of the organism. Such studies are referred to as chemotaxonomy. Finally, cytotaxonomists examine the chromosome structure of organisms to detect genetic similarities and differences in order to classify organisms.

The classification and naming of organisms by individual taxonomists would not be very useful if everyone had their own system and terminology. Fortunately, there is a single system of taxonomic classification that is generally accepted by all biologists and biogeographers. The development of our present taxonomic system is at the foundation of biology and biogeography. It is a history that can be traced back to ancient Greece. The roots of the modern taxonomic system began with Aristotle (384–322 BCE[1]). Aristotle, a student of Plato, is best known as a philosopher but was also active in biology, physics, astronomy, and psychology. He developed an early scientific system for classifying animals into groups that shared similar features. Aristotle believed that the specific form and behavior of individual plants and animals were inherited and immutable. He considered that all individuals belonged to groups, or **species** (from the Greek word *eidos*), of taxonomically similar individuals. Aristotle believed that the form and behavior of these species did not change from generation to generation. He taught that dogs form one species and cats another. Aristotle also argued that plant and animal species formed a hierarchy ranging from simple organisms, such as worms, to the most complex organism, which he considered to be humans. In the Middle Ages, European scientists, influenced by Aristotelian logic, grouped organisms that appeared to be generally, but not exactly, similar to each other into taxonomic units called **genera** (the singular form is **genus**). A genus would include, for example, the Scots pine trees of Scandinavia and the Mediterranean pines of southern Europe. Both the Scots pines and the Mediterranean pines belong to the genus called *Pinus*.

Before we proceed, let's examine the words *genus* and *species* and consider the continuing role of Latin in biology and biogeography. Genus is the Latinized form of the Greek word *genos*. In the Middle Ages the language of scholarship in Europe was Latin, and the names of the genera came from that language or are latinized versions of words from other languages. For example, in Latin the genus for pines is called *Pinus*. In comparison, the genus for cats is *Felis*. Although Latin is no longer widely spoken or understood, the formal names of organisms are still written in that language. The benefit of this convention is that no matter which language the scientist is working in, the names of the organisms are always presented in Latin using the Latin alphabet. Even papers and books written in Russian Cyrillic or in Chinese characters will always present the scientific names of organisms in the Latin alphabet. Biologists and biogeographers always know which organisms are being discussed, even if they can read nothing else in the document. The use of the proper scientific names for organisms is also important for avoiding confusion among scientists who speak the same language. Take pines, for example. In North America we have many different trees that are closely related and belong to the genus *Pinus*. In the South Pacific we might encounter a tree that is commonly called the Norfolk Island pine. This plant, however, is unrelated to our North American pines and belongs instead to the genus *Araucaria* (Fig. 2.1). In many instances the same plant or animal will have several different common names, but every organism has only one accepted scientific name. In addition, the Latin names of organisms often contain descriptions that can be understood by people with only a limited knowledge of Latin.

So what is the difference between a species and a genus? It is recognized that genera of plants and animals contain organisms that are related but are consistently distinguishable on the basis of their morphology. In addition, many of these taxa, though members of the same genus, cannot interbreed. These different members of a genus are species. For example, people in eastern North America will be familiar with the eastern white pine, and those in the western part of the continent might know the lodgepole pine. Both species belong to the genus *Pinus* and share certain broad similarities, such as long thin needles, cones, and upright growth (Fig. 2.2). There are, however, clear differences

---

[1] BCE stands for Before Common Era and is equivalent to the previously used notation BC. CE (Common Era) is equivalent to the previously used notation AD. Both notation systems are based on the Gregorian calendar which was introduced by Pope Gregory XIII in 1582 and used throughout the world today. The year 322 BCE is exactly equivalent to 322 BC, and 2019 CE is equivalent to 2019 AD. In this book, years that are within the Common Era will be cited without adding the CE notation; for example, 2019 = 2019 CE.

FIGURE 2.1   White pine trees of the genus and species *Pinus strobus* growing in the Pocono Mountains of Pennsylvania (*a*) and so-called Norfolk Island pines of the genus and species *Araucaria heterophylla* from Norfolk Island in the South Pacific (*b*). Although both species are called pines, they are unrelated, and only white pine is actually a member of the pine genus.

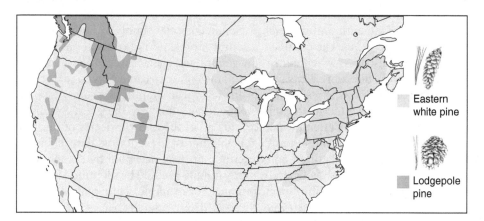

FIGURE 2.2   The needles, cones, and shapes of a mature eastern white pine *(Pinus strobus)* and a western lodgepole pine *(Pinus contorta)*. Notice that both pine species share a general resemblance but possess clear differences in terms of their needles, cones, and mature form.

between the trees. White pines can grow to well over 30 meters in height, whereas most lodgepole pines achieve heights of only 20 meters. The white pine carries its needles in bundles of five. In contrast, the lodgepole pine has bundles of two needles. The white pine has large cones that open immediately upon ripening and release their seeds. Lodgepole pines have small cones, many of which remain closed and retain their seeds until a fire causes the cone scales to open. In addition, eastern white pine and lodgepole pine cannot interbreed. The two trees are recognized as different species within the genus *Pinus*. Following the concepts of Aristotle, the English scholar John Ray (1627–1705) attempted to systematically define a species. He concluded that if seeds came from the same plant, the seedlings must be related and similar in character. He recognized that variations might occur between seedlings from the same parent, but he considered these to be "accidents." Ray's conception of species can be paraphrased as "Like begets like."

Since the time of Ray, species have formed the basic unit of the modern taxonomic system, but there remains much debate among scientists as to just how a species should be defined. Three main definitions are in use today. The phylogenetic species concept identifies a species as a group of sexually reproducing organisms that share at least one diagnostic character that is present in all members of the species but absent in other organisms. Unique behavioral traits are accepted as defining characters. This concept does not adequately take into account natural variations within species, such as differences in human hair color. If this concept were widely applied by biologists, far more species would be identified than is now the case. In addition, the phylogenetic species concept does not consider reproductive interactions between members of the species or their evolutionary history. The biological species concept defines a species as a group of organisms that can interbreed freely under natural conditions. This definition was articulated by the great evolutionary biologist Ernst Mayr in 1942, although its roots extend into the early part of the twentieth century. It remains the most widely used definition of a species in biology. Finally, paleontologists who study the fossil record often use the evolutionary species concept. Evolutionary species are organisms that have a direct ancestor-descendant relationship that is traceable in the fossil record. Usually, such a relationship is inferred from morphological similarities such as the size or shape of the fossils. This is because morphological features are generally all that is available from the fossil record. New species are differentiated when there are clear divisions of one evolutionary line into two or more new lineages. The evolutionary species concept allows for morphological changes in the species over time as evolution occurs.

In practice, biogeographers combine these concepts and view species as organisms that are morphologically similar and can interbreed freely under natural conditions. It is assumed that such organisms must share a common ancestor. Although this sounds pretty straightforward, we still encounter problems in defining species due to natural variability in the morphology and reproductive behavior. In addition, species are generally thought of in terms of sexually reproducing organisms. Organisms that reproduce asexually raise difficulties when the common phylogenetic and biological species definitions are used.

Until the eighteenth century, scientists named species simply by adding some descriptive words to the name of the genus. The eighteenth century was a period of great geographical and biological exploration and discovery when many new genera and species were found and described. To differentiate these newly discovered species, their names often became polynomials containing a dozen or more Latin words. Clearly, this system was unwieldy. The simplified system of naming organisms that we use today was invented in the mid-eighteenth century by the great Swedish scientist Carolus Linnaeus (1707–1778). He attempted to catalog all of the known species of the world. Linnaeus was trained as a medical doctor but inherited a keen interest in plants and natural history from his father. Many of the species he identified and described are still accepted by taxonomists today. Linnaeus introduced the binomial system whereby every organism could be identified by a unique combination of a generic (genus) and specific (species) name. For example, white pine is known as *Pinus strobus,* while lodgepole pine is known as *Pinus contorta.* These related species share the same genus, *Pinus,* but are differentiated by their species names, *strobus* and *contorta.* Species with the same generic names are assumed to be related. With the binomial system, different species can also share the same specific name. This occurs frequently because the specific names often describe some feature of the organism. As an example, the scientific name for white oak is *Quercus alba,* whereas the scientific name for a seabird known as the snowy sheathbill is *Chionis alba.* In this case, the word *alba* refers to the light coloring of the organisms and in no way implies a genetic relationship.

Because different, unrelated species can share the same specific name, a species is never referred to by its specific name alone—the genus must be indicated. If it is clear which genus is being discussed, you can use the first initial of the generic name instead of spelling the complete name (i.e., *Quercus alba* can be referred to as *Q. alba*). By convention, the generic and specific names are always italicized, and the generic name starts with a capital letter. The specific name starts with a lower-case letter. In this book you will find the scientific names of important genera and species in parentheses following the first usage of the common name. Sometimes the scientific name of a species is followed by the full name, abbreviated name, or initials of the taxonomist who first described it. A scientific name followed by the letter L, such as *Quercus alba* L., means that the species was first described by Linnaeus himself.

So what does the modern taxonomic hierarchy consist of? As we have seen, species form the lowest level of the hierarchy, although some species may be further subdivided into distinctive sub-species and races. Collections of species that are similar to one another, and presumably genetically related, are grouped together in the next level, the genera. Genera that are morphologically similar and likely possess evolutionary linkages are grouped together as **families.** Oaks, for example, belong to the family of trees called Fagaceae. This family also includes the genus beech *(Fagus)* and several other genera of trees. How higher taxonomic groups are classified remains a subject of revision and debate. According to a widely used system, families that are related are grouped together in taxonomic **orders.** The order for oaks is Fagales. In turn, orders are grouped into **classes.** All flowering plants are grouped into one of two classes. The oaks are members of the class **Magnoliopsida** (sometimes called **dicotyledons**). Among other traits, plants in this class have seeds that produce two initial leaves, or cotyledons, when they germinate. Many of the familiar trees of the deciduous forest, such as maples *(Acer),* birch *(Betula),* elm *(Ulmus),* willows *(Salix),* and garden flowers such as roses *(Rosa)* and daisies *(Aster)* are Magnoliopsida. The other class for flowering plants is **Liliopsida** (also called **monocotyledons**). Some examples of Liliopsida include orchids *(Orchidaceae)* and grasses *(Poaceae).* When the seeds of these plants germinate, they produce one cotyledon.

Classes of plants are grouped together into **divisions** (in the case of animals, the term **phyla** is used for this level of the hierarchy—the singular being **phylum**). The division for the Magnoliopsida and Liliopsida is **Magnoliophyta.** The Magnoliophyta are also referred to as the flowering plants or the **angiosperms. Conifers** (cone-bearing plants) such as the pines *(Pinus),* the spruces *(Picea),* and the firs *(Abies)* are neither Magnoliopsida nor Liliopsida. They do not belong to the Magnoliophyta but are instead members of the division **Pinophyta** and are also known as **gymnosperms.** Finally, at the highest level, plants are all grouped into the kingdom **Plantae.** Both the Magoliophyta and the Pinophyta have vascular systems to transport water and nutrients and are grouped together in a subkingdom called the **Tracheobionta.** Life is often grouped into **kingdoms** and sometimes a lesser number of higher-level **domains.** Debate continues on how to classify organisms at many levels, including these highest levels. According to one commonly used system, the kingdoms include Monera, Protista, Fungi, Plantae, and Animalia. The Monera are simple prokaryotic organisms such as bacteria. The Protista are single-celled eukaryotic organisms such as amoeba. You can probably guess who the members of the Fungi, Plantae, and Animalia kingdoms are. Thus, we can trace the affinity of every organism, including humans (Table 2.1), up through the taxonomic hierarchy to a kingdom. However, the exact number of kingdoms continues to be debated and numbers up to eight have been proposed in recent years. For example, the Monera are often subdivided into Bacteria and Archaebacteria kingdoms.

Notwithstanding continuing debate on specific aspects of it, the taxonomic hierarchy provides a convenient and generally accepted way to classify life. It should be remembered, however, that this

TABLE 2.1   A Systematics (Taxonomic Hierarchy) of Eastern White Pine *(Pinus strobus)* and Humans *(Homo sapiens sapiens)*

	White Pines		Humans
Species	*Pinus strobus*	**Species**	*Homo sapiens*
Genus	*Pinus*	**Genus**	*Homo*
Family	Pinaceae	**Family**	Homonidae
Order	Pinales	**Order**	Primates
Class	Pinopsida	**Class**	Mammalia
Division	Pinophyta	**Phylum**	Chordata
Kingdom	Plantae	**Kingdom**	Animalia

system is not static. Terminology associated with its subdivisions has evolved over time. New species are continuously being discovered and added. At times, new genera and families are added. The status of some of the kingdoms remains hotly debated. In addition, taxonomists may change the genus or family of older recognized species when new information comes to light. It sometimes happens that several taxonomists identify the same species and give it different names. These names are referred to as synonyms, and the earliest published name generally takes precedence.

Ecological Hierarchy

The taxonomic hierarchy provides for a division and ranking of life based on the morphology and the genetic relationship of organisms. It is the foundation for many aspects of biological research. There is, however, a need to consider a different hierarchy used to organize and interpret observations for the ecologist or biogeographer who goes out into the field to study and categorize how organisms are affected by the environment or how they interact with other species. It would be an intractable task to study the relationships between every organism on earth and the environment, or the interrelationships among all the organisms that populate the planet. Therefore, ecologists and biogeographers often concentrate on specific spatial and taxonomic scales of study. These different levels of ecological study can be organized into a hierarchy that reflects the increasing geographic and taxonomic scale being examined.

The lowest scale of study that is of interest to the biogeographer is the **individual** organism. For example, scientists might put a radio tracking collar on one Arctic fox *(Alopex lagopus)* and monitor its movements. At the next level, they might consider studying a **population** of Arctic foxes. A population is defined as all individuals of a given species in a prescribed area. Usually, members of the same population are assumed to be in close enough proximity to be able to interact and interbreed frequently. Of course, in many cases, members of the same species are found in different locations and do not interact on a regular basis. Interbreeding and other interactions between these separated populations of the same species may only occur infrequently. Such populations are referred to as **metapopulations.** The term was coined by the biologist and philosopher Richard Levins in 1969 to describe this concept of studying populations of separated populations of the same species. Metapopulations are found both in terrestrial and marine settings. Two broad types of metapopulations can be recognized. Loose metapopulations consist of subpopulations of the same species that live in different locations and interact with other subpopulations very infrequently. The distance between subpopulations is farther than most individuals travel during their life spans. Tight metapopulations consist of subpopulations that are close enough for individuals to travel between them and thus interact more frequently. The Arctic fox populations living on different islands in the Canadian Arctic separated by large expanses of water are an example of a loose metapopulation. Birds that live and nest in different woodlots that are separated only by agricultural fields are an example of a tight metapopulation.

We might also consider examining the interaction of our fox population with all the other species of organisms in its environment. In this case, we are studying an ecological **community.** A community can be broadly defined as all populations of organisms that live and interact within a prescribed area. Ecological research that focuses on one species is sometimes referred to as **autecology,** while research that focuses on the interactions between species in communities is referred to as **synecology.** Some ecologists and biogeographers limit their studies of communities to subsets of the total community. For example, phytogeographers might restrict their study to flowering plant species only. Some biogeographers would refer to this as the flowering plant community. However, other ecologists and biogeographers feel that the term community should only be used to refer to all species of organisms in an ecosystem. Subsets of a community, such as the flowering plant species, are often referred to as **assemblages.** One important subset of the complete community is the **guild.** Guilds are groups of animal species within a community that have similar forms, habitat, and resource requirements. The insectivorous bird species form one guild in a community, while the seed-eating bird species form another.

If we were to consider the relationship between the species of our community and the physical factors of the environment, particularly when we examine flows of energy and matter through this biophysical system, we would be conducting research at the scale of the **ecosystem.** It could be argued that the spatial boundaries of any ecosystem extend over the whole earth. Even the smallest area is linked to the world at large by physical processes such as global rainfall patterns and biological processes such as bird migrations or the input of tiny airborne microorganisms such as fungal spores. In practice, the boundaries of ecosystems are often defined by the researcher and can vary from very small areas such as the trunk of a decaying tree to large areas such as the tundra of Baffin Island in Arctic Canada. Very large areas of the earth's surface that have a similar climate and vegetation are referred to as **biomes.** The biomes are an important area of research in biogeography, and we will discuss them in detail later. Finally, all of life on the planet is collectively referred to as the **biosphere**—the highest and broadest level of ecological research. The other realms of the earth are physical ones of the **atmosphere, hydrosphere,** and **lithosphere.** The atmosphere includes all the components of the air; the hydrosphere includes all the water in the oceans, lakes, streams, and ground; and the lithosphere is the solid earth of rock and sediment. Although the research of biogeographers may focus on smaller levels such as ecosystems, the end goal of biogeography is to combine the results from all of these more specific studies and draw general conclusions about how the biosphere functions today, how it developed, and where it might be headed in the future. Remember, our future and that of the biosphere are really one and the same.

Trophic Hierarchy

A third way of ordering the biosphere is to examine the flow of energy through ecosystems. The various levels through which energy flows from its initial capture by the biosphere until its dissipation as waste heat are called **trophic levels.** The biosphere functions through the acquisition of energy by organisms and the flow of energy from one organism to another. With very few exceptions, all life is dependent on the sun to provide the energy that is consumed. Notable exceptions are the amazing ecosystems that have developed around undersea volcanic vents (Technical Box 2.1).

Most of the solar energy that reaches the earth is visible light, which is a form of short-wave radiation. The electromagnetic waves that make up visible light range in wavelength from approximately 0.4 microns to 0.6 microns. Of this incoming energy, approximately 32% is reflected back into space by the earth's atmosphere. Roughly 18% is absorbed by the earth's atmosphere. The surface of the earth absorbs about 50% of the incoming radiation, and about 20% of this is used in evaporation. Visible light energy from the sun is absorbed by the atmosphere and the earth's surface and then

TECHNICAL BOX 2.1

Deep-sea crabs and other marine life near a "black smoker" hydrothermal vent in the deep ocean. Illumination for the picture is provided by artificial lighting from a deep-sea submersible vehicle. *Source:* W.R. Normark, Dudley Foster / Wikimedia Commons.

Chemosynthesis and the Ecosystems of Oceanic Hydrothermal Vents

In 1977 scientists discovered that hydrothermal vents located 2500 meters below the ocean surface near the Galapagos Islands supported unexpectedly dense concentrations of organisms. Over 100 similar vents have been discovered throughout the world's oceans. The biomass of these vents is astounding and can range as high as 20 to 30 square kilometers. The vent organisms include giant red worms *(Riftia pachyptila)*, large clams *(Calyptogena magnifica),* and mussels *(Bathymodiolus thermophilus)*. In addition, various species of crabs, shrimps, and sea anemones are found at some vents. In all, some 700 new species, previously unknown to science, have been discovered near undersea hydrothermal vents. How can such fauna be supported so far from the upper ocean waters where photosynthesis occurs? These vents emit mineral-laden waters at temperatures ranging up to 450° C. When mixed with the cold ocean waters, this produces temperatures in the vent area of 8° C to 23° C. Hydrogen sulfide (H_2S) is particularly abundant in these plumes. Studies have revealed that these vents support a food chain based entirely on geothermal energy. At the base of the food chain are bacteria that oxidize the sulfur from the vents to form carbohydrates. In contrast to photosynthesis, this process is called **chemosynthesis** and proceeds as follows:

$$CO_2 + H_2S + O_2 + H_2O \rightarrow CH_2O + H_2SO_4$$

The CO_2 and O_2 in the process come from the sea water. The bacteria are then eaten by primary consumers such as limpets, mussels, and clams. Chemosynthetic bacteria utilizing hydrogen and aluminum have also been found in environments such as the hydrothermal vents, cold marine vents, and sunless caves.

radiated as long-wave (3-micron to 25-micron wavelength) thermal energy. This long-wave radiation is what we sense as heat. It is clear that almost 100% of the incoming solar radiation is reflected back into space or used in heating and evaporation. What then powers the biosphere? The entire biosphere, including the human race, is supported by a tiny fraction of the incoming solar radiation. The amount of solar energy captured for direct use by the biosphere is only 0.1% to 0.3% of the total input!

The solar energy used to power the biosphere is captured through the process of **photosynthesis.** For most plants, the radiation used for photosynthesis falls in the red through blue visible light (approximately 0.4-micron to 0.6-micron wavelength) portion of the electromagnetic spectrum. This light is referred to as photosynthetically active radiation, or PhAR. Most of the solar energy that the earth receives is PhAR, so it is no coincidence that plants evolved to use this portion of the electromagnetic spectrum in photosynthesis. It is also not surprising that our eyes are adapted to detecting radiation in these wavelengths. We can see visible light spectrum energy but not ultraviolet or infrared energy.

During photosynthesis, atmospheric carbon and water vapor are transformed into sugar, water, and oxygen. The energy for photosynthesis comes from PhAR. The process takes place in chloroplasts, which are small, green bodies within plant cells. The chloroplasts hold chlorophyll, which is the primary light capturing the pigment of plants. The overall process can be summarized as follows:

$$6CO_2 + 12H_2O \xrightarrow{\text{Light}} C_6H_{12}O_6 + 6H_2O + 6O_2$$

Carbon Dioxide	Water		Sugar	Water	Oxygen

The CO_2 enters the leaves of plants through openings created by specialized sets of cells called **stomata.** The stomata also allow the release of oxygen and water vapor from the interior of the leaf.

There are three different biochemical pathways that green plants use in photosynthesis. Most plants capture energy using the **C_3 pathway** described by Melvin Calvin of the University of California at Berkeley. In C_3 plants, the CO_2 from the atmosphere is converted into a 3-carbon molecule called 3-phosophoglyceric acid. In the 1960s it was discovered that sugar cane converts CO_2 into two 4-carbon molecules: malic and aspartic acid. This process became known as the **C_4 pathway.** Finally, some plants, such as the prickly pear cactus *(Opuntia),* use a modified form of photosynthesis called **crassulacean acid metabolism (CAM).** In CAM plants, CO_2 is absorbed at night and stored as malic acid. During the light of day, photosynthesis is conducted by the C_3 pathway. The term C_2 *photosynthesis* is also encountered when discussing photosynthesis and relates to the process of photorespiration whereby high light intensity and temperatures lead to a process with decreased photosynthetic efficiency. This will be discussed further in Chapter 3. In general, C_4 plants have the highest rates of photosynthesis, while CAM plants display the lowest rates. Interestingly, however, all plants are relatively inefficient in terms of energy fixation through photosynthesis. Only about 1% to 3% of the light hitting a leaf is transformed into chemical energy in the form of simple carbohydrates such as the sugars glucose and fructose.

Photosynthetic plants are the foundation of the trophic hierarchy almost everywhere except places like hydrothermal vents or sunless caves (Fig. 2.3). Plants are referred to as primary producers because they fix the energy of the sun into chemical energy used to power the biosphere. This term is slightly misleading, for plants cannot actually produce energy but merely transform it from one state to another. Plants are also referred to as **autotrophs** or phototrophs because of their ability to fix energy through photosynthesis rather than derive it from the consumption of other organisms. Organisms that eat plants to obtain energy are called primary consumers. Species that eat the primary

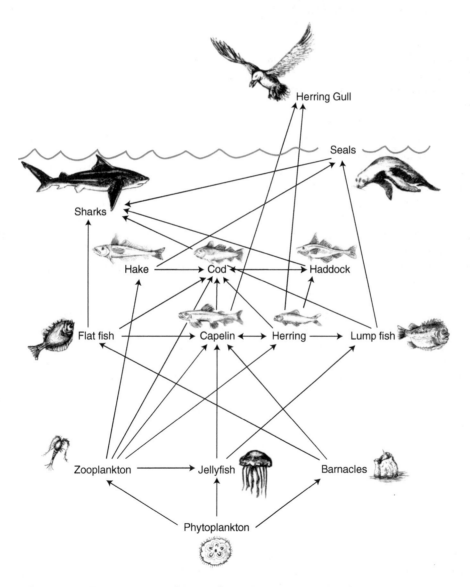

FIGURE 2.3 The complexity of energy flow within an ecosystem illustrated by a simplified and small part of the food web for the northwestern Atlantic Ocean (after Lavinge, 1992). Despite its complexity, a trophic hierarchy with planktonic plants at the base and sharks, mammals, and birds at the top is still apparent.

consumers are referred to as secondary consumers, while species that eat secondary consumers are referred to as tertiary consumers. Plant-eating species are also called herbivores, and meat-eating species are called carnivores. Animals that eat both meat and vegetable matter, such as most humans, are omnivores. When plants and animals die, decomposers consume them. Decomposition is the ultimate fate of all trophic levels. Herbivores, carnivores, omnivores, and decomposers are collectively known as **heterotrophs** because they rely on other organisms to provide energy.

The trophic levels extending from primary producers to the highest level of consumers are sometimes called food chains. However, the idea that energy flows in a linear fashion up the trophic hierarchy is simplistic and misleading. Generally, there is a skipping of levels and a back-and-forth exchange between different levels of consumers and decomposers. As a result, the actual flow of energy in an ecosystem is more like a **food web** than a linear chain (Fig. 2.3). The English ecologist Charles Sutherland Elton (1927) was one of the earliest scientists to identify the importance of ordering ecosystems according to energy flows. He also recognized that the flow of energy was more complex than a simple linear chain. The work of American ecologist Raymond Lindeman in 1942 was the first attempt to formally conceptualize and quantify the flow of energy within food webs. The study of food webs is the basis of modern systems ecology. It has been suggested that understanding food web structure is fundamental to understanding basic ecosystem structure, function, and response to disturbance and change.

Relatively complex trophic hierarchies have been identified from both terrestrial and marine ecosystems. However, tracking and quantifying the flow of energy in such systems are not easy. The three approaches commonly used to investigate and define food webs are direct observation of feeding habits, inspection of the stomach contents of dissected organisms, and use of radioactive tracers such as phosphorus-32 (^{32}P), which are introduced through injection into the vegetation and can be easily traced in tissue samples from animals.

The average length of time during which energy captured through photosynthesis is held by living plants in food webs varies from a few days in the open ocean to 3 years in grasslands and 22 to 25 years in some forests. Energy captured by photosynthesis can also be held in plant litter for significant periods of time before being used by decomposers. In the tropics, the energy held in plant litter may be released within 3 months, while it can be held for over 100 years in the litter of temperate and northern forests. However, studies using radioactive tracers show that, despite the potential long residence time of energy in plants and litter, much of the energy captured by photosynthesis moves very quickly through the food web. In both terrestrial and freshwater ecosystems, some of the energy captured by photosynthesis is passed up to the highest trophic levels within a matter of weeks.

Elton noted that trophic hierarchies appeared to be limited to only four or five levels. More recent examinations of the number of trophic levels in many food webs ranging from invertebrate communities living in dung to birds in forest stands to fish, birds, and mammals of the Antarctic pack-ice have found that most ecosystems support only three to six trophic levels. Why is this? It has been argued that since the transfer of energy between different trophic levels is relatively inefficient, the energy available to higher trophic levels will rapidly decrease. All organisms use energy to function through the process of **respiration,** which is the oxidative reaction that breaks the high-energy bounds of carbohydrates to release energy for the organism's metabolism. Thus, respiration is the reverse of the process of photosynthesis, with metabolic energy being produced, as shown here:

$$\text{Metabolic Energy}$$
$$\text{Energy} + 6CO_2 + 12H_2O \quad \leftarrow \quad C_6H_{12}O_6 + 6H_2O + 6O_2$$

Carbon Water   Sugar Water Oxygen
Dioxide

On average, only about 10% of the energy of any trophic level is passed on to the next trophic level. This generalization is known as the 10% rule. As a result of the 10% rule, each trophic level is expected to have about 90% less energy available to support it than the preceding trophic level. There is a high degree of variation in the actual energy efficiencies of different organisms in food webs. Birds and mammals are the least efficient because they use energy to maintain constant body heat. They assimilate only about 3% of the energy they receive in food. Insects have an energy efficiency of around 39%, while fish have efficiencies of around 10%. The decrease in available energy with each trophic level is extremely important in ordering ecosystems. When you consider how little energy is passed up through each trophic level, it is easy to understand why the bulk of all living things on earth are plants (perhaps greater than 90% by mass) and very large carnivores are rare, occurring at much lower numbers than their prey species. In theory, it takes a minimum average of 10 prey organisms to support one bird or mammal predator of equal mass.

The total number of consumers that any ecosystem can support is limited by the productivity of the plants. The productivity of vegetation is therefore a key focus of study for understanding the structure of ecosystems. The productivity of vegetation in fixing energy is known as **primary productivity.** Primary productivity can be viewed in two different ways. Gross primary productivity is the total energy fixed in an ecosystem by photosynthesis. However, since some of this energy is used through respiration to support the metabolic activities of the plants, not all of the photosynthetic energy is available to produce plant matter and be passed onto higher trophic levels. The gross primary productivity minus the energy lost by respiration is referred to as net primary productivity. So, how is net primary productivity measured? Many different techniques may be used, ranging from the measurement of carbon fixation and respiration losses of individual plants to satellite-based estimates of the photosynthetic activity of the world's oceans. A relatively direct way is to harvest and weigh a representative sample of the plant matter produced during a given time period in a given area. In terrestrial environments, the plants can be harvested from both above and below the ground to produce an estimate of aerial and root production. The plant material is dried and weighed. The mass of living material is referred to as **biomass.** The plant biomass is used to calculate primary productivity as follows:

$$dB = B_1 - B_2$$

where

dB = the change in biomass between time period 1 (t_1) and time period 2 (t_2)

B_1 = biomass at time t_1

B_2 = biomass at time t_2

However, we must also account for losses in biomass due to death and decay of plants *(L)* and losses to consumers *(G)*, so that the true net primary productivity = $dB + L + G$. Net primary productivity can be measured in terms of biomass as grams per square meter per year (g/m^2/yr). Alternatively, the caloric content of the biomass can be measured and net primary productivity reported as calories per square meter per year. Since the rate of vegetation growth varies in different ecosystems, net primary productivity is highly variable from place to place. In general, plant biomass and primary productivity are highest in the warm and wet tropics and lowest in cold polar ecosystems and dry deserts. The mean net primary productivity of a tropical rainforest ecosystem may be some 2200 grams per square meter per year. In desert environments, the net primary productivity averages from 90 to 3 grams per square meter per year. The net primary productivity of the open ocean averages 125 grams per square meter per year. We will discuss the global distribution of net primary production further when we examine the world's biomes.

When we look at the distribution of the standing biomass of different trophic levels, the impact of the 10% rule quickly becomes apparent. These distributions often have a pyramidal shape in which biomass can decrease by 90% or more with each increase in trophic level. However, the shape of such **pyramids** can vary widely. In Panamanian rainforests, it has been shown that it takes approximately 40,000 grams per square meter of vegetation biomass to support 14 grams per square meter biomass of primary consumers, secondary consumers, and decomposers. In coral reef systems, 703 grams per square meter of primary producer biomass can support 132 grams per square meter of herbivores and 11 grams per square meter of carnivores.

It would therefore seem that we should be able to predict the number of trophic levels that a given ecosystem will support. The food webs of insects, which have high-energy efficiencies, living in ecosystems with high primary productivity should contain more trophic levels than other types of communities. Interestingly, this has not been found to be the case. Most terrestrial food webs, despite differences in primary productivity and the thermal efficiency of the organisms, tend to have three to four trophic levels. This means that energy flow cannot be the only factor limiting the number of trophic levels. It is also interesting to note that despite the low primary productivity of the oceans, many marine communities have five trophic levels, and carnivores are much more abundant compared to the number of carnivores found in terrestrial systems. It is possible that the transfer of energy between the marine primary producers, which are mostly microscopic plants that float in the water, and primary consumers, which are mainly microscopic crustaceans called copepods, is much more efficient than the transfer of energy between land plants and terrestrial herbivores. Indeed, in some oceans the standing biomass of photosynthetic plankton may be low relative to the biomass of other trophic levels because of the efficiency with which the copepods harvest the phytoplankton and incorporate the energy captured by them. In the English Channel, 21 grams per square meter of herbivore standing biomass are supported by only 4 grams per square meter of primary producer standing biomass. Inverted biomass pyramids have also been identified in freshwater plankton-based communities and coral reefs. It has also been proposed that such inverted biomass pyramids may occur where the life spans of predators greatly exceed that of the food species or significant levels of immigration of prey occurs.

As a final thought on trophic levels, we might reflect on the place of humans in the food webs of the world's ecosystems. Estimates suggest that as a global average humans consume 0.7 tonnes of dry weight biomass per year. North Americans consume on average 1.9 tonnes. Since we are omnivores, we function as both primary consumers and higher-level consumers. It is easy to see that when we eat the meat of herbivores such as cows or sheep, we are secondary consumers. We thus are dependent on the larger primary productivity required to support the primary consuming animals we eat. Based on the 10% rule, a kilogram of meat would require 10 kilograms of plant matter to produce it. However, humans also function as tertiary or quaternary consumers and eat other carnivores. Most people eat fish, for example. Important food species such as tuna and salmon are carnivorous tertiary and quaternary consumers. Thus, in terms of marine food webs, we function as quaternary consumers or even higher in the trophic hierarchy. In addition to food, humans utilize biomass for things like building material and fuel. One estimate suggests that, in total, humans currently consume about 25% of the earth's total annual primary productivity each year. Measuring **human appropriation of net primary production (HANPP)** is one way of analyzing the increasing magnitude of humanity's demands on the biosphere and our role in the global ecosystem.

PHYSICAL GEOGRAPHY AND THE FUNCTIONING OF THE EARTH

The science of physical geography is concerned with the physical patterns and processes that occur at the earth's surface. Physical geographers conduct research on phenomena ranging from local landforms caused by glacial activity to large-scale variations in the world climate. The most important areas of physical geography for biogeographers are climatology, which is the study of the earth's atmosphere and climate; pedology, which is the study of soils; limnology, which is the study of fresh water; and oceanography, which is the study of the world's oceans. It is worthwhile to review a few important concepts from these areas in order to investigate how features of the earth's physical environment interact with the biosphere to influence the distribution of life.

Global Climate

Before we begin our examination of the earth's climate, we should note the difference between the terms *weather, climate,* and *atmosphere*. These are three very different concepts and are sometimes confused. Weather is the condition, or physical state, of the atmosphere at a given place at a given time. The weather changes from day to day or even hour to hour. Climate is the average condition of the atmosphere at a given place based on statistics from a long period of weather observations. When we say it is raining in Boston today, we are describing the weather. When we read that the average high temperature in Los Angeles in July is 28° C, we are reading a description of climate. By convention, such climatic averages are usually based on 30 years of observation.

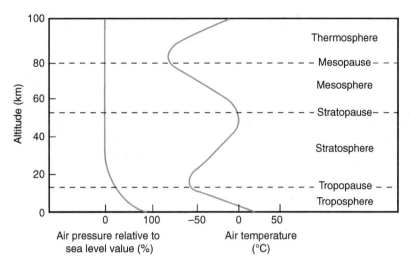

FIGURE 2.4   The atmospheric stratification of the earth and the variation in temperature and air pressure within the troposphere (*Source:* Strahler and Strahler, (1997) / John Wiley & Sons).

The atmosphere is the layer of gas that surrounds the earth. The atmosphere extends over 100 kilometers above the surface of the earth and is divided into several different layers based on temperature, pressure, and chemical composition (Fig. 2.4). However, all life exists in the lowest level of the atmosphere, called the **troposphere.** The troposphere extends some 9 to 17 kilometers above the earth and contains over 90% of the total mass of the atmosphere. The composition of pure dry air in the troposphere is roughly 78% nitrogen, 21% oxygen, 0.93% argon, and 0.036% carbon dioxide and a small amount of other gases. The air in the troposphere can be relatively dry or contain up to 5% water vapor in places such as the humid tropics. Both air pressure and temperature decrease with height in the troposphere. For every 275-meter rise in elevation, the air pressure decreases by about 3.3%, while temperature decreases at a rate of approximately 6.5° C for every 1000-meter increase in altitude. This is why the air is thinner and temperatures are cooler at the tops of high mountains.

As you recall from our discussion of photosynthesis, most of the sunlight received by the earth is in the visible light portion of the electromagnetic spectrum (approximately 0.4-micron to 0.6-micron wavelength). This energy is absorbed at the earth's surface and then emitted as heat energy at longer wavelengths of approximately 5.0 microns to 20.0 microns. The gases of the atmosphere allow most of the solar energy in the visible light wavelengths to pass through the atmosphere but tend to absorb the longer wavelength heat energy that is emitted from the earth's surface. Carbon dioxide and water vapor in the atmosphere are important in trapping this heat. This phenomenon of the transmission of incoming solar light energy and the trapping of outgoing heat energy is known as the **greenhouse effect.** Without this atmospheric effect, the earth would be too cold to support life.

The distribution of climatic conditions on the earth is not random but follows a predictable pattern that is controlled mainly by latitude, elevation, and the distribution of continents and oceans. As we shall see later, if we can predict the climatic conditions at a certain place, we can also make certain predictions about the type of organisms we will find there. Latitude is the most important control on the overall distribution of climatic conditions. Latitude determines the amount of solar energy that is intercepted over a given area of the earth's surface (Fig. 2.5). The earth orbits the sun at an average distance of about 150 million kilometers. The orbit is elliptical, so that in January the earth is about $2^{1}/_{2}$ million kilometers closer to the sun. This is referred to as the perihelion. The aphelion occurs in July when the earth is an additional $2^{1}/_{2}$ million kilometers away from the sun. At these distances, the electromagnetic energy from the sun can be assumed to be traveling as parallel waves with a constant energy of 1400 watts per square meter. During March and September, at the equator, these rays strike the earth at a 90° angle at noon, so that a square meter of incoming solar radiation is intercepted by a square meter of earth's surface. At higher latitudes, the angle of incidence of the sun's rays decreases. As a result, a square meter of solar energy is spread over an area much greater than a square meter of the earth's surface. Due to the decrease in the angle of insolation, temperatures should be cooler at higher latitudes because they receive less solar energy. However, that is only part of the story.

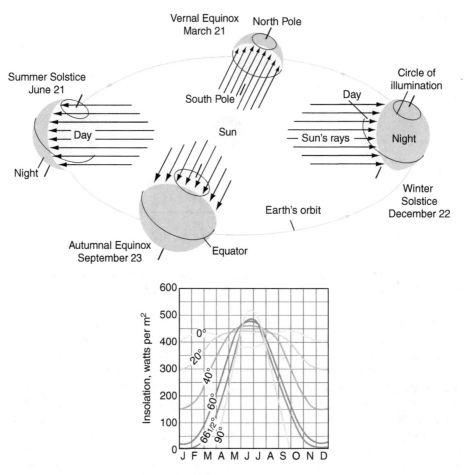

FIGURE 2.5 The seasonal orbital characteristics of the earth and the seasonal distribution of insolation at different latitudes (*Source:* Strahler and Strahler, (1997) / John Wiley & Sons). Notice how at the poles (90° latitude) there is less total insolation and greater seasonal variation than at the equator (0° latitude).

The latitudinal distribution of solar energy varies with season as well as latitude. Because the earth's axis is tilted about $23^1/_2$° relative to the plane of the orbit around the sun (Fig. 2.5), the amount of solar energy received at different latitudes varies throughout the year. At the December solstice, the northern hemisphere is tilted away from the sun so that the angle of incidence of solar energy is decreased and areas north of $66^1/_2$° (the Arctic Circle) do not receive any direct sunlight. The point of the earth's surface where the sun's rays strike the earth at a 90° angle lies at $23^1/_2$° south latitude (the Tropic of Capricorn). During June and the summer months, the northern hemisphere is tilted toward the sun, and the angle of incidence in the north is increased. North of the Arctic Circle the sun never sets. The sun's rays strike the earth at a 90° angle at the Tropic of Cancer ($23^1/_2$° north latitude). The tropics are defined as the region lying between $23^1/_2$° north and south latitude. During the equinoxes, on March 21 and September 23, neither hemisphere is tilted toward or away from the sun. The sun's rays strike the earth at a 90° angle at the equator, and there are 12 hours of daylight at all latitudes.

The seasonal distribution of insolation has important implications for temperatures and life on earth. The long days of insolation during May, June, July, and August at the high latitudes of the northern hemisphere provide large amounts of solar energy. This energy results in temperatures warm enough for ice and snow to melt and plant life to exist. In the southern hemisphere, the long days of high insolation are experienced in the months of November, December, and January. Thus, during the 24 hours of sunlight in June, the North Pole actually receives more insolation than the equator (Fig. 2.5). However, over the course of the year, the average insolation at the equator is much higher than at polar regions. A very important additional aspect of insolation distribution is the way that seasonal differences increase with latitude (Fig. 2.5). Organisms existing in the Arctic and Antarctic polar regions not only have to contend with less total insolation, but they have to adapt to large seasonal differences between winter and summer. A final interesting point to note, is that the high and midlatitudes experience one insolation peak during the summer, while the equator experiences two peaks in spring and fall.

If incoming insolation was the only control on temperatures at the surface of the earth, we would see surface temperatures decreasing in a smooth fashion, parallel with latitude, from the equator to the poles. However, if we look at maps of mean January and July temperatures, we see that temperatures do not decrease smoothly in parallel with latitude (Fig. 2.6). Some of this can be explained by differences in elevations. As we discussed earlier, high-elevation sites are cooler on average than low-elevation sites. Upon inspection of temperature maps, it is clear that summer temperatures are higher over land surfaces than over the sea. Winter temperatures are lower on land relative to the sea. This phenomenon is particularly noticeable in interior Alaska and central Siberia where average January temperatures range from −25° C to −50° C, while January temperatures in the adjacent North Pacific range from 0° C to −15° C. In fact, central Siberia is the coldest place in the northern hemisphere, with winter temperatures commonly far below those encountered at the North Pole. The coldest recorded temperature in North America occurred in the village of Snag in the Yukon Territory of Canada. In July the average daily temperatures in Siberia and Alaska rise to 10° C to 15° C, while the air temperatures over the oceans increase to only 5° C to 10° C. The reason for these differences in surface temperatures of continents and oceans is related to the differential rates of heating and cooling of land and water. The solid land surface heats and cools more rapidly than the liquid sea where energy is dispersed in depth. Thus, the land responds more dramatically to the seasonal increases and decreases in insolation. In addition, the oceans circulate heat laterally through currents. Such transfers do not occur on solid land surfaces. Sites that are located in the interior of landmasses and are subject to dramatic seasonal differences in temperature are said to experience continental climates, whereas islands and coastal sites, that have climates moderated by the ocean, are said to have marine climates. In the case of Siberia, the difference between the mean January and July temperatures can be as high as 60° C. In the interior of Alaska and Canada, this difference in seasonal mean temperatures is about 45° C. In contrast, the difference between January and July temperatures on islands and at coastal sites in the tropics is less than 3° C.

Now, let's consider the geographic distribution of precipitation. The term *precipitation* is used here to denote moisture that is deposited from the atmosphere to the land or water surface as rain, sleet, hail, or snow. Precipitation occurs when the water vapor in the atmosphere exceeds the moisture-holding capacity of the air. The amount of water that can be held as vapor increases rapidly with air temperature. Air with a temperature of 0° C can hold only about 5 grams of water per kilogram of air, whereas air with a temperature of 30° C can hold over 25 grams of water per kilogram of air. When air cools, its ability to hold moisture as vapor decreases and eventually precipitation forms. As we discussed, air generally cools with increasing altitude in the troposphere. The average rate of temperature decrease is about 6.5° C for every 1000 meters of altitude, although this varies. Air that is dry cools at a rate of about 10° C per 1000 meters. This is called the dry adiabatic lapse rate. Air in which condensation is occurring cools at the wet adiabatic lapse rate, which is approximately 5° C per 1000 meters. So, as air rises it cools, and if it cools enough, condensation and precipitation occur.

Three processes cause air to rise and generate precipitation. The first process is convective precipitation, which occurs when the warming of air at the surface causes the air to rise until it becomes cool enough for precipitation to occur. This phenomenon causes the thunderstorms and rain showers typical of summer weather in many parts of the world, particularly the tropics. The second process is frontal precipitation, which occurs when warm air masses rise up and over denser masses of cool air. An example of this phenomenon are the winter storms that occur in the middle latitude regions. Finally, air may be forced to rise when it encounters physical barriers such as mountains. Rain generated in this fashion is referred to as orographic precipitation. The high rainfall encountered on the western slopes of the Coast Ranges from northern California to Alaska is influenced by orographic conditions.

The global distribution of precipitation is controlled by surface conditions, such as the location of mountains and oceans, and by the general circulation of the atmosphere. The general circulation of the atmosphere is the large-scale patterns of winds that develop in response to differences in radiation and heat at the equator and the poles. Precipitation, due to convection, is typical of the equatorial

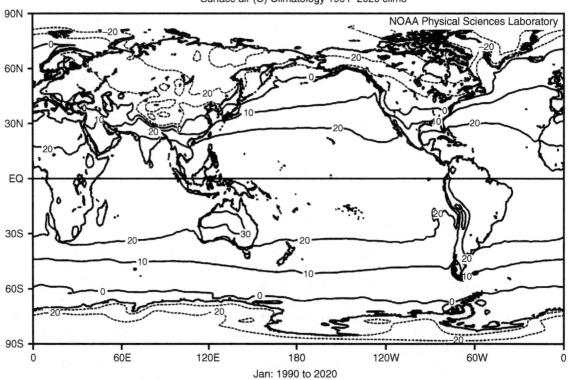

Jan: 1990 to 2020

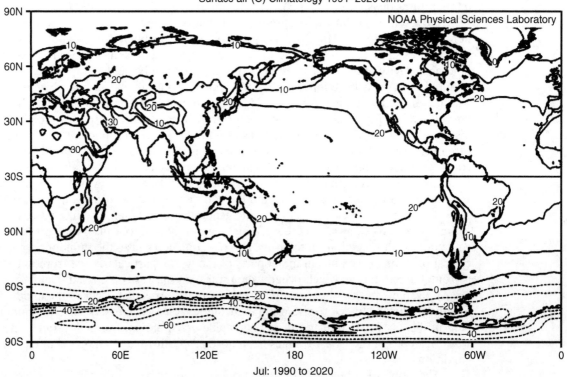

Jul: 1990 to 2020

FIGURE 2.6 Average monthly temperatures (°C) at the surface of the earth in January and July based on the 1990–2020 long-term mean (adapted from NOAA/ESRL Physical Sciences Division).

regions which, due to high insolation, experiences vigorous heating at the earth's surface. This surface heating causes air to warm and rise. High rates of evaporation from the warm sea surface provide abundant moisture to the atmosphere. At approximately 30° north and south latitude, the tropical air begins to subside and warm as it descends (Fig. 2.7). As the air warms, it can carry more water as vapor. Thus, areas that are typified by descending air are dry. The descending air causes a band of calm and dry climate in the subtropics at around 30° north and south latitude. These regions are referred to as the doldrums or horse latitudes because sailing ships often were becalmed in these latitudes and the sailors would jettison horses and other livestock to conserve water. At the surface, air flows from the region of the horse latitudes to replace the rising air in the tropics (Fig. 2.7). The flow of the air is deflected to the right due to the Coriolis effect imparted by the rotation of the earth. As a result, there is a zone of winds from the northeast and the southeast in the equatorial region that are referred to as the tropical trade winds. The region where the trade winds from the northern and southern hemisphere converge is called the Inter-Tropical Convergence Zone (ITCZ). The converging winds at the ITCZ replenish moisture for convective precipitation in the tropical zone. North and South of the horse latitudes, the surface air flows away from the zone of subsidence in an easterly direction, and this is referred to as the midlatitude westerlies. These westerlies dominate the temperate areas of the earth between 30° and 60° north and south latitude. Storms develop along the belt of westerlies as the atmosphere picks up moisture over the oceans and produces frontal precipitation. North and south of the midlatitude westerlies, polar air masses dominate. The polar air masses are generally very cold and very dry.

Elevation also plays a role in the distribution of precipitation. Because of orographic lifting and cooling of air, high elevations typically have higher amounts of precipitation than adjacent low-elevation sites. In areas such as coastal Oregon, Washington, and British Columbia, the frequent occurrence of midlatitude frontal storms brought ashore by the westerlies, coupled with orographic precipitation due to high mountain ranges along the coast, produces large annual totals of precipitation. Air also heats as it descends down the lee side of mountains. This promotes dry conditions, which are referred to as rainshadows. In the region of westerlies, rainshadows are often found on the east side of mountain ranges. The dry eastern slopes and the adjacent deserts of the Sierra Nevada Mountains of California are a result of the rainshadow effect. The warm and dry winds, experienced as air flows down mountains, are called Chinook Winds in the northern Rocky Mountains and Fohn Winds in Europe.

The general distribution of annual precipitation on the earth reflects the processes outlined above (Fig. 2.7). The greatest amounts of precipitation generally occur in the tropics where over 500 centimeters of rainfall may fall in a given year. Similar magnitudes of rainfall are experienced on the western slopes of coastal mountains in western North America, Norway, and western South America. A band of low precipitation is centered on 30° north and south latitude. This includes some of the world's great

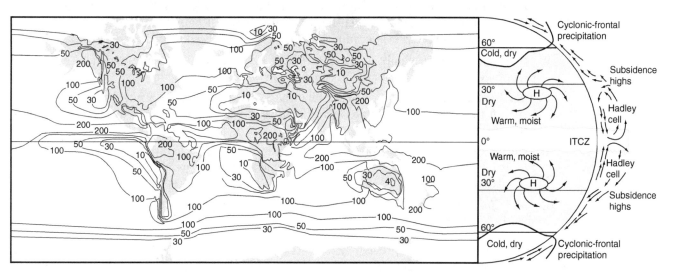

FIGURE 2.7   General circulation of the atmosphere and average annual precipitation (*Source:* Strahler and Strahler, (1997) / John Wiley & Sons).

deserts such as the Sahara in northern Africa, the Namibian in southern Africa, the Arabian Desert, the coastal deserts of Chile and Peru, the Australian Outback, and the deserts of the U.S. Southwest and adjacent Mexico. Average annual rainfall in some of these regions may be less than 2.5 centimeters per year. In some areas, such as the Sahara Desert of southern Sudan, 100 years may pass between occurrences of measurable rainfall. In the midlatitudes, very dry conditions and deserts may extend as far as 45° to 50° north and south latitude in the rainshadows of mountain ranges. The Great Basin Desert of California, Nevada, and Utah is an example. The Arctic and Antarctic regions also experience low annual precipitation. However, we generally do not think of these regions as deserts. Low precipitation is not the only factor in creating a desert. In true deserts, the rate of evaporation greatly exceeds precipitation, and there is a deficit in soil moisture at most times. The Arctic and Antarctic regions have low precipitation, but because of low temperatures, they also have low evaporation rates. Therefore, these polar regions cannot be considered to be true deserts in most cases.

In addition to geographic variations due to latitude and orographic features, seasonal differences in precipitation are important influences on the distribution of life. The ITCZ does not stay fixed at the equator but changes seasonally. During June and July, the ITCZ moves north toward $23\frac{1}{2}°$ north latitude. During December and January, it moves toward $23\frac{1}{2}°$ south latitude. As the ITCZ moves north and south, it brings precipitation to the arid subtropical regions near 30° north and south latitude. In addition, the increased warming of the land surface in the subtropical regions during the summer produces a pattern of enhanced convective uplift over the land and surface flow of moist air from the oceans. This creates a rainy season called the Summer Monsoon, which is important for bringing moisture to areas such as India, Southeast Asia, and the southwestern United States. Thus, the areas bordering the tropics experience their highest rainfalls during the summer months and often have very dry winters. In the midlatitudes, the westerly storm tracks are also displaced. In the summer, westerly storms do not generally travel as far south as California or Spain, and this brings dry summers to these regions. In contrast, during December and January, the storm tracks move south and bring winter rains.

Because of geography's role in controlling temperature and precipitation, the earth can be divided into broad climatic zones based on temperature and precipitation regimes. One such scheme, often used by biogeographers, is based on the 1918 work of Wladimir Köppen as later modified by Geiger and Pohl. Köppen was a knowledgeable biogeographer as well as a climatologist, so his system was designed to reflect the relationship between world climate and vegetation zones. The climatic zones he proposed are based strictly on long-term climatic averages. In determining the boundaries of the climatic zones, Köppen also used vegetation boundaries as guides. He reasoned that large-scale vegetation boundaries were likely related to climate. The complete Köppen system has 5 major climate zones and 13 climate types, with a mechanism for further subdivisions. A simplified global version of the Köppen system excluding highland areas is presented in Fig. 2.8. The five major climate zones are as follows:

A. *Tropical Rainy Climates* found in the equatorial regions. This zone has monthly average temperatures of 18° C or higher, with little seasonal variation. Rainfall is abundant throughout the year and always exceeds evaporation.

B. *Dry Climates* found mainly in the subtropical zone. Temperatures are generally warm, and evaporation exceeds precipitation throughout all or most of the year.

C. *Mild Humid Climates* found in the midlatitudes. The temperature of the coldest month falls between 18° C and −3° C, with a clear difference in winter and summer seasons. Precipitation exceeds evaporation.

D. *Snowy Climates* found in the mid- to high latitudes. The coldest month has an average temperature below −3° C, but the average for the warmest month is greater than 10° C. Precipitation exceeds evaporation.

E. *Polar Climates* found in the polar regions. The average temperature of the warmest month is less than 10° C. Precipitation is low, but evaporation is very limited.

Before we conclude this review of climate, we will consider some interesting properties of local climate related to mountains. Under normal conditions, temperatures decrease with altitude. Thus, high elevations are generally cooler than low elevations. The impact of this can be seen with particular clarity when we consider how the high elevations of mountains such as the Appalachians or Alps experience lower temperatures than the adjacent lowlands during both the summer and winter.

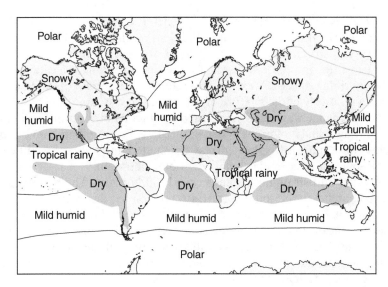

FIGURE 2.8   Simplified Köppen climate classification (*Source:* Strahler and Strahler, (1997) / John Wiley & Sons).

However, in some localized instances, valley bottoms in the mountains can experience colder temperatures than the surrounding slopes and peaks. The enhanced cooling of the valleys is caused by dense cold air which drains downward from higher elevations, particularly at night and in the early morning. The effect is known as cold air drainage. In some instances, valley bottoms may be too cold to support trees, which grow well on adjacent slopes.

Microclimate

In our discussion we have looked at relatively large-scale climatic conditions. However, significant differences in climate can occur at very small scales. Such very local conditions are referred to as microclimates. For example, aspect is the direction that a slope faces. A south-facing slope in North America can experience daytime temperatures that are over 10° C warmer than an adjacent north-facing slope. Evaporation rates are also higher on the warm south-facing slope, so soils are drier. Thus, the south-facing slope may support plants and animals that are typical of hot and dry climates, such as desert species, while the adjacent north-facing slope may be wooded. South-facing slopes in the Yukon Territory of Canada sometimes support plants typical of the northern Great Plains, while adjacent north-facing slopes support cool subarctic forest species. Sites close to the ground are generally warmer than elevated and exposed sites. In the Arctic, summer daytime temperatures can decrease by over 5° C from the soil surface to 30 cm above the surface. Temperatures decrease rapidly below the soil surface. The air temperature in rodent burrows, which are only a few tens of centimeters below the surface, may be over 10° C cooler than the exposed top soil.

Forests create profound differences in microclimate. The amount of sunlight received at the floor of a dense forest may be less than 10% of the amount received at the top of the forest canopy. Temperatures at the forest canopy can be well over 10° C higher than those found on the forest floor. Decreased evaporation rates and the transpiration of leaves cause forest floors to be much damper than at the canopy level. Finally, wind speed can decrease by as much as 90% between the top of the canopy and the forest floor.

World Soils

Soil is the uppermost layer of mineral and organic matter found on the earth's surface. Soil supports and sustains almost all plant life on which terrestrial ecosystems depend. In addition, many organisms, ranging from bacteria to mammals, live within the soil. Soil structure, texture, mineralogy, water content, and chemistry can exert a strong control on the distribution of organisms. Soil is formed by the physical and biological weathering of rock and the addition of organic matter. The formation of soils is dependent on many factors and can take thousands of years. The type of regolith (weathered rock and mineral matter) from which soil is developed is important. Plants and animals may depend on soils, but they also modify regolith to create soils. In this manner soil and life on the earth have evolved together. Climate is also extremely influential in determining the type of soil that will develop in a given locale. Both temperature and moisture conditions can have a great influence on soil development. Thus, soils vary in structure, texture, and chemical properties both locally and globally. A gradient in soil characteristics and types across landscapes is referred to as a catena.

One means of describing soils is by texture. This refers to the amount of sand, silt, and clay in the soil. Loam soils, with a relatively equal mixture of silt, sand, and clay, are often best for plant growth because they hold and exchange water easily and provide mineral nutrients in a form useful to plants. Water runs quickly through sandy and rocky coarse soils. In addition, plants cannot obtain nutrients from coarse particles. Water runs slowly through clay-dominated fine soils and is difficult for plants to acquire. Clay and silt soils that lack sand can impede root penetration.

Soils can be divided into stratigraphic units called soil horizons. The boundaries between the horizons generally run parallel to the soil surface. Each horizon has distinctive physical and chemical properties. Often, color can be a useful means of discriminating soil horizons and different soil types in the field. An objective means of describing soil color is provided by the use of a standardized color chart provided by a Munsell Color Book. A vertical section through the various horizons is called a soil profile, and comparing profiles provides another means of classifying soils. The principal horizons of a soil profile are generally designated as O, A, E, B, and C (Fig. 2.9).

The C horizon is the unconsolidated rock at the base of the soil profile. It can retain traces of the structures and bedding of the original rock. The C horizon has undergone little chemical

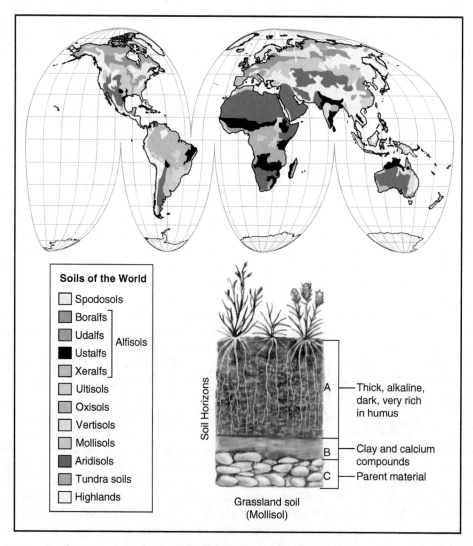

FIGURE 2.9 The world's major soil regions of the world and an idealized soil profile (after Stahler and Strahler, 1997; see also USDA Global Soil Regions Map, 2015; USDA Soil Gallery, 2005).

transformation and most closely resembles the chemical and mineral composition of the original rock. However, silica, carbonates, and salts, weathered from the overlying horizons, can accumulate in the C horizon. The base of the C horizon is often difficult to determine exactly as it grades into the unweathered underlying rock.

The B horizon overlies the C. It is highly weathered and does not retain any traces of the original bedding or structures of the bedrock. The B horizon is typified by illuviation (the accumulation of material) and contains substances such as clays, calcite, aluminum and iron oxides, and humus (organic matter) that have been weathered and leached downward from the overlying horizons. Some of these substances bind particles together, imparting structures of granules, plates, blocks, and columns to the B horizon. Under certain conditions the accumulation of iron oxides, gypsum, salts, silica, and clay can form impermeable layers that block the movement of water and nutrients in the soil profile. When such barriers consist of uncemented clays, they are called fragipans or claypans. When they consist of cemented material, they are called duripans or hardpans.

The A and E horizons are formed by eluviation (loss of material) and are depleted in clays, aluminum, and iron oxides relative to the B horizon. When leaching is particularly intense, the E horizon can be bleached of color, while organic matter in the A horizon makes it darker. Because biological activity such as rooting and soil-dwelling organisms is important, the A horizon contains more humus than the E. The A horizon can consist of up to 30% humus. The decay of humus releases organic acids that promote weathering and leaching in the soil. The replacement of cations such as calcium, magnesium, potassium, and sodium in the A and E horizons by hydrogen can reduce the ability of the soil to support plant growth. In some cases, an extremely organic-rich layer of plant matter, the O horizon, overlies the A horizon.

As is the case with climate, the soils of the world can be classified into broad categories to produce world soil maps. The pioneers in the area of soil description, classification, and mapping were the Russian scientists V. Dokuchaiv and K. Glinka and Americans such as E. Hilgard and C. Marbut who worked in the late nineteenth and early twentieth centuries. Broad classification systems take into account the structure, texture, chemical properties, and profiles of the soils. Unlike biological taxonomy, however, there is no international standard system for the classification of soils. Even in North America different systems are used. In the United States the USDA has produced a soil taxonomy system that is widely applied here and abroad. That system undergoes frequent revision and improvement and its eleventh iteration was published in 2010. Canada has a different system, the Canadian System of Soil Classification. Many other countries have developed their own systems or use terms from world soil classification schemes developed by Russian or British pedologists. The Food and Agriculture Organization (FAO) of the United Nations has also developed an international World Soil Classification map of the world that is widely used.

The current USDA taxonomy divides soils into 12 soil orders. The different orders are discriminated on the basis of criteria such as soil composition and texture, degree of horizon development, presence or absence of diagnostic horizons, and degree of weathering of soil minerals. The global distribution of these soil types has been mapped by the U.S. Department of Agriculture (Fig. 2.9). The distribution of some soil types does correspond with Köppen's climatic regions. The soil orders are as follows:

Entisols are mineral soils that have undergone little to no alteration. They lack distinct soil horizons and are often found on fresh deposits from rivers, glaciers, and sand dunes. Because of low weathering, it may be difficult for plants to obtain nutrients from some entisols. Many tundra and desert soils are entisols.

Inceptisols are young soils that possess enough soil moisture to support plants for at least 3 months of the year, have one or more discernible horizons, are fine to loamy in texture, and contain weathered minerals and associated plant nutrients. Some tundra soils are inceptisols. Cryaquept inceptisols are typically found in polar climate regions and high mountains. Aquept inceptisols are found in bogs and marshes throughout the world. Inceptisols are also found on floodplains where significant sedimentation is no longer occurring.

Histosols are generally moist and have up to 30% organic matter in the upper portion of the soil. The organics can come from mosses and other plants growing in water, such as in bogs, and from deep forest litter. These soils are commonly acidic and low in nutrients. High water content may produce reducing conditions and an abundance of iron oxides in the mineral soil beneath the A horizon. Histosols are common in the coniferous forests of the snowy climate region. They can also be found in other regions on wet, vegetated sites.

Gelisols have permafrost within 100 centimeters of the soil surface and/or have mineral or organic matter with evidence of frost churning (cryoturbation) and permafrost within 200 centimeters of the surface. Permafrost is a condition wherein the soil does not reach temperatures above freezing even in the summer.

Oxisols are characterized by the deep weathering of most minerals except quartz, aluminum, and iron. The soils have water, but the nutrient content is often low. They generally present little evidence of soil horizons, except that the surface is darker owing to the presence of organics. Hard concretions (plinthites) and layers (laterites) of iron concretions form in oxisols. Oxisols are found in the warm and moist areas of the tropical moist climate zone.

Ultisols have a prominent E horizon and a B horizon in which clay minerals accumulate and can form plinthite. These soils are low in plant nutrients. The mean soil temperature for formation of ultisols is 8° C, and they are associated with climates that produce an abundance of moisture in one season and dry conditions during another.

Vertisols have a high content of clay and shrink and swell depending on water content. They have cracks and other evidence of soil movement. Soil horizons may not be apparent. These soils have good nutrient availability but are difficult to work for agriculture. Vertisols form under grassland and savanna vegetation in the tropical and subtropical regions.

Alfisols have a gray to reddish horizon close to the surface that lack darkening by humus and an underlying layer of clay accumulation. Alfisols have good nutrient compositions for plants. Alfisols form under forests (Boralfs and Udalfs) and mixed forest-grassland, shrub cover (Xeralfs and Ustallfs) at sites ranging from the conifer forest in the snowy climate region to seasonally dry sites in the midlatitudes and subtropics.

Spodosols often have a bleached and light-colored E horizon that overlies a dark layer of illuviated humus, aluminum, and iron in the B horizon. Spodosols are generally acidic and low in nutrients. In many cases these soils are relatively young, despite the clear development of horizons. They are found under coniferous forests in the snowy climate zone.

Mollisols have a very thick and dark A horizon with a soft structure. Vertical cracks form in the soils due to cycles of wetting and drying out. These soils have high amounts of calcium and generally high nutrient content. Mollisols are perhaps the most fertile soils in the world. They form under grasslands in the dry climate regions of the midlatitudes.

Aridisols are dry soils that have low humus content but clear horizons. They can have high amounts of calcium and salt layers. They form in the arid climate zone where water content in the soil is too low to support much plant growth.

Andisols are soils formed on fresh deposits of volcanic ash. They have a small geographic distribution and can be very fertile.

The Physical Environment of Lakes

Lakes cover about 2% of the earth's surface and contain only about 0.01% of the world's water. However, lakes provide habitats for many types of plants, invertebrates, fish, amphibians, birds, and mammals and are an important focus of biogeographic research. Lake basins can be formed by a number of natural processes, including glacial activity, volcanic activity, tectonism, landslides, changes in river channels, and changes in ocean shorelines. Differences in light, temperature, acidity, and the chemical composition of water produce different physical environments in different lakes and within the same lake. These differences can have a profound impact on the geographic distributions of lake-dwelling organisms.

Aquatic organisms are sometimes classified by their habitat. Plankton are organisms that exist by passively floating in the waters of a lake or ocean. These include many microscopic plants (phytoplankton) and small invertebrate animals (zooplankton). Organisms such as fish that can propel themselves through the water are referred to as *nekton*. Finally, benthic plants and animals are those that exist on the bottom of lakes and oceans. The water environment is called the pelagic zone, while the bottom is called the benthos.

Plants that live in lake waters require sunlight for photosynthesis. When sunlight from the PhAR portion of the electromagnetic spectrum strikes the surface of a lake, some of it is reflected directly back skyward and does not enter the water. When light strikes the water at a 90° angle, only a small percentage of the energy is reflected. However, at a 10° angle, about 40% of the light is reflected off the lake waters. Once the light enters the water, its energy is dispersed by absorption and scattering. The transmission of PhAR is optimal in clear distilled water. In such water, light in the blue wavelength is transmitted best. Longer wavelength light, from the red portion of the spectrum, transmits poorly. About 53% of the light is absorbed and emitted as heat in the first meter or so of the water. However, lakes do not contain pure distilled water. Instead, their waters contain suspended and solid organic and inorganic substances and organisms such as plankton. The more matter and organisms in the lake water, the less light that can be transmitted through the water. In small, productive lakes with high amounts of suspended matter and organisms, the water is murky, and little light is transmitted below the first meter of water. In large, low-productivity lakes, such as Lake Tahoe in California or Crater Lake in Oregon, significant amounts of light can be transmitted tens of meters through the clear water. However, at greater depths, most of the light that is transmitted is in the blue portion of the spectrum. The portion of the lake water column that receives sufficient sunlight for photosynthesis is called the euphotic zone. Depths beneath this that receive enough light for fish and other organisms to see, but not enough for sustained photosynthesis, are referred to as the disphotic zone. Depths that are totally dark are called the aphotic zone. Photosynthetic plants can only live in the photic zone. Sometimes limnologists classify parts of lakes according to their ability to support photosynthetic plants. The shallow margins of the lake where plants can root and receive adequate light for photosynthesis is called the littoral zone, while the deep dark portion where rooted plants cannot exist is called the profundal zone (Fig. 2.10).

The impact of solar radiation coupled with warm air temperatures in the summer often produces a thermal stratification where temperatures are higher in the upper waters. Mixing of the surface waters by winds and currents generates isothermal conditions in the upper few meters of the lake waters. Below this, there is a gradient of decreasing temperature called a thermocline. At depth, the lake waters are cool and isothermal (Fig. 2.10). The difference in temperature between the surface and bottom waters can be as much as 20° C in relatively deep lakes. If you have dived and swum in small northern lakes, you may have experienced this thermal gradient as a feeling of increasing chill below a certain depth. These three thermal zones are called the epilimnion, metalimnion, and hypolimniom, respectively. The metalimnion produces a thermal barrier, and there is little mixing of

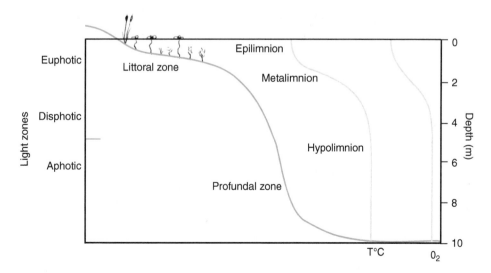

FIGURE 2.10 Life zones, light, temperature, and oxygen distributions in a small lake during summer.

waters from the epilimnion with the hypolimnion. Among other things, this phenomenon affects the distribution of oxygen. There are relatively high amounts of dissolved oxygen in the epilimnion and low amounts in the hypolimnion (Fig. 2.10). In some cases, the bottom waters are anoxic and cannot support organisms that require oxygen for respiration. This stratification of temperature and oxygen breaks down seasonally in lakes located in cold regions and allows oxygenation of bottom waters. Some lakes that are deep relative to their surface area never experience oxygenation of bottom waters.

The degree of thermal stratification in lakes can be related to the relationship between lake area and depth and climatic conditions—particularly air temperatures. This provides a means of lake classification linked to climate, as follows:

Amictic Lakes never lose their cover of ice and do not stratify. They are found only in the Antarctic, a few places in the Arctic, and on very high mountains.

Cold Monomictic Lakes never exceed 4° C water temperature and do not usually stratify. These lakes are found in the polar regions and on high mountains.

Dimictic Lakes are the stratified type described in the preceding discussion. They stratify in summer and mix in spring and fall.

Warm Monomictic Lakes never have water temperatures below 4° C. They stratify directly in summer. They are found in warm temperate areas and subtropical mountains.

Oligomictic Lakes have water much warmer than 4° C and irregular circulation events. There may be little temperature difference between the epilimnion and hypolimnion. These lakes are found in tropical regions.

Two categories of lakes can be found in both warm and cold climates:

Polymictic Lakes are shallow and circulate continuously.

Meromictic Lakes are deep lakes in which the bottom waters remain unoxygenated, even during lake overturn.

Lakes can also be classified in terms of chemistry, nutrient status, and productivity. Many lakes have fresh water; however, some saline lakes contain salt concentrations that are as high or higher than that of the ocean. Such saline lakes generally develop in arid regions and basins that lack outlets. Water is lost from these lakes mainly through evaporation, resulting in the accumulation of salts. Deep, cold lakes, with high amounts of oxygen and low amounts of phosphorus and other nutrients, typically have low primary productivity. These are referred to as oligotrophic lakes. In contrast, shallow, warm lakes with high amounts of nutrients have high primary productivity. These are called eutrophic lakes. However, eutrophic lakes generally have low amounts of oxygen due to high rates of respiration and aerobic decomposition. In some cases, the oxygen levels can decrease to the point where catastrophic die-offs occur for fish and other organisms.

The Physical Environment of Oceans

About 71% of the earth's surface is covered by ocean. This, coupled with the depth of the ocean, provides an area of habitat for marine life that is 300 times greater than the area provided by the terrestrial surface of the earth. The biogeographical importance of the world's oceans cannot be overstated.

As is the case with fresh water, light does not penetrate far down in ocean waters. In marine waters near the shore, sunlight may only reach the upper few meters of the water column. In clear midocean sites, light can penetrate many tens of meters, and 1% of the incident light may reach depths of 150 meters. Photosynthetic plants such as microscopic phytoplankton, or large plants such as kelp, can exist only in this upper photic zone. So, despite its great depth and volume, the marine environment must depend on a thin surface layer for its primary productivity.

As is the case with lakes, the ocean environment can be divided into pelagic and benthic zones (Fig. 2.11). In addition, in the sea the pelagic environment is subdivided into a nearshore neritic zone that extends to the edges of the continental shelves and a deepwater oceanic zone. The pelagic

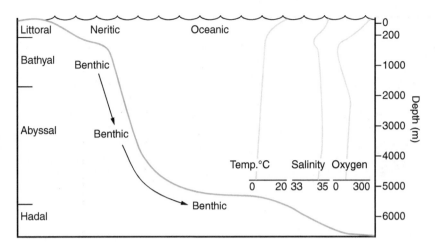

FIGURE 2.11   Marine life zones, salinity, temperature, and oxygen distributions in the ocean.

zone is also subdivided by depth into five regions (Fig. 2.11). All photosynthesis takes place in the uppermost epipelagic environment. Because the water is clearer in the midocean areas, the photic zone is deeper there than it is near the continents. Light sufficient for photosynthesis may penetrate for well over 10 meters in the midocean. The benthos of the ocean is also subdivided into a number of subzones extending from the intertidal supralittoral to the very deep hadal. The substrate of the benthos is very important in determining the organisms that can live there. Substrate varies by depth and proximity to land. Shallow areas near the shore may have substrates ranging from rocks, gravel, sand, and living coral. Areas in the deep sea away from land and the continental shelves often have substrates dominated by very fine silt and clay sediments.

A major difference between the marine environment and the freshwater environment of lakes and rivers is salinity. In general, the salinity of the ocean is about 35 grams of inorganic salts per 1 kilogram of sea water, the major salt being sodium chloride. However, the concentration of salts varies with both latitude and depth. In general, the salinity of surface waters is greatest in the regions between 20° and 40° north and south latitude where evaporation rates on the ocean are high relative to precipitation (Fig. 2.12). This corresponds with the dry climatic zone of terrestrial deserts discussed previously. In addition, in the low and midlatitudes, salinity decreases with depth between the surface and about 1000 meters. Below 1000 meters of depth, the salinity remains fairly constant.

The temperature of the ocean is an important control on planktonic and shallow nektonic organisms. As we might suspect, sea surface temperatures (SSTs) are highest in the low-latitude regions and decrease poleward (Fig. 2.12). Water temperature also varies with depth, and where the surface waters are relatively warm, a thermocline is usually found between the surface and 1000 meters of depth. Deep waters are not prone to seasonal changes in temperature. Waters below 2000 to 3000 meters never rise above about 4° C, and in the deepest portions of the ocean water temperatures can be below 3° C. In deep ocean waters, there is little temperature difference between the equator and the poles.

Oxygen in the oceans is highest in the surface waters and declines to minimum values at 500 to 1000 meters of depth. In some cases, oxygen concentrations may drop close to zero. The oxygen minimum zone is attributed to high rates of biological activity such as respiration and decomposition coupled with a lack of oxygen replenishment by photosynthesis because of the absence of light at this depth. Below the oxygen minimum zone, oxygen then increases with depth (Fig. 2.11). However, in some cases, such as off the coast of Santa Barbara, California, the sea bottom may be permanently anoxic. In other cases, such as the Gulf of Mexico, seasonally anoxic conditions may develop on the ocean floor.

The oceans display complex surface current systems that are largely driven by winds and the impact of the Coriolis effect (Fig. 2.12). Such systems work to bring warm waters poleward and moderate climates in areas such as northern Europe and western Alaska. These currents are also an important means for dispersing plankton and aiding in the migration and dispersal of nektonic

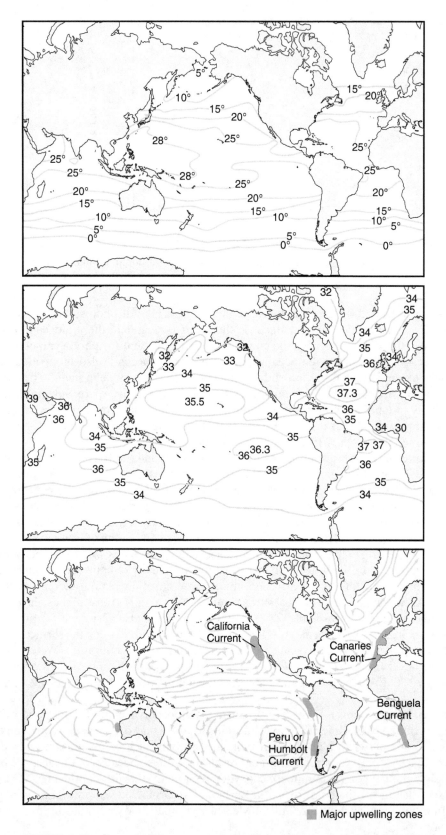

FIGURE 2.12   Ocean surface temperatures, salinity, currents, and major upwelling zones (information from a number of sources including Lalli and Parsons, 1997; Ross, 1992; Thurman, 1990).

organisms. In addition to surface currents, subsurface currents produce upwelling zones where cool nutrient-rich waters rise from depth to provide zones of high oceanic productivity (Fig. 2.12). Fisheries in areas such as coastal California owe their high catches to such upwelling.

A final important physical factor controlling the distribution of life in the oceans is pressure. Because the oceans are so deep, there are great differences between the ambient pressure that is exerted on organisms at the surface and pressure in the hadal zone. Pressure is determined by the weight of the overlying water. Ocean pressure is measured in atmospheres (atm units). One atmosphere equals the average pressure of the atmosphere at sea level. The pressure of the ocean increases by 1 atmosphere for every 10 meters of depth. So, in the deepest parts of the ocean, at 11,000 meters, the weight of water is equal to 1100 times the atmospheric pressure we experience at sea level. These pressures present special challenges for organisms that contain gas-filled sacks such as the swim bladders of fish or the lungs of marine mammals. In addition, pressure makes the exploration or collection of the biota of these depths extremely difficult. This is why the depths of the ocean represent one of the great unexplored frontiers of biogeography.

KEY WORDS AND TERMS

Angiosperms
Assemblage
Atmosphere
Autecology
Autotrophs
Biomass
Biome
Biosphere
C_3 pathway
C_4 pathway
Chemosynthesis
Class
Community
Conifer
Crassulacean acid metabolism (CAM)
Division
Domain
Ecosystem
Family

Food web
Genus, genera
Greenhouse effect
Guild
Gymnosperm
Heterotrophs
Hierarchy
Human appropriation of net primary production (HANPP)
Hydrosphere
Individual
Kingdom
Liliopsida (monocotyledon)
Lithosphere
Magnoliopsida (dicotyledon)
Magnoliophyta
Metapopulations
Order
Photosynthesis

Phyla
Pinophyta
Plantae
Population
Primary productivity
Pyramid
Respiration
Soil
Species
Spermatophyta
Stomata
Synecology
Systematics
Taxon, Taxa
Taxonomy
Tracheobionta
Trophic level
Troposphere

REFERENCES AND FURTHER READING

Aguado, E., and Burt, J. E. (2015). *Understanding Weather and Climate*, 7th edition. Pearson, London.

Begon, M., Mortimer, M., and Thompson, D. J. (2009). *Population Ecology: A Unified Study of Animals and Plants*. Wiley, Hoboken, NJ.

Bjorkman, O., and Berry, J. (1973). High efficiency photosynthesis. *Scientific American* **229**, 80–93.

Bowman, W. D., Hacker, S. D., and Cain, M. L. (2017). *Ecology*. Sinauer Associates, Sunderland, MA.

Brady, N. C. (2010). *Elements of Nature and Properties of Soils*, 3rd edition. Prentice Hall, Upper Saddle River, NJ.

Bridges, E. M. (1970). *World Soils*. Cambridge University Press, Cambridge.

Carpenter, S. R. (1988). *Complex Interactions in Lake Communities*. Springer-Verlag, Berlin.

Cavalier-Smith, T. (2004). Only six kingdoms of life. *Proceedings of the Royal Society B: Biological Sciences* **271**, 1251–1262.

Cole, G. A., and Weihe, P. E. (2016). *Textbook of Limnology*, 5th edition. Waveland Press, Long Grove, IL.

Committee of the Canadian Soil Survey (1978). *The Canadian System of Soil Classification*. Department of Agriculture, Ottawa.

Cracraft, J. (1983). Species concepts and speciation analysis. In *Current Ornithology*, vol. 1 (ed. R. F. Johnston), pp. 159–187. Plenum Press, New York.

De Angelis, D. L. (1992). *Dynamics of Nutrient Cycling and Food Webs*. Chapman and Hall, New York.

DeLeon-Rodriguez, N., Lathem, T. L., Rodriguez-R, L. M., Barazesh, J. M., Anderson, B. E., Beyersdorf, A. J., Ziemba, L. D., Bergin, M., Nenes, A., and Konstantinidis, K. T. (2013).

Microbiome of the upper troposphere: species composition and prevalence, effects of tropical storms, and atmospheric implications. *Proceedings of the National Academy of Sciences* **110**, 2575–2580.

Elton, C. (1927). *Animal Ecology*. Sidgwick and Jackson, London.

FAO-UNESCO (1988). *Soil Map of the World: Revised Legend (with Corrections and Updates)*. World Soil Resources Report 60. FAO, Rome.

Futuyma, D. J. (1979). *Evolutionary Biology*. Sinauer Associates, Sunderland, MA.

Gage, J. G., and Tyler, P. A. (1991). *Deep-Sea Biology: A Natural History of Organisms at the Deep-Sea Floor*. Cambridge University Press, Cambridge.

Garvey, J. E., and Whiles, M. (2016). *Trophic Ecology*. CRC Press, Boca Raton, FL.

German, C. R., Ramirez-Llodra, E., Baker, M. C., Tyler, P. A., and ChEss Scientific Steering Committee (2011). Deep-water chemosynthetic ecosystem research during the census of marine life decade and beyond: a proposed deep-ocean road map. *PLoS One* **6**, e23259.

Hanski, I. (1999). *Metapopulation Ecology*. Oxford University Press, Oxford.

Hutchinson, G. E. (1967). *A Treatise on Limnology*, vol. 2: *Introduction to Lake Biology and the Limnoplankton*. Wiley, New York.

Hutchinson, G. E. (1975). *A Treatise on Limnology*. Wiley, New York.

Ingmanson, D. E., and Wallace, W. J. (1973). *Oceanology: An Introduction*. Wadsworth, Belmont, CA.

Kinne, O. (1971). *Marine Ecology: A Comprehensive, Integrated Treatise on Life in Oceans and Coastal Waters*. Wiley Interscience, New York.

Krebs, C. J. (1994). *Ecology*, 4th edition. Harper-Collins College Publishers, New York.

Lalli, C., and Parsons, T. R. (1997). *Biological Oceanography*, 2nd edition. Pergamon Press, New York.

Larsen, C. P. S., and MacDonald, G. M. (1993). Lake morphometry, sediment mixing and the selection of sites for fine resolution palaeoecological studies. *Quaternary Science Review* **12**, 781–792.

Lindemann, R. L. (1942). The trophic-dynamic aspect of ecology. *Ecology* **23**, 399–418.

Lipscomb, D. (1996). A survey of microbial diversity. *Annals of the Missouri Botanical Garden* **83**, 551–561.

Maitland, P. S. (1990). *Biology of Fresh Waters*, 2nd edition. Chapman and Hall, New York.

Mayr, E. (1942). *Systematics and the Origin of Species*. Columbia University Press, New York. (Reprint edition: Dover Publications, New York, 1964.)

McKenney, C. (2014). *Introductory Plant Science*, 14th edition. Kendall/Hunt, Dubuque, IA.

Mittelbach, G. G. (2012). *Community Ecology*. Sinauer Associates, Sunderland, MA.

Molles, M. C., Cahill, J. F., and Laursen, A. (2015). *Ecology: Concepts and Applications*. McGraw-Hill, New York.

Nakagawa, S., and Takai, K. (2008). Deep-sea vent chemoautotrophs: diversity, biochemistry and ecological significance. *FEMS Microbiology Ecology* **65**, 1–14.

Nybakken, J. W. (1997). *Marine Biology: An Ecological Approach*, 4th edition. Addison Wesley Longman, Menlo Park, CA.

Odum, E. P., and Kuenzler, E. J. (1963). Experimental isolation of food chains in an old-field ecosystem with use of phosphorus-32. In *Radioecology* (ed. V. Schultzard and A. W. Klement), pp. 113–120. Van Nostrand, Reinhold, New York.

Pessarakli, M. (2005). *Handbook of Photosynthesis*. CRC Press, Boca Raton, FL.

Pimm, S. L. (1982). *Food Webs*. Chapman and Hall, London.

Polis, G. A., and Winemiller, K. O. (2013). *Food Webs: Integration of Patterns & Dynamics*. Springer Science & Business Media, Berlin.

Raven, P. H., Evert, F., and Eichhorn, S. E. (1999). *Biology of Plants*, 6th edition. W. H. Freeman, New York.

Raven, P. H., and Johnson, G. B. (1999). *Biology*, 5th edition. McGraw-Hill, Boston.

Ricklefs, R. E. (1990). *Ecology*, 3rd edition. W. H. Freeman, New York.

Roger, A. J., and Simpson, A. G. B. (2009). Evolution: revisiting the root of the eukaryote tree. *Current Biology* **19**, R165–167.

Ross, D. A. (1982). *Introduction to Oceanography*, 3rd edition. Prentice-Hall, Englewood Cliffs, NJ.

Ruggiero, M. A., Gordon, D. P., Orrell, T. M., Bailly, N., Bourgoin, T., Brusca, R. C., Cavalier-Smith, T., Guiry, M. D., and Kirk, P. M. (2015). A higher level classification of all living organisms. *PLoS One* **10**, e0119248.

Service, R. S. (1997). Microbiologists explore life's rich, hidden kingdoms. *Science* **275**, 1740–1742.

Soil Classification Working Group (1998). *The Canadian System of Soil Classification*, 3rd edition. Agriculture and Agri-Food Canada Publication 1646. NRC Research Press, Ottawa.

Staff of the Soil Survey (1999). *Soil Taxonomy: A Basic System of Soil Classification for Making and Interpreting Soil Surveys*, 2nd edition. U.S. Department of Agriculture Handbook 436

Strahler, A., and Strahler, A. S. (1997). *Physical Geography: Science and Systems of the Human Environment*. Wiley, New York.

Thurman, H. V. (1990). *Essentials of Oceanography*, 3rd edition. Merrill Publishing, Columbus, OH.

Trujillo, A. P., and Thurman, H. V. (2016). *Essentials of Oceanography*, 6th edition. Prentice Hall, Upper Saddle River, NJ.

Van Dover, C. L., German, C., Speer K. G., Parson, L. M., and Vrijenhoek, R. C. (2002). Evolution and biogeography of deepsea vent and seep invertebrates. *Science* **295**, 1253–1257.

Wang, H., Morrison, W., Singh, A., and Weiss, H. H. (2009). Modeling inverted biomass pyramids and refuges in ecosystems. *Ecological Modelling* **220**, 1376–1382.

Wetzel, R. G. (1983). *Limnology*, 2nd edition. Saunders College Publishing, New York.

Whittaker, R. H. (1975). *Communities and Ecosystems*, 2nd edition. Macmillan, New York.

CHAPTER 3

THE PHYSICAL ENVIRONMENT AND THE DISTRIBUTION OF LIFE

There are certain plants and animals that we automatically associate with specific environments. For example, we link elephants and palm trees with hot tropical regions. In contrast, we picture polar bears in cold Arctic habitats. These associations are typically based on the real geographic distributions of these species. We don't find palms and elephants living in the Arctic or polar bears in a tropical savanna. Why not? For one reason, plant and animal species require certain physical conditions related to temperature, moisture, chemistry, and light in order to survive. Exposure to temperatures that are too high or too low, or to conditions with too much or too little moisture, can prove lethal. As we have seen in Chapter 2, the physical environment of the earth varies greatly according to location, and there are large-scale geographic patterns in this variation that can be explained by physics and geography. The temperature regime experienced by organisms dwelling in the equatorial tropics is very different from the conditions experienced by organisms in the high-latitude Arctic. It is logical to assume that the differences in the physical environment must have a role in determining where certain species can be found and where they do not exist. An important goal for biogeographers is to understand how these geographic differences in the physical environment influence the geographic distributions of species. In this chapter we will investigate the role of the physical environment in controlling the geographic distributions of species. We should, however, keep in mind that physical conditions are just one factor in determining where a species is found and where it is absent. Both biological factors and history play a role in controlling geographic distributions.

One way of embarking on our investigation of the physical environment and the geography of life is to simply compare the mapped distributions of species with mapped distributions of physical factors such as average July temperatures. We would discover that the white spruce tree *(Picea glauca)* is typically not found in regions where average July temperatures are cooler than 10° C (Fig. 3.1). On this basis, we could speculate that July temperature controls the distribution of spruce trees. However, unless we have a good understanding of the physiological impact of temperature on plants and animals, we cannot be certain there really is a cause-and-effect relationship between July temperature and the distribution of a particular species such as white spruce. The similarity in the geographic distribution of the physical environmental variable and the species could simply be a coincidence. In addition, in many cases, a geographic correlation may exist between the range limits of the species and more than one environmental variable. This is particularly true for variables related to climate. Quite often, there is a high degree of cross-correlation between factors such as mean July temperature, maximum daily summer temperature, mean annual temperature, potential evaporation rates, and so on. In one example, it was shown that the eastern limits of holly *(Ilex aquifolium)* in Europe corresponded with the geographic distribution of at least 23 different climatic variables. Many of these aspects of climate may not, in fact, play any role in controlling the geographic distribution of the species. As we will see later, the geographic distributions of some species may also be influenced by biological factors such as the needs for special prey or competition for resources with other species. Finally, the present geographic distributions of some species may actually be in a state of change as the species disperse into new territory. Relations between the geographic distribution of such species and climatic conditions are transitory. Consequently, biogeographers cannot be content with merely identifying correlations between the spatial distributions of species and the spatial distributions of

Biogeography: Introduction to Space, Time, and Life, Second Edition. Glen M. MacDonald.
© 2025 John Wiley & Sons, Inc. Published 2025 by John Wiley & Sons, Inc.
Companion website: www.wiley.com/go/macdonald/biogeography2e

FIGURE 3.1   A stunted white spruce *(Picea glauca)* at the northern treeline in the Northwest Territories of Canada. Although such plants are often less than 1 meter tall, they can be over 100 years old. A combination of low summer temperatures, short summer growing season, and damage by blowing snow in winter determines the northern range limits of spruce at the northern treeline. Summer temperature and growing season length appear to be the most important factors, and the northern limit of spruce in Canada occurs where average July temperatures lie between 10° C and 12.5° C.

environmental variables. Biogeographers must base their analysis of these geographic distributions on an understanding of the impact of the physical environment on the physiological functioning of plants and animals.

In this chapter we will examine how physical conditions such as light, temperature, moisture, and chemistry impact the physiological functioning of plants and animals. We will also look at some of the remarkable adaptations that allow organisms to cope with stresses produced by the physical environment. Some examples of close links between the geographic distributions of species and the physical environment will be presented. We will then examine some examples of situations where two or more physical controls act together to determine the geographic distributions of plants and animals. Finally, we will consider how biogeographers and ecologists have developed conceptual frameworks for organizing, describing, and analyzing the relationships between the physical environment and the distribution and abundance of organisms.

LIGHT

Light in the PhAR portion of the electromagnetic spectrum provides the energy for photosynthesis. The amount of available light is a key control on the distribution of plants. Although the amount of sunlight received at the surface of the earth varies with latitude, most research on the impact of differences in light intensity on plants has focused on local conditions such as the distribution of light under forest canopies. Plants may be classified into two categories with respect to their light

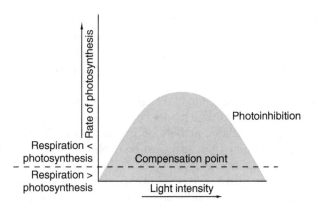

FIGURE 3.2   A generalized Photosynthesis Light Response Curve representing the relationship between light intensity and rate of photosynthesis.

requirements. **Heliophytes** grow best in full sunlight, and **sciophytes** grow best in shade. Plants that can only grow on sunny sites are referred to as obligate heliophytes and are also said to be shade intolerant. Heliophytes that can survive on shaded sites are referred to as facultative heliophytes. These plants are shade tolerant but do best in full sunlight. Sciophytes can similarly be divided into obligative and facultative categories. Most house plants are sciophytes from the tropics and subtropics. This allows them to live under the low light conditions found indoors.

The response of photosynthesis to light intensity can be graphed as a **Photosynthesis Light Response Curve** with a roughly hyperbolic shape (Fig. 3.2). Photosynthesis increases as light intensity increases until it reaches a maximum. After that, increasing light can cause declines in the rate of net photosynthetic gains. At low illuminations, there is an almost linear increase in the rate of photosynthesis in response to increasing fluxes of radiant energy. The point at which light intensity and photosynthesis are enough to offset the loss of energy through respiration is called the compensation point. The influence of increased radiation on photosynthesis declines at higher fluxes of radiant energy as the leaf reaches its saturation point for light relative to its capacity to absorb and use light for photosynthesis. For many plants, maximum net photosynthesis occurs when light intensities reach one-third to two-thirds the strength of full sunlight. However, there are large differences in the light intensities at which plants reach saturation. Herbs found on forest floors and shade-dwelling mosses often experience saturation at light intensities as low as 5% full sunlight or less. Some shade tolerant trees such as red maple *(Acer rubrum)* and American beech *(Fagus grandifolia)* may also reach light saturation at levels as low as 5% to 10% of full sunlight. The low light saturation level of sciophytes allows them to survive on very shady sites where other plants would not be able to conduct enough photosynthesis to balance energy lost through respiration. In fact, when most true sciophytes are exposed to full sunlight, they cannot conduct photosynthesis at nearly the same rate as heliophytes. High light intensities can actually cause decreased rates of photosynthesis in shade tolerant plants. This phenomenon, called photoinhibition, can be due to various factors including damage to the biochemical mechanisms through which chlorophyll absorbs light for photosynthesis, shrinking of chloroplasts, loss of chlorophyll, and loss of CO_2 due to photooxidation. Photooxidation is a particular challenge for C_3 plants and produces a condition called photorespiration, where CO_2 may be released without the production of chemical energy for the plant. Photorespiration is also referred to as the C_2 pathway. The process is related to high light intensities and high temperatures and uses O_2 while releasing CO_2 as cellular respiration. This produces decreased photosynthetic efficiencies. The rate of photorespiration is positively correlated with light intensity and temperature.

Plants have various adaptations to limit photoinhibition. Some C_3 plants, such as members of the wood sorrel genus *(Oxalis)*, have leaves that fold or turn, or produce red leaves through concentrations of anthocyanins, all to decrease photosynthetically active light exposure. In contrast to C_3 plants, C_4 plants have developed attributes to limit photorespiration; they thus have higher rates of photosynthesis at high light intensities and temperatures. However, C_4 plants are less efficient at photosynthesis under low light intensities and temperatures than C_3 plants. Indeed, C_4 plants are generally obligate heliophytes and are restricted geographically to areas of high insolation and heat. It is interesting to note that C_4 plants make up only about 3% of all land-plat species, but C_4 grasses dominate warm grasslands in the tropics and subtropics where light intensities and temperatures are high.

Aquatic vascular plants and algae are adapted to the low light intensities of their watery environment. Some only need low levels of light (approximately 1%–3% of full sunlight) to produce rates of photosynthesis that equal energy losses due to respiration. Many aquatic plants and planktonic algae reach the saturation point at 5% of full sunlight. Despite this relatively low saturation point, there is not enough light to support photosynthetic plants beyond the relatively thin upper zone of the oceans. In general, maximum photosynthetic activity occurs in the top 20 to 100 meters of the ocean. The absence of light at depth is one of the strongest controls on the vertical distribution of photosynthetic organisms in the ocean and lakes.

In addition to physiological adaptations, certain morphological (physical) adaptations take place in plants that make them uniquely suited to living in either high or low light intensities. For example, the leaves of plants adapted to live on sites with high light intensity tend to be small and thick and possess a thick cuticle (outer covering). In addition, the leaves of some high light intensity plants are slightly curled and possess a reflective surface layer. These plants also tend to have a high number of stomata, which often are found on both the upper and lower surface of the leaf. The stomata allow for high rates of CO_2 intake from the atmosphere while cooling the leaf by transpiring water vapor. Leaves growing under conditions of very high light conditions, such as the manzanita bushes *(Arctostaphylos patula)* of the Southwest are often oriented so that they do not receive high rates of insolation. In contrast, the leaves of plants living in low light intensity tend to be large and soft and have high amounts of chlorophyll. The leaves of low-insolation plants also live longer than those on plants adapted to high light intensity. Leaf adaptation to light is not restricted to land plants. For example, species of the Caribbean seaweed genus *Sargassum* that live in deeper waters with low light intensities have wider leaflike blades to capture more light than species that grow in shallower waters where light intensity is high. Some plants have heliotropic leaves and flowers that change their angle during the day to either increase or reduce insolation. The shifting flower heads of sunflowers, which seem to turn and follow the sun during the course of the day, are one example. The changing position of heliotropic leaves or flowers are caused by light creating differential rates of cell growth on stems.

In addition to morphology, the life history and reproductive behavior of plants can sometimes be related to light conditions. Shade tolerant plants are almost always **perennials** that live for more than one year. The rate of photosynthesis in shady spots is not enough to support the high-energy requirements of **annual** plants which must germinate from seed, grow, flower, and produce seed within a single year. The local biogeographic distribution of such plants can be tightly controlled by light availability. Grasses, for example, are often not able to grow in areas where trees produce shading of the ground surface. Plants often require high light intensities in order to produce flowers. In many temperate and northern forest stands, only the trees in the canopy are able to flower profusely, for the low light intensities beneath the canopy restrict flowering of understory trees. The production of cones by common North American coniferous trees such as Douglas fir *(Pseudotsuga menziesii)* and white spruce *(Picea glauca)* has been shown to be promoted by high light intensity. In addition, the seeds of many plant species will not germinate unless exposed to relatively high light intensities. For example, only about 3% to 4% of Virginia pine *(Pinus virginiana)* seeds will germinate in dark conditions, while 92% to 94% of the seeds will germinate when exposed to light.

The duration of daylight often serves as a stimulus for different seasonal changes in plant growth, such as flowering in spring or shedding of leaves in the fall. This relationship between **phenology** (the timing of changes in the growth of the plant) and light is called photoperiodism. A familiar example of this phenomenon in eastern North America is the late persistence in the fall of leaves on silver maple trees *(Acer sacharinum)* planted near streetlights compared to the earlier shedding of leaves from trees planted away from artificial light sources. Seasonal photoperiodism becomes more important at the mid- and high latitudes where there are pronounced changes in day length during the course of the year.

A classic example of the impact of light requirements on the distribution of a plant comes from the forests of southeastern North America. Loblolly pine *(Pinus taeda)* is found growing from eastern Texas to southern New Jersey. It is a common tree and an important timber species. Anyone familiar with the woods of the Southeast will know that the pine seedlings are numerous in open areas and at the edge of stands and occur less frequently within closed forests. In contrast, the seedlings of some deciduous trees such as eastern red oak *(Quercus rubra)* can be found within the forest. Loblolly pine has lower rates of photosynthesis than oak at all light intensities, but this difference is most pronounced under the low to moderate intensities typical of forested sites. Low light intensities under the forest canopy restrict reproduction by loblolly pine to sunny sites.

TEMPERATURE

Temperature exerts a strong physiological impact on both plants and animals. It has been said that "with regard to life on earth, temperature is—next to light—the most potent environmental component" (Kinne, 1970). Accordingly, a great deal of research has been conducted on the impact of temperature on the geographical distribution of species. Since the physiologies of plants and animals are so vastly different, it makes sense to divide our discussion of temperature into separate sections for plants and animals.

Plants

For plants, temperature can limit geographic distribution by causing metabolic difficulties and physical damage. On the whole, plants are **poikilotherms,** which means that they assume the temperature of their environment. It should not be assumed, however, that the temperature of a plant is the same as the air temperature. Because of the transpiration of water, leaves can be as much as 15° C cooler than air temperature, while portions of plants that intercept high amounts of sunlight may be 30° C warmer than the air. There are three amazing plants that have been shown to regulate their temperature biochemically. They are philodendron *(Philodendron selloum)*, skunk cabbage *(Symplocarpus foetidus)*, and sacred lotus *(Nelumbo nucifera)*. These plants can raise the tissue temperatures of their flowers up to 34° C above the air temperature for periods ranging from 18 hours to 2 weeks. The biochemical generation of this heat is not yet completely understood. It is thought that the heating of the floral parts may aid in the process of attracting and keeping pollen-carrying insects in contact with fertile flowers.

Plants depend on photosynthesis for food production. In turn, the rate of photosynthesis is dependent on temperature. If the temperature is too high or low for effective photosynthesis, the plant essentially starves to death. In addition to starving the plant, low rates of photosynthesis restrict the rate of growth and reproductive functions. There is wide variability among different plant species in the optimum temperature for photosynthesis. No higher plants are known to conduct photosynthesis at freezing temperatures. Some Arctic and alpine plants can conduct photosynthesis at temperatures as low as 5° C above freezing, while tropical plants often require temperatures greater than 15° C to 20° C for efficient photosynthesis. In general, tropical plants reach peak rates of photosynthesis at higher temperatures than plants from cooler regions (Fig. 3.3).

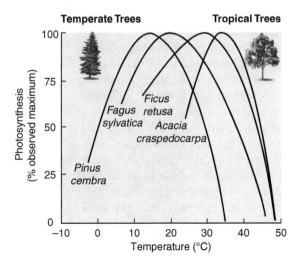

FIGURE 3.3 The relationship between temperature and rate of photosynthesis (Larcher, 1969; Kramer and Kozlowski, 1979).

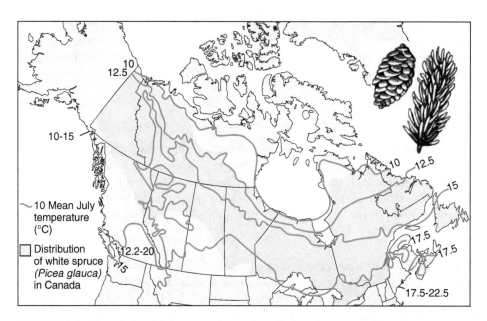

FIGURE 3.4   The relationship between the northern limits of spruce and July temperatures in Canada.

Low temperatures cause slow metabolic activity due to the kinetics of enzymes such as rubisco and depress the rate of photosynthesis. Low temperatures can also decrease rates of development of cells and tissue and thus decrease growth even when photosynthesis might support it. The circumarctic treeline of the northern hemisphere is a classic example of how low temperatures during the growing season appear to be responsible for determining the geographic range limits of plant species, particularly coniferous trees. The northern limit of trees such as white and black spruce *(Picea glauca* and *Picea mariana)* in North America, and Scots pine *(Pinus sylvatica),* Norway spruce *(Picea abies),* Siberian spruce *(Picea obovata),* Siberian larch *(Larix sibirica),* and Dahurian larch *(Larix dahurica)* in northern Eurasia corresponds to the 10° C to 12.5° C July temperature isotherm (Fig. 3.4). It has long been proposed that north of this region the summer temperatures are too low and the growing season too short to support the growth of trees. It has been demonstrated that the rate of photosynthesis for trees such as black spruce declines rapidly at temperatures less than 10° C. Growth of tissue such as radial wood formation is also observed to decrease under such low temperatures. In addition, the low temperatures during the summer appear to be insufficient for the development of fertile pollen and seeds. Finally, the low rates of photosynthesis and growth may make the trees more susceptible to damage due to desiccation and mechanical damage by blowing snow and ice during the winter. A similar correspondence has been noted between summer temperatures and high-elevation alpine treelines.

Cold temperatures can physically damage plants through chilling stress and freezing stress. Plants do not have to actually freeze to suffer severe damage due to low temperatures. Chilling stress takes place at temperatures that are above 0° C. Some tropical plants have been shown to experience chilling stress at temperatures as high as 20° C to 15° C. Chilling can cause reduced plant growth, leaf discoloration, curling, crinkling and lobing, surface lesions on fruit, splitting and dieback of stems, and death. In some tropical plants, temperatures that are cool but not cold enough to damage leaves and other tissue may disrupt reproductive functions, such as the formation of pollen grains. Young plants are often more susceptible to chilling damage. In addition, plants are more prone to chilling stress when they are not dormant and during the daylight hours when they are most physiologically active. The primary causes of chilling injury are a matter of debate. Cool temperatures can cause damage to plant membranes, decreased rates of nucleic acid and protein synthesis, and decreased photosynthesis.

It is not just tropical land plants that are susceptible to damage or death due to chilling stress. Tropical seaweeds are extremely sensitive to water temperature, and the geographic distributions of certain species can be linked to this. For example, the tropical seaweeds *Botryocladia spinulifera* and *Haloplegma duerreyi* have northern limits that coincide with the 20° C sea surface isotherm. It has been shown that both these species suffer severe damage when exposed to water temperatures of 18° C and die after 2 weeks in water that has a temperature of 15° C. A clear linkage has been found between their inability to survive cool waters and the relationship between their distribution and the 20° C isotherm.

Freezing stress occurs at temperatures below 0° C. Tropical plants that are susceptible to chill stress usually die when temperatures drop below freezing. In contrast, some Arctic plants can withstand temperatures as low as −70° C without damage. Aside from causing death, freezing can result in depressed growth rates, discoloration of leaves, fissuring of woody stems, malformed leaves and flowers, dieback of roots, stems, leaves, and flowers, and delays in sprouting and flowering. Damage to plants from freezing is caused primarily by the intracellular freezing of water. When ice crystals form inside cells, they cause ruptures to the cell wall and this is almost always lethal to the cell. The formation of extracellular ice can also occur in plant cells. As this process draws water out of cells, it can cause cellular damage due to dehydration and shrinkage of the cell walls. However, extracellular ice formation is not always harmful to the cells or plant tissue of many temperate and Arctic plants as long as the freezing occurs in the dormant period.

There are many interesting adaptations that allow plants to survive in cool environments. One obvious strategy is dormancy during the cold season. Most of the common tree genera in the forests of northeastern North America, western Europe, and eastern Asia, such as the maples, oaks, beech, birches, and ashes, are **deciduous** trees that lose their frost-sensitive leaves during the cold winter season. In most of these genera, the leaves suffer damage at temperatures of freezing or just below. The new leaves arise in the spring from winter buds that can remain viable at colder temperatures.

Most of the needle-leaved conifers of the northern and alpine forests, such as pines, spruces, and firs, do not lose their leaves during the winter. How do such evergreen plants escape intracellular freezing and tissue destruction when temperatures may drop to −40° C or colder? In these plants, the onset of cool temperatures causes physiological changes that allow plant tissue to either avoid freezing or restrict freezing to extracellular areas. For plants to avoid freezing, they must chemically alter their liquids into a form that is analogous to antifreeze in automobiles. The liquids in these plants can be cooled far below 0° C and will not freeze. This process is called supercooling and is achieved by the metabolic synthesis of sugars and other molecules which, when in solution in the plant's tissue, lower the temperature for ice formation to far below 0° C. Supercooling seems to be the prevalent mechanism of frost resistance in herbs. For woody plants, supercooling is augmented by declines of cellular water content, greater cellular accommodation to deformation, and processes that allow water to accumulate and freeze in extracellular spaces. The transference of water from the cells to extracellular areas increases the solute content of the remaining cell water, making it more resistant to freezing. The cell walls can accommodate the deformations caused by water freezing on the exterior of the cell. For northern and alpine evergreens such as pines and spruces, both supercooling and extracellular ice formation play a part in allowing the plants to withstand extremely cold temperatures. One interesting facet of these physiological adaptations to freezing is that most of these plants will still be damaged by cold temperatures if they do not have a period of cooling in which to adjust to the onset of winter. This process of physiological preparation for the onset of winter cold is called frost hardening. The cell lipids (fatty acids or their derivatives) may also change with developing low temperature conditions and this plays a role in plant's cells ability to withstand cool temperatures and eventual freezing temperatures. Plants typically are much more susceptible to freezing damage in spring and early fall than during winter when they have been frost-hardened.

Some members of the cactus family (Cactaceae) appear to resist freezing during cool nights by radiating heat stored during the day in their thick, moist tissue. The greater the mass of the cactus, the more heat it can store and the less prone it will be to freezing damage during the night. How is it then that these cacti can survive cold temperatures when they are young and small? The giant saguaro cactus (*Carnegiea gigantea*) is perhaps the most well-known symbol of the southwestern desert (Fig. 3.5). In the popular lore of North America, the distinctive "stovepipe" shape of the multistemmed saguaro is a universally recognized icon that is used to represent deserts in movies, television, and comics. Yet this distinctive and widely recognized plant is actually only found in the Sonoran Desert of California, Arizona, and adjacent Mexico. The saguaro cactus is damaged or killed if exposed to prolonged freezing temperatures. Desert climates in the northern Sonoran Desert are typified by warm days but sometimes experience nighttime temperatures that are below freezing in

FIGURE 3.5   Saguaro cacti *(Carnegiea gigantea)* growing in the Organ Pipe Cactus National Monument of Arizona. Water stored in stems and CAM photosynthesis help these large succulents to survive the aridity of the Sonoran Desert. The saguaro cannot survive more than 12 to 24 hours of air temperatures that are below freezing, and this restricts their northern distribution to areas such as southern Arizona and adjacent Mexico.
Source: Gregory Ochocki / Science Source.

the winter. Young saguaro that survive are found sheltered beneath more frost-tolerant desert shrubs. The cover of these shrubs acts as a thermal blanket, capturing heat radiated from the ground and keeping the microclimate of the small saguaro warm at night. As the cactus grows, it eventually rises above the cover of the protective shrub. The radiation of heat from the stalk of the large mature cactus prevents freezing. This strategy works up to a point. The range of the saguaro is restricted to areas that do not experience more than about 12 to 24 continuous hours of air temperatures below 0° C. It appears that after 24 hours of freezing air temperatures, not enough heat reserve is left in the saguaro to keep the tissue from freezing. Not all members of the cactus family are limited to regions with relatively mild winters. A small prickly pear cactus *(Opuntia fragilis)* grows as far north as central Alberta in Canada and can withstand temperatures as low as −50° C.

The broad geographic ranges of many plants can in part be explained by winter temperatures. For instance, most tropical and subtropical plant species are sensitive to temperatures at or below freezing and are restricted to low-latitude regions where air temperatures remain warm throughout the year. One example of this comes from a very familiar plant—the palm. Species of the palm family

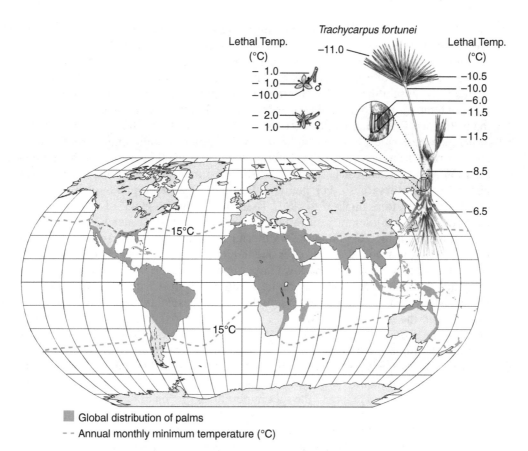

Lethal Temp.
(°C)

Trachycarpus fortunei

−11.0

Lethal Temp.
(°C)

− 1.0
− 1.0 ♂
−10.0

− 2.0
− 1.0 ♀

−10.5
−10.0
−6.0
−11.5

−11.5

−8.5

−6.5

■ Global distribution of palms
- - Annual monthly minimum temperature (°C)

FIGURE 3.6 The natural global distribution of palms and the susceptibility of different parts of the palm plant *Trachycarpus fortunei* to cold temperatures (Larcher and Winter, 1981; Sakai and Larcher, 1987). The relative resistance to freezing by the fronds and stems as well as sensitivity to freezing by the flowers are typical of palms.

(Arecaceae) are found throughout the world in the tropics and subtropics (Fig. 3.6). Despite this wide longitudinal distribution, they rarely grow north or south of the subtropics. The cabbage palmetto *(Sabal palmetto)* is common in the coastal regions of the southeastern United States from North Carolina to Florida. A careful analysis of the distribution of palms shows that they are restricted to areas where temperatures never drop below −10° C to −15° C. Palms have only one growing point, and if it is damaged, the palm fails to grow. All palm species that have been tested experience freezing damage at temperatures between −3° C and −13.5° C. This incapacity to survive freezing temperatures restricts their geographic range to warm regions. Some organs on the palms are more prone to freezing damage than others. Leaves and the inner stems of mature plants appear to be the most resistant, while reproductive organs are the least resistant (Fig. 3.6). For example, the fronds of some adult palms can survive temperatures of less than −10° C, while the flowers are killed by temperatures of −1° C. Thus, mature palms can be planted and survive in such unlikely places as the mild coastal areas of the British Isles or the Pacific Northwest of North America but are unable to reproduce in such regions.

Many plants in both the northern and southern hemispheres have geographic limits that are controlled by winter temperatures. Some well-known coniferous trees, including the coastal redwood of California *(Sequoia sempervirens)*, suffer frost damage at temperatures as high as −9.5° C. This intolerance to cold appears to keep the coastal redwood from growing north of southernmost Oregon or at higher elevations in the coastal ranges of California. Similarly, the buds of most midlatitude deciduous tree species such as the tulip tree *(Lirodendron tulipifera)* of North America, the gingko tree of China *(Gingko biloba)*, or the southern beech *(Nothofagus antarctica)* of Chile suffer damage at temperatures below −25° C to −35° C. There are exceptions. For example, the buds of the big-leafed oak *(Quercus macrocarpa)* of eastern North America can withstand winter temperatures of less than −60° C. Not surprisingly, this tree is distributed farther north in eastern North America than many other temperate deciduous trees such as the tulip tree.

Most plants also suffer from declining rates of photosynthesis once critically high temperatures are reached. The temperature at which plant species reach maximum photosynthetic rates often depends on the latitude at which the plant species is found. For some Arctic, alpine, and temperate plants, rates of photosynthesis decline with temperatures above 10° C. Tropical and desert plants, particularly those utilizing the C_4 photosynthetic pathway, may not attain peak rates of photosynthesis until temperatures reach 30° C. The decline in photosynthetic gains at high temperatures is due to many factors. Respiration rates increase more rapidly than photosynthesis at high temperatures. Some enzymes critical to photosynthesis are intolerant of high temperatures and break down. Cell structure may be altered, and cell membranes may become thin and leak at high temperatures. Finally, high temperatures can cause stomatal closure and inhibit the exchange of CO_2, water, and oxygen needed for photosynthesis. For many midlatitude and Arctic plant species, heat injury and death occur at temperatures of 30° C to 40° C. Tissue temperatures greater than 60° C to 70° C are lethal to even the most heat-tolerant plant, although cacti can withstand such heat for short periods of time.

It has been suggested that the southern range limits of many Arctic plants might be caused by the depletion of carbohydrate reserves due to high respiration rates generated by warm temperatures. In order for plants to survive the low temperatures and short duration of the Arctic growing season, they must be able to grow rapidly in the spring using stored reserves of carbohydrates. When subjected to long and warm growing seasons, however, these plants will deplete all reserves of carbohydrates over the summer and die. For example, the Arctic shoreline plant called Scots lovage (*Ligusticum scoticum*) reaches its southern limits on the Scottish and Irish coastlines where mean July temperatures reach about 14.4° C. Comparative physiological studies show that the respiration rate of Scots lovage is much higher than that of species of shoreline plants found to the south. This difference in respiration rates increases as temperature climbs. Scots lovage can lose 50% of its carbohydrate reserves in just 17 days if temperatures remain at 25° C or higher.

Adaptations to withstand heat include features that decrease leaf temperature such as the presence of reflective coatings and small hairs on leaf surfaces. Leaves may also be oriented vertically to reduce insolation or be heliotropic. Where water is plentiful, leaves may cool down through high rates of transpiration. When seasonal conditions are dry and hot, some plants lose their leaves during those times of the year. An interesting adaptation to high temperatures in dry regions is the development of **dimorphic leaves.** In hot and dry periods, plants with dimorphic leafing patterns are typified by small leaves, while in the cool season, they support larger leaves. A number of plants from the deserts of the southwestern United States and adjacent Mexico, such as honey mesquite (*Prosopis glandulosa),* have dimorphic leaves.

Animals

Like plants, the so-called cold-blooded animals such as fish, reptiles, amphibians, and insects are poikilotherms. Their body temperature generally reflects the ambient temperature of the environment. In contrast, birds and mammals are **homeotherms** that maintain a relatively steady body temperature through the metabolic generation of heat. Animals that are able to metabolically produce heat are also known as endotherms, while cold-blooded animals are referred to as ectotherms. Both homeotherm and poikilotherm animals can be damaged and killed by exposure to high and low temperatures. When they are in an active state, insects, fish, reptiles, amphibians, birds, and mammals must keep their body temperatures within a relatively narrow temperature range. The lowest body temperatures for active animals are found in fish from the polar regions. Species such as the Antarctic icefish (*Trematomus borchgrevinki)* can remain active at body temperatures of −1.9° C. In contrast, the desert pupfish (*Cyprinodon nevadensis),* found in hot springs near Death Valley, California, can withstand prolonged temperatures of up to 42° C. Ants of the species *Ocymyrmex barbiger* and spiders of the genus *Seothyra* from the Namib Desert of southern Africa have been shown to survive body temperatures of 52° C to 55° C for very short periods. However, these are exceptional species. The vast majority of animals cannot tolerate prolonged temperatures that are even close to freezing or above about 40° C. Temperatures that are too high or low for the organism can cause decreases in metabolic activity, physical movement, growth, and reproduction. Temperatures that are lethal and the range of temperatures that are optimum vary widely for different species. Species that can tolerate a wide range of temperature conditions are said to be **eurythermic.** Species with restricted temperature ranges are referred to as **stenothermic.**

In both homeotherms and poikilotherms, many metabolic functions are temperature dependent and increase exponentially with temperature. The increased rates of respiration and oxygen consumption generally result from increasing temperatures. However, at some point on the temperature scale the relationship between temperature and metabolic functioning becomes asymptotic. The physiological functioning of the organism no longer increases with temperature, and eventually further warming causes death. High body temperatures produce several different effects that can lead to death. Proteins may be broken down or caused to coagulate. Enzyme activity may be reduced and enzyme formation slowed. The rates of different temperature-controlled metabolic functions may be driven out of synchronization at high temperatures. Rates of oxygen intake may not equal respiration needs. Membranes may be distorted or destroyed. Finally, behavior such as feeding may be negatively altered at high temperatures. Salmon, for example, exhibit significantly decreased rates of feeding in warm waters.

For homeotherms, lethal core temperatures range from 37° C to 47° C. This is generally less than 10° C higher than the normal core temperatures of the animals. The core temperatures and lethal temperatures are higher for birds than for mammals. In contrast, the lethal temperatures for many cold-water fish are very low. For example, the lethal high temperature for Antarctic icefish is only 6° C. Thus, the geographic range of the icefish is tightly restricted to the cold waters around Antarctica. The exact causes of such low lethal temperatures in fish are not understood.

There are many ways in which animals in warm environments regulate their body temperatures to escape death through overheating. The metabolism of heat-tolerant fish such as desert pupfish or desert-dwelling rodents has been shown to differ from related organisms living in less extreme environments. For example, the overall metabolic rate is lower for desert rodents than for nondesert species. Large mammals, such as humans and horses, can cool themselves through the release of sweat. The evaporation of sweat dissipates heat from the body and cools the animal. The reason why humid nights in the summer seem so unbearable is that the high amount of water vapor present in the air decreases the rate of sweat evaporation and precludes us from cooling through sweating. Smaller mammals do not sweat because of their low ratio of body mass to surface area. Small animals would have to lose too great a proportion of their moisture to cool effectively by sweating. Some mammals also pant in order to release heat through the evaporation of moisture from the upper respiratory tract and to meet oxygen demands. Marsupials, such as Australian kangaroos, cool themselves through the production of saliva, which they distribute on their bodies by licking.

Small mammals may seek cool microclimates in the shelter of burrows and rock crevasses when temperatures are too high. In this case, small body size can be an advantage. The antelope ground squirrel *(Ammospermophilus leucurus)* of western North America has been shown to maintain sublethal core temperatures by running back into its burrow when its body temperature exceeds about 42.4° C. These squirrels' small bodies rapidly lose heat in the cool microclimate of the burrows, allowing them to quickly return to above-ground foraging activities. The lack of suitable soils for burrowing can be an effective control on the local geographic distribution of such animals. Birds do not sweat but are known to pant and rapidly oscillate the mouth and upper throat in an action called gular fluttering in order to cool themselves. Birds such as the ostrich *(Struthio camelus)* make their plumage go erect when they are in danger of overheating in order to insulate their skin from heat and solar radiation.

Reptiles and insects often rely on favorable microclimates in burrows and crevasses in order to avoid overheating. In many cases, the behavior of the organisms in selecting and frequenting the burrows is highly developed. If the animal stays in the shelter too long, it does not have sufficient time to meet its food needs. In addition, if the shelter is too cool, the low temperature will slow down the metabolism of the reptile or insect and hinder it. Reptiles have an amazing ability to use body positioning and differences in microhabitat to keep their body temperature within a very narrow optimum range for their metabolism. Reptiles can be very selective in choosing sites for thermoregulation. Snakes in California have been known to select intermediate-sized rocks that are 20 to 40 centimeters thick to hide under. Thinner rocks do not adequately protect the snakes from high temperatures, while thicker rocks provide too cool an environment for the snakes to maintain proper body temperature.

Fish are also cold blooded and often must move to different water depths to maintain optimal body temperature. In eastern North America, the optimal temperature for the growth of northern pike *(Esox lucius)* and walleye *(Stizostenodon vitreum vitreum)* is about 23° C. These species are found in the upper warm waters and will preferentially swim to such warm sites. In contrast, lake trout

(Salvelinus namaycush) and lake whitefish *(Coregonus clupeaformis)* have optimal growth at approximately 13° C and will actively seek out deep, cool waters.

The temperature requirements of adult animal species are often different from those of their developing embryos or young, which are often more sensitive to high temperatures than the adults. For example, the eggs of frogs from the genus *Rana* develop abnormally or die when exposed to temperatures of 25° C and above, although the adult frogs can tolerate these temperatures.

It is impossible to conclude a discussion of animal adaptations to heat without mentioning two well-known animals with very distinctive traits that aid in their thermoregulation—the chameleon and the African elephant. The true chameleons belong to the genus *Chamaeleo* and are found in semidesert regions of Africa. They are not related to the anole lizards of the genus *Anolis,* which are found in the New World and often sold as "chameleons" by pet shops. Like anole lizards, the true chameleons can change the color of their skin. Although this might help camouflage the lizard from prey and predators, research suggests that the color changes also facilitate thermal regulation. When the chameleons are cool, they disperse melanin pigment and become darker in color. This increases the amount of solar energy they absorb and warms them. The process reverses when they are hot. Similarly, certain tropical tree frogs become almost white when exposed to high heat and light intensities. Large ears may seem to be an adaptation for increased hearing. However, in the case of the African elephant *(Loxodonta africana),* the large thin ears appear to be an important mechanism to disperse body heat and cool the animal. The large ears contribute several meters of surface area for cooling, with very little additional body mass.

Cold temperatures serve to decrease rates of metabolic functioning and can prove lethal to both homeotherms and poikilotherms. In the morning, reptiles and amphibians sun themselves to bring their temperatures up to optimum levels. This can be very effective. Peruvian mountain lizards of the genus *Liolaemus* can be found living at altitudes of 4000 meters in the Andes. Even in the summer, temperatures at such high altitudes are often below freezing for part of the day. By sunning themselves, these lizards have been shown to raise their body temperature from 2.5° C to 33° C, even though the air temperature was only 1.5° C. Flying insects require relatively warm thoraxes in order for flight muscles to function. The decrease in mosquito activity that occurs when temperatures drop is a welcome result of this requirement. In bees and moths, it has been shown that minute rapid muscular motion, similar to shivering, allows some species to raise their thorax temperatures. The sphinx moth *(Manduca sexta)* is often active during cool evenings. Studies indicate that by rapid muscular motion the moth can raise the temperature of its thorax to 35° C even when the night air temperature is only 10° C.

The only way for most insects, amphibians, and reptiles to survive prolonged freezing temperatures is through avoidance and dormancy. Insects may find refuge underground or in rotting wood. Species in cold climates can often tolerate very cold temperatures during winter dormancy. For example, Alaskan beetles of the species *Pterosticus brevicornis* have been shown to survive temperatures of −35° C during winter hibernation. In many cases, a process of frost hardening through exposure to progressively cooler temperatures in the fall is required for poikilotherms to develop such freezing tolerance. Alaskan beetles will freeze to death at temperatures of just −6.6° C if exposed to such conditions during the summer. Amphibians, such as the frog species *Rana sylvatica,* and reptiles, such as the garter snake, can inhabit cold regions as far north as Alaska and the Canadian Northwest Territories because they settle down for winter dormancy. During this time, frogs seek soft mud while snakes take residence in rock crevasses. In the case of garter snakes, many thousands can be found aggregating in rock shelters in the fall and emerging again in spring in cold northern locations such as Wood Buffalo National Park and the Manitoba Snake Pits in Canada. Despite sheltered locations, dormant insects, amphibians, and reptiles often experience freezing temperatures in cold climates. The formation of intra- and extracellular ice would be lethal to many species even during dormancy. A number of species are able to supercool without the formation of ice because they produce glycerol, which acts as an antifreeze. Dormant reptiles and amphibians have been shown to supercool to temperatures as low as −8° C without damage.

Warm-blooded animals subject to cold temperatures must adapt by maintaining body temperature while minimizing their expenditure of energy. Going into hibernation allows the animal to minimize its energy loss by becoming inactive and slowing down its metabolism to decrease energy use. However, hibernating mammals can still freeze to death under extremely cold conditions or starve to

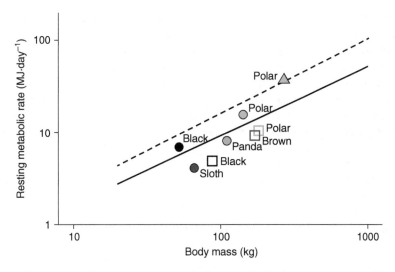

FIGURE 3.7   The high resting metabolic rate of polar bears compared with species from warmer climates— brown bears *(Ursus arctos)*, black bears *(Ursus americanus)*, sloth bears *(Melursus ursinus)*, and pandas *(Ailuropoda melanoleuca)*. The triangle and circles are active animals, and the squares represent hibernating animals (Pagano et al., 2018 / with permission of American Association for the Advancement of Science).

death if reserves of food energy are insufficient. Two primary adaptations exist to cope with cold for birds and mammals that remain active at low temperatures. The first is increased metabolic rates to generate body heat. This maximum cold-induced metabolic rate is referred to as the **Summit Metabolic Rate** and is positively correlated with cold tolerance in birds and mammals. As might be expected, the metabolic rates for both birds and mammals are generally higher for Arctic than for tropical species (Fig. 3.7). This suggests a general relationship between temperature, the metabolic functioning of the animals, and their geographic distribution. The second strategy involves increasing insulation to reduce heat losses. Insulation can take the form of fat, fur, or plumage. Larger birds and mammals can physically support thicker insulation; they depend on this strategy to counteract cold more than is the case for smaller animals. The blubber layer in marine mammals such as whales and seals is an important means by which they insulate their cores from cold water temperatures.

The polar bear *(Thalarctos maritimus)* is an example of a large mammalian predator well adapted to a very cold environment. Polar bears spend much of their lives in and out of the water hunting seals on the edges of Arctic sea ice. The Arctic climate is severely cold, and polar bears must contend with staying warm both when dry and when wet. The bears are large and present a small surface area for cooling relative to their mass. However, the bear's fur is coarse and open. It does not provide as much insulation in the air or in the water as a dense coat of fine fur would. It does, however, dry quickly. This is an advantage given polar bears' ice-shelf habitat and the severe wind-chill factor of the Arctic. In addition to fur, the bears have a layer of blubber that provides enhanced insulation in air and particularly in water. It may seem odd that polar bear fur is white. A dark coat would absorb more solar radiation and aid in heating. However, the white coat provides important camouflage capability, allowing the bear to approach its prey undetected. As might be expected, the metabolic rate of polar bears is high compared with species from warmer climates such as brown bears *(Ursus arctos)*, black bears *(Ursus americanus)*, pandas *(Ailuropoda melanoleuca)*, and sloth bears *(Melursus ursinus)*.

Small birds and mammals have two disadvantages when dealing with cold. First, because of their small ratio of body mass to surface area, they lose heat at a faster rate than larger animals. This places more demands on energy consumption in order to fuel the high metabolic rates needed to maintain body heat. Second, because of their small size, they cannot physically support enough fat or fur for insulation to counteract these demands. It would seem that the biogeographic implication of the thermal inefficiency of small animals is that Arctic animals should be larger than related temperate and tropical species. Within the same species or closely related species there is evidence that some populations living at higher latitudes tend to have larger bodies than those living in warmer lower latitudes. This general proposition has been referred to as **Bergmann's Rule.** It is named for the German biologist Carl Bergmann who described this phenomenon in 1847. However, as forcefully argued by the Canadian biologist Valerius Geist, this proposition cannot be accepted as universally true. It is possible that available food energy in the Arctic is insufficient to support greatly larger body masses in many cases. In addition, many birds and small mammals avoid the rigors of cold winters by migrating to warmer climates in the case of birds or by finding relatively warm microhabitats under the snow and other protected locations in the case of small mammals. Generally, however, animal species that live in cold environments have shorter extremities, such as limbs or ears, than related species in warm environments. The shorter the extremities are relative to body mass, the lower the rate of heat loss. This generality is known as **Allen's Rule** and is named after the American biologist Joel Asaph Allen who published a paper on this phenomenon in 1877. Desert rabbit and hare species, for example, typically have longer ears than Arctic species. It has also been found that bird species in hot regions often have larger bills than do similar species living in cold regions. The larger bills are effective in dissipating heat and help the birds maintain healthy core temperatures. Analysis of the body mass and wing length of bird species in the British Isles indicates a general correspondence to both Bergmann's and Allen's rules, but again with clear exceptions. Allen's Rule has also been extended to human populations that live in contrasting cold or hot climatic regions. Some studies suggest that human populations from the cold high latitudes of the Northern Hemisphere tend to have shorter arms and legs, stockier hands, and larger relative body mass than populations from the tropics.

It is not surprising that cold climate has been shown to limit the broad geographic distributions of many animals. Small birds often have winter range limits that are apparently dictated by low temperatures. The eastern phoebe *(Sayornis phoebe)* is one example. Its northern range in eastern North America corresponds fairly well with the −4° C isotherm for average minimum January temperatures (Fig. 3.8). Temperatures lower than this likely require increased metabolic rates that cannot be met by the feeding rates of the eastern phoebe at its northern limits. At least 14 other bird species in North America have northern range boundaries that coincide with the point at which the metabolic rate of the birds would have to be about 2.5 times higher than their basal metabolic rate in order to maintain body heat during cold winter conditions.

Cold temperatures frequently limit the ranges of organisms by affecting reproduction rather than through lethal cooling of adults. Corals, for example, are only found in regions with sea surface temperatures of 20° C or greater. Although adult coral can survive in cooler water, it is likely that they cannot breed. The same is true for other marine organisms. The northern limit of the barnacle species *Balanus amphitrite* coincides with the winter sea surface isotherm in summer of 18° C. Although the adults can

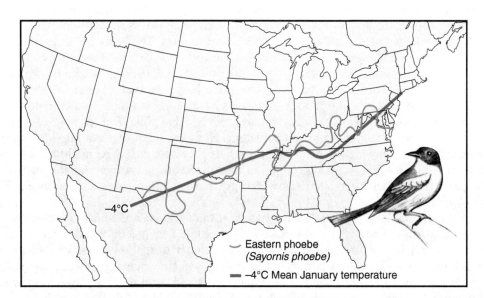

−4°C

— Eastern phoebe
(Sayornis phoebe)

— −4°C Mean January temperature

FIGURE 3.8 The relation between January temperature and the northern limits of the eastern phoebe *(Sayornis phoebe)*. North of the −4° C January isotherm, the birds cannot obtain food in sufficient quantities to support the metabolic activity required to maintain their body temperature above lethal levels (Root, 1993).

survive temperatures as low as 7.2° C, they require a mean monthly summer temperature of at least 18° C to reproduce. The North American distribution of different species of the frog genus *Rana* also illustrates the role that temperature-restricted reproduction plays in limiting the geographic distributions of species. The species *Rana sylvatica* can be found living and reproducing at sites near the Arctic Circle. This is much farther north than other frog species. It has been shown that *Rana sylvatica* can tolerate much lower temperatures during breeding and in the embryonic stage than other species of frogs.

As a final point, it is interesting to consider humans in terms of adaptations to Arctic versus tropical conditions. Humans do not have fur or high amounts of insulative body fat. In addition, our metabolic rates and lower critical thermal limits are similar to those of tropical mammals. On the basis of our natural cold tolerance abilities, we, like corals and palms, are truly creatures of the tropics and subtropics. It is only by our ability to manipulate our immediate environment through clothing, artificial heat, and specialized shelter that we can survive in the colder portions of the planet.

MOISTURE

As with temperature, the impact of moisture on the physiological functioning and geographical distribution of animals is of extreme importance and has long been studied by biologists and biogeographers. When examining the role of moisture in the distribution of life, it is again sensible to divide our discussion into separate considerations for plants and animals.

Plants

Water is essential to plants for a number of reasons. It is, of course, required for photosynthesis as well as other physiological processes such as cell growth and protein synthesis. The cells of plants also require sufficient water to maintain turgor (rigidity). The release of water from leaves helps to cool plants. Finally, in vascular plants water is essential to the movement of nutrients. Plants are often classified by their water demands. Those that can exist in dry environments are called **xerophytes.** Plants that require moderately moist conditions are referred to as **mesophytes,** while **hydrophytes** are those plants found growing in water or very wet soils.

In some sense, vascular plants may be thought of as conduits whereby water is taken up through absorption from the soil by fine roots and flows upward through the vascular system. The specialized vascular tissues that conduct water are called **xylem,** while those that transfer nutrients are called **phloem.** Eventually, water that is not retained in the plant or used directly in photosynthesis is released to the air through the stomata. Stomata are most commonly found on the underside of leaves but can also occur on the top of leaves, on stems, and on flower parts. The number of stomata on a leaf can be very high. The North American basswood tree *(Tilia cordata)* has about 13,000 stomata per square centimeter of leaf area, while the red oak *(Quercus rubra)* has 68,000. It is the difference between the low vapor pressure due to evaporation at the stomata and the high vapor pressure at the water-absorbing roots that causes water to move through the plant. The release of water to the atmosphere by plants is referred to as **transpiration.** The rate of water loss through transpiration is dependent on water availability, stomata shape and behavior, air temperature, humidity, and wind conditions. Hot, dry, and windy conditions increase the rate at which water evaporates from the leaf surface. In some regions, the loss of water to the air through transpiration is much greater than the amount of water lost through direct evaporation from the soil surface. Climatologists use the term **evapotranspiration** when considering the contribution of both sources of water to the atmosphere.

When the rate of water loss is high relative to absorption through the roots, water stress occurs. If the water deficit is great enough, the cells and tissues of the plant lose turgor and exhibit the soft and droopy conditions characteristic of the wilting point. In cases of severe water loss, the death of cells and tissue is the inevitable result. However, water stress can have many impacts besides wilting. Most important, when plants are exposed to water stress they experience decreased rates of photosynthesis. The inhibition of photosynthesis may occur because the stomata of many plants will close when dehydration causes a lack of cell turgor. This cuts off the supply of CO_2 for photosynthesis. There is also evidence that severe water deficits cause decreases in other physiological functions and further depress photosynthesis. Support for this theory has been found in tree species such as loblolly pine and Engleman spruce *(Picea englemannii)*. Analysis of widely different species such as periwinkle

(Catharanthus roseus), cotton *(Gossypium),* sunflowers *(Helianthus annuus),* and European blueberry *(Vaccinium myrtillus)* has shown water deficits produce decreases in chlorophyll upon which photosynthesis is dependent.

A clear biogeographic relationship exists between the ability of plant species to conduct photosynthesis at low levels of water availability and the places where the plant species are found growing. Trees that are typical of relatively moist forest sites in eastern North America, such as alder *(Alnus),* ash *(Fraxinus),* or loblolly pine, experience reduced rates of photosynthesis at moisture deficits of -1 megapascal (MPa; $1 MPa = 10^6$ Pa; pascals are a unit of pressure). Shrub species of the extreme desert in Death Valley, such as the creosote bush *(Larrea tridentata),* do not exhibit reduced rates of photosynthesis until water deficits of -2.0 to -2.9 megapascals are experienced.

Water stress can also cause the shedding of leaves and even stems. The bean caper *(Zygophyllum dumosum)* of the Negev Desert of Israel may lose up to 90% of its transpiring leaf and stem surface at the onset of the dry season. In a less extreme climate, the black sage *(Salvia mellifera)* of California can lose all but a few of its terminal leaves during the summer dry season but still maintain a net positive rate of photosynthesis. Such loss of leaves is not restricted to plants in dry regions. Well-known trees, such as the buckeye *(Aesculus glabra)* and black walnut *(Juglans nigra)* of the eastern North American forests, lose leaves relatively quickly when subjected to dry conditions during the growing season. Some oak species living in the same zones as black walnut and buckeye appear to lose their leaves less frequently during dry spells. Interestingly, they are found farther west on the dry fringes of the Great Plains than trees such as walnut and buckeye.

Dry conditions can be particularly lethal to seeds and seedlings. The seeds of some trees, particularly those associated with moist sites, such as willows, bald cypress, and elms, are sensitive to desiccation and can be killed when dried. In contrast, seeds of pines and eucalyptus are often quite tolerant of desiccation. The seeds of *Eucalyptus siberi* can survive several periods of desiccation and wetting. This ability is not surprising in view of the dry conditions in which most eucalyptus grow in Australia. When taken as a whole, the vast majority of terrestrial plant species produce seeds that can withstand some degree of desiccation.

In the southwestern United States and northern Mexico, the desert vegetation of the lowlands gives way to coniferous forest cover as elevation increases on mountains. This lower treeline is a biogeographic boundary that can be traced to moisture stress. With increasing elevation, temperatures decline. This produces lower evaporation rates in the upper elevations of the mountains. In addition, the mountains receive higher amounts of rainfall due to orographic precipitation. Thus, the mountains provide sufficient moisture to support tree growth, while the lower elevations do not. The pinyon pines *(Pinus edulis* and *P. monphylla)* occur in the southwestern United States and adjacent Mexico. The mature trees can survive extremely dry conditions. The seedlings, however, can generally survive no more than 12 days of wilting water stress, and this helps to exclude the pinyons from the drier lowland deserts of the region (Fig. 3.9).

A wide variety of adaptations allows plants to cope with dry conditions. Through these strategies plants can be grouped into three broad categories: water stress escapees, water stress avoiders, and water stress tolerators. Many smaller plants of seasonally dry environments survive the dry season as dormant seeds and are known as water stress escapees. The seeds of many desert plants can survive very long periods of desiccation and remain viable. There are accounts of seeds germinating in the Sahara Desert of Sudan after more than 100 years of dormancy. Annual plants in the Sahara may germinate, grow, flower, and release seeds in periods as short as 8 days following rainfall. These drought-escaping annual plants often have very high rates of photosynthesis and conductance of water from leaves during their short lives. Most are not adapted to function well when water is in short supply. Perhaps the most amazing story of long seed dormancy is the germination and healthy growth of a 2000-year-old seed of date palm *(Phoenix dactylifera)* recovered from an archaeological site in the desert of Israel.

Many other plants from dry regions are water stress avoiders that employ a number of strategies that allow individuals to survive during seasonally dry conditions. Certain tropical and subtropical regions experience dry winters, and many trees and large shrubs lose their leaves and become dormant during the dry season. The ocotillo plant *(Fouquieria splendens)* of the southwestern desert of the United States and Mexico may lose its leaves several times a year in response to episodes of drought and precipitation. Some smaller plants, such as the sedge species *Carex pachystylis* of the Negev Desert, allow their above-ground portions to die back while their underground portions

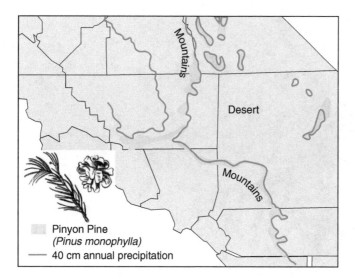

FIGURE 3.9 The distribution of the pinyon pine species *Pinus monophylla* and annual precipitation in the desert region of southeastern California. The pines are restricted to uplands where moisture availability is high compared to that in the dry lowlands.

remain alive and dormant in the dry season. A similar adaptation is found among grasses of dry sub-tropical regions. Plants that retain their leaves in the dry season often have hard and waxy cuticles that decrease moisture loss. Such leaves are called **sclerophyllous leaves.**

Another strategy to avoid water stress is to possess roots deep enough to obtain groundwater when the surface soil is dry. Even relatively small shrubs can possess very deep roots. For example, the chamise shrub *(Adenostoma fasciculatum)* of southern California has roots that can extend to 8 meters in depth, while the roots of the mesquite shrubs *(Prosopis)* of the southwestern desert can extend an amazing 53 meters. Another interesting adaptation for water stress avoidance is the ability of some plants to store water in their tissues. Several types of trees found in arid regions have enlarged trunks that can store water. These include the baobab *(Adansonia digitata)* of Africa and the bottle tree *(Brachychiton rupestris)* of eastern Australia. Baobabs may have a circumference of more than 20 meters and store up to 120,000 liters of water (Fig. 3.10). Some water stress avoiders use very little water in photosynthesis. The agave *(Agave deserti)* transpires only about 25 grams of water for every gram of carbon that is fixed during photosynthesis. In contrast, many plants from more mesic environments may transpire 250 to 600 grams of water for every gram of carbon they fix.

Perhaps the most interesting water stress avoiders are the members of the cactus family which are found in the deserts and dry regions of North and South America. These plants possess a number of adaptations to help them survive water stress and have been extensively studied by the American botanist Park Nobel and his colleagues. Cacti possess extensive fine root systems near the soil surface that quickly intercept any rainfall. The stems of the cacti are enlarged and contain tissue to store water. The barrel cactus *(Ferocactus acanthodes)* can store enough water for 50 days of photosynthetic activity. The stems of cacti are photosynthetic, and the leaves have been reduced to hard thorns. Some cacti have thick coverings of light-colored thorns that protect against the sun and drying wind. The covering of spines on the cholla cactus *(Opuntia bigelovii)* can cool its surface by as much as 11° C. In other cases, the thorns serve to defend the moist and tender stems against herbivores. The exteriors of cactus stems have a waxy coating and deeply inset stomata that help further reduce water loss. Finally, cacti use the CAM photosynthetic pathway, whereby the stomata open at night to take in CO_2. The CO_2 is converted to malic acid, which is decarboxylated during the day when C_3 photosynthesis converts the CO_2 to sucrose and starch. By keeping the stomata closed during the heat of the day, evaporative water loss through the stomata can be reduced. Despite these adaptations, cacti are not found in the very driest parts of the New World deserts. To survive, cacti must have sufficient rainfall to replenish their water supplies from shallow groundwater. In addition, CAM plants tend to have slow growth rates and are limited by the fact that they have the lowest photosynthetic rates of the flowering plants.

FIGURE 3.10 A large baobab tree *(Adensonia digitata)* growing in Africa. Baobabs may have a circumference of more than 20 meters and can store up to 120,000 liters of water. The stored water helps them cope with the dry subtropical winter.

All of these adaptations allow plants to survive and conduct photosynthesis when water is in short supply in the environment. However, very few plants can tolerate desiccation of tissue to the point of being near completely dry. True water stress tolerators are able to tolerate severe dehydration down to less than –13 megapascals. Some tropical C_4 grasses and members of the pea family can withstand this degree of desiccation. More amazingly, individuals of some mosses, lichen, algae, and fungi, such as the so-called resurrection plants *Selaginella lepidophylla* and *Craterostigma plantagineum,* can actually survive almost complete desiccation and begin functioning rapidly following rehydration. These are referred to as **resurrection plants** due to this amazing ability to seemingly come back from the dead.

Too much water can also be problematic for plants. The diffusion of oxygen through water-saturated soils is about 10,000 times slower than through dry soils. Plants that grow in water-saturated soils may not be able to obtain sufficient oxygen through their root systems to survive. Plants that grow in water-logged soils, however, often develop longitudinal air spaces within their roots. These spaces are called aerenchyma and aid in the storage and transfer of oxygen. Some plants have the ability to develop new roots from their stems if they are partially buried by wet soils following flooding. These roots are referred to as adventitious roots and are present in herbs such as the common dock weed *(Rumex)* as well as many trees and shrubs including the coastal redwood. Even plants that are normally associated with moist soils can differ markedly in their ability to tolerate very wet conditions. Both sycamore *(Platanus occidentalis)* and water tupelo *(Nyssa aquatica)* grow best on moist soils in southeastern North America. These trees are found along waterways and in damp areas. However, as soil moisture increases beyond 50% saturation, the growth rate of sycamore begins to decline while the water tupelo reaches maximum growth under water-saturated conditions. The water tupelo is, therefore, found on wetter sites than the sycamore.

Animals

Here we will consider the impact of moisture on terrestrial animals. Water is vital to the metabolic functioning of all animals. It has been said that "living animals can be described as an aqueous solution contained within a membrane." Most animals, including insects, are composed of two-thirds water by weight and must maintain this high volume in order to survive. Mammals can withstand a loss of about 10% of their body water, while losses of 15% to 20% generally prove fatal. In contrast, some frogs can lose as much as 40% of their body water and still survive. Water can be lost directly through evaporation from the skin. Evaporative water loss can be particularly high for animals with permeable skin such as worms or amphibians. Water can also be lost through respiration, excrement, and other secretions.

There is often a relationship between the environmental conditions that animals live in and their rate of water loss through evaporation. In general, insects and reptiles from dry environments have lower rates of water loss through skin evaporation than through respiration. For example, water snakes of the genus *Natrix* lose approximately 88% of their body moisture through evaporation and 12% through respiration. In contrast, desert-dwelling gopher snakes of the genus *Pituophis* lose only 64% of their body water through evaporation and 36% through respiration. In addition, total water loss rates tend to be lower in animals found in arid environments. Water snakes lose moisture at a rate that is three times faster than that for gopher snakes when subjected to the same dry conditions. Pond-dwelling turtles lose moisture at a rate that is eight times greater than the rate of moisture loss for desert tortoises. In desert rattlesnakes, evaporation and respiration combined account for only a 0.5% loss of their water body weight per day. At this rate, these reptiles could survive 2 to 3 months of water loss.

Water is generally replenished by drinking, and many animals are limited to areas where free water exists in lakes, streams, and springs and as dew on leaves. Water can also be obtained through food and, in the case of some insects, it is directly absorbed through the skin. In fact, a number of insects and some other animals never drink water but obtain sufficient moisture from food. Succulent leaves and fruits may contain up to 90% water. Some insects can subsist on dry leaves that contain only 5% to 10% water. Even if free water is limited in food, many animals can obtain water through the oxidation of organic matter. In general, the gastrointestinal system of desert mammals generally has more absorptive area for water than is the case for nondesert mammals.

The kangaroo rat *(Dipodomys spectabilis)* is a classic example of an animal adapted to an arid environment. This animal must maintain a body water concentration of about 66% while living in the dry deserts of southwestern North America. The kangaroo rat can live entirely on dry food without ever drinking free water. When eating 100 grams of dried seed, the rat can produce 54 milliliters of water through oxidation—even though the amount of free water in the grain may only be 6 milliliters. In order to avoid water losses through excrement, the rats produce very small quantities of very dry feces and urine which contain very highly concentrated amounts of urea relative to the water content. The rats do not sweat, and they lose only small quantities of water through evaporation. Despite these adaptations, the kangaroo rat still faces difficulty in compensating for moisture loss. In hot and dry atmospheric conditions, typical of desert days, the rats will lose more water through respiration and evaporation than they gain in eating dry food. These potential losses are mitigated by the fact that the rats are nocturnal and spend the daylight hours in cool and relatively moist burrows. In addition, female rats have increased water demands when they are in their reproductive period and producing milk. At this time, they must be able to increase their uptake of free water by eating green and moist plants. Despite these limitations, the kangaroo rat can survive in much drier environments than many other rodents. For example, packrats *(Neotoma)* also occur in the southwestern deserts but are restricted to moister regions than is the case for kangaroo rats. In contrast to kangaroo rats, packrats cannot obtain sufficient moisture through the oxidation of dry foods and are unable to produce urine which has the same high concentrations of urea to water.

OTHER PHYSICAL FACTORS

Plants and animals are sensitive to many other aspects of the physical environment besides light, temperature, and moisture. Plants require chemical nutrients in the soil such as nitrate, phosphorus, and potassium in order to grow. Aquatic organisms may be sensitive to the acidity of the water in which they live. Such physical factors are often extremely important at a local scale but less apparent in controlling broad global distributions. Detailed investigation of these physical factors falls more into the realm of the ecologist and physiologist rather than that of the biogeographer. Nonetheless, several of these physical properties of the earth's environment are important for controlling large-scale geographic distributions of organisms. We will examine a few examples of these physical factors in this section.

The most important chemical factor controlling the global geographic distribution of aquatic organisms is salinity. Most species of aquatic plants, crustaceans, fish, and other life can only be found in either fresh or salt water. How do differences in salinity control the distributions of aquatic organisms?

Organisms must maintain a relatively stable cellular chemistry. Increases or decreases in the salt content of the cells can cause changes in the reactions of proteins, enzymes, and other metabolic functions. This disruption can lead to death. If water salinity changes, an osmotic pressure gradient is formed. The osmotic pressure gradient causes water to flow through the permeable membrane of the organism. During this flow, pure water moves toward the more saline water. If a saltwater worm is placed in fresh water, the purer water will flow across the body wall into the worm. This leads to an imbalance in the salinity of the worm's cells. The opposite flow occurs for a freshwater organism placed in salt water. Under these circumstances, the freshwater species loses water and suffers desiccation. For this reason, it is impossible for freshwater amphibians, such as most frogs, to cross saltwater barriers. Indeed, most amphibians are restricted to fresh water. There is, however, a crab-eating frog in Southeast Asia *(Rana cancrivora)* that spends long periods in the ocean. It maintains an osmotic gradient against a lethal absorption of salt water by having a high urea content in bodily fluids.

Of course, fish and shelled crustaceans do not have the water-permeable skin of worms. In these organisms, osmotic exchange occurs through exposed permeable areas such as gills. Surprisingly, most marine fish have osmotic concentrations (concentrations of solutes) that are one-third to one-quarter that of sea water. They will lose body water through areas such as their gills and must avoid desiccation by taking in large amounts of sea water and evacuating salts through their gills and in urine. Sharks also have low concentrations of salts but maintain high osmotic pressure because they have high amounts of urea in their body fluids and cells.

Because of their sensitivity to osmotic pressure changes, the vast majority of marine plants and animals cannot survive in fresh water. Similarly, most freshwater species will die if placed in salt water. There are, however, species that spend part of their lives in salt water and part of their lives in fresh water. Salmon, as well as certain eels and sharks, found in Lake Nicaragua and in Malaysia, have this ability. In addition, organisms that live in estuarine environments where freshwater rivers enter the sea must deal with daily to seasonal swings in salinity as tides and river flow vary. These species have a special capacity to regulate their cellular chemistry and avoid the development of lethal osmotic pressure gradients. Such organisms are said to be **euryhaline.**

As a final point, there are certain lakes that have salinities much higher than those of the ocean. For example, the Great Salt Lake in Utah has salt concentrations that are more than seven times higher than concentrations in the ocean. Few aquatic organisms can tolerate such high salinities. The Great Salt Lake cannot support fish. Only two macroscopic invertebrates, a brine shrimp *(Artemeria salina)* and a benthic brine fly *(Ephydra cincera)* are able to tolerate the high salinity levels.

Both freshwater and marine animals obtain oxygen from the water. A lack of free oxygen in the water is lethal. In some lakes, the bottom waters become anoxic (low in oxygen) either seasonally or permanently. This condition may occur in deep lakes where current activity is not effective in mixing bottom and surface waters to replenish oxygen. Under such conditions, bottom-dwelling organisms such as worms and mollusks are excluded. A similar situation occurs in certain small deep ocean basins. Two well-known examples of such anoxic ocean basins are the Santa Barbara Basin off California and the Carioco Basin off the coast of Venezuela. The bottoms of these basins are devoid of the worms, mollusks, crustaceans, and ground fish typical of the ocean floor. Some organisms have evolved to accommodate life in waters with relatively low oxygen content. The so-called freshwater blood worms are actually the larval stage of midge flies (Chironomidae). The red color of these freshwater organisms is due to the extremely high hemoglobin content in their blood. This allows for a more efficient transfer of oxygen and gives blood worms the ability to live in lake waters that do not have enough free oxygen to support most other types of organisms.

Plants are often very sensitive to the chemical conditions of the soil and water. High concentrations of certain chemical elements may be deleterious to plants. Saline soils are lethal to most plants, and few terrestrial species can tolerate irrigation by sea water. High concentrations of magnesium are found in certain soils, particularly those derived from dolomite and serpentine rocks. The growth of many plants, from grasses and herbs to coniferous trees, has been shown to decrease in soils with high concentrations of magnesium. Vegetation cover on such soils is often sparse and may be dominated by species that are particularly adapted to tolerating high concentrations of magnesium.

Acidic precipitation and fog caused by the burning of fossil fuels can directly injure leaves and also lead to the mobilization of aluminum and other trace elements in some soils, and this can prove toxic to plants. It is estimated that between the years 1960 and 2000 about half of the high-elevation red spruce *(Picea rubens)* trees in the Adirondack Mountains of New York died due to acid precipitation.

In freshwater ecosystems, particularly lakes and ponds, the acidity of the water exerts a strong control on plankton and larger plants. Lakes that have become acidified due to air pollution experience great changes in plant species and animal species. Severely acidified lakes are often almost lifeless. Such lakes were becoming more common in portions of eastern North America and Europe. In the 1990s it was estimated that 15% of all lakes in New England were suffering from human-caused acidification. In the Adirondack Mountains, approximately 40% of all lakes were suffering from acidification. About 24% of the lakes in the Adirondacks had become so acidic that fish could no longer survive in them. Most of the acidic precipitation affecting vegetation and lakes in the northeastern United States was due to sulfur and nitrogen released into the atmosphere by coal-burning power plants and industry in the Ohio River Valley.

The good news is that the Canada–United States Air Quality Agreement, signed in 1991, has led to significant decreases in the airborne release of the pollutants that cause acid rain. Now, more than two decades after the agreement, there is evidence of the recovery of surface waters and soils. In studies at various sites over the period of 8 to 24 years following the agreement, reductions of wet SO_4^{2-} deposition of 5.7% to 76% have been observed. Both lakes and the soils in the watersheds around them have been showing increases in pH. However, the recovery of the lake biota has lagged behind these increases in pH. Many of the impacted lakes remain fishless, but others have been recolonized by fish. Hopefully, over time, there will be a full biological recovery. However, a renewed reliance on coal, as has been suggested by some policy makers in the United States, could reverse the gains against acid precipitation in the absence of safeguards.

INTERACTING PHYSICAL CONTROLS ON GEOGRAPHIC DISTRIBUTIONS

In this chapter, we have explored how individual physical factors such as light, temperature, or moisture affect the physiological functioning of organisms and limit the geographic distributions of species. Of course, frequently more than one physical factor may be responsible for determining whether a plant or animal can live at a given location. For example, in desert environments, the combination of low water availability, high temperatures, and high light intensities may all be important factors that work in combination to preclude the survival of most plant species.

The geographic distribution of plant species that employ the C_4 photosynthetic pathway provides a good example of how several different physical factors contribute to the geographic limits of species. The percentage of dicot species that use the C_4 pathway ranges from almost 5% in the southwestern United States to less than 1% in the Canadian grasslands. The percentage of C_4 species in the North American grass flora ranges from over 80% in the Southwest to 0% in northern Canada and Alaska. The C_4 plants can make up an even higher proportion of the flora in tropical and subtropical regions. In northern tropical Australia, C_4 species account for 80% to 100% of the grass flora in the dry interior. In the cooler and moister temperate portions of southern Australia and Tasmania, C_4 species make up only 2% to 10% of the grass flora (Fig. 3.11). In North America, there is a very high positive correlation between summer evaporation rates and the abundance of C_4 species in the flora. There is also a strong positive relationship between maximum daily temperature and the abundance of C_4 plants in the grass flora. C_4 plants appear to be better adapted to warm, dry, and sunny conditions for several reasons. First, C_4 plants are more efficient in CO_2 bonding at high light intensities and temperatures than are C_3 plants. Thus, through their stomata, C_4 plants can take in more CO_2 relative to the water they lose than C_3 plants. Second, C_4 plants can benefit from the higher light intensities in

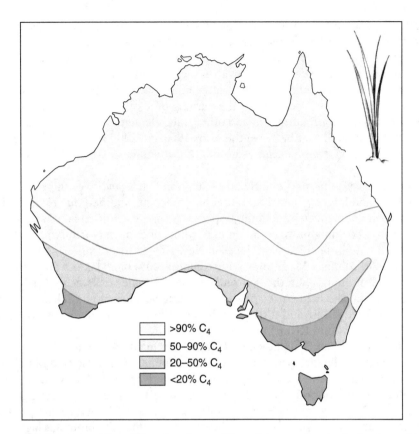

FIGURE 3.11 The relative abundance of C_3 and C_4 grass species in the flora of Australia (adapted from Henderson et al., 1993). The C_4 species are most abundant in the warm tropical and subtropical zones of the continent. A similar situation is found in the North American grasslands.

the lower latitudes because they do not lose carbon through photorespiration. This explains, in part, why C_4 plants can survive and dominate in sunny and dry subtropical and tropical grasslands. Why, however, are C_4 plants not abundant in the higher latitude vegetation of Canada or southern Australia? It appears that they are restricted from the higher latitudes because they require greater light intensities and higher temperatures for optimum photosynthesis than are required by C_3 grasses. Thus, there is a tradeoff between the ability of C_4 plants to tolerate hot, dry, sunny conditions and their ability to survive in the cooler and less sunny conditions found at high latitudes.

When we expand our examination of physical controls on species distributions to include different geographic areas—southern and northern range limits, for example—we almost always find that different factors are important at different geographic locations. Consider the northern and southern range distributions of spruce. In the north there appears to be a good correlation between low summer temperatures and the range limits of the species. However, when we examine the southern range limits in western Canada, we find little evidence that low summer temperatures control the distribution of the spruce species. Instead, it appears that moisture stress, caused by a combination of high summer temperatures and low precipitation, excludes spruce from growing on the prairies.

ENVIRONMENTAL GRADIENTS AND SPECIES NICHES

As we have seen, many different factors in the physical environment can limit the geographic distributions of plant and animal species. Are there some general models that biogeographers use to try and organize, conceptually at least, the relationships between different species and the physical environmental factors that affect them? In this section we will look at some ideas from ecology that provide useful conceptual frameworks to organize our observations on how the physical environment

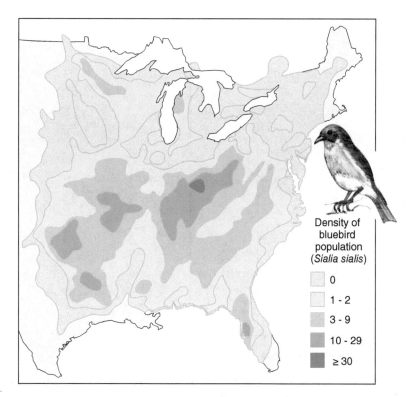

FIGURE 3.12   The range and population density of the eastern bluebird *(Sialia sialis)* in North America. Notice how population density is greatest in patches near the center of the geographic range (Bystrak, 1979; Brown and Gibson, 1983).

controls the distributions of species. First, let us consider how we can describe the geographic distributions of species. Two basic fundamental measures for describing the geographic distribution of a species are range and density. In the preceding sections, we used the term *range limits* to describe the geographic distribution of species. The **geographic range** of a species is the entire area where the species can be found regardless of whether it is common or rare. For example, the range of the eastern bluebird *(Sialia sialis)* extends from roughly the U.S.-Canadian border southward to the Gulf Coast and southern Florida. In the west, it extends to the Great Plains (Fig. 3.12). Within its range, a species may be found to be common in some areas and rare in others. These differences in the abundance of the species within its range are referred to as its **density.** Density is a measure of population abundance per unit of area and is often reported as the number of individuals per square meter or hectare on land and the number of individuals per cubic meter in aquatic systems.

The highest densities for species often occur in the central portions of their geographic ranges. Lowest densities are usually found toward the edges of species ranges. We can see that this is generally true for the eastern bluebird. The density of the species is greatest where the environment provides the best conditions to support individuals of the species. This often occurs in the central portions of the geographic range of the species. In contrast, at the range limits of the species, the environment is unable to support a large number of individuals and the density decreases. The number of individuals that the environment can support per meter or hectare is called its **carrying capacity.**

Now let's link these observations on geographic ranges and density to the physical environmental controls we have discussed. We have seen many examples of how the physiological functioning of species can be related to environmental conditions along the environmental gradients of light, temperature, and moisture. The physiological functioning of the organism increases along the gradient to an optimal level and then declines. The increase in physiological functioning along these physical environmental gradients frequently occurs in the form of a Gaussian curve (Fig. 3.13). In many cases, the environmental conditions that promote optimum physiological functioning produce the highest carrying capacities for the species. For example, a hypothetical plant species might reach its highest capacity for photosynthesis at about 20° C. At this

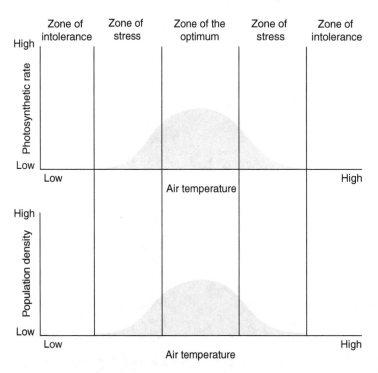

FIGURE 3.13 The photosynthetic rate and population density of a hypothetical plant species along an environmental gradient of low to high air temperature.

temperature the high rate of photosynthesis produces rapid and robust growth, allowing plants to mature quickly and produce more viable seeds, thereby providing better survival potential for the seedlings. The population size of the species is therefore higher in geographic regions with a temperature of 20° C than in areas where the temperature is warmer or cooler than 20° C. The geographic differences in species density can be related to increasing and decreasing physiological functioning along the environmental gradient of temperature. In general, the density of a species along an environmental gradient also takes the form of a Gaussian curve (Fig. 3.13). The study of how species distributions relate to the environment, including how species ranges and densities are distributed along environmental gradients, is called **ordination analysis.**

The environmental gradients for the physiological functioning and density of a species can be divided into several subzones. The zone of the optimum represents the best conditions for the species. In this zone, physiological functioning and density are at a maximum. The zones of stress provide conditions that allow the species to persist but at low carrying capacities. In the zones of intolerance, environmental conditions are too harsh for the species to survive. For our hypothetical plant species, average growing season temperatures above say 30° C or below 10° C are lethal and define the zones of intolerance (Fig. 3.13). The zones of tolerance for species may shift depending on the life stages of the plant or animals. For example, the seeds of a plant might tolerate temperatures of below freezing or above 40° C with no apparent damage. In contrast, seedlings frequently have much more restricted tolerances than seeds or adult plants.

In the preceding section, we have seen that the geographic distribution of a species may be controlled by more than one physical factor. Ecologists and biogeographers recognize that many environmental gradients, representing factors such as temperature, precipitation, and light intensity, can control the distribution of species. Species exist in a multidimensional space defined by all the different physical environmental factors that affect their physiological functioning and ultimately their abundance. If we again consider our hypothetical plant species, we might find that its range and abundance are influenced both by growing season temperature and annual precipitation. These two gradients form the axes of a two-dimensional environmental space in which the species can be found growing (Fig. 3.14). This two-dimensional representation of the environment in which the species can survive can be referred to as its **niche.** Of course, there are always more than two environmental

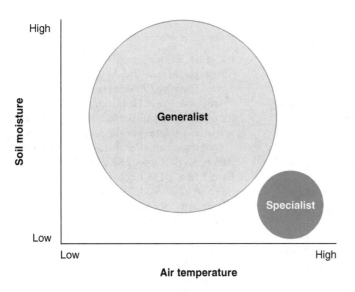

FIGURE 3.14   The niches of a hypothetical generalist plant species and a specialist plant species as defined by two gradients, air temperature and soil moisture.

variables that control the survival of an organism. So the real niches of species include many different gradients and are formally defined as *n*-dimensional hypervolumes that are defined by all *(n)* of the physical and biological environmental conditions the species requires to survive. The term *niche* was used by J. Grinnell in 1917 and C. Elton in 1927. G. E. Hutchinson, however, formulated its widely accepted definition in 1957. It is the Hutchinsonian niche concept that is used here. The Grinnellian and Eltonian niches are somewhat different conceptualizations of the interactions of species with environment and habitat. This will be discussed further in the next chapter. The Hutchinsonian niche should not be considered as synonymous with the habitat of the species. **Habitat** is the explicit spatial environment in which species can be found. For example, the damp surface soil beneath a rock is a habitat.

Different species have different niches. The niche provides a means of characterizing and comparing the general relationship between species and environment. Species, which can tolerate a wide range of environmental conditions, have wide ranges of tolerances on environmental gradients and wide niche breadths. Such species are referred to as **generalists.** Species, which have very narrow environmental tolerances, have more restricted gradient distributions and niche breadths and are referred to as **specialists** (Fig. 3.14). Since they can tolerate a wide variety of physical environmental conditions, generalists often have large geographic ranges compared to specialists. Although the concept of the ecological niche is easy to envision, determining all of the environmental factors that define the real niches of species is extremely difficult.

KEY WORDS AND TERMS

Allen's rule	Geographic range	Photosynthesis Light Response Curve
Annual	Habitat	Poikilotherm
Bergmann's Rule	Heliophytes	Resurrection plants
Carrying capacity	Homeotherm	Sciophytes
Deciduous	Hydrophyte	Sclerophyllous leaves
Density	Mesophyte	Specialist
Dimorphic leaves	Niche	Stenothermic
Euryhaline	Ordination analysis	Summit Metabolic Rate
Eurythermic	Perennial	Transpiration
Evapotranspiration	Phenology	Xerophyte
Generalist	Phloem	Xylem

REFERENCES AND FURTHER READING

Archibold, O. W. (1995). *Ecology of World Vegetation*. Chapman and Hall, London.

Arnott, S. E., Jackson, A. B., and Alarie, Y. (2006). Distribution and potential effects of water beetles in lakes recovering from acidification. *Journal of the North American Benthological Society* **25**, 811–824.

Arseneault, D., and Payette, S. (1997). Landscape change following deforestation at the Arctic tree line in Quebec, Canada. *Ecology* **78**, 693–706.

Atkinson, D., and Sibly, R. M. (1997). Why are organisms usually bigger in colder environments? Making sense of a life history puzzle. *TREE* **12**(6), 235–239.

Barbour, M. G., Berg, N. H., Kittel, T. G. F., and Kunz, M. E. (1991). Snowpack and the distribution of a major vegetation ecotone in the Sierra Nevada of California. *Journal of Biogeography* **18**, 141–149.

Bartholomew, G. A., and Lasiewski, R. C. (1965). Heating and cooling rates, heart rate and simulated diving in the Galapagos marine iguana. *Comparative Biochemical Physiology* **16**, 575–582.

Bauwe, H., Hagemann, M., and Fernie, A. R. (2010). Photorespiration: players, partners and origin. *Trends in Plant Science* **15**, 330–336.

Begon, M., Harper, J. L., and Townsend, C. R. (1996). *Ecology: Individuals, Populations and Communities*, 3rd edition. Blackwell Science, Oxford.

Begon, M., Townsend, C. R., and Harper, J. L. (2006). *Ecology: From Individuals to Ecosystems*. Blackwell Publishing, Malden, England.

Berry, J. A., and Bjorkman, O. (1980). Photosynthetic response and adaptation to temperature in higher plants. *Annual Review of Plant Physiology* **31**, 491–543.

Betti, L., Lycett, S. J., von Cramon-Taubadel, N., and Pearson, O. M. (2015). Are human hands and feet affected by climate? A test of Allen's rule. *American Journal of Physical Anthropology* **158**, 132–140.

Bjorkman, O. (1981). Responses to different flux densities. In *Encyclopedia of Plant Physiology*, vol. 12A (ed. P. S. Nobel, O. L. Lange, C. B. Osmond, and H. Ziegler), pp. 57–107. Springer-Verlag, Berlin.

Bjorkman, O., and Berry, J. (1973). High efficiency photosynthesis. *Scientific American* **229**, 80–93.

Bjorkman, O., Nobs, M., Mooney, H., Troughton, J., Berry, J., Nicholson, F., and Ward, W. (1974). Growth responses of plants from habitats with contrasting thermal environments: transplant studies in Death Valley and the Bodega Head experimental garden. *Carnegie Institute Year Book* **73**, 748–757.

Brooks, R. R. (1987). *Serpentine and Its Vegetation: A Multidisciplinary Approach*. Croom Helm, London.

Brower, L. P. (1969). Ecological chemistry. *Scientific American* **220**, 22–29.

Brown, J. H., and Gibson, A. C. (1983). *Biogeography*. C. V. Mosby, St. Louis.

Bystrak, D. (1979). The breeding bird survey. *Sialia* **1**, 74–79.

Chabot, B. F., and Bunce, J. A. (1979). Drought stress effects on leaf carbon balance. In *Topics in Plant Population Biology* (ed. S. T. Jain, O. T. Solbrig, G. B. Johnson, and P. H. Raven), pp. 338–355. Columbia University Press, New York.

Chapin III, F. S., Jefferies, R. L., Reynolds, J. F., Shaver, G. R., Svoboda, J., and Chu, E. W. (Eds.) (2012). *Arctic Ecosystems in a Changing Climate: An Ecophysiological Perspective*. Academic Press, Cambridge, MA.

Chappell, M. A., and Bartholomew, G. A. (1981). Activity and thermoregulation of the antelope ground squirrel *Ammospermophilus leucurus* in winter and summer. *Physiology Zoology* **54**, 215–223.

Chase, J. M. (2011). Ecological niche theory. In *The Theory of Ecology* (ed. S. M. Scheiner and M. R. Willig), pp. 93–107. University of Chicago Press, Chicago.

Cloudsley-Thompson, J. L. (1977). *Man and the Biology of Arid Zones*. Edward Arnold, London.

Cole, P. G., and Weltzin, J. F. (2005). Light limitation creates patchy distribution of an invasive grass in eastern deciduous forests. *Biological Invasions* **7**, 477–488.

Crawford, R. M. M. (1989). *Studies in Plant Survival: Ecological Case Histories of Plant Adaptation to Adversity*. Blackwell Scientific Publications, Oxford.

Crawford, R. M. M., and Palin, M. A. (1981). Root respiration and temperature limits to the North-South distribution of four perennial maritime plants. *Flora* **171**, 338–354.

Danner, R. M., and Greenberg, R. (2015). A critical season approach to Allen's rule: bill size declines with winter temperature in a cold temperate environment. *Journal of Biogeography* **42**, 114–120.

Daubenmire, R. (1974). Taxonomic and ecologic relationships between *Picea glauca* and *Picea englemannii*. *Canadian Journal of Botany* **52**, 1545–1560.

Daubenmire, R. F. (1943). Vegetational zonation in the Rocky Mountains. *Botanical Review* **9**, 325–393.

De Lucia, E. H. (1986). Effect of low root temperature on net photosynthesis, stomatal conductance and carbohydrate concentration in Engelmann spruce (*Picea engelmannii Parry ex Engelm*) seedlings. *Tree Physiology* **2**, 143–154.

Downton, W. J. S., Berry, J. A., and Seeman, J. R. (1984). Tolerance of photosynthesis to high temperatures in desert plants. *Plant Physiology* **74**, 786–790.

Downton, W. J. S., Loveys, B. R., and Grant, W. J. R. (1988). Stomatal closure fully accounts for the inhibition of photosynthesis by abscisic acid. *New Phytologist* **103**, 263–266.

Drezner, T. D. (2006). Regeneration of *Carnegiea gigantea* (Cactaceae) since 1850 in three populations in the northern Sonoran Desert. *Acta Oecologica* **29**, 178–186.

Drezner, T. D. (2014). The keystone saguaro (*Carnegiea gigantea*, Cactaceae): a review of its ecology, associations, reproduction, limits, and demographics. *Plant Ecology* **215**, 581–595.

Ehleringer, H. R., and Mooney, H. A. (1984). Photosynthesis and productivity of desert and Mediterranean climate plants. In *Encyclopedia of Plant Physiology,* New Series (ed. P. S. Nobel, O. J. Lange, C. B. Osmond, and H. Ziegler), pp. 205–231. Springer Verlag, Berlin.

Elliott-Fisk, D. L. (1983). The stability of the northern Canadian tree limit. *Annals of the Association of American Geographers* **73**, 560–576.

Elliott-Fisk, D. L. (2000). The taiga and boreal forest. In *North American Terrestrial Vegetation* (ed. M. G. Barbour and W. D. Billings), pp. 41–73. Cambridge University Press, Cambridge.

Elton, C. (1927). *Animal Ecology*. Sidgwick and Jackson, London.

Evanari, M., Shanan, L., and Tadmor, N. (1982). *The Negev— The Challenge of a Desert*, 2nd edition. Harvard University Press, Cambridge, MA.

Farrant, J. M., and Moore, J. P. (2011). Programming desiccation-tolerance: from plants to seeds to resurrection plants. *Current Opinion in Plant Biology* **14**, 340–345.

Geist, V. (1987). Bergmann's rule is invalid. *Canadian Journal of Zoology* **65**, 1035–1038.

Giller, P. S. (1984). *Community Structure and the Niche*. Chapman and Hall, New York.

Graham, D., and Patterson, B. D. (1982). Responses of plants to nonfreezing temperatures: proteins, metabolism and acclimation. *Annual Review of Plant Physiology* **33**, 347–372.

Grinnell, J. (1917). The niche-relationship of the California thrasher. *Auk* **34**, 427–433.

Hammond, K. A., and Diamond, J. (1997). Maximal sustained energy budgets in humans and animals. *Nature* **386**, 457–462.

Hattersley, P. W. (1983). The distribution of C_3 and C_4 grasses in Australia in relation to climate. *Oecologia* **57**, 113–128.

Hay, M. E. (1986). Functional geometry of sea-weeds: ecological consequences of thallus layering and shape in contrasting light environments. In *On the Economy of Plant Form and Function* (ed. T. J. Givnish), pp. 635–666. Cambridge University Press, Cambridge.

Heinrich, B., and Bartholomew, G. A. (1971). An analysis of pre-flight warm-up in the sphinx moth, *Manduca sexta*. *Journal of Experimental Biology* **55**, 223–239.

Henderson, S., Hattersley, P., von Caemmer, S., and Osmond, B. (1995). Are C_4 pathway plants threatened by global climatic change? In *Ecophysiology of Photosynthesis* (ed. E. D. Schulze and M. Caldwell), pp. 529–549. Springer-Verlag, Berlin.

Hengeveld, R. (1990). *Dynamic Biogeography*. Cambridge University Press, Cambridge.

Hengeveld, R., and Haeck, J. (1982). The distribution of abundance. I. Measurements. *Journal of Biogeography* **9**, 303–316.

Howe, H. F., and Westley, L. C. (1988). *Ecological Relationships of Plants*. Oxford University Press, New York.

Husic, D. W., Husic, H. D., Tolbert, N. E., and Black Jr., C. C. (1987). The oxidative photosynthetic carbon cycle or C2 cycle. *Critical Reviews in Plant Sciences* **5**, 45–100.

Hutchinson, G. E. (1957). Concluding remarks. In *Cold Spring Harbour Symposium on Quantitative Biology*, vol. 22, pp. 415–427. Cold Spring, NY.

Hutchinson, G. E. (1967). *A Treatise on Limnology*, vol. 2: *Introduction to Lake Biology and the Limnoplankton*. Wiley, New York.

Hutchinson, G. E. (1975). *A Treatise on Limnology*. Wiley, New York.

Jaleel, C. A., Manivannan, P. A., Wahid, A., Farooq, M., Al-Juburi, H. J., Somasundaram, R. A., and Panneerselvam, R. (2009). Drought stress in plants: a review on morphological characteristics and pigments composition. *International Journal of Agricultural Biology* **11**, 100–105.

Kinne, O. (1971). *Marine Ecology: A Comprehensive, Integrated Treatise on Life in Oceans and Coastal Waters*. Wiley Interscience, New York.

Körner, C. (2007). Climatic treelines: conventions, global patterns, causes. *Erdkunde* **61**, 316 –324.

Körner, C. (2015). Paradigm shift in plant growth control. *Current Opinion in Plant Biology* **25**, 107–114.

Kozlowski, T. T., Kramer, P. J., and Pallardy, S. G. (1991). *The Physiological Ecology of Woody Plants*. Academic Press, San Diego.

Kramer, P. J., and Decker, J. P. (1944). Relation between light intensity and rate of photosynthesis of loblolly pine and certain hardwoods. *Plant Physiology* **19**, 350–358.

Kramer, P. J., and Kozlowski, T. T. (1979). *Physiology of Woody Plants*. Academic Press, New York.

Krebs, C. J. (1994). *Ecology*, 4th edition. Harper-Collins College Publishers, New York.

Kreyling, J., Schmid, S., and Aas, G. (2015). Cold tolerance of tree species is related to the climate of their native ranges. *Journal of Biogeography* **42**(1), 156–166.

Lalli, C., and Parsons, T. R. (1997). *Biological Oceanography*, 2nd edition. Pergamon Press, New York.

Larcher, W. (1969). The effect of environmental and physiological variables on the carbon dioxide exchange of trees. *Photosynthetica* **3**, 167–198.

Larcher, W. (1980). *Physiological Plant Ecology*, 2nd edition. Springer-Verlag, Berlin.

Larcher, W. (2003). *Physiological Plant Ecology: Ecophysiology and Stress Physiology of Functional Groups*. Springer, Berlin.

Larcher, W., Kainmüller, C., and Wagner, J. (2010). Survival types of high mountain plants under extreme temperatures. *Flora-Morphology, Distribution, Functional Ecology of Plants* **205**, 3–18.

Larcher, W., and Winter, A. (1981). Frost susceptibility of palms: experimental data and their interpretation. *Principes* **25**, 143–153.

Lawrence, G. B., Hazlett, P. W., Fernandez, I. J., Ouimet, R., Bailey, S. W., Shortle, W. C., Smith, K. T., and Antidormi, M. R. (2015). Declining acidic deposition begins reversal of forest-soil acidification in the northeastern US and eastern Canada. *Environmental Science & Technology* **49**, 13103–13111.

Loach, K. (1967). Shade tolerance in tree seedlings. I. Leaf photosynthesis and respiration in plants raised under artificial shade. *New Phytologist* **66**, 607–621.

Lowe, C. H., Larnder, P. J., and Halpern, E. A. (1971). Supercooling in reptiles and other vertebrates. *Comparative Biochemical Physiology* **39A**, 125–135.

MacMahon, J. A. (2000). Warm deserts. In *North American Terrestrial Vegetation* (ed. M. G. Barbour), pp. 285–322. Cambridge University Press, Cambridge.

MacNally, R. C. (1995). *Ecological Versatility and Community Ecology*. Cambridge University Press, Cambridge.

Maitland, P. S. (1990). *Biology of Fresh Waters*, 2nd edition. Chapman and Hall, New York.

Malanson, G. P., Butler, D. R., Fagre, D. B., Walsh, S. J., Tomback, D. F., Daniels, L. D., Resler, L. M., Smith, W. K., Weiss, D. J., Peterson, D. L., and Bunn, A. G. (2007). Alpine treeline of western North America: linking organism-to-landscape dynamics. *Physical Geography* **28**, 378–396.

Marquard, R. D., and Hanover, J. W. (1984). The effect of shade on flowering of *Picea glauca*. *Canadian Journal of Forest Research* **14**, 830–832.

McCollin, D., Hodgson, J., and Crockett, R. (2015). Do British birds conform to Bergmann's and Allen's rules? An analysis of body size variation with latitude for four species. *Bird Study* **62**, 404–410.

Meyer, B. S., Anderson, D. B., Bohning, R. H., and Fratianne, D. G. (1973). *Introduction to Plant Physiology*. Van Nostrand Reinhold, Princeton, NJ.

Miller, L. K. (1969). Freezing tolerance in an adult insect. *Science* **166**, 105–106.

Moellering, E. R., Muthan, B., and Benning, C. (2010). Freezing tolerance in plants requires lipid remodeling at the outer chloroplast membrane. *Science* **330**, 226–228.

Mooney, H. A. (1988). Lessons from Mediterranean-climate regions. In *Biodiversity* (ed. E. O. Wilson and F. M. Peter), pp. 157–165. National Academy Press, Washington, DC.

Mooney, H. A., Cushman, J. H., Medina, E., Sala, O. E., and Schulze, E. (1996). *Functional Roles of Biodiversity: A Global Perspective*, p. 493. Wiley, Chichester, England.

Morrow, P. A., and Mooney, H. A. (1974). Drought adaptations in two California evergreen sclerophylls. *Oecologia* **15**, 205–222.

Murata, N., Takahashi, S., Nishiyama, Y., and Allakhverdiev, S. I. (2007). Photoinhibition of photosystem II under environmental stress. *Biochimica et Biophysica Acta (BBA)-Bioenergetics* **1767**, 414–421.

Nagy, K. A., and Gruchacz, M. J. (1994). Seasonal water and energy metabolism of the desert dwelling kangaroo rat *(Dipodomys merriami)*. *Physiological Zoology* **67**, 1461–1478.

Nielsen, S. L., and Simonsen, A. M. (2011). Photosynthesis and photoinhibition in two differently coloured varieties of Oxalis triangularis—the effect of anthocyanin content. *Photosynthetica* **49**(3), 346.

Nobel, P. S. (1977). Water relations and photosynthesis of a barrel cactus, *Ferocactus acanthodes,* in the Colorado desert. *Oecologia* **27**, 117–133.

Nobel, P. S. (1980). Leaf anatomy and water use efficiency. In *Adaptations of Plants to Water and High Temperature Stress* (ed. N. C. Turner and P. J. Kramer), pp. 43–55. Wiley, New York.

Nobel, P. S. (1980). Morphology, nurse plants, and minimum apical temperatures for young *Carnegiea gigantea*. *Botanical Gazette* **141**, 188–191.

Nobel, P. S. (1980). Morphology, surface temperatures, and northern limits of columnar cacti in the Sonoran Desert. *Ecology* **61**, 1–7.

Nobel, P. S. (1999). *Physicochemical & Environmental Plant Physiology*. Academic Press, Cambridge, MA.

Nobel, P. S. (2003). *Environmental Biology of Agaves and Cacti*. Cambridge University Press, Cambridge, England.

Nybakken, J. W. (1997). *Marine Biology: An Ecological Approach*, 4th edition. Addison Wesley Longman, Menlo Park, CA.

Osakabe, Y., Osakabe, K., Shinozaki, K., and Tran, L. S. P. (2014). Response of plants to water stress. *Frontiers in Plant Science* **5**, 86.

Pagano, A. M., Durner, G. M., Rode, K. D., Atwood, T. C., Atkinson, S. N., Peacock, E., Costa, D. P., Owen, M. A., and Williams, T. M. (2018). High-energy, high-fat lifestyle challenges an Arctic apex predator, the polar bear. *Science* **359**, 568–572.

Parker, K. C. (1989). Nurse plant relationships of columnar cacti in Arizona. *Physical Geography* **10**, 322–335.

Parker, K. C. (1993). Climatic effects on regeneration trends for two columnar cacti in northern Sonoran Desert. *Annals of the Association of American Geographers* **83**, 452–474.

Pearson, O. P. (1954). Habits of the lizard *Liolaemus multiformis multiformis* at high altitudes in Southern Peru. *Copeia* *1954* 2, 111–116.

Pocheville, A. (2015). The ecological niche: history and recent controversies. In *Handbook of Evolutionary Thinking in the Sciences* (ed. T. Heams, P. Huneman, G. Lecointre, and M. Silberstein), pp. 547–586. Springer, Dordrecht, Germany.

Rapoport, E. H. (1982). *Areography: Geographical Strategies of Species*. Pergamon Press, Oxford.

Rascio, N., and Rocca, N. L. (2005). Resurrection plants: the puzzle of surviving extreme vegetative desiccation. *Critical Reviews in Plant Sciences* **24**, 209–225.

Raven, P. H., Evert, F., and Eichhorn, S. E. (2005). *Biology of Plants*, 7th edition. W. H. Freeman, New York.

Rezende, E. L., Bozinovic, F., and Garland Jr., T. (2004). Climatic adaptation and the evolution of basal and maximum rates of metabolism in rodents. *Evolution* **58**, 1361–1374.

Richardson, A. D., and Friedland, A. J. (2009). A review of the theories to explain Arctic and alpine treelines around the world. *Journal of Sustainable Forestry* **28**, 218–242.

Root, T. (1988). Energy constraints on avian distributions and abundances. *Ecology* **69**, 330–339.

Sage, R. F. (2004). The evolution of C4 photosynthesis. *New Phytologist* **161**, 341–370.

Sakai, A., and Larcher, W. (1987). *Frost Survival of Plants*. Springer-Verlag, Berlin.

Sakai, A., and Larcher, W. (2012). *Frost Survival of Plants: Responses and Adaptation to Freezing Stress*. Springer Science & Business Media, Berlin.

Sallon, S., Solowey, E., Cohen, Y., Korchinsky, R., Egli, M., Woodhatch, I., Simchoni, O., and Kislev, M. (2008). Germination, genetics, and growth of an ancient date seed. *Science* **320**, 1464.

Schmidt-Nielsen, K. (1997). *Animal Physiology: Adaptation and Environment*, 5th edition. Cambridge University Press, Cambridge.

Scholander, P. F., Hock, R., Walters, V., Johnson, F., and Irving, L. (1950). Heat regulation in some Arctic and tropical mammals and birds. *Biological Bulletin* **99**, 237–258.

Scott, F. M. (1964). Lipid deposition in intercellular space. *Nature* **203**, 164–165.

Seymour, R. S. (1997). Plants that warm themselves. *Scientific American* **March,** 104–109.

Silen, R. R. (1973). First and second season effect on Douglas-fir cone initiation from a single shade period. *Canadian Journal of Forest Research* **3**, 428–435.

Stowe, L. G., and Teeri, J. A. (1978). The geographic distribution of C₄ species of the *Dicotyledonae* in relation to climate. *American Naturalist* **112**, 609–623.

Sutherland, J. W., Acker, F. W., Bloomfield, J. A., Boylen, C. W., Charles, D. F., Daniels, R. A., Eichler, L. W., Farrell, J. L., Feranec, R. S., Hare, M. P., and Kanfoush, S. L. (2015). Brooktrout Lake case study: biotic recovery from acid deposition 20 years after the 1990 clean air act amendments. *Environmental Science & Technology* **49**, 2665–2674.

Swanson, D. L., and Garland Jr., T. (2009). The evolution of high summit metabolism and cold tolerance in birds and its impact on present-day distributions. *Evolution* **63**, 184–194.

Teeri, J. A., and Stowe, L. G. (1976). Climatic patterns and the distribution of C4 grasses in North America. *Oecologia* **23**, 1–12.

Teskey, R. O., Fites, J. A., Samuelson, L. J., and Bongarten, B. C. (1986). Stomatal and non-stomatal limitations to net photosynthesis in *Pinus taeda L.* under different environmental conditions. *Tree Physiology* **2**, 131–142.

Toole, V. K., Toole, E. H., Hendricks, S. B., Borthwick, H. S., and Snow Jr., A. B. (1961). Responses of seeds of *Pinus virginiana* to light. *Plant Physiology* **36**, 285–290.

Walter, H. (1985). *Vegetation of the Earth*, 3rd edition. Springer-Verlag, New York.

Woodward, F. I. (1987). *Climate & Plant Distribution*. Cambridge University Press, Cambridge.

Wyse, S. V., and Dickie, J. B. (2017). Predicting the global incidence of seed desiccation sensitivity. *Journal of Ecology* **105**, 1082–1093.

Yamori, W., Hikosaka, K., and Way, D. A. (2014). Temperature response of photosynthesis in C3, C4, and CAM plants: temperature acclimation and temperature adaptation. *Photosynthesis Research* **119**, 101–117.

CHAPTER 4

BIOLOGICAL INTERACTIONS AND THE DISTRIBUTION OF LIFE

Sex is not the first thing that comes to most people's minds when they watch a hummingbird move from flower to flower. Yet that is precisely what the hummingbird is engaged in—on behalf of the flowers, that is. With each sip of nectar, the visiting bird brushes against the flower's anthers and takes away some pollen, which it then transports to the next flower. Some of the pollen picked up by the hummingbird will be transferred to the pistil of a flower of the same species where the male sperm nucleus within the pollen grain will fertilize the female ovule to form a seed. The nectar produced by the flowers feeds pollinating birds or insects. Sexual reproduction would be impossible for many flower species without the intervention of specific pollinators such as the hummingbird. If the pollinating species is not present, the flower species that depends on it will also be absent.

We have explored how variations in the physical environment of the earth affect the physiological functioning of organisms and ultimately influence the geographic distributions of species. However, plants and animals also interact with other species, and these biological interactions can under certain circumstances control the geographic ranges and population densities of species. In addition to flowers and their pollinators, there are many forms of such interaction. One species of insect may prey upon another, or two bird species may compete for the same nesting sites. Some interactions between species can be important factors in controlling geographic ranges. The study of interspecies interaction has long been a focus for both ecologists and biogeographers. In this chapter we will explore different types of interactions between species and how these interactions influence geographic distributions. We will examine how geographic distributions can be controlled by a combination of biological interactions and physical factors. We will then consider how biological interactions are accommodated within the concepts of gradient distributions and species niches.

PREDATION

Predation occurs when one organism consumes another. In this section we will consider predation to include both herbivory (animal-plant predation) and carnivory (animal-animal predation), although many ecologists treat these as separate relationships.

If a predator depends on only one particular species for its prey, the geographic distribution of the predator is necessarily limited to the geographic distribution of the prey species. Some predators are indeed highly selective in terms of their prey species. Predators that have a very narrow range of prey species are called **stenophagous,** or selective predators. In California, the geographic distribution of some races of the checkerspot butterfly *(Euphydryas editha)* are geographically limited to areas where their only food plant, a plantain species called *Plantago hookeriana,* is found. Interestingly, this plantain species itself is restricted to soils formed from serpentine rock. At a continental scale, the northern limits of the monarch butterfly *(Danaus plexippus)* corresponds exactly with the northern range limits of its food plant (Fig. 4.1), milkweed *(Asclepias* spp.). How common is such restricted stenophagy? If we look at all butterfly species in the Los Angeles area, we find that about one-third of the species are restricted to one species of food plant or several closely related species of the same genus. Within North America, about 80%

Biogeography: Introduction to Space, Time, and Life, Second Edition. Glen M. MacDonald.
© 2025 John Wiley & Sons, Inc. Published 2025 by John Wiley & Sons, Inc.
Companion website: www.wiley.com/go/macdonald/biogeography2e

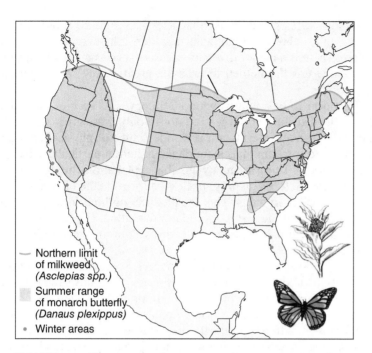

FIGURE 4.1 The correspondence between the northern range limits of the monarch butterfly *(Danaus plexippus)* and the northern range limits of its food plant, milkweed *(Asclepias* spp.) (adapted from Brower and Malcolm, 1991).

of all butterfly species are relatively stenophagous and restricted to a single family of plants for food. Worldwide, it appears that about 90% of all plant-eating insects (phytophagous species) are restricted to food plants from one or two plant families. Other insects are oligophagous, meaning they are restricted to several different food species.

So, for plant-eating insects, some degree of stenophagy is extremely common and must be considered an important factor in controlling geographic distribution. Is this also true for other herbivorous animals and carnivores? There are some famous instances of mammals that are highly stenophagous. Two of the best known are the Australian koala bear *(Phascolartus cinereus),* an arboreal (tree-dwelling) marsupial that eats only the leaves of the bluegum eucalyptus. The second, the Chinese giant panda bear *(Ailuropoda melanoleuca),* eats mainly the foliage of certain bamboo species. In North America, Albert's squirrel *(Sciurus alberti)* generally subsists on seeds from the ponderosa pine *(Pinus ponderosa).* However, the vast majority of terrestrial birds and mammals are relatively **euryphagous** (using many different plant species for food; also referred to as polyphagous). The degree of food selectivity appears to be less restrictive for large herbivores than for small ones. The decrease in selectivity for large herbivores likely reflects their need for a broad resource base to acquire enough energy to maintain themselves. A large herbivore has a better chance of meeting its energy requirements if it has a large foraging area and can eat many different species of plants. In general, carnivores display a much lower degree of stenophagy than do herbivores. This is to be expected, since carnivores are generally at the top trophic level and must have a relatively broad food resource base to acquire enough energy to persist. In addition, large carnivores will usually transect many different habitats when foraging, and it benefits them to take advantage of different species of prey that occupy these various environments.

We can therefore see a biogeographic relationship between the feeding practices of large herbivores and carnivores, the range of their foraging activities, and the number of different environments that they use for feeding. Such animals are sometimes referred to as fine-grained because their foraging is made up of resources from an array of many fine-scale environments within their range. In contrast, some insects may spend their entire life on one leaf. Animals that have very small ranges and restricted diets can be called coarse-grained. These organisms live in a world that is coarsely divided between the limited areas that they can live in and areas that they cannot.

A lower prevalence of stenophagy is also observed in smaller predators, such as spiders. In the case of spiders, the degree of stenophagy is low compared to herbivorous insects. One study examined the diets of over 500 spider species and found that less than 30% of those spider species were stenophagous. Interestingly, the degree of stenophagy in spiders is higher in the tropics and subtropics than in temperate regions. There is a higher diversity of insect species upon which spiders feed upon in the tropics and potentially this has helped to promote greater prey specialization by spiders in those regions.

Stenophagy would appear to be a relatively poor strategy for a species. The reliance on one or two species of prey limits the distribution of the predator to only those areas where the prey species can exist. The predator may perhaps tolerate a much wider range of physical environmental conditions than the prey. So why are some animals stenophagous, and in particular why is stenophagy so common in phytophagous insects? There are many possible explanations, but here we will outline just a few. First, it benefits predators to focus on prey species that provide the highest ratio of food energy relative to the energy required for foraging. This is called the **optimal foraging theory.** The theory assumes that food choices represent the result of predator evolution. The focus of herbivores on specific plants that provide high nutrition is the outcome of an evolutionary adaptation for using only the most nutritious food sources.

Second, many plants produce toxins to deter herbivores. Some herbivores, particularly insects, evolve resistance to the specific toxins of just one plant species. This allows the insect to feed on that plant with little competition from other herbivores. In addition to foraging benefits, ingesting high levels of such plant toxins can also make the insect unpalatable to carnivorous predators. In one experiment, the caterpillars of 70 different butterfly species from Costa Rica were offered as food to predatory ants. The ants avoided caterpillars that were stenophagous and ate plants containing toxic and unpalatable chemicals. Euryphagous caterpillars were more palatable to the ants and were selectively eaten.

Third, small insects may find that all of the microenvironmental conditions they require during their life cycle are provided by one individual plant, and it is efficient to restrict foraging to that plant. Finally, insects have relatively simple neural capabilities and may be behaviorally limited to identifying and consuming a very restricted number of plant species.

It is easy to see how the presence or absence of prey species can control the geographic distribution of a predator. Can the presence of a voracious euryphagous predator restrict the geographic distribution of prey species? In some instances, the introduction of a generalist predator by humans has led to the extinction or near extinction of prey species. In eastern North America, lake trout *(Salvelinus namaycush)* are driven toward extinction in lakes that contain lampreys *(Petromyzon marinus)*. Lampreys are fearsome predators that spend much of their life at sea but come into fresh waters to breed. They use rasping mouths to attach themselves to prey fish and literally suck the life out of them. Formerly, sea lampreys could not enter the Great Lakes beyond Lake Ontario. They were prevented from moving farther up the Great Lakes system by the presence of Niagara Falls. However, with the building of shipping canals to Lake Erie, lampreys were able to enter the upper Great Lakes. The result was the almost complete eradication of lake trout *(Salvelinus namaycush)* in the upper Great Lakes (Fig. 4.2). In the late 1950s, native lake trout were essentially extirpated from lakes Ontario, Erie, and Michigan. Populations of lake trout now have to be sustained by lamprey control programs and the breeding of trout in fish hatcheries for release into the lakes.

Although intensive management efforts are helping lake trout recover after lamprey introduction into the Great Lakes, the same positive news cannot be found for many bird species that have been subjected to predation by the introduction of domesticated cats *(Felis silvestris catus)* by humans. The impact of domestic cats in terms of causing extinctions and geographic range reductions of birds on islands has been particularly severe. Cats are nonselective predators and are estimated to kill one billion birds every year in the United States alone. It is thought that the extirpation of bird species on islands around the world by feral cats introduced by humans is responsible for almost 14% of global bird extinctions over the past few hundred years.

Experimental data also supports the contention that predation by euryphagous predators can reduce geographic distributions of prey species. An interesting set of long-term experiments has been conducted on the impact of predation on orb spiders that live on small islands in the Bahamas. Predatory lizards of the genera *Anolis* and *Leiocephalus* were introduced to some of the islands. Subsequent monitoring showed that islands with lizards had about one-tenth the spider population density and one-half the spider species as comparably sized non-lizard islands.

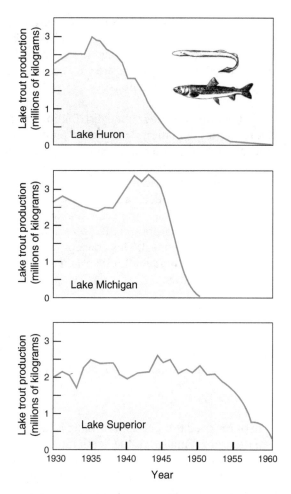

FIGURE 4.2   The devastating impact of the accidental introduction of sea lampreys *(Petromyzon marinus)* on lake trout *(Salvelinus namaycush)* in Lake Michigan. At present, lamprey control programs and the breeding of trout in hatcheries are used to maintain lake trout populations in the Great Lakes (Baldwin, 1964; Krebs, 1994).

Outside of these instances of human-introduced predators, is there any evidence to suggest that predators are important in restricting the geographic distributions of species in nature? The answer is yes—under certain circumstances. For some aquatic organisms such as large dragonflies (Order Odonatathe), the presence of predatory fish excludes the dragonfly nymphs from some lakes and ponds, while the dragonflies are found in nearby waters in which such fish are absent. Some species of damselflies of the genus *Enallagma* are found in waters where large dragonflies live, but are absent in waters with predatory fish. For other damselfly species, the opposite is true. Some species of *Enallagma* appear to be adapted to dragonfly predation and not fish predation, and vice versa. Studies show that damselfly species that swim vigorously to actively evade predation can survive in waters where dragonfly are predators, but are excluded where fish are present. Those damselfly species that remain motionless to evade predation can survive where fish are their predators, but not in the presence of large dragonflies.

Experimental studies show that rocky shorelines along the Pacific Coast of North America and coral reefs off Hawaii experience increased numbers of algae species following the artificial exclusion of animals that graze on algae. These species of algae are apparently not able to survive at the sites when grazers are present. However, these and other experiments, particularly in marine and freshwater ecosystems, have also shown that the long-term impact of excluding predators can cause the decline and extinction of many prey species. This result is counterintuitive but makes perfect sense if we consider it further. In some cases, predators may be **keystone species** that significantly influence the whole composition of ecosystems by controlling the population sizes of prey species. When the predators are removed, certain prey species are able to outgrow or outcompete other species and eventually eradicate some species. In some instances, the presence of generalist predators may allow species of prey to grow in areas they would be excluded from by competition with other prey species. From these studies we can derive an important lesson. Ecosystems are very complex. Seemingly simple changes that we might make, such as the removal of a predator species, can have unexpected and undesirable results.

Aside from examining the impact of predation on geographic distributions, much thought has gone into how predation might impact long-term population dynamics of predators and their prey. The population size of stenophagous and oligophagous predators can be highly dependent on the abundance of a single or few prey species. If predation pressure becomes intense, the population size of the prey will decrease, and this in turn causes a decrease in predators. Due to lack of prey, the predator population will decline because of starvation and decreased fertility. The decline in the predator population relieves the pressure on the prey, and the prey species is able to increase again. We would logically expect populations of stenophagous predators and their prey species to undergo periodic fluctuations in population size as exemplified in Fig. 4.3. This concept of linked oscillations in predator-prey populations is called the **Lotka-Volterra model** after its originators. The model can be expressed mathematically for prey populations as follows:

$$dN/dt = rN - a'PN$$

In this equation, dN/dt refers to the rate of change in the prey population (N) over time (t), r is the natural rate of growth of the prey population (N), a' is the hunting efficiency of the predator, and P is the size of the predator population. For predators, the model is expressed as follows:

$$dP/dt = fa'PN - qP$$

In this equation, dP/dt is the rate of growth of the predator population, and f is the efficiency by which the predator turns captured prey into new offspring. Finally, q is the death rate for the predator populations (P). The model produces a set of sinusoidal curves in which the curve for the prey population leads the rises and declines of the predator population. It is a simple and elegant model that explains why selective predators can cause changes in prey population size but are less likely to cause extinction. When the prey population size gets too low, the predator population size crashes.

Do Lotka-Volterra cycles occur in nature? Laboratory experiments using populations of mites and beetles have produced cyclic variations in predator and prey populations. Some short-term observational studies in nature suggest that such population cycles in prey and predators can exist for both carnivores and herbivores. For example, it has been shown that populations of English cinnabar moths *(Tyria jacobaea)* climb and fall one year behind increases and decreases in the abundance of their prey plant, the ragwort *(Senecio jacobaea)*. However, the changes in the abundance of the

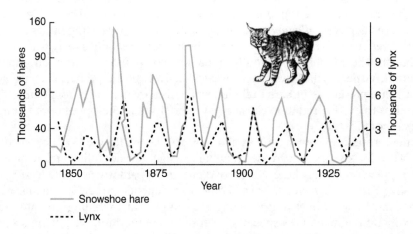

FIGURE 4.3 Historical cyclic variations in the population sizes of the snowshoe hare *(Lepus americanus)* and Canada lynx *(Lynx canadensis)* (MacLulick, 1937; Begon et al., 1996).

ragworts mainly reflected changes in seed germination conditions, and not pressure by predation from the cinnabar moths. A classic study of Lotka-Volterra oscillations between predator and prey comes from the fur trading records of the Hudson Bay Company in Canada. These records provide data on the number of pelts of Canada lynx *(Lynx canadensis)* and its main prey, snowshoe hare *(Lepus americanus),* that were obtained by the Hudson Bay Company from the early nineteenth century through the mid-twentieth century (Fig. 4.3). It is assumed that the number of pelts can be reliably related to the natural abundance of the lynx and hare. These records show an 8- to 11-year cycle in the lynx and hare population size. The lynx populations do indeed follow the increases and decreases in the hare population size. Interestingly, these cycles seem to be synchronous across northern Canada. For many years, this record of lynx and hare populations was used as evidence in support of the classic Lotka-Volterra model.

Experimental manipulations of Yukon hare populations show that both food availability and the presence of predators influence the development of cyclic increases and decreases in hare populations. If predators are excluded, but food supplies are allowed to vary naturally, periodic cycles in the size of the hare population remain. It has been shown that high populations of hares quickly deplete the supply of early easy-to-digest twigs of willow and birch. The subsequent regrowth shoots contain toxins that make them less palatable. So the hare-lynx cycle may also be an echo of the plant-hare cycle. More recent analysis of the empirical data on hare and lynx oscillations show that climatic variations can also influence these cycles. Interestingly, these reanalyses also suggest that while hare abundance seems to directly promote population growth in the lynxes, there is a lag between lynx population peaks and hare decline. This may be due to the longer-term effects of stress placed on hares, particularly potentially breeding females, by being chased by predators. It may also be due to increased toxicity of birch twigs the year following heavy hare browsing. However, taken as a whole, current thought is that although other physical and biological factors clearly influence the hare-lynx cycle, predator-prey interactions and the resulting population increases and declines are a necessary force in producing the observed cycles. Despite its shortcomings, the Lotka-Volterra model is a theoretical model that helps us understand why selective predators do not seem to be important in restricting the geographic ranges of prey species.

COMPETITION

Interspecific competition can be broadly defined as the interaction between individuals of two or more species in which the growth and/or fertility is decreased or the mortality is increased for both species. In the case of **intraspecific competition,** competing individuals are within the same species. In this section, we focus on the role of interspecific competition in determining the range limits of species. It has long been a topic of intense interest to biogeographers. The great evolutionist Charles Darwin placed particular emphasis on competition and its potential importance in limiting the distributions of species. Extensive reviews of field data indicate that some degree of interspecific competition between species is a common occurrence in all ecosystems. Interspecific competition is increasingly recognized as an important factor in limiting the geographic distribution of species based upon field studies, experiments, and theoretical modeling of species range limits and their controls.

There are two general forms of competition. The first is resource exploitation in which both species compete for the same resources without direct contact or interaction. For example, two species of birds that eat the same insect species compete for prey, although the birds might never encounter each other during foraging. The second form, interference competition, occurs when the competitors actually interact either physically or chemically. For example, two plants might establish as seedlings very close together. One may have a faster growth rate and physically exclude the other in the rooting zone and in the canopy. In some competitive interactions, one organism exudes a chemical that is deleterious to another organism. This is referred to as **allelopathy.** In animals there are some suspected instances of interference competition through allelopathy. For example, it is thought that algae that are found in the feces of the tadpoles of common frogs *(Rana temporaria)* can inhibit the growth and survival of competing natterjack toad *(Bufo calamita)* tadpoles.

Some biogeographers believe allelopathy is a major feature of interference competition for plants. It is known that many plants produce chemical substances that are toxic to other plants

(Technical Box 4.1). However, the true importance of allelopathy in the spatial distributions of plants continues to be debated. Classical research into the role of allelopathy in plant competition comes from studies of California chaparral (shrubland) and grassland communities. Halos of bare ground are found around many chaparral shrub species. Careful field experiments have demonstrated that factors such as root competition or canopy shading cannot explain these areas of bare ground. Common chaparral plants such as purple and black sage *(Salvia leucophylla* and *S. mellifera),* coastal sagebrush *(Artemisia californica),* and holly-leafed cherry *(Prunus ilicifolia)* all produce terpenes that may have phytotoxic (poisonous to plant) properties. These substances have been shown to be present in soils around these species. It is thought by some that these phytotoxins inhibit seed germination and thus limit competing plants from the vicinity of established plants. This is particularly important in chaparral where water can be a limiting resource for plants. Some experiments have shown that the soils around chaparral shrubs may not be as toxic to other species as was first thought. Experiments that exclude rodents and rabbits around chaparral plants show that removing such herbivores can lead successful plant growth within some halo areas. On the other hand, morphological and biochemical analyses on purple sage show that certain compounds associated with the plant negatively affects DNA synthesis in the roots of competing species and can inhibit root growth. Although the causes of the chaparral halos remain debated, many other studies and experiments provide evidence for the importance of allelopathy in other vegetation communities. For example, research conducted in Oklahoma showed that several competing grass species had low seed germination rates and grew poorly on soil taken from areas dominated by the grass species *Sporobolus pyramidus.* It is known that *Sporobolus pyramidus* produces p-coumeric and ferulic acid, which can act as phytotoxins.

Competition also occurs between individuals of the same species (intraspecific competition). In theory, competition should be most intense between individuals of the same, or closely related, species because these related organisms will likely have the greatest similarities in the resources they require to survive. The impact of both interspecific and intraspecific competition on population size has been demonstrated in many laboratory experiments. One such experiment was conducted by growing two diatom species in water enriched with silica. Diatoms are unicellular photosynthetic organisms that build protective shells out of silica. When the species *Asterionella formosa* was grown

TECHNICAL BOX 4.1 ALLELOPATHY AND CONCERNS ABOUT A WIDELY PLANTED TREE SPECIES: THE *EUCALYPTUS*

Trees of the genus *Eucalyptus* are native to Australia and surrounding islands. Because many *Eucalyptus* tree species grow rapidly and can survive in a variety of environments, including drought-prone areas, the genus has been widely introduced to areas such as China, California, and South America over the past 150 years. *Eucalyptus* was introduced to China in the 1890s and there are now about 300 varieties growing there and occupying some 3,680,000 hectares of land. In California several dozen species have become established since *Eucalyptus* was introduced in 1853. So common are they along coastal regions and portions of central California that many people assume *Eucalyptus* is native to the state.

Despite the beauty of many *Eucalyptus* species, a walk beneath a dense grove in California typically reveals a ground surface that is almost completely devoid of any other plant cover. The same phenomenon is experienced in other parts of the world where trees have been introduced. The genus is associated with strong allelopathy through volatilization of allelochemicals in the air, permeation of soils with allelochemicals, and a dense leaf litter that contains phytotoxins. These chemicals and related forest-floor conditions have been shown to be toxic to germination, root growth, and the establishment of many other plant species. This produces significantly decreased plant biodiversity within the *Eucalyptus* groves. Native plant species are often particularly vulnerable to the effects of *Eucalyptus* allelopathy in areas where the genus has been introduced. Aside from decreasing plant competition, some of the chemical compounds produced by *Eucalyptus* serve as a defense against herbivores and pathogens. Taken together with the decreased understory plant biodiversity, the result can be decreased insect, amphibian, reptile, bird, and mammal diversity. The visually beautiful and aromatic groves of introduced *Eucalyptus* trees can be an allelopathic killing ground for many other species. This has spurred widespread concern in areas outside of its native Australia where trees of the genus have become abundant.

alone, the population initially grew rapidly. The growth then tapered off as silica abundance declined. Finally, a stable population size was reached and maintained. A similar pattern occurred when the competing diatom species *Synedra ulna* was grown alone in the silica-enriched water. What happens when we observe the populations of competing species that are grown together in the laboratory? When the two diatom species are grown together, the rate of population increase for both species is slower. Eventually, *Asterionella formosa* is driven to extinction by *Synedra ulna*. In this case *Synedra ulna* is a superior competitor. Such interactions have been observed in many laboratory experiments using competing species.

Examination of maps showing the range limits of some species of plants and animals provides evidence in support of competition as an important biogeographic force. For example, the distribution of different species of kangaroo rats (*Dipodomys* spp.) in the desert of the southwestern United States and adjacent Mexico shows little geographic overlap in the ranges of these closely related species (Fig. 4.4). A similar pattern of nonoverlapping ranges is apparent in the geographic distributions of subspecies of chuckwalla lizards *(Sauromalus obesus)* in the same region or subspecies of rat snakes *(Elaphe obsoleta)* in the southeastern United States. It is possible to suggest, from such mapped distributions, that competitive exclusion of one species by another species may be an important factor in determining range limits. However, in many cases closely related species do overlap in range limits. One example is provided by the South American distribution of species of caiman *(Caiman)*, a relative of alligators and crocodiles. In the case of the caiman, only two of the six South American species have very sharp range limits with each other, whereas in most cases species overlap significantly in their geographic distributions. It is possible that the species with sharp range boundaries with no overlap exclude each other through competition for similar resources such as prey or nesting sites. The overlapping species may avoid competition by making use of different prey or habitats, thereby lessening competition pressure.

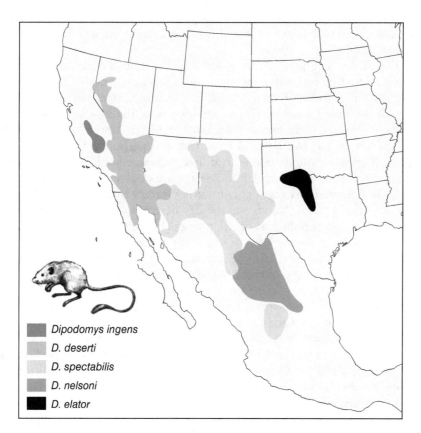

Dipodomys ingens
D. deserti
D. spectabilis
D. nelsoni
D. elator

FIGURE 4.4 The distribution of kangaroo rat species in the southwestern United States and Mexico shows sharp range boundaries that suggest the importance of competition with other species in limiting the geographic ranges of each species (Bowers and Brown, 1982; Brown and Lomolino, 1998).

Even when two closely related species do have sharp range limits and no geographic overlap in distribution, a number of factors other than competition may still influence their range limits. For example, the winter climate in the northern regions where the kangaroo rat species *Dipodomys deserti* and *D. ingens* are found is much colder than the climate in Mexico where the species *D. nelsoni* and *D. spectabilis* live. It is possible that the northern species are better adapted to cold temperatures. Cold temperatures could be the factor that ultimately controls the northern range limits of some kangaroo rats.

The distribution of the closely related cuckoo-dove species *Macropygia mackinlayi* and *M. nigrirostris* on small islands off New Guinea provides the opportunity to study competition in a region where such large differences in climate do not affect the species. It has been found that the two species of cuckoo-doves never occur together on the same island. This observation suggests that competitive exclusion keeps the two species from inhabiting the same islands. However, it has also been proposed that random chance alone could lead to the observed distributions of the two bird species. All in all, we can conclude that analysis of species range limits from maps cannot be considered unequivocal evidence that competition has a primary importance in determining geographic distributions.

Human introductions of alien plant and animal species do provide evidence that interspecific competition can cause changes in the geographic distribution of species. Perhaps one of the most striking results of such competitive displacement occurred through the introduction of Eurasian annual grass species to California by the Spanish and other European colonists.

Prior to European contact, the grassland in the low and mid-elevations of California was dominated by native annual and perennial grass and herb species. Europeans introduced a variety of annual grass species from Eurasia that are able to occupy the same habitat as the native herbs and grasses. The Eurasian annual grasses grow rapidly with the onset of the winter rains, flower and set seed, and then die during the waterless summer months. The next winter, the seeds germinate and begin the cycle again. These aggressive competitors are well adapted to California's seasonally wet and dry climate, and the native species have been displaced throughout most of the state's grasslands. The golden hills associated with summers in California are largely an artificial landscape dominated by competitively superior alien invaders.

Although the California grasslands are a dramatic example of competitive displacement, they represent a situation created by human intervention. Direct evidence of the competitive interactions controlling natural distributions is more difficult to find. One classic study that provides such evidence is that of the distribution of chipmunk species in the mountains of the southwestern United States. There are some 20 species of chipmunks in North America. Like the caiman in South America, some species do not overlap and have very distinct range boundaries, whereas others have overlapping distributions and appear to utilize different habitats within those ranges. The isolated mountains in the southwestern United States are cool and moist compared with the surrounding desert. These mountains support coniferous forests that provide a suitable habitat for chipmunks. The chipmunks cannot survive in the dry desert, so the mountains serve as islands of habitat that support isolated populations of chipmunks (Fig. 4.5). The distribution and behavior of four chipmunk species, *Eutamias dorsalis, E. minimus, E. quadrivittatus,* and *E. umbrinus,* have been studied in detail. Many mountains in the region have the right environment to support all four species. However, in many cases one or more of the species is absent. It is thought that competition may play a role in excluding some species from the mountains. When the two species *E. dorsalis* and *E. umbrinus* occur together, *E. umbrinus* is found only in the dense forests of the upper elevations and *E. dorsalis* only in the open woods at low elevations. A relatively narrow zone separates the ranges of the two species. However, if one of these species is present on a mountain and the other is absent, the species that is there can be found at both lower and higher elevations. This indicates that the two species can tolerate the range of physical and biological conditions found at both high and low elevations but are excluded from certain elevations when the competing chipmunk is present. The elevational distributions of *E. quadrivittatus* and *E. dorsalis* also appear to be strongly influenced by competition between these species (Fig. 4.5). Finally, when other species of chipmunks are present, *E. minimus* is always restricted to the low-elevation shrubby habitat. Observations suggest that interference competition plays a direct role in these elevational distributions. *E. dorsalis* is very aggressive in physically excluding other chipmunks within its range. This strategy is particularly

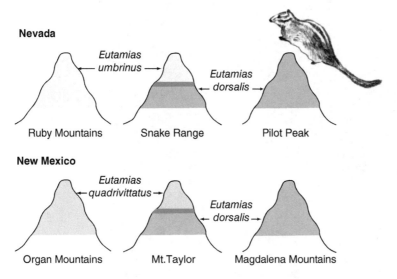

FIGURE 4.5 The distribution of three different chipmunk species on isolated mountains in the Great Basin of the western United States. Where *Eutamias umbrinus* or *E. quadrivittatus* are present, they competitively exclude the species *E. dorsalis* from highelevation sites (adapted from Brown and Lomolino, 1998).

effective in the open and scattered woods of lower elevations. At higher elevations, *E. dorsalis* expends much energy trying to defend habitat in relatively continuous forest. In contrast, the two species *E. umbrinus* and *E. quadrivitattus* do not waste as much energy in defending territory in the closed forest, and this gives them a survival advantage over *E. dorsalis*. The species *E. minimus* is less aggressive and behaviorally subordinate to all of the other species. As a final point, although competition appears to be clearly important in the elevational distribution of the chipmunk species, it must be remembered that habitat conditions in the form of patchy versus continuous tree cover ultimately determine which species wins the competition.

SYMBIOSIS: MUTUALISM, COMMENSALISM, PARASITISM, AND MIMICRY

Symbiosis is the close association between two species that generally develops through coevolution. In many cases, the symbiotic association is obligatory for the survival of one or both species. There are several different types of symbiosis. Different species may interact closely with each other in a manner that benefits both species, called **mutualism;** or benefits one species and has no impact on the other, called **commensalism.** In the case of **parasitism,** one species benefits at the expense of the other. **Parasites** are organisms that are wholly dependent on other organisms for nutrients and, in many cases, microhabitat. They do not immediately kill their hosts. Because of the obligatory nature of most symbiotic relationships, this type of biological control determines the geographic distribution of one or both of the species involved.

Some examples of mutualistic relationships are very obvious. For example, many plants require insects, birds, or mammals to transfer their pollen from flower to flower. Many plants produce nectar that provides food for the pollen-transferring organism. When, as we have already noted, a humming-bird places its beak within the flower, some pollen rubs onto it and is carried to the next flower that the bird visits. Pollinating relationships can tightly control the geographic distributions of the organisms involved. For example, in the desert of the Southwest, moths of the genus *Tegeticula* feed only on the desert plant *Yucca*. In turn, the *Yucca* depends on the moth to transfer their pollen. In some mutualistic relationships neither species can extend its geographic range into areas that cannot support the other. Most hummingbirds, however, are able to feed from a number of different plant species and do not have geographic ranges that are constrained by the distribution of a single flower species. Indeed, the relationship between plant species and the animals that pollinate them or disperse seeds is often not sufficiently obligative to be the primary control on the geographic distribution of the species.

FIGURE 4.6   Some species of clownfish *(Amphiprion)* are not harmed by the poison of sea anemones. Some species of clownfish are immune to the poison of only one or two species of anemones and have a similar geographic distribution to those anemone species. Other clownfish species can coexist with a number of different anemone species and do not have a geographic distribution that is tied to just one anemone species. *Source:* National Park Service / Public domain / https://www.nps.gov/media/photo/view.htm?id=EF600348-155D-4519-3EF2-617B6CE1CBC5

In the ocean, clownfish of the genus *Amphiprion* benefit from a close and often commensalistic association with sea anemones without adversely affecting the anemone (Fig. 4.6). Although they look like plants, anemones are animals that use their poisonous tentacles to stun or kill fish. The fish are then drawn in for digestion. The clownfish are impervious to the poison of the anemone and feed on detrital material left from the anemone's meals. In addition, the clownfish can escape predators by staying within the tentacles of the anemone. Some clownfish appear to be immune to the poison of one species of anemone but are killed when coming into contact with other species. However, other species of clownfish can survive in a relationship with a number of different anemone species. In some cases, the clownfish might serve to attract prey to the anemone, in which case the relationship between clownfish and anemone is mutualistic.

Some forms of mutualism are extremely intimate. For example, lichens, which form the splotchy black and yellow growths on rocks and old stone structures, are a symbiotic organism composed of a fungus and a photosynthetic partner such as algae, cyanobacteria, or both. The fungus is the visible part of the lichen and provides a microenvironment in which the photosynthetic organism can grow, while the algae or cyanobacteria capture energy through photosynthesis. The algal partner would never be able to survive on the rocky substrates, favored by many lichens, if the fungus did not provide protection against desiccation. The fungus, in turn, absorbs nutrients from the photosynthetic partner.

Corals perhaps provide the most striking example of intimate mutualism. The reef-building organisms in corals are members of the class Anthozoa. They are related to sea anemones. Many of these species exude tough outer skeletons made of calcium carbonate that produces the hard reef structure. The Anthozoa generally harbor symbiotic coralline algae within their tissue. The Anthozoa provide protection for the algae and benefit from the photosynthate produced. Such intimate symbionts must, of course, share identical geographic ranges.

The majority of parasites are restricted to one host species. The highly specific relationship between parasites and hosts is the result of an evolutionary process whereby hosts evolve chemical or physiological defenses against parasites and parasites in turn evolve specialized adaptations that circumvent these defenses. Upon reflection, one can see how stenophagous herbivorous insects are perhaps more akin to parasites than predators. Conversely, some ecologists consider parasitism to be a form of predation.

Almost all multicellular organisms, including humans, harbor parasites. Fortunately, many of these are not lethal. Microparasites include viruses, bacteria, and protozoa. Macroparasites include parasitic worms, ticks, fleas, lice, and mites. There are also some interesting plants that are macroparasites. These are plants that have no photosynthetic tissue and live by securing water and nutrients from host plants. One example from the forests of North America is the Indian Pipe *(Monotropa uniflora)*. This small white plant lacks chlorophyll and obtains its nutrients from the roots of other plants.

Since almost all parasites are very tightly restricted to certain host species, their geographic ranges are controlled by the distribution of their hosts. In some cases, the geographic distribution of the parasite may be smaller than the range of the host. There are several good examples of this among human parasites. Humans are host to the parasitic protozoans that cause malaria and sleeping sickness. However, the distribution of these parasites is largely confined to tropical regions. The reason for this distribution is that both parasites are transferred to humans by a second host. In the case of malaria, the other hosts are certain species of mosquito. Sleeping sickness is hosted by the tsetse fly. The mosquito species and tsetse flies that commonly host malaria or sleeping sickness cannot tolerate low temperatures and are therefore restricted to warm regions.

Parasites must have hosts that live long enough for the parasite to complete its life cycle and transfer its progeny to new hosts. Given this dependence on a living host, parasites may not be key factors in directly determining the geographic distributions of host species. However, parasites are probably one of the most important factors in determining mortality and natality rates, and thus the abundance of natural populations of plants and animals. Interestingly, cyclic variations in the abundance of parasites and hosts, similar to predator-prey cycles, have been observed in nature. Red grouse *(Lagopus scoticus)* in northern England and Scotland are afflicted with a parasitic nematode worm *(Trichostrongylus tenuis)* that causes death or decreased fertility. It has been observed that populations of the grouse and the nematode rise and fall on a 5-year cycle.

Prior to concluding our discussion of parasites, we should consider one important, and extremely sad, case of geographic range being controlled by the presence of a parasite. The native bird fauna of the Hawaiian Islands is dominated by a family of birds called honeycreepers (Drepanididae). They are finchlike birds, of which some 30 species are known from the islands. The Hawaiian honeycreepers evolved in isolation and are found nowhere else in the world. During the end of the twentieth century, naturalists noticed a steady decline in the honeycreeper population. On the island of Oahu, six species became extinct by 1900. It was noted that species found on mountains above 600 meters of elevation persisted, while the lowland honeycreepers became extinct. Eventually, it was shown that the honeycreepers were being killed by a form of avian malaria carried by the mosquito species *Culex pipens fatigans*. The mosquito had likely been introduced to Hawaii in the 1820s in water carried on European sailing ships. Not having evolved with this parasite, honeycreepers have no defense against it. Fortunately, the mosquito is a tropical species and cannot physiologically tolerate the cool temperatures of the Hawaiian mountains. Some honeycreepers have therefore escaped extinction but are limited in their geographic distribution to elevations above 600 meters. Most species of honeycreeper that were restricted to the lowlands have been lost forever.

One particularly interesting form of biological interaction is mimicry, in which one species evolves the appearance or behavior of another species. In many cases, the mimic species may not interact directly but may share common predators with which they both interact. **Müellerian mimicry** occurs when a species that is poisonous or unpalatable to predators possesses the same coloring or shape as another species that is also poisonous or unpalatable. Both species benefit since predators avoid either species. The term is named after the German-Brazilian naturalist Fritz Müller, who described the phenomenon in butterflies from South America in the nineteenth century. A striking example of the role of Müellerian mimicry in geographic distributions is provided by the ranges of different races of the butterfly species *Heliconius melpomeme* and *Heliconius erato* from South America. Both species are unpalatable to predators. Each species has a number of different races that are discernible by differences in wing coloration. The different races of the two species have almost completely overlapping ranges with the similarly colored races of the other species. In this case, both species benefit since predators, recognizing the coloring of either species, avoid eating them. The geographic distributions of the races appear to reflect this.

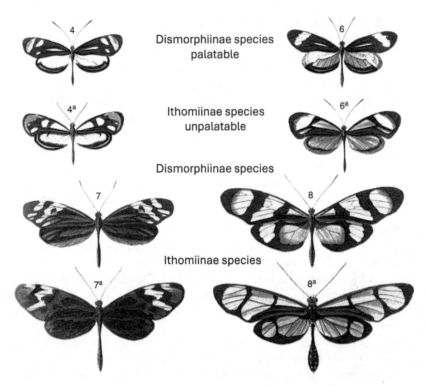

FIGURE 4.7   Batesian mimicry for the South American butterfly subfamilies—Dismorphiinae and Ithomiinae (based on illustrations by Bates [1861], 1981).

Batesian mimicry occurs when one species that is not poisonous or unpalatable has the same coloring or shape as a species that is poisonous or unpalatable. The palatable species benefits from coexisting with the unpalatable species because predators mistake it for the other species and avoid it. This can be considered a form of commensalism. A good example of the potential influence of Batesian mimicry on geographic distributions comes from the biogeography of poisonous coral snake *(Micrurus)* species and nonvenomous colubird snake *(Pliocercus)* species in Central America. The different colubird species possess coloring that mimics the markings of the coral snakes that live in the same geographic region. The term is named after the nineteenth-century English scientist and explorer Henry Walter Bates. He described this type of mimicry as also being prevalent in butterflies, of which there are numerous examples of palatable species having similar wing shapes and coloring to nonpalatable species (Fig. 4.7).

COMBINED PHYSICAL AND BIOLOGICAL CONTROLS ON DISTRIBUTION

Here and in Chapter 3, we have focused on cases where individual physical or biological factors appear to be important in controlling the geographic distributions of certain species. However, for the vast majority of nonsymbiotic species, the actual geographic range limits are probably controlled by a combination of several physical and biological factors. This is illustrated by a classic study of life on the rocky intertidal zone of Scotland. Two species of barnacles, *Balanus balanoides* and *Chthamalus stellatus,* are common on rocky intertidal shorelines in northwestern Europe. The adult *Chthamalus stellatus* is always found in the uppermost portion of the intertidal zone when the two species occur together, although the larvae can also be found at greater depths (Fig. 4.8). When *Balanus balanoides* is removed from the rocks, it has been found that *Chthamalus stellatus* can occupy both the upper and lower parts of the intertidal zone. *Balanus balanoides* grows very quickly and smothers out *Chthamalus stellatus* when the two species occur together. However, *Balanus balanoides* cannot tolerate the dry conditions of the upper intertidal zone. *Chthamalus*

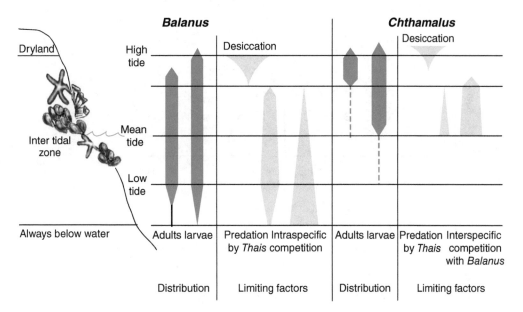

FIGURE 4.8 Distribution of the barnacle species *Balanus balanoides* and *Chthamalus stellatus* in the intertidal zone of the Scottish seacoast. The relative effects of physical factors (desiccation) and biological factors (competition and predation) on the distributions of the two barnacle species are indicated (adapted from Connell, 1961).

stellatus can survive at higher and drier sites than *Balanus balanoides* but dies from desiccation at sites above the mean spring high-water level. In the lower intertidal zone, competition with algae and predation by the snail *Thais lapillus* determines the lower range limits of *Balanus balanoides*. Thus, the simple spatial distribution of these two barnacle species reflects a combination of physical control through intolerance of desiccation and biological controls through competition and predation.

The determination of the geographic range limits of species through such combinations of physical and biological constraints is increasingly recognized, including at larger spatial scales. A study of the winter range limits of an insectivorous bird, the yellow-browed leafwarbler *(Phylloscopus humei)*, demonstrated that the northern limit was determined by the occurrence of climatic conditions cold enough to produce leaf loss in deciduous plants, and this resulted in the absence of the insect prey that the leafwarbler depended upon. The southern limit of the yellow-browed leafwarbler appears to be determined by the presence of the related species *Phylloscopus trochiloides*, which is much larger in body mass and outcompetes the smaller *Phylloscopus humei* for habitat. However, *Phylloscopus trochiloides* appears to be limited in its northern population size and distribution by a deficit of large-bodied insect prey in the north.

INTERACTIONS, GRADIENTS, AND NICHES

Based on the examples we have looked at in this chapter, we can conclude that certain types of biological interactions, such as symbiosis, are extremely important in controlling geographic ranges, while other interactions, such as predation and competition, are more variable in their impact on distributions. To conclude this chapter, we look at how these biological interactions are incorporated into the environmental gradient and niche concepts.

In the cases of predation and symbiosis, we can consider the abundance of prey or hosts as being part of the environmental variability that determines the carrying capacity of a site. Just as we might have a gradient for soil moisture and incorporate this gradient as an axis in defining the niche of a species, so too we might have a gradient and niche axis representing prey density. In some cases, we might also consider incorporating competition in a similar manner. However, it is very difficult to define the actual impact of competition in a numerical manner and make an appropriate gradient or

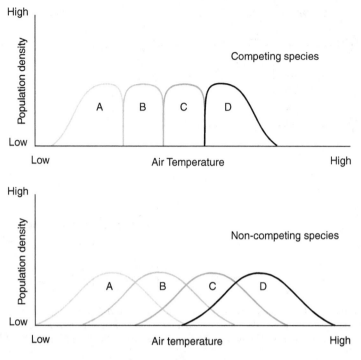

FIGURE 4.9 Hypothetical gradient distribution of four competing species and four noncompeting species.

niche axis to represent it. The impact of competition has been an area of special interest and accommodation within theories of gradient distributions and the niche.

The impact of competition on the distribution of species is often studied by examining the abundance of different species along environmental gradients. If we graph the population density of several species along an environmental gradient and find that the distribution of one species is abruptly truncated by another species (Fig. 4.9), we might conclude that competition between the two species leads to an abrupt replacement of one species by a competitor. Alternatively, we might find a high degree of overlap between the distributions of the species. We could then assume that competition between the species is not an important factor in controlling their distribution. Analyses of the elevational distribution of dominant tree species in the Great Smoky Mountains of eastern North America has been used to study competitive interactions of tree species along an environmental gradient. In this case, elevation represents a gradient of increasing moisture and decreasing temperature from low to high altitudes in the mountains. The data suggest that there is a gradual replacement of tree species along the elevational gradients and that competition does not play an important role in these distributions. Some ecologists, particularly those working with plant communities, feel that there is not much evidence that competition is a strong force in the modern ranges of plant species. They consider that because competition has a negative impact on both species involved, organisms should evolve in such a way as to avoid or reduce direct competition for resources. The perceived lack of competition in modern communities has been poetically referred to as the "ghost of competition past." However, not all ecologists agree with the conclusion that competition is unimportant in controlling natural distributions.

The impact of competition on the niches of species is often examined by portraying competing species within the same niche space (Fig. 4.10). Competition should be manifest as a reduction in the niche breadth of one or both species where they overlap. The broad niche of a species in the absence of competition is referred to as its **fundamental niche.** The more restricted niche that occurs if competition excludes the species from certain portions of niche space is called the **realized niche.** Within the natural environment, the fundamental niche would define all the environmental conditions and resources used if competition did not exist. In theory, no two species should have completely overlapping niches. This is called the **competitive exclusion principle** and can be paraphrased as "complete competitors cannot coexist." That is, if two species have absolutely identical requirements and

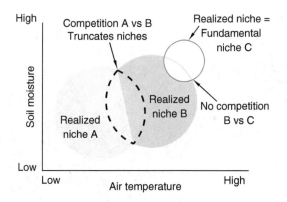

FIGURE 4.10 The realized and potential niches of three hypothetical plant species.

relationships to the environment, the high degree of competition between them will drive one species to extinction. Ecologists have long sought to understand how much differentiation in niche is required for species to coexist.

An interesting study of competitive exclusion between pine species has been conducted in the Sierra Nevada Mountains of California. The study looked at the distributions of closely related pines on sites that had similar slope, exposure, and soil type. By controlling for these habitat factors, the researcher was able to study how the pines were distributed along an elevational gradient, without local habitat differences influencing the study. In this case, increasing elevation represented a gradient of increasing moisture and decreasing temperature. The study found that when one looked only at sites with similar slope, exposure, and soil type, there was an abrupt replacement of pine species by other pine species along the elevational gradient. However, if the distributions of pines on different slopes, exposures, and soil types are included in the analysis, the actual elevational ranges of the pines are seen to overlap greatly. The elevational ranges of the pines overlap because they are able to coexist by taking advantage of different habitat. In the context of geographic distributions, we expect that when species face significant competition, they will have more restricted or fragmented distributions than when they do not face competition. One potential example of this is the distribution of the yellow warbler *(Dendroica petechia)*. In North America this insectivorous bird is found in a wide variety of habitats and geographic locations. However, the bird also ranges into the tropics of Central and South America. In these tropical regions, the yellow warbler is restricted to coastal sites and islands formed by mangrove trees. It is possible that competition from the large number of tropical insectivorous bird species excludes the yellow warbler from other habitat types within the tropical portion of its range.

It might be concluded that competition can exert a control on geographic distributions, particularly among closely related species or members of the same guild. However, it is also likely in other cases that small-scale heterogeneity in the environment, such as differences in slope, exposure, and soil type might allow for potentially competing species to coexist geographically by taking advantage of different habitat types. In addition, species may display different behaviors or phenotypic attributes when living together with closely competing species than when each species is living in separate areas. This is referred to as **character displacement** and results in **niche divergence.** An example of this is provided by the northern ranging Canadian lynx *(Lynx canadensis)* and the closely related, but more southerly ranging bobcat *(Lynx rufus)*. The two species are similar morphologically (Fig. 4.11) and utilize similar prey. In geographic regions where these two feline predators are found living together, the Canadian lynxes are limited to sites that are cooler and have greater snow depths compared to situations where lynxes are present and bobcats are absent. Thus, as a result of competition, the realized niche of the lynxes diverges from the fundamental niche and is more specialized in regions where bobcats are present. It is possible that competition by bobcats in areas that lack cold and snowy habitat is one of the factors controlling on the southern range limits of lynxes in the United States. The southern range limits of bobcats in Mexico appear to be in part limited by competition from a number of other felid predators, including puma *(Puma concolor)*, jaguar *(Panthera onca)*, and ocelot *(Leopardus pardalis)*.

FIGURE 4.11 Canadian lynx *(Lynx canadensis)* and the closely related, but more southerly ranging bobcat *(Lynx rufus)*. Competition by bobcats may be an important factor in determining the southern range limit for lynxes *(Source:* US Fish and Wildlife Service / Public Domain).

The Hutchinsonian niche concept has formed the basis of much of our discussion. It differs from the Grinnellian and Eltonian niche conceptualizations in various nuanced ways. The Grinnellian niche focuses on the habitat a species lives in and the adaptations, particularly behavioral, that habitat promotes. The Eltonian niche as originally conceived focused on the biotic environment and the relationship of the species to prey, predators, and competitors. That interaction is not static, but is altered by species over time, so the niche is created and altered by these. In recent years, there has been increasing focus on the process of **niche construction,** in which the interplay of ecology, genetics, and evolutionary changes and the way in which species themselves alter their biological and physical environments produce species niches and allow their alteration. Over medium to long time spans, niches are much more dynamic than might be appreciated based on a simple analysis of current relationships between a species and the environment.

KEY WORDS AND TERMS

Allelopathy	Intraspecific competition	Optimal foraging theory
Batesian mimicry	Keystone species	Parasites
Character displacement	Lotka-Volterra model	Parasitism
Commensalism	Müellerian mimicry	Predation
Competitive exclusion principle	Mutualism	Realized niche
Euryphagous	Niche construction	Stenophagous
Fundamental niche	Niche divergence	Symbiosis
Interspecific competition	Oligophagous	

REFERENCES AND FURTHER READING

Baldwin, N. S. (1964). Sea lamprey in the Great Lakes. *Canadian Audubon Magazine*, 2–7.

Bates, H. W. (1981). "Contributions to an insect fauna of the Amazon valley (Lepidoptera: Heliconidae)." *Biological Journal of the Linnean Society* **16**(1), 41–54. (Revised version of paper published in 1861)

Begon, M., Harper, J. L., and Townsend, C. R. (1996). *Ecology: Individuals, Populations and Communities*, 3rd edition. Blackwell Scientific Publications, Oxford.

Bowers, M. A., and Brown, J. H. (1982). Body size and coexistence in desert rodents: chance or community structure. *Ecology* **63**, 391–400.

Brower, L. P. (1977). Monarch migration. *Natural History* **86**, 40–53.

Brower, L. P., and Malcolm, S. B. (1991). Animal migrations: endangered phenomena. *American Zoologist* **31**, 265–276.

Brown, J. H., and Lomolino, M. V. (1998). *Biogeography*, 2nd edition. Sinauer Associates, Sunderland, MA.

Buss, L. W. (1986). Competition and community organization on hard surfaces in the sea. In *Community Ecology* (ed. J. M. Diamond and T. J. Case), pp. 517–536. Harper and Row, New York.

Case, T. J., Holt, R. D., McPeek, M. A., and Keitt, T. H. (2005). The community context of species' borders: ecological and evolutionary perspectives. *Oikos* **108**, 28–40.

Caughley, G., and Sinclair, A. R. E. (1994). *Wildlife ecology and management*. Blackwell Science, Cambridge, MA.

Chappell, M. A. (1978). Behavioral factors in the altitudinal zonation of chipmunks (*Eutamias*). *Ecology* **59**, 565–579.

Chu, C., Mortimer, P. E., Wang, H., Wang, Y., Liu, X., and Yu, S. (2014). Allelopathic effects of Eucalyptus on native and introduced tree species. *Forest Ecology and Management* **323**, 79–84.

Cockburn, A. (1991). *An Introduction to Evolutionary Ecology*. Blackwell Scientific Publications, London.

Connell, J. H. (1961). The influence of interspecific competition and other factors on the distribution of the barnacle *Chthamalus stellatus*. *Ecology* **42**, 710–723.

Connell, J. H. (1980). Diversity and the coevolution of competitors, or the ghost of competition past. *Oikos* **35**, 131–138.

Connell, J. H. (1983). On the prevalence and relative importance of interspecific competition: evidence from field experiments. *American Naturalist* **122**, 661–696.

Connor, E. F., and Simberloff, D. (1979). The assemblage of species communities: chance or competition. *Ecology* **60**, 1132–1140.

DeAngelis, D. L., Bryant, J. P., Liu, R., Gourley, S. A., Krebs, C. J., and Reichardt, P. B. (2015). A plant toxin mediated mechanism for the lag in snowshoe hare population recovery following cyclic declines. *Oikos* **124**, 796–805.

Dempster, J. P., and Lakhans, K. H. (1979). A population model for cinnabar moth and its food plant, ragwort. *Journal of Animal Ecology* **48**, 143–164.

Diamond, J. M. (1975). Assembly of species communities. In *Ecology and Evolution of Communities* (ed. M. L. Cody and J. M. Diamond), pp. 342–444. Belknap Press of Harvard University Press, Cambridge, MA.

Dobson, A. P., and Hundson, P. J. (1992). Regulation and stability of a free-living host-parasite system: *Trichostrongylus tenuis* in red grouse. II. Population models. *Journal of Animal Ecology* **61**, 487–498.

Ehrlich, P. R. (1961). Intrinsic barriers to dispersal in the checkerspot butterfly. *Science* **134**, 108–109.

Ehrlich, P. R. (1965). The population biology of the butterfly *Euphydryas editha*. II. The structure of the Jaspar Ridge colony. *Evolution* **19**, 327–336.

Elton, C. S., and Nicholson, M. (1942). The ten year cycle in numbers of the lynx in Canada. *Journal of Animal Ecology* **11**, 215–244.

Gaston, K. J. (2009). Geographic range limits: achieving synthesis. *Proceedings of the Royal Society of London Series B: Biological Sciences* **276**, 1395–1406.

Gulland, F. M. D. (1995). The impact of infectious diseases on wild animal populations—a review. In *Ecology of Infectious Diseases in Natural Populations* (ed. B. T. Grenfell and A. P. Dobson), pp. 20–51. Cambridge University Press, Cambridge.

Hanna, E., and Cardillo, M. (2014). Island mammal extinctions are determined by interactive effects of life history, island biogeography and mesopredator suppression. *Global Ecology and Biogeography* **23**, 395–404.

Hanski, I., Hansson, L., and Henttonen, H. (1991). Specialist predators, generalist predators, and the microtine rodent cycle. *Journal of Animal Ecology* **60**, 353–367.

Katti, M., and Price, T. D. (2003). Latitudinal trends in body size among over-wintering leaf warblers (genus *Phylloscopus*). *Ecography* **26**, 69–79.

Keith, L. B. (1983). Role of food in hare population cycles. *Oikos* **40**, 835–895.

Kingsland, S. E. (1985). *Modeling Nature: Episodes in the History of Population Ecology*. University of Chicago Press, Chicago.

Krebs, C. J. (1985). Do changes in spacing behaviour drive population cycles in small mammals? In *Behavioural Ecology* (ed. R. M. Sibly and R. H. Smith), pp. 295–312. Blackwell Scientific Publications, Oxford.

Krebs, C. J. (1994). *Ecology*, 4th edition. Harper-Collins College Publishers, New York.

Krebs, C. J. (2011). Of lemmings and snowshoe hares: the ecology of northern Canada. *Proceedings of the Royal Society of London B: Biological Sciences* **278**, 481–489.

Krebs, C. J., Boonstra, R., Boutin, S., Dale, M., Hannon, S., Martin, K., Sinclair, A. R. E., Smith, N. M., and Turkington, R. (1992). What drives the snowshoe hare cycle in Canada's Yukon. In *Wildlife 2001: Populations* (ed. D. R. McCullough and R. H. Barrett), pp. 886–896. Elsevier, New York.

Levins, R. (1968). *Evolution in Changing Environments*. Princeton University Press, Princeton, NJ.

Lotka, A. J. (1932). The growth of mixed populations: two species competing for a common food supply. *Journal of the Washington Academy of Sciences* **22**, 461–469.

MacLulick, D. A. (1937). Fluctuations in the number of the varying hare (Lepus americanus). *University of Toronto Studies, Biology Series* **43**, 1–136.

MacNally, R. C. (1995). *Ecological Versatility and Community Ecology*. Cambridge University Press, Cambridge.

Mattoni, R. (1990). *Butterflies of Greater Los Angeles*. Center for the Conservation of Biodiversity/Lepidoptera Research Foundation, Beverly Hills, CA.

May, F. E., and Ash, A. E. (1990). An assessment of the allelopathic potential of Eucalyptus. *Australian Journal of Botany* **38**, 245–254.

McGill, B. J., Enquist, B. J., Weiher, E., and Westoby, M. (2006). Rebuilding community ecology from functional traits. *Trends in Ecology & Evolution* **21**, 178–185.

McPherson, J. K., and Muller, C. H. (1969). Allelopathic effects of *Adenostoma fasciculatum*, "chamise" in the California chaparral. *Ecological Monographs* **39**, 177–198.

Medina, F. M., Bonnaud, E., Vidal, E., Tershy, B. R., Zavaleta, E. S., Josh Donlan, C., Keitt, B. S., Le Corre, M., Horwath, S. V., and Nogales, M. (2011). A global review of the impacts of invasive cats on island endangered vertebrates. *Global Change Biology* **17**, 3503–3510.

Morin, P. J. (2011). *Community Ecology*, 2nd edition. Wiley-Blackwell, Hoboken, NJ.

Muir, A. M., Krueger, C. C., and Hansen, M. J. (2012). Re-establishing lake trout in the Laurentian Great Lakes: past, present, and future. *Great Lakes Fishery Policy and Management: A Binational Perspective*, 2nd edition, pp. 533–588. Michigan State University Press, East Lansing.

Muller, C. N., Muller, W. H., and Haines, B. L. (1964). Volatile growth inhibitors produced by aromatic shrubs. *Science* **143**, 471–473.

Nishida, N., Tamotsu, S., Nagata, N., Saito, C., and Sakai, A. (2005). Allelopathic effects of volatile monoterpenoids produced by *Salvia leucophylla*: inhibition of cell proliferation and DNA synthesis in the root apical meristem of *Brassica campestris* seedlings. *Journal of Chemical Ecology* **31**, 1187–1203.

Peers, M. J., Thornton, D. H., and Murray, D. L. (2013). Evidence for large-scale effects of competition: niche displacement in Canada lynx and bobcat. *Proceedings of the Royal Society of London B: Biological Sciences* **280**, 20132495.

Peterson, A. T. (2003). Predicting the geography of species invasions via ecological niche modeling. *Quarterly Review of Biology* **78**, 419–433.

Peterson, A. T. (2006). Uses and requirements of ecological niche models and related distributional models. *Biodiversity Informatics* **3**, 59–72.

Petren, K., and Case, T. J. (1997). A phylogenetic analysis of body size evolution and biogeography in chuckwallas *(Sauromalus)* and other iguanines. *Evolution* **51**, 206–219.

Pimm, S. L. (1991). *The Balance of Nature?* University of Chicago Press, Chicago.

Pocheville, A. (2015). The ecological niche: history and recent controversies. In *Handbook of Evolutionary Thinking in the Sciences* (ed. T. Heams, P. Huneman, G. Lecointre, and M. Silberstein), pp. 547–586. Springer, Dordrecht, Germany.

Price, P. W., and Clancy, K. M. (1983). Patterns in number of helminth parasite species in freshwater fishes. *Journal of Parasitology* **69**, 449–454.

Price, T. D., and Kirkpatrick, M. (2009). Evolutionary stable range limits set by interspecific competition. *Proceedings of the Royal Society B: Biological Sciences* **276**, 1429–1434.

Quammen, D. (1996). *Song of the Dodo: Island Biogeography in an Age of Extinctions*. Touchstone, New York.

Rice, E. L. (1984). *Allelopathy*. Academic Press, New York.

Roughgarden, J. (1974). Niche width: biogeographic patterns among *Anolis* lizard populations. *American Naturalist* **108**, 429–442.

Roughgarden, J. (1986). A comparison of food-limited and space-limited animal competition communities. In *Community Ecology* (ed. J. Diamond and T. J. Case), pp. 492–516. Harper and Row, New York.

Sanchez-Cordero, V., Stockwell, D., Sarkar, S., Liu, H., Stephens, C. R., and Gimenez, J. (2008). Competitive interactions between felid species may limit the southern distribution of bobcats *Lynx rufus*. *Ecography* **31**, 757–764.

Savidge, J. A. (1987). Extinction of an island forest avifauna by an introduced snake. *Ecology* **68**, 660–668.

Schoener, T. W., and Spiller D. A. (1996). Devastation of prey diversity by experimentally introduced predators in the field. *Nature* **381**, 691–694.

Silvertown, J. W., and Doust, J. L. (1993). *Introduction to Plant Population Biology*. Blackwell Scientific Publications, Oxford.

Simberloff, D. (2004) Community ecology: is it time to move on? *American Naturalist* **163**, 787–799.

Sinclair, A. R. E., Krebs, C. J., Smith, J. N. M., and Boutin, S. (1988). Population biology of snow-shoe hares. III. Nutrition, plant secondary compounds, and food limitation. *Journal of Animal Ecology* **57**, 787–806.

Soberón, J. (2007). Grinnellian and Eltonian niches and geographic distributions of species. *Ecology Letters* **10**, 1115–1123.

Turner, N. C. (1986). Crop water deficits: a decade of progress. *Advances in Agronomy* **39**, 1–51.

Volterra, V. (1926). Fluctuations in the abundance of a species considered mathematically. *Nature* **118**, 558–560.

Warner, R. E. (1968). The role of introduced diseases in the extinction of the endemic Hawaiian avifauna. *Condor* **70**, 101–120.

Waser, N. M., Chittka, L., Price, M. V., Williams, N. M., and Ollerton, J. (1996). Generalization in pollination systems, and why it matters. *Ecology* **77**, 1043–1060.

Whittaker, R. H. (1956). Vegetation of the Great Smoky Mountains. *Ecological Monographs* **26**, 1–80.

Whittaker, R. H. (1960). Vegetation of the Siskiyou Mountains, Oregon and California. *Ecological Monographs* **30**, 279–338.

Williamson, G. B. (1990). Allelopathy, Koch's postulates, and the neck riddle. In *Perspectives on Plant Competition* (ed. J. B. Grace and D. Tilman), pp. 143–162. Academic Press, New York.

Willmer, P. G., and Stone, G. N. (1997). How aggressive ant-guards assist seed-set in *Acacia* flowers. *Nature* **388**, 165–167.

Yan, C., Stenseth, N. C., Krebs, C. J., and Zhang, Z. (2013). Linking climate change to population cycles of hares and lynx. *Global Change Biology* **19**, 3263–3271.

Yeaton, R. I. (1981). Seedling morphology and the altitudinal distribution of pines in the Sierra Nevada Mountains of central California: a hypothesis. *Madrono* **28**, 487–495.

Zhang, C. L., and Fu, S. L. (2009). Allelopathic effects of *Eucalyptus* and the establishment of mixed stands of *Eucalyptus* and native species. *Forest Ecology and Management* **258**, 1391–1396.

CHAPTER 5

DISTURBANCE

In 1988 Yellowstone National Park experienced its most active wildfire season in recorded history. Despite the $120 million spent to fight the blazes, approximately 320,915 hectares of forest were burned, affecting some 36% of the total area of the park. Countless animals were either killed or driven from their original habitat. It is estimated that the toll included 300 large mammals such as elk, bison, mule deer, and moose. Despite the apparent damage fires cause in places such as Yellowstone National Park, biogeographers recognize that fires are a natural part of many ecosystems. By the 2020s, the areas of Yellowstone that were burned and apparently left lifeless in 1988 were alive with hundreds of species of wildflowers, shrubs, and small trees. Aspens and lodgepole pines grew quickly after the blaze and new forests of young trees have regrown across the park. The bird and mammal populations of the park are thriving. Rather than long-term destruction, the effect of the 1988 fires and subsequent fires has led to rejuvenation of many ecosystems. The 1988 fires were not the first nor the last to burn in Yellowstone. Over the past 50 years, Yellowstone National Park has experienced an average of 26 fires and 2368 hectares burned each year. About 78% of these fires are sparked by lightning. Biogeographers recognize that we cannot understand the functioning and distribution of organisms and ecosystems in Yellowstone or any other region without understanding the role of natural disturbances such as fires.

Ecosystems are prone to many forms of periodic disturbances. In many cases, ecosystems depend on disturbance for long-term functioning. The term **disturbance** is used to describe short-term physical or biological events that significantly alter ecosystems. These events include things like fires, wind storms, landslides, floods, avalanches, and outbreaks of diseases or pests. Rather than being thought of as unnatural events that act to destroy ecosystems, disturbances are generally natural and vital parts of the environment and important for the long-term functioning of ecosystems. The frequency and impact of disturbances vary widely between and within ecosystems. Giant sequoia (*Sequoiadendron giganteum*) forests of the southern Sierra Nevada Mountains of California might experience fires in the low vegetation of the ground surface as frequently as once every 7 to 9 years. These fires kill ground vegetation and seedlings but rarely damage mature sequoia trees. In contrast, hurricanes may only impact the forests of the southeastern Great Lakes region less than once every century but can destroy most of the tree canopy in their paths when they do occur.

A useful formal definition of disturbance and its impact is provided by White and Pickett (1985): "A disturbance is any relatively discrete event in time that disrupts ecosystem, community, or population structure and changes resources, substrate availability, or the physical environment" (p. 7). They suggest that disturbances can be described in terms of cause, frequency, predictability, spatial extent, magnitude of impact on ecosystems, synergism with other disturbances, and seasonal timing. Some disturbances, such as small fires generated by lightning, may frequently occur at almost random intervals, while other disturbances, such as outbreaks of pathogenic insects, appear to occur at more regular intervals. Volcanic eruptions and other catastrophic disturbances may be very rare and unpredictable. Some disturbances are linked together in synergistic relationships. For example, in many mountainous regions, fires destroy vegetation and produce unstable soils that lead to landslides and further vegetation disturbance.

The adjustments and changes that occur in ecosystems following disturbances have long been topics of great interest to biogeographers and ecologists. Let's start by considering some of the classic terms and concepts applied to the recovery of an ecosystem following disturbance. We will then reexamine these concepts in light of more recent research. **Succession** is the term used to describe the changes in physical and biological conditions that follow disturbances. There are two major types of succession. **Primary succession** occurs when a previously lifeless surface is first colonized by plants and animals. Specifically, it takes place when new islands or land surfaces are formed by volcanic activity, when sea-level changes or shoreline deposition create new land surfaces, or when glaciers melt and expose new areas of land. **Secondary succession** occurs when an existing ecosystem recovers from a disturbance such as fire or flood.

In some cases, species that colonize a disturbed site change the environment and allow the establishment of yet other species. For example, the establishment of a tree canopy during succession provides a cool, moist, and shady environment that supports forest floor species. The process by which the establishment of one species changes the environment and allows the subsequent establishment of other species is called **facilitation.**

The process of succession is often divided into different stages. A typical sequence of these stages for a given vegetation type is called a **sere.** Seral stages can be observed when agricultural fields in the southern Great Lakes region are allowed to return to a forested state (Fig. 5.1). The initial successional vegetation consists of annual grasses and herbs such as quackgrass *(Agropyron repens)* and ragweed *(Ambrosia artemisiifolia)*. Within the first 10 years, perennial herbs and grasses such as Kentucky bluegrass *(Poa pratensis)* dominate the site. By 40 years, there is a marked increase in dominance by woody shrubs such as roses *(Rosa* spp.) and bramble *(Rubus* spp.). The seedlings and saplings of white pine *(Pinus strobus)* and red and white oak *(Quercus rubra* and *Q. alba)* become more dominant during this time. If no further disturbance occurs, a mixed oak and pine forest will occupy the former agricultural field within 100 years. The final vegetation type developed during succession is sometimes referred to as a **climax** community.

A special type of succession is recognized when ponds and wetlands fill in with sediment and become terrestrial environments. This process is referred to as *hydroseral* or *hydrarch succession*. As water depths in ponds and lakes decline due to natural infilling by mineral sediment and organic debris, plants such as water lily *(Nuphur* spp.) and cattail *(Typha* spp.) can root in the mud, trap additional sediment, and produce increased organic debris. Eventually, the level of sediment rises to the point where it can be colonized by terrestrial plants such as willows *(Salix* spp.) and alders *(Alnus* spp.). Over time, the site will become dry enough to support forest cover. Hydroseral successions are not true primary successions because there is a preexisting aquatic ecosystem at the site. Nor are they secondary successions because the process of infilling is a long-term progressive event, and not the result of a discrete disturbance.

The seral successional patterns outlined above follow the classic model of succession as proposed by the American ecologists Henry Cowles and Frederic Clements at the beginning of the

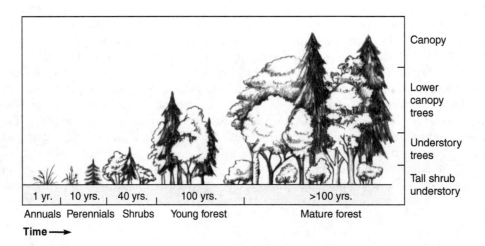

FIGURE 5.1 An idealized "classical" secondary succession on an abandoned agricultural field in the North American Midwest.

twentieth century. Cowles conducted much of his research on the primary succession that occurs on newly stable beaches. Clements worked with secondary succession in deciduous forests and grasslands. Clements was much influenced by the writings of Darwin on evolution and the potential role of the coevolution of species in the development of modern flora. According to Clements' model, seres are highly predictable sequences that are controlled by facilitation. The organisms in the later seres depend on the environmental changes caused by the establishment of the earlier invading plants and animals. The climax community is the predictable end product of succession and is self-perpetuating in the absence of further disturbance. In a sense, the climax community is the norm, and the seral stages are anomalies in nature caused by the destructive impact of disturbances. The exact nature of the climax community was thought to be determined largely by regional climate. A deciduous forest was the natural climax in northeastern North America, whereas a coniferous forest was the natural climax of mid-elevation sites in the Rocky Mountains.

As more data have been collected on disturbance and succession, some original tenets of the Clementsian model of facilitative succession followed by climax conditions have been questioned and often refuted. Various alternative models of succession have been proposed. Some of the most common include the following:

Tolerance Model   In many instances, species of both early and late seral stages are established very quickly following disturbance. The plants that dominate the early successional period simply are those that grow fastest. The slower growing plants that dominate the later stages eventually displace the early dominants. All of the species can tolerate the early successional environments, but as the succession progresses, some of the early dominants can no longer tolerate the increased competition for light and moisture.

Inhibition Model   In some ecosystems, particularly following small disturbances, it is very difficult to predict which species will invade and eventually dominate a site. It appears that proximity to seed sources and chance factors related to seed dispersal play a major role in determining which species occupy the site. Once the site is occupied, the presence of the early invaders inhibits establishment of other species.

Initial Floristic Composition Model   The species composition of the site following disturbance both during the early and late stages of the succession is determined mainly by which species are present at the site immediately following disturbance. The different successional stages largely reflect the different life cycles and morphology of the plants as they grow and mature.

Random Model   Finally, it has been suggested that succession can be typified simply as the random arrival and survival of species at a disturbed site. This model assumes that there is no facilitation or inhibition.

Many studies have shown successional pathways to be much more complicated and unpredictable than the unidirectional seres leading to a stable climax. For example, much work has been conducted on the primary succession that occurs as glaciers have melted and exposed land surfaces at Glacier Bay, Alaska. Since the rate of glacial retreat is known, the ages of different land surfaces can be estimated. By comparing the vegetation on young surfaces, intermediate-aged surfaces, and old surfaces, the temporal pattern of succession may be reconstructed. This technique of using different aged patches of the landscape to reconstruct a general pattern of succession is called the *chronosequence approach* and is widely used. The chronosequence approach is often referred to as a "space for time" method of inferring succession patterns.

Based on chronosequence data, it was initially thought that the sites on Glacier Bay represented a classical case of facilitative succession through predictable seres reaching stable climax. On sites deglaciated within the past 20 years, the vegetation on the rocky surfaces includes rock mosses *(Racomitrium canescens* and *R. lanuginosum),* herbs such as fireweed *(Epilobium latifolium)* and mountain avens *(Dryas drummondii),* and small shrubs such as Arctic willow *(Salix arctica).* Slightly older sites are dominated by larger species of willow and mountain alder *(Alnus crispa).* The alder may form large, almost pure thickets on sites that are 50 years old. On still older sites, alder thickets are replaced by cottonwood trees *(Populus trichocarpa).* Sitka spruce *(Picea siitchensis)* eventually comes to dominate the vegetation on sites that have been ice-free for 120 years. By 200 years, a forest dominated by sitka spruce and western hemlock *(Tsuga heterophylla)* is established. The importance of alders in succession at Glacier Bay has been of particular interest because their roots contain nodules that host symbiotic bacteria. The bacteria have the ability to transform nitrogen from the atmosphere into nitrate, which is required by plants. It has been argued that the nitrogen-fixing ability of the alders made them important facilitators of further succession by coniferous trees.

Later research shows that the unidirectional successional model developed from the chronosequence studies at Glacier Bay is oversimplistic. Research has been conducted using tree-ring analysis to examine the actual ages and year of establishment of woody plants on the different sites around Glacier Bay. Contrary to earlier conclusions, the tree-ring-based study shows that sitka spruce invaded the older sites very quickly after deglaciation and did not appear to require facilitation by the nitrogen-fixing alders. In fact, alder appears to have been important in the successional vegetation of only some of the sites. The early invasion of sites by spruce appears to be positively correlated with the proximity of spruce stands. Sites that were near stands of spruce received high seed falls and were more quickly dominated by spruce following deglaciation. Cottonwood is only an important component in the seral vegetation at some of the sites, notably those deglaciated in the past 100 years. Finally, although spruce and hemlock form a climax community on some sites, on poorly drained locations Sphagnum moss and shore pine *(Pinus contorta* ssp. *contorta)* eventually replace this forest. Thus, the primary succession at Glacier Bay follows multiple pathways and can lead to very different seral and climax communities. Because stochastic factors such as proximity of seed sources and environmental factors such as slight differences in drainage or soil type can be very important in determining successional communities on different sites, the results of chronosequence studies must be interpreted with great caution. For this reason, tree-ring analysis of the long-term history of individual sites has become increasingly important for reconstructing vegetation disturbance and succession in forested ecosystems (Technical Box 5.1).

Glacier Bay represents a relatively simple system. An example of greater complexity in multi-path successional sequences is provided by primary succession on sand dunes along the shores of Lake Michigan. The level of the lake relative to the land has been falling over the past 12,000 years. As a result, chronosequence studies of ancient dune surfaces and newly created ones can provide evidence of successional and climax communities of many different ages. At a typical site, there is a sequence of vegetation succession that starts with marram grass *(Ammophila breviligulata)* colonization of freshly stabilized dunes. Marram grass is a very interesting species that is particularly adapted to primary succession on dunes. It seldom reproduces by seeds, and it spreads mainly through underground rhizomes and pieces of leaf and stem. It aggressively colonizes and stabilizes dune surfaces within a few years after colonization. If further sand accumulates, marram grass grows rapidly upward and keeps pace with deposition. Interestingly, the grass seems to die out naturally after the dune's surface is stabilized and after deposition of new sand ceases. The stabilized dunes are eventually occupied by jack, red, and white pine *(Pinus banksiana, P. resinosa, P. strobus)* forest within 100 to 200 years. Finally, deciduous forest dominated by black and white oaks *(Quercus velutina, Q. alba)* and/or sugar and red maple *(Acer saccharum, A. rubrum)* occupy the site. However, because of very small differences in initial topography and drainage conditions, a number of other successional pathways can be identified, and at least six different "climax" vegetation communities can come to dominate individual sites (Fig. 5.2).

Research on post-fire succession in Australian mountain ash *(Eucalyptus regnans)* forests in the Central Highlands of Victoria also show that there are multiple successional pathways that may be followed after disturbance and that these can lead to different "climax" communities as a result. For example, if a second fire occurs at a mountain ash site in less than 25 years after the initial fire, the newly established ash trees will not have had time to have reached sexual maturity and the resulting forest will become dominated by *Acacia* trees instead.

After considering both the plethora of successional concepts and theories that have developed over the past century, and considering their experience with Australian mountain ash ecosystems, Stephanie Pulsford, David Lindenmayer, and Don Driscoll suggested that the development of vegetation following disturbance can be attributed to five main mechanisms: (1) pulse events when sequential establishment of different plant species produces rapid and pervasive shifts in the vegetation; (2) stochastic events related to chance occurrences such as climatic conditions following disturbance that may favor some species, but not others; (3) facilitation as described previously; (4) competition such as described previously as inhibition; and (5) the influence of the species composition of the earliest

TECHNICAL BOX 5.1

Tree-Ring Analysis, Stand History, and Fire Reconstruction

Tree-ring analysis is one of the most widely used techniques to study forest disturbance and succession in nontropical regions. Many biogeographers use tree-ring analysis. The science of dating past events by tree rings is called dendrochronology. Many coniferous and dicotyledonous trees growing in areas with seasonally variable climates produce annual growth rings. Monocots, such as palm trees, do not produce annual rings. In most conifers, the rings are composed of couplets of light and dark wood. Wood consists of the xylem cells (water-conducting tissue) of the tree. The light-colored wood is produced by rapid xylem growth at the start of each year's growth season. The light-colored wood is composed of thin-walled and large xylem cells and is called early wood. The start of the growth season might be the spring in forests that experience cold winters. As the tree grows during the summer, additional xylem cells form and add to the width of the ring. As the autumn approaches, xylem growth slows, and the cells are smaller and have thicker walls. These dense, late-growth cells are generally dark in color and are called late wood. With the onset of winter, the tree becomes dormant, and xylem growth ceases until the next spring when a new ring couplet begins to grow. Somewhat similar couplets of early wood and late wood can be detected in dicotyledonous trees, but the transition from early wood to late wood is often not as clear. The only living cells in the trunk of a tree are found in this narrow tree-ring-forming zone just beneath the bark. This living zone of the trunk is called the cambium. The phloem cells that form each year at the outside edge of the cambium are crushed outward by the growing xylem formed at the inside edge and are incorporated into the bark.

The simplified anatomy of the wood of a tree typical of those used in dendrochronological studies. The oldest rings are near the pith, and the youngest are near the bark.

If you cut down a tree with annual rings and look at the trunk, you will see concentric circles of couplets of early wood and late wood. The innermost point of wood is called the pith. The pith represents the tree's first year of growth. The outermost ring represents the most recent year of growth. So as to not kill or injure trees, special small corers are generally used to obtain tree-ring samples. By counting the rings from the pith to the outermost ring, the age of the tree in years can be estimated. Dendrochronologists use a number of checks and statistical techniques to verify that the rings they count are annual. The process by which dendrochronologists measure, count, and assign an age to a ring is called cross-dating. By cross-dating the rings from many trees in a stand, one can estimate the age of the patch that the stand comprises. The age of the trees can provide a minimum estimate of the last time a major flood, avalanche, wind storm, fire, or other disturbance created the gap in which the present trees of the patch became established.

Obtaining tree-ring cores from spruce trees for dendrochronological analysis at a site in Utah.

Some types of trees, such as many species of pines, can preserve evidence of past fires within their tree rings. Surface fires often burn away portions of bark and wood and kill a part of the cambium. This can leave a long and deep vertical scar on the lower portion of the trunk. The dead cambium does not recover but can be eventually overgrown by living cambium at its edges. These areas of dead cambium produce fire scars that are preserved in the tree-ring record. Dendrochronologists can cross-date the rings prior to the scar and the rings formed by undamaged cambium around the scar to estimate the age of the fire scar. Trees such as pines *(Pinus)* and redwoods *(Sequoia)* from western North America can preserve dozens of scars from individual surface fires. By sampling a large number of trees for fire scars, spatially and temporally detailed histories of past fire activity can be reconstructed.

Fire-scarred coastal redwood tree in California.

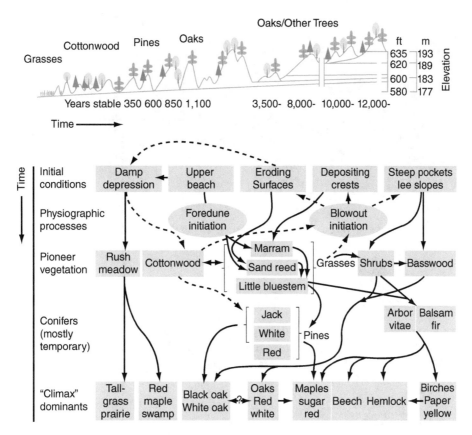

FIGURE 5.2 The multiple successional pathways evident for succession on sand dunes along the shoreline of Lake Michigan (Olson, 1958; Krebs, 1959; Krebs, 1994).

arriving/reproducing plants in defining the post-disturbance community as described by the Initial Floristic Composition Model. These researchers make the important point that these processes can all be operating concomitantly.

As a final point regarding classical succession theory, many studies have now shown that so-called climax communities cannot in fact propagate themselves in the absence of disturbance. This is particularly true in forests, where shading by the mature canopy has been shown to exclude both the early seral species and many of the so-called climax trees. Regeneration in these instances often requires the formation of canopy gaps due to external disturbances or the senescence and mortality of stand dominant trees. In short, although the Clementsian model of facilitative succession and climax is elegant, it is far too great a simplification and too often misleading.

Over the years, many properties have been attributed to the various seral stages that occur during succession. It has been suggested that the process of succession leads invariably to higher biomass, higher growth rates, higher diversity, or greater stability. Such generalizations cannot really be applied to a broad spectrum of ecosystems. For example, plant species diversity in the coniferous forests of Alaska is often highest during the early stages of succession. During this time, annual and perennial grasses, herbs, and shrubs are intermixed with tree seedlings and small saplings. The biomass of small mammal communities is highest in these early successional communities and decreases as shrubby vegetation and coniferous forest are established. In contrast, the species diversity of small mammals is highest during the period when shrubby vegetation is dominant.

Biogeographers are often interested in disturbances because the landscapes they study consist of a mosaic of patches that represent sites at different stages of successional recovery from disturbance. When you stand atop a mountain and look out upon the forest below, you are seeing a patchwork of vegetation made up of units of different sizes and ages that were created by disturbance-generated gaps (Fig. 5.3). Some patches, such as those created by large fires, are huge in size, while others, perhaps created by the wind toppling a single tree, are quite small. The physical and biological environments

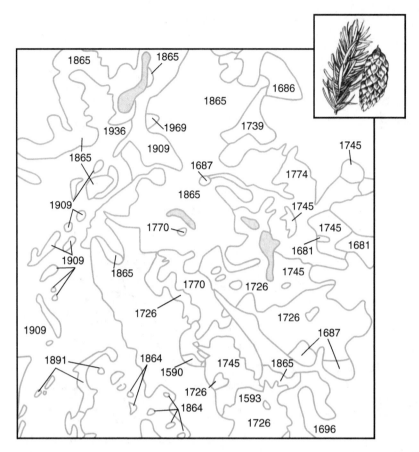

FIGURE 5.3 Different-aged patches that make up a portion of the Rocky Mountain coniferous forest in Alberta. The map shows how the environment is made up of patches of different ages and sizes caused by disturbances. In this case, most of the patchiness is imparted by different forest fires that have occurred over the period from 1590 to 1969. The outline of each patch and the age of the fire that created it are indicated on the map. The vegetation and microclimate of a recent patch are very different from those of a site that was burned hundreds of years ago (adapted from Johnson and Larsen, 1991).

within patches of varying ages and sizes can be quite different. For example, a large area that has been recently burned experiences significantly higher light intensities, higher air temperatures, and lower humidity than adjacent forested sites. Some disturbances and the patch environments they create are detrimental to certain species and thus limit geographic distributions. Frequently, disturbances create patches that allow some species to persist in areas where they might not otherwise be found. A closed forest canopy does not provide enough insolation at ground level for many light-demanding herbs and grasses. A recently opened gap in the forest provides the proper physical environment for such plants and allows them to persist. Thus, biogeographers who try to understand the distribution of species without considering the impact of disturbance and succession do so at great risk. In this chapter we will examine some common forms of physical and biological disturbance, associated successional patterns, and the influence of disturbance on geographic distributions.

FIRE

Fire is the most pervasive and well-studied form of disturbance in terrestrial environments. Wildfires have been recorded at places ranging from Arctic Alaska and Canada to the depths of the South American tropical rainforests. Wildfires have been an important part of the earth's ecology for more than 400 million years. Satellite remote sensing provides an important tool for mapping the earth's wildfires even in the most remote areas. Satellite observations indicate that approximately 200 to

500 million hectares of the earth's land surface burns annually. The highest amounts of annual area burned by wildfires are found in tropical seasonal forest, savanna, and grassland systems, but fires are also common in many other ecosystems, including temperate grasslands and boreal forest. Despite expenditures of billions of dollars on fire prevention and suppression, fire remains a potent form of disturbance even where intensive prevention and suppression efforts are employed. For example, Canada is a world leader in the development and application of equipment and techniques to fight forest fires. However, according to one estimate, each year in Canada thousands of recorded fires occur, mainly in coniferous forests, and the average area burned each year was over 1.6 million hectares for the period from 1985 to 2015. Over this time period, 51 million hectares burned in total, impacting more that 6.5% of Canada's forests. However, the annual totals are quite variable; in some years, over 7 million hectares have burned. The forests of Canada may seem remote and in many cases are lightly inhabited. However, at the other end of the continent, a single fire in April 1999 burned approximately 28,000 hectares in the Everglades region of south Florida, despite the proximity of the blaze to major cities and a large fire-fighting infrastructure. Across the United States from 2011 to 2020 there was an annual average of approximately 62,000 wildfires and on average some 3 million hectares impacted annually. As in Canada, the annual area burned by wildfires in the Unites States can be quite variable. In 2020 wildfires burned about 4 million hectares, largely in the western United States.

The occurrence of fire is so widespread and its ecological impacts so important that its geographical study can be identified as a special subfield of biogeography. The term **pyrogeography** is used to denote the study of the past, present, and projected future distribution of wildfire and its impacts. Evidence suggests that despite advanced fire-fighting capacity, in many parts of the world wildfires are becoming larger and more frequent, making pyrogeography research all the more critical. In this section we will look at the behavior, natural occurrence, and impact of fire. We will also examine the interesting fire-related adaptations that some plants possess and consider the positive impacts that fire has on certain environments.

The occurrence of fire requires adequate fuel and an ignition source. The fuel for wildfires is provided by vegetation, while the ignition source for most natural fires is lightning. Physical factors such as topography, climate, and the type of vegetation that serves as fuel all influence the behavior of fire. Different types of vegetation have inherently different fuel characteristics. Both plant structure and composition determine the fuel characteristics of vegetation. Trees and shrubs, which maintain dead lower branches, provide fire ladders that allow flames to climb from the ground to the canopy. Many species of pine possess dead lower limbs that serve as fire ladders. The lack of dead branches on self-pruning trees such as Douglas fir (*Pseudotsuga* spp.) is one reason the adult trees of this genus are not killed by the frequent surface fires that burn through their stands. Some plants, including pines and eucalyptus, contain high amounts of resins or other volatile oils and ignite much more easily and burn at higher temperatures than other plants. Pines, for example, typically burn at temperatures of up to 800° C. Jack and lodgepole pines *(Pinus contorta* and *P. banksiana)* of the northern forests of North America have been found to be much more flammable than associated deciduous trees such as aspen *(Populus tremuloides)* and paper birch *(Betula papyrifera)*. As a result, the chance of high-intensity fires in stands dominated by pine is generally higher than in deciduous or mixed stands. In addition to structure and chemical composition, the flammability of vegetation can be enhanced by climatic factors. Low precipitation coupled with strong winds causes drying of living and dead biomass and greatly increases flammability.

The total amount of flammable biomass available for burning is important in determining fire occurrence and intensity. For example, a very direct relationship has been shown between maximum temperature of fires in Texas grassland and grass biomass (Fig. 5.4). Areas with patchy vegetation are unable to support extensive or intense fires. Many deserts do not experience large fires because plants are small and sparsely distributed. Dead biomass, in the form of leaf litter or dead branches, is an important fuel for fires in forested regions. Deadwood can make up one-third of the standing biomass in some California shrub communities. Removal of living and dead biomass by fires decreases fuel load and reduces the probability of subsequent severe burning. The chance of a severe fire increases over time as vegetation regrowth leads to an accumulation of living and dead fuel. In annual grasslands, such as those found in California, high amounts of fuel can accumulate within one growing season after a fire. The fire-prone chaparral and coastal sage shrublands of southern California

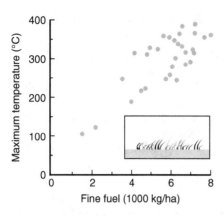

FIGURE 5.4 The positive relationship between fuel biomass and fire temperature in Texas grasslands (adapted from Stinson and Wright, 1969).

recover quickly and regain pre-fire biomass within 10 to 20 years following an intense burn. Post-fire accumulation rates of fuel in forests are often slower than in grasslands and shrublands. However, in Australian eucalyptus forests, it takes only 10 to 20 years for the highly flammable leaf litter to build up to pre-fire levels.

The principal natural ignition source for wildfires is lightning. At any one time as many as 1800 lightning storms may be occurring on earth. Lightning is generated by unstable atmospheric conditions associated with convective storms and the uplift of warm air along cold frontal systems. In tropical regions, convective storms can occur throughout the year, whereas in the Arctic they are extremely rare at any time. In North America lightning storms are most common in the warm and humid Southeast and least common in the cool Arctic and along the Pacific Coast. There are roughly 10 times as many thunderstorms per year in the southeastern United States as there are in the Pacific Northwest. Many of the southeastern storms occur in the summer. In coastal California lightning can develop with winter frontal storms, but there are few convective storms during the summer. Lightning storms can develop in southern California when warm, monsoonal air flows northward, usually in late summer.

A lightning strike does not always ignite a fire; fuel conditions must be right for a lightning strike to create a fire. The conditions that favor ignition include prolonged dry conditions, high air temperatures, low humidity, and high winds. A lightning strike will not ignite a fire in a damp forest. Highly resinous pines are unlikely to ignite if they have moisture contents that are greater than 35% of their dry weights. Even in summer, the relatively moist conditions typical of the eastern United States preclude ignition of large fires by lightning. Therefore, the number of fires does not necessarily correspond to the number of lightning strikes. The southwestern United States receives frontal precipitation in the winter and monsoonal precipitation in the summer months. The late winter and early spring tend to be extremely dry. Most fires occur in the months of May and June, despite the fact that the highest thunderstorm activity occurs in July and August. The winter months experience some lightning but almost no fires because of the cool and damp conditions.

Humans are the leading ignition source in inhabited areas. An analysis of over 100 years of data for California showed that 95% of recorded wildfires were started by humans. In Canada approximately 50% of all forest fires are started by humans. If ignition does occur, what determines whether a fire will scorch the ground over a few hectares or destroy all vegetation over thousands of square kilometers? The impact of a fire is determined by its rate of spread and intensity. The rate of spread of a fire is simply the distance it advances during a given amount of time. Fires spread through the progressive advance of the burning front and through the ignition of spot fires by flying embers. Fire fronts can advance at rates of several kilometers per hour, while embers can generate spot fires kilometers beyond the main fire front. Fire intensity is sometimes described in terms of its impact on the vegetation. In forested settings, low-intensity **surface fires** destroy ground vegetation while leaving trees intact. Fires that are sufficient to cause tree canopies to ignite produce high-intensity **crown** or **canopy fires.** Fire intensity can also be measured in terms of kilowatts of energy generated per meter along the fire front. Surface fires reach intensities of less than 260 kilowatts per square meter, whereas crown fires may reach intensities of more than 2800 kilowatts per square meter. Several factors govern rates of fire spread and intensity. First, the vegetation has to provide adequately flammable and closely spaced fuel to carry a fire. Fuel flammability is increased by long periods of dry, hot weather. Second, fire spread is promoted by topography that is without natural fire breaks such as lakes, rocky ridges, and rivers. Finally, dry winds provide an important mechanism to increase the rate of spread and intensity. Winds drive the fire forward and bring the flame front down into contact with flammable ground fuels. For the same reasons, fires can also spread more quickly and burn with greater intensity when moving upslope.

Because fire spread is affected by fuel conditions, topography, and weather, there is wide variation in the average size of fires from region to region and year to year. In some damp tropical forests, it is not unusual for most fires to be restricted to individual trees that are hit by lightning. In contrast, wildfires in the coniferous forests of western Canada can frequently exceed 100,000 hectares in size. Although only 3% of the fires in Canadian forests are greater than 10,000 hectares in size, these large fires account for 90% of the total area burned each year. Such extensive fires are generally associated with particularly dry and hot summers. During certain hot, dry years or following prolonged droughts, very high fire activity can be experienced across a broad belt of Canada's forested regions. During the dry summer of 1995, for example, 8,302 fires burned over 7 million hectares in Canada. The hot and dry summer of 2023 saw Canadian fires extend over a record-breaking 16 million hectares. In contrast, during the cool moist summer of 1978, less than 1 million hectares were consumed by fire. Interestingly, during 2020, Canada experienced a very low annual area burned total of about 227,477 hectares during a time of relatively high precipitation in some regions and restrictions on travel, intentional fires, and off-road vehicle use in some forested regions due to COVID-19 pandemic policies. The latter possibly decreased human ignition events. In contrast, the western United States experienced record-setting aridity and saw over 2.5 million hectares burned.

The frequency of fire is also influenced by fuel and climate. There are large differences between the frequency of fires in moist tundra and eastern deciduous forests (where fires may occur less than once every 300 to 1000 years) and seasonally dry areas found in western North America (where surface fires can burn through pine stands and grasslands almost annually). Even within subcontinental regions, there are significant differences in fire frequency. For example, in the northwestern United States, a clear relationship exists between fuel type, temperatures, and moisture stress, and the frequency of fires in forest and woodland ecosystems (Fig. 5.5). Cool-moist coastal forests experience little fire activity, while ponderosa pine *(Pinus ponderosa)* and oak-juniper woodlands in the interior experience fires every 1 to 25 years. Interestingly, when fires do occur in the coastal and high-elevation forests of the Pacific Northwest, they are often much more destructive to the tree cover than the frequent surface burns experienced at the low-elevation and inland sites.

From this discussion we may draw some geographic generalizations regarding the role of fire as an agent of disturbance. Areas such as subtropical grasslands and savanna (mixtures of grasses and woody plants)—which have low topographic relief; are subject to prolonged periods of aridity, low humidity, high temperatures, high winds, and frequent lightning storms; and possess fast-growing vegetation—will experience frequent large fires. In portions of Africa, Asia, Australia, and South America, savanna ecosystems experience fires on an average of every 1 to 5 years. In contrast, the Arctic tundra (low herb and small shrub vegetation) has many lakes and rivers that serve as topographic fire breaks; the tundra is also cold and moist, has few thunderstorms, and supports sparse, slow-growing vegetation. Southern tundra areas burn perhaps once every 1000 years, and fires are generally quite small. Most of the northern Arctic almost never experiences fires.

Plants that exist in regions that are frequently disturbed by fire often exhibit adaptations that allow them to cope with the effects of burning. Some tree species possess fire-resistant bark. Mature trees can be killed by direct burning of foliage or by lethal heating of the cambium layer beneath the bark. A tree with a bark thickness of 0.6 centimeters can survive external heating of 500° C for about 1 minute, whereas a tree with a bark thickness of 2.6 centimeters can survive the same temperatures for 20 minutes. Douglas fir *(Pseudotsuga menziesii* and *P. macrocarpa),* western larch *(Larix occidentlis),* and ponderosa pine *(Pinus ponderosa)* stands in western North America experience frequent fires. All of these trees have bark that can grow to over 10 centimeters in thickness. These species survive significantly longer periods of heating than trees with thinner bark. Fires can also kill root

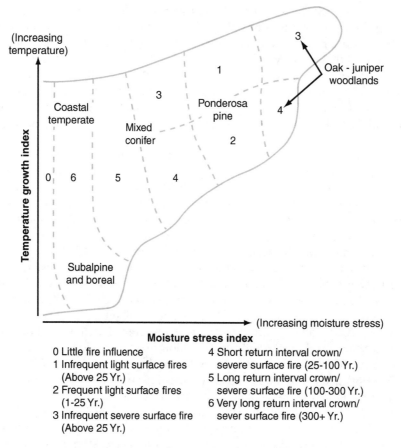

(Increasing temperature)

Temperature growth index

3

Oak - juniper woodlands

1

3

Coastal temperate

Ponderosa pine

Mixed conifer

2

4

0 6   5   4

Subalpine and boreal

(Increasing moisture stress)

Moisture stress index

0 Little fire influence
1 Infrequent light surface fires (Above 25 Yr.)
2 Frequent light surface fires (1-25 Yr.)
3 Infrequent severe surface fire (Above 25 Yr.)

4 Short return interval crown/ severe surface fire (25-100 Yr.)
5 Long return interval crown/ severe surface fire (100-300 Yr.)
6 Very long return interval crown/ sever surface fire (300+ Yr.)

FIGURE 5.5 An idealized representation of the relationship between temperature, moisture stress, fuel type, and fire frequency in the Pacific Northwest of North America (Agee, 1981).

systems, and all four of these species have relatively deep root systems that are insulated by soil cover against high temperatures. Temperatures during fires decline quickly with soil depth. A surface fire burning at a temperature of 600° C may produce a temperature of only 80° C at the soil surface and 33° C at 10 centimeters of depth.

A number of tree and shrub species display epicormic sprouting that allows the regrowth of foliage from trunks and branches after fire. Epicormic sprouting occurs from latent buds that lie behind the bark. In some cases, only buds protected by the thick bark of the trunk may survive, and the new foliage will appear as a *fire column* growing from the standing trunk. Coniferous trees, such as pitch pine *(Pinus resinosa)* in eastern North America and coastal redwood *(Sequoia sempervirons)* in western North America, and angiosperms, such as live oak *(Quercus agrifolia)* in California and *Eucalyptus* species in Australia, display epicormic sprouting following fires.

Some plants can produce new growth from lignotubers, which are swellings that occur at the interface between the roots and shoots. The lignotubers contain buds and reserves of food. Soil cover protects the lignotubers from lethal temperatures during fires. Moderate heating by the fire may even stimulate the buds to grow. A number of shrubs found in the shrublands, or chaparral, of California and adjacent Mexico, such as chamise *(Adenstoma fasiculatum)*, sprout vigorously from lignotubers following fires. Other plants, such as trembling aspen *(Populus tremuloides)* and paper birch *(Betula papyrifera)*, sprout suckers from root buds. Most savanna and prairie grasses resprout from root systems following fires.

A number of plants produce seeds that remain dormant in the soil until **scarification** by heating causes them to germinate following fires. Although the parent plants may be destroyed by the fire, the newly emerged seedling can take advantage of the high soil nutrients and competition-free environment that exists following the burn. For example, the seeds of deerbrush shrubs *(Ceanothus* spp.)

FIGURE 5.6 A dense stand of young lodgepole pines *(Pinus contorta)* established following a fire in Glacier National Park, Montana. Fire in this forested ecosystem is a natural event. The destruction of the forest canopy allows light-demanding species such as lodgepole pine to become established and grow (*Source:* Reimar / Adobe Stock).

germinate best after experiencing temperatures of 75° C to 100° C. The deerbrush seeds can remain dormant and viable in the soil for several decades. The store of dormant seeds in the soil is referred to as a *soil seed bank*. Some plants, especially pines, possess **serotinous cones** that remain closed and hold their seeds until heated by a fire. Jack and lodgepole pine cones can be held on the trees and contain viable seeds for 25 to 75 years. The resin keeping the cone scales closed melts at temperatures of about 60° C, and this releases seeds during and after the fire. The seeds can remain viable when exposed to temperatures as high as 370° C for short periods. The pine seedlings take advantage of the favorable soil conditions and lack of competition that results from fires. It is not unusual to find jack and lodgepole pine seedling densities of over 1 million individuals per hectare following severe fires (Fig. 5.6).

Although we may think of fire as a destructive and negative aspect of the environment, it is clear that fire plays a beneficial role in many ways. Fires serve to release nutrients from dead and living vegetation back into the soil. Fires also clear away leaf litter and decadent vegetation cover. Most plants cannot germinate and grow under the shade of a dense forest canopy. Destruction of the canopy by fire allows shade-intolerant plants to regenerate and take advantage of soils enriched by the minerals in the ash from the fire.

Patterns of post-fire succession have been intensively studied in many forests. For example, stands dominated by white spruce *(Picea glauca)* in the forests of Alaska and northwestern Canada burn every 80 to 400 years (Fig. 5.7). Within 2 to 5 years following a fire, the site is usually dominated by light-seeded plant species, plants with seeds that lie dormant in the soil until stimulated by heat and chemical changes associated with fire, and plants that can sprout from surviving roots and corms. These early dominants include grasses, herbs such as geranium *(Geranium bicknellii)*, fireweed *(Epilobium angustifolium)*, roses such as *Rosa acicularis,* and willows such as *Salix bebbiana* and *S. scouleriana*. However, even at this very early stage, the seedlings of deciduous and coniferous trees are often present in the vegetation. The tree seedlings are nevertheless very small and grow too slowly to dominate the early vegetation. During the next 25 years, large shrubs such as willows and birches *(Betula nana)* and the saplings of deciduous trees such as aspen *(Populus tremuloides)* and paper birch *(Betula papyrifera)* come to dominate the vegetation. Over the next 50 years, deciduous trees dominate the site, and the shade-intolerant grasses, herbs, and willows are replaced by more shade-tolerant species. Spruce, which has been present all along, becomes increasingly important over time, until after 100 years it dominates the vegetation. After 200 years, a dense moss cover forms on the ground, and most of the deciduous trees have died and disappeared from the forest.

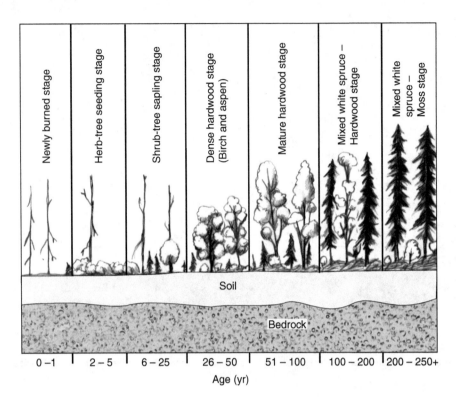

FIGURE 5.7   Idealized post-fire succession in a boreal white spruce *(Picea glauca)* forest (adapted from Van Cleve and Viereck, 1981).

The herbs, shrubs, and deciduous trees that regenerate following northern fires are important sources of food for animals such as moose, beaver, and snowshoe hare. In turn, these herbivores serve as food for predators such as wolf and lynx. Fires are therefore important in supporting the mammalian fauna. The burned stands also provide a number of different microhabitats encompassing fallen and dead trees, older surviving trees, and new vegetation. These microhabitats are used by many different animal species. As one example, the *Melanophila* beetle found in North America deposits its eggs under the bark of freshly burned trees. There the larvae feed on the dead cambium of the trees. The adult beetles have specialized organs to sense heat and find freshly burned stands. The beetles and larvae serve as food for a number of birds such as woodpeckers, which also use standing dead trees as nesting sites.

As biogeographers, we are interested in the role of fire in controlling the geographic distributions of species. One example of this role is provided by the northern distribution of black ash *(Fraxinus nigra)* and some species of maple and oaks in Canada. The trees are typically found in eastern deciduous forests of North America. The trees are almost never found in the more northern boreal forests, which are dominated by coniferous trees such as spruce, fir, pine, and larch. In contrast to the deciduous forest, the northern coniferous forests are subject to frequent intense fires. Ash trees, for example, are thin-barked and do not appear to be particularly well adapted to frequent fires. However, the northern range of ash extends far into the boreal forest in moist areas and along river valleys. Among other things, these areas provide habitat in which fire frequencies are lower than in the surrounding boreal forest. It is possible that the ash trees are able to tolerate the colder northern climates of the southern boreal zone as long as they are protected from frequent fires.

European land-use practices in North America have led to decreased fire activity in a number of areas. The response of vegetation to this decrease in fire provides further evidence of the role of burning in controlling geographic distributions of plants. The area of grassland in southern Texas has declined sharply since 1860 (Fig. 5.8). Much of the grassland first encountered by American settlers has been replaced by mesquite *(Prosopis)* shrubland and woodland. The reduction in grassland and expansion of shrubland and woodland can be partially attributed to fire suppression following American settlement. Cattle ranchers view the expansion of mesquite as a menace to rangelands and

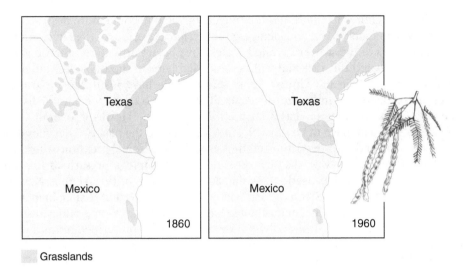

Grasslands

FIGURE 5.8 The expansion of mesquite *(Prosopis)* shrubland at the expense of grassland between 1860 and 1960. Decreases in burning due to human land-use change and fire control are thought to be a major contributor to the decline in grassland cover (Johnston, 1963; Brown and Lomolino, 1998).

undertake various measures, including setting fires to battle it. However, cattle grazing itself may be an additional contributing factor to shrub expansion. Prescribed burns are one tool used to maintain grasslands. Research in the semiarid grasslands of southern Arizona shows that the natural fire frequency there is about 4 to 8 years. Decreased frequency of fire due to grazing and fire suppression has allowed the advance of pine and oak forest at the expense of grassland. Similar recent increases in woodland at the expense of grassland are evident on the Canadian and American plains and in dry portions of the Pacific Northwest of the United States. For example, poplar and spruce have invaded the periphery of the northern prairies in Canada, pines have invaded grasslands in Colorado, while in the Pacific Northwest juniper *(Juniperus* spp.) woods have invaded upland grasslands. The exclusion of shrubs and trees by fire is crucial in maintaining grassland and savanna in many other settings throughout the world. Suppression of fires can also change the species dominance in wooded sites. Studies in northern Florida indicate that the Ocala sand pine *(Pinus clausa)* regenerates poorly beneath its own canopy. Frequent fires serve to open the canopy and allow new pines to establish. Fire control measures over the past century have allowed the pine stands to be invaded by shrubs and evergreen oaks which were formerly excluded by fires. The shade cast by these plants makes it even more difficult for the sand pines to regenerate.

On the other hand, not only has fire been an important agent of disturbance over geologic time, but fire has been used as a land management tool by many cultures for thousands of years. Archaeological evidence for controlled use of fire appears as early as 700,000 to 800,000 years ago. Many forested and nonforested vegetation covers and landscapes around the world have been significantly shaped by these practices. Changes in human population density and land use can drastically alter fire regimes and even change large-scale relationships between climate and fire. In tree-ring-based studies of fire history in the Sierra Nevada Mountains of California, the geographer Alan Taylor and his colleagues found that prior to the arrival of Europeans, Native American use of fire in the forests promoted open stands and decreased forest sensitivity to climate variations in terms of fire activity. Following the crash in Native populations due to introduced diseases and direct exclusion by Europeans, the fire regime was altered by the lack of Native American–set fires. Due to increased fuel loads because of the absence of Native American–set fires and the introduction of twentieth-century fire suppression practices, these Sierra Nevada forests became more susceptible to climatically driven variations in fire behavior.

Although large fires are still capable of burning beyond our control, fire suppression methods were relatively successful in the United States and Canada over the twentieth century. One consequence has been the buildup of a dangerously high level of fuel in some forest ecosystems. These fuels, such as sapling-sized trees and brush, increase the flammability of the ecosystems and allow

low-intensity ground fires to climb into the tree canopy and become more destructive. As a result of this artificially high fuel loading and changing fuel architecture, recent fires in coniferous forests that dominated portions of the western United States have been particularly intense and large. Climate change has exacerbated this risk and will be discussed later in the book. To mitigate this situation, government authorities in the United States and Canada have instituted programs of prescribed burns that are set under carefully controlled conditions in order to imitate natural disturbance regimes and decrease fuel loads. In many cases, such efforts are useful. However, overall, there has been an increase in large fires in much of western North America over the past two decades despite fuel management programs and improved fire-fighting capacity. On some occasions these prescribed fires have gotten out of control, with disastrous results. For example, a prescribed fire that went out of control in the spring of 2000 caused severe damage to the town of Los Alamos, New Mexico, despite intensive efforts to control it. Given the artificially high fuel loads that remain in many western North American forests, coupled with climate change, similar near catastrophes and catastrophes will likely continue into the foreseeable future. It will take many years of prescribed burns and other fuel management techniques to remedy the situation in North America. A study in the Boundary Waters Canoe area of Minnesota concluded that it would take 50 to 75 years of prescribed burning to return the forest structure of that area to its pre-fire-suppression state. Only about 20% of the forested land of California's Sierra Nevada Mountains have been receiving sufficient fuel-thinning treatments, and at that rate will never see significant restoration of current stands to pre-suppression fuel conditions.

WIND

After fire, catastrophic wind is the most frequent and widespread agent of terrestrial ecosystem disturbance (Fig. 5.9). Although fires dominate disturbance regimes in grassland, savanna, and coniferous forests, wind is a more important agent of disturbance in many temperate deciduous forests and tropical forests. Wind is also a major cause of disturbance in many high-altitude coniferous forests. Exposed alpine sites generally experience much higher winds than surrounding lowlands. Research near alpine treeline in western Oregon suggests that between 20% and 70% of the fir stands were established after windthrow events that occurred in the past 60 years. Documentary evidence shows that at least eight severe wind storms occurred during that period.

Forests are more sensitive to wind damage than other vegetation types. Winds in forests strip foliage, break branches, shear trunks, and uproot trees (Fig. 5.10). The uprooting of trees causes soil disturbance and the exposure of mineral soils. Winds of over 100 to 130 kilometers per hour typically

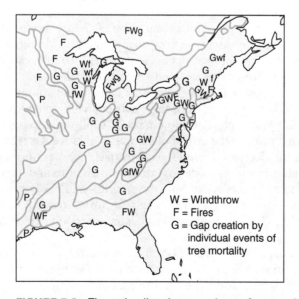

FIGURE 5.9   The major disturbance regimes of eastern North American forest types (Runkle, 1990).

FIGURE 5.10   A windblast destroyed the coniferous tree canopy and allows light-demanding grasses, herbs, shrubs, and tree saplings to become established at this site in Oregon. *Source:* Flickr/R6, State & Private Forestry, Forest Health Protection.

strip trees of foliage and topple them. Tornado-induced winds of 160 kilometers per hour have been shown to be responsible for reducing tree cover by 68% in affected stands in Minnesota. Wind speeds as low as 60 kilometers per hour can still produce significant destruction under certain conditions. Trees are more susceptible to being uprooted by winds when the wind event follows or is contemporaneous with high precipitation—which leaves soils soggy and decreases root stability.

Catastrophic winds, with gusts over 200 kilometers per hour, are generated by hurricanes and tornadoes. Wind microblasts, with velocities reaching 400 kilometers per hour, cause small linear patches of intense tree mortality. Tornadoes generally affect very small areas. However, they can be frequent occurrences in the plains and adjacent forests of interior North America. In contrast, individual large hurricanes cause severe damage to forests over thousands of kilometers. For example, in August 1992 Hurricane Andrew created a 20-kilometer-wide swath of wind damage that impacted over 1600 square kilometers of southern Florida and disturbed 80% of the vegetation in Everglades National Park (Fig. 5.11). It is estimated that Hurricane David (August 1979) killed 5 million trees on the island of Dominica. The impact of Hurricane Katrina (August 2005) on the forests of the Gulf Coast of the United States was estimated using ecological field studies, satellite remote sensing data, and modeling. It was concluded that the hurricane caused severe structural damage or death to approximately 320 million large trees. Hurricane Katrina produced sustained wind speeds of up to 280 kilometers per hour. Hurricanes and associated tropical storms are categorized by their wind speeds. Powerful class 4 and 5 storms, like Hurricane Andrew and Hurricane Katrina, have sustained winds of over 200 kilometers per hour and cause the greatest damage to forest cover. Hurricanes develop in late summer and early fall over warm tropical seas. The convective energy that drives hurricanes dissipates quickly over land or cool oceans. The frequency of ecosystem disturbance by hurricanes is greatest in the tropics and subtropics. It has been estimated that 83% of the forested area of Puerto Rico has been disturbed by hurricanes over the past 100 years. A study in Sri Lanka reports that 73% of the forest there has been similarly disturbed by hurricanes (called cyclones in that region) over the past century. The frequency of hurricanes diminishes relatively quickly as one moves away from the tropics. Nonetheless, the Gulf of Mexico and Atlantic coastlines of the United States also experience hurricanes relatively frequently. Remarkably, during the twentieth century, hurricanes have traveled as far inland and north as the province of Ontario in eastern Canada. Thus, hurricanes

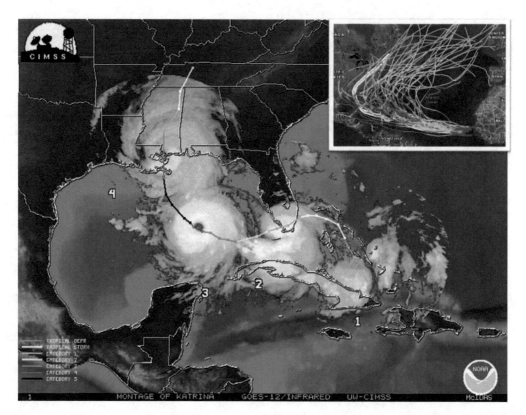

FIGURE 5.11   Time-lapse satellite imagery of the path of Hurricane Katrina (2005). The inset map shows the paths of Category 4 and 5 hurricane tracks over the period 1851–2016 in the East Atlantic Basin (*Source:* NOAA / Public domain).

may be an important force of vegetation disturbance as far north as the forests of New England and adjacent Canada.

Although catastrophic wind events such as hurricanes can impact very large areas of land, the actual size of the gaps (contiguous areas where large numbers of trees have been toppled) created by wind disturbance is highly variable. Average gap size created by wind disturbances generally ranges from the size of individual trees to around 30 hectares. In some cases, gaps of over 3000 hectares have been reported for hurricanes. In contrast, in the coastal forests of the Pacific Northwest of North America, 46% of the windthrow gaps are created by the felling of single trees. Most wind events create a number of different gap sizes. The percentage of trees actually toppled in such gaps can range from 1% to over 90%.

The geographic pattern of tree destruction during catastrophic wind events is highly variable. Tornadoes tend to destroy relatively narrow and discontinuous swaths. Hurricanes and large cyclonic storms cause damage over wider areas. Exposed stands on high ground, at the edges of forests, or on shallow soils are most susceptible to windthrow. Within continuous forests, tall canopy trees are the most prone to destruction. They have bigger areas of foliage relative to trunk size and are more exposed to the shear stresses caused by wind. Data from the impact of Hurricane Andrew on Florida showed that canopy trees suffered up to 70% mortality, while smaller trees suffered less than 20% mortality. When large canopy trees fall, they may crush smaller trees and shrubs. One study in palm forests showed that 70% of the tree mortality during wind storms was due to large canopy trees crushing smaller subcanopy individuals.

Not all wind-generated gaps are associated with catastrophic events. Many small gaps are caused when old and senescent trees are felled by relatively mild winds. Such events are part of the **background mortality** that occurs naturally in forests and causes losses of perhaps 5% of the canopy each year. Interestingly, one study in Puerto Rico showed that hurricanes could be expected to kill about 472 trees per hectare per century. In contrast, background mortality likely causes the death of 1973 to

2650 trees per hectare per century. Thus, in terms of general mortality, hurricanes cause devastating short-term effects but are not the primary cause of tree mortality in this tropical forest regime.

As is the case with fire, wind disturbance plays a crucial role in the functioning of ecosystems. The clearing of canopy trees increases the light received by the lower vegetation layers. In many cases, this facilitates the germination and growth of seedlings that would have been unable to grow under the shade of a dense canopy. The dead vegetation supplements soil nutrients. The churning action of uprooted trees can produce areas of exposed mineral soil which certain plants require for seedling establishment. Soil temperature generally increases in gaps, while humidity decreases. These changes aid in the establishment and growth of plants that cannot survive under dense forest canopies.

There are key differences between the impact of fires and wind disturbance. Surface fires may destroy low vegetation but leave canopy trees intact. Winds always impact the canopy more than the subcanopy layer. Although falling trees can damage the subcanopy vegetation, the damage to the lower vegetation layers during wind events is much less than during fires. Finally, while fires can clear the litter layer and return nutrients to the soil as ash, winds often add undecomposed organic matter to the litter layer.

Tropical and subtropical areas, which are prone to hurricane disturbance, appear to show remarkable rates of succession and recovery. Closed forest canopies at sites in Central America damaged by Hurricane Joan in 1988 developed within 5 years. The height of the canopy, however, was still much lower than its pre-hurricane condition. Hurricanes also kill wildlife and cause severe damage to nesting sites. Of particular concern is damage to nesting sites for birds. However, studies in the Everglades following Hurricane Andrew showed that populations of many wading birds remained stable or increased slightly following the storm. Bald eagles have been shown to have recovered quickly despite the loss of about 44% of their nesting sites in areas of South Carolina impacted by Hurricane Hugo.

Wind is certainly a factor in controlling the geographic distribution of trees in alpine environments. High winds cause limb breakage and the destruction of foliage and buds by blowing snow and ice crystals during the winter. Cold and dry winter winds can also cause the death of needles and buds by desiccation. Winds may also remove soils and seeds. There are many examples where the elevation of the treeline on exposed and windy slopes is considerably lower than on protected sites. In these cases, wind is more important than temperature in determining the local position of the treeline.

The gaps caused by wind are undoubtedly important in allowing shade-intolerant plant species to exist in regions of continuous forest cover. Wind fragmented forests provide a wider range of habitats and may therefore support a larger number of different species. Tree species such as Sitka spruce (*Picea sitchensis*) require the relatively continuous creation of gaps of at least 800 to 1000 square meters in order to have sunlight sufficient to regenerate and remain codominant with coastal hemlock (*Tsuga heterophylla*) in the Pacific Northwest. Hemlock is more successful than Sitka spruce on shady sites and in very small gaps. In the absence of large wind gaps, the spruce would be excluded in many areas. The importance of wind gaps in creating diverse habitat and promoting greater biodiversity is an ubiquitous feature in forested ecosystems.

FLOODING

Flooding is an important form of disturbance impacting low-lying areas near rivers and shoreline ecosystems along oceans and lakes. A flood is formally defined as a high flow of water that overtops normal confinements and covers land that is normally dry. In areas of flat topography, river flooding can expand over thousands of square kilometers. Despite many control measures, widespread river flooding still occurs throughout the world. One example was the destructive flooding of communities and large areas of farmland and forest along the Red River in North Dakota, Minnesota, and adjacent Manitoba during the spring thaw of 1997. In total, the flood waters covered over 20,000 square kilometers. Even this pales in comparison with the Mississippi River Flood of 1927, which inundated some 70,000 square kilometers from Missouri to Louisiana.

Three types of floods are recognized. First, river and associated lake flooding take place when high seasonal precipitation and/or spring snowmelt exceeds soil moisture-holding capabilities and produces high amounts of runoff. River flooding can also be caused by channel blockages owing to factors such as river ice. The area of low-lying land that is typically inundated when a river floods is referred to as the floodplain. Second, flash floods are short-duration events that occur when individual storms produce more runoff water than can be accommodated by stream and river channels. In general, steep mountainous rivers in semiarid regions experience flash floods. Third, coastal floods occur along the ocean margin and are often associated with storm surges that coincide with high tides and hurricanes.

The season, frequency, and severity of flooding vary greatly. Low-lying areas along many rivers are flooded annually. In Brazil some 300,000 square kilometers of land surface are prone to annual flooding by the Amazon and its tributaries. When such regular cycles of dry and wet conditions exist, the term *flood* may be replaced by a term such as *seasonal inundation*. Where long records of river flow are available, it is clear that many flood-prone rivers are subject to frequent low-magnitude floods, with very high-magnitude flooding recurring at widely spaced intervals. The flood frequency of the Red River, which was responsible for the 1997 flooding of North Dakota and Manitoba, shows this pattern. Fargo, North Dakota in most years experiences peak flows of about 1000 cubic meters of water per second or less. The flow during the 1997 flood was three times this amount. It is estimated that such high flows and flooding occur only once every 200 years in that system. For planning and conservation purposes, many resource managers use a measure called the 100-year flood as a worst-case scenario for anticipating flood impacts. The 100-year flood is the peak discharge that is expected to occur once each century. For the Red River at Fargo, the 100-year flood discharge is 2300 cubic meters of water per second.

Flooding impacts terrestrial ecosystems in a number of ways. High water can drown animals and kill plants by restricting oxygen uptake through roots. Decreased root activity can also kill plants by causing cessation of carbohydrate production. In effect, the flooded vegetation starves to death. Flood waters from the ocean carry salts that are lethal to many plants. Evaporation of fresh water can also produce salt crusts that inhibit plant establishment and growth. Anoxic conditions in water-logged soils also lead to the production of toxins such as ferrous and manganese ions and sulfides. Flood waters carry suspended sediment loads that are deposited on the floodplain. The addition of large amounts of sediment to the soil surface can kill plants by depriving root systems of oxygen. Floods also kill plants by eroding soils and removing the rooting surface for plants. In cold climates, flooding can damage vegetation by pushing large slabs of ice onto the shoreline. Because flood events can cause disturbance of ecosystems not just due to inundation by water, but through geomorphological processes such as erosion and sediment deposition, floods are best considered to be a **hydrogeomorphic disturbance.**

Ecosystems in humid, low-lying regions often must be able to endure long periods of inundation by low-energy flood waters. In contrast, floodplain ecosystems in arid and hilly regions often experience high-energy floods that produce short periods of inundation. In these systems, the erosion and deposition of sediment are the most important factors in ecosystem disturbance. It has been calculated that a complete reworking of sediment along the floodplains of rivers on the Great Plains near the Rocky Mountains occurs every 200 to 300 years. In most floodplains, both inundation by water and erosion and deposition caused by floods are active forms of disturbance.

In addition to their destructive potential, floods play a positive role in many ecosystems. High water associated with flooding increases soil moisture and allows forests to grow in regions that are normally too dry to support trees. Such woods are referred to as **riparian forests,** gallery forests, or riverine forests. Examples of riparian trees found on very frequently flooded sites include plains poplar *(Populus deltoides)* stands found on the North American Great Plains, *Eucalyptus camaldulensis* forests in arid portions of Australia, and *Populus ilicifolia* stands in the savanna of Kenya. Floodplain trees such as those mentioned above are usually fast growing, reproduce at an early age, and can withstand soil inundation. In addition to providing moisture, floods can replenish soil nutrients in areas where sediment is deposited. The irrigation and fertilization of drylands by floods have played an important role in human history in many regions. Perhaps most notable is the development of early Egyptian civilization along the floodplain of the lower Nile River. Floods also serve as an important mechanism for seed dispersal. The seeds of many common floodplain species, such as poplar and willow, can survive transport in water. A number of Amazon trees use fish to disperse seeds. In fact, some Amazon plants require that seeds be chewed by fish in order to release them from their hard outer coats. Most relatives of the flesh-eating pirhana fish *(Serrasalmus rhombeus)* of South America are fruit eaters that rely on flooded forests for habitat and food. Similarly, in Africa, the butter barbel fish *(Schilbe intermeddins)* is important in the dispersal of seeds from the sycamore fig *(Ficus sycomorus)*.

Aside from seeds that are adapted for transport by water, floodplain plants possess other adaptations that allow them to cope with floods. The hollow and tubular stems of bullrush *(Schoenoplectus lacustris)* serve to oxygenate water-covered roots. Other plants have long cavities in their wood and roots that allow downward oxygen flow. The coastal variety of lodgepole pine *(Pinus contorta* ssp. *contorta),* which grows on wet sites in the Pacific Northwest and Alaska, produces such cavities. Many floodplain plants have metabolic and physical adaptations that allow them to persist for long periods of time with their roots in anoxic soils. Floodplain plants generally have high tolerances for ferrous and manganese ions or root systems that are inefficient at extracting these potentially toxic ions from the soil. The swamp cypress *(Taxodium distichum)* of southeastern North America produces elongated vertical structures from their roots. These structures, called **pneumatophores,** extend above flood level and allow the root systems to obtain oxygen. Many floodplain plants, such as cattails *(Typha latifolia),* possess **rhizomes** (horizontal stems and associated root systems) that contain large reserves of carbohydrates and can tolerate submergence and oxygen deprivation for months at a time. When detached from the parent plants by floods, these rhizomes can grow and produce a complete new plant. Following floods, new growth can sprout from the toppled stems of riparian willows and poplars (willows and poplars that grow along rivers or streams). The California coastal redwood *(Sequoia sempervirons)* produces new roots from its trunk when buried by flood deposits. These **adventitious roots** develop close to the new soil surface where oxygen is more plentiful.

Flooding controls the geographic distribution of plant species at a number of different spatial scales. At the larger scale, the additional soil moisture provided by flooding allows some tree species to extend their range into arid regions. At a smaller scale, flooding is important in determining local plant distributions. Slight differences in elevation produce large changes in floodplain vegetation. Examples of such local vegetation zonation are apparent from most river systems. As river channels shift, they leave behind new terrain for colonization by terrestrial plants. Succession on these abandoned river channels is also an important factor in floodplain vegetation zonation.

A good example of the local geographic vegetation zonation that develops in floodplains comes from the Tana River of Kenya. Here the floodplain supports gallery forests of *Acacia,* poplar, and a number of evergreen tropical trees (Fig. 5.12). The channel of the Tana shifts often, and the sediments of the floodplain are completely reworked by erosion and deposition on a 150-year cycle. The area above the floodplain is dominated by arid shrubland. Lower levels of the floodplain support a mosaic of bare soil, small shrubs, and saplings. The sparse vegetation on these sites results from frequent erosional and depositional events, as well as long periods of inundation. Forest is restricted to higher areas that experience less than 28 days of continuous inundation. The most diverse forest is found on sites that do not experience flood erosion. In the case of the Tana River, flooding allows trees to extend their range into a relatively arid region but restricts forest development to very specific sites along the floodplain.

Patchy floodplain zonation and succession are typical of forested regions throughout the world. Along the Tanana River in Alaska, recently stabilized river deposits are colonized within a few years by a shrub-dominated vegetation consisting of willow *(Salix alaxensis)* and alder *(Alnus tenuifloia).* If the site is high enough above the current floodplain, the shrubby vegetation is followed by balsam poplar *(Populus balsamifera)* forest, and then mixed balsam poplar, aspen *(Populus tremuloides),* and white spruce *(Picea glauca)* forest. On the oldest and highest land surfaces the forest is dominated by white spruce.

One of the most interesting examples of how flooding controls the very large-scale geographic distribution of a species is provided by bald cypress *(Taxodium distichum)* in the southeastern United States (Fig. 5.13). Bald cypresses are large, striking trees well adapted to flood-prone areas and swamp. They have trunks that spread out at the base in buttress fashion to provide support in water-logged soils. They are also noted for their large and abundant pneumatophores. The seeds of the bald cypress survive immersion in water and are typically transported by floods. The trees are found in the floodplains of slow-moving streams with gentle gradients, in swamps that experience inundation by water for a large portion of the year, and in small shallow lakes called oxbows. Oxbow lakes are formed when the meandering movement of a river or stream channel experiences a rapid shift that leaves a portion of the channel cut off from regular flow. These cutoff channels form small lakes that eventually fill with sediment and disappear.

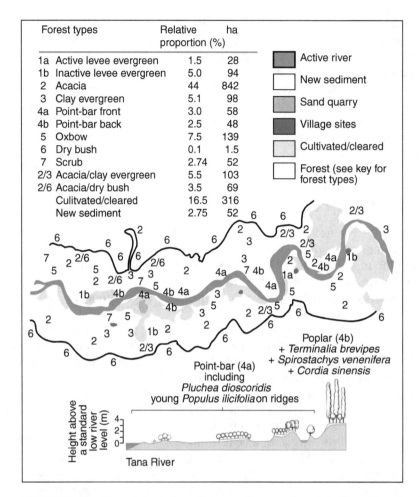

FIGURE 5.12   The patchy soil conditions and vegetation of the Tana River floodplain near the village of Pumwani in Kenya. The vegetation typical of different successional stages along the river is illustrated (adapted from Hughes, 1990).

FIGURE 5.13   Bald cypress *(Taxodium distichum)* trees in parts of the southeastern United States require regular, low-velocity flood events in order to successfully reproduce. *Source:* National Park Service / Public domain / https://www.nps.gov/places/teel-road-cypress-swamp.htm

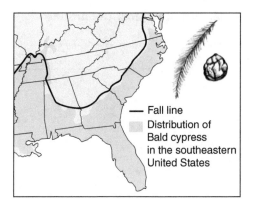

FIGURE 5.14 The correspondence between the northern limits of bald cypress *(Taxodium distichum)* and the boundary of the Coastal Plain. Seasonal flooding on the Coastal Plain appears to be crucial for bald cypress regeneration (Shankman and Kortright, 1994).

The geographic distribution of bald cypress in the southeastern United States has a relatively sharp border called the fall line that coincides with the physiographic boundary between the flat Coastal Plain of the Southeast and the steeper terrain of the surrounding Appalachian Highlands (Fig. 5.14). Bald cypress trees require regular, low-velocity flood events in order to successfully reproduce. Bald cypress seeds germinate after being transported by floods and deposited on moist surfaces. They will not germinate on dry soils. The seedlings, however, are killed if they are covered by water for more than a few days following germination. The floodplains of streams on the uplands are narrow and do not provide much habitat for bald cypress seeds to settle and germinate. Most important, incidences of flooding are more frequent and erosive in the narrow upland floodplains than on the broad floodplains of the Coastal Plain. The Coastal Plain, which experiences low-energy floods followed by long dry intervals, is ideal for bald cypress reproduction. In addition, the flood regime prohibits the establishment of inundation-intolerant tree species and thus reduces competition for the bald cypress trees. Research by geographers David Shankman and Justin Hart indicates that the Coastal Plain–Appalachian Highlands fall line is an important hydrogeomorphic and pedological control on the geographic distribution of a number of tree species and forms an important biogeographic boundary (Table 5.1).

Flooding can create different habitat conditions over time or space and be a promoter of higher biodiversity. Research on the Prosser and Little Swanport rivers in Tasmania provided evidence of the role of flooding in producing habitat diversity important for biodiversity. Species richness for herbs was highest in areas eroded by floods, whereas shrub and grass diversity tended to be high in area where floods deposited sediment. Human activity can alter flood regimes by changing the frequency or geographic extent of flooding and dramatically alter habitat conditions and diversity. Reservoirs built along river systems can serve as modulators of river flow by

TABLE 5.1 **Some Tree Species Found Growing Below or Above the Atlantic Fall-Line Hydrogeomorphic and Pedological Boundary**

Flood-Adapted Species Found Below the Fall Line

Bald cypress *(Taxodium distichum)*
Water tupelo *(Nyssa aquatic)*
Water hickory *(Carya aquatica)*
Water elm *(Planera aqua)*

Xeric Species Found Above the Fall Line

Chestnut oak *(Quercus montana)*
Scarlet oak *(Quercus coccinea)*
Virginia pine *(Pinus virginiana)*

Source: Adapted from Shankman and Hart, 2007.

TABLE 5.2   **Impact of Dams on the Physical and Biological Conditions of Downstream Rivers**

Physical Changes

Water temperature

Water sediment content

Water chemistry

Magnitude of flow

Timing of high and low flow

Variability of flow

Changes in river channel morphology

Changes in bank morphology

Changes in floodplain surface and soil water

Biological Changes

Fragmentation of aquatic populations

Death of individuals at spillways and intakes

Disruption of aquatic migration routes

Increased colonization of banks

Decreased abundance or extinction of flood-dependent species

Increased abundance or invasion off flood-intolerant species

Alterations of food webs

Changes in biodiversity

capturing and storing water during periods of natural peak flow and then releasing it during periods of natural low flow. If such reservoir systems are extensive, they can virtually remove any chance of flooding. Over time, this can cause severe problems for floodplain ecosystems as soils dry out, mineral soil cover ceases to be replenished or eroded, and vegetation and organic matter build up on the land surface and in pools (Table 5.2).

One example of the damaging impact of such flood control on vegetation is provided by the lower Colorado River in the Grand Canyon. The completion of the Glen Canyon Dam above the Grand Canyon in 1963 caused the cessation of the large annual floods that were a feature of the river's hydrology. In addition, the fine sediment that normally was carried into the Grand Canyon by the Colorado River became trapped in the still waters behind the dam. Following the construction of the dam and the cessation of large floods, woody vegetation began to thicken on the lower banks of the river. Many of the successful plants were nonnative to the canyon. On higher areas, the loss of water from floods caused the native mesquite vegetation to die back. In addition, sandy bars no longer received fresh inputs of sediment, and they began to disappear. Within the river, trout, which had not been able to live in the previously warm and sediment rich waters, began to flourish. In 1996 the government and scientists, recognizing the changes that the dam was creating in the natural ecology of the Grand Canyon, began to experiment with controlled flooding. In March 1996 a total of 900 million cubic meters was released from Glen Canyon Dam over a one-week period to simulate a natural flood. The rate of discharge was, however, less than half of a natural flood on the Colorado. The experiment resulted in the creation of new sandy bars and other changes in the river channel. However, it did not have a significant impact on the woody vegetation that now lines much of the lower Colorado.

Modification of river drainage basins can also serve to increase river flows and the frequency and magnitude of flooding. Removal of vegetation and soil cover causes increased rates of water runoff into rivers. Some scientists argue that increased agricultural land clearance in the mountainous regions above Bangladesh has led to more frequent and intense flooding along the lower Ganges River. The building of levees and dikes along rivers can constrict their channels and keep high flows from flooding adjacent low-lying areas. Much of the lower Mississippi River is constrained by artificial levees and controlled by water diversion channels. Decreased incidents of flooding by the Mississippi are causing a loss of bald cypress swamps along the lower portions of the river.

Not all human modifications of riparian vegetation make stream banks more unstable. When compared to natural forest vegetation, the establishment of pasture grasses has been shown to cause decreased channel erosion rates in some small stream systems. The dead branches and trunks of trees that fall into forested streams are important in causing ripples and pools for aquatic life and in directing erosive river flow. In some instances, clearing forest cover and replacing it with grass may artificially stabilize floodplains and be destructive to aquatic ecosystems and riparian vegetation.

OTHER PHYSICAL DISTURBANCES

Avalanches and landslides are important agents of disturbance in mountainous regions. Avalanches are flows of snow and debris that travel with great speed and force. Avalanches are capable of destroying entire stands of trees. Avalanches often occur along the same relatively narrow tracks year after year (Fig. 5.15). They may affect areas of ten to over a hundred hectares. The occurrence of frequent avalanches precludes the growth of mature trees, and avalanche tracks are typically vegetated by grasses, herbs, shrubs, and saplings. Within forested regions, avalanche tracks provide important open habitat for shade-intolerant plants. The herbaceous and shrubby vegetation of avalanche tracks is important feeding ground for deer, moose, and bear within forested mountain areas. Research in the Alps has shown that more plant species are found growing in active and inactive avalanche tracks than in adjacent undisturbed spruce *(Picea abies)* and larch *(Larix decidua)* dominated subalpine forest. The greater number of species likely reflects indicated a wider range of ecological niches in the avalanche tracks than in mature closed forest. On sites where avalanche activity has been artificially suppressed by barriers, the plant species composition can begin to change markedly after a few decades.

The term *landslide* encompasses a number of different types of slope failure. Landslides can be defined as sudden mass movements of rock, soil, and organic material downslope along sheer surfaces. These movements can range from rockfalls off of steep cliffs, to slides and slumping of large coherent blocks of wet or dry earth, to debris flows of water-saturated material. Landslides destroy vegetation by removing soil, shearing stems and trunks, and burying plants. The frequency and

FIGURE 5.15 An avalanche path has cleared a swath through coniferous forest. Avalanche tracks can provide habitat for light-demanding grasses, shrubs, and small trees. They can also serve as natural fire breaks in the forest. *Source:* Mario / Adobe Stock.

magnitude of landslides are controlled by many factors, including slope angles, bedrock type, soil properties, rainfall, and vegetation cover. Vegetation-landslide relations are often a two-way street. Certain vegetation covers, such as deep-rooted forests, can increase slope stability and inhibit landslide activity. This decrease in landslide disturbance in turn allows for mature forest to be maintained. On the other hand, studies of big node bamboo *(Phyllostachys nidularia)* on landslide-prone slopes in Sichuan, China, suggest that this shallow-rooted species can enhance landslide propensity as it outcompetes other deeper-rooted plants on unstable slopes. Frequent landslides allow the bamboo to escape competition from trees that, if established, would inhibit landslides. It has been argued that landslides have not received the attention they deserve as an important biogeographic disturbance agent. In costal British Columbia. debris-flow landslides and snow avalanches have been identified as the dominant agent of vegetation disturbance. The habitat diversity created by landslides there is in turn an agent promoting greater plant and animal diversity. Landslides are a disturbance factor that can be readily altered by humans. Human activities, such as deforestation, grading of slopes, and increasing groundwater through upslope irrigation, can produce increased landslide activity. A study of landslides in northeast Puerto Rico found that 53% of the landslides recorded between 1964 and 1989 were associated with road building.

Landslides are a particularly important form of disturbance in steep tropical landscapes where deeply weathered rock and high rainfall promote slope instability. Hurricanes are important in promoting tropical landslides as they strip away protective vegetation cover and produce extremely intense rainfall. It is estimated that 0.8% to 7.5% of tropical forests are affected by landslides in the Caribbean region each century. Although huge tropical landslides do occur, most are small in area. Studies in Jamaica indicate that 80% of the recent landslides were less than 0.08 hectares in area, while a similar study in Puerto Rico showed that 77% of recent landslides were less than 0.4 hectares in size.

Succession patterns following tropical landslides appear to be highly dependent on the proximity of seed sources and the degree to which the original soil is disturbed by the landslide. The initial vegetation on tropical landslides is usually dominated by shade-intolerant plants such as tree ferns, climbing ferns, grasses, large herbs, shrubs, and vines. Completely denuded and unstable surfaces may be colonized by mosses and other nonvascular plants. Establishment of fern and shrub thickets leads to decreased soil evaporation, increased moisture availability, and more soil stability. Increasing moisture and soil stability eventually facilitate tree establishment. As the trees grow, they shade-out the pioneer vegetation. In Puerto Rico, the vegetation on landslides can revert to forest within 50 years. However, on sites where no residual soil remains, it may take several hundred years before a mature forest cover is established.

In cold mountain regions, the paths of debris flows often coincide with avalanche tracks. The frequency of both forms of disturbance can keep such areas in permanent states of disruption when compared to surrounding areas. However, not all landslides in cold regions are restricted to avalanche paths. Deep unstable soils, slope angles greater than 30°, and a lack of deeply rooted vegetation cover contribute to landslides. Debris flows are promoted by the combination of moist soils due to melting snow and intense rainfall. Analysis of satellite images from a small section of Glacier National Park in Montana produced evidence of 73 recent debris flows associated with avalanche paths, 48 associated with slope failures beneath large snow patches, and 28 flows down ravines that were not avalanche paths. The widespread occurrence of landslide activity in the northern Rocky Mountains has been shown to depress the elevation of the treeline by excluding the growth of forest trees in protected ravines and other sites that would otherwise provide favorable habitat for trees. Studies of landslides in the Canadian Arctic show that, even in relatively simple plant ecosystems, 50 years are required after a landslide for disturbed areas to achieve plant species composition similar to pre-landslide conditions and the vegetation cover may remain discontinuous.

Fires can promote landslides by stripping away vegetation cover and destabilizing slopes. In the semiarid chaparral vegetation of California, large fires in the late summer are often followed by severe landslide activity the following winter. Aside from removing the vegetation cover, the fires also decrease the moisture-repelling properties of the soil surface, leading to greater absorption of water and increased soil instability. Interestingly, research in the Canadian Rocky Mountains indicates that the shrubby vegetation associated with avalanche tracks and debris flow paths form important fire breaks that restrict the spread of fires in the surrounding conifer forests.

Volcanic activity is one of the most spectacular forms of disturbance. Volcanoes can create disturbances through several different processes. Lava flows cover existing ecosystems and provide new landscapes for primary succession. During volcanic eruptions, large amounts of fine ash, called **tephra,** are created and deposited. Tephra can bury existing ecosystems and provide a new surface for primary succession. Explosive eruptions can denude the landscape over large areas. Hot gases, lava, and ejected material can cause fires in the vicinity of the volcanic eruption. Finally, unstable land surfaces left by volcanic eruptions are prone to landslides. The eruption of Mount Saint Helens in Washington provides a good example of volcanism as an agent of disturbance. On May 18, 1980, Mount Saint Helens was rocked by an explosive eruption that removed 400 meters off the top of its cone. Due to the explosive nature of the eruption, a larger area of devastation called the blast zone extended some 18 kilometers north of the former mountain peak. Some 500 square kilometers of forest were destroyed by the searing blast. The blast zone was typified by bare soil surfaces and the broken and burned stumps of millions of dead trees. Debris flows of wet sediment came down the south side of the mountain and extended considerable distances along adjacent river valleys. In all, more than 60 square kilometers of landcover was destroyed by pyroclastic flows of scorching gases and pumice. In the first years after the eruption, the windborne seeds of many plants were deposited on the bare ash and lava surfaces. Most of these seeds were unable to establish plants on the dry and nitrogen-deficient soils. Lupines *(Lupinus lepidus),* however, were able to establish. They have seeds that contain starch reserves for the seedlings and roots that possess nodules that allow nitrogen fixation. These features allowed lupines to dominate the early vegetation on many of the bare sites. The patches of lupines were in turn invaded by plants such as asters *(Aster ledophyllus)* and fireweed *(Epiliobium angustifolium),* which apparently relied on the more favorable soils created by the lupines (Fig. 5.16). In this case, the lupines are referred to as nurse plants because their presence facilitates the establishment and growth of other species. In most cases, the lupines were replaced by other plant species after about five years. Now that several decades have passed since the eruption, vegetation cover and species numbers in both terrestrial and aquatic ecosystems in the blast zone have increased remarkably. Perhaps the most remarkable evidence of succession and revegetation comes from satellite remote sensing (Fig. 5.17) which shows the inexorable spread of life in the form of herbaceous and shrubby vegetation over the once seeming lunar landscape. In some places, scattered tree saplings can be seen emerging above the herbs, grasses, and shrubs. The process of succession is still under way on Mount Saint Helens, and it will take decades before coniferous forest cover and associated ecosystems are fully reestablished over formerly wooded areas.

FIGURE 5.16   In the years after the 1980 eruption, fireweed *(Epilobium angustifolium)* and later, young coniferous forest were established on parts of the Mount Saint Helens blast zone. The remains of trees killed by the blast remain on the slopes. *Source:* khomlyak / Adobe Stock.

FIGURE 5.17   Vegetation succession on Mount Saint Helens from 1985 to 2016 as seen from space. The dark areas are vegetated surfaces (*Source:* NASA / Public Domain).

PATHOGENS

The outbreak of pathogens, especially plant-destroying insects, is an important and widespread form of disturbance. In the early twenty-first century, tree diseases and destructive insects destroyed some 60 million cubic meters of Canadian timber each year. This is second only to fire that destroyed an average of 80 million cubic meters of timber each year. An outbreak of spruce bark beetles *(Dendroctonus rufipennis)* in Alaska from 1989 to 2003 infested over 275,000 hectares of forests in that state alone. Further south in Utah, spruce bark beetles infested more than 49,371 hectares and killed more than 3 million trees in the 1990s. The bark beetles weaken and kill trees by eating the cambial layer beneath the bark and can girdle the tree, cutting off the supply of water from roots to foliage (Fig. 5.18). One of the most destructive insects in the forests of northern North America is the spruce budworm *(Choriostoneura fumiferana)*. During a peak outbreak in 1978, over 72 million hectares of forest were affected by this insect. The larvae eat the foliage and young cones of spruces and

FIGURE 5.18   Engelmann spruce trees *(Picea engelmannii) (left)* killed by spruce bark beetle *(Dendroctonus rufipennis)* in southern Utah and the galleries *(right)* carved by the beetle larvae under the bark. The destruction of vascular tissue by the galleries eventually weakens and kills the trees.

firs, but the principal host is balsam fir *(Abies balsamea)*. After three to four years of infestation, even mature trees are killed. Severe outbreaks of spruce budworm occur approximately every 30 years and may take 7 to 15 years to subside. In Quebec the 1978 event was part of a longer-term major outbreak that extended to 1991. Tree-ring analysis suggests the following major outbreak events in that province: 1976–1991, 1946–1959, 1915–1929, 1872–1903, 1807–1817, 1754–1765, 1706–1717, 1664–1670, and 1630–1638.

During bark beetle and budworm outbreaks, individual trees and whole stands may be killed, opening many different-sized gaps in the forest. It has been suggested that the dry, dead foliage produced during budworm outbreaks promotes the ignition and spread of high-intensity fires. However, there are some controversies about spruce budworm and similar pathogen outbreaks and their impacts on forests. It has been suggested that outbreaks have been more synchronized and thus widespread in the twentieth and twenty-first centuries. Research in Quebec, however, suggests that this may not be the case for spruce budworm outbreaks there. In addition, research in the Rocky Mountains of Colorado suggests that climatic variations have a far more prominent role in producing large fires than do outbreaks of tree mortality linked to insect pathogens. In addition, the development of overly dense stands of trees due to human suppression of forest fires coupled with climate warming and drying may make stands and individual trees more susceptible to insect infestation in the first place. It has been estimated by the United States Forest Service that between 2010 and 2018 some 129 million trees have died in California's national forests due to conditions caused by climate change, unprecedented drought, bark beetle infestation, and high tree densities. Teasing out the relative impact of each of these factors can be difficult though. Regardless of this all, the patchwork of living and dead stands produced by spruce budworm and similar tree pathogens such as the spruce beetle can also produce diverse ranges of habitat in forested ecosystems and promote increased species diversity.

An interesting example of how disturbance by a pathogen can be transmitted through different levels of an ecosystem is provided by outbreaks of anthrax among grazing and browsing mammals in Africa. Anthrax is a bacterium that causes death in mammals. Following outbreaks, anthrax can remain dormant in soils for years. In Tanzania, epidemics of anthrax among impalas *(Aepyceros melampus)* caused widespread mortality in 1961, 1977, 1984, and 1998. The declines in the impala population led to decreased browsing on shrubs and the invasion of some grasslands by shrubby species and umbrella thorn trees *(Acacia tortilis)*. In this manner, the disturbance provided by outbreaks of anthrax among mammals contributes to maintaining the mosaic of grassland, shrubs, and trees that makes up the savanna landscape. Recent work suggests that these population fluctuations in impalas produced by anthrax are of central importance in woodland dynamics and the generation of habitat diversity.

MARINE DISTURBANCES

Marine ecosystems are also subject to physical and biological disturbances. Physical disturbances include abrupt changes in water temperature and salinity, submarine landslides, and episodes of rapid erosion and deposition. The impact of the 1982–1983 El Niño event demonstrates how different marine communities are impacted by rapid changes in sea surface temperatures. During El Niño events, there is a rapid warming of surface waters in the eastern equatorial region of the Pacific Ocean. The 1982–1983 event produced unusually strong warming of 3° C to 4° C off the coast of Central and South America. The higher temperatures rendered reef-building corals more sensitive to ultraviolet radiation and caused them to expel their symbiotic algae. Within 2 to 4 weeks the corals died. The mortality extended to corals growing as deep as 20 meters below the sea surface. In some reefs, up to 97% of the coral died. Off the Galapagos Islands temperatures increased by 5° C. This warming caused a decline in the upwelling rate of deep, nutrient-rich waters. The warmer water temperatures and decreased upwelling led to declines in marine plant productivity and a southward movement of many fish populations. Marine iguanas *(Amblyrhynchus cristatus)* of the Galapagos Islands are dependent on seaweed for food. Due to decreased productivity during El Niño, the iguanas suffered population declines of 60% to 70%. Since El Niño events recur every 5 to 11 years, the iguana population has adapted to such disturbances. In the year following an El Niño, the age for first breeding by females decreases and the average egg clutch size increases. Thus, the iguana population appears to respond to this periodic disturbance and persists on the Galapagos Islands through increased fertility following El Niños.

Although such climatic extremes may be rare, their occurrence can lead to range reduction and extinction. In contrast to the recovery of the marine iguana, the corals have suffered much more long-lasting damage. Erosion by physical processes such as storms and bioerosion, caused by coral-eating fish and burrowing mollusks, has led to complete destruction of some reef systems. Several species of *Millepora* coral may have been driven to extinction by the 1982–1983 El Niño.

An example from coral ecosystems in the Caribbean demonstrates how biological disturbances can impact marine environments. In 1983 an unknown pathogen killed more than 95% of long-spined sea urchins *(Diadema* spp.). The loss of the sea urchins decreased grazing pressure on seaweeds, which then overgrew corals at many sites. The corals that were covered by seaweed died. The sea urchins were a keystone species in the coral ecosystem, and their decline caused the loss of over 90% of corals surrounding Jamaica.

As in many terrestrial ecosystems, succession in coral reef systems is often controlled by the species that reach the site first. In general, large filamentous algae establish quickly upon bare surfaces and dead coral. Grazing pressure from urchins and fish generally limits algal growth and allows corals to grow and regain dominance. The long-term decrease in sea urchins that graze on algae may not only destroy mature coral, but also inhibit its reestablishment. The exact species of coral that eventually dominates the patch in the latter stages of succession is often largely dependent on the species of coral larvae that first settle there. The hard surfaces and nooks and crannies produced by mature coral subsequently facilitate occupancy of the reef by a wonderful assortment of sea anemone, mollusks, worms, crustaceans, and fish. Unfortunately, human activity has increased the sensitivity of corals to physical and biological disturbance. Increased deposition of fine sediments due to land clearance has choked and weakened many Central American coral communities. The recent destruction of corals in the Caribbean also has a human component. Extreme overfishing off Jamaica removed many fish species that feed on seaweed. This made the explosive growth of seaweed possible after the sea urchins were killed by the unknown pathogen. As the 2000s have progressed, there is evidence that coral decline following natural and human disturbance is global in scale. However, there is also evidence that some coral systems have a high degree of resilience to disturbance. Coral reefs in Moorea, French Polynesia, have been subjected to a number of large-scale disturbances since the 1990s. These disturbances include coral bleaching, large cyclones, and outbreaks of the crown-of-thorns starfish *(Acanthaster)*. The starfish kills corals by devouring the hard coral polyps. Study of these reef systems indicates that within 4 years after disturbance and coral decline, the abundance of juvenile and adult colonies can reach or even exceeded pre-disturbance levels. In addition, no replacement of corals by algae was observed. Unfortunately, this apparent resilience does not seem to be ubiquitous in coral reefs around the world.

KEY WORDS AND TERMS

Adventitious roots
Background mortality
Canopy fire
Climax
Crown fire
Disturbance
Facilitation
Hydrogeomorphic disturbance

Inhibition Model
Initial Floristic Composition Model
Pneumatophores
Primary succession
Pyrogeography
Random Model
Rhizome
Riparian forests

Scarification
Secondary succession
Sere
Serotinous cones
Succession
Surface fires
Tephra
Tolerance Model

REFERENCES AND FURTHER READING

Adjeroud, M., Kayal, M., Iborra-Cantonnet, C., Vercelloni, J., Bosserelle, P., Liao, V., Chancerelle, Y., Claudet, J., and Penin, L. (2018). Recovery of coral assemblages despite acute and recurrent disturbances on a South Central Pacific reef. *Scientific Reports* **8,** Article 9680.

Agee, J. K. (1993). *Fire Ecology of Pacific Northwest Forests.* Island Press, Washington, DC.

Allen, J. L., Wesser, S., Markon, C. J., and Winterberger, K. C. (2006). Stand and landscape level effects of a major outbreak of spruce beetles on forest vegetation in the Copper River Basin, Alaska. *Forest Ecology and Management* **227,** 257–266.

Ansley, R. J., Kramp, B. A., and Jones, D. L. (2015). Honey mesquite *(Prosopis glandulosa)* seedling responses to seasonal timing of fire and fireline intensity. *Rangeland Ecology & Management* **68,** 194–203.

Baker, W. L. (1989). Landscape ecology and nature reserve design in the Boundary Waters Canoe Area, Minnesota. *Ecology* **70,** 23–35.

Baker, W. L. (1990). Climatic and hydrologic effects on the regeneration of *Populus angustifolia* James along the Animas River, Colorado. *Journal of Biogeography* **17,** 59–73.

Baker, W. L. (1992). Effects of settlement and fire suppression on landscape structure. *Ecology* **73,** 1879–1887.

Baker, W. L. (1993). Spatially heterogeneous multi-scale response of landscapes to fire suppression. *Oikos* **66,** 66–71.

Baker, W. L. (1994). Restoration of landscape structure altered by fire suppression. *Conservation Biology* **8,** 763–769.

Baker, W. L. (2009). *Fire Ecology in Rocky Mountain Landscapes.* Island Press, Washington, DC.

Baker, W. L., and Walford, G. M. (1995). Multiple stable states and models of riparian vegetation succession on the Animas River, Colorado. *Annals of the Association of American Geographers* **85,** 320–338.

Barry, John M. (2007). *Rising Tide: The Great Mississippi Flood of 1927 and How It Changed America.* Simon and Schuster, New York.

Beatty, S. W. (1989). Fire effects on soil heterogeneity beneath chamise and redshanks chapparal. *Physical Geography* **10,** 44–52.

Beatty, S. W. (1991). Colonization dynamics in a mosaic landscape: the buried seed pool. *Journal of Biogeography* **18,** 553–563.

Bebi, P., Kulakowski, D., and Rixen, C. (2009). Snow avalanche disturbances in forest ecosystems—state of research and implications for management. *Forest Ecology and Management* **257,** 1883–1892.

Begin, Y., and Filion, L. (1995). A recent downward expansion of shoreline shrubs at Lake Bienville (subarctic Quebec). *Geomorphology* **13,** 271–282.

Belcher, C. M., Collinson, M. E., and Scott, A. C. (2013). A 450-million-year history of fire. In *Fire Phenomena and the Earth System: An Interdisciplinary Guide to Fire Science* (ed. C. M. Belcher), pp. 229–249. Wiley, New York.

Bendix, J. (1997). Flood disturbance and the distribution of riparian species diversity. *Geographical Review* **87,** 468–483.

Birkeland, G. H. (1996). Riparian vegetation and sandbar morphology along the lower Little Colorado River, Arizona. *Physical Geography* **17,** 534–553.

Blais, J. R. (1983). Trends in the frequency, extent, and severity of spruce budworm outbreaks in eastern Canada. *Canadian Journal of Forest Research* **13,** 539–547.

Boulanger, Y., Arseneault, D., Morin, H., Jardon, Y., Bertrand, P., and Dagneau, C. (2012). Dendrochronological reconstruction of spruce budworm *(Choristoneura fumiferana)* outbreaks in southern Quebec for the last 400 years. *Canadian Journal of Forest Research* **42,** 1264–1276.

Bowman, D. M., Balch, J. K., Artaxo, P., Bond, W. J., Carlson, J. M., Cochrane, M. A., D'Antonio, C. M., DeFries, R. S., Doyle, J. C., Harrison, S. P., and Johnston, F. H. (2009). Fire in the earth system. *Science* **324,** 481–484.

Bowman, D. M., Balch, J., Artaxo, P., Bond, W. J., Cochrane, M. A., D'Antonio, C. M., DeFries, R., Johnston, F. H., Keeley, J. E., Krawchuk, M. A., and Kull, C. A. (2011). The human dimension of fire regimes on earth. *Journal of Biogeography* **38,** 2223–2236.

Brown, P. M. (2013). Dendrochronology: fire regimes. *Encyclopedia of Scientific Dating Methods* **10,** 978–994.

Butler, D. R. (1994). Physical geography and alpine treeline: an introduction. *Physical Geography* **15,** 101–103.

Chambers, J. Q., Fisher, J. I., Zeng, H., Chapman, E. L., Baker, D. B., and Hurtt, G. C. (2007). Hurricane Katrina's carbon footprint on US Gulf Coast forests. *Science* **318,** 1107.

Chuvieco, E., Mouillot, F., van der Werf, G. R., San Miguel, J., Tanase, M., Koutsias, N., García, M., Yebra, M., Padilla, M., Gitas, I., and Heil, A. (2019). Historical background and current developments for mapping burned area from satellite earth observation. *Remote Sensing of Environment* **225,** 45–64.

Clements, F. E. (1916). *Plant Succession: An Analysis of the Development of Vegetation.* Carnegie Institute, Washington, DC.

Clements, F. E. (1936). Nature and structure of the climax. *Journal of Ecology* **24,** 252–284.

Cohen, W. B., Yang, Z., Stehman, S. V., Schroeder, T. A., Bell, D. M., Masek, J. G., Huang, C., and Meigs, G. W. (2016). Forest disturbance across the conterminous United States from 1985–2012: the emerging dominance of forest decline. *Forest Ecology and Management* **360,** 242–252.

Collier, M. P., Webb, R. H., and Andrews, E. D. (1997). Experimental flooding in Grand Canyon. *Scientific American* **January,** 82–89.

Conedera, M., Tinner, W., Neff, C., Meurer, M., Dickens, A. F., and Krebs, P. (2009). Reconstructing past fire regimes: methods, applications, and relevance to fire management and conservation. *Quaternary Science Reviews* **28,** 555–576.

Congressional Research Services (2021). *Wildfire Statistics.* https://sgp.fas.org/crs/misc/IF10244.pdf. Library of Congress, Washington, DC.

Connell, J. H., and Slatyer, R. O. (1977). Mechanisms of succession in natural communities and their role in community stability and organization. *American Naturalist* **111,** 1119–1144.

Cook, J. E., and Halpern, C. B. (2018). Vegetation changes in blown-down and scorched forests 10–26 years after the eruption of Mount St. Helens, Washington, USA. *Plant Ecology* **219,** 957–972.

Coops, N. C., Hermosilla, T., Wulder, M. A., White, J. C., and Bolton, D. (2018). A thirty year, fine-scale, characterization of area burned in Canadian forests shows evidence of regionally increasing trends in the last decade. *PloS One* **13,** e0197218.

Cordes, L. D., Hughes, F. M. R., and Getty, M. (1997). Factors affecting the regeneration and distribution of riparian woodlands along a northern prairie river: the Red Deer River, Alberta, Canada. *Journal of Biogeography* **24,** 675–695.

Covington, W. W., Fule, P. Z., Moore, M. M., Hart, S. C., Kolb, T. E., Mast, J. N., Sackett, S. S., and Wagner, M. R. (1997). Restoring ecosystem health in ponderosa pine forest of the southwest. *Journal of Forestry* **95,** 23–29.

Cowles, H. C. (1901). The physin graphic ecology of Chicago and vicinity. *Botanical Gazette* **31,** 73–108.

Craig, M. R., and Malanson, G. P. (1993). River flow events and vegetation colonization of point bars in Iowa. *Physical Geography* **14,** 436–448.

Cross, W. F., Baxter, C. V., Donner, K. C., Rosi-Marshall, E. J., Kennedy, T. A., Hall, R. O., Kelly, H. A. W., and Rogers, R. S. (2011). Ecosystem ecology meets adaptive management: food web response to a controlled flood on the Colorado River, Glen Canyon. *Ecological Applications* **21,** 2016–2033.

Dale, V. H., Crisafulli, C. M., and Swanson, F. J. (2005). 25 years of ecological change at Mount St. Helens. *Science* **308,** 961–962.

Del Moral, R., Saura, J. M., and Emenegger, J. N. (2010). Primary succession trajectories on a barren plain, Mount St. Helens, Washington. *Journal of Vegetation Science* **21,** 857–867.

Dennison, P. E., Brewer, S. C., Arnold, J. D., and Moritz, M. A. (2014). Large wildfire trends in the western United States, 1984–2011. *Geophysical Research Letters* **41,** 2928–2933.

Dyer, J. M., and Baird, P. R. (1997). Wind disturbance in remnant forest stands along the prairie-forest ecotone, Minnesota, USA. *Plant Ecology* **129,** 121–134.

Egler, F. E. (1954). Vegetation science concepts I. Initial floristic composition, a factor in old-field vegetation development. *Plant Ecology* **4,** 412–417.

Elsner, J. B., Liu, K., and Kocher, B. (2000). Spatial variations in major U.S. hurricane activity: statistics and a physical mechanism. *Journal of Climate* **13,** 2293–2305.

Everham III, E. M. (1996). Forest damage and recovery from catastrophic wind. *Botanical Review* **62,** 113–185.

Franke, M. A. (2000). *Yellowstone in the Afterglow: Lessons from the Fires.* National Park Service, Mammoth Hot Springs, Wyoming.

Fritts, H. C. (1976). *Tree Rings and Climate.* Academic Press, New York.

Geertsema, M., and Pojar, J. J. (2007). Influence of landslides on biophysical diversity—a perspective from British Columbia. *Geomorphology* **89,** 55–69.

Giglio, L., Csiszar, I., and Justice, C. O. (2006). Global distribution and seasonality of active fires as observed with the Terra and Aqua MODIS Sensors. *Journal of Geophysical Research—Biogeosciences* **111,** G02016.

Glenn-Lewin, D. C., Peet, R. K., and Veblen, T. T. (1992). *Plant Succession: Theory and Prediction.* Chapman and Hall, London.

Goldammer, J. G. (1990). Fire in the tropical biota: ecosystem processes and global challenges. In *Ecological Studies,* vol. 84, p. 497. Springer-Verlag, Berlin.

Goldblum, D., and Veblen, T. T. (1994). Fire history of a ponderosa pine/Douglas fir forest in the Colorado Front Range. *Physical Geography* **13,** 133–148.

Grime, J. P. (2006). *Plant Strategies, Vegetation Processes, and Ecosystem Properties*. Wiley, Hoboken, NJ.

Grissino-Mayer, H. D., Baisin, C. H., and Swetnam, T. W. (1995). Fire history in the Pinaleno Mountains of southeastern Arizona: effects of human-related disturbances, in biodiversity and management of the Madrean-Archipelago. *USDA Forest Service General Technical Report RM-GTR-264*, 163–171.

Grissino-Mayer, H. D., and Swetnam, T. W. (2000). Century-scale climate forcing of fire regimes in the American Southwest. *The Holocene* **10**, 213–220.

Guiterman, C. H., Margolis, E. Q., and Swetnam, T. W. (2015). Dendroecological methods for reconstructing high-severity fire in pine-oak forests. *Tree-ring Research* **71**, 67–77.

Hadley, K. S., and Savage, M. (1996). Wind disturbance and development of a near-edge forest interior, Marys Peak, Oregon Coast Range. *Physical Geography* **17**, 47–61.

Hadley, K. S., and Veblen, T. T. (1993). Stand response to western spruce budworm and Douglas-fir bark beetle outbreaks, Colorado Front Range. *Canadian Journal of Forest Research* **23**, 479–491.

Hampson, K., Lembo, T., Bessell, P., Auty, H., Packer, C., Halliday, J., Beesley, C. A., Fyumagwa, R., Hoare, R., Ernest, E., and Mentzel, C. (2011). Predictability of anthrax infection in the Serengeti, Tanzania. *Journal of Applied Ecology* **48**, 1333–1344.

Higuera, P. E., and Abatzoglou, J. T. (2021). Record-setting climate enabled the extraordinary 2020 fire season in the western United States. *Global Change Biology* **27**, 1–2.

Holtmeier, F. K., and Broll, G. (2010). Wind as an ecological agent at treelines in North America, the Alps, and the European subarctic. *Physical Geography* **31**, 203–233.

Honer, T. G., and Bickerstaff, A. (1985). *Canada's Forest Area and Wood Volume Balance 1977–81: An Appraisal of Change under Present Levels of Management*. Canadian Forestry Service, Pacific Forestry Centre BC-X-272.

Hope, A., Tague, C., and Clark, R. (2007). Characterizing post-fire vegetation recovery of California chaparral using TM/ETM+ time-series data. *International Journal of Remote Sensing*, **28**, 1339–1354.

Horn, S. P. (1997). Postfire resprouting of *Hypericum irazuense* in the Costa Rican paramos: Cerro Asuncion revisited. *Biotropica* **29**, 529–431.

Hughes, F. M. R. (1990). The influence of flooding regimes on forest distribution and composition in the Tana River floodplain, Kenya. *Journal of Applied Ecology* **27**, 475–491.

Jackson, J. B. C., and D'Croz, L. D. (1997). The ocean divided. In *Central America* (ed. A. C. Coates), pp. 38–71. Yale University Press, New Haven, CT.

Johnson, E. A., and Larsen, C. P. S. (1991). Climatically induced change in fire frequency in the southern Canadian Rockies. *Ecology* **72**, 194–201.

Johnson, E. A., and Miyanishi, K. (2010). *Plant Disturbance Ecology: The Process and the Response*. Elsevier, Amsterdam.

Johnson, W. C., Dixon, M. D., and Simons, R. (1995). Mapping the response of riparian vegetation to possible flow reductions in the Snake River, Idaho. *Geomorphology* **13**, 159–173.

Johnston, M. C. (1963). Past and present grass-lands of southern Texas and northeastern Mexico. *Ecology* **44**, 456-466.

Jolly, W. M., Cochrane, M. A., Freeborn, P. H., Holden, Z. A., Brown, T. J., Williamson, G. J., and Bowman, D. M. (2015). Climate-induced variations in global wildfire danger from 1979 to 2013. *Nature Communications* **6**, 7537.

Kaib, M., Baisan, C. H., Grissino-Mayer, H. D., and Swetnam, T. W. (1996). Fire history in the gallery pine-oak forests and adjacent grasslands of the Chiricahua Mountains of Arizona. In *Effects of Fire on Madrean Province Ecosystems*, vol. *General Technical Report RM-GTR-289* (ed. P. F. Folliott, L. F. DeBano, M. B. Baker, G. J. Gottfried, G. Solis-Garza, C. B. Edminster, D. G. Neary, L. S. Allen, and R. H. Hamre), pp. 253–264. U.S. Department of Agriculture, Forest Service, Washington, DC.

Keeley, J. E., Fotheringham, C. J., and Baer-Keeley, M. (2005). Determinants of postfire recovery and succession in Mediterranean-climate shrublands of California. *Ecological Applications* **15**, 1515–1534.

Keim, R. F., Dean, T. J., and Chambers, J. L. (2013). Flooding effects on stand development in cypress-tupelo. In *Proceedings of the 15th Biennial Southern Silvicultural Research Conference* (ed. J. M. Guldin), pp. 431–437. U.S. Department of Agriculture, Forest Service, Southern Research Station, Asheville, NC.

Kipfmueller, K. F., and Swetnam, T. W. (2001). Using dendrochronology to reconstruct the history of forest and woodland ecosystems. In *Historical Ecology Handbook* (ed. D. Egan and E. A. Howell), pp. 199–228. Island Press, Washington, DC.

Kitzberger, T., Veblen, T. T., and Villalba, R. (1997). Climatic influences on fire regimes along a rain forest-to-xeric woodland gradient in northern Patagonia, Argentina. *Journal of Biogeography* **24**, 35–47.

Knapp, P. A. (1997). Spatial characteristics of regional wildfire frequencies in intermountain West grass-dominated communities. *Professional Geographer* **49**, 39–51.

Kozlowski, T. T., and Ahlgren, C. E. (1974). *Fire and Ecosystems*, p. 542. Academic Press, New York.

Krawchuk, M. A., Moritz, M. A., Parisien, M. A., Van Dorn, J., and Hayhoe, K. (2009). Global pyrogeography: the current and future distribution of wildfire. *PloS One* **4**, e5102.

Krebs, C. J. (1994). *Ecology*, 4th edition. Harper-Collins College Publishers, New York.

Kupfer, J. A., and Runkle, J. R. (1996). Early gap successional pathways in a *Fagus-Acer* forest preserve: pattern and determinants. *Journal of Vegetation Science* **7**, 247–256.

Lafon, C. W. (2005). Reconstructing fire history: an exercise in dendrochronology. *Journal of Geography* **104**, 127–137.

Larsen, C. P. S. (1997). Spatial and temporal variations in boreal forest fire frequency in northern Alberta. *Journal of Biogeography* **24**, 663–673.

Larsen, C. P. S., and MacDonald, G. M. (1995). Relations between tree-ring widths, climate, and annual area burned in the boreal forest of Alberta. *Canadian Journal of Forest Research* **25**, 1746–1755.

Likens, G. E. (1989). *Long-Term Studies in Ecology*, p. 214. Springer-Verlag, New York.

Lite, S. J., Bagstad, K. J., and Stromberg, J. C. (2005). Riparian plant species richness along lateral and longitudinal gradients of water stress and flood disturbance, San Pedro River, Arizona, USA. *Journal of Arid Environments* **63,** 785–813.

Liu, K. B. (2007). Uncovering prehistoric hurricane activity: examination of the geological record reveals some surprising long-term trends. *American Scientist* **95,** 126–133.

Liu, K. B., and Fearn, M. L. (1993). Lake-sediment record of late Holocene hurricane activities from coastal Alabama. *Geology* **21,** 793–796.

Lugo, A. E. (1988). Ecological aspects of catastrophes in Caribbean islands. *Acta Cientifica* **2,** 24–31.

Lugo, A. E., and Scatena, F. N. (1996). Background and catastrophic tree mortality in tropical moist, wet and rainforests. *Biotropica* **23,** 585–599.

MacDonald, G. M. (2023). Global wildfire. In *International Encyclopedia of Geography: People, the Earth, Environment and Technology* (ed. D. Richardson, N. Castree, M. F. Goodchild, A. Kobyashi, and R. A. Marston), https://doi.org/10.1002/9781118786352 .wbieg2152. Wiley, Hoboken, NJ.

Madin, J. S., and Madin, E. M. (2015). The full extent of the global coral reef crisis. *Conservation Biology* **29,** 1724–1726.

Malanson, G. P. (1993). *Riparian Landscapes*. Cambridge University Press, Cambridge.

Marston, R. A. (2010). Geomorphology and vegetation on hillslopes: interactions, dependencies, and feedback loops. *Geomorphology* **116,** 206–217.

Mast, J. N., Veblen, T. T., and Hodgson, M. E. (1997). Tree invasion within a pine/grassland ecotone: an approach with historic aerial photography and GIS modeling. *Forest Ecology and Management* **93,** 181–194.

McArthur, A. G., and Cheney, N. P. (2015). The characterization of fires in relation to ecological studies. *Fire Ecology* **11,** 3–9.

Medley, K. E., and Hughes, F. M. R. (1996). Riverine forests. In *East African Ecosystems and Their Conservation* (ed. T. R. McClanahan and T. P. Young), pp. 361–383. Oxford University Press, Oxford.

Mertes, L. A. K., Daniel, D. L., Melack, J. M., Nelson, B., Martinelli, L. A., and Forsberg, B. R. (1995). Spatial patterns of hydrology, geomorpology and vegetation on the flood-plain of the Amazon River in Brazil from a remote sensing perspective. *Geomorpology* **13,** 215–232.

Mietkiewicz, M., and Kulakowski, D. (2016). Relative importance of climate and mountain pine beetle outbreaks on the occurrence of large wildfires in the western US. *Ecological Applications* **26,** 2523–2535.

Millspaugh, S. H., and Whitlock, C. (1995). A 750-year fire history based on lake sediment records in central Yellowstone National Park, USA. *The Holocene* **5,** 283–292.

Minnich, R. A., and Bahre, C. J. (1995). Wildland fire and chaparral succession along the California-Baja California boundary. *International Journal of Wildland Fire* **5,** 13–24.

Mitchell, S. J. (2013). Wind as a natural disturbance agent in forests: a synthesis. *Forestry: An International Journal of Forest Research* **86,** 147–157.

Moreno, J. M., and Oechel, W. C. (1994). Fire intensity as a determinant factor of postfire plant recovery in southern California chaparral. In *The Role of Fire in Mediterranean-type Ecosystems* (ed. J. M. Moreno and W. C. Oechel), pp. 26–45. Springer-Verlag, New York.

Naito, A. T., and Cairns, D. M. (2011). Patterns and processes of global shrub expansion. *Progress in Physical Geography* **35,** 423–442.

National Park Service. (2008). *The Yellowstone Fires of 1988.* https://www.nps.gov/yell /learn/nature/upload/firesupplement.pdf.

New, T., and Xie, Z. (2008). Impacts of large dams on riparian vegetation: applying global experience to the case of China's Three Gorges Dam. *Biodiversity and Conservation* **17,** 3149–3163.

North, M., Collins, B. M., and Stephens, S. (2012). Using fire to increase the scale, benefits, and future maintenance of fuels treatments. *Journal of Forestry* **110,** 392–401.

O'Kane, C. A., Duffy, K. J., Page, B. R., and Macdonald, D. W. (2012). Heavy impact on seedlings by the impala suggests a central role in woodland dynamics. *Journal of Tropical Ecology* **28,** 291–297.

Oliver, C. D., and Larson, B. C. (1996). *Forest Stand Dynamics*. Wiley, New York.

Olson, J. S. (1958). Rates of succession and soil changes on southern Lake Michigan sand dunes. *Botanical Gazette* **119,** 125–170.

Owens, M. K., Mackley, J. W., and Carroll, C. J. (2002). Vegetation dynamics in mixed mesquite/acacia savannas. *Journal of Range Management* **55,** 509–516.

Paquin, P. (2008). Carabid beetle (Coleoptera: Carabidae) diversity in the black spruce succession of eastern Canada. *Biological Conservation* **141,** 261–275.

Parker, K. C., Parker, A. J., Beaty, R. M., Fuller, M. M., and Faust, T. D. (1997). Population structure and spatial pattern of two coastal populations of Ocala sand pine (*Pinus clausa* [Chapm. Ex Engelm.] Vasey ex Sarg. Var. *clausa* D. B. Ward). *Journal of the Torrey Botanical Society* **124,** 22–33.

Parker, T. J., Clancy, K. M., and Mathiasen, R. L. (2006). Interactions among fire, insects and pathogens in coniferous forests of the interior western United States and Canada. *Agricultural and Forest Entomology* **8,** 167–189.

Pausas, J. G. (2015). Evolutionary fire ecology: lessons learned from pines. *Trends in Plant Science* **20,** 318–324.

Pausas, J. G., and Keeley, J. E. (2019). Wildfires as an ecosystem service. *Frontiers in Ecology and the Environment* **17,** 289–295.

Perry, D. A. (1994). *Forest Ecosystems*. Johns Hopkins University Press, Baltimore, MD.

Peterken, G. F., and Hughes, F. M. R. (1995). Restoration of floodplain forests in Britain. *Forestry* **68,** 187–202.

Pickett, S. T. A., Cadenasso, M. L., and Meiners, S. J. (2009). Ever since Clements: from succession to vegetation dynamics and understanding to intervention. *Applied Vegetation Science* **12,** 9–21.

Prach, K., and Walker, L. R. (2011). Four opportunities for studies of ecological succession. *Trends in Ecology & Evolution* **26,** 119–123.

Prichard, S. J., Stevens-Rumann, C. S., and Hessburg, P. F. (2017). Tamm review: shifting global fire regimes: lessons from reburns and research needs. *Forest Ecology and Management* **396,** 217–233.

Pulsford, S. A., Lindenmayer, D. B., and Driscoll, D. A. (2016). A succession of theories: purging redundancy from distur-bance theory. *Biological Reviews* **91,** 148–167.

Rebertus, A. J., Kitzberger, T., Veblen, T. T., and Roovers, L. M. (1997). Blowdown history and landscape patterns in the Andes of Tierra del Fuego, Argentina. *Ecology* **78**, 678–692.

Richards, P. W. (1996). *The Tropical Rainforest,* 2nd edition. Cambridge University Press, Cambridge.

Rixen, C., Haag, S., Kulakowski, D., and Bebi, P. (2007). Natural avalanche disturbance shapes plant diversity and species composition in subalpine forest belt. *Journal of Vegetation Science* **18**, 735–742.

Roman, C. T., Aumen, N. G., Trexler, J. C., Fennema, R. J., Loftus, W. F., and Soukup, M. A. (1994). Hurricane Andrew's impact on freshwater resources. *Bioscience* **44**, 247–255.

Romme, W. H., Boyce, M. S., Gresswell, R., Merrill, E. H., Minshall, G. W., Whitlock, C., and Turner, M. G. (2011). Twenty years after the 1988 Yellowstone fires: lessons about disturbance and ecosystems. *Ecosystems* **14**, 1196–1215.

Rood, S. B., Goater, L. A., Mahoney, J. M., Pearce, C. M., and Smith, D. G. (2007). Floods, fire, and ice: disturbance ecology of riparian cottonwoods. *Botany* **85**, 1019–1032.

Rowe, J. S. (1983). Concepts of fire effects on plant individuals and species. In *The Role of Fire in Northern Circumpolar Ecosystems* (ed. R. W. Wein and D. A. MacLean), pp. 135–154. Wiley, Chichester, England.

Schulte, L. A., and Mladenoff, D. J. (2005). Severe wind and fire regimes in northern forests: historical variability at the regional scale. *Ecology* **86**, 431–445.

Schweingruber, F. H. (2012). *Tree Rings: Basics and Applications of Dendrochronology*. Springer Science & Business Media, Berlin.

Seidl, R., Schelhaas, M. J., Rammer, W., and Verkerk, P. J. (2014). Increasing forest disturbances in Europe and their impact on carbon storage. *Nature Climate Change* **4**, 806–810.

Shakesby, R. A., and Doerr, S. H. (2006). Wildfire as a hydrological and geomorphological agent. *Earth-Science Reviews* **74**, 269–307.

Shankman, D., and Hart, J. L. (2007). The fall line: a physiographic-forest vegetation boundary. *Geographical Review* **97**, 502–519.

Shankman, D., and Kortright, R. M. (1994). Hydrogeomorphic conditions limiting the distribution of bald cypress in the southeastern United States. *Physical Geography* **15**, 282–295.

Shinneman, D. J., and Baker, W. L. (1997). Non-equilibrium dynamics between catastrophic disturbances and old-growth forests in ponderosa pine landscapes of the Black Hills. *Conservation Biology* **11**, 1276–1288.

Shroder Jr., J. F., and Bishop, M. (1995). Geobotanical assessment in the Great Plains, Rocky Mountains and Himalaya. *Geomorphology* **13**, 101–119.

Smith, J. A., and Baeck, M. L. (2015). Prophetic vision, vivid imagination: the 1927 Mississippi River flood. *Water Resources Research* **51**, 9964–9994.

Stinson, K. J., and Wright, H. A. (1969). Temperatures of headfires in the southern mixed prairie of Texas. *Journal of Range Management* **22**, 169–174.

Stoffel, M., and Wilford, D. J. (2012). Hydrogeomorphic processes and vegetation: disturbance, process histories, dependencies and interactions. *Earth Surface Processes and Landforms* **37**, 9–22.

Stokes, A., Lucas, A., and Jouneau, L. (2007). Plant biomechanical strategies in response to frequent disturbance: uprooting of Phyllostachys nidularia (Poaceae) growing on landslide-prone slopes in Sichuan, China. *American Journal of Botany* **94**, 1129–1136.

Strahler, A., and Strahler, A. (1997). *Physical Geography: Science and Systems at the Human Environment*. Wiley, New York.

Swetnam, T. W., Allen, C. D., and Betancourt, J. L. (1999). Applied historical ecology: using the past to manage for the future. *Ecological Applications* **64**, 1189–1206.

Syphard, A. D., and Keeley, J. E. (2015). Location, timing and extent of wildfire vary by cause of ignition. *International Journal of Wildland Fire* **24**, 37–47.

Taylor, A. H. (1990). Disturbance and persistence of sitka spruce (*Picea sitchensis* [Bong] carr.) in coastal forests of the Pacific Northwest, North America. *Journal of Biogeography* **17**, 47–58.

Taylor, A. H., Trouet, V., Skinner, C. N., and Stephens, S. (2016). Socioecological transitions trigger fire regime shifts and modulate fire–climate interactions in the Sierra Nevada, USA, 1600–2015 CE. *Proceedings of the National Academy of Sciences*, **113**, 13684–13689.

Teague, W. R., Grant, W. E., Kreuter, U. P., Diaz-Solis, H., Dube, S., Kothmann, M. M., Pinchak, W. E., and Ansley, R. J. (2008). An ecological economic simulation model for assessing fire and grazing management effects on mesquite rangelands in Texas. *Ecological Economics* **64**, 611–624.

Teltscher, K., and Fassnacht, F. E. (2018). Using multispectral landsat and sentinel-2 satellite data to investigate vegetation change at Mount St. Helens since the great volcanic eruption in 1980. *Journal of Mountain Science* **15**, 1851–1867.

Thom, D., and Seidl, R. (2016). Natural disturbance impacts on ecosystem services and biodiversity in temperate and boreal forests. *Biological Reviews* **91**, 760–781.

Tiegs, S. D., O'Leary, J. F., Pohl, M. M., and Munill, C. L. (2005). Flood disturbance and riparian species diversity on the Colorado River Delta. *Biodiversity & Conservation* **14**, 1175–1194.

Trimble, S. W. (1997). Stream channel erosion and change resulting from riparian forests. *Geology* **25**, 467–469.

Tutin, C. E. G., White, L. J. T., and Mackanga-Missandzou, A. (1996). Lightning strike burns large forest tree in the Lope Reserve, Gabon. *Global Ecology and Biogeography Letters* **5**, 36–41.

Vale, T. R. (1982). *Plants and People: Vegetation Change in North America*. Association of American Geographers, Washington, DC.

Van Cleve, K., and Viereck, L. A. (1981). Forest succession in relation to nutrient cycling in the boreal forest of Alaska. In *Forest Succession: Concepts and Application* (ed. D. C. West, H. H. Shugart, and D. B. Botkin), pp. 185–211. Springer-Verlag, New York.

Vandermeer, J., Boucher, D., Perfecto, I., and Granzow de la Cerda, I. (1996). A theory of disturbance and species diversity: evidence from Nicaragua after Hurricane Joan. *Biotropica* **28**, 600–613.

Veblen, T. T., Hadley, K. S., Reid, M. S., and Rebertus, A. J. (1989). Blowdown and stand development in a Colorado subalpine forest. *Canadian Journal of Forest Research* **19**, 1218–1225.

Veblen, T. T., Hadley, K. S., Reid, M. S., and Rebertus, A. J. (1991). Stand response to spruce beetle outbreak in Colorado subalpine forests. *Ecology* **72,** 213–231.

Veblen, T. T., and Lorenz, D. C. (1991). *The Colorado Front Range: A Century of Ecological Change.* University of Utah Press, Salt Lake City.

Vetaas, O. R. (1994). Primary succession of plant assemblages on a glacier foreland—Bodalsbreen, southern Norway. *Journal of Biogeography* **21,** 297–308.

Wallace, J. B., Eggert, S. L., Meyer, J. L., and Webster, J. R. (1997). Multiple trophic levels of a forest stream linked to terrestrial litter inputs. *Science* **277,** 102–104.

Walsh, S. J., Butler, D. R., Allen, T. R., and Malanson, G. P. (1994). Influence of snow patterns and snow avalanches on the alpine treeline ecotone. *Journal of Vegetation Science* **5,** 657–672.

Wein, R. W., and MacLean, D. A. (1983). *The Role of Fire in Northern Circumpolar Ecosystems,* p. 322. Wiley, New York.

West, D. C., Shugart, H. H., and Botkin, D. B. (1981). *Forest Succession: Concepts and Application,* p. 517. Springer-Verlag, New York.

West, D. C., Shugart, H. H., and Botkin, D. F. (Eds.) (2012). *Forest Succession: Concepts and Application.* Springer Science & Business Media, Berlin.

Westerling, A. L., Hidalgo, H. G., Cayan, D. R., and Swetnam, T. W. (2006). Warming and earlier spring increase western US forest wildfire activity. *Science* **313,** 940–943.

Whelan, R. J. (1995). *The Ecology of Fire.* Cambridge University Press, Cambridge.

Whited, D. C., Lorang, M. S., Harner, M. J., Hauer, F. R., Kimball, J. S., and Stanford, J. A. (2007). Climate, hydrologic disturbance, and succession: drivers of floodplain pattern. *Ecology* **88,** 940–953.

Whitlock, C., Higuera, P. E., McWethy, D. B., and Briles, C. E. (2010). Paleoecological perspectives on fire ecology: revisiting the fire-regime concept. *Open Ecology Journal* **3,** 6–23.

Wintle, B. C., and Kirkpatrick, J. B. (2007). The response of riparian vegetation to flood-maintained habitat heterogeneity. *Austral Ecology* **32,** 592–599.

Wolf, J. J., and Mast, J. N. (1998). Fire history of mixed conifer forests of the North Rim, Grand Canyon National Park, Arizona. *Physical Geography* **19,** 1–14.

Wright, H. A., and Bailey, A. W. (1982). *Fire Ecology: United States and Southern Canada.* Wiley, New York.

Zimmerman, J. K., Willig, M. R., Walker, L. R., and Silver, W. L. (1996). Introduction: disturbance and Caribbean ecosystems. *Biotropica* **28,** 414–423.

CHAPTER 6

COMMUNITIES, FORMATIONS, AND BIOMES

Imagine you are taking a road trip across North America. You start driving southwest from the border between New Brunswick and Maine. At first, you are surrounded by dark forests of coniferous trees such as pine, spruce, and fir (Fig. 6.1). For hours you drive through stands dominated by these same trees. Long before you cross into Pennsylvania, the forest has become quite different. In place of pines and spruces, the dominant trees are broadleaved deciduous genera such as maple, oak, ash, and hickory. Crossing central Kansas, you find that the deciduous forest has been replaced by open grasslands. West of Denver you climb into the Rocky Mountains and discover that the vegetation is again dominated by coniferous forests of pine, spruce, and fir. For a short time, near the crest of the Rockies, you see treeless alpine vegetation of low herbs and grasses. After you have descended from the mountains and drive across southern Nevada, you pass through a vast expanse of desert with small shrubs and much bare ground. Approaching the end of your trip at the Pacific Coast of central California, you find dry grassy meadows dotted with scattered oaks. As an astute biogeographer, you have noticed how certain combinations of the same species and vegetation types seem to dominate the landscape over large areas and then disappear altogether. You might wonder what controls such associations of species and causes large geographic changes in vegetation? Biogeographers have marveled at these geographic changes in vegetation and considered these same questions for hundreds of years. Finally, you have noticed that much of the natural vegetation along your route is restricted to areas such as parks and reserves. In many cases, urban and agricultural land use has led to the removal of natural forests or grasslands. The causes and extent of such human impacts have also long been a topic of concern to biogeographers.

In the preceding chapters, we have seen how physical and biological factors influence the geographic distributions of individual species. However, it is often observed that some groups of species are almost always found living together. Species may occur together because they are dependent on each other or because they have very similar tolerances and requirements from the environment. Some biogeographers concentrate their studies on identifying these recurring groupings of species and understanding why such associations occur. At the smallest spatial scale, such studies focus on single communities described from one location. At the largest spatial scale, the terrestrial biosphere can be subdivided into a number of supercontinental biogeographic units called biomes. In this chapter we will look at some basic principles of community ecology that all biogeographers should be familiar with. We will then consider how vegetation structure can be systematically described and used to subdivide the earth's biosphere into biomes. Since an understanding of the biome concept and an appreciation of the different biomes are fundamental knowledge for every biogeographer, we'll conclude with a tour of the world's biomes.

COMMUNITIES

The modern concept of biogeographical communities can be partially traced to the work of the British ecologist Edward Forbes in the mid-nineteenth century. He described the species of animals found in British coastal waters and contrasted this assemblage to those found in the Mediterranean Sea. Today ecologists and biogeographers define the community as an assemblage of all organisms

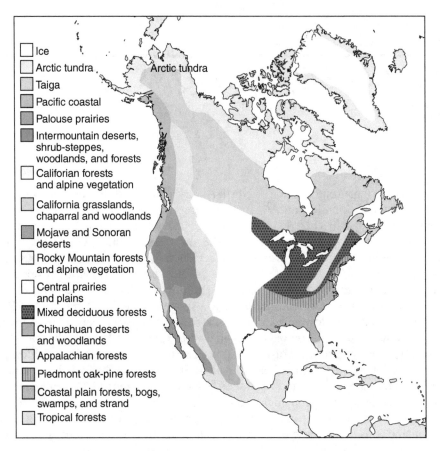

FIGURE 6.1 The major vegetation communities (formations) of North America (*Source:* Babour and Billings, (2000) / CAMBRIDGE UNIVERSITY PRESS).

living in a prescribed place or habitat. We could also consider the entire biosphere as one large community. In practice, the area encompassed by an individual community is usually defined by the researcher. The spatial boundary between adjacent communities is called an **ecotone.** The term **community type** is used to describe a group of species that is typically found in a specific type of habitat. For example, we could study all of the organisms living under many different damp stones, note which species are generally found in such habitat, and describe this collection of recurring coexisting species as the "damp stone community type."

In many cases, researchers are interested only in the plants, animals, or microbes within the community. Plant ecologists employ the term **association** when referring to groups of plant species commonly found in similar habitats. The forest dominated by sugar maple (*Acer saccharum*), red maple (*Acer rubrum*), white oak (*Quercus alba*), red oak (*Quercus rubra*), ash (*Fraxinus*), and hickory (*Carya*), found on mesic sites in northeastern North America, is an example of a plant association. Forest plant ecologists generally refer to the individual sites as **stands.** To simplify descriptions, plant biogeographers and foresters often describe associations in terms of the dominant plant species and call such associations **dominance types.** Central America has forested areas that contain a number of different plant species, but Caribbean pine (*Pinus caribaea*) is one of the region's more common canopy-forming trees. This assemblage can be referred to as the Caribbean pine dominance type.

A number of properties can be measured or described for a community. The properties can relate to either static features of community structure or dynamic properties of community function. Some of the most important properties of community structure include (1) the species that occur, (2) overall species diversity (the number of different species in the community), (3) the species that is most abundant, (4) the species that is most dominant in terms of biomass or structure, (5) **species evenness** (i.e., the degree to which the number of individual organisms are evenly divided between the different species of the community), (6) the general growth form or structure of the organisms or vegetation that makes up the community, (7) the number of trophic levels, and (8) primary productivity. Properties of ecosystem function include features such as rates of energy cycling, rates of nutrient cycling, or exchange rates of gases between the community and the atmosphere. In addition, researchers might examine the genetic diversity within the species that make up the community.

Although the concept of subdividing the geographic distribution of life into community types seems straightforward and useful, it has been one of the most contentious problems in biogeography and ecology. The nature of communities has been a particularly keen subject of debate among plant ecologists. The problem can be illustrated by examining the two classical and diametrically opposed views of natural communities. One view was initially outlined by Stephen Forbes in 1887 in the United States and forcefully developed by the American ecologist and biogeographer Frederic Clements in 1916. From the last chapter we recall that Clements was also a founder of the classical succession theory and the concept of the climax community. Clements argued that because of interactions between species during the climax phase of succession and because of facilitation during seral stages, plants and animals are bound together into tight communities that behave like superorganisms. In this view, all of the species in a community are like the different organs of a larger organism. All the species depend on the functioning of the other species in the community and cannot persist outside of this association. Clements was much influenced by the evolutionary theory propounded by Charles Darwin, and implicit in Clements' theory is the assumption that the species in these communities coexisted and evolved (see Chapter 9) together over very long periods of time. The Clementsian view of community was embraced by many famous ecologists of the mid-twentieth century such as Arthur Tansley in England and Josias Braun-Blanquet in France. One attraction of the superorganism model is that it allows biogeographers to assume that the community types that are described and mapped are stable features of the biosphere.

The opposing view of the community was developed by the American ecologist Henry Gleason in the early twentieth century. He argued that what we perceive as biological communities are simply areas of similar habitat where species coexist because they have somewhat comparable environmental tolerances and resource demands. Gleason went on to argue that each of these species has its own individual sets of tolerances and demands and do not always occur together. Since the species do not always have to occur together, communities cannot be thought of as superorganisms. Gleason's model is called the **individualistic community concept.**

If the **superorganism community concept** is correct, it has important implications. Let's make an extreme case. First, communities should be easy to define in the field. They should be homogeneous in terms of species composition, dominance, structure, and so on, and quite distinctive from adjacent communities developed in a different habitat. Second, from the view of conservation, the removal of one or more species from a community should have a significant impact. If the individualistic concept is correct, then communities should be relatively difficult to define on the basis of species composition. Species from one community will likely be found in several other habitats. Instead of sharp ecotones, communities will blend into each other and be more difficult to discern. The overall community should be less sensitive to the removal of one species.

A convenient way of summarizing these two opposing views is to consider the graphical representation of several plant species along an environmental gradient—soil moisture, for example (Fig. 6.2). The examination of the distribution of species and communities relative to environmental conditions and along environmental gradients is known as ordination analysis (see Chapter 3). On such ordinations, species might be distributed in three different ways. First, they might be distributed in tight groupings with sharp boundaries between different communities. This configuration will occur if there is strong dependency between species within the community and competition with species from outside of it (Fig. 6.2a). Second, species might be segregated into tight community groupings, but the boundaries between communities may be less distinct because of low competition with species outside of the community (Fig. 6.2b). All of the species may be individualistically distributed along the gradient but have sharp boundaries in distribution due to competition with other species (Fig. 6.2c). Finally, the species may be individualistically distributed with no apparent competition or interaction with other species (Fig. 6.2d).

In most cases, the natural distribution of species along environmental gradients appears to fit an individualistic model. Some species are widely distributed in a number of communities and others are more restricted, but distributions where many species have similar limits and only occur together are uncommon. A classic ordination study along a moisture gradient in the Siskiyou Mountains of Oregon

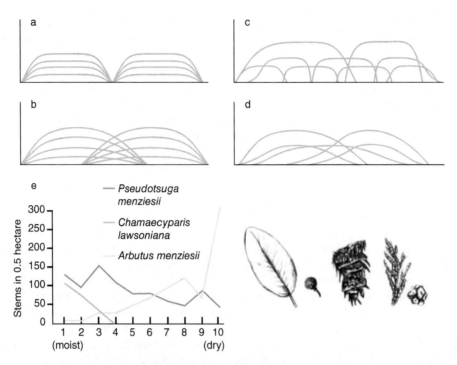

FIGURE 6.2   The ordination of plant species along a hypothetical environmental gradient. (*a*) Species with strong dependency between species within the community and competition with species from outside of it. (*b*) Species with strong dependency but low competition with species outside of the community. (*c*) Individualistically distributed species exhibiting competition with other species. (*d*) Individualistically distributed species with no apparent dependency or competition. (*e*) The actual distribution of dominant shrubs and trees along an altitudinal gradient in the Siskiyou Mountains of Oregon and California (adapted from Whittaker, 1960).

and California illustrates this idea. Based on dominance types, the vegetation of the mountains can be divided into a Pacific Madrone (*Arbutus menziesii*) shrub association on dry sites, a Douglas fir (*Pseudotsuga menziesii*)–dominated forest association on intermediate sites, and a Port Orford cedar (*Chamaecyparis lawsoniana*)–Douglas fir forest association on moist sites. However, Pacific Madrone is relatively common on all but the very wettest sites, and Douglas fir is common on all sites (Fig. 6.2). When the distributions of these species is graphed along a moisture gradient, it is difficult to detect sharp boundaries between the communities. The boundaries between supposed communities can become even less clear if nondominant plants such as small shrubs and herbs, which straddle both communities, were included on the gradient diagram.

The spatial mapping of the local distributions of plant communities sometimes does seem to reveal clear boundaries between communities. In some cases, this may be because the landscape does not really provide a continuum of different environmental conditions, but rather is composed of sites that offer highly contrasting habitat. Perhaps there are two very different soil types with a sharp boundary between them, or slopes that have different degrees of sun exposure. For example, a landscape consisting of small rolling hills could provide either very dry sites on hill slopes or very moist sites in the hollows. We observe two very different plant associations for these two sites, but the differences are exaggerated because there are no intermediate semimoist sites available to provide a true continuum of environmental conditions.

Ecological modeling suggests that competition between species may also play a role in producing sharp boundaries between different communities along ecological gradients. In cases where strong competition exists between the constituent species of the two communities, sharp geographic boundaries can be produced. If competitive interactions between the species are weak, then the boundaries between the communities are more indistinct.

Perhaps the most damaging evidence against the superorganism concept comes from studies of past vegetation change. Research on the past distributions of plant species based on fossil evidence indicates that most modern plant associations have been in existence for only the past 10,000 years.

Many communities of plants have existed in their current state for only relatively short periods. For example, hemlock trees (*Tsuga canadensis*) are important species in the northern forests of the midwestern United States. Hemlock trees have been a part of these forests for only the past few thousand years. As the noted American paleoecologist Margaret Davis argued, the relative youthfulness of many of our present plant associations means that they have not evolved together as superorganisms over long periods of time. As climate has changed during ice ages and warm periods like today, the species composition of plant communities has altered in response.

It is true that there is much individualism in the distributions of species along gradients or on the landscape. However, anyone hiking in the forests of southeastern North America can readily see that there is a difference in the plant species and vegetation structure found in a wet bald cypress (*Taxodium distichum*) swamp versus a loblolly pine (*Pinus taeda*) stand on adjacent dryland. Similarly, a hiker in southern California cannot help but be struck by the difference in vegetation as one ascends from the dry chaparral shrubs at low elevation to the stands of huge big cone Douglas fir (*Pseudotsuga macrocarpa*) and Jeffrey pine (*Pinus jefferyi*) found at higher elevations. Clearly, different associations of species do exist on different sites. Differing environmental requirements between the constituent species as well as competition may play a role. Even if they change over time in response to climatic changes, and their spatial or environmental boundaries may be fuzzy, identifiable communities, particularly dominance-type communities, can often be delineated at many different spatial scales and provide a useful way to describe and study the geographic distribution of life.

Given the sometimes indistinct nature of community boundaries, much work by biogeographers has gone into identifying communities and associations and mapping their spatial boundaries. In Europe, many biogeographers use the Braun-Blanquet **releve** approach to describe and classify plant communities. Using this approach, the researcher makes a number of qualitative observations on the species observed in the community. The observations fall into three major areas: (1) the Sociability Classes relate to the spatial grouping of the plants belonging to the species, (2) the Vitality Classes represent the reproductive ecology of the species, and (3) the Abundance Classes measure how common the species is in the vegetation. The Braun-Blanquet technique and similar releve approaches are relatively rapid to use in the field but suffer from their subjective nature.

Another approach that is widely used is to quantitatively survey the plant species present and then to mathematically compare the similarity and dissimilarity of stands. When a quantitative sampling is conducted for a number of stands, we can calculate a number of measures to describe the community. For any single species, these measures include density (the number of plants per square meter or hectare), frequency (the percentage of samples or stands that contain the species), cover (the amount of ground surface occupied by the plant), dominance (which can be calculated in many ways, including the summing of density, frequency, and cover, for example), and basal area (the area per hectare occupied by the cross-section trunks of tree species). For forest trees, there is a widely used set of quantitative measures of community structure. For a given species (x), these include the following:

$$\text{Relative density:} \frac{\text{Individuals of species } x}{\text{Total individuals of all species}} \times 100$$

$$\text{Relative frequency:} \frac{\text{Frequency of species } x}{\text{Sum of frequency of all species}} \times 100$$

$$\text{Relative dominance:} \frac{\text{Basal area of species } x}{\text{Total basal area of all species}} \times 100$$

Importance value: Relative density + Relative frequency + Relative dominance of species x

In conducting surveys of stands, it is important to examine a large enough area to provide a valid sampling of the true species composition. Samples of species composition are often made by laying out square subsampling areas called quadrats. If the quadrat is too small, it will not capture all of the true species diversity of the stand. If the quadrat is too large, it takes too long to sample and yields no additional information on species diversity compared to a properly sized smaller quadrat. When sampling large areas, or examining dominant tree species only, line or belt transects are often used. A line through the stand is surveyed and all trees touching it, or growing within a certain distance of the line, are identified and recorded. For very large stands, where only the dominant trees are being studied, point-quarter transects are used. For this technique, very long line transects are established and divided into widely separated points. The trees growing nearest to each point, in each of the four cardinal directions, are identified and recorded. In addition to the size of the quadrat or transect, the placement of the sampling points is critical in community analysis. In many cases, researchers use random number tables to generate geographic sampling coordinates for quadrats, transects, and nodes.

Once quantitative information is obtained from stands, a number of mathematical techniques are used to compare the species compositions of the samples and determine whether there are differences between different communities and whether two communities representing the same supposed habitat are really similar. The degree of difference in the species composition of two different habitats or community types is often referred to as **beta diversity.** The degree of difference between stands of the same community type is called **alpha diversity.** One of the simplest mathematical measures of beta or alpha diversity is the index of similarity. The index is based on the presence or absence of species in two different stands and has the following form:

$$\textbf{Index of similarity:} \quad \frac{2 \times \text{the number of species that occur in both stands}}{\text{Total species in stand}\,1 + \text{Total species in stand}\,2}$$

If there is no similarity between the stands (i.e., no shared species), the index will equal 0. If the species present in the two stands are exactly the same, the index will equal 1. There are many other mathematical forms of this general approach using species counts from two stands to calculate similarity.

The problem with the index of similarity is that it does not consider the differences between the two stands in terms of the abundance of the different species. One stand may be dominated by a single tree species and contain a few shrubs, while the other might be dominated by shrubs and contain one or two trees. The species present in the two stands could be identical, but the structure of the stands and habitat would be completely different. In view of this consideration, some researchers use more sophisticated mathematical measures that include the abundance of individuals of each species.

The classification schemes that are used to group sites into associations or community types have two general forms: hierarchical and reticulate. For the hierarchical form, we assume that there is one major community type—deciduous forest, for example. This large community can be subdivided into smaller units that are generally similar to each other but show consistent differences in their species composition. One example might be the grassland association of the midwestern United States. The extensive grasslands can be broken down into a tall grass prairie in the East and a short grass prairie in the West. The two grassland types share some species but also have clear differences in species composition and structure. In an hierarchical classification, we would consider the two prairie types to be subsets of the midwestern grassland association.

Alternatively, we might instead consider that the communities are joined in a reticulate classification network in which each identifiable community is considered a separate entity and not a subset of a larger association. An example might be the forest, tundra, and peatland patches found at the northern treeline. These associations occur in close proximity, may share some species, but are generally distinct from each other in terms of species composition and structure. A number of mathematical techniques are used to classify stands and identify formations of recurring species associations. In addition, some of these techniques can help determine whether the relationship between formations should be considered hierarchical or reticulate.

The impact of disturbance adds to the difficulty in sampling and describing communities. Most of the community types and associations that are described are implicitly assumed to represent "climax" states. The species composition, abundance, and dominance of a recently disturbed site may be significantly different from that of the pre-disturbance conditions. Over time, succession will lead to marked changes in the species composition and structure of the stand. If we compare communities along environmental gradients and include stands at different successional stages, the results can obscure evidence of community structure.

PLANT PHYSIOGNOMY, VEGETATION STRUCTURE, AND FORMATIONS

In addition to species composition, communities are sometimes defined on the basis of plant physiognomy (growth form and function) and vegetation structure. In such schemes, the physiognomy of the plants and the structure of the vegetation are used to describe the community. Units that are defined on the basis of similar vegetation structure are called **formations.** Our imaginary drive across North America took us through a number of the various vegetation formations recognized on the continent (Fig. 6.1). The major North American formations range from treeless tundra in the far north of Alaska and Canada to tropical forest in southern Mexico. The formal use of the term *formation* to characterize vegetation units can be traced back to the work of the German ecologists August Grisebach and Johannes Warming in the nineteenth and early twentieth centuries. Of course, people have been informally characterizing landscape on the basis of vegetation structure since the earliest times. Informal terms such as *meadow* or *woodland* are common ways to denote different vegetation structures. The value of physiognomic and structural classifications of vegetation is threefold. First, physiognomic and structural classifications do not require a specialized knowledge of plant species. Thus, these classifications can be applied quickly and with little technical expertise. Second, the physiognomy of the plants and structure of the vegetation can provide insight into the climatic and soil requirements of the plants, the habitat provided by the plants, and the primary productivity of the vegetation. Finally, physiognomic and structural classifications allow us to compare and group vegetation communities on different continents where we would not find many shared species.

Just as scientific names help to ensure a standardized way to describe species present in communities, there are standardized methods for describing the physiognomy of plants and the structure of vegetation. The Raunkiaer System developed in the early twentieth century by the Danish botanist Christen Raunkiaer is a widely used method of systematically describing plants on the basis of physiognomy and life history. The Raunkiaer system divides plants into six main categories called **life forms:**

Phanerophytes: Perennial land plants that are trees and very large shrubs with persistent above-ground stems and buds. These plants do not require protected microsites in order to grow or shield buds during cold winters or other dormant periods. The phanerophytes are broken down into five subcategories that include megaphanerophytes, which are over 30 meters tall, and nano-phanerophytes, which are less than 2 meters tall, such as vines (called climbing phanerophytes).

Chamaephytes: Perennial land plants that grow as small shrubs. They have persistent above-ground stems and buds. These plants do depend on protected microsites to protect buds in winter or during dormant periods.

Hemicryptophytes: Perennial land plants that are small shrubs and herbs. They have persistent stems and buds located at the soil surface.

Cryptophytes: Perennial land plants that are small shrubs and herbs. They depend on subsurface organs such as bulbs and corms to survive winter or dormant periods.

Therophytes: Annual land plants that survive winter or other periods, such as summer drought, as seeds only.

Hydrophytes: Water plants.

By using sampling techniques similar to the strategies used for studying plant species composition, we can typify stands by the relative dominance of different Raunkiaer life forms. For example, tropical forest stands typically have over 70% phanerophytes and only 1% therophytes. In contrast, herb-dominated tundra in the high Arctic typically has over 50% hemicryptophytes and no phanerophytes.

Although the life form classes developed by Raunkiaer have proven a useful method for describing plant physiognomy and characterizing different vegetation formations, they have two drawbacks. First, they require some knowledge of the life history of the plant. Second, they do not truly capture the great diversity of plant forms that can be seen in nature. An alternative is to classify stands and define formations on the basis of the plant's **growth forms.** The Canadian ecologist Pierre Dansereau developed one system of growth form description that is widely used. Dansereau's method divides plants into six main classes: Trees, Shrubs, Herbs, Bryophytes (nonvascular plants), Epiphytes (plants that grow on other plants but do not tap the host plant's vascular system), and Lianas (woody vines). These main classes are then subdivided on the basis of their leaf shape, size, texture, and function. The system includes symbols for each physiognomic type and can portray plant growth form and vegetation structure graphically. The symbols make the Dansereau method very easy to use in the field.

A final way in which plants may be classified and communities described is in terms of their functional requirements from the environment. The Swiss botanist Augustin de Candolle, working in the early nineteenth century, subdivided plants into different groups based on heat and moisture requirements. The main plant functional types that he defined were as follows:

Megatherms: Plants that require high temperatures and abundant moisture.

Xerophiles: Plants that require high temperatures and tolerate drought.

Mesotherms: Plants that require moderate temperatures and moisture availability.

Microtherms: Plant that have relatively low heat and moisture requirements.

Hekistotherms: Plants that tolerate extremely cold conditions.

In recent work, biogeographers have attempted to define plant functional types on the basis of more refined quantitative relationships between plant form and climatic variables such as tolerances to low temperatures and soil moisture requirements. Perhaps one of the most sophisticated of these studies is the series of "BIOME" models by I. Colin Prentice and his colleagues. These models produce biome estimates from a set of climatic input data.

Vegetation structure can be described in many ways. Most schemes start by dividing vegetation into formations with trees and formations without trees. Formations with trees are then subdivided into those that have continuous canopy cover and those with discontinuous canopy cover. Tree formations may also be subdivided on the basis of the number of strata or layers in the canopy. These might include a canopy layer of dominant trees, a subcanopy layer of smaller trees and saplings, a shrub layer, and a ground layer. Sites without trees might be divided into shrub-dominated and grass-dominated formations. Brown and Lomolino (1998) list six basic vegetation structural types recognized by biogeographers:

Forest: Sites dominated by trees and a generally continuous canopy.

Woodland: Sites typified by widely spaced trees allowing for substantial areas dominated by shrubs, grasses, or herbs.

Shrubland: Sites dominated by a relatively continuous cover of shrubs.

Grassland: Sites dominated by grasses and herbs.

Scrub: Sites dominated by widely spaced shrubs.

Desert: Sites dominated by sparse xerophytic plant cover with mostly bare ground.

In practice, scientific descriptions of vegetation structure often include information on the physiognomy of the dominant plants, degree of canopy closure, and number of vegetation strata.

ECOLOGICAL EQUIVALENTS, LIFE ZONES, AND THE BIOMES

Biogeographers have long been interested in vegetation classification systems that embrace the entire earth. To this end, much work has been done on defining and describing the world's biomes. The modern biome concept of subdividing the biosphere into intercontinental formations of similar climate and vegetation structure can be traced back to phytogeographers working in the late eighteenth and early nineteenth centuries. At the heart of the biome concept is the premise that similar regional climates produce similar plant physiognomic adaptations and vegetation structure. The early phytogeographers noticed that similar plant growth forms and vegetation formations could be found in many parts of the world, even in cases where the similar formations were far distant from each other and did not share any of the same species. For example, the life forms, growth forms, and the vegetation structure of the Amazon rainforest is very similar to tropical forested sites in Africa and Southeast Asia. However, the species of plants, and indeed in many cases the genera and families found in the Amazon, Africa, and Southeast Asia, are very different. Such widely separated, but physiognomically and structurally similar, species and vegetation formations are called **ecological equivalents.**

The scientific study of ecologically equivalent vegetation and the causes of this phenomenon date back over 200 years. The naturalist Johann Forester sailed around the world with Captain James Cook in order to study the natural history of the globe. In 1778 he published his observations on how the earth could be divided into latitudinal belts of similar vegetation structure and similar climate. The German geographer Alexander von Humboldt is often described as the father of biogeography and the originator of the modern ecological equivalence concept. Interestingly, he is also considered to be a founder of the science of geophysics. Humboldt was a young nobleman, 30 years of age, when he sailed from Europe in 1799 and spent the next 5 years on an expedition to the Americas (Fig. 6.3). In his travels, he explored the Amazon, Orinoca, and Magdalena rivers of South America. He also climbed to the 5800-meter summit of Mount Chimborazo in the Andes. During his explorations he took measurements of the climates he encountered, studied ocean currents, and noted the plant and wildlife. He was particularly interested in the relationship between altitude and temperature. Upon his return to Europe in 1804, von Humboldt combined his observations of altitudinal vegetation zonation in South America with Forester's latitudinal evidence and deduced that as one progresses from high elevation to low elevation the climate and vegetation structure change in a manner similar to the changes encountered when one travels from the polar regions to the equator. We may recall from Chapter 2 that air cools in the troposphere at a rate of about 6.5° C for every 1000-meter increase in elevation. Thus, if the sea-level air temperature at an equatorial site is 31° C, it is likely to be a chilly 5° C at the top of an adjacent 4000-meter-high mountain. Humboldt's scientific

FIGURE 6.3 The German geographer Alexander von Humboldt (1769–1859), who conducted some of the earliest scientific work linking the geographic distribution of vegetation formations with climatic gradients. He is considered a founding figure in biogeography and geophysics (*Source:* caifas / Adobe Stock).

explorations did not conclude with his travels in the Americas. In 1829, he set off on an expedition to the Ural and Altai mountains of Russia. His scientific work culminated in the publication of a five-volume work called the *Kosmos*. The *Kosmos* presents Humboldt's accumulated concepts on geography and geophysics.

In North America, during the nineteenth century, Clinton Merriam compared the distribution of desert, coniferous forest, and tundra found at different elevations in the San Francisco Mountains of Arizona, with similar changes in vegetation that occur as one goes northward from the southwestern United States to the Canadian Arctic. He deduced that the climate and vegetation at progressively higher altitudes were equivalent to the climate and vegetation at progressively higher latitudes. He correlated the conifer forest vegetation found at mid-elevations in the mountains with conifer-dominated forests found growing at low elevations 1200 miles north in Canada. He used the term **life zones** to refer to these elevational and latitudinal bands of similar climate and vegetation (Fig. 6.4). Merriam estimated that a 1-mile increase in elevations in the Southwest was roughly equivalent to an 800-mile-northward displacement in latitude.

By the mid-twentieth century, much information had been gathered on natural vegetation formations from around the world. It was clear that formations with similar plant physiognomy and vegetation structure could be observed on different continents. In addition, detailed data on climate was being gathered from meteorological stations all over the world. From this information, it was possible to analytically compare the relationship between large vegetation formations and climatic conditions on a global scale. In 1936 Frederic Clements and Victor Shelford coined the term **biome** to describe the dominant vegetation formation associated with specific climatic conditions. Based on the premise that plant physiognomy and vegetation structure are directly controlled by climatic conditions, the 1936 Köppen climatic subdivision of the earth (discussed in Chapter 2) used vegetation ecotones to help infer the location of climatic boundaries for mapping the world's climatic regions. Other scientists have devised similar divisions of the earth based on large climatic zones and associated vegetation types. In 1947 the English botanist Leslie Holdridge published a subdivision of the

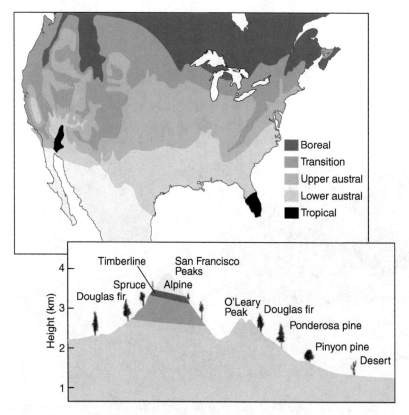

FIGURE 6.4   Merriam's life zones for the classification of North American vegetation based on similarities in structure and climate (based on Merriam, 1890, 1894; Bailey, 1996; Brown and Lomolino, 1998).

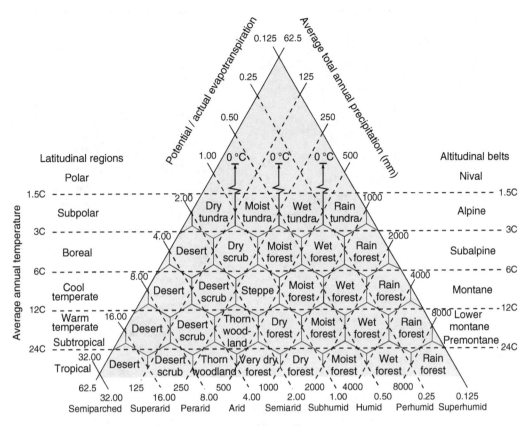

FIGURE 6.5   The Holdridge System for classification of the world's vegetation on the basis of climate or geographic location (Holdridge, 1947; Holdridge et al., 1971).

earth based on the relationship between vegetation formations and mean annual temperature, mean annual precipitation, and mean annual potential evapotranspiration (Fig. 6.5). Modified versions of the Holdridge system are particularly popular today for use by climate modelers interested in the changes in vegetation that may result from future climate change. Many formation and biome schemes have been produced for individual continents or for the entire globe. They all have in common the use of some measure of heat (such as temperature or potential evaporation) and some measure of moisture (such as annual precipitation or soil moisture deficit). Each biome is typified by similar climate and vegetation structure. Although terminology and details on geographic boundaries and number of formations may differ, the general patterns of vegetation and vegetation-climate relationships are similar from system to system.

In 1975 the American ecologist Robert H. Whittaker published a classification of global vegetation formations and their relationship to mean annual precipitation and mean annual temperature (Fig. 6.6). One shortcoming of Whittaker's approach is that it does not incorporate any measure of seasonality. The magnitude of seasonal differences in temperature and precipitation can be extremely important in determining dominant plant physiognomy and vegetation structure. However, this simple scheme remains widely used as a basic introductory model for the description of the world's biomes. Whittaker's system divides the biosphere into a number of distinctive biomes, ranging from tropical rainforest to Arctic and alpine tundra. The geographic distribution of the biomes reflects general global climatic zones and more regional influences on vegetation such as the presence of cool conditions in high-altitude mountain ranges or the presence of rainshadows. Estimates of the total area, net primary productivity, and mean biomass for the biomes are also available from work by Whittaker (Table 6.1). The biomes exhibit large differences in net primary production, which range from 2000 grams per square meter for tropical rainforests to 71 grams for deserts. Each biome has characteristic soil conditions that form due to the

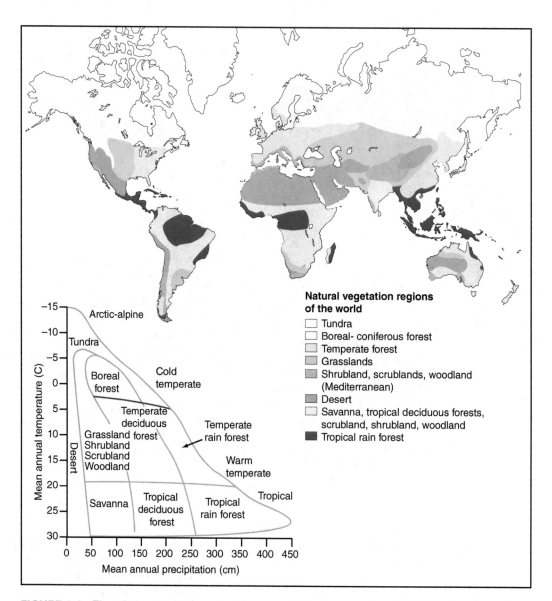

FIGURE 6.6   The relationship between the biomes and climate according to Whittaker's system and a general map of the world's biomes (Whittaker, 1975 / CAB International).

TABLE 6.1   Primary Productivity of the Major Biomes

Biome	Area (million square kilometers)	Primary Production (grams per meter per year)	Total Biomass (kilograms biomass per meter)
Tropical rainforest	17	2000 – >3000	44
Tropical deciduous forest	7.5	1500	36
Savanna	15	400 – 700	4
Desert	18	0 – 90	0.67
Mediterranean and other shrublands	8	300 – 600	6.8
Temperate grassland	9	400 – 1000	1.6
Temperate deciduous forest	7	600 – 1500	30
Temperate rainforest	5	1300	36
Boreal forest	12	200 – 800	20
Tundra	8	100 – 200	0.67

Source: Data from Whittaker and Likens, 1975; Ricklefs, 1990; Forseth, 2010.

particular climatic and vegetation conditions associated with the biome. Using Whittaker's system as a general basis for our exploration, let's take a tour of the world's biomes, starting at the steaming tropical rainforests of the equator and moving poleward to the cold, dry Arctic tundra (Fig. 6.6). We will look at the geographic distribution and climate of each biome. We will investigate the unique features of each biome's vegetation and make a few observations on the fauna. During our journey we will also consider the impact of direct human actions and land use on the vegetation of the biomes. Later in this book we will examine the current and projected effects of human-caused climate change on important aspects of the world's biomes.

Tropical Rainforest

Tropical rainforest occurs in the equatorial regions between about 25° north and south latitude. About 50% of the total area of tropical rainforest is found in Central and South America, 30% in southeastern Asia and eastern Australia, and the remainder in central Africa. The total area of the biome has been estimated at 17 million to 19 million square kilometers. Tropical rainforest is found in areas that have an average annual temperature of greater than 20° C and receive more than 200 to 250 centimeters of precipitation per year. In some areas average annual rainfall exceeds 400 centimeters. This tropical biome has little daily or seasonal variation in temperature (less than 5° C) and never experiences freezing conditions. Precipitation occurs throughout the year. Although dry seasons may occur, the rainfall during the driest month in the tropical rainforest biome averages greater than 12 centimeters. On an annual basis, moisture stress is not a problem as potential evapotranspiration does not exceed actual evapotranspiration.

The warm, moist conditions of the tropical climate lead to deep weathering and rapid leaching of soluble nutrients from soils. The result is a dominance by oxisols. These soils often have a deep red color due to the accumulation of iron oxides. Oxisols are typically very low in plant nutrients. In tropical areas of Central America and Southeast Asia, the dominant soils are developed on fresh volcanic deposits and have greater nutrient content. Over time, these soils will weather and become similar to the more mature oxisols.

The physiognomy of tropical plants and the structure of tropical forest vegetation reflect several key factors. Temperatures are relatively warm, so adaptations to periods of freezing are not required. Soil water is plentiful, so adaptations to seasonal dry soil conditions are not needed. Thus, photosynthesis and plant growth can continue throughout the entire year. However, plants do face several environmental challenges that are reflected in the tropical vegetation. In addition to heat and water, the photosynthetic process requires light. The floor of tropical rainforests typically receives only 1% of the photosynthetically active radiation that is received at the top of the canopy. Much of the physiognomy and structure of rainforest vegetation can be understood in terms of competition for light. Although water is plentiful in the rooting zone, the high temperatures and evaporative stress on top of the canopy cause high rates of water loss from foliage through evapotranspiration. In contrast to the canopy, the forest floor may be 2° C to 4° C cooler and maintain a relative humidity of close to 100% at all times. Such high humidities can hinder the evaporation of excess water from leaf surfaces and decrease rates of photosynthesis. In oxisolic regions, soil nutrients are in short supply. Finally, the diversity of phytophagous insects is extremely high in the tropical rainforest, and plants must be adapted to herbivory by these insects.

As a result of the year-round growth, relatively plentiful moisture, and competition for light, the tropical rainforest is dominated by tall evergreen trees with an understory of perennial plants. It has been estimated that trees can make up 70% of the plant species found in a rainforest stand. Annual plant species are almost exclusively restricted to forest gaps caused by windthrow, fires, or landslides. The stratification of the rainforest vegetation is the most complex of all biomes. In many cases there are five distinct strata (Fig. 6.7). The A stratum consists of scattered supercanopy trees that have trunks rising above the continuous canopy cover. The A stratum trees typically grow to 30 to 50 meters but can reach heights of over 70 meters. An 84-meter-tall *Koompassia excelsa* tree growing on Sarawak Island in Indonesia is reputed to be the tallest tree in the tropics. This is roughly the height of a 15- to 20-story building! The B stratum is the continuous canopy of dominant trees that typically reach heights of 18 to 27 meters. The C stratum is made up of subdominant trees and immature trees that are 8 to 14 meters tall. The D stratum consists mainly of smaller trees, such as palms and tree ferns, and immature canopy trees. On the forest floor, the E stratum contains small herbaceous plants and low shrubs but is often dominated by tree seedlings. There is a common perception that tropical rainforests are a tangled mass of dense vegetation. This is only true around openings and gaps where light

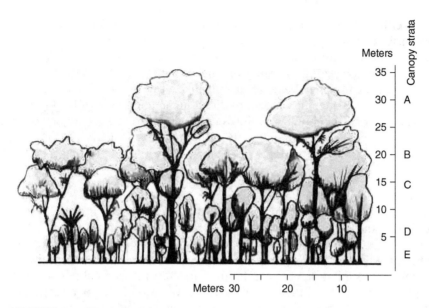

FIGURE 6.7   Tropical rainforest structure and stratification (adapted from Richards, 1996).

is plentiful. The low light levels beneath the canopy of an undisturbed rainforest do not support much vegetation, and the understory is often quite open.

The tropical rainforest contains the greatest amount of standing biomass of any biome. In addition, about 20% of the total biomass of the tropical rainforest is found underground in roots, debris, and other soil organics. The complexity of the tropical forest vegetation and the many different microenvironments in the forest is reflected in the high species diversity typical of this biome. Although the exact number is unknown, it has been estimated that there may be 260,000 to 375,000 different flowering plant species in the tropical rainforest biome. That means that up to 75% of all plant species on earth might be found in this one biome. Over 170 different tree species have been recorded on single hectares of rainforest in both Indonesia and South America. African rainforests have lower diversities, but even there, some 60 to 130 different tree species can be found on a hectare of rainforest land. In parts of the Amazon Basin over 300 species of trees have been found on 1 hectare.

Tropical plants have developed some interesting physiognomic adaptations in response to the low light conditions and different microclimates found within the rainforest. The high growth rates and continuous growing season mean that tree species must reach and dominate the canopy quickly if they are to survive. Canopy and supercanopy trees commonly have very slender trunks supported by spreading lower trunks called buttresses (Fig. 6.8). Plants in the intermediate canopy sometimes are supported by stilt roots. The buttresses and stilts provide stability to the tall slender trunks. A number of plant species reach canopy and subcanopy heights by using the physical support of other plants (Fig. 6.8). Woody stemmed lianas and soft vines are diagnostic plants of the tropical rainforest. They sprout in the soil and use tendrils, adhesive surfaces, thorns, and twining to climb the trunks and branches of trees. The advantage for vines and lianas is that they do not have to invest energy and nutrients into the formation of large trunks in order to reach light. The climbing palms of the genus *Calamus* can reach lengths of 165 meters. This is the plant from which rattan furniture and baskets are made.

Epiphytes sprout and grow on the trunks and branches of trees. They usually never have contact with the soil zone. Epiphytes receive all their water and nutrients from rainfall. Unlike parasitic plants, epiphytes do not tap into the vascular tissue of their hosts. In the tropical rainforest, precipitation is plentiful, and annual rainfall may contribute as much as 21 kilograms per hectare of nitrogen, 16 kilograms of phosphorus, 18 kilograms of potassium, and 16 kilograms of calcium. Epiphytes are dominated by relatively small perennials such as orchids, bromeliads, and ferns. Many ferns and bromeliads grown as houseplants throughout the world are tropical epiphytes. Epiphytes put mechanical strain on their hosts and compete with the host's foliage for light. However, nutrients in the humus produced by epiphytes can benefit tree hosts by adding to the soil nutrient balance. The strangler fig trees of the genus *Ficus* combine aspects of epiphyte and liana growth with a deadly result for the host tree (Fig. 6.9). The seeds of the strangler figs are deposited on the branches of host trees in bird droppings. The young fig begins life as an epiphyte. However, it also develops woody roots that grow

FIGURE 6.8 Buttress trunk (*left*) and a mass of epiphytes and lianas (*right*) on trees in the tropical rainforest of Guatemala.

FIGURE 6.9 A large strangler fig (*Ficus virens*) in the rainforest of Queensland, Australia. The fig begins as an epiphyte living in the upper branches of a host tree. It sends down long tendrils that establish roots. Eventually, the fig kills the host tree. A number of other epiphytic plants have established on the upper branches of the strangler fig.

downward along the trunk of the host tree. These roots eventually reach the soil. Constriction by the fig roots and shading by the expansive fig canopy lead to the death of the host tree. In most cases, the fig tree remains standing, supported by its woody root structure. Plants such as the strangler fig, which combine liana and epiphyte characteristics, are classified as hemiepiphytes. The bark of most tropical trees is very smooth, and this may be an adaptation that decreases the ability of epiphyte seeds to adhere to and germinate on them.

With its trees, lianas, vines, epiphytes, and hemiepiphytes, the upper branches and foliage of the continuous B stratum canopy provide a huge habitat that can stretch for thousands of kilometers. This surface is broken only by tree-fall gaps, disturbed areas, and the courses of larger streams and rivers. Many of the plants and animals of the rainforest exist on this leafy surface and may never have contact with the ground. The hot sunny world of the canopy is very different from the cooler and wetter depths of the forest floor. Rainforest plants display many adaptations that result from the different microclimatic conditions found in the upper canopy and in the lower strata. Leaves in the upper strata are often sclerophyllous in form, being small and having stiff, thick cuticles with waxy coatings and stomata that are sunk in pits. These features protect the leaves against short-wave radiation and help decrease water loss. Upper stratum leaves are often held at high angles relative to incoming sunlight, which decreases radiation and heating of the leaf surface. Leaves on lower strata plants often have sharp elongated tips called drip tips. The presence of drip tips significantly increases water drainage from the surface of the leaf, and this allows for better photosynthesis. The drainage of water from the leaf surface also inhibits growth of fungi, algae, and lichens.

Low windspeeds within the canopy make wind dispersal of pollen and seeds inefficient in the rainforest. Instead, more than 90% of the plant species in some tropical forests depend on insects and animals for pollination and seed dispersal. The shape, color, and smell of many tropical flowers attract specific pollinators. Beetles typically pollinate bow-shaped flowers, which are greenish or off-white and emit strong odors. Hummingbirds are associated with tubular flowers that are brightly colored but scentless. Some species of tropical bats feed only on nectar and pollen and are associated with flowers that bloom at night. The Indonesian corpse flower (*Rafflesia arnoldii*) produces a meter-tall flower that is colored brown-red and emits an odor like decaying meat in order to attract flies as pollinators. Studies in the African rainforest indicate that through selective eating of soft fruit tissue, droppings, and regurgitation, monkeys disperse about 96% of seeds they encounter while feeding and birds disperse 88%. As discussed in Chapter 4, fruit-eating fish are important in dispersing seeds in seasonally flooded portions of the tropical rainforest. The low seasonal variability in climate and the dependence of rainforest plants on different animal and insect pollinators and seed dispersers result in flowers of different species blooming at different times throughout the year. A number of tree species exhibit cauliflory—the development of flowers and fruits directly on the trunk.

Although soils may be poor in nutrients, decay rates of soft tissue and wood are extremely high in the rainforest. This allows for quick cycling of nutrients and for support of the forest vegetation. Most rainforest plants have extensive shallow root systems that are more efficient at taking up nutrients brought in by rainfall and released from decaying litter. In addition, many plants have internal and external fungi associated with their root systems. The hypha of the fungi produce extensive absorbent surfaces that increase rates of nutrient uptake by the host plants.

A unique type of forest found in the tropical regions are mangroves. These forests are found growing in calm ocean waters and estuaries along tropical coastlines. Mangroves are often dominated by tress of the genera *Rhizophora* or *Avicennia*. Like many tropical forest canopy trees, mangroves have stilt roots and sclerophyllous leaves. However, mangrove trees are generally only 10 meters or less in height. Many mangrove trees have pneumatophores (see Chapter 3) that allow them to obtain sufficient oxygen while growing in flooded soils. The interesting facet of mangroves is their ability to grow in saline water. The salinity of soil water in mangrove stands equals that of sea water. To cope with these high salinities, mangroves have a number of interesting adaptations (Fig. 6.10). Mangrove roots can filter out a certain amount of salt. *Avicennia* trees have special glands on the undersides of their leaves that exude salt to remove it from the tree. Excess salt is

FIGURE 6.10 Mangrove stilt roots (*left*) and pneumatophores (*right*) from a site on the Pacific Coast of Mexico.

also stored in leaf tissue and eliminated when old leaves drop from the trees. Trees of the genus *Rhizophora* produce seeds that germinate while still on the tree. These are called viviparous seedlings. When the seedlings are attached to the mother tree, they receive nonsaline water from the parent. Eventually, the seedlings fall from the parent and root. At this point they appear capable of using saline water to grow.

Tropical rainforests contain the highest diversities of insects, birds, and some mammals, such as primates and bats, in the world. There may be as many as 30 to 50 million different species of rainforest insects and even 180 species of primates ranging from gorillas (*Gorilla gorilla*) to the tiny mouse lemur (*Microcebus murinus*). Many of the insects and other animals are phytophagous. Rainforest plants have a number of strategies that combat herbivory. Some vines such as *Calumus* have thorny stems that assist in climbing and protect against larger herbivores. Many plants contain chemicals that are toxic to herbivores. For example, manioc (*Manihot ultissima*) contains cyanogenic glycosides that release cyanide when ingested.

The tropical rainforests of the world have been cleared and destroyed at a rapid pace. Satellite remote sensing provides a powerful tool to monitor rainforest clearances at local to global scales (see Technical Box 6.1). By the 1990s, the rate of rainforest clearance averaged 170,000 square kilometers per year. That represents a loss of almost 1% of the total rainforest of the world each year. In areas such as Central America, less than half of the original rainforest cover remained by the end of the 1990s (Fig. 6.11). Tropical rainforest clearance has occurred in all parts of the biome. Many of the countries with the highest rates of rainforest loss also have rapidly growing populations and limited economic resources. For many people in such countries, exploitation of the rainforest is seen as one of the only means of survival. Even Australia, however, which is an affluent country and contains only a tiny area of rainforest along its eastern edge, has transformed a significant area to pasture and farming. Although still high, rates of tropical rainforest clearance have decreased in South America, South Asia, and Southeast Asia between 2000 and 2018 as conservation priorities have increased. Let's explore this a little more below.

Until this century, the human impact on some tropical rainforests was relatively limited. Small bands of hunter-gatherers lived in some areas, while a shifting pattern of slash-and-burn agriculture was practiced in other regions. Such agriculture, also known as swidden farming, consists of clearing small plots of forest and growing crops on them for 1 to 5 years. After this time, the soil fertility is generally reduced, and the plot is abandoned to forest succession. This type of farming is sustainable

TECHNICAL BOX 6.1

Satellite Remote Sensing of Forest Clearance

Given the massive size of the tropical rainforest biome and the fact that many regions of the biome are remote and relatively inaccessible, monitoring the global loss of these forested regions presents a challenge. The launch of the first Earth Resources Technology Satellite (ERTS-1; later known as LANDSAT) in 1972 was an important milestone in the monitoring of land-cover change related to forest clearance. The satellite was equipped with a Multispectral Scanner (MSS) that detected visible light spectrum energy and infrared energy useful in observing the alteration of land cover through the removal of forest vegetation. The satellite circled the earth at an elevation of 917 kilometers and completed 14 orbits each day. It would reimage a location on the earth's surface every 18 days. The MSS had a ground resolution of roughly 80 meters. Changes in near-infrared bands are particularly indicative of vegetation alteration. This provided a tool to monitor changes. A combination of visible and near-infrared bands, called the Normalized Difference Vegetation Index (NDVI), has been particularly useful in vegetation studies.

$$NDVI = \frac{(\text{Near infrared} - \text{Red})}{(\text{Near infrared} + \text{Red})}.$$

Since 1972 there have been a number of different satellites launched that use various sensors to capture signals related to land-cover changes such as deforestation. Some of these platforms offer much greater spatial resolution than the early satellites. Many have passive optical systems such as expanded and refined versions of the early MSS, and some offer active systems such as synthetic aperture radar (SAR) imaging. Literally thousands of such vegetation change studies from the earth's different biomes have been published.

One of the most popular satellite sensor systems for assessing changes in vegetation, including rainforest clearance, is the Moderate Resolution Imaging Spectroradiometer (MODIS) provided by the NASA Terra satellite system. This provides vegetation indices at 250-meter resolution with a 16-day repeat cycle. This is ideal for regional to global vegetation monitoring. The Terra satellite with MODIS was launched in 1999. Below are two MODIS images of tropical forest clearance from the state of Rondônia in western Brazil. This is one of the most highly deforested parts of the Amazon basin. The first image is from 2000 and the second is from 2012. Forested areas are dark and the cleared areas are light in shade. The massive extent of deforestation is evident from the images.

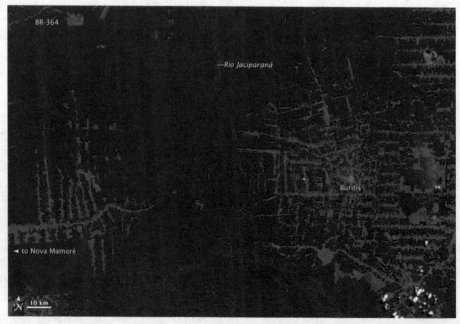

2000

2012 Satellite image of forest clearance in Brazil (NASA / https://earthobservatory.nasa.gov/world-of-change/Deforestation / last accessed December 26, 2023).

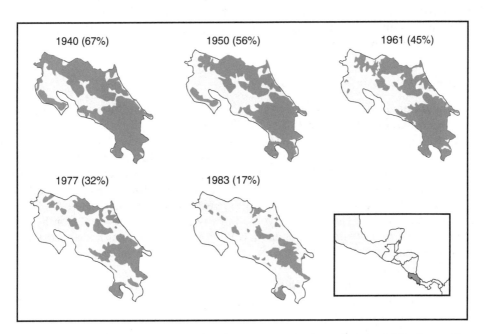

FIGURE 6.11   The loss of primary rainforest in Costa Rica from 1940 to 1983 (Sader and Joyce, 1988; Whitmore, 1990; Richards, 1996).

in the rainforest but can only support small populations of people. It has been estimated that the population density of the Yanomami people of the Amazon, who practice hunting, gathering, and shifting agriculture, is only one person per every 777 hectares. Increasing population pressure and access to the rainforest in areas such as the Amazon Basin have led to attempts to establish permanent farm plots on cleared rainforest land. In many cases, the soil fertility is quickly reduced, and the farmland is colonized by tropical grasses such as *Imperata cylindrica* and other light-demanding herbs and ferns. Introduction of livestock to such areas maintains poor-quality grasslands at the expense of

rainforest succession. Rainforest around the world has also been cleared or altered by large plantations growing commercial crops such as coffee (*Coffea* spp.), bananas (*Musa* spp.), and rubber trees
(*Hevea brasiliensis*).

Logging has also had a dramatic impact on rainforests. Even selective logging can cause widespread damage due to road construction, log felling, and log transport. Removal of teak (*Tectona
grandis*) and other valuable hardwoods, which make up about 10% of Southeast Asian rainforests,
has been shown to lead to damage to about 65% of the forest. In countries such as Indonesia and
Malaysia, hardwood logging almost quadrupled between 1965 and the 1990s. Once logging roads are
established, they are often used for access by farmers, and this leads to permanently cleared land.
Other human disturbances include clearance for crops and plantations, road building, mining, and
dam construction. Between 1997 and 2015, Indonesia lost approximately 90,000 square kilometers of
forest. In recent decades, Indonesia and Malaysia have seen widespread conversion of natural forest
to oil palm (*Elaeis guineensis*) plantations. Over 90% of the world's palm oil is produced on the
islands of Borneo and Sumatra and on the Malay Peninsula. It has been estimated that 50% of the
tropical forest cover on the Malaysian portions of Borneo have been transformed to oil palm plantations. Tropical mangrove forests are not immune to degradation and deforestation. Clearance for
aquaculture, the use of wood, and other forces have resulted in a global rate of removal estimated to
be 2% to 8% per year.

There is mixed news on the current state of tropical deforestation. According to the 2015 UN
FAO Global Forest Resource Assessment, although there had been a net loss of over 1.95 million
square kilometers of tropical forest area, the annual rate of loss decreased by almost 55% from 2000
to 2015. Rapid second-growth tropical forests have produced a phenomenon of reforestation in some
regions of the tropics in the early portion of the twenty-first century. For example, from 2000 to 2010,
forest vegetation increased over some 360,000 square kilometers in Latin America and the Caribbean.
Unfortunately, in many parts of the region, that trend reversed by 2014, and sustained forest loss
became more common than regrowth. In the Amazon about 50% of the secondary forest regrowth
was re-cleared within 8 years.

Tropical Seasonal Forest

Tropical seasonal forest, also called tropical dry forest, occurs as broad bands to the south and north
of the tropical rainforest biome. It is generally restricted to the tropics but can be found as far as 30°
north latitude in India. The total area of the biome is around 7.5 million square kilometers.

The mean annual temperature of the tropical seasonal forest region ranges from 20° C to 30° C
with rainfall averaging 150 to 250 centimeters per year. However, it is the seasonal variability in rainfall that most strongly differentiates the tropical seasonal forest from the rainforest. The tropical
seasonal forests experience a monsoon rainfall pattern, with the vast majority of precipitation falling
in the summer months. The winter is typified by dry conditions for 4 to 7 months. Seasonal variations
in temperature are not large. Frost generally does not occur but has been reported in portions of the
tropical seasonal forest in southern Africa, for example. Potential evaporation exceeds precipitation
in the dry winter months.

The soils in the seasonal tropical forest can vary widely by region, but many tend to be deeply
weathered, leached oxisols. However, owing to the dry season, leaching is often not as extreme as in
the tropical rainforest, and nutrient retention can be higher than in the rainforest.

As a result of the seasonal differences in precipitation, the dry tropical forest differs in many
important ways from the adjacent rainforest. First and foremost, most of the canopy trees are deciduous. They become dormant during the winter dry season and lose their leaves (Fig. 6.12). Some understory small trees and shrubs may maintain their leaves during the winter, and these can be
sclerophyllous. In many cases, shrubs, vines, and herbs go dormant in the dry season. The leaves of the
canopy trees tend to be larger than those found in the rainforest canopy. The deciduous leaves must
develop quickly at the start of the wet season, and then they die when moisture stress becomes acute
at the start of the dry season. The canopy trees often bloom just before the onset of the wet season.
This allows pollinators to find blooms before canopy leaves obscure them. Fruit and seeds can then
be formed during the wet summer months and are often dispersed during the subsequent dry season.

FIGURE 6.12   Dry tropical deciduous trees in northern Queensland, Australia. These trees cope with the rainless winter season by releasing their leaves and going dormant. When rains return in the summer, the canopy will become leafy and green.

There is less cauliflory than in the tropical rainforest. There is a greater proportion of wind-dispersed plant species than is the case in the tropical rainforest.

The seasonal tropical forest has a much simpler structure than the rainforest, with only three dominant strata—canopy, shrub and sapling, and ground cover. The height of the canopy tends to be less than the tropical rainforest, generally averaging 20 to 30 meters. Vines and lianas are found in the tropical seasonal forest, but they are not as common or as diverse as in the rainforest. Epiphytes are rarer and less diverse because they often cannot obtain sufficient water or nutrients to persist. The ground layer vegetation may be denser than in the tropical rainforest because of greater light penetration.

The standing biomass and species diversity in tropical seasonal forests are second to tropical rainforests. In terms of insect and animal fauna, some tropical seasonal forests have diversities close to those of adjacent rainforests. An 11,000-hectare tropical seasonal forest in Costa Rica is estimated to host some 130,000 species of insects, 175 species of birds, 115 species of mammals, and 75 species of reptiles and amphibians. The floral diversity of tropical seasonal forests is often less than 50% of the diversity found in nearby rainforests. In particular, the number of tree species and epiphytes found in the dry forest is much less than that of the rainforest. A total of 700 angiosperm species is estimated to inhabit the 11,000-hectare Costa Rican dry tropical forest.

As is the case for tropical rainforests, tropical seasonal forests are gravely threatened by human activity. The biome is generally found in areas of the world where increasing human population size and persistent poverty make conservation difficult. The forests are easy to traverse in the dry season and easy to clear using fires. Some forest is lost to intentional fires, while large areas are also destroyed by accidental burning. Once the trees and shrubs are cleared, the long dry season makes it easy to use fire to preclude forest reestablishment. In many cases, the soils are better for crops or grazing land than those found in the rainforest, and permanent agriculture has been established in place of forest. It has been estimated that only 2% to 10% of the original tropical seasonal forest in Central America remains intact. Only a paltry 480 square kilometers of the original 550,000 square kilometers of seasonal tropical forest in Central America has official conservation status. In all of the Americas, less than 5% of the tropical deciduous forest biome was protected by law as of 2010. Of this, 65% is located in just two countries, Brazil and Bolivia. In Southeast Asia, teak (*Tectona grandis*) logging negatively impacted large areas of the forest. Globally, the tropical deciduous forest is considered by some scientists to be the most threatened biome on earth. Despite its high productivity, biomass, species diversity, and threatened status, the seasonal tropical forest has not received nearly the same level of scientific research, public attention, and conservation efforts as the rainforest.

Tropical Savanna

Tropical savannas and associated thorny woodlands occur in tropical and subtropical regions, centered on about 30° north and south latitude. The tropical savanna borders tropical forest areas, and often there is a diffuse ecotone between savanna and tropical seasonal forest. Over half of all tropical savanna is found in Africa, where this biome comprises 65% of the total land cover. Tropical savannas and thorny woodlands have many different regional names; in Africa the savanna may be called *bushveld,* in Australia it can be called *brigalow,* and in South America it is known as *campos* or *llanos.* The total area of the biome is about 15 million square kilometers. The tropical savanna is typified by average annual temperatures of between 20° C and 30° C and annual precipitation of between about 50 and 150 centimeters. Seasonal differences between winter and summer temperatures average 10° C to 15° C. Most tropical savanna sites experience prolonged winter dry seasons. The dry season can persist as long as 7 months. Frosts are uncommon.

Reddish oxisols are found on moist savanna sites. Drier portions of the biome are dominated by alfisols with brownish to red surface horizons. These soils are higher in nutrients than oxisols. Shallow and sandy entisols develop in very dry areas. The entisols are low in nutrients and may also be relatively saline. Because of the contrasting wet and dry conditions, bases, clay minerals, and sesquioxides accumulate in the B horizon of savanna soils. In some cases, saline horizons or layers of calcium carbonate develop. Iron and aluminum can accumulate to form hardpan layers known as laterite. These layers are impenetrable by water and block drainage, upward movement of groundwater, and nutrient flow. The organic content of all savanna soils is very low (less than 3%). One contributor to the low organic content of the soils is the rapid decomposition of plant matter by termites. The termite population of a savanna site in Zaire has been estimated to be 17.5 million individuals per hectare. They can make up 80% of the soil organism biomass. In Africa, termites are responsible for 90% of the decomposition of dead grass. Termites and their mounds are common in all tropical savanna regions.

Tropical savanna consists of grasslands with scattered trees and woody shrubs. The relative abundance and the spatial distribution of woody plants and grassland can differ markedly. In some areas, such as the *Curatella* llanos of Venezuela, the *Eucalyptus* savanna of eastern Australia, or the *Mopane* savanna of southern Africa, scattered individual trees dot the grasslands. In other cases, the savanna has a parkland configuration in which areas of open or closed canopy forest are found on well-drained uplands and along rivers. These forest stands are surrounded by extensive areas of treeless grassland. In some cases, trees are rare, and the woody vegetation is dominated by often thorny shrubs.

The trees and woody shrubs of the savanna can be either evergreen, often with sclerophyllous leaves, or deciduous. Deciduous tree species are most common in Africa, and evergreen species are more common in other parts of the world. Deciduous leaves and dormancy do not allow trees to completely escape moisture stress during the dry winter months. Water can be lost through bark and branches, causing lethal desiccation. The thick bark of some savanna trees reduces this water loss. Perhaps the most interesting adaptation to dry season moisture loss is found in species of the Bombaceae family that grow in seasonal tropical forests and savanna of Africa and South America. These trees such as the baobab (*Adensonia digitata*) of Africa and the silk cotton tree (*Ceiba pentadra*) of South America have enlarged trunks where they store water in spongy tissue. Baobabs can reach a circumference of more than 20 meters and store up to 120,000 liters of water (Fig. 6.13).

The evergreen trees of the savanna have deep tap root systems that penetrate through laterite layers and provide access to the water table. The *Curatella americana* tree of the South American savanna reaches a height of only 6 meters, but it develops tap roots as long as 20 meters. Some shrubs, such as the palm species *Attalea exigua* in Brazil, have large woody organs called xylopdia that lie beneath the soil surface. During the dry season or following fires, the above-ground portion of the plant may die back. The plants resprout from xylopdia when the rains return. The height of trees and forest canopies is generally less than 20 meters and often less than 10 meters. Vines are not common, and epiphytes are extremely rare. The grasses of the savanna tend to be perennial species. The above-ground portions of the grasses die back during the dry season, and the plants resprout from underground organs, such as rhizomes and stolons, in the wet season. Most grasses are of the C_4 variety (see Chapter 2 for a description of C_4 photosynthesis) and can conduct photosynthesis at high temperatures and illumination. For example, many temperate C_3 grass species have maximum rates of

FIGURE 6.13   Baobab tree (*Adensonia digitata*) in the savanna region of Mali (*left*). Grassy vegetation dominating wet, poorly drained low spots with woody vegetation occupying better drained and more nutrient-rich soil on old termite mounds in the savanna region of Mali (*right*).

photosynthesis at around 24° C. In contrast, tropical C_4 grass species such as *Paspalum dilatum* reach peak net photosynthesis at 35° C. In general, the percentage of woody species and the height of trees and shrubs decrease as one moves from moister to drier portions of the savanna. In the very driest areas of the savanna biome, annual grasses become increasingly important.

What controls the spatial distribution of woody plants and grasslands in tropical savannas has long been a topic of great interest. Why do neither the woody plants nor the grasses come to dominate the entire landscape (Fig. 6.13)? It is thought that this may be due to a fine balance between the two types of plants. The grasses have shallow and extensive root systems (Fig. 6.14). They depend on efficient interception of rainwater close to the surface. Tree seedlings often cannot compete with the grass roots for moisture. In addition, the grasses grow quickly during the rainy season and can attain heights of greater than 2 meters. In these circumstances, the slower-growing seedlings of tree and shrub species are shaded out. Grazing and browsing by animals also likely favors grasses. The perennial grasses can endure relatively close grazing and still resprout and grow quickly. Recovery of trees and shrubs after intense browsing is slower. Close grazing will often kill tree and shrub seedlings. Elephants uproot and destroy small trees when browsing and often destroy several trees in the course of a day. Thorns and tough leaves are common in woody plants of the savanna. These offer some protection against browsing. Finally, lightning strikes are common, and the dry savanna grasses burn easily. Fires can kill tree and shrub seedlings.

When trees are established, they produce deep intensive roots to obtain groundwater. Trees tend to limit grassland cover by shading out the grass understory, which decreases competition for water and the amount of ground fuel for fires. In addition, many trees and shrubs have adaptations to fire, such as thick bark or the ability to resprout from underground organs. Establishment of trees also increases soil nutrients and can promote fast growth of the trees themselves. In some cases, the location of trees and parkland groves appears dependent on small differences in topography and soils (Fig. 6.14). Tree roots may not be able to penetrate to the water table in areas with laterite, and this restricts the survival of woody species. Low-lying sites, which are underlain by laterites, often flood during the rainy season, precluding the establishment and growth of trees. Old termite mounds

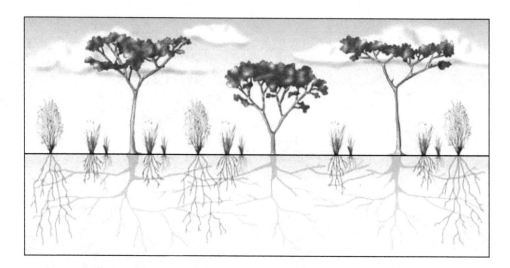

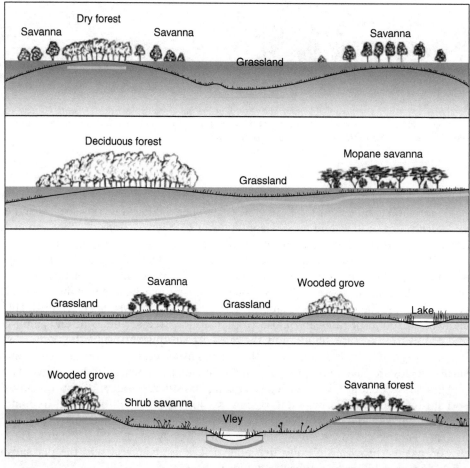

Hard pan soil layers

FIGURE 6.14 Small differences in topography and differences in drainage caused by hardpans help control the distribution of trees, shrubs, and grasses on the tropical savanna. The grasses have shallow, extensive root systems, while trees have deep, intensive systems (Walter, 1985 / Springer Nature).

sometimes form small raised areas that are occupied by woody vegetation, while adjacent seasonally wet areas are occupied by grasses sedges and herbs. In addition to providing better drainage, the mounds sometimes have higher concentrations of potassium, calcium, and magnesium than adjacent soils. The increased nutrient availability helps to support the woody plants (Fig. 6.14). Areas with coarse, rapidly drained soils may not hold enough water at the surface to support shallow rooted grasses but can support deeply rooted trees and shrubs.

The standing biomass of the tropical savanna biome is only a fraction of that found in the tropical rainforest. The species diversity for plants, particularly woody plants and vines, is also much lower. The tree flora over large areas is dominated by a handful of species. Savannas do support a considerable diversity of birds. The Cerrado savanna of Brazil contains 850 bird species, while 794 bird species are known from the African savanna. In total tropical savannas are home to at least 2,951 bird species. That represents almost 30% of all known bird species on earth. Large flightless birds including the ostrich (*Struthio camelus*) from Africa, the emu (*Dromaius novaehollandia*) from Australia, and the rhea (*Rhea americana*) from South America are all found in the savanna. The mammal fauna of the African savanna is distinctive for its great number of large grazers, browsers, and their predators. Over 90 different species of ungulates (hoofed mammals) alone are found there. Some of the more well-known African savanna mammals include elephant, giraffe, rhinoceros, hippopotamus, zebra, gazelle, lion, cheetah, and hyena. Such diversity of large mammals is not found on the savanna of any other continent (Fig. 6.15). The mammal population sizes, particularly for grazers, can be quite high on the African savanna. It has been estimated that the Serengeti-Mara National Parks of Tanzania and Kenya support some 3 million ungulates in an area of 25,000 square kilometers.

Human impact on the savanna ecosystem includes the introduction of domesticated grazing animals such as sheep and cattle, decimation of populations of native mammals such as elephant, and changes in fire regime. These changes have occurred in almost all savanna regions that are not protected as parks and preserves. Large savanna preserves are particularly well developed in Africa. In some areas, control of fire and declines in native grazing populations have led to the invasion of grassland by woody species. In other areas, increased rates of burning to spur grass production have led to declines in woodland cover. The African elephant (*Loxodonta africana*) has faced the threat of extinction in its native savanna at the hands of ivory poachers. In recent years, rigorous protection from poachers and an international boycott of new ivory have led to a resurgence of African elephant populations in some regions, such as portions of Tanzania. However, this is not universal. A census

FIGURE 6.15   The grasslands, short woodlands, and wide diversity of large grazing mammals on the tropical savanna of Kenya are well illustrated in this picture. The various mammal species on the African savanna take advantage of different types and layers of vegetation for feeding and also migrate seasonally to take advantage of different feeding grounds (*Source:* Nikolay N. Antonov / Adobe Stock).

conducted in Angola in 2015 concluded that there was a 21% decline in elephant population size since 2005. Both poaching and displacement of elephants by the expansion of human populations and land-use practices are contributing to the decline in Angolan elephant populations. Globally, it is estimated that 50% of the tropical and subtropical savannas have been sequestered for human land-use practices, and it is possible that 20% more could be lost by 2050.

Desert

Arid regions of desert and semidesert cover some 40 million square kilometers and comprise about 30% of the earth's land surface. Most of the world's great deserts are centered on latitudes 20° to 30° north and 20° to 30° south, where descending air from the tropical regions produces warm and dry conditions. The desert biome extends beyond 40° north and south latitude in areas such as central Asia, the western United States, and South America where large mountain ranges produce intense rainshadows. The largest single desert in the world is the Sahara of North Africa, which covers some 9 million square kilometers.

Some tropical portions of the desert experience average annual temperatures of 30° C with little seasonal variation. Cold desert regions in North America, central Asia, and South America can have average annual temperatures of less than 15° C, with differences of as much as 35° C between July and January temperatures. Many deserts found in higher latitudes experience freezing temperatures in winter. Diurnal variations in temperature may also be quite high. The defining characteristic of desert climate is severe moisture deficit. Annual precipitation in the desert biome is less than 50 centimeters per year and is often less than 10 centimeters per year. Average annual precipitation in the Atacama Desert of Chile averages 0.4 centimeters per year, while portions of the eastern Sahara Desert may experience no significant rainfall for over 100 years. Low-latitude deserts generally receive their rainfall in the summer when the Inter-Tropical Convergence Zone shifts to higher latitudes and monsoon precipitation occurs. Some northern deserts, such as the Great Basin and Mojave deserts of the United States (Fig. 6.16), experience greatest precipitation in winter when they are crossed by midlatitude storm tracks. In the Great Basin and Mojave deserts of California and Nevada, precipitation can come as rain or snow. In some cases, such as the Sonoran Desert of southern Arizona and northern

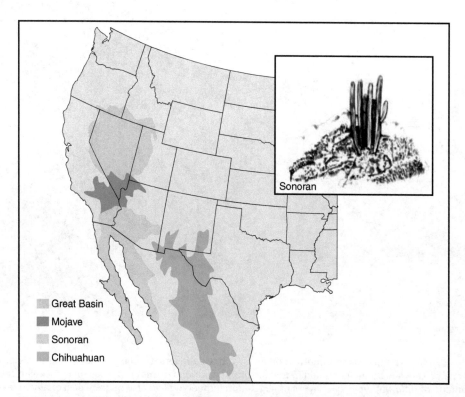

FIGURE 6.16   The desert regions of North America (Brown and Lomolino, 1998; MacMahon, 2000).

Mexico, rainfall occurs during the summer due to monsoon activity and in the winter due to midlatitude storms. The Chihuahuan Desert of New Mexico and Mexico is dominated by summer rain. Freezing temperatures or snow is extremely rare in the Sonoran Desert but occurs in other North American deserts. The Great Basin Desert and portions of the Mojave can experience particularly cold temperatures. In all cases, annual, and generally monthly, potential evapotranspiration rates significantly exceed precipitation. In tropical and subtropical deserts, that ratio of annual potential evapotranspiration to annual precipitation generally exceeds 10:1.

Aridisols dominate desert regions. Due to dry conditions, these soils are little weathered. They are generally shallow and coarse and contain negligible organic matter. They are often alkaline and can contain bands of calcium carbonate or salts. Entisols are found in areas of active sand dune activity. These soils are very poorly developed and contain very little organic matter of accessible plant nutrients. Substrate and soil type are very important for local differentiation of arid land vegetation, and deserts are sometimes classified on the basis of edaphic (soil) conditions. Hamadas are rocky deserts where fine soil particles have been washed or blown away. Regs are deserts with a ground cover predominantly made of gravel. Sandy deserts, with their characteristic shifting sands and dunes, are called ergs. Dry valleys in desert regions are sometimes referred to as wadis. Playas or salt pans are the beds of ancient lakes, or ephemeral desert lakes. They consist of fine sediment, such as clays and silts, and are generally very rich in salts and other evaporites. They may have salt crusts at the surface. Oases are areas where groundwater rises to the surface, or at least close enough for access by the roots of plants. Owing to the availability of water, plants found at oases are often not true desert species.

The structure of desert vegetation is characterized by scattered shrubs, herbs, or grasses. There is usually more bare ground than vegetation. Because of the range in precipitation and temperature regimes experienced by different deserts, there is a wide variety of plant physiognomic features and vegetation densities (Fig. 6.17). In general, the relative amount of vegetation cover is positively linked to precipitation (Fig. 6.18), but soil conditions are also extremely important. The heavy and saline soils of playas frequently do not support vascular plants. Similarly, few plants can establish or persist on actively moving sand. In deserts with extensive vegetation cover, the dominant plant species often appear to be more or less regularly spaced with little growth between them. Some work has shown that this is due to root competition for moisture and nutrients. In the most arid deserts, plants are extremely rare and may only be found in very sheltered sites along valleys and in rock crevasses. Trees are generally absent except at oases and along river courses. Some desert plants, however, such as the Joshua tree (*Yucca brevifolia*) found in the Mojave Desert, the saguaro cactus (*Carnegiea gigantea*)

FIGURE 6.17   The relatively diverse succulent and nonsucculent vegetation of the Sonoran Desert is well represented at Organ Pipe Cactus National Monument in Arizona. The two species of large cacti are organ pipe (*Stenocereus thurberi*) in the foreground, with saguaro (*Carnegiea gigantea*) in the background. Despite the seemingly dense vegetation, much of the ground is actually bare of any plant cover (*a*). In contrast to the Sonoran Desert, the drier and colder Mojave Desert and Great Basin Desert have much lower plant species diversity. The green plants shown here are the ubiquitous creosote bush (*Larrea tridentata*). Areas such as Death Valley National Park in California have very low amounts of vegetation cover (*b*). *Source:* (*a*) reisegraf / Adobe Stock and (*b*) Dan Suzio / Photo Researchers.

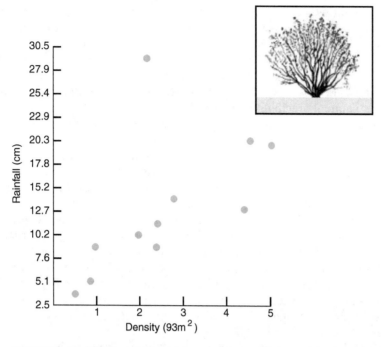

FIGURE 6.18   The correlation between annual precipitation and the density of creosote bushes in the Mojave Desert of California (adapted from Woodell et al., 1969).

of the Sonoran Desert, and the kokerboom (*Aloe dichotoma*) of South Africa can grow to heights of over 4 meters.

Desert vegetation generally contains both perennial and annual plants. In hyper-arid regions, however, with very rare rainfall events such as the interior Sahara, annual plants may dominate. In these cases, the plants persist for decades as dormant seeds that germinate, mature, and set seed within days or weeks after rainfall events. Both evergreen and deciduous plants are found in deserts. In most cases, flowering occurs during the wet season or following individual precipitation events. The beautiful coulter hibiscus (*Coulter hibiscus*) of the southwestern desert can bloom several times a year whenever a sufficiently heavy rainfall event occurs. Seedling establishment can be difficult in deserts, and many of the perennial plants can resprout from below-ground organs if their leaves and stems are destroyed. The ubiquitous creosote bush (*Larrea tridentata*) of the Southwest does this readily, and it has been estimated that some patches of this plant may represent individuals that have persisted for 11,000 years or more. A number of plants can regenerate from portions of detached leaf or stem. The jumping cholla cactus (*Opuntia fulgida*) of the Sonoran Desert has spiny stem joints that are very loosely attached to the parent plant. The joints break easily, produce roots, and grow into full-sized plants. This species appears to rely more on this mode of regeneration than on seedlings.

The physiognomies and life histories of desert plants generally reflect adaptations to moisture stress. There are many different ways in which such desert plants survive in their arid environments.

Drought-escaping annuals germinate, mature, and set seed when soil moisture is sufficient to support them. The growing plants have no physiological mechanism that copes with severe drought. Seeds can remain dormant in the upper soil layers or, in some cases, in dried fruiting bodies on the dead parent plant. Plants have appeared from the soil seed bank in areas that have not received rain for almost 100 years. Not all drought-escaping seeds are stored in the soil. The seeds of the southwestern desert plant *Chorizanthe rigida* are released from the parent plant over periods of years as the flower and fruiting structure that holds them disintegrates. After germination, the growth and seed set by drought-escapees happen very quickly, generally within 6 to 8 weeks. The annual plant *Boerhaavia repens* of the Sahara can germinate and set seed within 8 days. Many desert annuals can self-pollinate so that viable seeds are produced during their short life spans even if they do not come into reproductive contact with other members of their species.

Some perennial species evade the impact of aridity by remaining dormant until there is sufficient moisture to trigger renewed activity. As in the case of drought-escaping annuals, these drought-avoiding plants have no physiological adaptations that cope with arid conditions during their growth periods. Some of these plants persist as bulbs and rhizomes after the above-ground portion of the plant dies off. They become active very quickly once the soil is moist. The sedge species *Carex pachystylis* of Israel's Negev Desert produces new rootlets and begins spreading within 12 hours after precipitation. A number of woody species lose their leaves and become dormant during dry periods. The wolfberry (*Lycium berlandieri*) of North American deserts can lose its leaves several times each year and survive for many months until the next rainfall. The white thorn acacia (*Acacia constricta*) of the American Southwest loses its leaves in response to drought and frost. As a result, the plant can spend much of the year in a leafless state.

Perennial species that continue functioning even during the driest portions of the year can be divided into three broad classes based on leaf structure. First, *stenohydric xerophytes*, such as members of the genus *Euphorbia* found in deserts of Africa and Asia, shut their stomata to stop water loss during drought and persist for long periods by relying on stored carbohydrate reserves. However, because photosynthesis ceases when the stomata are closed, the plant can eventually starve from lack of CO_2. This strategy is therefore not very useful in regions where very long periods occur between precipitation.

Second, sclerophyllous xerophytes, such as agritos (*Berberis trifoliolata*), found in southwestern American deserts, have stiff sclerophyllous leaves with waxes, resins, and other coatings that decrease evaporative loses. On agritos leaves, the light gray color of the waxy coating also helps to reflect sunlight and keep the leaf surface cool. Other features typical of sclerophyllous leaves, such as stomata in pits and inrolled leaf margins, are also found on many desert plants. Some plants, such as wild buckwheat (*Eriogonum fasciculatum*), also have a covering of small hairs on the leaf surface to help reflect sunlight and decrease evaporative water loss. Leaves may also be held with their thin edges toward the sun to keep temperatures and water losses low. The jojoba plant (*Simmondsia chinensis*) found in Arizona and California has such high-angle leaves. This is the same plant used today in shampoos, soaps, and waxes. The ground jojoba seeds were long used by Native Americans for shampoo.

Interestingly, sclerophyllous xerophyty does not appear to be a suitable strategy for extremely arid deserts. In such areas, the leaf adaptations are not enough to counter water losses. In more arid settings, plants lose all or most of their leaves during drought. Such plants belong to the third broad class, malakophyllous xerophytes. The term *malakophyllous* means "soft leafed." However, many plants do not go dormant. They have green photosynthetic tissue in stems and remain active. There are many examples of such plants in the deserts of the southwestern United States and Mexico, including the paloverde plants (*Cercidium* spp.) common in the Sonoran Desert. Photosynthetic stems have a low ratio of surface area and stomata relative to tissue. Some desert plants, such as desert broom (*Baccharis sarothroides*), have dispensed with leaves altogether and rely solely on photosynthetic stems. Finally, some perennials, such as mesquite (*Prosopis* spp.), typical of Texas deserts, have extremely deep root systems in order to reach water during dry periods.

Some perennials store water and have other physiological adaptations that cope with dry periods. Succulent plants with thick water-storing stems and leaves are the most common example of this strategy. The cacti of North, Central, and South America and the euphorbs typical of African and Asian deserts are well-known examples of succulents. The saguaro and barrel cactus (*Ferocactus wislizenii*) of the Sonoran Desert have fleshy stems that are composed of 70% water. Cacti and euphorbs also have waxy coatings on leaves and stems, and many have spines. Both these features decrease tissue temperatures and evaporative water loss. The cacti also use the CAM pathway and obtain CO_2 at night (see Chapter 2), so they can keep their stomata closed during the day and limit evapotranspiration.

Although highly adapted to relatively dry conditions, the cacti are not as dominant in American deserts as one would at first suspect. Cacti have shallow root systems, and despite their water storage capabilities must receive relatively frequent infusions of moisture. The cacti cannot rely on the transpiration of water from their stomata to cool leaf and stem surfaces during the day. In addition, their high water content makes many species of cacti extremely intolerant of freezing temperatures. It is not surprising that in the Southwest the greatest diversity of cacti species and their greatest dominance is found in the Sonoran Desert, with its combination of warm temperatures throughout the year and the addition of soil moisture during both winter and summer wet seasons.

The standing biomass of the world's deserts is low compared with other biomes. In addition, over half of the total biomass may be found below ground in the roots and associated structures. Productivity and standing biomass are positively related to precipitation. Plant species diversity in this harsh biome is also low. Large areas may support only one or two species of plant. The entire desert biome in Africa supports approximately 3000 flowering plant species, but individual desert subregions within the continent typically have fewer than 400 plant species. The Tomboctou Desert of Mali supports only 134 plant species. The Australian desert regions have some 1000 different plant species. In North America the deserts support well over 3000 different plant species. The most diverse is the Sonoran Desert, which has approximately 2700 species of plants. In contrast, Death Valley supports only about 279 plant species. Animal diversities are similarly low in desert regions.

Human impact on deserts varies greatly. In some areas, such as the central Sahara, the desert's inhospitable conditions preclude human occupation except at oases. It has been suggested that overgrazing in areas such as the southern fringe of the Sahara Desert in Africa may be leading to an increase in desert at the expense of adjacent shrubland and savanna. This conclusion is not universally accepted, and climate change is also considered a driving force in changing the extent of the Sahara Desert. It has been estimated that the Sahara expanded southward by some 130 kilometers between 1980 and 1990. This represents an increase in the Sahara of some 7%. However, during this time there were also northward shifts by as much as 110 kilometers owing to climatic variations. More recent research combining satellite observations, climatic data, and modeling suggests a net southward expansion of the Sahara Desert of 100 kilometers over the period 1950–2015. However, most of this occurred up through the 1980s, and this trend has reversed in places. Climate was found to be a key factor driving these shifts. Increased salinization of soils due to improper irrigation can also cause expansion of desert regions. When large amounts of irrigation water are allowed to evaporate from fields, they leave behind salts that can make the soil infertile. In the United States, rapid development of desert regions has taken place in areas such as Las Vegas, Nevada; Phoenix and Tucson, Arizona; and Lancaster, California. This urban growth depends largely on imported water. The construction of buildings and infrastructure for large cities can directly destroy local desert sites, while road building, recreational activity, and other land uses at the fringes of these metropolitan areas cause significant deterioration of desert environments and declines in native plant populations' species richness. Replacement of native desert plant species by introduced species is also a challenge (Technical Box 6.2). One difficulty that faces efforts to preserve desert environments is that, despite their beauty and the many fascinating adaptations of their biota, deserts are often perceived as wastelands.

TECHNICAL BOX 6.2

Invasive Buffelgrass in the Sonoran Desert

Buffelgrass (*Pennisetum ciliare* L.) is a perennial bunch grass that is native to parts of Africa and southern Asia. It was intentionally introduced to southern Arizona in the early through mid-twentieth century in an attempt to improve range conditions for cattle grazing. The species is now established in desert and grassland regions from southern California to southern Texas and adjacent Mexico. There are also scattered occurrences throughout the southern states to Florida. Buffelgrass grows quickly and can reach a height of over 1 meter. Once it is established, it is difficult to eradicate. The plants create a large amount of vegetative fuel that burns easily and at high temperatures in the dry season. Buffelgrass fires have produced recorded temperatures of 871° C. Buffelgrass is able to quickly resprout from underground roots and rhizomes after fires. Patches typically can withstand annual burns. Buffelgrass produces large amounts of small fluffy seeds which are easily spread by the wind and water and through transport on animals and people. It can also spread via rhizomes.

The advance of buffelgrass in the Sonoran Desert poses a threat to the cactus-rich native vegetation. Buffelgrass grows quickly into dense tall stands and can outcompete native species that are low in height or at the seedling stage. Due to the dispersed spatial distribution of the individual native plants and the fact that many are succulents, fire is not a frequent source of disturbance in the Sonoran Desert. Buffelgrass promotes fire activity within the Sonoran vegetation by producing a continuous carpet of highly flammable vegetative fuel. Although buffelgrass can withstand frequent burning, many native plants in the Sonoran Desert cannot.

Buffelgrass can establish itself in the Sonoran Desert and spread very rapidly. A small patch that was detected in the Santa Catalina Mountains of southern Arizona in the 1980s had grown to 66 hectares by 2008. The invasive patch was capable of doubling in size in as short a period as 2.26 years. Eradication of buffelgrass has proven difficult. Burning the patches is largely ineffective and can be harmful to native vegetation. Mechanical removal may leave behind roots and rhizomes, which will regenerate. Typically, resource managers have to resort to chemical herbicides to eradicate buffelgrass.

Invasive buffelgrass resprouting after fire (*left*) and spreading into Sonoran Desert habitat (*right*) in southern Arizona.

The Mediterranean Biome

Next we will explore a biome that is not well accommodated within Whittaker's simple biome model. This is the Mediterranean or winter rain biome. The Mediterranean is the smallest of all the biomes we will consider. It occupies only about 1.8 million square kilometers and is found as narrow strips of land along the Mediterranean Sea; the southeastern coasts of Africa, Australia, and South America; and the coasts of California and closely adjacent Mexico. More than half of the total area of the biome is found in the Mediterranean region of southern Europe, western Asia, and northern Africa. In terms of temperature and annual precipitation, the biome would fall into Whittaker's Woodland-Grassland-Shrubland category. The biome experiences average annual temperatures of 15° C to 20° C and total annual precipitation of 30 centimeters to 100 centimeters. Summers are warm and winters mild. Frosts are rare. What makes the biome distinctive is the fact that precipitation is limited to the fall, winter, and spring months. For 4 months or so during the summer, there usually is no precipitation.

The characteristic soils of Mediterranean regions are formed from the deep weathering of bedrock and are rich in clay and poor in nutrients. The organic content of the Mediterranean soils is generally low. In areas of calcareous rocks, a deep red color is imparted to the soil by oxidized iron. These distinctive soils are known as terra rosa.

The typical vegetation structure of the Mediterranean biome includes a mosaic of valley forests, open woodlands, shrublands, and grassland. The woodlands in both the Mediterranean Sea region and in California are typically dominated by oaks (*Quercus* spp.). In Australia the dominant tree is eucalyptus. Southern beech (*Nothofagus obliqua*) and acacia (*Acacia*) are found in Chilean woodlands. Conifers also occur in some woodlands. These include pines (*Pinus*), firs (*Abies*), and cedars (*Cedrus*) in the Mediterranean Basin, and pine and bigcone Douglas fir (*Pseudotsuga macrocarpa*) in California. The South African portion of the biome is dominated by shrubs and grasses. The spatial distribution of woodland, shrubland, and grassland can reflect a number of factors, including regional rainfall differences, slope aspect, substrate, and disturbance. In regions where annual precipitation is less than 40 centimeters, there is often insufficient deep percolation of water to support deep-rooted shrubs. In these areas small shrubs, annual plants, and grasses predominate. Fire is common during the long dry summers and can restrict shrub and tree dominance. In addition, it is thought that some of the extensive shrublands around the Mediterranean Sea are the result of overgrazing by goats and other livestock, followed by erosion of the topsoil. The shrublands of the biome are known by many different regional names: *chaparral* and *coastal sage scrub* in California (Fig. 6.19), *matorral* in Chile, and *fynbos* in South Africa. In Europe the Mediterranean shrublands are divided into two main categories: *maquis*, which are evergreen shrubs and small trees over 2 meters in height, and *garrique*, which consists of shrubs between 0.6 meters and 2 meters in height.

Three environmental factors are most important in the physiognomy and life history of Mediterranean vegetation: the long summer drought, fires, and low soil nutrients. Annual plants grow during the moist and mild winters and then survive the dry summers as seeds. Annuals often account for 40% to 50% of the plant species found in Mediterranean regions. In general, annuals dominate on very dry sites and recently burned areas. Perennial herbs are less common because they must face severe summer drought on dry sites and must compete with the roots and canopies of small trees and shrubs that dominate moister sites.

Interestingly, many trees and shrubs in the Mediterranean biome are evergreen rather than summer deciduous. It has been postulated that, despite the summer drought conditions, evergreen leaves are favored because they allow the plant to take advantage of any soil moisture that is available during the late spring and early fall. This may be an advantage over deciduous species, which have no capacity to benefit from late- or early-growing season moisture. In a compromise strategy, the black sage (*Salvia mellifera*) of California loses all but its outermost leaves in the summer. Other species of small shrubs, such as *Teucrium polium*, possess large leaves during the winter and replace them with much smaller leaves during the summer. Nutrient demands are also higher for species that must

FIGURE 6.19   Late summer picture of semi-deciduous coastal sage scrub (*foreground*) and evergreen chaparral (*background*) of the California Mediterranean biome covering a hillside in southern California. The dry grasses are dominated by invasive annual species from Eurasia.

produce new leaves each year; this is important in the poor soils that often typify sites in the biome. Interestingly, some trees such as California valley oak (*Quercus lobata*) actually lose their leaves in the moist winter season and must rely on deep root systems to provide water during their summer growing season when the climate is dry.

Because of the dry conditions of the summer, many Mediterranean shrubs and trees possess sclerophyllous leaves. Some well-known examples include the olive (*Olea sativa*) and the cork oak (*Quercus suber*) from Europe, manzanita (*Arctostaphylos* spp.) from California, and eucalyptus from Australia. The small size and stiff cuticles of these leaves cut water loss by reducing the evaporative surface area and diffusion rates. Stomata also tend to be small and recessed in pits. Species such as the California lilacs (*Ceanothus* spp.) have stomata that respond quickly to moisture stress by closing. Some plants, such as California chamise (*Adenstoma fasciculatum*), have leaves reduced to needlelike spines, while Spanish broom (*Spartium junceum*) has sclerophyllous photosynthetic stems instead of leaves. Some plants have small leaves with dense vein systems, which may protect them from the physiological effects of water stress.

Many evergreen shrubs produce aromatic organic substances that are deposited on the soil around the plant and decrease rates of water infiltration. This characteristic may seem odd in a water-stressed environment. However, these substances also decrease rates of water evaporation from soils. In addition, the morphology of stems on shrubs causes water to flow toward the base of the plant stem. Studies have shown that water infiltration rates are higher at the base of chaparral shrubs and that this provides water directly to the root system.

The root systems of many shrubs and trees in the Mediterranean biome are also adapted to the summer dry conditions. Plants such as black sage, which typify coastal sage scrub, are almost dormant in the summer and typically have shallow and extensive root systems to capture winter moisture. Evergreen sclerophyllous plants, such as chamise, which typify chaparral, have very deep root systems that can extend 6 meters to 8 meters and supply moisture during the summer. However, even deep-rooted plants tend to have an extensive network of fine roots near the soil surface to maximize water absorption during winter rains.

Adaptations to the low level of nutrients in Mediterranean soils are common. A number of species have root nodules that contain nitrogen-fixing bacteria. Some plants are parasitic and tap into the root systems of adjacent plants with subsurface organs called *haustoria*. The haustoria of the Australian plant *Nytusia floribunda* can reach over 100 meters in length. Many shrubs in the California chaparral store nitrogen in old leaves during the winter and then translocate it to new growth during the spring and summer.

Fire is common in the Mediterranean biome. Under natural conditions, many sites will burn every 10 to 30 years. Plant species display strong adaptations to this frequent disturbance by burning. Some trees such as the cork oak and bigcone Douglas fir have thick bark that shields the cambium from heat damage. Cork used in wine bottles is made from the bark of this Mediterranean oak species. Many plants can resprout from underground organs or display epicormic sprouting following fires. A number of species of oak and eucalyptus share both capacities for sprouting. The aleppo pine (*Pinus halepensis*) has serotinous cones that open only after fire and distribute the seeds of the species onto the newly burned soil surface. A number of other species of plants produce seeds that require high temperatures prior to germination. These seeds may remain dormant in the soil until activated by fire. Examples include acacias in Australia and California lilac. Some ecologists believe that the shrublands of the Mediterranean biome are largely a fire-dependent system. The vegetation structure and the volatile oils found in many shrubs promote burning. After long periods of time without fire, the species diversity, vegetation productivity, and amount of live vegetation cover decrease. A number of species seem unable to reproduce without fire, even if open gaps are created by the removal of other plants.

The species diversity of the Mediterranean biome is high considering its small spatial extent. It has been estimated that the Mediterranean Basin has over 23,000 different vascular species. In other parts of the world, the number of plant species ranges from 8500 in the Cape Province of South Africa to 2100 in Chile, 8000 in Australia, and 4300 in California. Many of these plant species are found only in the Mediterranean biome. For example, about 75% of the native vascular plant flora of the South African portion of the Mediterranean biome are found only in that one region.

Human impact on areas of the biome in Europe and California has been particularly high. Much of the area in Europe has been impacted by grazing and agricultural and urban development that extends back thousands of years. Similar pressures are increasing in other parts of the biome but have not resulted in the same degree of vegetation change. Although agricultural conversion of Mediterranean biome lands has decreased, and even reversed, in some portions of the biome, urban areas have expanded and human populations have grown across the biome. The wildland-urban interface (WUI) where human development interfaces with wildland vegetation has grown greatly. This can produce direct replacement of native vegetation by buildings, roads, and other infrastructure. Increased development in the WUI can also alter fire regimes.

The introduction and naturalization of exotic plant species are observed throughout the biome and are particularly problematic in California where more than 1000 alien plant species have been introduced. Over 70 are naturalized and could be invasive of native habitat. Introduced Eurasian annual grasses have virtually replaced native perennial grasses in California. Large pampas grasses (*Cortaderia selloana* and *Cortaderia jubata*) have expanded into many parts of coastal California, displacing native plants. Eucalyptus trees introduced from Australia have established in woodlands and grasslands. Shrubs and annuals, such as black mustard (*Brassica nigra*), Spanish broom, and cardoon thistle (*Cynara cardunculus*) from Europe, have expanded into shrublands and grasslands, further displacing native plants. Interestingly, when all parts of the biome are considered, the Mediterranean Basin itself appears to be most resistant to plant invasions. This may be because of its long history of human landscape and vegetation modification.

Changes in fire regime have also impacted the biome. Decreases in burning due to fire suppression in some local settings have allowed shrubs such as coyote bush (*Baccharis pilularis*) to invade some grasslands in California. However, invasive annual grasses and plants, such as black mustard, produce highly flammable fuels in the summer that promote fires. Large human populations in the WUI offer greater risks of human-caused ignition of wildfires, more frequent fires, and the loss of life and property. In California, a shortening of the time between fires to intervals of less than 10 to 15 years has the potential of eliminating reproduction of chaparral shrubs and the replacement of native chaparral by invasive grasslands. The expansion of the WUI and projected changes in climate and fire-prone weather conditions over the twenty-first century are a cause for concern in California's Mediterranean biome region and elsewhere.

Temperate Grassland

Extensive temperate grasslands, dominated by species of the family Poaceae, are found in the southern and central interior of North America, small portions of east-central South America, and southern Africa. Temperate grasslands also occur as a long band extending from southeastern Europe through central Asia. Limited areas of grassland are found in New Zealand and portions of Australia. Grasslands are referred to as *prairie* in North America, *pampas* in South America, *grassveld* in South Africa, and *steppe* in Eurasia. The total area of the temperate grassland biome is around 9 million square kilometers.

The mean annual temperature in the temperate grassland biome ranges from over 15° C in areas such as South Africa or northern Mexico, to anywhere from 2° C to 3° C in Canada and Siberia. Freezing temperatures are rare in the grassland areas of South Africa and South America. In contrast, average January temperatures in grassland areas of western Canada and Siberia are generally less than −10° C. Annual precipitation averages between 30 centimeters and 100 centimeters. The season of highest precipitation varies regionally. Potential evapotranspiration exceeds precipitation during the growing season. Grasslands in North America and central Asia also suffer from periodic severe droughts.

Mollisols are the common soil type in the grassland biome. These soils have a thick dark brown to black surface horizon, high nutrient content, and a loose granular structure. Mollisols are excellent for farming, being particularly suitable for cereal crops such as wheat or barley.

The structure of grassland vegetation is relatively simple. Grasses and herbs dominate. Grasses make up about 90% of the biomass at most grassland sites. In terms of species diversity, only about 20% of the plant species found on grasslands are grasses; the other 80% are mainly herbs. What at first glance looks like a monotonous cover of a few species of grass may be found to contain a wonderful assortment of herbaceous wild flowers lurking amongst the dominant grasses (Fig. 6.20). Some trees and shrubs are found in protected valley sites and as gallery forests along rivers. In some cases,

FIGURE 6.20 A remnant of tall grass prairie in Michigan. Although the above-ground biomass in grasslands is dominated by grasses, there is often a good diversity of colorful flowering herbs in the vegetation. (*Source:* Dean Pennala / Adobe Stock)

trees such as limber pine (*Pinus flexilis*) and ponderosa pine (*Pinus ponderosa*) can be found growing on rocky outcrops in the northern prairie of central North America. Scotts pine (*Pinus sylvestris*) can be found on similar sites in portions of the Eurasian steppes.

The physiognomy of the plant species reflects the influence of three main pressures: aridity, fire, and grazing. Prior to human intervention, grasslands experienced frequent fires. Some grasslands in the southwestern United States appear to have burned almost every year. Native grazing animals found in temperate grasslands include buffaloes (*Bison bison*) in North America, horses (*Equus caballus*) in Eurasia, and many smaller mammals and insects such as grasshoppers. Large grazing animals appear to have been a critical element in the development and maintenance of temperate grassland in areas such as northern Europe. Similarly, buffalo have been identified as a keystone grazer in maintaining grassland diversity in the plains of North America.

Most temperate grasslands are dominated by perennial grasses. One exception is in North America in the Central Valley of California, where introduced species of European annual grasses have displaced the native perennial grasses and other vegetation types. The perennial bunch grasses, such as needle grass (*Stipa* spp.), have very short main stems from which arise photosynthetic stems and leaf blades. The photosynthetic stems and blades of grass are called tillers (Fig. 6.21). In sod-forming grasses, such as wheat grass (*Agropyron* spp.), tillers arise from lateral underground rhizomes. Some grasses, such as buffalo grass (*Buchloe dacyloides*), produce above-ground stolons that also grow laterally and produce new tillers. Rhizomes and stolons allow the sod grasses to spread over considerable areas and produce a dense mat of vegetation. Such grasses can spread over 20 square meters in as little as 2 years. Following heavy browsing or fires, most perennial grasses can resprout from their roots or rhizomes. Rhizomes and stolons also store carbohydrates and help grasses survive during periods of drought. Grassland herb species such as golden rod (*Solidago* spp.) also possess rhizomes.

In some cases, the root biomass of grasslands can be three times larger than the living biomass found above ground. On average, over 70% of the biomass in grasslands is found below ground in the roots and associated structures. The root systems of grassland plants reflect two main strategies for coping with arid conditions. Short grass species (with tillers usually less than 50 centimeters in height) found on arid sites typically have intensive networks of dense fine roots within the top 150 centimeters of the soil surface (Fig. 6.21). Examples include buffalo grass and wheat grass. Some short bunch grasses have as much as 90% of their root biomass concentrated in the top 15 centimeters of soil. Such

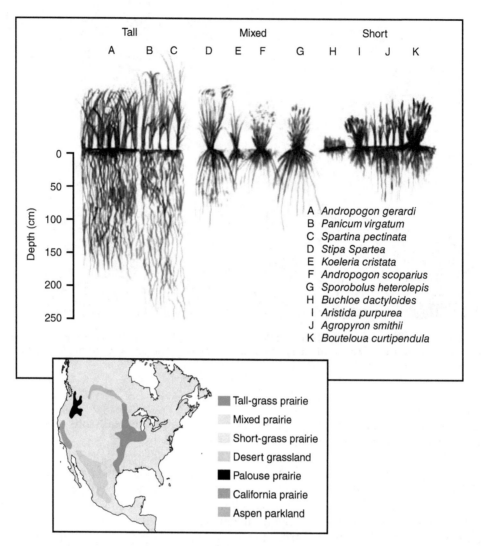

FIGURE 6.21   Major grassland formations of North America (Sims and Coupland, 1979) and the tillers, stolons, and root systems of some common grasses (Weaver, 1968; Archibold, 1995).

species depend on efficient capture of rainfall at the surface, leaving little water to percolate down to lower levels in the soil. This strategy is advantageous in dry areas. Tall grass species (tillers up to 1 meter in height), such as switch grass (*Pancium vigatum*), grow on moist sites and have long roots that can extend over 2 meters below the surface (Fig. 6.21). These species take advantage of deeper groundwater and are particularly suited to areas where rainfall is sufficient for water to percolate to depth in the soil. Many grassland herb species have similar deep roots that can extend more than 3 meters in depth. Some species, such as blue-stem grass (*Andropogon gerardi*), have a combination of dense and fine extensive root systems in the top 30 to 50 centimeters of the soil and coarse intensive roots that penetrate more than 2 meters (Fig. 6.21).

The root systems of grassland species may also be important in determining the position of the ecotone between grassland and adjacent forest. The concentration of fine roots near the surface allows the grasses to capture precipitation before it can percolate downward and provide moisture to the deeper root systems of trees and shrubs. The dense root mat also makes it difficult for tree and shrub seedlings to establish. Finally, sprouting from roots and rhizomes allows perennial grasses to survive fires that kill tree and shrub seedlings.

The wide range in temperatures, and to a lesser extent precipitation regimes, experienced within the grassland biome produce regional differences in vegetation composition and structure. Within

North America, the grassland biome is often divided into several different formations (Fig. 6.21). The relatively moist eastern portions of the central plains support tall grass prairie dominated by species such as eastern blue stem (*Andropogon gerardi*) and switch grass (*Panicum virgatum*). Some of these grasses can grow to heights of 2 meters. The central portion of the plains supports mixed-grass prairie with some tall grasses, medium-sized species such as little bluestem (*Andropogon scoparius*), and short grass species such as buffalo grass. The height of vegetation in the mixed-grass prairie generally does not exceed 1.5 meters for the taller grasses, with an understory of short grasses that grow to 30 centimeters in height. The western edge of the plains is composed of short grass prairie that includes species such as buffalo grass and blue grama (*Bouteloua gracilis*). The height of the vegetation is generally from 30 centimeters to 50 centimeters. The California prairie originally supported more perennial bunch grasses from genera such as three awn grass (*Aristida* spp.), bunch grass (*Poa* spp.), and needle grass (*Nassella* spp.). Starting with settlement by the Spanish in the late eighteenth century, European annual grasses of the genera wild oats (*Avena* spp.), brome (*Bromus* spp.), and fescue (*Festuca* spp.) have been introduced and have come to dominate. The Palouse prairie of the Pacific Northwest was also occupied by more bunch grasses but has come to be increasingly dominated by introduced annual grasses.

Net primary productivity in grasslands is positively correlated with rainfall in the grassland biome. Annual productivity may range from 200 grams per square meter where annual precipitation is 40 centimeters to over 400 grams per square meter where annual precipitation is 80 centimeters. The diversity of plant species found in grasslands is not as high as in more structurally complex systems, such as tropical and temperate forests. However, in the central plains of North America, over 100 different species of grasses and herbs can be found in areas of 1 hectare to 2 hectares.

Human impact on the grassland biome has been considerable. It is arguably the biome most altered by human activity. Temperate grasslands are extremely important for growing crops and grazing livestock. For example, grain production from former temperate grasslands in the United States and Canada is on the order of almost 500 million tonnes per year. It has been estimated that 45.8% of the world's temperate grasslands have been converted to human-dominated uses. Most of the area of natural grassland vegetation in North America and Eurasia is now used for agriculture. In North America the area covered by grasslands in the Great Plains decreased from 370 million hectares to 125 million hectares between 1830 and 2000. In some cases, such as the tall grass and mixed-grass regions of the Great Plains, fields of crops such as wheat and corn have almost completely replaced native prairie. Less than 10% of the tall grass prairie vegetation in the United States remains intact. Within the short grass prairie, extensive grazing has changed the nature of the vegetation. Heavily grazed fields have less dense vegetation cover and a much higher proportion of invasive nonnative plant species. In portions of Texas, Oklahoma, and Nebraska, nonnative plant species can make up over 90% of the grass cover on overgrazed fields. Introduced species of grasses and herbs are found in all North American grassland formations. The California grasslands are completely dominated by such invasive species. Even in a protected area such as Badlands National Park in South Dakota, some 28% of the grass species found there are nonnatives.

In some areas, such as the southern and northern edges of the central plains of the United States, the exclusion of fire coupled with heavy grassing has allowed shrubs and trees to invade former grasslands. In areas of the southern Great Plains, such as Texas, the honey mesquite tree (*Prosopis glandulosa*) has come to dominate large areas of former grassland. This phenomenon is not restricted to the United States. Nonnative species of mesquite have also replaced grass-cover in parts of Australia, for example. The replacement of grassland by the expansion of populations of native and nonnative woody species has been reported from many other regions. Overgrazing, transport of woody plant seeds by livestock, introduction of nonnative woody plant species, suppression of natural fire regimes, and climate change have all been identified as driving forces in this transition. Temperate grasslands have been greatly reduced in extent by past human activity and are still under threat today. It has been estimated that during the period 2018–2019 approximately 1 million hectares of grassland in the United States and Canada were plowed up and largely converted to row cropping. We might ask ourselves, what is the right balance between the importance of temperate grassland regions in feeding the world through the growth of crops and grazing, and preserving the original vegetation of the biome?

Temperate Forests

Temperate forests are found in eastern North America, portions of Europe, eastern Asia, southern South America, southeastern Australia, and New Zealand. The biome occupies some 7 million square kilometers. The mean annual temperature of the biome ranges from 20° C to 5° C. Average annual precipitation is between 50 centimeters and 250 centimeters. A distinguishing characteristic of the biome is the occurrence of freezing temperatures in the winter. Some regions, including northeastern North America and portions of Asia, experience long cold winters with average January temperatures of less than –10° C. Areas of the biome in the southeastern United States and New Zealand have milder winters, with January mean temperatures that are above freezing. The annual distribution of precipitation is highly variable from region to region. Areas of eastern Asia experience a monsoonal regime, with most precipitation occurring as summer rainfall. In contrast, western Europe has generally high precipitation throughout the year.

Soils in the temperate forest biome include alfisols, ultisols, and entisols. The alfisols, common in cooler areas of eastern North America, Europe, and portions of Asia, have a brown surface horizon that is rich in humus. They are generally quite fertile. Ultisols are associated with warmer portions of the biome in southeastern North America, Asia, Australia, and New Zealand. The ultisols are generally red to yellow in color. Most nutrients are found in decaying organic matter at the top of the soil and are quickly recycled by the forest vegetation.

Within the northern hemisphere, the key characteristic of the temperate forest biome is the dominance by winter deciduous trees including maples (*Acer* spp,) and a number of other genera (Fig. 6.22). Winter temperatures in the portion of the biome that lies within the southern hemisphere are milder. Most of the temperate forest regions there are dominated by broadleaf evergreen trees such as mountain southern beech (*Nothofagus solandri*) in New Zealand and eucalyptus species in Australia. Evergreen conifer species including matai (*Podocarpus spicatus*) and the beautiful kauri tree (*Agathis australis*) in New Zealand also occur. The only southern hemisphere forest that is dominantly deciduous is found in the drier and colder portions of Patagonia. Here the forest is dominated by deciduous species of southern beech such as lengue (*Nothofagus pumilo*).

The deciduous temperate forests of the northern hemisphere are distinctive for the winter loss of leaves from broadleaved trees and shrubs. During the winter, temperatures are too cold for photosynthesis, and freezing conditions damage exposed soft tissues. As a response to cold conditions, deciduous plants lose their leaves and go dormant. It is thought by some that the deciduous trait may

FIGURE 6.22   Autumn foliage in an area of deciduous forest in New Hampshire. The loss of leaves and dormancy allows the broadleaved tree species to survive the long and cold winter climate.

have developed in plants in response to winter drought in the tropical seasonal forest region in the geologic past. Others believe that the deciduous leaf trait observed in temperate deciduous forests developed in situ in response to long-term climatic cooling. The environmental cue that leads to leaf mortality and release varies for different tree species. In some cases, tree species respond to decreased daylight in the autumn. In other cases, increased moisture stress due to decreased rates of evapotranspiration and water conductance leads to leaf mortality. At the onset of dormancy, the leaves lose their chlorophyll and turn red or yellow as pigments such as anthocyanins are exposed. The walls of abscission cells at the base of the leaves disintegrate and release the leaf from the tree. Many tree species require a period of chilling followed by sustained warm temperatures in order to stimulate bud opening and the development of new leaves. Red maple (*Acer rubrum*) from northeastern North America requires at least a month of freezing temperatures followed by warm temperatures of 15° C or more in order to initiate leafing out. The chilling requirement keeps trees from breaking dormancy before the passing of winter and the onset of warm spring conditions. For the deciduous strategy to be effective, the summer growth period must be long and warm enough for the trees to produce and form buds for the next year and to produce sufficient reserves of carbohydrates to sustain the plant during the winter and spring when new leaves begin to develop. It can take several weeks for the new leaves to mature and reach peak photosynthetic activity. Some trees such as European hornbeam (*Carpinus betulus*) obtain a competitive advantage by producing leaves early in the spring before trees such as beech and oak do. Early leafing out of saplings of other tree species has been observed, in advance of the overhead canopy leaf development, and may be important in providing sufficient growing season light and photosynthesis for the persistence and growth of these young trees.

The timing for leaf abscission and subsequent leafing out varies from species to species and from year to year. In general, most tree species within a stand will lose their leaves during the same 14- to 30-day period. Leafing out in the spring will also occur at a similar time frame. Flowering of the canopy trees and many of the understory plants occurs early in the spring before full leaf development. As a result, it is easier for windborne pollen to be transported through the canopy, and insects and other animal pollen vectors have a better opportunity to find flowers.

The structure of deciduous forests in North America, Europe, and Asia is quite similar. In addition, the same genera of broadleaved and coniferous trees dominate the forests on the three continents. There is a canopy formed by large deciduous trees such as beech (*Fagus* spp.), maple (*Acer* spp.), basswood (*Tilia* spp.), hickory (*Castanea* spp.), elm (*Ulmus* spp.), ash (*Fraxinus* spp.), and oak (*Quercus* spp.). The height of the mature canopy is generally between 20 meters and 30 meters. Beneath the canopy, there is a layer of subdominant trees that may include saplings of the canopy-dominant trees, shade-tolerant subdominant trees, and large shrubs such as dogwood (*Cornus* spp.) and hazel (*Corylus* spp.). Finally, there is a ground layer formed by tree and shrub seedlings, small shrubs, herbs, and ferns. Vines occur in the temperate deciduous forests but are not as common or as diverse as in the rainforest. Most temperate deciduous forests also contain evergreen conifers from genera such as pine (*Pinus* spp.) and hemlock (*Tsuga* spp.) and a few evergreen shrubs such as holly (*Ilex* spp.).

Deciduous forest grows across a wide latitudinal range in eastern North America, extending from northern Florida to southern Canada. Within this broad expanse, differences in seasonal temperatures and precipitation produce differences in the forest composition. The North American deciduous forest is often subdivided into a number of associations or formations (Fig. 6.23). From south to north these include a southern mixture of deciduous trees, evergreen broadleaf trees such as southern magnolia (*Magnolia grandiflora*), and conifers such as loblolly pine (*Pinus taeda*) and shortleaf pine (*Pinus echinata*); an association dominated by a number of species of oaks and pines; an association dominated by oaks and formerly by chestnut (*Castanea dentata*); an area of mixed mesophytic deciduous trees, including species of maples, beeches, elms, basswoods, and the tulip tree (*Lirodendron tulipifera*) (this mixed mesophytic association has the highest diversity of broadleaf trees in North America); an association in the southern Great Lakes region typified by maples and beech; a forest bordering the Great Plains dominated by oaks and hickory; and a mixed wood in the north typified by deciduous species such as maple and oak in association with conifers such as white pine (*Pinus strobus*) and eastern hemlock (*Tsuga canadensis*).

During the winter, about 70% of the light received at the top of the leafless canopy is transmitted to the ground surface. However, during mid-summer, only about 10% of the incident light might be transmitted to the forest floor. This light on the forest floor is more than that in tropical

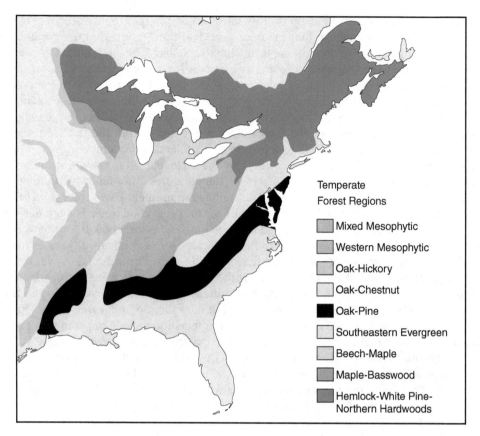

FIGURE 6.23 Eastern North American temperate forest formations (*Source:* Archibold, 1995 / Springer Nature).

rainforests; as a result, the understory vegetation in the deciduous forest can be relatively lush (Fig. 6.24) Some trees, such as oaks, have different leaves on canopy branches as compared to lower limbs. The sun leaves found at the canopy level are smaller in area, are thicker, and have more chlorophyll and more stomata than the lower shade leaves. Certain tree genera, including birch (*Betula* spp.), cherry (*Prunus* spp.), and poplar (*Populus* spp.), that occur in the deciduous forest do not grow as high as canopy dominants and are intolerant of shade. They require gaps in the forest in order to establish and grow. Due to natural mortality, windthrow, and other disturbances, it is estimated that approximately 10% of the deciduous forest consists of canopy gaps at any given time. Shade-intolerant subcanopy trees such as cherry or poplar can persist 30 to 80 years in such gaps before being shaded out by canopy-dominant trees such as maple or beech. When undisturbed, the canopy-dominant trees generally live 300 to 400 years. Disturbances create a complex forest structure of different ages and sizes of trees and some deadwood. It is thought that such complex stands in the temperate deciduous forest are important in maintaining forest diversity and can make the forest more resistant and resilient to new disturbances.

In terms of total plant biomass, about 20% can be found below ground in the roots. The plant species diversity in the temperate deciduous forests is relatively high. The tree species diversity in North American deciduous forests ranges from 180 species in the Southeast to 60 species in the western Great Lakes region. Even in the Great Lakes region, it is not unusual to find 10 or more different species of trees on a 1-hectare study site. Within the southeastern deciduous forest, some 2000 species of vascular plants are known. As many as 14 different herbaceous species can be found within a single square meter of forest floor. Asia also has a diverse deciduous flora. Over 50 genera of broadleaved trees and 12 coniferous genera are known from China, with over 50 individual species of maple alone. The diversity of tree species is not nearly as great in Europe as in North America or Asia, but it is still considerable. Interestingly, the temperate forests of New Zealand have high tree species diversity. There are 215 tree species found there despite the small total land area of the country. However,

FIGURE 6.24 Temperate deciduous forest understory in Quebec, Canada.

a high proportion of these are relatively small in stature (less than 15 meters in height) compared to the northern hemisphere temperate tree flora.

Because of their temperate climate and fertile soils, much of the world's temperate forests have been cleared for agriculture and settlement. In addition, extensive logging has long been practiced in these forests. Significant clearing of European forests for agriculture commenced some 5000 to 8000 years ago and progressed northwestward from Turkey and Greece. Between 2000 and 4000 years ago, most of the arable lands throughout Britain were cleared of forest. Clearing of deciduous forests in China for agriculture began some 5000 years ago. Indigenous peoples in eastern North America were clearing small sections of forest for swidden agriculture 2000 years ago and perhaps modified larger areas by burning. However, the arrival of Europeans caused massive clearing of North American forests. By the early twentieth century, only a tiny fraction of original forest remained in the eastern United States. Stands that have not been cleared by human activity are called **primary forest.** If the primary forest has not been significantly altered by recent natural disturbance, it is sometimes referred to as **old growth forest.** The most recently cleared deciduous forest in the United States is found in the western Great Lakes region. During 1850–1995, almost 99% of that forest was cleared or altered. In many other areas of the temperate forest biome, primary forest has been reduced to only 1% to 2% of its natural extent.

Invasive plants are found in the temperate deciduous forest of North America, but they largely occur in the understory and vines. Some common invasive species include Common invasive trees or shrubs include honeysuckles (*Lonicera* spp.), English ivy (*Hedera helix*), kudzu (*Pueraria montana*), glossy buckthorn (*Rhamnus cathartica*), privet (*Ligastrum* spp.), autumn olive (*Eleagnus umbellata*), burning bush (*Euonymus alatus*), and Japanese barberry (*Berberis thunbergii*). Native trees continue to make up about 98% of the canopy in the eastern North American temperate deciduous forest. Intact forest stands appear to be more resistant to nonnative plant species than disturbed stands.

Invasive plants are not restricted to only the North American portion of the temperate deciduous forest biome. The North American black cherry (*Prunus serotina*) has become established as a widespread invasive species in European forests, for example. Invasive plants can change the forest by outcompeting native plants, altering habitat conditions for animals and changing soil conditions.

The accidental introduction of insect species and other pathogens has been an additional anthropogenic pressure on the deciduous forest biome in North America. Populations of beech, chestnut, and elm have all declined due to introduced blights and other pathogens. For example, in the early twentieth century, the chestnut fungus *Cryphonectria parasitica* was introduced to the eastern deciduous forest from Eurasia. Old World relatives of the North American chestnut have a resistance to the fungus. By 1950, all North American chestnut trees were infected by the blight, and the species only persists as small suckers that emerge from the root system and are killed before reaching tree form. Thus, a dominant tree has been essentially eliminated from the forest. Widespread mortality of hemlock trees (*Tsuga canadensis* and *T. caroliniana*) caused by hemlock woolly adelgid (*Adelges tsugae*), an aphid-like insect from Japan, has further altered large areas of the eastern North American temperate deciduous biome.

In recent times, two factors have led to some recuperation of deciduous forests. First, parks and conservation areas have been created around the world to protect remaining forests. The preservation of deciduous forest patches has its origins several hundred years ago with the establishment of game parks used for hunting by European aristocracy. Second, in areas such as New England, the northern Great Lakes region, and portions of northern Europe, it is not economically viable to practice farming on some formerly forested sites. Farms have been allowed to revert back to forest. Such reforested sites are called **second growth forest.** In general, they do not have the same species diversity or structural diversity of large primary forest stands that include closed canopy and gap sites. Many parts of northern Europe and northeastern North America have far more deciduous forest today than they did 100 years ago. In fact, it is estimated that owing to the increase in deciduous forest, the total rate of reforestation exceeded deforestation in the United States in the twentieth century. Europe has experienced a similar increase in forested land. It is estimated that forest cover in Europe, largely temperate deciduous forest, has increased by 25% to 30% since the 1950s. However, the reforestation trajectory may be reversing in the eastern United States. Although conversion of abandoned agricultural land to forest has continued, the loss of forest to urbanization, timber-harvesting, and other land uses has accelerated and one estimate suggests a 4% decline in regional forest cover in the eastern United States in recent decades.

Temperate Rainforest

The temperate rainforest is a small and scattered biome. The largest conterminous area of temperate rainforest is found on the northwest coast of North America, extending from northern California to the panhandle of Alaska. Smaller strips of temperate rainforest are found in southeastern Australia, Tasmania, New Zealand, and the southwestern coast of South America. Annual rainfall exceeds 200 centimeters, and precipitation occurs throughout the year. In some regions total rainfall approaches 400 centimeters. All areas of temperate rainforest are found close to the oceans and have mild temperature regimes. Average annual temperatures range from just less than 20° C to about 5° C. Winters may experience freezing temperatures but are generally mild. Summers are cool.

Along the California coast, the temperate rainforest vegetation extends quite far south into the dry summer climatic region of the Mediterranean biome. This extension of the rainforest is due to the prevalence of summer fogs in this region. Condensation along the branches and needles of the redwood trees (*Sequoia sempervirens*) that dominate the forest in coastal California falls as large droplets at the base of the canopy and provides an important source of moisture during the dry summer months (Fig. 6.25). In addition, redwood trees and many of the other plant species associated with them can absorb moisture from the fog through their needles and stems via a process called *foliar uptake*. In addition, large amounts of winter orographic rainfall are generated by air rising over the coastal mountains, and this produces high soil moisture that helps trees and associated plants persist over the drier summer.

Soils in the temperate rainforest biome are generally spodosols, which are acidic and low in nutrients. They often contain a B horizon composed of reddish mineral matter overlain by a light gray E horizon that has lost iron and aluminum through intensive eluviation to the B horizon. In poorly

FIGURE 6.25   Summer fog delivering important moisture to a stand of California redwood trees (*Sequoia sempervirens*).

drained areas, histosols, made up of poorly decomposed organic matter, are common. These soils are extremely acidic and contain almost no nutrients.

The temperate rainforest biome is dominated by evergreen trees. In the southern hemisphere, the dominant trees include karri (*Eucalyptus diversicolor*) and other eucalyptus species in Australia, southern myrtle beech (*Nothofagus cunninghammi*), southern mountain ash (*Eucalyptus regnans*), and the huon pine (*Dacrydium franklinii*) in Tasmania; and southern beeches (*Nothofagus* spp.), Patagonian cypress (*Fitzroya cupressiodes*), and monkey puzzle tree (*Araucaria araucana*) in South America. The North American portion of the biome is dominated by evergreen conifers, including the giant coastal redwood in California. The redwood can reach heights of over 110 meters. Douglas firs (*Pseudotsuga menziesii*), which are commonly over 50 meters tall, grow from California to Alaska. Other massive, dominant trees in the northwestern rainforests include red cedar (*Thuja plicata*), sitka spruce (*Picea sitchensis*), amabalis fir (*Abies amabalis*), and yellow cypress (*Chamaecyparis nootkatensis*). In addition to great size, the dominant trees of the temperate rainforests often have very long life spans. Douglas firs, yellow cypress, red cedar, and huon pine all can typically reach ages of 1000 years or greater. California redwoods may reach ages of over 2000 years.

Light penetration in the mature canopy of the temperate rainforest is very limited (Fig. 6.26). In addition, the litter from the dominant conifers makes the soils quite acidic. As a result, the structure of the forest is very simple, usually consisting of a continuous conifer canopy and sparse subdominant layer made up of early successional trees such as California bay (*Umbellularia californica*) and immature conifers. A lower layer of shrubs includes huckleberry (*Vaccinium* spp.), together with various herbs and ferns. Thick carpets of moss are also encountered. Many epiphytes, such as leather leaf fern (*Polypodium scouleri*) and red elderberry (*Sambucus callicarpa*), are found on redwood trees. The canopy of the forest generally ranges from 40 to 60 meters in height. The world's tallest tree is a California redwood with a height of over 114 meters. Trunks of the dominant trees are often massive, reaching 20 meters in circumference. Buttressed trunk systems occur in a number of species

FIGURE 6.26 The cool and moist floor of the coastal temperate rainforest in Oregon. The dense closed canopy formed by large conifers with buttress trunks allows little light to penetrate to the ground. Vascular plant vegetation on the forest floor is relatively sparse and there is an abundance of mosses (*Source:* aiisha / Adobe Stock).

(Fig. 6.26). Tree species diversity is not as high as in the temperate deciduous forest. In addition, because of the low light conditions and acidic soils, the shrub and herb diversity, particularly within mature stands, is also relatively low.

Due to their moist conditions, fire is often not a major source of disturbance in temperate rainforests. However, fires do occur in redwood stands, and redwood trees have a number of adaptations to persist despite burning. These include thick bark, epicormic sprouting, and the ability to resprout from the below-ground tissue following fires. Windthrow and geomorphic forces such as landslides are common agents of disturbance in many temperate rainforests. Such disturbances may be quite small in spatial scale. A Canadian study of coastal temperate rainforest in British Columbia found that over the past 140 years only about 3.1% of the forest had been affected by natural disturbances.

The large trees of the temperate rainforest have made this region particularly attractive to logging operations and forest clearance. For many years, the rugged and inaccessible nature of much of the Pacific Northwest restricted logging activities. However, the increasing value of quality wood and the advent of practices such as helicopter slinging to remove timber has changed this situation. In addition, logging in the coastal rainforest has tended to use the clear-cut approach in which essentially all trees are removed from large swaths of forests. Approximately 76% of the Olympic National Forest of Washington was cut between 1940 and 1990. In Oregon and California, perhaps only 4% to 5% of the coastal rainforest remains unlogged to some extent. British Columbia has seen over 50% of its rainforest logged, whereas Alaska has had about 11% logged. Worldwide, it is estimated that by the end of the twentieth century about 55% of the temperate rainforest had been logged, and only 17% was protected in parks or reserves. Destruction of large portions of the forest has led to the endangerment of bird and mammal species adapted to the old growth stands. In North America, both the spotted owl (*Strix occidentalis*) and marbled murrelet (*Brachyramphus marmoratus*) have been threatened with extinction due to loss of old growth habitat. Efforts to protect the owl led to large areas of old growth forest being closed to logging in the United States. In some areas, the remaining old growth stands are too small to support the industry for more than a few years in any case. The debates and battles between environmentalists and logging interests over preservation of California redwoods were particularly heated, and it has been argued that they changed the course of environmental protests and politics in the United States. Today about 75% of the remaining old growth coastal redwood stands are protected. However, such stands are a tiny proportion of the current forest. Only about 26% of the entire coastal redwood forest is currently protected in California. Ongoing battles between conservationists and logging and other development interests continue in the temperate rainforest biome.

Coniferous Boreal (Taiga), Subalpine, and Montane Forests

The coniferous boreal, subalpine, and montane forests are found only in the northern hemisphere. The boreal forest, known as *taiga* in Eurasia, extends as a broad band centered on 60° north latitude (Fig. 6.27). The boreal forest extends across northern Eurasia from Norway to the Pacific Coast of Russia and the northern islands of Japan. In North America the boreal forest extends from coastal Alaska to the island of Newfoundland. It is the largest conterminous forest in the western hemisphere. Montane coniferous forests are found at lower to mid-elevation sites on mountains, while subalpine coniferous forests extend from higher elevations to the treeline. The largest subalpine/montane portion of the coniferous biome extends down the Rocky Mountain region of North America into the highlands of Mexico. Small portions are also found at higher elevations in the Appalachian Mountains in the East. In Eurasia, areas of subalpine/montane coniferous forest are found at higher elevations in alpine regions such as the Alps or the Himalayas. The total area of the biome is around 12–15 million square kilometers. It has been estimated that the boreal forest constitutes one-quarter to one-third of all forest cover on the planet. Plant biomass varies widely depending on climate and local site conditions. About 20% of total biomass may be found in the roots.

The coniferous boreal and subalpine/montane forests are found in regions with a mean annual temperature of between 5° C and –5° C. Mean annual precipitation varies between about 20 centimeters and 200 centimeters. In general, precipitation is lowest in the boreal portions of the biome. Summers are generally mild, with average July temperatures of 10° C to 20° C. Winters are generally cold with average January temperatures reaching as low as –50° C in central Siberia. The seasonal distribution of precipitation is regionally variable. Central Alaska receives its greatest amount of precipitation in the summer months, while Scandinavia has its highest precipitation in the winter.

Why are the boreal and coniferous subalpine/montane forest biome regions restricted to the northern hemisphere? A similar broad band of cool summer and cold winter boreal climate does not occur in the southern hemisphere owing to the absence of continental land in the vicinity of 60° south latitude. The mountains of the southern hemisphere do not have conifer-dominated subalpine/montane forest that is

FIGURE 6.27   Winter of the boreal forest in the Yukon Territory of Canada. The diversity of tree species is very low in this biome. All of the coniferous trees in the picture are either white spruce (*Picea glauca*) or black spruce (*Picea mariana*). *Source:* Scalia Media / Adobe Stock.

floristically similar to that found in the northern hemisphere. For example, eucalyptus trees are found at high elevations of the Snowy Mountains in southern Australia, and southern beech forests are found in the mountains of New Zealand and Patagonia. In tropical Africa, woodlands are formed on the subalpine zone of Mount Kilimanjaro by a relative of the ragwort, *Senecio keniodendron*, and a relative of the lobelia, *Lobelia keniensis*. These remarkable landscapes bear no resemblance to the coniferous forests developed on similar cool alpine sites in the northern hemisphere.

Spodosols are often found in coniferous forest regions, particularly in the boreal zone. In addition, alfisols may be found in southern portions of the boreal biome and in some montane regions. Histosols are common as extensive peatlands, and fens are found throughout the boreal zone and on poorly drained sites in other portions of the biome. Subalpine regions often have poorly developed rocky soils, which are sometimes referred to as *regosols*. High-latitude sites and high-elevation sites may have permafrost. In general, the southern border of continuous permafrost lies at the northern edge of the boreal forest zone in North America. The boreal forest here is underlain by discontinuous to sporadic permafrost in places. In Siberia continuous permafrost underlies almost all of the boreal biome.

The coniferous forests in North America and Europe are generally dominated by evergreen species from the genera pine (*Pinus* spp.), spruce (*Picea* spp.), and fir (*Abies* spp.). In Siberia, much of the boreal zone is dominated by the deciduous conifer species Siberian larch (*Larix sibirica*) and dahurian larch (*Larix dahurica*). The larch trees, though related to the evergreen conifer genera, turn a beautiful golden color in the fall and then lose all of their needles during the winter. It is possible that this adaptation allows them to dominate in Siberia because of the extremely cold winter temperatures experienced in this continental region. Despite adaptations to the cold, exposed needles of evergreen species can be susceptible to frost damage and desiccation in the extreme cold of the northern Siberian winter. Recently disturbed sites and areas along rivers are sometimes dominated by deciduous broadleaf genera such as birch (*Betula* spp.) or aspen and poplar (*Populus* spp.).

The structure of the conifer forests is relatively simple. There is a tree layer typically dominated by conifers, and the conifer canopy is generally no more than 30 to 40 meters in height, and often much less. In the case of the giant sequoia (*Sequoiadendron giganteum*) of the Sierra Nevada mountains of California, the dominant trees can reach heights of close to 100 meters. The giant sequoia is the largest tree in the world in terms of volume. The conifers from this biome are the oldest trees in the world. Bristlecone pines (*Pinus aristata*) from the White Mountains of California reach ages of over 4000 years. Other species such as giant sequoia and limber pine (*Pinus flexilis*) can reach ages of 1000 to 2000 years. Ages of 500 years are not uncommon for many other coniferous tree species in the biome.

The forest canopy is closed in the southern and middle portions of the boreal biome. Similarly, mid-elevation sites in coniferous montane forests generally have closed canopies. Toward the northern limits of the boreal forest and the upper treeline in the subalpine zone, the forest canopy begins to open. The trees also become smaller in stature. At their extreme limits, trees are often restricted to protected sites where they grow as low shrubs called krummholtz. The dominance of the conifer biome by a few families of trees with the same physiognomy and stand structure produces very similar landscapes in widely separated regions such as Siberia, Finland, Alaska, and Canada. Some refer to the conifer forests, particularly the boreal portion, as monotonous.

The underlying vegetation in the coniferous forest often includes evergreen shrubs such as many members of the heath family (Ericaceae) and junipers (*Juniperus* spp.). Deciduous shrubs such as willow (*Salix* spp.), alder (*Alnus* spp.), and shrub birch are also common. In addition, stunted saplings of the dominant conifers may be present. In many cases, however, there is little regeneration of dominant trees beneath their own canopies. Due to dense shade and acidic needle litter, the ground cover can be relatively sparse and consists of a few herb species. In some regions, the ground surface is covered by feather moss (*Pleurozium schreberi*), peat moss (*Sphagnum* spp.), and other mosses. Black spruce (*Picea mariana*) and larch (*Larix laricina*) stands in the North American boreal forest are often found on thick organic soils composed of peat mosses. Similarly, the larch forests of Siberia often grow on layers of moss that can be up to 10 meters thick. Peat is "mined" in many parts of the world as a fuel for fire and a soil additive. The bags of peat moss that are purchased in local nurseries generally come from the boreal forest. In portions of eastern North America, thick growths of *Cladonia* lichens form a dry yellow crust on the ground. In some dry conifer forests, such as those at lower elevations in the Southwest, frequent burning and lack of soil moisture preclude understory

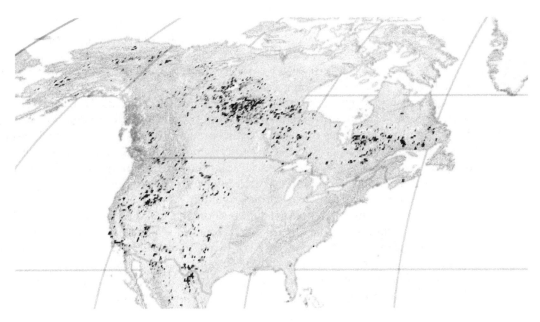

FIGURE 6.28   Fires detected by satellite remote sensing, mainly in the boreal subalpine/montane biome, over the period January 1, 2003, to December 31, 2016 (from the *NASA Global Fire Atlas*; courtesy of NASA, https:// earthobservatory.nasa.gov/images/145417/a-new-global-fire-atlas).

growth. Here the ground can be relatively bare, or in some cases it can support sparse grass cover if the canopy is open enough to allow sufficient light.

Fire is the most common agent of vegetation disturbance in coniferous forests (Fig. 6.28), although wind, avalanches, permafrost activity, insects, and other pathogens can also be important causes of local to regional vegetation disturbance. It has been estimated that 5 million to 20 million hectares of boreal forest burn each year, representing up to 2% of the upland boreal forest. From 2004 to 2016, Canada saw an average of 2.26 megahectares burned annually. Even in areas where fire suppression has been widely practiced, fires in the boreal zone are numerous and burn very large areas. The most frequent fires in coniferous forests are very small, consuming a few tens of hectares. However, it is the less frequent large fires (greater than 10,000 hectares) that are often responsible for most of the area burned. In the boreal forest of Canada over the period 1959–2015, large fires, making up only 3.2% of all reported fires, were responsible for 98.8% of the annual area burned. Similar relationships are observed in subalpine/montane conifer forests. In California, fires of greater than 150 hectares accounted for 3% of burns, but represented 64% of the total burned area.

Two important factors influence the number and size of conifer fires: high rates of ignition events and high rates of spread. Fire ignition may be caused by human activity and lightning. Lightning-caused fires account for 91% of the area burned in the boreal forest of Canada. However, in less remote regions such as montane forests of California, human-caused ignition can be responsible for over 90% of all fires and play a larger role in contributing to annual area burned. As the occurrence of summer frontal activity and lightning storms decreases toward treeline, so too does the number of ignition events. Conifer-dominated stands provide high levels of fuel in the form of dry litter, fine needles, twigs, bark flakes, and resinous products. In addition, the conical shape of conifers promotes crown fires and rapid spread. Understory shrubs and lichens are highly flammable and carry fire between trees.

The average frequency of fires in the biome varies according to climatic setting and site type. Low-elevation ponderosa pine (*Pinus ponderosa*) stands in the southwestern United States experience very dry summers and frequent lighting storms. They have ground fires every few years. In contrast, lightning storms are rare at the northern treeline of the boreal forest, and cool temperatures produce low rates of evaporation and keep the landscape moist. It may be thousands of years between fires at treeline. In the central boreal forest, areas with natural fire breaks such as lakes and sharp topography may not burn for 100 to 400 years. Dry areas with sandy soils and little topographic relief can have average fire-return intervals as short as 28 years. In many regions, the high number of fires coupled with their variability in size imparts a mosaic pattern to coniferous forest in which patches of various ages, some large and some quite small, typify the landscape. For example, about 10% of the boreal forest stands are less than 20 years old, while 75% are between 20 and 75 years old. The old growth forest in the boreal region is very small.

Many plants of the coniferous forest are adapted to disturbance by fire. The lodgepole pine (*Pinus contorta*) of western North America produces serotinous cones that open following heating. The seedlings of this species thrive on freshly burned soils and require the high insolation provided by an open canopy in order to grow. Aspen (*Populus tremuloides*) and birch trees, as well as many other deciduous species, resprout vigorously following fires. Conifers such as Douglas fir (*Pseudotsuga menziesii*), ponderosa pine, and many larches have extremely thick bark that protects the vascular tissue of the mature tree from heat during ground fires.

Insect pathogens are important agents of disturbance in conifer forests. From the 1990s into the present, outbreaks of bark beetles have caused widespread mortality of trees in montane forests of western North America. By 2012, over 750,000 hectares of forest in the southern Rocky Mountains had been impacted by mountain pine beetle (*Dendroctonus ponderosae*). The main defoliator of conifers in the North American boreal forest is spruce budworm (*Choristoneura fumiferana*), potentially affecting over 90% of spruce- and fir-dominated regions through cyclic outbreaks.

The impact of direct human activity on the coniferous forests varies from region to region. In large areas of the boreal forest, particularly in Russia, Alaska, and Canada, there are few settlements or roads, and so human impact has been minimal. In contrast, humans have used the forests of the Alps for timber, firewood, and pasture land for thousands of years. Many of the conifer stands in this region are carefully managed resources. In Canada and Russia, harvesting of the boreal forest for timber and paper is not insignificant. About two-thirds are managed for wood production and provide 37% of global wood supplies. The average area of the Canadian boreal forest that is clear-cut each year is roughly 400,000 hectares. Increasing population size has caused increased pressure on some forested regions. In North America, montane coniferous forests are often seen as prime real estate for recreational development. In Nepal, increased demands for fuel wood have led to high rates of deforestation. Fire management practices, particularly in the United States and Canada over the past century, have led to a decrease in small, low-intensity fires in many areas. This has allowed fuels to build up to high levels. Such areas may experience large fires that will be difficult, if not impossible, to control. In addition, controls on natural fires have allowed fire-sensitive tree and shrub species to increase at the expense of fire-resistant species in certain regions. For example, many old growth ponderosa pine stands in the southwestern United States are becoming closed-canopied or are being invaded by newly established spruce and fir. In montane conifer forests of California, comparison of current conifer forest tree-stem density and structure with conditions in the early twentieth century shows decreases in gaps between trees, more closed canopies, more small trees, and less diversity in structure. These stands burn with greater severity and appear less resilient to disturbance. Prior to fire control measures, frequent low-intensity ground fires that killed seedlings and small trees kept the canopies of pines and other conifers open and restricted other species from occupying these sites. Controlled burns and fuel-thinning efforts have been implemented in some areas to cope with these problems. However, even with such efforts it can take many decades for conifer forests to return to their pre-fire-suppression state.

Tundra

The final biome we will consider is the tundra. This sparse herb- and shrub-dominated vegetation is found beyond the treeline in the Arctic and on high mountains. Tundra covers some 8 million square kilometers of the earth's surface. The largest areas of conterminous tundra are found in Arctic portions of Eurasia and North America. Large areas of continuous alpine tundra are found in the Rocky Mountains, the Andes, and the Himalayas. Antarctica is largely covered by ice and supports little vegetation. The vascular plant flora of that continent is dominated by only two species: a grass called *Deschampsia antarctica* and a small cushion-form plant called *Columbanthus quitensis*.

The key climatic characteristic of the tundra biome is low temperature. Mean annual temperatures in the biome are less than −15.0° C. Average temperatures of the warmest month of the year are generally less than 10° C, whereas average temperatures during the coldest month of the year are typically less than −30° C. The growing season is quite short, averaging under 3 months. Diurnal variations in temperature during the growing season in Arctic regions are relatively low, owing to the long hours of daylight in the summer. In contrast, during the summer, alpine regions can experience relatively warm days and below-freezing temperatures at night. Radiation is also higher during the day in high-elevation alpine regions. Alpine regions near the equator have little seasonal variation in temperature. Precipitation in the Arctic is generally quite low, ranging from about 40 centimeters to less than 10 centimeters per year. In contrast, some alpine tundra regions, such as the coastal mountains of Washington and British Columbia, can receive well over 200 centimeters of precipitation each year.

The very low temperatures of the tundra biome restrict soil development. In many areas, the substrate consists of regolith with low organic content and low amounts of available nutrients. In the very high Arctic, vegetation cover often increases in the vicinity of the bones of dead animals due to the increased availability of nutrients. Histosols are found in some regions and can be several meters in depth. Permafrost is continuous in the Arctic and in very high alpine regions. Areas such as the Rocky Mountains may have sporadic permafrost. Frost heaving and other cryogenic processes associated with permafrost can cause the development of sorted stone polygons in regolithic areas and small hummocks and depressions in histosol-dominated sites. The patterned ground surface in such regions is often very active and important in determining local vegetation cover.

The structure of tundra vegetation is extremely simple (Fig. 6.29). In southern portions of the Arctic and at lower elevations in the mountains, small shrubs such as willow (*Salix* spp.), shrub birch (*Betula nana*), alder (*Alnus* spp.), and heaths form a canopy that is up to 2 meters tall. The ground cover in such areas includes sedges, grasses, diverse herbs, and mosses. This vegetation is sometimes referred to as low Arctic tundra. Further north and at higher elevations, the shrub cover becomes sparser, shrubs become more prostrate, and plant height generally decreases. Areas of poorly drained soils are often dominated by the sedge species called cotton grass (*Eriophorum vaginatum*). Patches of unvegetated ground become increasingly common away from the treeline zone. In the far north and at very high elevations, vegetation cover is extremely sparse. In general, the climate is very cold, precipitation is low, and soil nutrients are very limited. In the Arctic, such areas are referred to as polar desert or high Arctic tundra. In some cases, vascular plants can only survive in moist microsites in valleys, along wetlands, and adjacent to summer snow fields. Only a few very small prostrate plants, lichens, and mosses exist in the polar desert. The vascular plants often include low creeping species such as Arctic willow (*Salix arctica*), which is related to willow trees but grows as a low woody plant that rises only a few centimeters above the ground. Other plants, such as the purple saxifrage (*Saxifraga oppositifolia*), grow as compact cushions 10 centimeters across and a few centimeters high. The low stature of the prostrate plants and the cushion form allows plants to grow in a relatively warm microclimate close to the soil surface (Fig. 6.30). In addition, low stature allows the plants to be insulated from cold winter temperatures by a covering of snow. In some cases, 70% of the plant biomass in the tundra can be found protected below the ground in roots and associated structures. Some tundra plants have heliotropic flowers that rotate as the stems elongate so that the flower head always faces the sun. Heliotropism keeps the flower warm and aids in pollen, ovule, and seed production. In addition, the warm microclimate of the flower is beneficial for insect pollinators.

Almost all tundra plants are perennials. There is simply insufficient time for annual plants to germinate, mature, and set seed during the short growing season. Most upright shrubs in the tundra are deciduous, but many of the low prostrate plants such as crowberry (*Empetrum nigrum*), Arctic heather (*Cassiope tetrgona*), and dryas (*Dryas integrifolia*) are evergreen. Plants with persistent leaves demand less time to reach maximum rates of photosynthesis and require fewer resources during the short summer than is the case for deciduous plants, which must produce new leaves at the start of each summer. The leaves of these evergreens are protected beneath snow cover during the winter.

The standing biomass of the world's tundra is extremely low. The species diversity of tundra flora is also extremely low. The entire Arctic tundra region of North America and Eurasia has only about 900 species of flowering plants. Many of the same species are found on all three continents and

FIGURE 6.29   The short vegetation cover of the Arctic tundra near Churchill, Manitoba, in Canada. The low stature of the plants provides them with optimum warmth in the summer and protection beneath snow cover in the winter.

in the floras of adjacent alpine regions. The highest regional diversity of flowering plants is found in Alaska where approximately 600 species are known. In contrast, the flowering plant flora of the northernmost Arctic islands of Canada has less than 100 species. Some plants, such as crowberry, are found in the Arctic and in alpine tundras of Eurasia and South America. In general, however, the plants found on alpine tundra sites in the southern hemisphere are not closely related to Arctic tundra species. For example, the prostrate shrub *Dracophyllum muscoides* and the cushion plant *Raoulia*

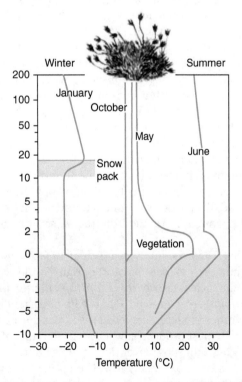

FIGURE 6.30   The microclimatic conditions of the Arctic tundra in northern Sweden during different months of the year (after Rosswell et al., 1975). The thermal advantage to low-growing plants is strongest in spring and winter.

youngii are common at alpine tundra sites in New Zealand, but these genera are not found in the northern hemisphere. In addition to low species richness, the phenotypic variability between tundra plants of the same species is relatively low, likely reflecting the selective pressures of the harsh conditions in which the plants must survive.

Direct human impact is limited in most areas of the tundra biome because of the harsh climatic conditions and inaccessibility. Reindeer herding is practiced extensively in northern Eurasia. Grazing pressure and trampling have caused local disruption of tundra vegetation. Intensive grazing by reindeer can cause shifts from moss- and shrub-dominated vegetation to grass-dominated. Similarly, grazing by cattle and sheep has long been practiced in the alpine meadows of the Alps and is common in many mountain areas of North and South America. Resource extraction for oil and minerals is increasing in many tundra regions. For example, some of the largest reserves of natural gas in the world lie in the Yamal Peninsula region of Arctic Russia. Large reserves of oil have been discovered in northern Alaska and northwestern Canada. Diamonds have been discovered and mined in the central Canadian Arctic. Mines to extract metals such as gold, nickel, tungsten, and platinum are in production at a number of Arctic and alpine tundra sites in North America and Eurasia. Emissions from smoke stacks have caused serious damage to tundra in portions of Russia. Buildings and road construction often cause long-term damage to permafrost soils. A rut created in permafrost by a wheeled or tracked vehicle will often be apparent for decades. In some cases, such compaction causes local melting of the permafrost and the subsequent development of wetlands. This process is called thermokarst erosion. Like all other biomes, even the sparsely populated tundra is facing increasing environmental pressure from human activity.

KEY WORDS AND TERMS

Alpha diversity	Ecotone	Primary forest
Association	Formation	Releve
Beta diversity	Growth form	Second growth forest
Biome	Individualistic community concept	Species evenness
Community type	Life form	Stand
Dominance type	Life zone	Superorganism community concept
Ecological equivalent	Old growth forest	

REFERENCES AND FURTHER READING

Aerts, R., Ewald, M., Nicolas, M., Piat, J., Skowronek, S., Lenoir, J., Hattab, T., Garzón-López, C. X., Feilhauer, H., Schmidtlein, S., and Rocchini, D. (2017). Invasion by the alien tree Prunus serotina alters ecosystem functions in a temperate deciduous forest. *Frontiers in Plant Science* **8**, 179.

Aide, T. M., Clark, M. L., Grau, H. R., López-Carr, D., Levy, M. A., Redo, D., Bonilla-Moheno, M., Riner, G., Andrade-Núñez, M. J., and Muñiz, M. (2013). Deforestation and reforestation of Latin America and the Caribbean (2001–2010). *Biotropica* **45**, 262–271.

Anderson, R. C. (2006). Evolution and origin of the Central Grassland of North America: climate, fire, and mammalian grazers. *Journal of the Torrey Botanical Society* **133**, 626–647.

Ansley, R. J., Boutton, T. W., and Hollister, E. B. (2021). Can prescribed fires restore C4 grasslands invaded by a C3 woody species and a co-dominant C3 grass species? *Ecosphere* **12**, e03885.

Archibold, O. W. (1995). *Ecology of World Vegetation*. Chapman and Hall, London.

Arianoutsou, M., Delipetrou, P., Vila, M., Dimitrakopoulos, P. G., Celesti-Grapow, L., Wardell-Johnson, G., Henderson, L., Fuentes, N., Ugarte-Mendes, E., and Rundel, P. W. (2013). Comparative patterns of plant invasions in the mediterranean biome. *PLoS One* **8**, e79174.

Augspurger, C. K. (2008). Early spring leaf out enhances growth and survival of saplings in a temperate deciduous forest. *Oecologia* **156**, 281–286.

Bailey, R. G. (1996). *Ecosystem Geography*. Springer, New York.

Barbour, M. G., and Billings, W. D. (2000). *North American Terrestrial Vegetation*, 2nd edition. Cambridge University Press, Cambridge.

Barbour, M. G., and Minnich, R. A. (2000). Californian upland forests and woodlands. In *North American Terrestrial Vegetation* (ed. M. G. Barbour and W. D. Billings), pp. 161–202. Cambridge University Press, Cambridge.

Begon, M., Harper, J. L., and Townsend, C. R. (1996). *Ecology: Populations and Communities*. Blackwell Science, Oxford.

Bergeron, Y., and Fenton, N. J. (2012). Boreal forests of eastern Canada revisited: old growth, nonfire disturbances, forest succession, and biodiversity. *Botany* **90**, 509–523.

Billings, W. D. (1974). Arctic and alpine vegetation: plant adaptations to cold climates. In *Arctic and Alpine Environments* (ed. J. D. Ives and R. G. Barry), pp. 403–443. Methuen and Co., London.

Billings, W. D. (1990). The mountain forests of North America and their environments. In *Plant Biology of the Basin and Range* (ed. C. B. Osmund, L. F. Pitelka, and G. M. Hidy), pp. 47–86. Springer-Verlag, New York.

Billings, W. D. (2000). Alpine vegetation. In *North American Terrestrial Vegetation* (ed. M. G. Barbour and W. D. Billings), pp. 537–572. Cambridge University Press, Cambridge.

Bjorkman, A. D., Myers-Smith, I. H., Elmendorf, S. C., Normand, S., Rüger, N., Beck, P. S., Blach-Overgaard, A., Blok, D., Cornelissen, J. H. C., Forbes, B. C., and Georges, D. (2018). Plant functional trait change across a warming tundra biome. *Nature* **562**, 57–62.

Bowers, J. E. (1984). Plant geography of southwestern sand dunes. *Desert Plants* **6**, 31–42, 51–54.

Brown, J. H., and Lomolino, M. V. (1998). *Biogeography*, 2nd edition. Sinauer Associates, Sunderland, MA.

Chapman, T. B., Veblen, T. T., and Schoennagel, T. (2012). Spatiotemporal patterns of mountain pine beetle activity in the southern Rocky Mountains. *Ecology* **93**, 2175–2185.

Clements, F. E. (1916). *Plant Succession: An Analysis of the Development of Vegetation*. Carnegie Institute, Washington, DC.

Clements, F. E., and Shelford, V. E. (1939). *Bioecology*. Wiley, New York.

Crawford, R. M. M. (1989). *Studies in Plant Survival: Ecological Case Histories of Plant Adaptation to Adversity*. Blackwell Scientific Publications, Oxford.

Dansereau, P. (1957). *Biogeography: An Ecological Perspective*. Ronald Press, New York.

Dark, S. J. (2004). The biogeography of invasive alien plants in California: an application of GIS and spatial regression analysis. *Diversity and Distributions* **10**, 1–9.

de Groot, W. J., Cantin, A. S., Flannigan, M. D., Soja, A. J., Gowman, L. M., and Newbery, A. (2013). A comparison of Canadian and Russian boreal forest fire regimes. *Forest Ecology and Management* **294**, 23–34.

Delcourt, H. R., and Delcourt, P. A. (2000). Eastern deciduous forests. In *North American Terrestrial Vegetation* (ed. M. G. Barbour and W. D. Billings), pp. 357–395. Cambridge University Press, Cambridge.

Delcourt, P. A., and Delcourt, H. R. (1987). *Long-Term Forest Dynamics of the Temperate Zone: A Case Study of Late-Quaternary Forests in Eastern North America*. Springer-Verlag, New York.

Dexter, K. G., Pennington, R. T., Oliveira-Filho, A. T., Bueno, M. L., Silva de Miranda, P. L., and Neves, D. M. (2018). Inserting tropical dry forests into the discussion on biome transitions in the tropics. *Frontiers in Ecology and Evolution* **6**, 104.

Dimson, M., and Gillespie, T. W. (2020). Trends in active restoration of tropical dry forest: methods, metrics, and outcomes. *Forest Ecology and Management* **467**, 118150.

Dong, C., Williams, A. P., Abatzoglou, J. T., Lin, K., Okin, G. S., Gillespie, T. W., Long, D., Lin, Y. H., Hall, A., and MacDonald, G. M. (2022). The season for large fires in Southern California is projected to lengthen in a changing climate. *Communications Earth & Environment*, **3**, 1–9.

Drummond, M. A., and Loveland, T. R. (2010). Land-use pressure and a transition to forest-cover loss in the eastern United States. *BioScience* **60**, 286–298.

Dryflor, Banda- R. K., Delgado-Salinas, A., Dexter, K. G., Linares-Palomino, R., Oliveira-Filho, A., Prado, D., Pullan, M., Quintana, C., Riina, R., and Rodríguez M., G. M. (2016). Plant diversity patterns in neotropical dry forests and their conservation implications. *Science* **353**, 1383–1387.

Dyer, J. M. (2006). Revisiting the deciduous forests of eastern North America. *BioScience* **56**, 341–352.

Edwards, E. J., Chatelet, D. S., Chen, B. C., Ong, J. Y., Tagane, S., Kanemitsu, H., Tagawa, K., Teramoto, K., Park, B., Chung, K. F., and Hu, J. M. (2017). Convergence, consilience, and the evolution of temperate deciduous forests. *American Naturalist* **190**, S87–S104.

Elliott-Fisk, D. L. (2000). The taiga and boreal forest. In *North American Terrestrial Vegetation* (ed. M. G. Barbour and W. D. Billings), pp. 41–73. Cambridge University Press, Cambridge.

Ewing, H. A., Weathers, K. C., Templer, P. H., Dawson, T. E., Firestone, M. K., Elliott, A. M., and Boukili, V. K. S. (2009). Fog water and ecosystem function: heterogeneity in a California redwood forest. *Ecosystems* **12**, 417–433.

Eyre, S. R. (1968). *Vegetation and Soils*. Aldine Publishing, Chicago.

FAO (2022). FRA 2020 Remote Sensing Survey. FAO Forestry Paper, No. 186. https://doi.org/10.4060/cb9970en. FAO, Rome, Italy.

Fischer, A., Marshall, P., and Camp, A. (2013). Disturbances in deciduous temperate forest ecosystems of the northern hemisphere: their effects on both recent and future forest development. *Biodiversity and Conservation* **22**, 1863–1893.

Foley, C. A., and Faust, L. J. (2010). Rapid population growth in an elephant Loxodonta africana population recovering from poaching in Tarangire National Park, Tanzania. *Oryx* **44**, 205–212.

Forseth, I. (2010). Terrestrial biomes. *Nature Education Knowledge* **3**, 11.

Franchin, A. G., de Freitas Juliano, R., Kanegae, M., Marçal Jr., O., DelClaro, K., Oliveira, P. S., and Rico-Gray, V. (2009). Birds in the tropical savannas. *Tropical Biology and Conservation Management* **10**, 160–197.

Franklin, J. F. (1988). Pacific Northwest forests. In *North American Terrestrial Vegetation* (ed. M. G. Barbour and W. D. Billings), pp. 103–130. Cambridge University Press, Cambridge.

Franklin, J. F., and Halpern, C. B. (2000). Pacific Northwest forests. In *North American Terrestrial Vegetation* (ed. M. G. Barbour and W. D. Billings), pp. 123–159. Cambridge University Press, Cambridge.

Fuchs, R., Herold, M., Verburg, P. H., Clevers, J. G., and Eberle, J. (2015). Gross changes in reconstructions of historic land cover/use for Europe between 1900 and 2010. *Global Change Biology* **21**, 299–313.

Furley, P. (2006). Tropical savannas. *Progress in Physical Geography* **30**, 105–121.

Gallery, R. E. (2014). Ecology of tropical rain forests. In *Ecology and the Environment*, vol. 8: *The Plant Sciences* (ed. R. Monson), pp. 1–22. Springer, New York.

Gauthier, S., Bernier, P., Kuuluvainen, T., Shvidenko, A., and Schepaschenko, D. (2015). Boreal forest health and global change. *Science* **349**, 819–822.

Ghazoul, J., and Sheil, D. (2010). *Tropical Rain Forest Ecology, Diversity, and Conservation*. Oxford University Press, Oxford.

Gibson, D. J. (2009). *Grasses and Grassland Ecology*. Oxford University Press, Oxford.

Gillespie, T. W., Grijalva, A., and Farris, C. F. (2000). Diversity, composition and structure of tropical dry forests in Central America. *Plant Ecology* **147**, 37–47.

Gleason, H. A. (1926). The individualistic concept of the plant association. *Torrey Botanical Club Bulletin* **53**, 7–26.

Goldammer, J. G. (1990). Fire in the tropical biota: ecosystem processes and global challenges. In *Ecological Studies*, vol. 84, p. 497. Springer-Verlag, Berlin.

Goode, J. D., Barefoot, C. R., Hart, J. L., and Dey, D. C. (2020). Disturbance history, species diversity, and structural complexity of a temperate deciduous forest. *Journal of Forestry Research* **31**, 397–414.

Grimm, W. C. (1967). *Familiar Trees of America*. Harper and Row, New York.

Hall, R. J., Skakun, R. S., Metsaranta, J. M., Landry, R., Fraser, R. H., Raymond, D., Gartrell, M., Decker, V., and Little, J. (2020). Generating annual estimates of forest fire disturbance in Canada: the National Burned Area Composite. *International Journal of Wildland Fire* **29**, 878–891.

Hanes, C. C., Wang, X., Jain, P., Parisien, M. A., Little, J. M., and Flannigan, M. D. (2019). Fire-regime changes in Canada over the last half century. *Canadian Journal of Forest Research* **49**, 256–269.

Hannah, L., Carr, J. L., and Lankerani, A. (1995). Human disturbance and natural habitat: a biome level analysis of a global data set. *Biodiversity and Conservation* **4**, 128–155.

Holdridge, L. R. (1947). Determination of world plant formations from simple climatic data. *Science* **105**, 367–368.

Holdridge, L. R., Grenke, W. C., Hatheway, W., H., Liang, T., and Tosi, J. A. (1971). *Forest Environments in the Tropical Life Zones*. Pergamon Press, Oxford.

Janzen, D. H. (1988). Tropical ecological and biocultural restoration. *Science* **239**, 243–244.

Keeley, J. E. (2000). Chaparral. In *North American Terrestrial Vegetation* (ed. M. G. Barbour and W. D. Billings), pp. 203–253. Cambridge University Press, Cambridge.

Keenan, R. J., Reams, G. A., Achard, F., de Freitas, J. V., Grainger, A., and Lindquist, E. (2015). Dynamics of global forest area: results from the FAO Global Forest Resources Assessment 2015. *Forest Ecology and Management* **352**, 9–20.

Klein, T., Vitasse, Y., and Hoch, G. (2016). Coordination between growth, phenology and carbon storage in three coexisting deciduous tree species in a temperate forest. *Tree Physiology* **36**, 847–855.

Kramer, H. A., Mockrin, M. H., Alexandre, P. M., and Radeloff, V. C. (2019). High wildfire damage in interface communities in California. *International Journal of Wildland Fire* **28**, 641–650.

Krebs, C. J. (1994). *Ecology*, 4th edition. Harper-Collins College Publishers, New York.

Lazzeri-Aerts, R., and Russell, W. (2014). Survival and recovery following wildfire in the southern range of the coast redwood forest. *Fire Ecology* **10**, 43–55.

Liautaud, K., van Nes, E. H., Barbier, M., Scheffer, M., and Loreau, M. (2019). Superorganisms or loose collections of species? A unifying theory of community patterns along environmental gradients. *Ecology Letters* **22**, 1243–1252.

Limm, E., Simonin, K., and Dawson, T. (2012). Foliar uptake of fog in the coast redwood ecosystem: a novel drought-alleviation strategy shared by most redwood forest plants. In *Proceedings of Coast Redwood Forests in a Changing California: A Symposium for Scientists and Managers. General Technical Report PSW-GTR-238* (tech. coord. R. B. Standiford, T. J. Weller, D. D. Piirto, and J. D. Stuart), pp. 273–281. U.S. Department of Agriculture, Washington, DC.

Liu, Y., and Xue, Y. (2020). Expansion of the Sahara Desert and shrinking of frozen land of the Arctic. *Scientific Reports* **10**, 1–9.

Lovett, G. M., Weiss, M., Liebhold, A. M., Holmes, T. P., Leung, B., Lambert, K. F., Orwig, D. A., Campbell, F. T., Rosenthal, J., McCullough, D. G., and Wildova, R. (2016). Nonnative forest insects and pathogens in the United States: impacts and policy options. *Ecological Applications* **26**, 1437–1455.

Lugo, A. E., Colon, J. F., and Scatena, F. N. (2000). The Caribbean. In *North American Terrestrial Vegetation* (ed. M. G. Barbour and W. D. Billings), pp. 593–622. Cambridge University Press, Cambridge.

Lydersen, J. M., North, M. P., Knapp, E. E., and Collins, B. M. (2013). Quantifying spatial patterns of tree groups and gaps in mixed-conifer forests: reference conditions and long-term changes following fire suppression and logging. *Forest Ecology and Management* **304**, 370–382.

MacMahon, J. A. (2000). Warm deserts. In *North American Terrestrial Vegetation* (ed. M. G. Barbour and W. D. Billings), pp. 285–322. Cambridge University Press, Cambridge.

McDonald, C. J., and McPherson, G. R. (2013). Creating hotter fires in the Sonoran Desert: buffelgrass produces copious fuels and high fire temperatures. *Fire Ecology* **9**, 26–39.

McGlone, P. S., Richardson, S. J., and Jordan, G. J. (2010). Comparative biogeography of New Zealand trees: species richness, height, leaf traits and range sizes. *New Zealand Journal of Ecology* **34**, 137–151.

McIntosh, R. P. (1985). *The Background of Ecology: Concept and Theory*. Cambridge University Press, Cambridge.

Meijaard, E., Brooks, T. M., Carlson, K. M., Slade, E. M., Garcia-Ulloa, J., Gaveau, D. L., Lee, J. S. H., Santika, T., Juffe-Bignoli, D., Struebig, M. J., and Wich, S. A. (2020). The environmental impacts of palm oil in context. *Nature Plants* **6**, 1418–1426.

Mendelssohn, I. A., and McKee, K. L. (2000). Saltmarshes and mangroves. In *North American Terrestrial Vegetation* (ed. M. G. Barbour and W. D. Billings), pp. 501–536. Cambridge University Press, Cambridge.

Merriam C. H. (1894). Laws of temperature control of the geographic distribution of terrestrial animals and plants. *National Geographic* **6**, 229–238.

Merriam, C. H., and Stejneger, L. (1890). Results of a biological survey of the San Francisco Mountain region and desert of the Little Colorado, Arizona. *North American Fauna* **3**, 1–136.

Mistry, J., and Beradi, A. (2014). *World Savannas: Ecology and Human Use*. Routledge, London.

Mittelbach, G. G., and McGill, B. J. (2019). *Community Ecology*. Oxford University Press, Oxford.

Moffat, A. S. (1998). Temperate forests gain ground. *Science* **282**, 1253.

Mooney, H. A. (1988). Lessons from Mediterranean-climate regions. In *Biodiversity* (ed. E. O. Wilson and F. M. Peter), pp. 157–165. National Academy Press, Washington, DC.

Morin, P. J. (2011). *Community Ecology*, 2nd edition. Wiley-Blackwell, Chichester, England.

Murphy, B. P., and Bowman, D. M. (2012). What controls the distribution of tropical forest and savanna? *Ecology Letters* **15**, 748–758.

Nardini, A., Gullo, M. A. L., Trifilò, P., and Salleo, S. (2014). The challenge of the Mediterranean climate to plant hydraulics: responses and adaptations. *Environmental and Experimental Botany* **103**, 68–79.

Navarro, L., Morin, H., Bergeron, Y., and Girona, M. M. (2018). Changes in spatiotemporal patterns of 20th century spruce budworm outbreaks in eastern Canadian boreal forests. *Frontiers in Plant Science* **9**, 1905.

Nunes, S., Oliveira, L., Siqueira, J., Morton, D. C., and Souza, C. M. (2020). Unmasking secondary vegetation dynamics in the Brazilian Amazon. *Environmental Research Letters* **15**, 034057.

Ocón, J. P., Ibanez, T., Franklin, J., Pau, S., Keppel, G., Rivas-Torres, G., Shin, M. E., and Gillespie, T. W. (2021). Global tropical dry forest extent and cover: a comparative study of bioclimatic definitions using two climatic data sets. *PloS One* **16**, e0252063.

Olofsson, J. (2006). Short-and long-term effects of changes in reindeer grazing pressure on tundra heath vegetation. *Journal of Ecology* **94**, 431–440.

Olsson, A. D., Betancourt, J. L., Crimmins, M. A. and Marsh, S. E. (2012). Constancy of local spread rates for buffelgrass (*Pennisetum ciliare* L.) in the Arizona Upland of the Sonoran Desert. *Journal of Arid Environments* **87**, 136–143.

Ornduff, R. (1974). *Introduction to California Plant Life*. University of California Press, Berkeley.

Palmero-Iniesta, M., Pino, J., Pesquer, L., and Espelta, J. M. (2021). Recent forest area increase in Europe: expanding and regenerating forests differ in their regional patterns, drivers and productivity trends. *European Journal of Forest Research* **140**, 793–805.

Pärtel, M., Bruun, H. H., and Sammul, M. (2005). Biodiversity in temperate European grasslands: origin and conservation. *Grassland Science in Europe* **10**, 1–14.

Pearson, A. F. (2010). Natural and logging disturbances in the temperate rain forests of the Central Coast, British Columbia. *Canadian Journal of Forest Research* **40**, 1970–1984.

Peet, R. K. (2000). Forests and meadows of the Rocky Mountains. In *North American Terrestrial Vegetation* (ed. M. G. Barbour and W. D. Billings), pp. 75–121. Cambridge University Press, Cambridge.

Perry, D. A. (1994). *Forest Ecosystems*. Johns Hopkins University Press, Baltimore, MD.

Pielou, E. C. (1979). *Biogeography*. Wiley, New York.

Podani, J. (2006). Braun-Blanquet's legacy and data analysis in vegetation science. *Journal of Vegetation Science* **17**, 113–117.

Poorter, H., Niklas, K. J., Reich, P. B., Oleksyn, J., Poot, P., and Mommer L. (2012). Biomass allocation to leaves, stems and roots: meta-analyses of interspecific variation and environmental control. *New Phytologist* **193**, 30–50.

Portillo-Quintero, C. A., and Sánchez-Azofeifa, G. A. (2010). Extent and conservation of tropical dry forests in the Americas. *Biological Conservation* **143**, 144–155.

Prentice, I. C., Cramer, W., Harrison, S. P., Leemans, R., Monserud, R. A., and Solomon, A. M. (1992). A global biome model based on plant physiology and dominance, soil properties and climate. *Journal of Biogeography* **19**, 117–134.

Raunkiaer, C. (1934). *The Life Forms of Plants and Statistical Plant Geography*. Clarendon Press, Oxford.

Richards, P. W. (1996). *The Tropical Rainforest*, 2nd edition. Cambridge University Press, Cambridge.

Ricklefs, R. E. (1990). *Ecology*. W. H. Freeman, New York.

Risser, P. G. (1988). Abiotic controls on primary productivity and nutrient cycles in North American grasslands. In *Concepts of Ecosystem Ecology* (ed. L. R. Pomeroy and J. J. Alberts), pp. 115–129. Springer-Verlag, New York.

Ritchie, J. C. (1987). *Postglacial Vegetation of Canada*. Cambridge University Press, Cambridge.

Robinson, T. P., Van Klinken, R. D., and Metternicht, G. (2008). Spatial and temporal rates and patterns of mesquite (*Prosopis* species) invasion in Western Australia. *Journal of Arid Environments* **72**, 175–188.

Rosswall, T., Flower-Ellis, J. G. K., Johansson, L. G., Jonsson, S., Ryden, B. E., and Sonesson, M. (1975). Stordalen (Abisko) Sweden. *Ecological Bulletin, Swedish National Science Research Council* **20**, 265–294.

Roughgarden, J. (2009). Is there a general theory of community ecology? *Biology & Philosophy* **24**, 521–529.

Rundel P. W. (2018). California chaparral and its global significance. In *Valuing Chaparral. Springer Series on Environmental Management* (ed. E. Underwood, H. Safford, N. Molinari, and J. Keeley), pp. 1–27. Springer, Cham, Switzerland.

Runkle, J. R. (1990). Gap dynamics in an Ohio *Acer-Fagus* forest and speculations on the geography of disturbance. *Canadian Journal of Forest Research* **20**, 632–641.

Russell, W., Sinclair, J., and Michels, K. H. (2014). Restoration of coast redwood (*Sequoia sempervirens*) forests through natural recovery. *Open Journal of Forestry* **4**, 106.

Saatchi, S. S., Harris, N. L., Brown, S., Lefsky, M., Mitchard, E. T., Salas, W., Zutta, B. R., Buermann, W., Lewis, S. L., Hagen, S., and Petrova, S. (2011). Benchmark map of forest carbon stocks in tropical regions across three continents. *Proceedings of the National Academy of Sciences* **108**, 9899–9904.

Sader, S. A., and Joyce, A. T. (1988). Deforestation rates and trends in Costa Rica. *Biotropica* **20**, 11–19.

Schlossberg, S., Chase, M. J., and Griffin, C. R. (2018). Poaching and human encroachment reverse recovery of African savannah elephants in south-east Angola despite 14 years of peace. *PLoS One* **13**, e0193469.

Schröder, J. M., Rodriguez, L. P. A., and Günter, S. (2021). Research trends: tropical dry forests: the neglected research agenda? *Forest Policy and Economics* **122**, 102333.

Schwartz, N. B., Aide, T. M., Graesser, J., Grau, H. R., and Uriarte, M. (2020). Reversals of reforestation across Latin America limit climate mitigation potential of tropical forests. *Frontiers in Forests and Global Change* **3**, 85. https://doi.org/10.3389/ffgc.2020.00085.

Scott, G. A. J. (1995). *Canada's Vegetation: A World Perspective*. McGill-Queen's University Press, Montreal and Kingston.

Shinneman, D. J., and Baker, W. L. (1997). Nonequilibrium dynamics between catastrophic disturbances and old-growth forests in ponderosa pine landscapes of the Black Hills. *Conservation Biology* **11**, 1276–1288.

Sims, P., and Coupland, R. T. (1979). Producers. In *Grassland Ecosystems of the World: Analysis of Grasslands and Their Uses* (ed. R. T. Coupland), pp. 49–72. Cambridge University Press, Cambridge.

Sims, P. L., and Risser, P. G. (2000). Grasslands. In *North American Terrestrial Vegetation* (ed. M. G. Barbour and W. D. Billings), pp. 323–356. Cambridge University Press, Cambridge.

Sodhi, N. S. (2008). Tropical biodiversity loss and people—a brief review. *Basic and Applied Ecology* **9**, 93–99.

Speece, D. F. (2017). *Defending Giants: The Redwood Wars and the Transformation of American Environmental Politics*. University of Washington Press, Seattle.

Steel, Z. L., Safford, H. D., and Viers, J. H. (2015). The fire frequency-severity relationship and the legacy of fire suppression in California forests. *Ecosphere* **6**, 1–23.

Thomas, H. J., Bjorkman, A. D., Myers-Smith, I. H., Elmendorf, S. C., Kattge, J., Diaz, S., Vellend, M., Blok, D., Cornelissen, J. H. C., Forbes, B. C., and Henry, G. H. (2020). Global plant trait relationships extend to the climatic extremes of the tundra biome. *Nature Communications* **11**, 1–12. https://doi.org/10.1038/s41467-020-15014-4.

Tinley, K. L. (1982). The influence of soil moisture balance on ecosystem patterns in southern Africa. In *Ecology of Tropical Savannas* (ed. B. J. Huntley and B. H. Walker), pp. 175–192. Springer-Verlag, Berlin.

Trammell, T. L., D'Amico III, V., Avolio, M. L., Mitchell, J. C., and Moore, E. (2020). Temperate deciduous forests embedded across developed landscapes: younger forests harbour invasive plants and urban forests maintain native plants. *Journal of Ecology* **108**, 2366–2375.

Tsujino, R., Yumoto, T., Kitamura, S., Djamaluddin, I., and Darnaedi, D. (2016). History of forest loss and degradation in Indonesia. *Land Use Policy* **57**, 335–347.

Underwood E., Safford H., Molinari N., and Keeley J. (Eds.) (2018). *Valuing Chaparral. Springer Series on Environmental Management*. Springer, Cham, Switzerland.

Underwood, E. C., Viers, J. H., Klausmeyer, K. R., Cox, R. L., and Shaw, M. R. (2009). Threats and biodiversity in the Mediterranean biome. *Diversity and Distributions* **15**, 188–197.

Utz, R. M., Slater, A., Rosche, H. R., and Carson, W. P. (2020). Do dense layers of invasive plants elevate the foraging intensity of small mammals in temperate deciduous forests? A case study from Pennsylvania, USA. *NeoBiota* **56**, 73–88.

Vale, T. R. (1982). *Plants and People: Vegetation Change in North America*. Association of American Geographers, Washington, DC.

Vellend, M., and Geber, M. A. (2005). Connections between species diversity and genetic diversity. *Ecology Letters* **8**, 767–781.

Vogiatzakis, I. N., Mannion, A. M., and Griffiths, G. H. (2006). Mediterranean ecosystems: problems and tools for conservation. *Progress in Physical Geography* **30**, 175–200.

Wang, P., Heijmans, M. M., Mommer, L., van Ruijven, J., Maximov, T. C., and Berendse, F. (2016). Belowground plant biomass allocation in tundra ecosystems and its relationship with temperature. *Environmental Research Letters* **11**, 055003.

Walker, J. S., Grimm, N. B., Briggs, J. M., Gries, C., and Dugan, L. (2009). Effects of urbanization on plant species diversity in central Arizona. *Frontiers in Ecology and the Environment* **7**, 465–470.

Walter, H. (1985). *Vegetation of the Earth*, 3rd edition. Springer-Verlag, New York.

Weaver, J. E. (1968). *Prairie Plants and Their Environment: A Fifty Year Study in the Midwest*. University of Nebraska, Lincoln.

Whitford, W. G., and Duval, B. D. (2019). *Ecology of Desert Systems*. Academic Press, London.

Whitmore, T. C. (1990). *An Introduction to Tropical Rainforests*. Clarendon Press, Oxford.

Whittaker, R. H. (1960). Vegetation of the Siskiyou Mountains, Oregon and California. *Ecological Monographs* **30**, 279–338.

Whittaker, R. H. (1975). *Communities and Ecosystems*, 2nd edition. Macmillan, New York.

Williams, C. B., and Sillett, S. C. (2007). Epiphyte communities on redwood (*Sequoia sempervirens*) in northwestern California. *The Bryologist* **110**, 420–452.

Wilson, E. O. (1988). *Biodiversity*, p. 521. National Academy Press, Washington, DC.

Wolf, J. J., and Mast, J. N. (1998). Fire history of mixed conifer forests of the North Rim, Grand Canyon National Park, Arizona. *Physical Geography* **19**, 1–14.

Woodell, S. R. J., Mooney, H. A., and Hill, A. J. (1969). The behaviour of *Larrea divaricata* (creosote bush) in response to rainfall in California. *Journal of Ecology* **57**, 37–44.

Woodward, F. I. (1987). *Climate & Plant Distribution*. Cambridge University Press, Cambridge.

Woodward, S. L. (2003) *Biomes of Earth*. Greenwood Press, Westport, CT.

World Wildlife Federation (2021). *Plowprint Report*. World Wildlife Foundation, Washington, DC. https://www.worldwildlife.org/publications/2021-plowprint-report.

PART II

TIME AND LIFE

CHAPTER 7

CHANGING CONTINENTS AND CLIMATES

The Latin term *terra firma* means solid earth, and indeed that is how we generally think about the continents—solid and unmoving. However, geological and paleontological evidence tells us that the continents and oceans have shifted and changed their geographic positions over time. For example, if we rearranged the continents to look as they did some 200 million years ago, the cities of Boston, Massachusetts, and Lisbon, Portugal, would have almost been neighboring communities with no Atlantic Ocean between them. Although changes in the geography of the continents and oceans have occurred at almost imperceptibly slow rates over millions of years, these changes explain many aspects of the modern distributions of species. Thus, it is important for biogeographers to understand and consider the past changes in the geography of the earth when trying to comprehend present-day plant and animal distributions.

As the continental landmasses shifted their geographic positions relative to the equator, poles, and each other, their climates changed. In addition to the slow movement of continents and oceans, there have been large-scale changes in the earth's climate driven by other factors. Many of these climatic changes have occurred on much shorter time scales. In the last 2 million years or so, climatic changes have generated repeated episodes of growth and retreat by glaciers at high latitudes and high elevations. Standing in the sweltering summer heat and humidity of Chicago, it seems impossible that the region was covered year-round by thick glacial ice only 20,000 years ago. The climatic and geographic changes that accompanied these glacial cycles have had a profound impact on the modern distributions of life. Over the past 11,000 years or so there have been changes in climate that, though smaller in magnitude than the glacial cycles, have also affected the modern distributions of species. Finally, we are now standing on the edge of a unique event in the climatic and biogeographic history of the earth. For the first time, human activity is changing the climate of the entire planet in a major fashion. The profound changes that humans are creating in the atmosphere and climate of the earth is one reason why many scientists believe that we are witnessing a new geological epoch—the Anthropocene.

Biogeographers recognize that the modern distribution of life reflects both present-day environmental conditions and the past history of the planet. Current and future climatic changes will also affect the distribution of life. In this chapter we will review the processes and patterns of past changes in the earth's geography and climate. We will then look at the role of human activity in altering the climate of the planet. To begin, however, we should review a little bit of basic geology.

LIFE AND THE GEOLOGIC TIME SCALE

For most people, events that occurred a few decades ago seem like ancient history. Events that go back thousands of years appear extremely remote, and the concept of dealing with changes and processes that occur over millions of years seems overwhelming. Biogeographers, however, know that the modern distributions of species are strongly influenced by events that occurred thousands to millions of years ago. Geologists, paleontologists, and biogeographers must routinely consider events of the distant past and deal with huge units of time. The **geologic time scale** is an internationally accepted

Biogeography: Introduction to Space, Time, and Life, Second Edition. Glen M. MacDonald.
© 2025 John Wiley & Sons, Inc. Published 2025 by John Wiley & Sons, Inc.
Companion website: www.wiley.com/go/macdonald/biogeography2e

system that serves as the standard means of subdividing the many millions of years of earth's history. The geologic time scale has its origins in the nineteenth century when the sciences of geology and paleontology were rapidly developing. Even today the geologic time scale continues to be refined in the light of new geological and paleontological discoveries, and some details of the subdivisions remain debated.

The exposures that geologists study consist of layers of rock with different compositions and textures. These layers are referred to as strata, and their analysis and classification is known as stratigraphy. Rocks can be classified into three broad categories: igneous rocks formed by the cooling and solidification of molten material called magma, intrusive igneous rocks formed when the cooling occurs deep underground, and extrusive igneous rocks formed when cooling occurs at the surface. Examples of intrusive and extrusive igneous rocks include granite and basalt. Over time, physical, chemical, and biological processes cause the igneous rocks to break down into fine particles and soluble chemicals. This process is referred to as weathering. Weathered material is transported away by water, wind, and gravity. The transport of these weathered materials is called erosion. Sedimentary rocks are formed when the weathered material accumulates. The accumulating substances might include sand grains that are deposited into lakes by rivers and form sandstone, or calcium carbonate ooze precipitated in the ocean that forms limestone. The accumulated material is buried and over long periods of time becomes compacted and cemented into solid rock. The physical and chemical process by which sediments are turned into rock is called lithification. Sandstone and limestone are just two examples of the wide range of sedimentary rock. Sometimes the nature of the sedimentary rock provides clues to the climate at the time of its formation. For example, glaciers develop under very cold conditions and by eroding underlying rock and soil produce sedimentary deposits called tillites (also called glacial tills), which contain a poorly sorted mixture of clay, silts, sand, and rock fragments. Metamorphic rocks are formed when igneous or sedimentary rocks are altered by high temperatures and pressures. Marble is a metamorphic rock that is formed when limestone is exposed to high heat and pressure. Geologic strata consist of bedded layers of different igneous, sedimentary, and metamorphic rock.

Some sedimentary strata contain the remains of plants and animals. In many cases, these remains were altered by chemical processes so that they became lithified. Such lithified remains are called fossils. Some researchers consider all preserved organic remains to be fossils, whereas others contend that the remains must be lithified. The presence of fossil remains of extinct animals in sedimentary rocks has been known for a long time. In ancient China it was thought that some of these remains were mythical beasts such as dragons. In Europe these fossils were thought to be the remains of animals that had not been able to get to Noah's Ark and were destroyed during the Great Flood described in the Judeo-Christian Bible.

During the nineteenth century, geologists and paleontologists began to classify strata on the basis of their physical characteristics and their fossil content. The geologists made two assumptions. First, they assumed that strata that were lower in a stratigraphic exposure were older than overlying strata. Thus, the fossils in lower strata were assumed to represent older forms of life than those in higher strata. This assumption is called the **law of superposition.** Second, they assumed that the processes that weathered, eroded, and deposited rock in the past were the same as those operative today. They therefore concluded that rates of erosion and deposition in the past were no faster or slower than those observed today. This assumption is called the **law of uniformitarianism.** From observations of rock and fossil stratigraphy, mainly in Europe and North America, the geologic time scale was developed and refined during the nineteenth and twentieth centuries.

Initially, geologists could only try to estimate the number of years that different strata represented and guess how old fossils were. Eighteenth- and nineteenth-century geologists such as James Hutton and Charles Lyell of Britain realized that the earth must be very ancient in order for the rocks of so many different strata to be formed, weathered, eroded, deposited, and lithified. By the end of the nineteenth century, geologists, paleontologists, and biogeographers concluded that the earth must be several hundred million years old. During the twentieth century, sophisticated means of dating rocks by using the radioactive decay of unstable isotopes and elements have allowed more reliable estimates of the age of the earth. Physicists have calculated how long it takes radioactive elements and isotopes to decay to daughter elements and isotopes. For example, it takes about 1.3 billion years for half of the total amount of the radioactive element potassium 40 to decay to calcium and argon. Physicists refer to the time needed for half of the radioactive element present in a material to decay as the half-life of the element. By examining the ratio of potassium 40 to calcium and argon in a mineral crystal, the age of the crystal can be estimated. Some of the oldest dated mineral crystals are zircons, which are 4.1 billion years old. It is estimated that the earth is about half a billion years older than the age of these zircons. By using radiometric dating techniques, we can affix ages to different strata and their fossils. The modern geologic time scale is based on the classification of strata by their lithology, relative position, and fossils, with ages provided by radiometric dating. The geologic time scale encapsulates, in a very broad manner, the history of earth and its life.

The modern geologic time scale extends back over 4 billion years and is subdivided into an hierarchy extending from eons to eras, periods, and epochs (Fig. 7.1). The names for the different periods can be quite confusing to the novice. Some, such as the Pennsylvanian Period, are named for geographic regions where diagnostic deposits of this age are found. Others such as the Quaternary Period are drawn from other classification schemes that are no longer in use. Some, such as the Silurian Period, are even more obscure in origin. The Silurian is named after an ancient tribe of Britain, the Silures.

We live in the Phanerozoic Eon, which began 570 million years ago. The first era of the Phanerozoic is the Paleozoic, which is typified by the appearance of complex forms of life. The appearance of complex, multicellular life appears to have occurred during the Cambrian between 570 and 505 million years ago. The rapid proliferation of multicellular exoskeleton (external hard covering) marine species is called the Cambrian Explosion and peaked about 530 million years ago. The ancestors of mollusks, crustaceans, and insects all appear in the oceans of the Cambrian Period. The first marine vertebrate also appears in the Cambrian. The Ordovician Period, extending from 505 to 438 million years ago, is noted for the development of vertebrate saltwater and freshwater fish. The rise of primitive land plants, land-dwelling invertebrates, and amphibians occurred during the Silurian and Devonian periods about 438 to 360 million years ago. The oceans of this time teemed with fish species and primitive sharks. The land plants likely started as semiaquatic algal seaweeds followed by primitive ferns. The first true land plants were only a few centimeters in height. By the end of the Silurian, tree-like plants with leaves over 10 centimeters had long existed, and the ancestral form of gymnosperm trees had arisen. Centipedes and spiders appeared on land in the Silurian. By the middle Devonian, lobe-finned fish were making forays to feed on the land. By the late Devonian, the first true amphibians appeared. Development of terrestrial vegetation and land animals continued through the Mississippian and Pennsylvanian periods. The Permian Period, some 286 to 245 million years ago, is characterized by the appearance and development of true reptiles and the ancestor of the crocodiles, caimans, and alligators.

The Mesozoic Era began 245 million years ago and is best known for the proliferation and dominance of dinosaur species. However, the ancestors of modern mammals also appeared at the same time that the dinosaurs rose to dominance in the Triassic Period. The evolution of birds appears to have begun in the Jurassic Period between 208 and 144 million years ago. A possible early ancestor of birds, *Archaeopteryx*, possessed some features of modern birds, such as feathers, but also had features of dinosaurs, including its basic skeletal structure and teeth. Conifers, cycads, and ferns dominated the vegetation, but flowering plants also appeared by the early Cretaceous Period, 144 million years ago.

The current era is the Cenozoic. It began 65 million years ago and is typified by the extinction of the dinosaurs and the dominance of mammals in the terrestrial megafauna. Flowering plants diversified and came to dominate the flora. One important group of flowering plants, the grasses, appeared in the mid-Tertiary Period around 24 million years ago. We live in the Quaternary Period, which began about 2 million years ago. In revised versions of the geologic time scale (Fig. 7.1), the Tertiary has been relegated to an informal period and is replaced by the Paleogene Period, which extends from 65 million to 24 million years ago, and the Neogene Period, which extends from 24 million years ago to the start of the Quaternary about 2 million years ago.

The Quaternary is characterized by the appearance of the species of plants and animals found living today, including our own species, *Homo sapiens*. We currently live in the Holocene Epoch. The Holocene represents the past 11,500 years and is typified by the presence of modern species of plants and animals. The rise of human civilization and the explosive growth of the human population are additional features of the Holocene. Some scientists believe that the impact of humans has been so profound that a new epoch, the Anthropocene, should be recognized to replace, subdivide, or conclude the Holocene. This proposition was rejected by the International Union of Geological Sciences in 2024 and remains a controversial matter of ongoing discussion. Despite the importance we place

RELATIVE GEOLOGIC TIME					ABSOLUTE TIME (Millions of Years)
EON	ERA		PERIOD	EPOCH	
Phanerozoic	Cenozoic (Recent life)		Quaternary	Holocene	
				Pleistocene	2
		Tertiary	Neogene	Pliocene	5
				Miocene	24
			Paleogene	Oligocene	37
				Eocene	58
				Paleocene	66
	Mesozoic (Middle life)		Cretaceous	Late	
				Early	144
			Jurassic	Late	
				Middle	
				Early	208
			Triassic	Late	
				Middle	
				Early	245
	Paleozoic (Ancient life)		Permian	Late	
				Early	286
			Pennsylvanian	Late	
				Middle	
				Early	
			Mississippian	Late	
				Early	360
			Devonian	Late	
				Middle	
				Early	408
			Silurian	Late	
				Middle	
				Early	438
			Ordovician	Late	
				Middle	
				Early	505
			Cambrian	Late	
				Middle	
				Early	570
	Precambrian				3800+

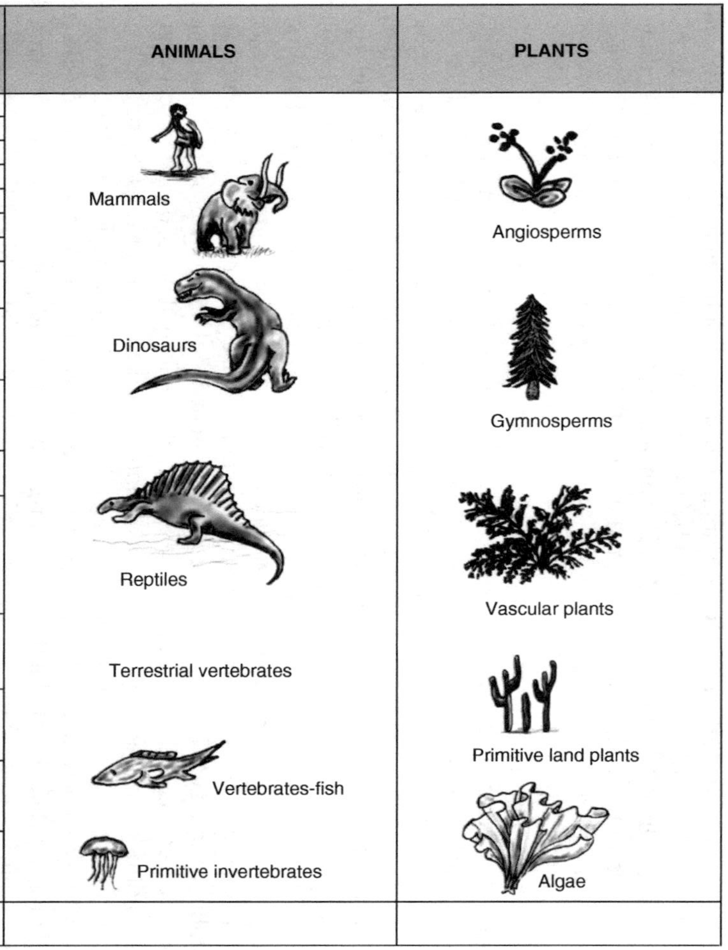

FIGURE 7.1 The geologic time scale with approximate ages of boundaries, and the types of animals and plants typical of the fossil record for different time periods (Birkeland and Larson, 1989; updated from Geologic Names Committee of the U.S. Geological Survey, 2007). The Tertiary, although widely used as an informal period, has been replaced by the Paleogene and the Neogene.

on humans and our environmental impacts, our actions over the Holocene Epoch represent but a brief moment in geologic time. Although we may think of the now extinct dinosaurs as failures, they did dominate the earth for a period of time that is thousands of times longer than the Holocene.

Charles Lyell recognized that since certain sedimentary rocks and fossils can be closely related to past climatic conditions, it is possible to reconstruct a record of climatic change over geologic time by carefully examining rock strata. He came to the conclusion that the earth must have experienced a number of large-scale swings in climate over its history. Continued work has documented the general timing and magnitude of such swings. On the broadest scale, long periods of time seem to have been dominated by generally warm conditions. However, during certain intervals, there is evidence for cold temperatures and widespread glacial conditions. Within the Phanerozoic, such glacial climates existed around 450 million years ago during the late Ordovician, around 286 million years ago, at the boundary between the Pennsylvanian Period and the Permian Period, and during the Pleistocene Epoch between 2 million and 11,500 years ago. Within some of these periods of glaciation, we also find evidence of alternating warm climates and glacial conditions. The Pleistocene, for example, is typified by periods of glaciation over large areas of the high latitudes in the northern hemisphere and on many mountain ranges and by other intervals when the climate is relatively warm like that of today. What changes could have occurred to the earth causing such dramatic shifts in climate over both long and short time spans? This question will be explored in the next two sections of this chapter.

SHIFTING CONTINENTS

A casual look at a globe reveals a curious match between the eastern coastlines of North and South America with those of western Europe and Africa. It would seem that South America could fit nicely into the concave coastline of western Africa, while northwestern Africa and Europe could be accommodated within the Caribbean Basin and adjacent portions of North America. This apparent fit between the Old and New World was noticed as early as 1598 by the cartographer Abraham Ortelius who pointed out that such a fit between these coastlines suggested a dislocation of the continents relative to each other. In 1858, the American Antonio Snider-Pelligrini published a detailed map showing how North and South America may have at one time been joined to Europe and Africa. He postulated that the biblical flood of Noah had somehow caused the continental masses to be sundered apart.

In addition to cartographic evidence, data from fossil finds, geological mapping, and the distributions of living species also pointed to a closer connection between North America and Europe. For example, as early as the 1750s, the Compte de Buffon pointed out the similarity between fossils found in Ireland and North America. Evidence of similarities in fossils and geology continued to accumulate. In 1872 another Frenchman, E. Reclus, argued for the existence of a closer connection between Europe and America on the basis of their similarity in geological structure. In 1853 the biogeographer J. D. Hooker noted the similarity in plant species, genera, and families on the widely separated continents and islands of the southern hemisphere and concluded that the modern flora must represent the remnants of a once continuous land vegetation that was broken up by geological and climatic changes. By the early twentieth century, a number of geologists and paleontologists had speculated on the possibility that the continental landmasses had been closer together in the geologic past. However, it remained for the German astronomer, meteorologist, and geophysicist Alfred Wegener to marshal compelling evidence for continental movement and make a strong case for this phenomenon (Fig. 7.2).

Wegener warrants some description before we continue with our examination of his ideas. He was born in Germany in 1880 and was educated at the universities of Heidelberg, Innsbruck, and Berlin. In 1906 he went to Greenland as an expedition meteorologist for the Danish government. Following that, he taught meteorology and astronomy at Marburgh University in Germany. He wrote that his first ideas about continental movement arose in 1910 while he was looking at a geographic atlas and noticed the match between the Atlantic coastlines of the Old and New Worlds. In 1912 he published his first account of his theory of the Horizontal Displacement of the Continents. He would publish a number of book editions and papers on this topic between 1912 and 1929. He had a more than passing acquaintance with contemporary biogeography: in 1913 he married the daughter of the famous climatologist and biogeographer Köppen.

FIGURE 7.2   The German geophysicist and Arctic explorer Alfred Wegener (1880–1930). Wegener was the father of the modern theory of continental drift. His ideas were not widely accepted during his lifetime. Wegener died on the Greenland Ice Cap during a scientific expedition (*Source:* Large Norwegian encyclopedia).

Wegener's ideas on the movement of the continents have come to be known as the **theory of continental drift.** In a treatise published between 1912 and 1929, he outlined the following arguments in support of continental drift. First, he pointed out that the crust of the continents was composed of relatively light rocks, or sial. He suggested that the ocean floors were largely composed of dense basalts, or sima. The lighter continental rocks could essentially float on the viscous mantle of the earth and override oceanic crust. Second, Permian glacial deposits in Brazil, South Africa, India, and Australia indicate that these subtropical areas were once glaciated, and this could only occur if those landmasses were formerly located close to the polar regions (Fig. 7.3). Third, similarities in the shape of continents, their geology, and fossils suggest closer proximity in the past. For example, a Permian fossil flora that is typified by the fern genus *Glossopteris* is found in areas of South America, Africa, and India (Figs. 7.3 and 7.4). Fourth, geodetic measurements of the latitude and longitude of Greenland taken in 1823, 1870, 1923, and 1927 indicated a continuing westward movement of as much as 9 to 32 meters per year. We now know that this movement is real but occurs at a slower rate of a few centimeters a year.

Based on the evidence he had marshaled, Wegener concluded that all of the world's continents had once been joined in one huge landmass. He thought that this landmass, which he called **Pangaea,** had existed until Permian-Pennsylvanian times. The southern portion of Pangaea, which included modern South America, Africa, Australia, Antarctica, and India, came to be known as **Gondwanaland,** while the northern area comprised of North America, Europe, and most of Asia came to be called **Laurasia.**

According to Wegener, the supercontinent of Pangaea broke apart in the Mesozoic. The Atlantic Ocean formed as a great rift between South America and Africa that grew northward. Wegener thought that Europe and North America had separated only in the last few million years. The problems with Wegener's theory included sparse information on the geology of the deep oceans, a scarcity of absolute dates and rates for many of the changes he envisioned, a lack of knowledge about continental locations prior to the Pennsylvanian-Permian, and unsatisfactory identification of the mechanism that caused the continents to move about on the mantle. Wegener suggested that radioactive heating and flow of the mantle coupled with other factors likely caused the continental blocks to move. In addition, in drawing together so many lines of evidence, Wegener made a number of factual errors regarding geology and paleontology in his publications.

Wegener's theories were not greeted with universal acceptance, and his later years were not happy ones. He and his theory of continental drift were harshly criticized by the Anglo-American

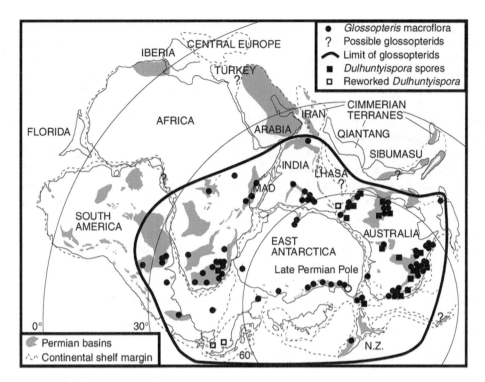

FIGURE 7.3   The distribution of *Glossopteris* floral macrofossils and the Permian geography of the continents (*Source:* McLoughlin, 2001 / CSIRO Publishing).

FIGURE 7.4   A fossil of a *Glossopteris* leaf from Permian-aged deposits in New South Wales, Australia. Fossils of this land plant have been found in South America, Africa, India, Australia, and Antarctica and provide evidence that all of these landmasses were joined as one continent in the Permian (*Source:* James St. John / Wikipedia Commons / CC BY 2.0.).

geological establishment. In 1926 a symposium on continental drift was sponsored by the American Association of Petroleum Geologists (AAPG), and a number of prominent American and British geologists attacked the theory. Wegener continued to have some supporters, such as the South African geologist Alexander du Toit, who coined the terms *Laurasia* and *Gondwana* for the northern and southern continental masses. In many ways, the AAPG symposium led to the almost complete disappearance of Wegener's theory for 25 years. For Wegener himself, the shunning of his theory would not matter for long. He died on the Greenland Ice Cap during an expedition in 1930. His body remains there still, entombed within the glacial ice.

Despite the widespread dismissal of Wegener and his theory, from the 1940s onward, new evidence began to accumulate that would lead to the vindication of continental drift. Some of the evidence simply supported the contention that the continents must have been in closer proximity in the past. For example, in the 1960s and 1970s, fossils of an early Triassic reptile called *Lystrosaurus* were unearthed in Antarctica. Fossils of this genus are also known from Africa, India, China, Mongolia, and Russia. The finds showed that Antarctica must have been considerably warmer in the past and suggest a closer connection to other continents. At the same time, biogeographers such as P. J. Darlington were showing that the modern distributions of some genera and families of plants and animals could best be explained by a closer connection between Africa, South America, Australia, and New Zealand in the past. Such information seemed to indicate that, despite his factual errors in geology and paleontology, Wegener was on the right track. More important perhaps, geophysical evidence from the oceans and continents was helping to solve the riddles of the history, rates, and causes of continental movement.

Much progress in oceanographic geophysics in the twentieth century was initiated by submarine warfare in World War II. Conditions and features beneath the sea became of great strategic interest. Work accelerated on devices such as sonar, which allowed the detection of enemy submarines but could also be used to map the sea floor. For example, the marine geologist H. H. Hess sailed on American troopships across the Atlantic and used an echo sounder to chart the sea floor. Once maps of the ocean floor were completed, it was clear that the fit between the submerged continental margins of areas such as South America and Africa was even better than the fit that was seen when the modern coastlines were compared. Over the 1940s and 1950s, it was discovered that a global system of midoceanic ridges bisected the South Pacific, Atlantic, and Indian oceans (Fig. 7.5). These ridges are centers of volcanic activity and the emergence of lava. Oceanic trenches, several kilometers deep, are found near the margins of continents. Volcanic activity and earthquakes are common near the oceanic trenches. The famous Ring of Fire, the zone of intense volcanic and earthquake activity along the edges of the Pacific Ocean, is associated with oceanic trenches around the Pacific. In addition, flat-topped mountains called guyots lie beneath the sea in some regions, often being found at greater depths toward the continental margins. Deep-sea cores revealed that the oceans were indeed underlain by basalt, and most interestingly, radiometric dating showed that this basalt was generally less than about 150 million years old. This is quite young compared to continental rocks, which can date back billions of years. In addition, basalts near the midoceanic ridges are generally younger than those found in the oceanic trenches.

Paleomagnetism is an important geophysical technique that was developed in the postwar period and has been applied to both marine and terrestrial geology. When igneous rocks that contain elements such as iron or titanium cool, the earth's magnetic field causes mineral grains to become oriented to the magnetic field. Once the rocks have solidified, this orientation is locked in. This remnant magnetization of the rocks can be measured in the laboratory. An igneous rock that is formed by a volcanic eruption today will have remnant magnetization very much like a compass needle, with the minerals preferentially oriented along an axis toward the north magnetic pole. Geophysicists can collect a sample of ancient igneous rock, carefully record its modern position and orientation, and then measure its remnant magnetic properties. If the magnetic orientation of the rock is different from the modern compass orientation, they know that there has been movement of the land surface relative to the earth's magnetic poles. By careful geometric analysis and radiometric dating of the igneous rocks, the past orientation of the landmass can be reconstructed over millions of years. These studies have shown huge changes in the latitude and longitude of continental regions over the past billion years. For example, paleomagnetic studies show that some 300 million years ago the continent of North America was located at the equator.

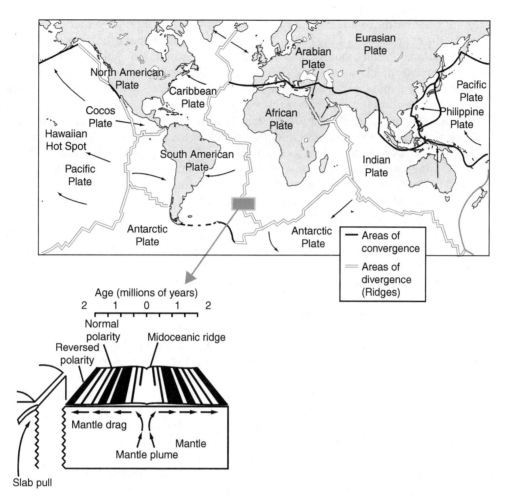

FIGURE 7.5 The earth's plates, zones of spreading and collision (adapted from Birkland and Larson, 1989), and the Hawaiian Islands associated with the Pacific hot spot. The insert shows a cross-section of the earth at the mid-Atlantic Ridge. The symmetrical magnetic banding found on both sides of the ridge is indicated, as is the general paleomagnetic stratigraphy for the past 160 million years. The factors contributing to plate movement are noted.

An important feature of the earth's magnetic field is its capacity to change polarity. If the magnetic field were to change polarity today, we would find all our compasses pointing south instead of north. The cause of such shifts is unclear, but they occur over periods of hundreds of thousands to millions of years. Today we live in a geomagnetic period called the Bruhnes Normal Chron. This magnetic period began about 780,000 years ago. The magnetic field of the earth was reversed most of the time between 2.58 million years ago and 780,000 years ago, and this period is called the Matuyama Reversed Chron. Paleomagnetic studies of ocean basalts reveal that these rocks are banded parallel to the midoceanic ridges and that the paleomagnetic banding on each side of the ridge is a mirror image of the other side (Fig. 7.5). The younger rocks near the ridges formed during the Bruhnes Normal Chron; older rocks away from the ridges formed during earlier magnetic chrons. The increasing age of the rocks away from the ridges is symmetrical on both sides of the ridge.

Taken together, the geophysical data collected from the continents and the oceans lead to four important conclusions. First, the continents are less dense than the rock underlying the oceans. Second, paleomagnetic evidence shows that the continents have indeed moved very significant distances during geologic time. Third, the oceans are much younger than the continents. Fourth, the age of the oceanic basalts increases symmetrically outward from the midoceanic ridges. These observations support Wegener's contentions that the continents had indeed moved, and they also provide some evidence of the mechanism that drives such movements.

The modern theory of how continental drift occurs is referred to as **plate tectonics.** Geophysicists now believe that the movement of the continents is ultimately driven by the dynamics of the inner earth. To understand the forces behind continental movement, we need to know a little about the composition of the earth. At the center of our planet is a dense inner core surrounded by a 1700-kilometer-thick outer core of molten liquid iron and nickel. This in turn is surrounded by a 2900-kilometer-thick mantle of extremely hot silica, oxygen, iron, magnesium, and other minerals. Above this is a thin layer of cool, solid rock that is 5 kilometers thick under the oceans and 60 kilometers thick beneath the continents. This outer layer is called the crust. The upper 50 to 125 kilometers of rock in the mantle and crust is relatively hard and brittle rock and is referred to as the lithosphere. Beneath the lithosphere, the hot and plastic rock of the mantle is called the asthenosphere. Volcanic activity occurs when the molten rock of the asthenosphere penetrates the crust. The plastic conditions of the asthenosphere extend some 300 kilometers into the mantle. The remainder of the mantle is called the mesosphere. The continents are less dense than the mantle and therefore float on top of the lithosphere and asthenosphere. The fluid nature of the underlying asthenosphere allows the plates to move across the surface of the earth. Several forces appear to propel such movements.

Part of the force causing the continents to move comes from ridge push. This occurs when convective flow in the mantle causes molten rock to come to the surface in areas such as the midoceanic ridge zones. The magma cools and forms new solid basalt. This new material pushes older basalt outward away from the ridge zone (Fig. 7.5). Hence, the basalt near the midoceanic ridges is younger than the basalt near the continental margins, and symmetrical paleomagnetic banding occurs parallel to the ridges. Ridge push also explains why guyots farther from the midoceanic ridges are older and more eroded than those close to the ridges. Over millions of years, this process causes sea floors to expand, separating continental plates on either side of the midoceanic ridge. Lateral convective flow of the mantle occurs below the crust, and friction between the crust and the flowing mantle also causes continental plates to move. This process, somewhat like a conveyor belt, is referred to as mantle drag. Spreading due to ridge push and mantle drag is not always confined to midoceanic ridges. Rift zones, such as the Red Sea Basin, are areas where spreading has occurred in continental areas. The Atlantic Ocean began as a rift zone in the ancient supercontinent of Pangaea. Along the continental margins, the lower density of the continental rocks allows the landmasses to ride over the expanding oceanic crust. Such areas are called subduction zones because the oceanic basalts are subducted beneath the continents and remelted. The deep oceanic trenches are formed in the subduction zones. The process of subduction leads to much friction and fracturing of the crust. This causes earthquakes and volcanic activity in the vicinity of the subduction zone and explains why the Ring of Fire exists around the Pacific Ocean. The process of subduction also helps pull the remaining oceanic crust toward the subduction zone. This process is called slab pull.

Today we can identify 16 major tectonic plates that are bordered by spreading zones, subduction zones, transform zones, and collision zones. Transform zones and collision zones are areas of convergence where two plates move laterally against each other or collide (Fig. 7.5). Volcanic activity and earthquakes are common near spreading zones and subduction zones. Earthquakes also occur near strike-slip fault zones. The San Andreas Fault of California represents the strike-slip zone between the Pacific and North American plates. This region is well known for earthquakes, including the 1906 quake that destroyed much of San Francisco. At the San Andreas Fault, the Pacific Plate is moving northward relative to the North American Plate. Deformation of rock strata by movement along the plate and volcanism is responsible for the development of mountain ranges from the west coast to the Rocky Mountains. One of the most dramatic collision zones occurs where the Indian Plate contacts the Eurasian Plate. This collision has produced the Himalayan Mountains, the tallest mountain range in the world. These major geologic disturbances that result in mountain ranges are called **orogenies.** Thus, plate tectonics not only explains the past movement of continents, but also helps us understand phenomena such as volcanic activity, earthquakes, and some episodes of mountain building.

With radiometric dating of rocks, paleomagnetism, and an understanding of plate tectonic mechanisms, it is possible to reconstruct the relative positions of the continents over the past 400 million years. The plate tectonic history of the past 250 million years is crucial (Fig. 7.6) to an understanding of the modern physical geography and biogeography of the earth. As we will see in later chapters, many aspects of the distributions of plant and animal families, genera, and species are explainable in light of past continental movements.

During much of the early to mid-Permian (286 million to 250 million years ago), Gondwanaland, made up of the plates that comprise modern Antarctica, Australia, India, South America, and Africa,

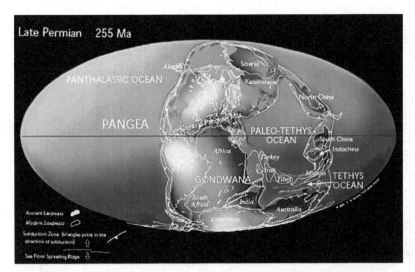

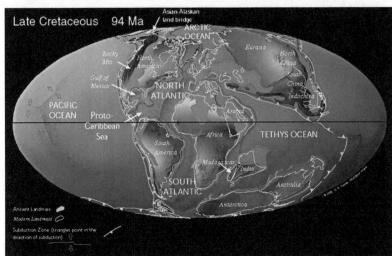

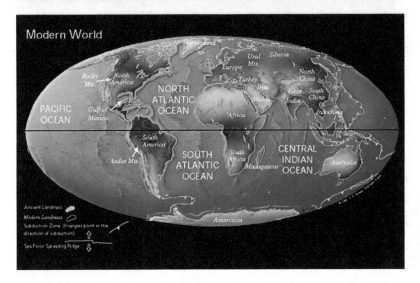

FIGURE 7.6 The movement of the major plates over the past 250 million years and 250 million years into the future (*Source:* Scotese and Baker, 1975; Scotese, 2021, from PALEOMAP website http://www.scotese.com).

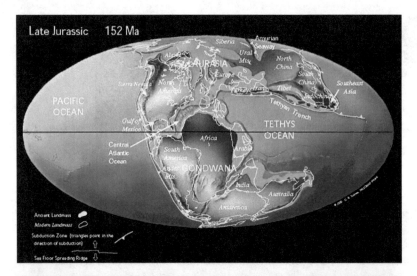

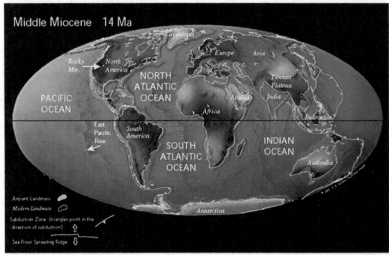

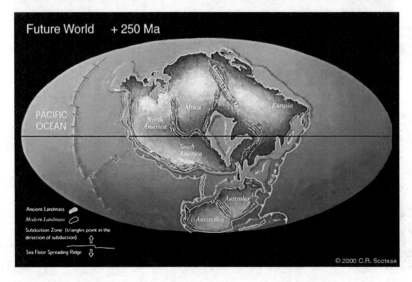

FIGURE 7.6   (Continued)

was located in the southern hemisphere. Portions of Gondwanaland nearest the South Pole experienced intensive glaciation during this time. The Laurasian landmasses, consisting of the North American and Eurasian plates, were located in the northern hemisphere and equatorial region. The plate that makes up modern North America had collided and joined with the African Plate about 350 million years ago, linking Gondwanaland and Laurasia. By the early Jurassic, the northward movement of Gondwanaland had led to the end of glaciation and more complete fusion with Laurasia. In addition, the Asian portions of Laurasia had fused together and joined with the European and North American plates. In this manner, all of the continental masses converged and become the supercontinent of Pangaea by 195 million years ago. The supercontinent of Pangaea stretched from near the South Pole to the North Pole and was surrounded by one large continuous ocean called Panthalassa.

Geologically speaking, the supercontinent of Pangaea was short-lived and began to split apart 180 million years ago. A rift that would become the midoceanic ridge of the Atlantic opened between the North American and European plates during the Jurassic. By the late Cretaceous, Asia and North America became linked together near present-day Alaska and Siberia. Rifting began to separate Africa and South America by 180 million years ago. By 150 million years ago, in the late Jurassic, continued sea-floor spreading severed the land connection between the North American Plate and Gondwanaland. During the late Jurassic, the subduction of the Pacific Plate by the North American Plate initiated the Cordilleran Orogeny. Mountains such as the Coast Ranges, Sierra Nevada, and Rocky Mountains of western North America were formed by the Cordilleran Orogeny.

By the late Cretaceous, some 90 to 100 million years ago, a seaway separated North America from the South American and African plates. North America itself was divided into western and eastern landmasses by a shallow sea. Rifting and subsequent sea-floor spreading continued to drive Africa and South America apart. In addition, India and Australia-Antarctica were now separating from each other and the rest of Gondwanaland. By about 90 million years ago, not only had Pangaea ceased to exist, but Laurasia and Gondwanaland were splitting apart.

During the Paleogene and Neogene (Tertiary), between approximately 65 million and 2 million to 3 million years ago, the modern configuration of our oceans and continents came into being. Northeastward movement of Africa and India brought Africa in contact with the European Plate and India into contact with the Asian Plate by the Miocene Epoch 14 million years ago. During the same period, Australia moved northeastward and separated from Antarctica. The northward movement of South America and formation of land by volcanic activity led to the creation of the Central American Isthmus and the joining of North America and South America by a land bridge 3.5 million years ago. By 2 million to 3 million years ago, the configuration of the earth's continents and oceans was essentially the same as it is today.

Prior to leaving the topic of crustal movement and plate tectonics, we might consider the special geological history of the Hawaiian Islands. The islands are the most isolated in the world and have never been part of any continental plate. They have a core of oceanic basaltic rock. The island chain extends from Hawaii northwestward to include Midway Island. One predominant theory is that the islands were created as the Pacific Plate moved over a volcanic hot spot extending into the mantle of the earth. Over the past 80 million years, this hot spot has produced volcanic activity that leads to the buildup of extruded magma. The magma eventually rises above sea level and forms islands. As crustal movement occurs at a rate of approximately 4 centimeters per century in this region, the islands formed by the hot spot gradually move northwestward and cease to experience volcanic activity and growth. The big island of Hawaii lies near the hot spot and is increasing in area due to active volcanism. Islands to the west, such as Oahu and Kauai, are no longer near the hot spot and are decreasing in surface area through erosion and subsidence. Northwestward, beyond Midway Island, lies a chain of guyots and seamounts called the Emperor Seamounts. These guyots were once islands much like the current Hawaiian chain. Cores of the guyots and seamounts show that they once supported terrestrial environments and shallow-sea reefs. Erosion and crustal subsidence have caused these former islands to sink hundreds of meters beneath the sea. Plate movement has carried them northwestward thousands of kilometers. There is a sharp bend in the line of islands and seamounts that suggests a shift in the direction of Pacific Plate motion about 43 million to 45 million years ago. The oldest seamounts are about 80 million years old and the oldest existing island in the chain is Kure Atoll in the west, which formed about 30 million years ago. The island of Hawaii at the eastern end formed less than a million years ago. Western islands such as Kauai formed about 5 million years ago. The Hawaiian Islands are both a creation of plate tectonics and striking evidence of the earth's crust in motion. Eventually, all of the current Hawaiian Islands will move westward and disappear beneath the waves. This process will take several million years. East of Hawaii, volcanic activity is already forming a new island, called Loihi, which will rise above sea level at some point in the future.

QUATERNARY CLIMATIC CHANGE

We live in the Quaternary Period of the geologic timescale and a key feature of the Quaternary has been the repeated alternation between cold conditions, when large areas on northern North America and Europe were covered by think glacial ice, and interceding warm periods with conditions similar to today. Today almost everyone knows that there have been past "ice ages" on earth. However, recognition of these past **glaciations** occurred only in the nineteenth century. The concept that glacial ice cover had once been much greater than it is today was developed by the Swiss-American scientist Louis Agassiz. While working as a young researcher in the Alps, he noticed that new glacial deposits being formed there by existing alpine glaciers were similar to older rock deposits found far away from the glaciers in low-lying areas around the periphery of the mountains. Agassiz concluded that these older deposits could only be explained by the theory that the alpine glaciers were once much larger than they are today and extended into the lowlands. In 1846 he came to North America and was appointed to Harvard University, which became his permanent home. Uniformitarianist geologists accepted Agassiz's theory because it explained the occurrence of glacial deposits in areas that are too warm to support glaciers today. However, others greeted the glacial hypothesis with skepticism. Some scientists subscribed to the theory of diluvialism and believed that the tills and other glacial features that Agassiz identified were really the result of massive water erosion and deposition caused by the biblical flood of Noah. In the end, Agassiz was correct about the existence of past ice ages. It has been argued that he not only changed the sciences of geology and geomorphology, but helped change the course of American science in general.

LOUIS AGASSIZ: A CONTRADICTORY AND TROUBLING LEGACY

Louis Agassiz was born in the canton of Fribourg, Switzerland, on May 28, 1807. He attended the universities of Zürich, Heidelberg, and Munich and obtained both a doctorate of philosophy and a doctorate of medicine. However, his real love was natural history. He moved to Paris and was mentored there by the great geographer Alexander von Humboldt. In Paris he also began to focus on ichthyology, which would remain one of his major scientific interests. In 1832 he was appointed professor of natural history at the University of Neuchâtel in Switzerland. In 1837 he publicly presented his groundbreaking idea that the earth had experienced geographically extensive ice ages in the recent geologic past and published his two-volume treatise, *Études sur les glaciers,* in 1840. After coming to the United States and taking up his position at Harvard in the 1840s, he became both a major scientific figure and a public intellectual until his death on December 14, 1873, in Cambridge, Massachusetts.

Although Agassiz's contribution in developing the theory of past glaciation and in overcoming religious objections to his theory is widely admired, his legacy is also contradictory and deeply marred. Agassiz adhered to a religious belief in polygenesism, by which he maintained that all plants and animals, and different races of humans, were created separately by God and for distinct geographic provinces. He vigorously opposed the new ideas of Charles Darwin and his adherents regarding evolution. He argued that such beliefs reduced the importance of God and must have felt that they also blurred the lines between different species and the different human races. It is ironic that someone whose work and energy overturned the biblical explanations invoking Noah's Flood to account for widespread glacial deposits would be so adamant in his disdain for similarly iconoclastic and mechanistic explanations for the diversity of life.

Much more troubling than Agassiz's scientific contradictions regarding evolution were his views on race. Although he appears to have been no supporter of slavery, he expressed strong opinions regarding the racial inferiority of Black slaves after he came to the United States and met African Americans for the first time. Agassiz did not accept that all humans belonged to the same polygenic province or species and he argued vehemently in public for the inferiority of non-white human groups. In this regard, he became a scientific "leader" in offering rationales for racist theories and for racists. This dark aspect of Agassiz has led to important reappraisals of his legacy and has resulted in actions such as the removal of his name from an elementary school near Harvard University.

Subsequent geological investigations in northern Europe and North America confirmed the presence of glacial deposits and led to the development of a fourfold model of glaciation. According to this model, there had been four major glaciations in the Quaternary. Each time, glacial ice cover extended over all of Scandinavia, most of the British Isles, and the Alps. Almost all of Canada, portions of the Great Lakes region, and New England were covered by ice, and the glaciation extended as far south as portions of Kansas in the plains. The model also postulated that these glaciations were short-lived and that the interglacial climate of the Quaternary was similar to today's climate. This model remained the accepted view on glaciation until the early 1970s. We now know that both temporal and spatial patterns of glaciation were far more complex than the fourfold model.

As the twentieth century progressed, new geological and paleontological evidence, along with new techniques to dating deposits, permitted a more complete picture of the extent of glaciation and its impact on the earth. In the late 1940s, the American physicist Willard Libby developed the radiocarbon technique for dating organic remains. Radiocarbon dating is based on measuring the amount of the radioactive isotope carbon 14 (^{14}C) found in plant and animal remains. Isotopes are forms of an element that have the same atomic number but differ in their number of neutrons. Carbon has three isotopes: carbon 12 (^{12}C) and carbon 13 (^{13}C), which are stable isotopes, and ^{14}C, which is a radioactive isotope that decays over time to nitrogen. ^{14}C is formed naturally in the atmosphere when cosmic rays bombard nitrogen. Plants acquire ^{14}C during normal photosynthesis and accumulate it in their tissue. When animals eat plants, they accumulate ^{14}C. When a plant or animal dies, it ceases to accumulate ^{14}C. The isotope decays to nitrogen at a rate of one half-life, which equals 5730 years. By measuring the amount of ^{14}C in a plant or animal remains, we can estimate the length of time that elapsed since the organism died. Radiocarbon dating of plant and animal remains provides good chronological control over events of the past 40,000 years. Samples older than 40,000 to 50,000 years in age typically have too little ^{14}C remaining to provide reliable age estimates. Thanks to abundant geological, geochemical, and paleontological evidence, coupled with many radiocarbon dates and other forms of absolute chronological control, we have a very good understanding of conditions during the last glacial maximum (LGM), during which global glacial ice reached its maximum extent about 20,000 years ago. We can consider these conditions to be a very general approximation of what the world was like during major glaciations earlier in the Quaternary, although each warm and cold period also likely experienced some unique conditions. In the classical fourfold model of glaciation, each of the four supposed past glacial intervals was given a name. The last glacial maximum is equivalent to the Wisconsin Glaciation in North America, the Devensian Glaciation in Britain, and the Würm Glaciation in central and western Europe. Although these names sometimes appear in the literature, as we shall see, the simple fourfold model of glaciation was wrong.

One of the most detailed records of climatic conditions during the last glaciation and some of the older glacial episodes comes from ice cores taken in Greenland and Antarctica. These cores extend from the surface of the modern ice caps down to depths of over 3000 meters. The upper portions of the cores contain annual layers of winter and summer snow accumulation. Like tree-rings, these annual layers can be counted to provide chronological control. Lower portions of the cores must be dated using the occurrence of volcanic ash layers, which have been radiometrically dated elsewhere, and correlations with other paleoclimatic evidence. The Greenland ice cores provide records dating back over 130,000 years, while Antarctic ice cores date as far back as 800,000 years ago. Samples of ice ranging as far back as 2.7 million years have recently been found in shallow deposits in the Allan Hills Blue Ice Area of Antarctica. There the bedrock topography and strong winds lead to the exposure of old ice that was once deep beneath the surface. Important records of past atmospheric conditions can be captured from the ice cores. The fresh snow layers at the top of the ice contain much air space. Over time, as more snow accumulates, the older snow is compressed to firn (ice crystals) and finally clear ice. During this process, the air is compressed into bubbles trapped in the ice. Geochemists and geophysicists can sample these tiny air bubbles and extract the trapped atmospheric gases. Analysis of the gases provides important clues about past atmosphere, climate, and the age at the ice cores.

One important constituent of the ice core records is the relative abundance of the ^{18}O and ^{16}O isotopes. The isotope ^{18}O is chemically similar to ^{16}O but is heavier due to the additional two neutrons. Variations in the relative amount of the isotopes ^{18}O and ^{16}O in ice cores appear to be related to the temperature of the atmosphere at the time of precipitation. The heavy isotope ^{18}O condenses and

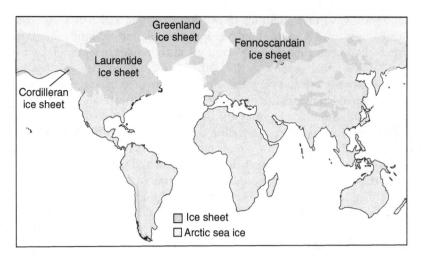

FIGURE 7.7 The location of major ice sheets and major areas of continental shelf exposed due to lower sea levels (eustatic sea-level changes) during the last glacial maximum (*Source:* Roberts, 2014 / Scientific Research Publishing / CC by 4.0.).

precipitates more readily at high temperatures than does the light isotope ^{16}O. The relative abundance of ^{18}O increases with increasing temperature during condensation and precipitation. Studying changes in the relative amounts of ^{18}O and ^{16}O in the ice cores allows scientists to estimate changes in temperature at the high latitudes. Such estimates are generally consistent from Greenland and Antarctic ice cores for the past 100,000 to 140,000 years. The temperature records from the ice cores suggest that between 110,000 and 140,000 years ago, the earth was as warm, or slightly warmer, as it is today. By 100,000 years ago, high-latitude temperatures had declined and were variable but remained somewhat depressed until about 12,000 years ago. The last glacial maximum at ~20,000 years ago coincides with a period of particularly low temperatures.

Globally, climate during the last glacial maximum was much cooler than present. In ice-covered areas such as Canada and Scandinavia, summer and winter temperatures were as much as 20° C to 28° C colder than today. Even in equatorial regions, the difference between glacial and modern temperatures was generally 2° C to less than 5° C. In many equatorial regions, precipitation was significantly less than today. In other areas, such as the Great Basin of the United States, evaporation was lower and precipitation was higher than today. Such prolonged periods of wetter conditions are called **pluvials.** Ice sheets that were 2 kilometers to 4 kilometers thick were centered over Scandinavia and Canada (Fig. 7.7). These sheets extended far south into northern Europe and the northern United States. A person standing at the location of present-day Winnipeg, Canada, or Stockholm, Sweden, 20,000 years ago would have experienced a climate and a view very similar to what someone standing in the middle of the Greenland Ice Cap would experience today. Interestingly, much of Alaska and Siberia were free of ice except in mountainous regions. It is likely that the climate was cold in these regions, but it was too dry to develop ice sheets. Montane ice caps developed over alpine regions, such as the Rocky Mountains, Alps, Andes, and Southern Alps of New Zealand, in both the northern and southern hemispheres. Alpine glacier complexes developed in mountains such as the Sierra Nevada of California and the high peaks of east Africa and New Guinea.

The growth of glacial ice had a profound impact on regional and global sea levels. At the regional level, the weight of the ice depressed the land surface beneath it and caused the sea to inundate adjacent coastlines. However, in many areas, the ice sheets also caused more distant land surfaces to be displaced upward. This caused land to rise above sea level. Such changes in sea level, caused by the depression and distortion of the earth's crust by glaciation, are called **isostatic sea-level changes.** Some regions, such as the Hudson Bay of Canada are still experiencing upward rebound following the end of the last glacial maximum. In addition, the growth of the ice sheets required great amounts of water. The growth of glaciers occurs when snow does not melt over the summer and is buried by snowfall during the following year. Over many years, the overlying snow builds up, and its weight causes compaction and deformation of the snow crystals until solid glacial ice results. Water evaporated from oceans, lakes, and rivers and soil fell as snow on the ice sheets and remained frozen there.

This water, therefore, was tied up in the growing ice and did not return to the sea as runoff. As the ice sheets grew, more and more of the world's water was tied up in them. The result was a drop in the level of the world's oceans by some 80 to greater than 130 meters during glacial periods. Because all the oceans are connected, this drop in sea level was the same everywhere. Such global changes in the level of the oceans are called **eustatic sea-level changes.** During the last glacial maximum, sea level likely dropped by around 134 meters.

Eustatic sea-level changes during glacial periods caused some very important changes in the geography of the earth (Fig. 7.7). The drop in sea level caused the Bering Strait between Alaska and Siberia to disappear and produced a large land bridge in **Beringia** that directly connected Asia and North America. Large islands such as Sumatra, Java, and Borneo in Southeast Asia became linked with the Asian mainland. The islands of New Guinea and Tasmania were linked with Australia. Many other islands, peninsulas, and isthmuses were greatly expanded in size. The linkages between land areas that form when sea level drops allow land species to move across them and can have important biogeographic implications.

In addition to drops in sea level, glaciation caused other significant hydrological changes. In the Great Basin region of the United States, the glacial climate was wetter and colder than the modern climate. The increased precipitation and decreased evaporation caused large lake systems to develop in what is now an extremely arid desert environment. Such water bodies, which appear with large-scale variations in climate, are called **pluvial lakes.** The modern Great Salt Lake of Utah is a small remnant of a much larger pluvial lake called Lake Bonneville. This lake stretched from modern-day Salt Lake City to the Utah-Nevada border. The famous Bonneville Salt Flats were formed by the evaporation of this once gigantic lake. Another pluvial lake even occupied portions of Death Valley, which is a region of extreme desert today. The relict shorelines and sedimentary deposits from these lakes are found throughout the Great Basin region of the United States.

The altered climate and geography of glacial periods produced profound changes in the geographic distributions of the world's vegetation. Studies of fossil pollen, also known as palynology, have been widely used to reconstruct past Quaternary vegetation (Technical Box 7.1). One recent study was able to combine the fossil pollen records from almost 600 sites to examine global changes in vegetation composition and structure from the last glacial to the present. Studies of fossil pollen, along with analysis of fossil wood, needles, and seeds, have allowed biogeographers to reconstruct many aspects of the earth's vegetation during the last glacial maximum. Pollen and other fossil plant remains are often recovered from lake sediments, peats, flood deposits, and wind-blown sediments.

TECHNICAL BOX 7.1

Palynology (Fossil Pollen and Spore Analysis)

Conifers and flowering plants produce pollen grains that contain male gametophytes and protect them during transfer to receptive ovules. Lower plants such as ferns and mosses produce spores that also aid in the dispersal of gametophytes. Pollen grains and spores are very small, generally ranging in size from 10 microns to 150 microns. This is about the size of fine silt. Pollen grains are transferred from flower to flower by many agents, including insects, birds, mammals, and water. However, many plants rely on wind to disperse pollen grains and spores. As wind dispersal is a relatively random process, most pollen grains will not reach a receptive ovule of the same plant species and will not fulfill their reproductive function. Thus, flowering plants and gymnosperms that use wind pollination must produce many millions of grains in order to ensure success in sexual reproduction with other plants of the same species. For example, in one year a single lodgepole pine *(Pinus contorta)* tree may produce over 25 billion pollen grains.

Many wind-dispersed pollen grains are deposited on soil and washed into lakes or the ocean. Many others are directly deposited into lakes, oceans, or peat bogs. Although the living cytoplasm in the interior of such grains quickly dies and decays, the external wall of the grain, called the exine, is constructed of a relatively inert polymer called sporopollenin. The sporopollenin is very resistant to decay, particularly in anaerobic environments such as lake bottoms or within fast-growing peat moss. Within such environments, pollen grains and spores can be fossilized and preserved indefinitely. Spores that are more than 500 million years old have been recovered from sedimentary rocks.

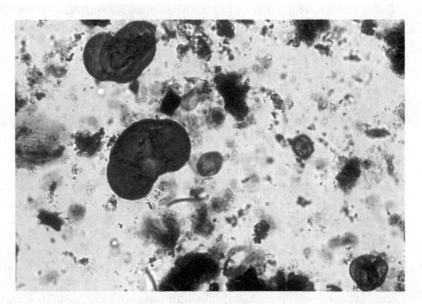

Fossil pollen grains recovered from the sediments of a small lake in the modern Arctic tundra zone of northern Canada. The pollen grains are approximately 5000 years old. The two large kidney-shaped pollen grains are from spruce trees *(Picea)*. Each spruce grain is about 100 microns long. The high numbers of spruce pollen grains in sediments of this age from the lake provide evidence that spruce trees once grew near the site. Biogeographers often use fossil pollen analysis to reconstruct past vegetation.

Most biogeographers work with fossil pollen from the Quaternary. They generally obtain sediment samples for pollen analysis from lakes, peatlands, and the ocean. Treatments of acids and bases are used to digest other organic and inorganic material in the sediment samples and recover the fossil pollen and spores. After the pollen grains and spores are recovered from the sediment samples, they are identified and counted using transmitted light microscopy at 400X to 1000X magnification. Pollen samples also contain other material such as fossil charcoal, which can be used to reconstruct past fire regimes, and plant remains such as stomata or algal fragments. Most Quaternary pollen grains and spores can be identified as to the genus or species of plant they belong to. Identifications are based on the shape, size, exine wall structure, openings in the exine (called apertures), and sculpturing on the exine.

The science of palynology is now over 100 years old, having been pioneered by the Swedish scientist Ernst Jakob Lennart von Post. By conducting many studies of the relationship between modern vegetation and the pollen and spores deposited in present-day lakes and peatlands, palynologists have demonstrated that a good relationship exists between the vegetation surrounding a site and the pollen and spores that are deposited. Since vegetation is often controlled by climatic conditions, good relationships have also been found between pollen deposition and climate. Palynology is used widely to reconstruct both past vegetation and climate for the Quaternary. Thousands of pollen records, extending for centuries to tens of thousands of years, have been recovered and provide crucial insights into Quaternary environmental history.

One of the most amazing repositories of larger plant remains is the middens of packrats (*Neotoma* spp.). The middens are collections of plant fragments that have been gathered by the packrats and accumulated in small rocky hollows. The urine of the packrats cements the plant remains together and protects them from deterioration. Many packrat middens dating from the last glacial maximum have been found in the desert regions of western North America. Thus far there are thousands of radiocarbon dates for plant fossils from sites across the arid western regions of Mexico, the United States, and Canada (Fig. 7.8).

Evidence from fossil pollen records, packrat middens, and other sources allows for the reconstruction of the vegetation of the last glacial maximum in a number of regions. During the glacial maximum, much of the area occupied by Arctic and alpine tundra today was covered by glacial ice. In turn, tundra and steppe vegetation replaced the coniferous boreal forest in Siberia and Alaska, as well as the boreal and deciduous forests of Europe. Imagine the city of Paris, France, surrounded not by forests and fertile fields, but by tundra and cold steppe. In North America, boreal forest replaced the

Packrat Midden Sample Localities

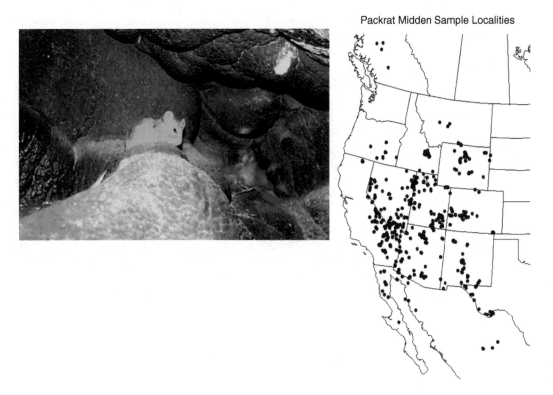

FIGURE 7.8   A packrat (*Neotoma* sp.) surrounded by urine-cemented middens of plant remains *(left)* and packrat midden sample sites from western North America *(right)* (*Source:* USGS / https://geochange.er.usgs.gov/midden / last accessed December 26, 2023).

grasslands on much of the Great Plains. In western North America, pine trees expanded along coastal regions of southern California and across inland areas that are desert or grassland today. Northern temperate rainforest trees such as sitka spruce *(Picea stichensis)* grew along the coast south of San Francisco. The tropics were much drier than today, and rainforest vegetation was often fragmented into patches and restricted to areas along rivers and other favorable localities. Tropical and subtropical savanna and desert expanded. Florida, for example, was dominated by sand dunes and dry scrub vegetation. Mountain vegetation in zones in the tropics and higher latitudes were depressed a thousand meters or more. In the Andes, vegetation zones may have dropped in altitude by close to 2000 meters. The Andean treeline occurred at about 2000 meters in elevation during the last glacial maximum. Today the treeline is found at 4000 meters in elevation. Packrat midden records show that the southwestern United States and adjacent Mexico were much cooler and moister than present during the last glacial maximum. Packrat middens provide evidence that coniferous trees grew in large areas of the modern Great Basin, Mojave, Sonoran, and Chihuahuan deserts.

Some vegetation types, such as the boreal forest of Eurasia, virtually disappeared. The Eurasian boreal trees grew as small populations in local areas of favorable habitat. The woodlands of the Mediterranean vegetation in Europe also became restricted to small fragments in protected locations. The deciduous forest of Europe was similarly reduced to small fragments. In other vegetation formations, significant changes took place in the relative dominance of plant species. In the eastern deciduous forest of North America, many tree species that are common today became rare. The fragmentation and rearrangement of major vegetation suggest that some of the broad regional climatic regimes that existed during the glacial maximum have no counterpart in the modern climates of the world. The glacial climates in many parts of the world consisted of combinations of seasonal temperatures and precipitation that do not occur today.

In addition to the changes in temperature and precipitation, evidence from Greenland and Antarctic ice cores tells us that CO_2 (carbon dioxide) levels in the atmosphere during the last glacial maximum were only 180 to 200 parts per million. This is some 80 to 100 parts per million less than average Holocene concentrations of around 280 parts per million. Decreased CO_2 in the atmosphere can result in decreased rates of photosynthesis. Under low amounts of atmospheric CO_2, many plants produce greater numbers of stomata. The increased stomata may have allowed the plants to exchange gases more freely with the atmosphere. However, the increase in stomata also raises rates of transpiration and increases the plant's sensitivity to water stress. Consequently, some plant species were more sensitive to water stress and drought during the last glacial maximum than they are today. It has been found that C_3 plants can experience up to 40% to 70% declines in rates of photosynthesis and production of biomass under low CO_2 conditions. Rates of photorespiration can also increase. Some plant species are more sensitive to these CO_2 changes than others. For example, experimental studies and paleoecological observations suggest that the California redwood *(Sequoia sempervirens)* to be very sensitive to CO_2 concentrations of less than or equal to 200 parts per million, similar to the low levels that occurred during the last glacial maximum. The end result of these differences between glacial and nonglacial climate and atmospheric CO_2 is that communities of plant species existed during glaciations that are not present today, and some modern vegetation types were absent or greatly reduced in area.

During the last glacial maximum, it was not only vegetation communities that were often greatly different from modern ones. Paleontologists have found evidence that groups of animals that are not found living together today actually formed communities during the glacial maximum. For example, species of collared lemmings, shrews, and ground squirrels that today inhabit very different portions of the North American continent were all living together in Pennsylvania during the last glacial maximum (Fig. 7.9).

FIGURE 7.9 The modern distributions of eastern shrew *(Sorex fumens),* eastern collared lemming *(Dicrostonyx hudsonius),* prairie ground squirrel *(Spermophilus tridecemlineatus),* and western collared lemming *(Dicrostonyx torquatus),* and a site in Pennsylvania where fossil evidence indicates that all four species coexisted during the last glacial maximum, although they clearly do not live together today *(Source:* Graham, 1986; Graham et al., 1996; Brown and Lomolino, 1998).

Identifying the cause of the ice ages has long been a topic of great interest and debate among scientists. The question of what caused the Quaternary ice ages can be divided into two main issues. First, there is no evidence for extensive glaciation or alternating glacial and nonglacial conditions during the rest of the Cenozoic Era. So why did large-scale glaciation begin in the Quaternary? Second, what caused the Quaternary climate to shift from glacial to nonglacial conditions? Let's explore each of these issues in turn.

Some of the most important evidence for the first onset of widespread glaciation comes from marine sediment cores. Some midoceanic cores from the high latitudes have small pebbles in them These start to appear in sediments that date to between 2 million and 3 million years ago. These pebbles are called drop stones and could only have been transported to the midocean by icebergs calving off of glaciers. Geologic, paleontological, and geophysical evidence all suggest that areas such as Greenland and Antarctica have been ice-covered throughout the Quaternary. In contrast, for much of the earlier Cenozoic these areas were free of ice. Fossil evidence shows that 40 million to 50 million years ago (Eocene), areas such as Greenland and the Canadian Arctic Islands supported swamp and deciduous forest. For example, the stumps, needles, and seeds of the Tertiary tree *Metasequoia occidentalis,* a relative of the redwood, can be found on Axel Heiberg Island and in Arctic Canada. Other fossil trees found in Eocene-aged deposits of the High Arctic include maidenhair tree *(Gingko)* and cypress *(Glyptostrobus)*. Mean annual temperatures in the Arctic may have been 30° C higher than today. During the same time, tropical rainforest and dry subtropical forest may have covered much of North America, southern Europe, and Siberia. By 25 million years ago, cooler temperate deciduous and boreal coniferous forests had spread across northern North America, Europe, and Siberia. By 3 million years ago, tundra was developing in the Arctic.

Clearly, the earth became cooler during the transition to the Quaternary. Although much of the story remains to be resolved, this cooling of the earth during the Late Cenozoic and the initiation of glaciation have roots in plate tectonics. As the time progressed in the Cenozoic, large landmasses, such as northern North America, Europe, and Asia, became concentrated around the north polar region at the same time Antarctica moved to its position at the South Pole. The positioning of land in the polar regions produced continental conditions with cold winters. The land also provided a base for ice accumulation. In addition, the landmasses inhibited the northward transport of heat by the oceans. The atmospheric transport of heat to the north polar regions was also hindered by the development of the Himalayas and the Tibetan Plateau over the last 10 million years. While Antarctica remained attached or very close to South America, warm oceanic currents moving along the east coast of South America brought heat to the southern polar region. Between 37 and 34 million years ago, the distance had increased to the point that a strong eastward flow of oceanic currents developed between the two continents. Warmer currents from the south could no longer reach Antarctica, and so glaciation commenced in the mountains and eventually its ice cap began to form. Finally, the development of the Isthmus of Panama about 3 million years ago further disrupted oceanic circulation. At this point permanent glaciers had developed in the high latitudes of the northern hemisphere.

Other factors likely influenced cooling of the planet and the development of glaciation. Declining atmospheric concentration of the greenhouse gas CO_2 between 7 million and 5 million years ago would have been conducive to cooling. As large areas of ice formed at the high latitudes, they would have produced a positive feedback promoting further cooling because their white surfaces reflected a considerable amount of solar energy back into space. In addition, drying conditions produced desertification in some regions, such as the Sahara Desert, and further increased the reflection of energy back into space, enhancing cooling.

Plate tectonic movements and associated orogenies in the Late Cenozoic help explain why the earth cooled and ice caps developed in areas such as Greenland, the Canadian Arctic, and Antarctica. Plate tectonics cannot, however, explain why the Quaternary has experienced alternations between cold conditions, like the last glacial maximum, and warm conditions like today. The rate of plate movement and orogenic mountain formation is simply too slow to explain these large variations in Quaternary climate. To understand what may have caused the alternations between glacial and nonglacial conditions, we must first determine the frequency and timing of such climatic changes during the Quaternary.

It is difficult to use terrestrial geologic deposits to determine how many times major glaciations have occurred in the Quaternary. As glacial ice advances at the start of a glaciation, it erodes deposits left by prior ice advances. As glacial ice retreats, meltwater also erodes tills or covers them with lake and stream deposits. As a result, the advance and subsequent retreat of glaciers destroys and covers evidence of prior glaciations. With the advent of radiometric dating techniques and more sophisticated stratigraphic analysis, it has become clear that the classic fourfold model of glaciation was based on an oversimplified classification of glacial and nonglacial deposits. However, no matter how sophisticated the analysis, the terrestrial record of Quaternary glaciation is problematic, especially for older glacial episodes. Surprisingly, the best evidence of how often Quaternary glaciations have occurred, and how extensive these glaciations were, comes from the oceans. Unlike deposits on land, deposits in deep ocean basins are not prone to erosion. Sediments deposited in the deep ocean accumulate slowly but remain undisturbed. Cores from the deep ocean can provide continuous sediment records that extend back millions of years.

Several types of evidence found in deep ocean cores can be used to reconstruct past variations in ocean temperature and even the relative amount of the world's water that is tied up in ice sheets. The sediment cores are often dated using radiocarbon analysis and paleomagnetic dating. Sediment cores from the deep ocean generally contain large numbers of shells from tiny planktonic and benthic protozoans called foraminifera. These organisms live in the water column and sediments of the world's oceans and produce distinctive calcium carbonate shells. Many species of foraminifera and other plankton are sensitive to water temperatures, and analysis of their fossil shells can be used to reconstruct past temperatures. For example, glacial-aged sediments from the North Atlantic Ocean contain high numbers of shells from the Arctic-subarctic species *Neogloboqadrina pachyderma*. It has also been found that the tropical species *Pulleniantina obliquiloculata* and *Sphaeroidinella dehiscens* are absent from equatorial waters of the Atlantic and Caribbean during the last glacial maximum. It has been suggested that the presence of northern foraminifera and the absence of equatorial species indicate a drop in sea surface temperatures of as much as 7° C to 8° C compared to modern conditions. The silica shells of planktonic zooplankton, such as radiolaria, and phytoplanktonic unicellular algae, such as diatoms, are also found in deep-sea sediments and have been used to infer past sea surface temperatures. In general, such biological evidence points to a cooling of the world's oceans during the last glacial maximum, with decreases in temperature ranging from 0.5° C to 9.0° C. Reconstruction of long-term changes in sea surface temperatures from plankton assemblages indicate that there have been at least six episodes of sea surface cooling similar to the last glacial maximum over the past 500,000 years.

Geochemical analysis, particularly of the calcium carbonate ($CaCO_3$) shells of foraminifera, has proven an especially powerful tool for reconstructing the timing and relative extent of Quaternary glaciations. When aquatic organisms form calcium carbonate shells, they obtain oxygen from the water in which they live to produce their shells. The oxygen atom of a water molecule (H_2O) can be either of the two isotopes ^{18}O or ^{16}O, although ^{18}O is much less abundant in the oceans than ^{16}O. As ice sheets build up during a glaciation, they grow by the addition of snow. Most of the water in the atmosphere that forms the snow crystals comes from evaporation at the surface of the oceans. During this process, the light isotope, ^{16}O, evaporates more readily than the heavy isotope, ^{18}O. The water that is tied up in the growing ice sheets does not flow back into the ocean to replenish the ^{16}O that is preferentially evaporated. Over thousands of years, the relative amount of ^{18}O to ^{16}O in the oceans increases as the glaciers grow. When the glaciation ends and meltwater enters the oceans, the ^{16}O is replenished and the relative amount of ^{18}O decreases. As a rough approximation, the volume of global glacial ice is reflected in the amount of $^{18}O/^{16}O$ in sea water. The oxygen isotope record from the world's oceans for the past 5 million years suggests that there was little to no glacial ice in the northern hemisphere prior to about 3 million years ago (Fig. 7.10). There was a general increase in ice cover from about 3 million years ago through the Quaternary. This coincides with the first appearance of drop stones in deep-sea cores. The Quaternary record is marked with a number of swings in the relative abundance of ^{18}O that suggest as many as 50 advances and retreats of glacial ice over the past 2.5 million years. However, very prolonged periods of large-scale glaciation, such as the last glacial maximum, have only occurred in the past 1 million years or so. Higher resolution analysis of deep-sea cores indicates that perhaps 20 large glaciations have taken place in the past 800,000 years. In addition, the isotopic records from the ocean indicate that very warm periods similar to the Holocene Epoch are relatively rare in the late Quaternary. One can also detect a shift from a 41,000-year periodicity in glaciation to a 100,000-year cycle that commenced about 1 million years ago. It is also notable that amplitude in

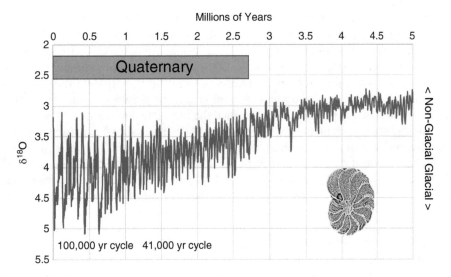

FIGURE 7.10 Stacked variations in $\delta^{18}O$ content values from benthic foraminifera *(inset)* from a global network of 57 ocean cores extending over the past 5 million years (adapted from Lisiecki and Raymo, 2005). The Quaternary Period and the 100,000-year and 41,000-year glacial cycles are indicated. Note that the scale for $\delta^{18}O$ is reversed, so that periods of high $\delta^{18}O$ (glacials) appear as depressions in the time-series.

the difference between glacial and non-glacial conditions has become larger in the late Quaternary (Fig. 7.10). Previous generations of geologists and biogeographers believed that the ice ages were short-term anomalies in global climate and that warm conditions, such as we experience at present, were the norm. We now know that the warm climate and resulting distribution of life that characterizes our world today is, instead, the exception. This becomes clear when we compare the sharp and brief excursions in ^{18}O that typify the Holocene and earlier interglacials with the overall record for the late Quaternary.

It is clear from the marine record that many episodes of climatic cooling and warming have occurred over the Quaternary. With the number and timing of glacial and non-glacial intervals documented, it is possible to search for the causes of the Quaternary climate shifts. Although there are several theories pertaining to this phenomenon, the most widely accepted hypothesis is the **Milankovitch orbital theory of glaciation** elaborated in detail in the 1920s by the Serbian meteorologist Milutin Milankovitch. He argued that three natural and periodic variations occur in the orbital geometry of the earth and cause changes in the seasonal and latitudinal distribution of incoming solar radiation. These changes in insolation trigger periodic glaciations. The concept that changes in the orbital geometry of the earth caused ice ages is actually quite old and dates back to the work of French mathematician Joseph Adhémar in the 1840s and the Scottish naturalist James Croll in the 1860s and 1870s. According to the Milankovitch Theory, periodic glaciations occur for three main reasons. First, the orbit of the earth around the sun is not circular but eccentric, and the degree of eccentricity increases and decreases on an approximately 95,800-year cycle. The time of the year when the earth is closest to the sun is called the perihelion; the time when it is farthest away is the aphelion. Today we experience an intermediate degree of eccentricity, and the perihelion occurs in January. This makes winter in the northern hemisphere slightly milder than it would be if the orbit was circular; second, the tilt of the earth's axis varies from 21.8° to 24.4° on an approximately 41,000-year cycle. If the tilt is increased, the difference between winter and summer insolation is increased. This makes summers warmer and winters cooler. Today the tilt is 23.5°. Third, the seasonal timing of the perihelion and aphelion (also called the precession of the equinoxes) varies on a 21,700-year cycle. Today the perihelion occurs in January; 11,000 years ago it occurred in June.

Milankovitch proposed that glaciation takes place when the orbital geometry of the earth causes the high latitudes of the northern hemisphere to receive low amounts of summer insolation. Conditions in the northern hemisphere are crucial because this region offers the greatest amount of land area for new ice sheets to form. The growth of northern hemisphere ice sheets influences global cooling by reflecting large amounts of sunlight back into space and disturbing ocean circulation.

Changes in southern hemisphere insolation do not appear to be very important. Summer insolation is important because heating in the summer causes the ice and snow to melt. Winters in the high latitudes of the northern hemisphere are always cold enough for snow to occur. If the summers are cool enough, the winter snow does not melt and ice sheets form. Minimum radiation input to the northern hemisphere occurs when the eccentricity of the earth's orbit is high, the tilt of the axis is low (21.8°), and the aphelion occurs in the summer. A comparison of glacial ice volume, global sea level, and changes in northern hemisphere insolation since the last glacial maximum shows a good relationship between minimum summer insolation 18,000 to 20,000 years ago and maximum ice volume (Fig. 7.11). The subsequent decay of the ice sheets and resulting infusion of meltwater in the world's oceans corresponds to increasing summer insolation in the northern hemisphere during the late Pleistocene and early Holocene. Detailed analysis of sedimentalogical records of earlier glaciations, such as the Permian, also suggests that Milankovitch cycles produced variations in glaciation during those times.

The Milankovitch Theory also casts some light on large-scale climatic changes during the Holocene. There is much evidence from a number of regions that summer temperatures were a few degrees warmer than present during the early to mid-Holocene. The timing of this warm period, known as the altithermal, climatic optimum or Holocene Thermal Maximum varies regionally, but ranges from about 8000 to 4000 years ago. For example, well-preserved tree stumps are found across the Arctic tundra of northern Eurasia and show that boreal forest grew right up to the present Arctic coastline between 8000 and 4000 years ago. The timing of this maximum Holocene warming lags behind the timing of maximum summer insolation, which occurred between 12,000 and 9000 years ago. It is likely that the presence of large areas of decaying ice in northern North America and the relatively slow warming of the oceans kept summer temperatures cool until the mid-Holocene. By 7000 to 6000 years ago, the ice had melted, and the sea surface had warmed. The cooling that has been experienced in the past 4000 years is probably the result of decreasing summer insolation caused by the changing orbital geometry of the earth (Fig. 7.11). This cooling is entirely consistent with the Milankovitch Theory.

Important questions remain to be resolved in regard to the detailed linkages between the subtle effects of orbital variations on the earth's energy budget and the glaciation history of the earth in the Quaternary. How do such small changes in insolation drive such large changes in ice-cover? Why were the early Pleistocene glaciations aligned with the 41,000-year cycle of the tilt of the earth's axis,

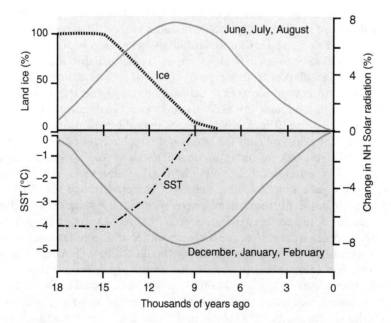

FIGURE 7.11   Changes in summer and winter insolation for the northern hemisphere as predicted by the Milankovitch Theory for the past 18,000 years. Changes in global glacial ice volume and sea surface temperatures (SSTs) are also shown (*Source:* Kutzbach and Street-Parrott, 1985; see also Kutzbach et al., 1998). Modern conditions of global sea surface temperatures and glacial ice extent were developed by 6000 to 7000 years ago.

while the cycle of glaciations over the past million years seems to have been dominated by a 100,000-year cycle? Why has the amplitude in the difference between glacials and interglacials increased? Although we have come a long way toward solving the mysteries of the ice ages, important riddles remain for scientists to ponder.

FUTURE CHANGES IN CONTINENTS AND CLIMATE

We have seen that the past has experienced profound changes in the geography and climate of the earth. It is natural to wonder what is in store for the future. With an understanding of the patterns and mechanisms of plate tectonics, it is possible to predict the future movement and configuration of the continental plates (Fig. 7.6). As the continents change position, global climate will also change, and the ranges of species will be split or brought together. Some areas that are cold today will warm and other regions will cool. In 150 million years from now, Antarctica and Australia will have merged again. Africa will have moved north and collided with the European landmass. The Mediterranean Sea will no longer exist. Indeed, the shrinkage of the Mediterranean Sea is already commencing. Eventually, the northern coast of Africa will lie at 60° north latitude. That is as far north as modern Scandinavia. In contrast, Siberia will have swung south and lie at about the same latitude as modern southern China. In 250 million years, the continental landmasses will have rejoined each other or will lie very close together. Pangaea will exist again.

In the shorter term, the natural changes in the earth's orbit have produced conditions of increasingly low summer insolation in the northern hemisphere over the late Holocene. Evidence abounds of a cooling climate in the late Holocene. Glaciological studies indicate that mountain glaciers around the world began to expand about 4000 years ago. Northern treeline and the boreal biome, along with some alpine treelines, began to retreat southward or to lower elevations. Since the 1960s and 1970s, as orbital forcing of glaciations became widely accepted and evidence of cooling over the late Holocene accumulated, some have wondered if the orbital geometry of the earth was placing the planet on the verge of a new ice age. The complexity of the linkages between orbital changes and the ultimate growth of continental ice sheets, and the additional complexity added by human activity as discussed below, makes this a challenging question to answer. Past interglacials have lasted between 10,000 and 30,000 years and there is no reason to suspect that under natural conditions the warmth of the Holocene Epoch would persist indefinitely.

Aside from shifting continents, changes in the orbital geometry of the earth, and other natural influences, a new and potent force is exerting its influence on the earth's present and future climate. That force is anthropogenic climate change caused by increasing amounts of greenhouse gases in the atmosphere. **Greenhouse gases** are molecules such as CO_2, methane (CH_4), and nitrous oxide (N_2O), which in the atmosphere are transparent to energy in the visible light portion of the electromagnetic spectrum, but absorb long-wave radiation (heat) rather than allowing it to escape into space. Almost all of the absorption of heat energy in the atmosphere takes place within the lower 4 kilometers of the troposphere. The most abundant and important of these three greenhouse gases is CO_2. The concentrations of these gases are very low, measured in parts per million and parts per billion of the total volume of the atmosphere. Despite their tiny concentrations, the absorption and reradiation of heat energy by greenhouse gases keep the atmosphere and surface of the earth significantly warmer than they would be without the presence of such gases. Today, the average temperature at the surface of the earth is approximately 15° C. If our atmosphere did not contain greenhouse gases, the average temperature would be −18° C. Without greenhouse gases, life would not exist on our planet.

Greenhouse gases such as CO_2, CH_4, and N_2O are natural constituents of the atmosphere. Carbon dioxide is released in volcanic eruptions, by the weathering of calcareous rocks, by respiration, by organic decay, and by the burning of biomass from wildfires. Methane is mainly produced by anaerobic decay in soils, wetlands, and peats. Insects, such as termites, and animals, such as cows, also produce significant amounts of CH_4 during digestion. Volcanic activity, biomass burning, respiration, and decomposition of organic matter are important sources of N_2O. In the natural carbon cycle, photosynthesis by terrestrial and aquatic plants removes CO_2 from the atmosphere and incorporates the carbon into plants and, eventually, animal tissue. In some instances, the carbon stored in the plant and animal tissue is quickly released back to the atmosphere following the death and decomposition of the organism. In other cases, such as the organic matter deposited in peatlands, lakes, and the ocean, the carbon can be stored for thousands to millions of years. Coal and oil deposits are the geological remains of past plants and animals and represent such stored carbon. Increased levels of atmospheric CO_2 serve as a fertilizer promoting increased photosynthesis and plant growth. The increases in photosynthesis and plant growth results in higher rates of CO_2 removal from the atmosphere and storage of the carbon in organic material. This feedback helps to regulate the concentration of CO_2 in the earth's atmosphere. Natural feedbacks within biogeochemical cycles have maintained relatively stable concentrations of CO_2 in the atmosphere of approximately 280 parts per million by volume (ppmv) during the late Holocene. However, as we shall see below, the amount of CO_2 in the earth's atmosphere has varied significantly during the Quaternary and over longer geologic time.

Several different human activities have led to increasing amounts of CO_2 and other greenhouse gases in the atmosphere. Fossil fuels, such as coal, oil, and natural gas, are formed from the biomass of organisms that died millions of years ago. The carbon in these remains has been stored underground in sedimentary deposits and has not been released back into the atmosphere. In burning fossil fuels, we release this stored carbon into the atmosphere at a rapid rate. Destruction of living vegetation can also release carbon into the atmosphere and decrease the rate at which carbon is taken up by photosynthesis. With increasing world population, industrialization, and land-use change starting in the late 1700s and early 1800s, there have been great increases in the amount of fossil fuels burned each year to produce energy. The rate of the destruction of natural vegetation has also increased as human population numbers have grown and additional land has been cleared for cities and agricultural uses. How much has this human activity changed the atmospheric concentrations of CO_2 and other greenhouse gases? The technology to measure the small concentrations of CO_2, CH_4, and N_2O in the atmosphere was first developed in the late 1950s by Charles David Keeling at the Scripps Institution of Oceanography at the University of California, San Diego. Direct measurements of CO_2 in the atmosphere are taken from remote regions such as the top of the Mauna Loa volcano in Hawaii; Point Barrow, Alaska; and the Vostok Station, Antarctica, because the atmosphere at such sites is not affected by nearby human activities that would locally release large quantities of CO_2 and other greenhouse gases. These measurements show that since the late 1950s, CO_2 has increased throughout the world from approximately 315 ppmv to 419 ppmv by 2022 (Fig. 7.12). By 2024 this had increased to about 424 ppmv.

To obtain longer records of atmospheric greenhouse gases, scientists have studied the concentration of CO_2 and other greenhouse gases that is preserved in air bubbles from Greenland and Antarctic ice cores. They have found that atmospheric CO_2 increased from about 280 ppmv at the start of the Industrial Revolution in the early 1800s to the more than 420 ppmv observed in 2023 (Fig. 7.13). Analysis of ice core records indicate that late Holocene levels of CO_2 were relatively stable at around 280 parts per million prior to the Industrial Revolution. Accelerating human population growth, industrialization, use of fossil fuels, deforestation, and land-use change have produced a rapid rise in atmospheric CO_2 over the past two centuries. It is very clear that rise in CO_2 from 280 ppmv in the early 1800s to 419 ppmv in 2022 is well outside the bounds of natural variability experienced over the past 800,000 years (Fig. 7.13). What is strikingly apparent from these long records is that at no time over the past 800,000 years have levels of CO_2 in the atmosphere been as high as they are today. Natural variability has ranged from lows of around 180 ppmv during glacials to peaks of around 280 ppmv during interglacials (Fig. 7.13). These changes in CO_2 during the Quaternary served to amplify the differences in climate between glacial and interglacials. Although the specific causes of these Quaternary variations in CO_2 are still being resolved, it seems clear that biogeochemical cycling by the oceans plays a major role. In any case, the level of CO_2 in the atmosphere today is about 50% higher than any of the previous natural peaks observed in the ice core records.

The net CO_2 emissions from human activity between 1850 and 2019 is 2400 gigatonnes (Gt). Emissions accelerated over this period (Fig. 7.12) and about 42% of this increase (1000 Gt) occurred between 1990 and 2019. It is estimated that as much as 85% of the increased levels of CO_2 in the atmosphere are derived from the burning of fossil fuels. Land clearance and other activities such as cement production contribute the next highest portion. In 2020 the burning of coal accounted for about 40% of the global anthropogenic CO_2 emissions related to fossil fuels. Oil accounted for about

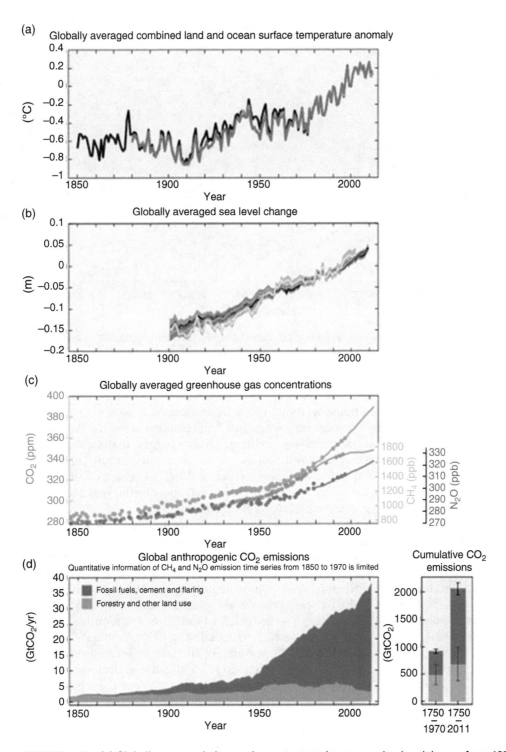

FIGURE 7.12 (*a*) Globally averaged observed temperature change over land and the sea from 1850 to 2010. (*b*) Globally averaged sea level change from 1900 to 2010. (*c*) Globally averaged atmospheric greenhouse gas concentrations. (*d*) Global atmospheric measurements from 1850 to 2010 (IPCC, 2014).

32%, and natural gas burning contributed 21%. Other greenhouse gases such as CH_4 and N_2O have also increased in a very similar manner over the past 200 years. In 2023 methane has risen from around 722 parts per billion by volume (ppbv) in the pre-industrial era and constitutes about 1919 ppbv. Nitrous oxide rose from around 275 ppbv and today accounts for 336 ppbv. It is important to remember that the burning of fossil fuels or the clearing of lands has been done to meet human demands. About 73% of current total greenhouse gas emissions are related to energy used for

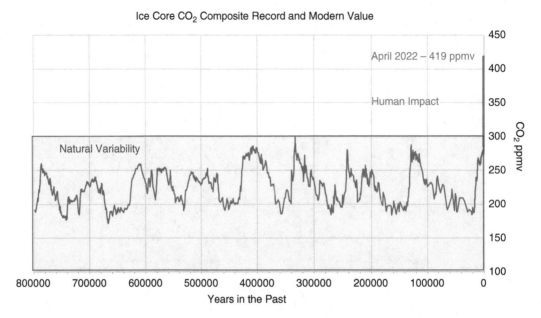

FIGURE 7.13   Changes in atmospheric CO_2 concentrations for the past 800,000 years as evident in a composite record of ice cores from Antarctica. Recent increases and level in atmospheric CO_2 are also indicated (adapted from Luthi et al., 2008; NOAA).

electricity, heating, and transportation. About 18% is related to agriculture, forestry, and other land uses. Analysis of human population growth, fossil fuel demands, land-use patterns, and current political, economic, and societal responses to climate change suggest that atmospheric concentrations of CO_2 and other greenhouse gases will continue to increase in the future. Under pessimistic scenarios developed by the United Nations Intergovernmental Panel of Climate Change (IPCC), it is projected that atmospheric CO_2 concentrations could exceed 900 ppmv by the year 2100. This is well over three times higher than the pre–Industrial Revolution concentration.

It is overwhelmingly accepted by climate scientists that human activity over the past two centuries has led to substantial increases in the amount of greenhouse gases in the atmosphere and that these increases have had, and will continue to have, a significant impact on world climate. This should not be surprising as it has long been postulated that increasing greenhouse gas concentrations will lead to the increased retention of heat energy in the atmosphere. This would produce higher temperatures in the troposphere and at the earth's surface. In 1856, the American scientist Eunice Foote, and in 1861, the Irish scientist John Tyndall, independently suggested that changes in atmospheric CO_2 could produce climatic changes. In the 1890s, Svante Arrhenius concluded that a doubling of CO_2 could lead to a 5° C warming of the atmosphere. In the mid-twentieth century, English engineer and amateur meteorologist Guy Callendar collected estimates of atmospheric CO_2 and compared these to surface temperature records to argue there had been a linked increase in both since the nineteenth century. In the 1950s the Canadian physicist Gilbert Plass performed calculations that suggested humans could potentially raise the global average temperature by 1.1° C per century through the greenhouse effect. Oceanographer Roger Revelle at the Scripps Institution of Oceanography began to argue forcefully that the oceans could not buffer all of the additional greenhouse gas additions and humans might indeed transform the climate. In the 1970s and 1980s, American climatologist Stephen Schneider and many others provided increasing evidence and warnings about the potential warming of the earth due to increasing greenhouse gases. Continued intense efforts have been made to estimate the impact of increasing CO_2 and other greenhouse gases on climate. This difficult problem has been tackled mainly through the use of meteorological observations, paleoclimatic records, and sophisticated computer models of the earth's climate. A number of such models are in use today. The Coupled Model Intercomparison Project 6 (CMIP6) climate modeling project includes more than 100 models from over 50 different research centers. **General Circulation Models** (GCMs) suggest that the increases in CO_2 and other greenhouse gases

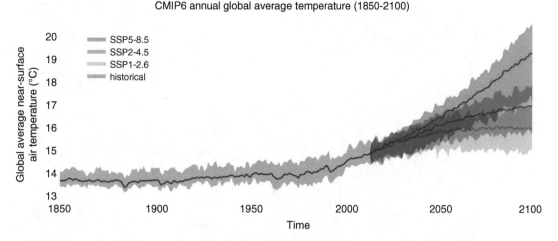

CMIP6 annual global average temperature (1850-2100)

FIGURE 7.14 The historical instrumental record of variations in global temperature and CMIP6 climate modeling estimates of increases in annual mean temperature (°C) that could be experienced by the year 2100 due to increasing greenhouse gases under different emissions scenarios (from *Source:* Copernicus Climate Change Service, ECMWF, https://climate.copernicus.eu/latest-projections-future-climate-now-available / ECMWF / last accessed on 28 December 2023).

can lead to an increase in global temperature of 1.5° C to 5° C in the twenty-first century depending upon what actions are taken to curb emissions. This would raise the global average temperature to over 19° C under the worst-case scenario (Fig. 7.14). Warming will not be even across the whole earth. For example, it is likely that particularly strong temperature increases will occur in the Arctic. The GCMs provide less consistent results in terms of changes in precipitation. It is likely that areas which already have high precipitation, such as the tropics, will get more precipitation, while some already dry regions may get less precipitation. However, even if precipitation remains the same in arid regions, the increased evaporation caused by the higher temperatures will lead to higher moisture stress on the vegetation. Additional features of twenty-first-century climate change will likely be an increase in the number or magnitude of extreme weather events, such as floods, and the development of regional climates that are unlike any current climates.

One important change is widely forecast for the cryosphere and hydrosphere. It is estimated that the melting of glaciers and thermal expansion of water in the ocean will produce a rise in sea level of 30 centimeters to 300 centimeters by 2100. This increase would be very problematic for present coastal ecosystems such as mangrove forests and saltwater marshes around the world. Low-elevation coastal cities and agricultural lands would be similarly threatened with inundation. About 50% of the city of New Orleans already lies below sea level and is protected by levees and pumps. This infrastructure would require significant improvements. Some agricultural areas, such as large portions of the Netherlands, that lie below sea level today would require additional dikes. Such expensive mitigation measures are available to rich developed nations. A country like Bangladesh, which has much of its land near sea level and is already prone to inundation, would be devastated. The nation of the Maldives Islands could simply cease to exist.

Is there any evidence that global warming due to increasing greenhouse gases has taken place already? Analysis and synthesis of instrumental temperature records from meteorological stations around the world show that the average global temperature increased by over 1° C since the end of the nineteenth century and continues to rise (Fig. 7.12 and Fig. 7.14). Glaciers have shrunk and retreated in many regions (Fig. 7.15) and the resulting meltwater, coupled with thermal expansion of ocean waters, has produced a rise in sea level of about 20 centimeters today compared to 1900 (Fig. 7.12). The Greenland Ice Cap lost an estimated 2900 Gt of mass in the two decades between 1991 and 2011 and this in itself likely contributed 8 millimeters to the global rise in sea level. Each year, more such evidence accumulates as anthropogenic greenhouse gases increase in the atmosphere and the climate continues to change.

Instrumental climate records extend back in time only 100 to 150 years in most cases. They do not provide long temporal views of natural climate variability to compare with current trends

FIGURE 7.15 The melting and retreat of the Athabasca Glacier portion of the Columbia Icefield in the Canadian Rocky Mountains, Alberta. A marker shows its former position in 2000. As can be seen, by 2010 when this picture was taken, it had retreated significantly. The glacial front has been retreating rapidly over the past 100 years. Such retreat is evident for a number of alpine glaciers as a result of global climate warming.

and projected future conditions. A number of biogeographers use various techniques, such as tree-ring analysis (dendroclimatology), to construct records of past climate that extend back centuries to millennia. As we discussed earlier, some of the greatest warming is expected to occur in the high latitudes of the northern hemisphere. Analysis of temperature records of thermometer readings from scattered northern weather stations provides some evidence of a warming trend during the twentieth century. However, such observations are very sparse for the Arctic and subarctic. Biogeographers have turned to paleoclimatological techniques to reconstruct past changes in temperature in northern North America and Eurasia. Tree-ring analysis is one of the most important paleoclimatic techniques used to reconstruct recent changes in the northern climate (Technical Box 7.2). Boreal trees such as white spruce *(Picea glauca)* in North America or Eurasian trees such as Scots pine *(Pinus sylvestris),* Norway spruce *(Picea abies),* and Siberian larches *(Larix sibirica* and *L. dahurica)* can live for over 400 years. Preserved wood that is thousands of years old has also been recovered in the north and can sometimes be cross-dated with living trees to extend the paleoclimatic records. The growth of these northern trees is often limited by cold climatic conditions, particularly during the summer. During warm summers, the annual rings of the trees tend to be wide; during cold summers, the rings are narrow. By counting the rings and measuring changes in their width, dendroclimatologists are able to reconstruct annual changes in temperature back hundreds to thousands of years. Tree-ring records and other paleoclimatic data from ice cores, lake sediments, and ocean sediments from many sites around the world show that the twentieth century experienced greater and/or more sustained high temperatures than at any other time over the past 400 to 2000 years. Evidence from the Arctic shows that climate warming has been particularly pronounced in the high latitudes of the northern hemisphere. Analysis of these records show that the trend toward warmer summer temperatures is the opposite of the cooling that natural orbital variations should be producing in the higher latitudes of the northern hemisphere. The more paleoclimatic records that are acquired, the clearer the evidence becomes that the climate of the earth is warming significantly and the primary cause for this warming is increased concentrations of greenhouse gases in the atmosphere due to human activity.

Dendroclimatology (Tree-Ring Analysis)

We have already seen how the analysis of tree-rings can be used to reconstruct the history of forest stands and fires (Chapter 5). In addition, tree-rings are one of the most important tools used by biogeographers to reconstruct recent past climatic changes. In many parts of the world, instrumental climate records from meteorological stations extend back in time 100 years or less. This span is too short to allow analysis of natural climate variability or to detect the impact of increasing greenhouse gases. In some areas, coniferous trees can be found that are 1000 to over 4000 years old. In addition, it is sometimes possible to find preserved wood from trees that lived and died thousands of years ago. By analyzing the rings of ancient living trees and preserved wood, it is often possible to construct long records of past climate that extend for hundreds to thousands of years.

The climate of the past is captured in tree-ring records because trees in cold regions or arid areas are often limited in terms of growth by temperature or precipitation. For example, a ponderosa pine *(Pinus ponderosa)* growing on a rocky ridge in the western interior of the United States will grow more vigorously during a wet year than during a dry year. When growth is vigorous, the size of the annual ring produced by the tree is large. During a dry and stressful year, the size of the ring is small.

Tree-ring samples are usually obtained using a very fine coring device that penetrates the tree but does not injure it. Dead wood is sampled by sawing a cross-section of the trunk. The samples are finely sanded and then mounted for counting using a stereomicroscope and special computerized measuring table. By carefully counting, cross-dating, and measuring the rings, dendroclimatologists can construct tree-ring records that extend far into the past. In addition to optical counts, dendroclimatologists also use X rays of wood samples to examine variations in wood density. Generally, many trees are sampled at each site, and the counts from these samples are combined into an indexed record for that location. Prior to amalgamation, mathematical detrending techniques are used to remove variations in the ring records that might arise due to biological factors such as tree-aging or competition with other trees. The resulting detrended annual variations in the rings are then statistically compared to annual variations in climate that have been recorded at nearby meteorological stations. From this statistical comparison a mathematical model is developed to link changes in ring widths to variations in climate. The variations in the rings are then used as input in a mathematical transfer function model that estimates past climate on the basis of the tree-rings. A number of statistical tests are performed to make sure the estimates of past climate are reliable. Tree-ring records of past climate are available from many tree species and many parts of the world, ranging from larches *(Larix dahurica)* at the Arctic treeline in Siberia to teak *(Tectona grandis)* trees at the equator in Southeast Asia. Tree-ring records from living and preserved oaks in Europe and the bristlecone pines *(Pinus aristata)* in California extend back through the entire Holocene. A number of tree-ring studies from the Arctic and elsewhere provide evidence that the twentieth century experienced longer and more prolonged warming than any time over the past 400 to 1000 years.

The author collects tree-ring samples from a larch tree *(Larix dahurica)* in northern Siberia. A small boat anchored on the lake is being used as a platform to recover sediment cores for pollen analysis. Analysis of Arctic tree-ring cores from Siberia and elsewhere provides evidence that the twentieth century has experienced unusually warm temperatures compared to the past 400 to 1000 years.

KEY WORDS AND TERMS

Beringia
Eustatic sea-level change
General Circulation Models
Geologic time scale
Glaciation
Gondwanaland

Greenhouse gases
Isostatic sea-level change
Laurasia
Law of superposition
Law of uniformitarianism
Milankovitch orbital theory of glaciation

Orogeny
Pangaea
Plate tectonics
Pluvials
Pluvial lakes
Theory of continental drift

REFERENCES AND FURTHER READING

Allen, J. R., Forrest, M., Hickler, T., Singarayer, J. S., Valdes, P. J., and Huntley, B. (2020). Global vegetation patterns of the past 140,000 years. *Journal of Biogeography* **47**, 2073–2090.

Annan, J. D., and Hargreaves, J. C. (2013). A new global reconstruction of temperature changes at the Last Glacial Maximum. *Climate of the Past* **9**, 367–376.

Bartoli, G., Sarnthein, M., Weinelt, M., Erlenkeuser, H., Garbe-Schönberg, D., and Lea, D. W. (2005). Final closure of Panama and the onset of northern hemisphere glaciation. *Earth and Planetary Science Letters* **237**, 33–44.

Basinger, J. F. (1991). The fossil forests of the Buchanan Lake Formation (early Tertiary), Axel Heiberg Island, Canadian Arctic Archipelago: preliminary floristics and paleoclimate. In *Tertiary Fossil Forests of the Geodetic Hills, Axel Heiberg Island, Arctic Archipelago* (ed. R. L. Christie and N. J. McMillan), pp. 39–65. Geological Society of Canada, Ottawa.

Bell, M., and Walker, M. J. C. (1992). *Late Quaternary Environmental Change: Physical and Human Perspectives*. Wiley, New York.

Bennett, K. D., Tzedakis, P. C., and Willis, K. J. (1991). Quaternary refugia of north European trees. *Journal of Biogeography* **18**, 108–115.

Benson, L., and Thompson, R. S. (1987). The physical record of lakes in the Great Basin. In *Geology of North America, Volume K–3: North America and Adjacent Oceans During the Last Deglaciation* (ed. W. F. Ruddiman and H. E. Wright Jr.), pp. 241–260. Geological Society of America, Boulder, CO.

Betancourt, J. L., VanDevender, T. R., and Martin, P. S. (1990). *Packrat Middens*. University of Arizona, Tucson.

Bhagwat, S., and Willis, K. J. (2008). Species persistence in northerly glacial refugia of Europe: a matter of chance or biogeographical traits? *Journal of Biogeography* **35**, 464–482.

Binney, H., Edwards, M., Macias-Fauria, M., Lozhkin, A., Anderson, P., Kaplan, J. O., Andreev, A., Bezrukova, E., Blyakharchuk, T., Jankovska, V., and Khazina, I. (2017). Vegetation of Eurasia from the last glacial maximum to present: key biogeographic patterns. *Quaternary Science Reviews* **157**, 80–97.

Birkeland, P. W., and Larson, E. E. (1989). *Putnam's Geology*, 5th edition. Oxford University Press, New York.

Birks, H. J. B., and Berglund, B. E. (2018). One hundred years of Quaternary pollen analysis, 1916–2016. *Vegetation History and Archaeobotany* **27**, 271–309.

Birks, H. J. B., and Willis, K. J. (2008). Alpines, trees, and refugia in Europe. *Plant Ecology & Diversity* **1**, 147–160.

Bowen, D. Q. (1978). *Quaternary Geology: A Stratigraphic Framework for Multidisciplinary Work*. Pergamon Press, Oxford.

Bradley, R. S. (2015). *Paleoclimatology: Reconstructing Climates of the Quaternary*, 3rd edition. Academic Press, Oxford.

Briggs, J. C. (1995). *Global Biogeography*. Elsevier, Amsterdam.

Brown, J. H., and Gibson, A. C. (1983). *Biogeography*. C. V. Mosby, St. Louis.

Brown, J. H., and Lomolino, M. V. (1998). *Biogeography*, 2nd edition. Sinauer Associates, Sunderland, MA.

Bryant, E. (1997). *Climate Process and Change*. Cambridge University Press, Cambridge.

Campisano, C. J. (2012) Milankovitch cycles, paleoclimatic change, and Hominin evolution. *Nature Education Knowledge* **4**, 5. https://www.nature.com/scitable /knowledge/ library/milankovitch-cycles-paleoclimatic-change-and-hominin-evolution-68244581/.

Clague, D. A., and Dalrymple, G. B. (1994). Tectonics, geochronology, and origin of the Hawaiian-Emperor Volcanic Chain. In *A Natural History of the Hawaiian Islands—Selected Readings II* (ed. E. A. Kay), pp. 5–40. University of Hawaii, Honolulu.

Claussen, M., Selent, K., Brovkin, V., Raddatz, T., and Gayler, V. (2013). Impact of CO_2 and climate on last glacial maximum vegetation—a factor separation. *Biogeosciences* **10**, 3593–3604.

Cole, K. L., Fisher, J. F., Ironside, K., Mead, J. I., and Koehler, P. (2013). The biogeographic histories of *Pinus edulis* and *Pinus monophylla* over the last 50,000 years. *Quaternary International* **310**, 96–110.

Cook, J., Nuccitelli, D., Green, S. A., Richardson, M., Winkler, B., Painting, R., Way, R., Jacobs, P., and Skuce, A. (2013). Quantifying the consensus on anthropogenic global warming in the scientific literature. *Environmental Research Letters* **8**, 024024.

Cook, J., Oreskes, N., Doran, P. T., Anderegg, W. R., Verheggen, B., Maibach, E. W., Carlton, J. S., Lewandowsky, S., Skuce, A. G., Green, S. A., and Nuccitelli, D. (2016). Consensus on consensus: a synthesis of consensus estimates on human-caused global warming. *Environmental Research Letters* **11**, 048002.

Dai, A. (2013). Increasing drought under global warming in observations and models. *Nature Climate Change* **3**, 52–58.

Dai, A., Zhao, T., and Chen, J. (2018). Climate change and drought: a precipitation and evaporation perspective. *Current Climate Change Reports* **4**, 301–312.

Dansgaard, W., Johnsen, S. J., Clausen, H. B., Dahl-Jensen, D., Gundestrup, N. S., Hammer, C. U., Hvidberg, C. S., Steffensen, J. P., Sveinbjornsdottir, A. E., Jouzel, J., and Bond, G. (1993). Evidence for general instability of past climate from a 250-kyr ice-core record. *Nature* **364**, 218–220.

Darlington Jr., P. J. (1965). *Biogeography of the Southern End of the World*. Harvard University Press, Cambridge, MA.

D'Arrigo, R. D., Jacoby, G. C., and Krusic, P. J. (1994). Progress in dendroclimatic studies in Indonesia. *Terrestrial, Atmospheric and Ocean Sciences* **5**, 349–363.

Davies, B. J., Hambrey, M. J., Smellie, J. L., Carrivick, J. L., and Glasser, N. F. (2012). Antarctic Peninsula ice sheet evolution during the Cenozoic Era. *Quaternary Science Reviews* **31**, 30–66.

Davis, M. B. (1981). Quaternary history and stability in forest communities. In *Forest Succession: Concepts and Applications* (ed. D. C. West, H. H. Shugart, and D. B. Botkin), pp. 132–153. Springer-Verlag, New York.

Delcourt, P. A., and Delcourt, H. R. (1987). *Long-Term Forest Dynamics of the Temperate Zone: A Case Study of Late-Quaternary Forests in Eastern North America*. Springer-Verlag, New York.

Dewbury, A. (2007). The American School and scientific racism in early American anthropology. *Histories of Anthropology Annual* **3**, 121–147.

Edwards, K. J. (2018). Pollen, women, war and other things: reflections on the history of palynology. *Vegetation History and Archaeobotany* **27**, 319–335.

Edwards, K. J., Fyfe, R. M., and Jackson, S. T. (2017). The first 100 years of pollen analysis. *Nature Plants* **3**, 17001. https://doi.org/10.1038/nplants.2017.1.

Elias, S. A. (2007). Rodent middens. *Encyclopaedia of Quaternary Science*, pp. 2356–2367. Elsevier, Oxford.

Erickson, J. (1995). *A History of Life on Earth: Understanding Our Planet's Past*. Facts on File, New York.

Faegri, K., Kaland, E., and Krzywinski, K. (1989). *Textbook of Pollen Analysis*, 4th edition. Wiley, New York.

Finney, S. C., and Edwards, L. E. (2016). The "Anthropocene" epoch: scientific decision or political statement? *GSA Today* **26**, 4–10.

Frenzel, B., Pecsi, M., and Velichko, A. A. (1992). *Atlas of Paleoclimates and Paleoenvironments of the Northern Hemisphere: Late Pleistocene, Holocene*. G. Fischer, Stuttgart.

Frisch, W., Meschede, M., and Blakey, R. C. (2011). *Plate Tectonics: Continental Drift and Mountain Building*. Springer, Heidelberg, Germany.

Fritts, H. C. (1976). *Tree Rings and Climate*. Academic Press, New York.

Funder, S., Kjeldsen, K. K., Kjær, K. H., and Cofaigh, C.Ó. (2011). The Greenland Ice Sheet during the past 300,000 years: a review. *Developments in Quaternary Sciences* **15**, 699–713.

Gale, S. J., and Hoare, P. G. (2012). The stratigraphic status of the Anthropocene. *The Holocene* **22**, 1491–1494.

Gehrels, R. (2010). Sea-level changes since the Last Glacial Maximum: an appraisal of the IPCC Fourth Assessment Report. *Journal of Quaternary Science* **25**, 26–38.

Geologic Names Committee of the U.S. Geological Survey (2007). *Divisions of Geologic Time: Major Chronostratigraphic and Geochronologic Units*. USGS, Reston, VA. https://pubs.usgs.gov/fs/2010/3059/pdf/FS10-3059.pdf.

Gerhart, L. M., Harris, J. M., Nippert, J. B., Sandquist, D. R., and Ward, J. K. (2012). Glacial trees from the La Brea tar pits show physiological constraints of low CO_2. *New Phytologist* **194**, 63–69.

Graham, R. W. (1986). Response of mammalian communities to environmental changes during the late Quaternary. In *Community Ecology* (ed. J. M. Diamond, and T. J. Case), pp. 300–313. Harper and Row, New York.

Graham, R. W., Lundelius, E. L., Graham, M. E., Schroeder, E. K., Toomey, R. S., Anderson, E., Barnosky, A. D., Burns, J. A., Churcher, C. S., Grayson, C. K., Guthrie, R. D., Harrington, C. R., Jefferson, G. T., Martin, L. D., McDonald, H. G., Morlan, R. E., Semken, H. A., and Webb, S. D. (1996). Spatial response of mammals to late-Quaternary environmental fluctuations. *Science* **272**, 1601–1606.

Grigg, R. W. (1988). Paleoceanography of coral reefs in the Hawaiian-Emperor Chain. *Science* **240**, 1737–1743.

Hallam, A. (1994). *An Outline of Phanerozoic Biogeography*. Oxford University Press, Oxford.

Hirabayashi, Y., Mahendran, R., Koirala, S., Konoshima, L., Yamazaki, D., Watanabe, S., Kim, H., and Kanae, S. (2013). Global flood risk under climate change. *Nature Climate Change* **3**, 816–821.

Houghton, J., Meira Filho, L. G., Callander, B. A., Harris, N., Kattenberg, A., and Maskell, K. (1996). *Climate Change 1995: The Science of Climate Change*. Cambridge University Press, Cambridge.

Houghton, J. T., Jenkins, G. J., and Ephraums, J. J. (1990). *Climate Change: The IPCC Scientific Assessment*. Cambridge University Press, Cambridge.

Houghton, J. T., Meira Filho, L. G., Bruce, J., Hoesung Lee, Callander, B. A., Haites, E., Harris, N., and Maskell, K. (1995). *Climate Change 1994: Radiative Forcing of Climate Change and an Evaluation of the IPCC 1992 Emission Scenarios*. Cambridge University Press, Cambridge.

Huntley, B., and Birks, H. J. B. (1983). *An Atlas of Past and Present Pollen Maps for Europe: 0–12000 Years Ago*. Cambridge University Press, Cambridge.

Imbrie, J., and Imbrie, K. P. (1979). *Ice Ages: Solving the Mystery*. Macmillan, London.

Imbrie, J., van Donk, J., and Kipp, N. G. (1973). Paleoclimatic investigation of a late Pleistocene Caribbean deep-sea core: comparison of isotopic and faunal methods. *Quaternary Research* **3**, 10–38.

IPCC (2014). *Climate Change 2014: Synthesis Report. Contribution of Working Groups I, II and III to the Fifth Assessment Report of the Intergovernmental Panel on Climate Change* (ed. R. K. Pachauri and L. A. Meyer). IPCC, Geneva, Switzerland.

Irmscher, C. (2013). *Louis Agassiz: Creator of American Science*. Houghton Mifflin Harcourt, Boston.

Jackson, R. (2020). Eunice Foote, John Tyndall and a question of priority. *Notes and Records* **74**, 105–118.

Jahren, A. H. (2007). The Arctic forest of the middle Eocene. *Annual Review of Earth and Planetary Science* **35**, 509–540.

Jouzel, J., Barkov, N. I., Barnola, J. M., Bender, M., Chappellaz, J., Genthon, C., Kotlyakov, V. M., Lipenkov, V., Lorius, C., Petit, J. R., Reynaud, D., Raisbeck, G., Ritz, C., Sowers, T., Stievenard, M., Yiou, F., and Yiou, P. (1993). Extending the

Vostock ice-core record of paleoclimate to the penultimate glacial period. *Nature* **364**, 407–412.

Kaufman, D. S., Schneider, D. P., McKay, N. P., Ammann, C. M., Bradley, R. S., Briffa K. R., Miller, G. H., Otto-Bliesner, B. L., Overpeck, J. T., Vinther, B. M., Arctic Lakes 2k Project Members (M. Abbott, Y. Axford, B. Bird, H. J. B. Birks, A. E. Bjune, J. Briner, T. Cook, M. Chipman, P. Francus, K. Gajewski, Á. Geirsdóttir, F. S. Hu, B. Kutchko, S. Lamoureux, M. Loso, G. MacDonald, M. Peros, D. Porinchu, C. Schiff, H. Seppä, and E. Thomas) (2009). Recent warming reverses long-term Arctic cooling. *Science* **325**, 1236–1239.

Koehler, P. A., Anderson, R. S., and Spaulding, W. G. (2005). Development of vegetation in the central Mojave Desert of California during the late Quaternary. *Palaeogeography, Palaeoclimatology, Palaeoecology* **215**, 297–311.

Kohfeld, K. E., Quéré, C. L., Harrison, S. P., and Anderson, R. F. (2005). Role of marine biology in glacial-interglacial CO_2 cycles. *Science* **308**, 74–78.

Kohfeld, K. E., and Ridgwell, A. (2009). Glacial-interglacial variability in atmospheric CO_2. *Surface Ocean: Lower Atmosphere Processes* **187**, 251–286.

Kukla, G. J., and Matthews, R. K. (1972). When will the present interglacial end? *Science* **178**, 190–191.

Kutzbach, J. E., Gallimore, R., Harrison, S. P., Behling, P., Selin, R., and Laarif, F. (1998). Climate and biome simulations for the past 21,000 years. *Quaternary Science Reviews* **17**, 473–506.

Kutzbach, J. E., and Street-Parrott, A. F. (1985). Milankovitch forcing of fluctuations in the lake level of tropical lakes from 10 to 0 kyr. *Nature* **317**, 130–134.

Lambeck, K., Rouby, H., Purcell, A., Sun, Y. and Sambridge, M. (2014). Sea level and global ice volumes from the Last Glacial Maximum to the Holocene. *Proceedings of the National Academy of Sciences* **111**, 15296–15303.

Lambert, F., Delmonte, B., Petit, J. R., Bigler, M., Kaufmann, P. R., Hutterli, M. A., Stocker, T. F., Ruth, U., Steffensen, J. P., and Maggi, V. (2008). Dust-climate couplings over the past 800,000 years from the EPICA Dome C ice core. *Nature* **452**, 616–619.

Lemieux-Dudon, B., Blayo, E., Petit, J. R., Waelbroeck, C., Svensson, A., Ritz, C., Barnola, J. M., Narcisi, B. M., and Parrenin, F. (2010). Consistent dating for Antarctic and Greenland ice cores. *Quaternary Science Reviews* **29**, 8–20.

Lowe, J. J., and Walker, M. J. (2015). *Reconstructing Quaternary Environments*, 3rd edition. Routledge, Abingdon, United Kingdom.

Lüthi, D., Le Floch, M., Bereiter, B., Blunier, T., Barnola, J. M., Siegenthaler, U., Raynaud, D., Jouzel, J., Fischer, H., Kawamura, K., and Stocker, T. F. (2008). High-resolution carbon dioxide concentration record 650,000–800,000 years before present. *Nature* **453**, 379–382.

MacDonald, G. M. (1996). Quaternary palynology. In *Stratigraphic Palynology* (ed. J. Jansonius and D. C. McGregor), pp. 879–910. American Association of Stratigraphic Palynologists, Houston.

MacDonald, G. M. (2011). Global climate change. In *Geographical Knowledge* (ed. J. A. Agnew and D. N. Livingstone), pp. 540–548. Sage, London.

MacDonald, G. M., Bennett, K. D., Jackson, S. T., Parducci, L., Smith, F. A., Smol, J. P., and Willis, K. J. (2008). Impacts of climate change on species, populations and communities: palaeobiogeographical insights and frontiers. *Progress in Physical Geography* **32**, 139–172.

MacDonald, G. M., Case, R. A., and Szeicz, J. M. (1998). A 538-year record of climate and tree-line dynamics from the Lena River region of northern Siberia. *Arctic and Alpine Research* **30**, 334–339.

MacDonald, G. M., and Cwynar, L. C. (1985). A fossil pollen based reconstruction of the late Quaternary history of lodgepole pine *(Pinus contorta* spp. *latifolia)* in the western interior of Canada. *Canadian Journal of Forestry Research* **15**, 1039–1044.

MacDonald, G. M., and Edwards, K. J. (1991). Holocene palynology I. Principles, palaeoecology and palaeoclimatology. *Progress in Physical Geography* **15**, 261–289.

MacDonald, G. M., Kremenetski, K. V., and Beilman, D. W. (2008). Climate change and the northern Russian treeline zone. *Philosophical Transactions of the Royal Society B: Biological Sciences* **363**, 2283–2299.

MacDonald, G. M., Velichko, A. A., Kremenetski, C. V., Borisova, O. K., Goleva, A. A., Andreev, A. A., Cwynar, L. C., Riding, R. T., Forman, S. L., Edwards, T. W. D., Aravena, R., Hammarlund, D., Szeicz, J. M., and Gattaulin, V. N. (2000). Holocene treeline history and climate change across northern Eurasia. *Quaternary Research* **53**, 302–311.

Mann, M. E. (2021). Beyond the hockey stick: climate lessons from the Common Era. *Proceedings of the National Academy of Sciences* **118**(39), e2112797118.

Marshak, S., and Repcheck, J. (2013). *Essentials of Geology*. Norton, New York.

McCoy, R. M. (2006). *Ending in Ice: The Revolutionary Idea and Tragic Expedition of Alfred Wegener*. Oxford University Press, Oxford.

McLoughlin, S. (2001). The breakup history of Gondwana and its impact on pre-Cenozoic floristic provincialism. *Australian Journal of Botany* **49**, 271–300.

McLoughlin, S. (2011). Glossopteris—insights into the architecture and relationships of an iconic Permian Gondwanan plant. *Journal of the Botanical Society of Bengal* **65**, 93–106.

Meinshausen, M., Smith, S. J., Calvin, K., Daniel, J. S., Kainuma, M. L., Lamarque, J. F., Matsumoto, K., Montzka, S. A., Raper, S. C., Riahi, K., and Thomson, A. G. J. M. V. (2011). The RCP greenhouse gas concentrations and their extensions from 1765 to 2300. *Climatic Change* **109**, 213–241.

Moore, P. D., Webb, J. A., and Collinson, M. E. (1991). *Pollen Analysis*, 2nd edition. Blackwell Scientific Publications, Oxford.

Morain, S. A. (1984). *Systematic and Regional Biogeography*. Van Nostrand Reinhold, New York.

Müller, U. C., and Pross, J. (2007). Lesson from the past: present insolation minimum holds potential for glacial inception. *Quaternary Science Reviews* **26**, 3025–3029.

Neall, V. E., and Trewick, S. A. (2008). The age and origin of the Pacific islands: a geological overview. *Philosophical Transactions of the Royal Society of London B: Biological Sciences* **363**, 3293–3308.

Neilson, R. P., and Marks, D. (1994). A global perspective of regional vegetation and hydrologic sensitivities from climatic change. *Journal of Vegetation Science* **5**, 715–730.

Nolan, C., Overpeck, J. T., Allen, J. R., Anderson, P. M., Betancourt, J. L., Binney, H. A., Brewer, S., Bush, M. B., Chase, B. M., Cheddadi, R., and Djamali, M. (2018). Past and future global transformation of terrestrial ecosystems under climate change. *Science* **361**, 920–923.

North Greenland Ice Core Project Members. (2004). High resolution climate record of the northern hemisphere reaching into the last interglacial period. *Nature* **431**, 147–151.

O'Dea, A., Lessios, H. A., Coates, A. G., Eytan, R. I., Restrepo-Moreno, S. A., Cione, A. L., Collins, L. S., De Queiroz, A., Farris, D. W., Norris, R. D., and Stallard, R. F. (2016). Formation of the Isthmus of Panama. *Science Advances* **2**, e1600883.

O'Hara, K. D. (2018). *A Brief History of Geology*. Cambridge University Press, Cambridge.

Orme, A. R. (2008). Pleistocene pluvial lakes of the American West: a short history of research. *Geological Society, London, Special Publications* **301**, 51–78.

Overpeck, J. T., Hughen, K., Hardy, D., Bradley, R., Case, R., Douglas, M., Finney, B., Gajewski, K., Jacoby, G., Jennings, A., Lamoureux, S., Lasca, A., MacDonald, G. M., Moore, J., Retelle, M., Smith, S., Wolfe, A., and Zielinski, G. (1997). Arctic environmental change of the last four centuries. *Science* **278**, 1251–1256.

PAGES2k Consortium (2017). A global multiproxy database for temperature reconstructions of the Common Era. *Scientific Data* **4**, 88.

Paillard, D. (2015). Quaternary glaciations: from observations to theories. *Quaternary Science Reviews* **107**, 11–24.

Past Interglacials Working Group of PAGES (2016). Interglacials of the last 800,000 years. *Reviews of Geophysics* **54**, 162–219.

Pielou, E. C. (1979). *Biogeography*. Wiley, New York.

Pielou, E. C. (1991). *After the Ice Age: The Return of Life to Glaciated North America*. University of Chicago Press, Chicago.

Plummer, C. C., and McGeary, D. (1985). *Physical Geology*, 3rd edition. Wm. C. Brown Publishers, Dubuque, IA.

Prentice, I. C., Harrison, S. P., and Bartlein, P. J. (2011). Global vegetation and terrestrial carbon cycle changes after the last ice age. *New Phytologist* **189**, 988–998.

Ramsey, C. B. (2008). Radiocarbon dating: revolutions in understanding. *Archaeometry* **50**, 249–275.

Rapp, D. (2019). *Ice Ages and Interglacials: Measurement, Interpretation and Models*, 3rd edition. Springer Nature, Cham, Switzerland.

Ray, N., and Adams, J. M. (2001). A GIS-based vegetation map of the world at the last glacial maximum (25,000–15,000 BP). *Internet Archaeology* **11**, 2.

Raymo, M. E. (1992). Global climate change: a three million year perspective. In *Start of a Glacial* (ed. G. J. Kukla and E. Went), pp. 207–223. Springer-Verlag, Berlin.

Raymo, M. E., and Huybers, P. (2008). Unlocking the mysteries of the ice ages. *Nature* **451**, 284–285.

Ritchie, H., and Roser, M. (2020). *CO_2 and Greenhouse Gas Emissions*. OurWorldInData.org. https://ourworldindata. org/co2-and-other-greenhouse-gas-emissions.

Roberts, N. (2014). *The Holocene*, 3rd edition. Wiley Blackwell, Oxford.

Romano, M., and Cifelli, R. L. (2015). 100 years of continental drift. *Science* **350**, 915–916.

Sauermilch, I., Whittaker, J. M., Klocker, A., Munday, D. R., Hochmuth, K., Bijl, P. K., and LaCasce, J. H. (2021). Gateway-driven weakening of ocean gyres leads to Southern Ocean cooling. *Nature Communications* **12**, 6465. https://doi.org/10.1038/s41467-021-26658-1.

Schuster, M., Duringer, P., Ghienne, J.-F., Vignaud, P., Mackaye, H. T., Likius, A., and Brunet, M. (2006). The age of the Sahara desert. *Science* **311**, 821–821.

Schweingruber, F. H. (1996). *Tree Rings and Environment: Dendroecology*. Paul Haupt, Berne, Switzerland.

Schwörer, C., Gavin, D. G., Walker, I. R., and Hu, F. S. (2017). Holocene tree line changes in the Canadian Cordillera are controlled by climate and topography. *Journal of Biogeography* **44**, 1148–1159.

Scotese, C. R. (2021). An atlas of Phanerozoic paleogeographic maps: the seas come in and the seas go out. *Annual Review of Earth and Planetary Sciences* **49**, 679–728.

Scotese, C. R., and Baker, D. W. (1975). Continental drift reconstructions and animation. *Journal of Geological Education* **23**, 167–171.

Scotese, C. R., Gahagan, L. M., and Larson, R. L. (1988). Plate tectonic reconstructions of the Cretaceous and Cenozoic ocean basins. *Tectonophysics* **155**, 27–48.

Shakun, J. D., and Carlson, A. E. (2010). A global perspective on Last Glacial Maximum to Holocene climate change. *Quaternary Science Reviews* **29**, 1801–1816.

Sigman, D. M., Hain, M. P., and Haug, G. H. (2010). The polar ocean and glacial cycles in atmospheric CO2 concentration. *Nature* **466**, 47–55.

Solomina, O. N., Bradley, R. S., Hodgson, D. A., Ivy-Ochs, S., Jomelli, V., Mackintosh, A. N., Nesje, A., Owen, L. A., Wanner, H., Wiles, G. C., and Young, N. E. (2015). Holocene glacier fluctuations. *Quaternary Science Reviews* **111**, 9–34.

Stanley, S. (1987). *Extinction*. Scientific American Books, New York.

Steffen, W., Leinfelder, R., Zalasiewicz, J., Waters, C. N., Williams, M., Summerhayes, C., Barnosky, A. D., Cearreta, A., Crutzen, P., Edgeworth, M., and Ellis, E. C. (2016). Stratigraphic and Earth System approaches to defining the Anthropocene. *Earth's Future* **4**, 324–345.

Steinthorsdottir, M., Wohlfarth, B., Kylander, M. E., Blaauw, M., and Reimer, P. J. (2013). Stomatal proxy record of CO2 concentrations from the last termination suggests an important role for CO2 at climate change transitions. *Quaternary Science Reviews* **68**, 43–58.

Stott, P. (2016). How climate change affects extreme weather events. *Science* **352**, 1517–1518.

Strahler, A., and Strahler, A. S. (1997). *Physical Geography: Science and Systems of the Human Environment*. Wiley, New York.

Strahler, A. N. (1998). *Plate Tectonics*. Geo Books Publishing, Cambridge.

Tanner, T., Hernández-Almeida, I., Drury, A. J., Guitián, J., and Stoll, H. (2020). Decreasing atmospheric CO2 during the late Miocene cooling. *Paleoceanography and Paleoclimatology* **35**, e2020PA003925.

Tarbuck, E. J., Lutgens, F. K., Tasa, D., and Tasa, D. (2005). *Earth: An Introduction to Physical Geology*. Pearson/ Prentice Hall, Upper Saddle River, NJ.

Tarduno, J., Bunge, H. P., Sleep, N. and Hansen, U. (2009). The bent Hawaiian-Emperor hotspot track: inheriting the mantle wind. *Science* **324**, 50–53.

Tennant, C., and Menounos, B. (2013). Glacier change of the Columbia Icefield, Canadian Rocky Mountains, 1919–2009. *Journal of Glaciology* **59**, 671–686.

Tzedakis, P. C., Channell, J. E. T., Hodell, D. A., Kleiven, H. F., and Skinner, L. C. (2012). Determining the natural length of the current interglacial. *Nature Geoscience* **5**, 138–141.

van den Broeke, M., Box, J., Fettweis, X., Hanna, E., Noël, B., Tedesco, M., van As, D., van de Berg, W. J., and van Kampenhout, L. (2017). Greenland ice sheet surface mass loss: recent developments in observation and modeling. *Current Climate Change Reports* **3**, 345–356.

Velichko, A. A., Wright Jr., H. E., and Barnowsky, C. A. (1984). *Late Quaternary Environments of the USSR*. University of Minnesota, Minneapolis.

Webb III, T. (1987). The appearance and disappearance of major vegetation assemblages: long-term vegetational dynamics in eastern North America. *Vegetatio* **69**, 177–187.

West, C. K., Greenwood, D. R., and Basinger, J. F. (2015). Was the Arctic Eocene "rainforest" monsoonal? Estimates of seasonal precipitation from early Eocene megafloras from Ellesmere Island, Nunavut. *Earth and Planetary Science Letters* **427**, 18–30.

Wicander, R., and Monroe, J. S. (2015). *Historical Geology*. Cengage Learning, Boston.

Williams, J. W., and Jackson, S. T. (2007). Novel climates, no-analog communities, and ecological surprises. *Frontiers in Ecology and the Environment* **5**, 475–482.

Williams, J. W., Jackson, S. T., and Kutzbach, J. E. (2007). Projected distributions of novel and disappearing climates by 2100 AD. *Proceedings of the National Academy of Sciences* **104**, 5738–5742.

Windley, B. F. (1977). *The Evolving Continents*. Wiley, London.

Woillez, M. N., Kageyama, M., Krinner, G., de Noblet-Ducoudré, N., Viovy, N., and Mancip, M. (2011). Impact of CO_2 and climate on the Last Glacial Maximum vegetation: results from the ORCHIDEE/IPSL models. *Climate of the Past* **7**, 557–577.

Wolff, E. W., Barbante, C., Becagli, S., Bigler, M., Boutron, C. F., Castellano, E., De Angelis, M., Federer, U., Fischer, H., Fundel, F., and Hansson, M. (2010). Changes in environment over the last 800,000 years from chemical analysis of the EPICA Dome C ice core. *Quaternary Science Reviews* **29**, 285–295.

Wright Jr., H. E., Kutzbach, J. E., Webb III, T., Ruddiman, W. F., Street-Perrott, F. A., and Bartlein, P. J. (1993). *Global Climates Since the Last Glacial Maximum*. University of Minnesota Press, Minneapolis.

Wu, H., Zhang, S., Hinnov, L. A., Jiang, G., Feng, Q., Li, H., and Yang, T. (2013). Time-calibrated Milankovitch cycles for the late Permian. *Nature Communications* **4**, 2452. https://doi.org/10.1038/ncomms3452.

Yan, Y., Bender, M. L., Brook, E. J., Clifford, H. M., Kemeny, P. C., Kurbatov, A. V., Mackay, S., Mayewski, P. A., Ng, J., Severinghaus, J. P., and Higgins, J. A. (2019). Two-million-year-old snapshots of atmospheric gases from Antarctic ice. *Nature* **574**, 663–666.

Yau, A. M., Bender, M. L., Blunier, T., and Jouzel, J. (2016). Setting a chronology for the basal ice at Dye-3 and GRIP: implications for the long-term stability of the Greenland Ice Sheet. *Earth and Planetary Science Letters* **451**, 1–9.

CHAPTER 8

DISPERSAL, COLONIZATION, AND INVASION

During the 1930s, a series of extremely severe droughts affected the Great Plains of the United States. These droughts destroyed crops and vegetation over large areas of the southern Great Plains. Lack of vegetative cover allowed the top soil to be removed by the wind. Fields were rendered infertile due to lack of water, loss of soil, or deep accumulations of dust. Areas of Oklahoma and adjacent states, which were once productive farmland, became known as the Dust Bowl. People had been migrating from the region in the 1920s, but now many more people living in the Dust Bowl region could no longer support themselves and their families. A number of those displaced packed what possessions they could and moved to other parts of the country that were not as deeply impacted by the drought, such as the central valley of California. The hardships of those times are recounted in family histories, songs, and novels such as John Steinbeck's *Grapes of Wrath*. As personally tragic as the 1930s drought was, it illustrates that all species, including humans, are vulnerable to environmental change and that one of the ways of surviving such changes is for organisms to alter their geographic distributions. The biogeographic history of plant and animal species is in many ways a story of changing distributions in the face of environmental change. In this chapter, we will explore how plants and animals facilitate such changes in geographic distributions. We will also explore the potential consequences that such changes in one species' range can have on other plant and animal species.

As we have seen, changes in the geography and environment of the earth have occurred in the past and will continue into the distant future. Such changes can be either negative or beneficial to individual organisms and species. Changing climate may make the current geographic area occupied by a species inhospitable. At the same time, new areas may become available for occupation. Species escape extinction due to unfavorable environmental conditions and take advantage of new areas that are available for occupation by having the capacity to disperse either themselves or their progeny to new geographic locations.

For many plant species, a single seed that germinates and matures can lead to the establishment of a viable population. In the case of some plants, such as marram grass *(Ammophila arenaria)* which is common on beach dunes, a single stem fragment can grow vegetatively and establish a new population. For most animal species, at least one male and one female are required to establish a viable population. There are, however, some exceptions. For many species of insects and birds, a single female carrying fertilized eggs can establish a viable population. Snail species can self-fertilize, and a single individual can produce a viable population. Some fish species have been shown to actually undergo spontaneous changes in gender, allowing two females to establish a reproductive population. In view of this diversity of strategies, biogeographers use the term **propagule** to refer to the stage in the life cycle (a plant seed, for example), the part of the organism (a piece of marram grass that can develop a new set of roots and grow into a full-sized grass plant), or the group of organisms (a male and female rabbit of mating age) that is required to establish a new reproducing population. Upon arrival at a new geographic location, the propagule must then be able to establish a viable reproducing population in order for the species to survive in that location. In many cases, propagules are dispersed to locations where they cannot establish a viable population due to the existing physical or biological conditions. In other cases, the establishment of a new species causes dramatic changes in the physical and biological environment of the newly occupied area. In many cases, propagules

Biogeography: Introduction to Space, Time, and Life, Second Edition. Glen M. MacDonald.
© 2025 John Wiley & Sons, Inc. Published 2025 by John Wiley & Sons, Inc.
Companion website: www.wiley.com/go/macdonald/biogeography2e

disperse and establish within the current geographic ranges of the species. Such processes are crucial for the survival of a species within its range. In other cases, dispersal and establishment occur at sites that are beyond the current geographic range of the species. Such events that lead to range expansions are called **colonizations** or **invasions.** The term *invasion* is often used to describe geographic range extensions that are caused by the introduction of exotic species by humans. The changes in range boundaries that occur due to dispersal and establishment are an important component of understanding the geographic distributions of species. In this chapter, we will examine the processes and patterns of dispersal, colonization, and range expansion events. We will also look at some examples of biological invasions and their impacts.

DISPERSAL

Although **dispersal** can be defined simply as the movement of an organism away from its point of origin, this definition is too vague to be very useful. From the perspective of the biogeographer and the ecologist, there are two main categories of dispersal events. The first category is *intra-range dispersal* or *ecological dispersal*, which results in the movement of a propagule from its place of origin to a new site within the current geographic range of the species. Intra-range dispersal is extremely important in maintaining species populations within their geographic ranges. In some instances, the local environment of a plant or animal's place of origin may not be suitable for survival of progeny. Many tree species provide good examples of this. Seedlings of light-demanding genera, such as pines, have low photosynthetic rates when shaded. These seedlings cannot survive when they germinate directly under the canopy of their parents. Dispersal away from the parent tree is crucial for the survival of progeny. The study of seedling distributions for trees and shrubs often shows a lack of regeneration directly beneath the parent. In many cases, dispersal of progeny decreases competition with parents and siblings for light, water, food, or nesting sites. In some instances, dispersal also decreases rates of predation upon offspring. Many seed predators such as rodents and squirrels concentrate their feeding in areas close to parent plants where seeds are most plentiful. In such cases, seeds that are dispersed away from the parent have a higher survival rate. Finally, changes in physical and biological conditions, caused by disturbance and subsequent succession, can make sites uninhabitable for species. Many early successional herb species cannot survive shading and competition by shrubs and trees that come later. In order to survive, the early successional species must be able to disperse seeds onto newly disturbed sites.

The second general form of dispersal is *extra-range dispersal* or *biogeographical dispersal.* This type of event results in the movement of the propagule away from its place of origin to a new site that lies outside the current geographic range of the species. Extra-range dispersal allows species to colonize new regions and is especially crucial when an environment becomes unfavorable since it helps the species avoid extinction. Obviously, these types of dispersal events are of great interest to biogeographers.

Evolutionary ecologists sometimes use the term *long-distance dispersal* to signify the movement of a propagule outside the parent population's geographic limits and beyond genetic contact with the parent population. New populations established in this manner are genetically isolated from the parent populations and over time may exhibit evolutionary traits that are not shared with the parent population. Most extra-range dispersal events could be considered long-distance dispersal in this context. However, dispersal of propagules over long spatial distances can occur within the general geographic range of a species and does not equate with a change in the species' overall geographic range.

Organisms disperse their propagules in a number of different ways. Dispersal mechanisms may be passive or active. **Passive dispersal** requires an outside force to move the propagule. The force can be a physical one, such as wind and water, or it can be in the form of biological agents, such as birds and mammals that transport seeds in their feathers, fur, or digestive system. Plants cannot move under their own power and are almost all obligate passive dispersers. **Active dispersal** relies on the propagule itself to provide motion. Species that propel themselves, such as birds, bats, and insects, have been known to cross 3000 kilometers of open ocean. An example is the dispersal of the cattle egret *(Bubulcus ibis)* to South America from Africa (Fig. 8.1). Either a breeding pair of egrets or a female with fertile eggs crossed the Atlantic Ocean from Africa and established a population of egrets

FIGURE 8.1 A cattle egret *(Bubulcus ibis)*. In the nineteenth century, the cattle egret spread from Africa to South America, and it has now spread throughout much of the New World. Cattle egrets that spread to New Zealand in the 1960s have now become established there. Part of the recent spread of the cattle egret can be attributed to increased areas of pasture land throughout the world *(Source:* Cavan / Adobe Stock Photos).

in South America in the late 1800s. The distance flown in this case was over 2000 kilometers. Even large terrestrial mammals, such as elephants, have been shown to swim distances of 10 kilometers and can disperse to nearby islands. In general, plants are passive dispersers. However, there are some interesting species that actively disperse seeds. One example is the dwarf mistletoe *(Areuthobium americanum),* which is a parasite on lodgepole pine *(Pinus contorta)* and jack pine *(Pinus banksiana).* The dwarf mistletoe carries its seed in its fleshy base from which pressure changes can eject the ripe seed to distances of several meters.

Many animals are active dispersers, but not all. For example, the larvae of many marine invertebrates, including many gastropods, bivalves, worms, and crustaceans, exist as free-floating plankton that are moved by water currents. Some planktonic invertebrate larvae found in the Pacific Ocean can survive 3 to 6 months and travel distances of 2000 to 4000 kilometers to new shallow-shore habitats. Such dispersal is responsible for the rich invertebrate fauna associated with reefs on even the most remote Pacific islands such as Hawaii. Mature barnacles, such as the common goosefoot barnacle *(Lepas fascicularis),* adhere to marine organisms such as turtles and whales, as well as ships, and are passively dispersed. The various modes and geography patterns of dispersal by marine organisms has generated much scientific interest and some general insights. Not unexpectedly, it has been found that fish and other species that produce free-floating pelagic eggs exhibit greater dispersal distances than species that produce demersal eggs, which rest on the sea floor. Larger fish species tend to exhibit greater dispersal distances than smaller fish. Interestingly, it has been found that fish species that live at the higher latitudes exhibited greater dispersal distances than those found in the topics.

Passive dispersal by animals is not restricted to aquatic settings. For example, a number of spider species utilize passive dispersal by a process called ballooning. Small spiderlings release silk strands that catch the wind and disperse the spiderlings on air currents. The black widow spider *(Latrodectus hesperus)* exhibits this mode of passive dispersal by its spiderlings. Ballooning spiderlings can be carried great distances. The famous evolutionary biologist Charles Darwin recorded the presence of hundreds of ballooning spiders 100 kilometers off the coast of Argentina.

Passively dispersing plants and animals are often classified according to the modes by which their propagules are dispersed. **Anemochores** are dispersed by the wind. For many plants, the wind disperses their seeds, which have specialized shapes and structures to aid in dispersal. The seeds of maple trees *(Acer* spp.) and many pines *(Pinus* spp.) possess winglike structures of thin, hard tissue attached to the seed coat. Seeds of this type are referred to as **samara seeds.** The name comes from the winged helmets worn by ancient Japanese warriors. When samara seeds fall, the wing causes the seed to rotate much like a helicopter blade. The spinning imparts a small amount of lift and decreases the settling velocity of the propagule. This allows the seed to remain suspended in the air and be transported away from the parent tree. Other seeds, such as the common dandelion *(Taraxacum officinale)* and aspen trees *(Populus tremuloides),* possess small tufts of hairlike material that catches the wind and helps keep the seeds aloft. Interestingly, some plants, such as birches, that live in environments with winter snow cover release a proportion of their seeds in winter when the seeds can be blown large distances across the smooth snow surface. Tumbleweeds exhibit another interesting form of anemochory. Adult tumbleweeds die and are rolled along by the wind. During this movement, the tumbleweed releases seeds. The Russian tumbleweed *(Salsola iberica),* originally from Eurasia but now found widely in North America, disperses some 20,000 to 50,000 seeds over great distances through this strategy.

Anemochory is not restricted to plants. We have already described ballooning by spiderlings. Even without such an adaptation, wind can carry insects very great distances. Insect fauna captured in air sampling traps over the central Pacific Ocean include spiders, mites, flies, butterflies, moths, beetles, and other insect groups in roughly the same proportion as these invertebrates are represented in the fauna of Hawaii. Ancestors of many of the native spiders, mites, and insects found on Hawaii likely made the 3000- to 4000-kilometer trip from Asia, Australia, or North America as anemochores. For these insects, and in many cases for birds and bats, long-distance dispersal is caused mainly by the organism being wafted along by strong winds rather than by the navigational efforts of the organism itself.

Hydrochores are dispersed by water. The adults, larvae, and eggs of many aquatic organisms are hydrochores. Even relatively large organisms such as crabs and starfish may have a juvenile stage that is dispersed as plankton by water currents. One of the most striking hydrochores among terrestrial plants is the coconut palm *(Cocos nucifera)*. The thick husk and shell of the coconut keep the seed afloat and protected from salt water for very long periods (Fig. 8.2). When deposited by storm waves

FIGURE 8.2 Coconut palm *(Cocos nucifera)* seeds germinating on a tropical beach. The widespread distribution of the palms on tropical coastal areas and islands owes much to the ability of palm seeds to remain viable after long transport on ocean currents. In addition, Polynesians brought coconut palms with them and introduced the species to many remote islands such as the Hawaiian Islands (*Source:* OlegD / Adobe Stock).

on suitable beaches, the coconuts germinate and grow. In this manner, coconuts have been able to spread to many tropical islands. Organisms dispersed by sea water are also called thalassochores. Such propagules must be able to withstand submergence in salt water. For the seeds of many plants, contact with salt water causes mortality. However, the seeds of other plants, like the coconut, can tolerate relatively long periods of contact with sea water. A study of seeds from plants growing on the Aldabra Atoll in the Indian Ocean showed that 81% of the native species had seeds that could survive up to 8 weeks of submergence in sea water.

Anemohydrochores are dispersed by wind or water. The tiny plumed seeds of aspen trees can be dispersed by both wind and water. This ability is particularly useful for floodplain species and plants that exist on islands.

Zoochores are dispersed by animals. Many plants have seeds that are adapted to cling to the coats of passing mammals. This form of transport is called exo-zoochory. For example, ragweeds *(Ambrosia spp.)* produce seeds with small spikes that hook into fur or clothing. Plants like burclover *(Medicago polymorpha)* produce barbed pods that fasten to fur or clothes and disperse a number of seeds as the transporting animal or person moves on. Other plants, such as the colorful garden vines of the genus *Plumbago,* produce seeds that are sticky and adhere to passing animals.

Dispersal of seeds through the feeding habits of animals is a common form of zoochory. Birds and squirrels, for example, feed on acorns from oak trees and often carry and store the acorns in the ground for later use. Some of the acorns are not retrieved by the animals and grow into new trees. Jays can bury as many as 4600 acorns in a single season and often transport caches of 5 seeds over 100 meters from parent trees. Other plants, such as fig trees, produce nutritious fruit that is eaten by birds and other animals. The seeds contained within the fruit pass through the digestive track of the animal and are eventually excreted. Some of these seeds come to rest at sites where they can germinate and grow. This form of transport is called endo-zoochory. As might be expected, plants that depend on smaller birds and mammals for dispersal tend to exhibit lower dispersal distances and lower probabilities of long-distance dispersal than plants that depend on larger birds and mammals.

Anthropochores are zoochores that are dispersed by humans. Aside from agricultural crops such as corn *(Zea maize),* which have been specifically bred and cannot effectively disperse seeds without human intervention, most of the anthropochores are dispersed by a number of different animals in addition to humans. The common weed ribwort *(Plantago lanceolata)* is a good example. The ribwort is native to Eurasia but is now found growing throughout the world. Its small brown seeds are sticky and are easily carried on clothing and fur. It spread wherever Europeans settled and introduced horses and cattle grazed. Native Americans sometimes referred to it as "white man's footprint."

Anemochory and zoochory are the two main dispersal strategies for plants. What are the relative advantages and disadvantages? A prime adaptation for wind dispersal is lightness. Wind-dispersed seeds contain very little endosperm to support the seedling after germination. The resistance to death by desiccation or by shading of germinating seedlings of plants such as oaks and chestnuts, which possess large seeds and much endosperm material, is often greater than for species such as birch, which have small light seeds (Fig. 8.3). In addition, the distribution of anemochorous seeds is relatively random depending on the wind. Such seeds can be deposited as easily on inhospitable sites as favorable sites. Many zoochorous seeds are transported from the parent plant to similar sites that are capable of supporting the plant species. However, each of these large seeds costs the parent plant more resources than is required to produce a small anemochorous seed. One result is that fewer seeds are produced for dispersal. The dispersal of zoochorous species is also dependent on the animal that transports the seeds. It has been suggested that the dispersal of North American beech seeds *(Fagus grandifolia)* was in part dependent on the passenger pigeon *(Ectopistes migratorius),* and the extinction of this bird in the early twentieth century hinders beech dispersal today. Deforestation and loss of beech trees may have contributed to the extinction of the passenger pigeon.

In terms of actual transport distance, the differences between anemochores and zoochores are sometimes not as great as might be expected. We can compare the average transport distances for the small seeds of a eucalyptus *(Eucalyptus regnans)* and the zoochorous acorns of oaks dispersed by the European jay *(Garrulus glandarius)* (Fig. 8.4). The vast majority of the eucalyptus seeds are deposited within 50 meters of the parent stands, and only a tiny portion are transported more than

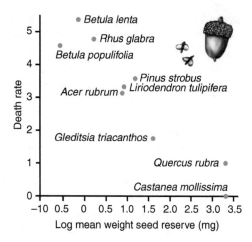

FIGURE 8.3 The relationship between seed size and mortality by shading for a number of tree species found in North American and European deciduous forests (adapted from Grime and Jeffery, 1965).

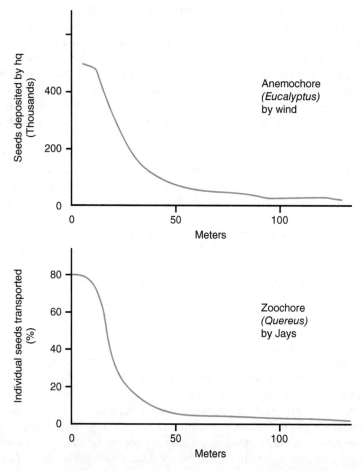

FIGURE 8.4 Seed dispersal curves for an anemochore (Eucalyptus) and the individual seeds of a zoochore (Quercus) carried by jays (Garrulus glandarius) (data from Cramer, 1965; Begon et al., 1996).

100 meters away. The same is true for the transport of individual oak seeds by jays. In general, jays do not transport single seeds more than 100 meters. However, when the jays transport groups of two to five seeds, they often carry these more than 100 meters. For both anemochores and zoochores, the majority of seeds are often dispersed very close to the parent plant and a few are dispersed distances

of hundreds to thousands of meters. We know from evidence of plant establishment on isolated islands, or the spread of newly introduced plant species, that both anemochorous and zoochorous seeds are occasionally dispersed hundreds to thousands of kilometers.

Biogeographers and ecologists recognize that simply transporting seeds does not itself ensure that the propagule performs its reproductive function and leads to the development of a mature plant capable of reproduction. It is also recognized that a combination of agents might be responsible for the dispersal event. For example, a bird may deposit a seed in water and that then delivers the seed to a place where it germinates. The agents of dispersal are more completely assessed using the framework of seed disperser effectiveness (SDE). Within this framework, both the dispersal of the seed and its probability of establishing an adult plant once dispersed are assessed. SDE is quantified as the number of seeds dispersed by a given dispersal agent, or combination of agents, multiplied by the probability that an individual seed will produce a new adult.

If we look more closely at plots of the distance that propagules are dispersed from parents, we see that the resulting dispersal curve for most species follows a generally similar pattern. High numbers of individuals are dispersed close to the parent, and an exponential or normal curve of declining numbers of individuals are dispersed farther away (Fig. 8.4). These types of dispersal curves are often described using the term **dispersal kernel.** The dispersal kernel is the mathematical function that describes the probability of dispersal of the propagule different distances. Most dispersal kernels are typified by distribution curves in which most of the seeds are dispersed short distances. The probabilities of dispersal more than a few meters or tens of meters is often extremely low. An exponential dispersal curve, with a very high proportion of offspring dispersing close to the parent, appears to be more common for passive anemochores, while a normal curve, with less offspring establishing very close to parents, is often observed for animals. The difference in dispersal kernels may be because animals can actively choose to disperse farther, thereby avoiding competition, while passive dispersal by wind is essentially a physical process dependent on winds and settling velocities of the propagules. Although only a tiny fraction of propagules may be dispersed long distances, biogeographers and ecologists are increasingly realizing the importance of such events in the ability of species to modify their ranges and occupy new regions. Long-distance dispersal events are also recognized as being very important for the evolutionary development of plant and animal species.

We have seen that dispersal is important for species to escape inhospitable conditions and reach new habitats. However, a propensity for distant dispersal can also be disadvantageous. The fact that an organism occurs in a specific place often indicates that conditions at that site are favorable for individuals of that species to survive and grow to reproductive maturity. In addition, having many propagules dispersed and established together on a nearby favorable site may allow the species to out-compete other species. Dispersal away from the site may bring the organism to a location in which conditions are inhospitable and competition from other species is high. For example, anemochorous seeds from island-dwelling plants are likely to be blown into the ocean and die. Flying insects and birds could potentially venture too far from the island and be lost. Interestingly, some plant species that have evolved on islands lack plumed or winged seeds that are typical for related species living on the mainland. Similarly, some island species of birds, flies, and beetles have lost the ability to fly. The dodo bird *(Raphus cucullatus),* an extinct relative of the pigeons, that lived on Mauritius Island in the Indian Ocean, is a good example. The ancestors of the dodo had dispersed to the islands by flying, but the dodo birds were incapable of flight. The typical distribution patterns of both anemochorous and zoochorous seeds, with the majority of propagules falling close to the parent and a few being transported great distances, naturally increases the chance of species survival. This form of dispersal curve means that the majority of propagules found close to the parent can take advantage of the probability that sites there will be hospitable. However, the smaller number of propagules that are dispersed to great distances may help to advance the range and long-term survivability of the species by taking advantage of potentially favorable conditions at new sites. Some plants, such as the telegraph weed *(Heterotheca latifolia),* produce both plumed seeds for air dispersal and nonplumed seeds that fall close to the parent plant. In general, the ability to disperse propagules great distances seems favored by organisms living in frequently disturbed environments, while it is not so favored by species in very stable environments or on islands. The growing evidence of the ecological and evolutionary importance of these relatively rare long-distance dispersal events is a topic of intense interest among biogeographers and ecologists,

COLONIZATION, SEASONAL MIGRATIONS, AND IRRUPTIONS

Colonization occurs when a propagule arrives in an area previously unoccupied by the species and establishes a reproducing population, thereby expanding the geographic range of the species. Being adept at dispersal does not benefit a species if it is unable to colonize its new location. Failure to establish a permanent population is common in widely dispersed organisms. Each year, about 100 species of birds from Asia and Europe are seen in North America, yet do not establish permanent populations. Before we discuss the process of colonization, we should consider two interesting phenomena that do not represent true dispersal-colonization events. These two phenomena are seasonal migrations and episodic irruptions.

Seasonal migrations are the annual movements of organisms from one regularly occupied geographic region to another for purposes of avoiding harsh conditions, feeding, and mating. Many species exhibit such behavior. Not only do they move in a regular annual fashion from one region to another, but they often follow the same routes between regions. The area occupied by the species during the summer is often referred to as its summer range to differentiate it from the area occupied in the winter, which is called its winter range. The regular paths taken repeatedly between winter and summer ranges are called migration routes. Technological advances over the past four decades have greatly expanded our ability to follow the migrations of individual animals and increased our understanding of these migration routes. The techniques and science of **biologging** to study animal migration is further discussed in Technical Box 8.1.

Seasonal migrations are very common for bird species. Many song birds in the eastern United States migrate to Mexico and Central and South America during the winter and return north to breed in the summer. In fact, about two-thirds of the bird species breeding in eastern North America migrate to the subtropics and tropics seasonally. Some shorebirds such as golden plover *(Pluviialis dominica)* and knots *(Calidris* spp.) breed in the tundra biome of Europe, Asia, or North America and spend the winter months in coastal regions of southern Africa, South America, Australia, and

---

TECHNICAL BOX 8.1

Biologging and Tracing Animal Dispersal and Migrations from Space

Direct observational studies and the passive tagging of larger organisms with things like bands on birds have long been used to trace animal dispersal and migration events. In the 1960s this even included techniques such as attaching lines with balloons onto marine mammals and physically observing movements. Over the past three decades, the science of *biologging*—the use of highly miniaturized active tags such as tiny recording devices or transmitters to track animal movements—has grown rapidly and allowed for much more detailed understandings of dispersal events and migrations of a much wider range of terrestrial and marine animal species. In addition to tracking locations, biologging devices can also be used to record information on the animal's physiology, the animal's behavior, and the ambient environmental conditions as the animals move.

The growth in biologging not only reflects the miniaturization of sophisticated electronic transmitting and recording devices, but also the development of satellite systems for receiving transmissions from even the most remote locations and the evolution of satellite systems, geographic positioning system (GPS) technologies, and analytic tools. Tiny neurologgers that measure less than 60 millimeters in their longest dimension and less than 1.5 grams can capture multiple channels of physiological or environmental information and easily be deployed on small birds. One such logger tracked the migration of an alpine swift *(Tachymarptis melba)* from Europe to west Africa and showed the bird remained airborne for a remarkable nonstop flight of 200 days. Some biologging requires the recovery of the electronic tag from the animal and this may not always be possible. Telemetric tracking uses a signal from a tag to geolocate the animal remotely. The Argos satellite program was initiated in 1978 mainly to collect locational and environmental data from marine locations. The system also has the capacity for animal telemetry tracking from remote locations around the world. Satellite tracking of animals expanded greatly during the 1980s using the Argos system. Early technologies provided animal locations to within several kilometers to several hundred meters. Current technologies can sometimes provide location within tens of meters. In addition, new analytic approaches, such as Fastloc GPS processing, allow positioning to be estimated from the very short periods in which some marine mammals and amphibians come to the surface for air. Satellite tracking using the Argos system and Fastloc GPS processing revealed that one green turtle *(Chelonia mydas)* being monitored undertook a migration of 3979 kilometers in the Indian Ocean. This is the longest yet recorded for this species.

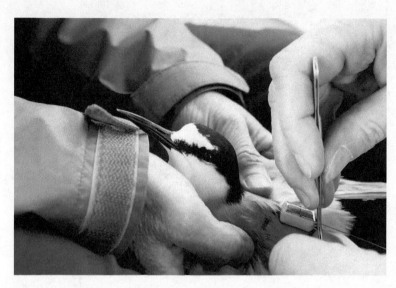

Satellite biotagging an Aleutian tern *(Onychoprion aleutica)*. This species migrates from Alaska to Southeast Asia on a seasonal basis (*Source:* Courtesy of US Fish and Wildlife Services).

New Zealand (Fig. 8.5). Insects may also undertake considerable seasonal migrations. One of the most celebrated instances of this pattern is the travel of the monarch butterfly *(Danaus plexippus)*, which migrates annually from a summer range in the northern United States and southern Canada to a winter range in coastal California and central Mexico. Monarch butterflies that have been tagged in Ontario, Canada, have been found over 2600 kilometers to the south in their Mexican winter range. As remarkable, it is also thought that the wandering glider dragonfly *(Pantala flavescens)* migrates in the fall from India and travels across the Indian Ocean to east Africa using islands such as the Maldives and Seychelles as waystations. Indeed, it is now recognized that extremely long-distance migration is widespread among high-flying insect species. Wildebeests *(Connochaetes* spp.), zebra *(Equus* spp.), gazelle *(Gazella* spp.), and other grazing animals of the Serengeti Plain in Africa also traverse hundreds of kilometers each year from feeding and breeding grounds in the south to feeding grounds in the north. Many marine animals also display seasonal migratory behavior. The Pacific gray whale *(Rhachianectes glaucus)* spends the summer in the North Pacific off the coast of Alaska and the winter off the coast of Mexico. Each year leatherback sea turtles *(Dermochelys coriacea)* travel from the waters of eastern Canada to overwinter as far away as South America and return to Canada in the summer. These are both roundtrip journeys of over 8000 kilometers.

What triggers animals to begin their seasonal migration and what guides them along the same routes and to the same seasonal ranges has long been a topic of research and debate. Cues to commence migration and to navigate along long migration routes can take many forms. This includes the availability of food, smells and other chemical cues, changes in daylight or other celestial phenomena, temperature, and sensing of the earth's magnetic field. Some migratory bird species have been shown to have unusual concentrations of iron minerals contained in the beaks that allows for magnetoreceptive alignment with the magnetic field or sensing magnetic intensity. Experiments with migratory species such as European robins *(Erithacus rubecula)* and reed warblers *(Acrocephalus scirpaceus)* suggest that changes in magnetism are sensed and influence behavior. In fact, sea turtles, fish, and amphibians have all displayed evidence of the use of magnetic properties such as polarity and

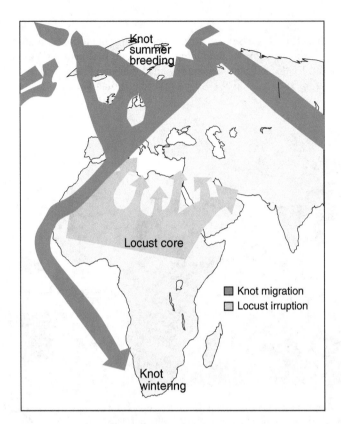

FIGURE 8.5 Examples of migration and irruption. Remarkable long-distance annual migration by different species of Eurasian knots *(Calidris)* (after Piersma and Davidson, 1992). Irruption expansion of the desert locust *(Schistocerca gregaria)* swarms from breeding areas in central Africa (Begon et al., 1996).

inclination in navigation. A number of insect species also display evidence similar sensitivity to the earth's magnetic field, as well as sensitivity to cues from celestial sources such as the sun. Monarch butterflies appear to use both solar cues and geomagnetic cues to guide them on their long migrations. Many birds also use celestial cues for navigation.

Irruptions are episodic explosions in the population size and geographic ranges of insects or animals. Locusts are the best-known example of species that experience dramatic irruptions. The desert locust *(Schistocerca gregaria)* has core areas in central Africa where the species usually lives and reproduces. Increases in population, often tied to increased rainfall, then followed by decreased food supplies in their home range, cause the locusts to swarm and disperse over huge areas of Africa and adjacent Asia (Fig. 8.5). The 2004 irruption was particularly large and eventually extended from the Canary Islands to west Africa, to the Red Sea, and northward into the Mediterranean basin. Some African locust swarms have even crossed the Atlantic to South America. However, typically irruptive species such as locusts cannot reproduce and permanently occupy this extended range. The range of the species eventually collapses back down to a core area. The desert locust has peaks in such irruptions roughly every 15 years. The desert locust is not the only such irruptive locust. The Moroccan locust *(Dociostaurus maroccanus)* is found in some Central Asian countries. Irruptions of the Italian locust *(Calliptamus italicus)* can cause extensive crop damage in areas of Russia and China. The Australian plague locust *(Chortoicetes terminifera)* experiences irruptions that lead it to swarm, expanding into that continent's agricultural regions in search of food. The effects of locust irruptions are so large in scale that satellite remote sensing is an important tool in monitoring locust irruptions and the damage created. The sensors used range from the Advanced Very-High-Resolution Radiometer (AVHRR), Landsat, the Moderate-Resolution Imaging Spectroradiometer (MODIS), and the Satellite Pour l'Observation de la Terre — Végétation (SPOT-VGT).

During a true colonization event, a species arrives in a new area and establishes a reproducing population. Population biologists have long been interested in the demography of such events. If an organism were to colonize a new area, which possessed unlimited resources to support the species,

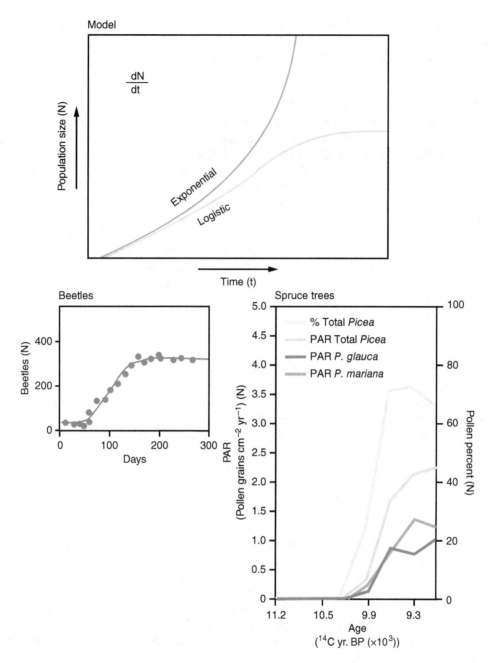

FIGURE 8.6 Exponential and logistic population growth curves. *(Lower left)* An example of logistic population growth observed in the laboratory for beetles *(Rhizopertha dominica)* (data from Crombie, 1945). *(Lower right)* An example of logistic population growth for spruce tree species *(Picea)* colonizing areas of western Canada during the Holocene (MacDonald, 1993). The population size of the spruce trees in the study region is inferred from the amount of spruce pollen deposited in lake sediments.

we would expect population growth to be exponential (Fig. 8.6). Imagine that a male and female finch are blown to a large island and reproduce. The next year the original breeding pair may have produced two male and two female offspring. The three pairs of finches that now exist on the island in turn produce a total of 12 more offspring, bringing the total population to 18 in a relatively short span of time. If we were to plot the growth of the finch population against time, we would observe an exponential curve (Fig. 8.6). The rate of **exponential population growth** can be described by the following simple equation:

$$dN / dt = rN$$

where dN = the change in population size (N), dt equals the time increment (one generation in our example), and r is the per capita rate of population growth (the number of finches that each parent produces minus the number that die prior to reproduction in our example). The rapidity of growth that is possible due to exponential increases in population explain why it is often so difficult to control invading species.

If we ponder the example of the colonizing finches a bit more, we must consider that the island that they have invaded does not really offer infinite resources for population growth. Eventually, all of the sites on the island that can support finch nesting and feeding will be occupied, and the population cannot continue to grow. Our hypothetical island may be able to support only 100 finches. We may recall from earlier chapters that the environmentally controlled limit to the population size of a species in a particular area is called the carrying capacity. In our example, the carrying capacity of our island for finches is 100. As population size approaches carrying capacity, intraspecific competition increases for resources. For example, two pairs of finches that live near each other may compete for nesting sites and food. In the end, only one pair might survive and reproduce. This increasing competition between offspring causes the rate of population growth to slow as carrying capacity is approached. The population growth of a colonizing species in a setting with finite resources has the S-shaped form of **logistic population growth** (Fig. 8.6). We can mathematically describe the logistic population growth of a colonizing species as follows:

$$dN / dt = rN\left(\frac{K - N}{K}\right)$$

where K is the carrying capacity. The equation simply expresses that as N approaches K, the rate of population increase will slow and there is no growth after $N = K$. Since all environments can support only a limited number of individuals of any species, the logistic model is a better general approximation of the growth of colonizing species than the exponential model.

One convenient way of comparing the growth rates for different populations that are increasing either exponentially or logistically is to calculate the population doubling time. This is simply the number of years it takes for the population to double in size. Once the per capita rate of population growth rate (r) is calculated, the population doubling time (t_d) is calculated from the following equation:

$$t_d = \ln 2 / r$$

The ln 2 approximately equals 70, and this rounded value is typically used in the equation.

The logistic equation for the population growth of colonizing species has been tested in many laboratory experiments on population growth of single-celled organisms such as yeast cells, insects such as fruit flies *(Drosophila melanogaster)*, water fleas *(Daphnia magna)*, beetles *(Tribolium* and *Rhizopertha)*, and simple plants such as aquatic duckweed *(Spirodela oligorrhiza)*. There is also evidence that natural populations grow in a logistic manner. For example, a study of the primary succession on the volcanic island of Surtsey near Iceland showed that the colonization of bare rock by moss followed a clear logistic pattern. Fossil pollen evidence of the population growth of some, but not all, tree species colonizing deglaciated landscapes (Technical Box 8.2) at the start of the Holocene also reveals logistic growth (Fig. 8.6). The population doubling times for different tree taxa ranged from less than 30 years to over 1000 (Table 8.1).

In addition to the smooth S-shaped pattern of simple logistic growth, some colonizing populations show evidence of a phenomenon that can be called carrying capacity overshoot. In these cases, the initial growth of the population continues beyond the average carrying capacity of the area, and the population experiences a catastrophic decline to the carrying capacity, or below it. One example comes from studies of reindeer *(Rangifer tarandus)* populations on the Pribilof Islands of the Bering Sea. In 1911, 4 male and 21 female reindeer were introduced to St. Paul Island in order to develop herds for food, clothing, and export. The initial growth of the population was exponential, and by 1938 the island supported more than 2000 reindeer. Then, with no prior evidence of logistic decline in population growth rate, there was a massive and continued decline until only 8 reindeer remained in 1950. The carrying capacity of reindeer on the islands has remained very low. The decline appears to have been prompted by overgrazing by the reindeer and sharp declines in lichens and plants. In a sense, the introduction and rapid population growth of the reindeer caused a catastrophic

Reconstructing Plant Colonization Using Fossil Pollen

Fossil pollen grains found in lake sediments and peatlands have often been used to reconstruct the history of plant colonization, particularly the northward spread of North American and European tree species that occurred at the end of the Pleistocene. The technique is based on the premise that when a plant species first arrives in the vicinity of a lake or peatland, it will be represented by very low amounts of pollen in sediments deposited at that time. As the population of the colonizing plant increases in the vicinity of the site, the amount of pollen from that species being deposited will increase in proportion to the size of the plant population. Radiocarbon dating is usually used to derive a time scale for the lake sediment and peatland stratigraphic records. The British palynologist Keith Bennett showed that by studying the increase in the accumulation rates of tree pollen over time at one site, palynologists could reconstruct the population growth rate for the tree species as it colonized the area around the site. Such studies have been carried out for a number of species and genera in England, Europe, Australia, and North America.

The American palynologist Margaret B. Davis pioneered the approach of mapping the post-glacial geographic expansion of tree species. She did this by using networks of many fossil pollen sites. She demonstrated that by collecting pollen stratigraphies from a number of sites and carefully radiocarbon dating those stratigraphies, it is possible to obtain one line of evidence of the northward colonization of tree species that occurred following the close of the Pleistocene. Reconstructions of post-glacial plant migrations have been carried out for a number of North American and European tree species and genera. The work revolutionized the understanding of plant migration and the changeable nature of plant community species compositions.

Pollen analysis provides one of the ways in which we can study the population dynamics and geographic patterns of naturally colonizing plant species. However, it is recognized that the fossil pollen record is an imperfect one. For example, some trees, such as pines, produce great amounts of pollen that can be carried far from the parent tree. Pine pollen can be found at sites that lay far beyond the range limits of the trees. Alternatively, some trees, such as firs, produce small amounts of heavy pollen and may not be detected in pollen records from sites around which the trees are present. Finally, it is not known if the relationship between tree population size and the amount of pollen deposited in lakes and bogs is linear. New approaches to tracing the past geography of plant species distribution in the Late Quaternary include simulation modeling and the analysis of genetic data from modern plant populations. In a number of cases, these lines of evidence suggest pollen-based estimates may miss small populations and overestimate rates of diffusive migration. Thus, reconstructions of population growth based on pollen records are best considered uncertain approximations, but can be important lines of evidence when combined with other indicators of past plant distributions.

Distinguished palynologist Margaret B. Davis (*Source:* Courtesy of the University of Minnesota).

TABLE 8.1   The Post-Glacial Population Growth Rates (as Expressed as Doubling Time) of Some Tree Genera at Various Sites in North America[a]

Tree Genus	Doubling Time (Years)
Fir (*Abies*)	408
Pine (*Pinus*)	80–1100
Douglas fir (*Pseudotsuga*)	52–365
Hemlock (*Tsuga*)	31–239
Beech (*Fagus*)	124–444

[a] Estimates are calculated from fossil pollen data.
Source: Adapted from MacDonald, 1993.

disruption of the entire island's ecosystem. Such exponential growth followed by a precipitous decline down to carrying capacity has been shown in a number of laboratory and field-based studies of colonizing organisms. The phenomenon seems prevalent in species that have significant overlaps in generation and where breeding aged individuals can survive the initial onset of resource decline.

Many factors can work to impede colonization. In some instances, unfavorable physical environmental conditions cause mortality or reproductive failure. For example, coconut palms cannot survive frost. Although on occasion coconuts may be carried to high-latitude beaches, they will not germinate, or the seedling will be killed at the first frost. The same fate awaits tropical birds or insects that are blown to high-latitude locations. Coral larvae are widely dispersed as plankton in oceanic currents; yet most corals do not grow in waters that fall below 18° C in temperature, and many of these larvae end up in waters in which they cannot survive. Biological factors can also work against successful colonization. The arriving species may be confronted by a superior competitor and not be able to survive. Many laboratory studies show that rates of population growth decline significantly in colonizing populations in the presence of a competitor. Plant-eating insects are often highly stenophagous and may not survive because their food plant is not present. Finally, some species may face predators or pathogens against which they have no defense.

What determines the general probability that an organism will be able to successfully colonize a new area? Some organisms seem particularly well suited for rapid dispersal and successful colonization. These species are sometimes referred to as **supertramps.** The term was used by the American evolutionary biologist and geographer Jared Diamond to describe bird species that seemed adept at dispersing to and colonizing recently disturbed islands off New Guinea. The characteristics that enhance the ability of species to disperse and colonize new habitat include widely dispersed propagules, rapid population growth rates, and the ability to survive and reproduce in a wide range of environmental conditions and with a wide range of resources (in other words, possessing a generalist niche). Many supertramp species also appear adept at colonizing disturbed sites. In many cases though, supertramps do badly in the face of heavy competition from other species. The Eurasian dandelion (*Taraxacum officinale*) is an example of a supertramp. This annual plant has wide climatic tolerance, thrives in a number of soil conditions, matures and sets seeds in one growing season, and has small plumed seeds that are easily carried by wind or affixed to fur and clothing. It is often inadvertently transported by humans. It can rapidly colonize disturbed and open canopy sites. Thanks in part to human dispersal, the dandelion is now found on every continent except Antarctica. Surprisingly, the propagules of many supertramps are spread by passive dispersal rather than active dispersal. An analysis of dispersal and colonization rates, compared to body size and maximum speed of active dispersal, shows that large animals, which have the potential for active dispersal over large distances, often have slower rates of dispersal and colonization than small passive dispersers.

DIFFUSION VERSUS JUMP DISPERSAL

In many cases, species expand their geographic ranges by dispersing into closely adjacent areas. Each generation expands the range of the species slightly. This form of range adjustment is sometimes referred to as *slow penetration* but should be more properly thought of as **diffusion.**

In animals, an example of slow diffusion is provided by the nine-banded armadillo *(Dasypus novemcinctus)*. Spreading naturally from Mexico into the southwestern United States and introduced into Florida, the armadillo has taken about 100 years to expand its range from the Mexican border to Arkansas. The diffusion of the armadillo is an example of slow penetration diffusion. During diffusion events, animal species with high dispersal abilities and short generational times can expand their geographic ranges extremely quickly. The European starling *(Sturnus vulgaris)* was introduced in New York in 1891. By 1980, the species had expanded its range in North America to the Pacific Coast of Alaska. The average rate of range expansion for the starling was greater than 40,000 meters per year. The house sparrow *(Passer domesticus)* is also a nonnative species that was introduced to North America from Europe in the nineteenth century. It spread across North America between 1852 and 1910 at a rate similar to that of the starling.

Diffusion is typical of plants that disperse seeds relatively close to the parent and rely on the germination and reproduction of each generation in order to expand into favorable new habitat. During the last glacial maximum, many of the tree species typical of the northern deciduous forest appear to have survived in glacial refugia in what is now the southeastern United States. With the warming of climate and the retreat of ice at the end of the Pleistocene, these species expanded their range northward. Using fossil pollen to reconstruct their northward expansion (see Technical Box 8.1), it has been estimated that some species expanded their ranges northward quickly, and some species, such as eastern hemlock *(Tsuga canadensis)*, experienced slow range expansions (Fig. 8.7). The rates of range expansion estimated for boreal and temperate tree species in eastern North America during

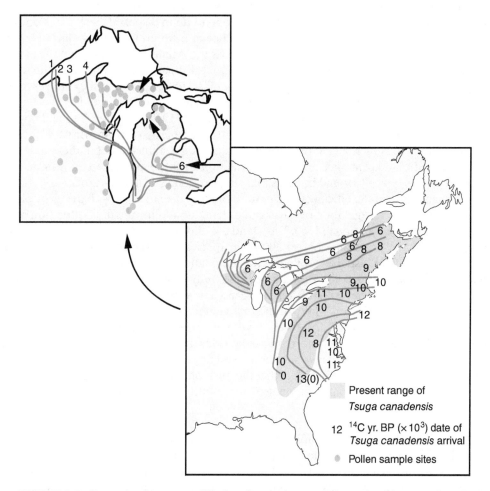

FIGURE 8.7 Example of apparent diffusion dispersal on a continental scale reconstructed using fossil pollen records. The natural spread of eastern hemlock *(Tsuga canadensis)* in North America during the post-glacial period (MacDonald, 1993; adapted from Davis et al., 1986). As noted in the text, such pollen-based reconstructions provide one line of evidence of past plant distributions, but are prone to potential problems.

the Holocene varied greatly and range from less than 100 meters per year to 400 meters per year. However, these estimates must be viewed with caution. Simulation modeling of range expansion rates for tree species, other fossil evidence such as macro-fossils or preserved conifer stomata, and the analysis of genetic data from modern populations often suggest that these pollen-based estimates of diffusion rates may be too fast. The pollen data may miss the presence of small populations in refugia or small populations established by long-distance dispersal and overestimate rates of diffusion. For example, pollen data from Europe for the post-glacial range expansion by silver fir *(Abies alba)* suggest Holocene expansion rates of 400 to 700 meters a year. When genetic evidence for potential refugia of small outlying populations of silver fir are considered and modeled, the maximum rate of expansion is revised to being 250 meters per year. A similar conclusion has been reached for the migration rates of beech *(Fagus grandifolia)* in eastern North America. Rather than the 169 to 200 meters per year calculated from pollen data, molecular genetics data suggest the existence of small refugial populations closer to the glacial ice-front and the revision of migration rates downward to less than 100 meters per year. In terms of the pollen-based reconstructions, what initially appears to be a traceable process of diffusion may be much more complex.

In some cases, dispersal and colonization events do not occur as gradual expansions along the range limits of a species. New populations can be established thousands of kilometers away from the range limits of the species. Such events are known as **jump dispersals.** If species did not possess the ability to disperse in this manner, remote islands such as Hawaii would be devoid of life. The importance and frequency of successful jump dispersal events are evident from the plants and animals that are found on remote islands. The probability of an organism spreading by jump dispersal is generally high for plant species with small wind-blown seeds and for flying animal species. Propagules that are carried by flying organisms are also prone to jump dispersal events. It is not surprising, then, that remote islands such as New Zealand or Hawaii have no native mammals except for bats. The dispersal of the cattle egret *(Bubulcus ibis)* across the Atlantic Ocean from Africa to South America is an example of a jump dispersal over a very great distance across a barrier. Interestingly, the jump dispersal across the Atlantic was followed by the diffusion dispersal of the cattle egret northward and westward into South and North America. More recently, the cattle egret has also expanded its range to New Zealand, becoming established there in the early 1960s. The New Zealand populations arrived from Australia. Although it appears that the cattle egret crossed from Africa to the New World and from Australia to New Zealand on its own, its successful establishment in these new regions is in part attributable to land clearance and the introduction of cattle and sheep by Europeans. Cattle egrets feed on insects typically found in fields that support domesticated livestock.

An example of the efficiency and pattern of jump dispersal across barriers by plants and animals is provided by the revegetation of the island of Krakatoa following the destruction of all life there by a massive volcanic eruption in 1883. The island lies about 40 kilometers from other islands. Yet, within 60 years after the eruption, close to 300 plant species and 30 bird species had colonized Krakatoa. An analysis of plant dispersal strategies for the plants on Krakatoa shows that sea-dispersed plants had the highest initial colonization rates and fully colonized the island within about 40 years. Anemochores had slower colonization rates, and new species were still being added to the island flora a century after the eruption. Many of the first anemochores were species of herbs, grasses, and ferns that grow in open conditions. As forest cover has formed on the island, other anemochores such as orchids, which require moist and shady conditions, began to colonize. Zoochores had the slowest initial colonization rates, and new species are still arriving on the island. The slow rate of colonization by zoochores is influenced by the fact that the birds and bats that carry these seeds will likely not visit the island for any appreciable amount of time until environmental conditions, such as vegetation cover, provide suitable habitat.

The increasing species diversity of the island for thalassochores, anemochores, and zoochores follows a logistic pattern with rapid initial increase in colonizing species and then a flattening out of colonization events. The logistic pattern of colonization probably reflects two factors. First, species with readily dispersed propagules and populations close to Krakatoa dispersed there very quickly. Subsequent arrivees were likely representative of less efficiently dispersed species or plants that had parent populations located farther away. Such species had a lower probability of reaching Krakatoa, and greater amounts of time elapsed before a chance event carried their seeds to the island. Second, the early arriving plant species probably faced little competition and were able to establish reproducing populations relatively easily. Some of the species that arrived later may not have been able to establish populations on the island because of competition from early colonizers. The revegetation of Krakatoa shows us that successful jump dispersal events require both the arrival of the propagule and suitable physical and biological environmental conditions for establishment.

Jump dispersal events are not always restricted to species crossing large expanses of water or other barriers. Sometimes rapid range expansion over large areas of suitable habitat can occur via jump dispersal. In these cases, propagules are dispersed over areas of suitable habitat and establish disjunct populations beyond the main diffusion front. Such events are important for colonization or invasions because these small disjunct populations serve as inoculation centers from which the colonizing or invading species can then expand.

The difference between diffusion and jump dispersal in environments that do not contain large barriers can often be more a matter of scale than anything else. A detailed analysis of the spread of many invading species generally shows that small disjunct populations are often established in advance of the main diffusion front. A good example of this comes from the spread of the house finch (Carpodacus mexicanus) in eastern North America. The birds are native to western North America. The first house finches in the east were released on Long Island in the 1940s. By the mid-1970s, the species had spread from Massachusetts to North Carolina and is now found throughout the eastern and central United States and closely adjacent portions of eastern Canada. A careful and spatially detailed analysis of bird count records shows how the spread of the finch was often facilitated by jump dispersal and the establishment of small disjunct populations in advance of the main diffusion front (Fig. 8.8).

In analyzing the diffusion and continental jump dispersal for a number of different species, some general patterns can be identified. In particular, when the geographic range expansion of a

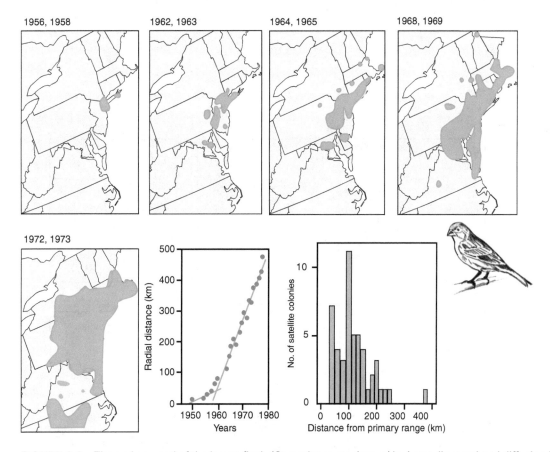

FIGURE 8.8 The early spread of the house finch (Carpodacus mexicanus) by jump dispersal and diffusion in eastern North America (Mundinger and Hope, 1982; Shigesada and Kawasaki, 1997). The mean jump distance for establishment of disjunct populations is 115 kilometers. The species is now found throughout the eastern and central United States and into portions of southern Canada.

colonizing species is graphed relative to time, the form of the relationship is often a logistic curve. This can be seen for both animals and plants and for diffusion and jump dispersals. The northwestward expansion of the American muskrat *(Ondatra zibethica)* following its accidental introduction near Prague, Czech Republic, provides an excellent example of slow initial spread followed by rapid expansion and then gradual slowing. Following its introduction in 1905, the range expansion of the muskrat in Europe was very slow diffusion. The rate of expansion increased between 1920 and the 1950s and then slowed greatly as the species approached its climatically controlled range limits in 1963. In addition, increased trapping programs likely contributed to the decreasing rate of expansion. In recent decades there has been a sharp decline in muskrat populations and the sites in which they are found in countries such as Poland. One factor in this decline may be predation by another invasive species, the American mink *(Neovison vison)*. Cheatgrass *(Bromus tectorum)* was accidentally introduced to western North America in the late nineteenth century. Through a series of jump dispersal events, aided by the transport of its seeds in feed and with livestock, it spread throughout the West by the 1930s. Although it continues to fill in portions of its western range, cheatgrass initially expanded its range in a logistic manner between 1900 and 1930 (Fig. 8.9).

The logistic pattern of range expansion has three phases. First, the initial rate of range expansion is often relatively slow because the small initial population size of the colonizing species limits propagule production and dispersal. As the size of the colonizing population increases, propagule numbers climb and expansion rates increase. The initial slow spread of the house finch in the vicinity of New York City likely reflects this principle. In actively dispersing animals, high densities at the point of introduction may further promote active dispersal as individuals move to avoid intraspecific competition. In some instances, the initial point of colonization may not be the optimal environment for the species. Rates of population growth and range expansion can increase when the species reaches areas that provide better conditions for growth and reproduction. The rapid spread of the armadillo across Texas may reflect the fact that Texas provided better habitat than the desert of

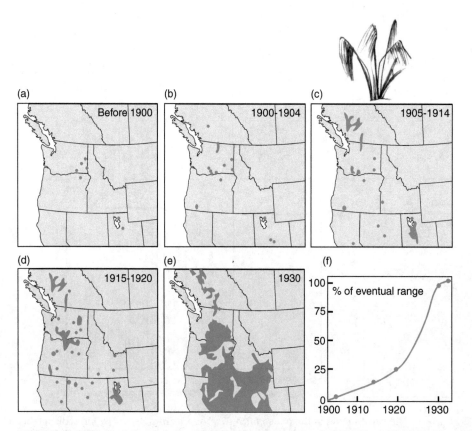

FIGURE 8.9 The logistic expansion of invading species spreading by either diffusion or jump dispersal as displayed by cheatgrass (*Bromus tectorum*) expansion in western North America from around 1900 to 1930 (a–e) and the observed increase in its total geographic range over the same period (f) (Mack, 1981; Shigesada and Kawasaki, 1997).

northern Mexico. Rates of range expansion generally increase after population size builds up and the species moves into favorable habitat. In many cases, rates of expansion decline as the environmentally controlled range limits of the species are approached. This decline probably reflects unfavorable environmental conditions near the range limits of the species that cause slower rates of population growth and propagule production. In addition, survival rates of offspring may be lower, and the chance of successful establishment and colonization decreases. The slow rate of range expansion for the starling in northwestern Canada and the armadillo in northeastern Texas potentially reflects the fact that these environments are not optimal for these species.

BARRIERS, CORRIDORS, FILTERS, STEPPING STONES, AND SWEEPSTAKES

Biogeographers often classify geographic features in terms of their impact on dispersal and colonization. Geographic features that block dispersal and colonization are called biogeographic **barriers.** In contrast, geographic features that promote dispersal and colonization are called biogeographic **corridors.** As we will see, the terms *barrier* and *corridor* are very relative in nature. For example, large water bodies may serve as barriers to land organisms, but may also act as corridors that allow the dispersal of aquatic species. When the term *barrier* is used in a biogeographic sense, it must be related to specific species or groups of species. A barrier for one species is often a corridor for another!

Large expanses of water are a barrier to the dispersal of most terrestrial organisms. Very few reptile or mammal species native to Europe and Asia are native to North America, despite similarities in the physical environments available on both continents. The tree species found in North America are also different from those found in Asia and Europe. Finally, even among land birds and bats, relatively few species are native to both Eurasia and North America. Oceans also serve as important barriers to the dispersal of freshwater fish and amphibians. Most species adapted to life in fresh water cannot tolerate prolonged contact with salt water. Few freshwater fish species ever disperse across ocean barriers and colonize islands and other continents. We know that in many cases, fish species will thrive in new regions if introduced by humans. North American trout species (*Salmo* spp.), for example, have been introduced to New Zealand as game fish and do very well there. However, even small expanses of salt water can pose insurmountable barriers to freshwater fish. Bass (*Micropterus* spp.) and sunfish (*Lepomis* spp.) are extremely abundant in lakes and streams in North America. These genera were not, however, able to cross the relatively short distances from mainland North America to the islands of the Caribbean Sea. Again, populations of these fish transplanted to the Caribbean Islands by humans have survived very well. Isolated islands, such as New Zealand and Hawaii, commonly have no native amphibian species. Even short expanses of salt water are difficult for amphibians to cross. The frog species found on islands in Indonesia are far less diverse than those found on nearby mainland Asia.

Land areas form barriers for the dispersal of marine and freshwater species. Many marine species found in the Pacific Ocean do not occur in the Atlantic. Even small land areas can pose insurmountable barriers to aquatic animals. The Isthmus of Panama is less than 100 kilometers wide in some places, yet many marine species are restricted to either the Pacific Ocean or the Caribbean Sea because of the physical barrier of the isthmus. Even the poisonous marine snake *Pelamis platurus*, which can survive for short periods out of water, is found naturally occurring only in the eastern Pacific Coast and has not been able to disperse overland to the nearby Caribbean Coast. In fact, some 70 species of sea snakes are found in the Indian Ocean and the Pacific Ocean but are absent from the Atlantic Ocean. Both land areas and regions of cold ocean water temperatures form barriers to the dispersal of the sea snakes. Recently, however, the marine snake *Pelamis platurus* has been reported in the Caribbean Ocean off Colombia. The species appears to have been inadvertently dispersed there by humans.

Environmental conditions vary across landmasses and oceans, and this variation can also pose a barrier to dispersal. For example, expanses of desert pose a barrier for the dispersal of mesic plants and moisture-demanding animals. The Sahara Desert in Africa is an important barrier between the circum-Mediterranean region and central and south Africa. The northern African flora and fauna of the Mediterranean coast have great affinity to Europe and much less to sub-Saharan Africa. The relatively moist areas of coastal and montane California are separated from the mesic habitat of the

Rocky Mountains and eastern North America by the Mojave and Great Basin deserts. Over 1500 species of plants found in California are not found in areas east of the state. Western tree species, such as Douglas fir *(Pseudotsuga menziesii)* grow very well when planted in eastern North America, but their populations cannot naturally cross the barriers posed by the dry deserts and grasslands in central North America, or the very cold winter climate and short summers of the boreal forest and tundra biomes to the north.

Within the Great Basin of the United States, a number of mountain ranges rise above the desert and provide mesic montane and alpine conditions. These isolated mountain environments are sometimes called sky islands. The small mammal fauna of these ranges includes species such as squirrels and chipmunks that cannot survive desert conditions. Analysis of the small mammal fauna of these mountains shows that desert distances of a few tens to hundreds of kilometers serve as a barrier to the dispersal of these species. It is thought that the montane and alpine mammal species dispersed to these isolated mountain ranges during glacial periods when the climate of the Great Basin was moister and cooler than it is today. A similar history of cooler temperatures during the last glacial maximum providing a biogeographic connection between currently isolated sky islands has been postulated for high mountain peaks in eastern Africa. Here it is thought that the treeline descended by 1000 meters in elevation and the alpine lifezone was eight times more expansive in size during the last glacial maximum than today. This effectively removed the topographic/climatic barrier that exists today between these now environmentally isolated mountains.

Mountains are also themselves a particularly interesting form of barrier. High-altitude locations have cooler temperatures than adjacent low-elevation sites. In addition, mountainous regions are often moister than nearby lowlands. Since many species of plants and animals are extremely sensitive to cold temperatures, high mountain ranges can be effective barriers to migration and dispersal. However, the degree to which mountains serve as a barrier is also related to latitude. In high-latitude locations, winter temperatures are cold in both the mountains and on the adjacent lowlands. Consequently, the flora and fauna at the higher latitudes must be adapted to withstand cold conditions regardless of whether they live in the mountains or the lowlands. In the tropics, however, there is much less seasonal variation in temperature, and lowlands are generally always warm. High elevations may warm up slightly during the summer but in general remain relatively cool throughout the year. Tropical lowland flora and fauna are often incapable of withstanding cold temperatures, and even in the summer the mountains may not be warm enough to allow survival of lowland organisms. Mountains in the tropics, therefore, are greater barriers than high-latitude mountain ranges.

In concluding our consideration of barriers, it is important to note that barriers to dispersal and migration can be biological as well as physical. Stenophagous predators cannot survive in areas where their prey species are lacking, even if the physical conditions are satisfactory for the predator. Intense competition from other species can also serve as a barrier to colonization. Animal behavior can present yet another barrier to dispersal. A number of tropical bird genera including cotingas, toucans, and ant birds have been shown to avoid crossing even very short distances of open water, despite the presence of suitable habitat only a few hundred meters away.

As mentioned earlier, areas that promote dispersal and colonization are called corridors. A river joining two lakes would be a corridor for the dispersal of aquatic organisms, while an isthmus joining two continents would serve a similar role for terrestrial plants and animals. According to its strict biogeographical definition, a corridor allows the unrestricted movement of all taxa from either side of it. Over time, dispersal and colonization along such a corridor lead to the harmonization of the flora and fauna on both sides. Biogeographical **harmonization** means that similar species of flora and fauna are found on both sides of the corridor. According to this definition, land corridors are continuous areas of similar climate and vegetation. Thus, some extensive biomes, such as the boreal forest of northern North America or northern Eurasia, form corridors. In the case of the North American boreal forest, the flora and fauna found in Alaska are indeed very similar to the flora and fauna found across the continent in eastern Canada. Mountain ranges can serve as corridors for alpine species. Many of the same species of plants and animals found in the Rocky Mountains of Canada occur as far south as the Rocky Mountains of Colorado.

Obviously, in many instances geographic connections, such as rivers or isthmuses, link two regions, but because of physical or biological conditions, all species cannot disperse with equal ease. Avenues of dispersal and colonization that are not equally favorable for all species are called **filters.**

When filters exist, they allow some species to cross and restrict others. Physical environmental conditions, such as temperature, may restrict some species. In addition, biological factors along the filter route, such as competition with other species, may also serve to restrict the dispersal and colonization of some species. The end result is that the flora and fauna on both sides remain distinctive. The Isthmus of Panama, linking North and South America, is an excellent example of a modern biogeographic filter. The formation of the Isthmus of Panama and the biogeographic impact of this landbridge between North and South America is one of the most interesting topics in dispersal biogeography and is well worth our detailed consideration.

Between about 15 million and 3 million years ago, the continents of North and South America lay at about their modern geographic positions. In place of the Isthmus of Panama, there was likely a series of islands set upon a relatively shallow sea of about 150 meters in depth. The lack of a continuous landbridge joining the two continents restricted, but did not completely preclude, the exchange of terrestrial fauna, particularly mammals and flightless birds. However, plant species appear to have been better able to migrate northward and southward by taking advantage of the islands as biogeographic **stepping stones,** a special type of dispersal route. The islands were formed by displacement and volcanic activity along the point of contact between the Caribbean Plate to the east and the Cocos Plate to the west. The ongoing collision between the two plates led to the continued rise of land in the area of the Isthmus until about 3 million to 2.7 million years ago when a continuous landbridge was formed between North and South America. Once the Isthmus of Panama was in place, it facilitated the dispersal of greater numbers of land animals and terrestrial plants north and south along its length. The movement of terrestrial fauna that occurred following the establishment of the isthmus is known as the **Great American Biotic Interchange** (GABI).

For the most part, mammal species in South America had evolved in isolation from other continents, while those in North America represented native families and colonizers from Eurasia. Exchange between North America and Eurasia was facilitated by the late separation of northern Europe and North America and the subsequent close proximity of Asia and North America at Beringia. The North American mammals were placental, whereas South America's mammals were dominated by marsupials. When the Isthmus was formed, there was a considerable exchange of mammals between the two continents (Table 8.2). Interestingly, mammals such as llamas and jaguars, which we typically think of as representatives of the modern South American fauna, are actually derived from colonizers from North America. The opossum, common in North America today, is a colonizer from the south. Aside from the opossum, there were many other interesting marsupials (mammals that carry their young in a pouch), such as giant anteaters *(Myrmecophaga)* and giant sloths *(Eremotherium),* and a giant carnivorous bird *(Titanis)* that stood about 4 meters tall that came northward across the landbridge. However, it appears that North American mammals were far more successful in colonizing South America than South American mammals and birds were in moving northward. It may be that the North American–Eurasian species were competitively superior to the South American species, which had evolved in relative isolation. Some of the North American mammal families also appear to have experienced rapid evolutionary diversification in South America. In contrast, South American mammals appear to have experienced higher extinction rates. As a result of all this, about half of the living mammal genera in South America are descendants from North American families, and living mammal lineages of South American origin are far less prevalent in North America.

When the isthmus first developed, it contained a landscape of both forest and open savanna and functioned, more or less, as a corridor allowing dispersal by many different species. However, during the Pleistocene, dense tropical forest developed in lowland Central America, and the isthmus became more of a filter that allowed dispersal by tropical forest species but restricted the southward migration of North American grazing fauna such as bison and antelope. The dense tropical forest of the modern southern Isthmus still serves as a filter to dispersal. Tropical forest animals can traverse it, but temperate, grassland, savanna, and desert species cannot. The more xeric savanna and desert conditions found in the northern portions of Central America and Mexico in turn serve as barriers to the northward movement of tropical species. The Isthmus also serves as a filter for the dispersal of plants. Pines, which are widely distributed across North America and Eurasia, reach their modern range limits in Nicaragua and appear to have been unable to cross the lowland tropical environment of the southern isthmus and colonize South America. Indeed, even humans are daunted by this tropical forest

TABLE 8.2   Some of the Most Commonly Known Mammal Families of the Great American Biotic Interchange Between North and South America via the Isthmus of Panama

Origin	Family	Common Name
North America		
	Leporidae	Rabbits
	Sciuridae	Squirrels
	Cricetidae	Field mice
	Felidae	Cats
	Mustelidae	Skunks and otters
	Canidae	Foxes
	Ursidae	Bears
	Equidae	Horses
	Camelidae	Camels and lamas
	Cervidae	Deer
South America		
	Didelphidae	Opossums
	Dasypodiadae	Armadillos
	Megatheriidae	Giant ground sloths
	Bradypodidae	Three-toed sloth
	Myrmecophagidae	Anteaters
	Cebidae	Monkeys
	Erethizontidae	Porcupines
	Caviidae	Guinea pigs

Source: Webb, 1997; Webb, 2006.

barrier—there is still no paved road connecting South and North America due to the difficulty of traversing the Darien Gap in southern Panama and adjacent Colombia.

Chains of closely distributed islands are known to form stepping stones. Stepping stones are always filters mainly because not all species of flora and fauna can cross the gaps between them. The ancient islands in the vicinity of the modern Isthmus of Panama served such a function. Today, islands of Indonesia and the Philippines provide similar stepping stones for the dispersal of flora and fauna from Asia. A number of bird species from Asia are found in the Philippines or eastern Indonesia. The amphibian and reptile fauna originating in Asia, however, is very poorly represented away from the mainland and adjacent islands. Similarly, the proportion of reptile and insect fauna from New Guinea and Australia decreases sharply westward toward the Asian mainland. New Guinea and the Melanesian Islands provide stepping stones for the dispersal of Asian and Australian fauna eastward into the Pacific. The distribution of mangrove tree species in New Guinea and Melanesia shows the filter effect of this stepping stone route (Fig. 8.10). A steady and marked decrease in the number of mangrove taxa is encountered westward from New Guinea. Clearly, either due to differences in dispersal and colonization abilities or due to chance, some species have been able to disperse across the Pacific island stepping stones while others have not.

Adjacent mountain ranges can also serve as stepping stone filters for the dispersal of montane and alpine flora and fauna. In the southern Rocky Mountains of the United States and the Sierra Madre Occidental of Mexico, elevations over 2000 meters provide relatively cool and moist montane and alpine environments. Lower elevations are dominated by desert, which forms a barrier for the montane species. The sky island mountains in between the southern Rockies and the northern Sierra Occidental provide a stepping stone migration route, but one that is a filter. The proportion of northern forest and woodland mammal species gradually declines southward, while the proportion of southern mountain species declines northward. The filter effect probably results from difficulties dispersing across the desert and competition from the already established northern and southern mammal species on the mountains.

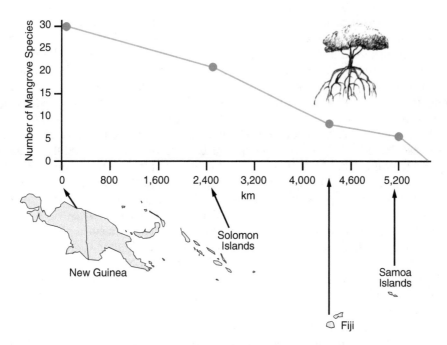

FIGURE 8.10   The impact of an island stepping stone filter on mangrove species in the South Pacific (adapted from Woodroffe, 1987). The number of species able to disperse to and colonize the most distant islands is a fraction of the genera found in New Guinea.

Some dispersal routes are typified by the long-distance transport of propagules in air or water currents, which provide a relatively small chance of successful colonization of new regions. An example is the release of coconuts into the sea. Most will perish, and a very small proportion may end up on a beach suitable for colonization. The planktonic larvae of coral reef fauna and flora depend on such dispersal, but also face the probability that most will not succeed in dispersing to new colonizable environments. When such long-distance dispersals are successful, they are sometimes called **waif dispersal events.** Waif dispersal is very important for the colonization of coral atolls and other shallow marine environments. Waif dispersal is also important for the colonization of islands, such as Hawaii, by small anemochorous insects and some plants. Waif dispersal routes are also filters, for not all organisms can survive the transport or colonize the new regions. Thus, the Hawaiian Islands have much smaller coral reef fauna and flora and smaller insect fauna than mainland areas of Asia or the Americas, or islands that lie closer to those continents. Such relatively impoverished biota are sometimes called waif flora or waif fauna due to the impact that isolation and dispersal constraints have on them.

Some routes of dispersal and migration only rarely allow successful dispersal and colonization. Crossing such routes occurs by chance and has a very low probability of success. Such dispersal pathways are called **sweepstakes routes.** The chances of an organism successfully dispersing by this route and colonizing a new area is similar to the odds of a person winning a lottery or sweepstakes. However, over thousands to millions of years, some organisms will disperse across sweepstakes routes by unlikely events and be able to colonize new areas. The dispersal of the cattle egret across the Atlantic to North America is an example of a sweepstakes route. The fact that even extremely remote islands, such as Hawaii in the Pacific and the Azores Islands in the Atlantic, have flora and fauna derived from a number of different families, not all of which are known waif dispersers, shows that, despite the low probability of success, sweepstakes routes are important modes of dispersal and colonization that have had clear impacts on the current distributions of plant and animal species.

RECENT INTRODUCTIONS AND INVASIONS BY NONNATIVE SPECIES

Among many of its other impacts, the voyage of Christopher Columbus to the islands of North America in 1492 marked the start of a period of unprecedented long-distance dispersal of plant and animal species between the world's continents by humans. This dispersal of plants and animals, and associated dispersal of people, cultures, and in some instances disease and human misery, is referred to as the **Columbian Exchange.** It has changed forever the biogeography of the earth. When a species is transported to a region outside its natural geographic range by humans, it is called an introduced species. Increasing world trade and travel typical of twenty-first-century globalization have served to further fuel the introduction of nonnative species to new regions. The magnitude of this ongoing biological exchange in terms of numbers of introduced species is staggering. Let's take plants as an example. Over 3400 species of nonnative vascular plants are known to have been introduced and now grow in the continental United States and Canada. This represents about 17% of the known vascular flora in the two countries. In Europe, over 2800 introduced plant species have been reported. Within the United States, California is host to approximately 1100 nonnative plant species. This is about 20% of the total California flora. Plants that are commonly associated with the California landscape, such as the eucalyptus (*Eucalyptus*—Australia), Mexican fan palm (*Washingtonia robusta*—Mexico), Canary Island pine (*Pinus canariensis*—Canary Islands), giant pampas grass (*Cortaderia selloana*—South America), and black mustard (*Brassica nigra*—Europe, Africa, and Asia), are all introduced alien species. The situation in the Hawaiian Islands is even more striking. The islands have a native flora of about 3000 plant species but host about 1200 additional species of alien plants. New Zealand has the highest number of nonnative plant species of any island group in the world. It supports approximately 1800 species of introduced plants, which now make up about 50% of the country's total flora. The introduction, establishment, and spread of nonnative plant species continues today. In Europe, approximately 6 new introduced plant species are discovered each year. Taken globally, at least 13,168 plant species have been introduced to new continents, islands, or waterways by human activity.

Large-scale introductions are certainly not limited to plants. In North America, there are about 100 introduced bird species that have become established. At least 9 nonnative mammal species have been introduced and established in the European country of Germany. In California, 35% of the freshwater fish species found today have been introduced from other parts of North America and other continents. In Hawaii, at least 40% of the bird species now found on the islands are introduced species, while 94% of the mammal species and 100% of the reptiles have been introduced to the Hawaiian Islands by humans. In many cases, insects make up a large proportion of introduced species. In the United States, over 450 introduced species of forest insects have been reported. In New Zealand, 1477 nonnative insect species have become established.

In some instances, such introductions of nonnative species were initially made deliberately. This is the case of ornamental garden plants, agricultural crops and animals, and game species that have been introduced to new regions by humans. In many cases, the introductions are purely accidental. Seeds may be carried in animal feed stocks or pets may escape and establish breeding populations. In New Zealand, for example, species introduced as ornamental garden plants make up the greatest number of the introduced plant species. These are followed in numbers by unintentional introductions for purposes such as forestry or hunting, and then agricultural plant species. Similarly, in Europe about two-thirds of the introduced plant species were imported for ornamental or agricultural purposes. Many plant species imported for gardens or agriculture may not be capable of establishing self-sustaining populations in the wild. However, from the evidence discussed above, many species can make this jump. When an introduced plant or animal species establishes self-sustaining populations, it is said to be established or naturalized. One rough rule of approximation is that about 10% of species introduced into a new region will become established.

The problems caused by introduced plant and animal species have become acute. The introduction of nonnative species can cause significant damage to native ecosystems, threats to native species, and problems for humans. Introduced species often cause considerable damage to agriculture, forestry, and other natural resources used by humans. Approximately 50% to 70% of the so-called weedy species that cause damage to agricultural crops in North America are introduced Eurasian plants. When an introduced species becomes established, spreads geographically, and causes measurable environmental, economic, or human health impacts, it is classified as an **invasive species.** Today, there are thousands of introduced species, including plants, mammals, birds, reptiles, amphibians, fish, and insects, that are classified as invasive. The numbers and impacts are seen around the world (Fig. 8.11). The continent of Europe alone has more than 300 terrestrial plants, 100 terrestrial vertebrates, and 600 terrestrial invertebrates that are classified as invasive species. There are also a great many invasive

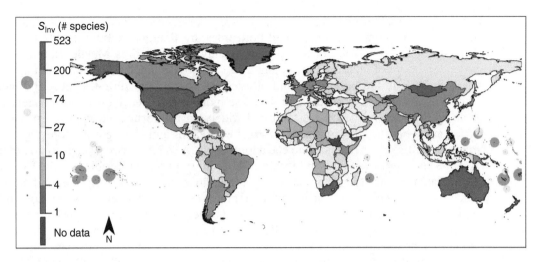

FIGURE 8.11 Numbers of invasive plant and animal species based on international databases (*Source:* Turbelin et al. 2017 / John Wiley & Sons / CC BY 4.0.).

microorganisms, including those that cause disease in humans. The annual global economic cost of all invasive species is incalculable. In the United States alone, it is estimated that the economic cost created by invasive species between 1960 and 2020 was at least $1.22 trillion, with a current annual cost of $21 billion.

The impact of introduced plants can be extremely disruptive to natural ecosystem functioning and to human land use. A classic example of plant invasion and its impact comes from Australia. Commencing in 1893, species of the prickly pear cactus *(Opuntia)* were brought from the Americas to Australia and planted to form hedges and other uses. The cacti form dense thickets that are over 1 meter high and impossible for humans or large animals to pass through. Prickly pear cacti, which are native to the Americas, lacked natural predators and parasites in Australia. In addition, the prickly pear was well adapted to thrive in the semiarid climate of Australia. Being free of predators and competitively superior to Australian plants, the prickly pear cactus increased dramatically in numbers and geographic distribution. In 40 years, it was recognized as a major pest that displaced native flora and destroyed grazing fields. By 1900, the prickly pear had spread over an area of 40,000 square kilometers in southeastern Australia. By 1925, the area infested with prickly pear cactus had increased to 243,000 square kilometers. The invasion of fields by prickly pear was a major economic catastrophe. It was not economically feasible to poison or remove the plants manually. In addition to the prohibitive cost of removing the cacti, such a strategy only brings short-term relief. Prickly pear can resprout from its roots, and its seeds remain viable in the soil for some 15 years. In some cases, scientists and resource managers turn to the use of predatory or pathogenic species from the native range of the invasive species to control the invader. This is referred to as **biocontrol.** Australian entomologists discovered that a moth species from Argentina called *Cactoblastis cactorum* produced larvae that feed exclusively on the cacti. The larvae burrow into the cactus, facilitating the entry of fungi and other pathogens. Starting in 1925, *Cactoblastis cactorum* were imported to Australia and introduced into areas where prickly pear were widespread. Similarly, the scale insect *Dactylopius*, which feeds on prickly pear, was imported as an additional biocontrol and released in 1935. By 1940, the prickly pear population had plummeted, and today the prickly pear cacti occur in Australia mainly as scattered individual plants. The control of *Opuntia* is considered the most successful biocontrol effort in Australia. Prickly pear has also been a destructive invasive plant in South Africa. Recent research there suggests that *Dactylopius* may be the most effective of the two biocontrol insects.

Australia has, in a way, repaid the United States for the damage caused by the prickly pear cactus. During the late nineteenth and early twentieth centuries, the Australian eucalyptus (*Eucalyptus* spp.) was introduced and widely planted in California. An incredible total of 374 different species of eucalyptus were introduced and 18 species have become naturalized. It has become the most common nonnative tree in California. At the same time, the Australian melaleuca tree (*Melaleuca* spp.) was introduced to Florida. The two genera were imported and planted as garden specimens, windbreaks, and timber trees. In both states, these introduced tree genera lacked serious predators and pathogens. Wildfire is common in Australia, and both trees are flammable, but adapted to frequent burning. They resprout easily from roots and grow quickly following fires. Melaleuca trees that survive fires produce millions of seeds, and the young trees can grow as much as 1 meter in a single year. The dense shade, thick leaf and bark litter, and oily residue found beneath eucalyptus and melaleuca stands often kill all understory plants and decrease biodiversity. Eucalyptus and melaleuca have escaped from cultivation and become established widely. In California, eucalyptus groves are particularly problematic because they replace oak-grass woodlands and some coastal forest and shrubland communities. Little grows beneath the eucalyptus canopy. *Melaleuca quinquenervia* has occupied at least 200,000 hectares of wetlands in Florida and become a particular problem in the Everglades National Park. The melaleuca trees were introduced in the region as a timber species and has become impossible to control. The trees have a profound impact on native vegetation. Areas covered by melaleuca typically support only two or three native plant species. Similar areas, which do not have melaleuca stands, generally support 60 to 80 different native plant species. Both the melaleuca and eucalyptus are particularly serious fire hazards. The trees contain high amounts of volatile oils and burn quickly at very high intensities. A 1992 fire in the eucalyptus-covered hills of Oakland, California, covered 650 hectares, destroyed over 3000 homes, and killed 25 people. In 1985, a melaleuca fire near Miami, Florida, consumed over 3500 hectares in a 35-day period.

Following the example of prickly pear in Australia, biocontrol methods are bring used to try and control melaleuca in Florida. In 1997, the Australian weevil *Oxyops vitiosa* was released in such an effort. The herbivorous insect *Boreioglycaspis melaleucae* was released in 2002. Such insect-based biocontrol approaches are credited with causing severe damage to melaleuca trees and the death of seedlings. Biocontrol along with herbicidal treatments and mechanical removal of melaleuca has achieved success on managed public lands in the Everglades, but populations on nearby untreated private lands provide for rapid reinvasion of treated sites.

One widespread, and troubling, invasive terrestrial plant in North America is cheatgrass *(Bromus tectorum)*. Some of the first reports of this Eurasian plant in North America came from British Columbia, Canada, in 1889. By 1930, cheatgrass was reportedly growing as far south as Utah, Nevada, and California (Fig. 8.12). By the 1990s, cheat grass had expanded into southern Arizona and occupied a total of some 40 million hectares. This invasive species is now found in all 50 states. Cheatgrass is easily spread by cattle and horses as these animals consume the plants and their resulting feces can contain viable seeds. Ironically, another introduced species, feral horses *(Equus ferus caballus)* have been shown to be an important dispersal agent for cheatgrass in the western United States. In the late twentieth century, it was estimated that cheatgrass was still expanding its footprint in western North America at a rate of over 1800 hectares per year.

Cheatgrass is spread by fire and in turn helps promote burning. Cheatgrass is an annual plant. In the summer, the plant dries and produces a continuous cover of fine fuel. Following burning, cheatgrass rapidly disperses seeds into burned areas. In areas formerly dominated by semiarid shrublands, the continuous cover of cheatgrass promotes frequent and widespread fires, which cause a decline in areas dominated by native shrubs such as sage brush (*Artemisia* spp.) and the replacement of species-rich shrubland by cheat grass–dominated grassland (Fig. 8.12). Cheatgrass is now considered a primary force that is transforming sagebrush-dominated ecosystems in regions such as the Great Basin into grasslands. The establishment of such grasslands in turn leads to a decrease in the species diversity of insects, birds, mammals, and reptiles. One species of particular concern is the sage grouse *(Centrocercus urophasianus),* which could see its populations decline by over 40% if cheatgrass-dominated grasslands continue to replace native shrublands. Cheatgrass has also spread into the understory of ponderosa pine *(Pinus ponderosa)* forests, potentially altering their fire regimes. In addition to its negative impact on native flora and fauna, cheatgrass is a common and destructive weed in wheat fields. Controlling cheatgrass is difficult, often requiring mechanical removal, herbicides, and reseeding with native species.

Another problematic invader in western North America is the saltcedar (*Tamarix* spp.). Saltcedar is an Asian desert shrub that grows quickly and produces copious amounts of seeds that are dispersed by wind and water. It was introduced to Texas in 1877. This invasive plant initially spread to southwest desert floodplains, sweeping as far north as central Utah and as far west as the Mojave Desert of California. The rate of expansion for the saltcedar in the desert regions has been greater

FIGURE 8.12 Cheatgrass *(Bromus tectorum),* an invasive annual plant from Eurasia, dominates the post-fire vegetation in a recently burned site in southwestern Utah. Cheatgrass dies and dries out in the summer, providing fuel for frequent fires that eventually exclude many native species of trees, shrubs, herbs, and grasses.

than 20,000 meters per year. It has now spread into the Great Plains region. By 2010, it was estimated that saltcedar occupied 364,000 hectares in the western United States. In 2007, naturalized saltcedar was discovered in the Canadian province of British Columbia. Ecological modeling suggests that saltcedar has the potential of becoming an invasive species throughout much of the Canadian Prairies. Aside from displacing native floodplain vegetation, each mature saltcedar can take 800 liters of water out of the soil each day. It has been estimated that saltcedars can remove as much water from southwestern soils as is used by all of the cities in southern California. The plant has proven difficult to control. Several saltcedar leaf beetle species *(Diorhabda* spp.) have been imported from Asia as a bicontrol agent and show some promise.

Invasive water plants are difficult to control and have caused significant problems in many parts of the world. The Amazonian water hyacinth *(Eichhornia crassipes)* was taken to Africa as an ornamental plant in the nineteenth century. It has beautiful floating foliage and purple flowers. It also has one of the highest growth rates of any vascular plant in the world. In 1990, escaped hyacinths were discovered in Lake Victoria. Victoria is the second largest lake in the world, and its fishery supports millions of people. By 1996, water hyacinth occupied 90% of the lake's shoreline, destroying important fish-breeding areas. The plants grow thick enough to choke off the entry to harbors for fishing boats. In addition, they clog water intake pipes for settlements and power plants. However, over the course of the late 1990s, the water hyacinth populations in Lake Victoria began to plummet and by the end of the twentieth century the species had disappeared from most areas. One factor contributing to this was the large-scale introduction of the South American species of weevils *(Neochetina* spp.) as a biocontrol agent in 1996. Weather conditions, ongoing lake eutrophication, competition from other emergent aquatic plants, and the use of herbicides may also have played a role. Interestingly, monitoring using satellite remote sensing and other means indicates a somewhat cyclical pattern of growth and then decline in water hyacinth populations and areal extent in Lake Victoria.

North America has been plagued by a number of invasive aquatic plant species. The water hyacinth was introduced to California in 1904 and has spread throughout the Sacramento–San Joaquin Delta (Fig. 8.13). Here it depresses water oxygen levels, competes with native plants, supports nonnative invertebrate communities, and can restrict travel by small boats. Satellite remote sensing indicates that water hyacinth covers some 1200 hectares of the delta. Invasive water primrose *(Ludwegia peploides),* native to South and Central America is another invasive plant now common in the delta.

FIGURE 8.13   A thick floating mat of the invasive plant species water hyacinth *(Eichhornia crassipes)* with a smaller cover of the invasive water primrose *(Ludwegia peploides)* in the Sacramento–San Joaquin River Delta of California.

The aquatic species hydrilla *(Hydrilla verticilatta)* was introduced to North America from Southeast Asia as an aquarium plant and escaped into the wild in the 1950s. This plant can grow as much as 10 centimeters per day and spreads through the dispersal of leaves and stems. It is now a very damaging weed in both the eastern United States and along the West Coast. By 1990, approximately 40% of Florida's waterways were infested by hydrilla. The species has now spread as far north as Connecticut. Hydrilla grows into dense mats that choke off light and cover submerged surfaces. The decay of the mats depletes oxygen from the water and can kill fish and other organisms. Aside from displacing native plants and disrupting aquatic ecosystems, hydrilla is a major culprit in clogging water intake valves, irrigation systems, and natural waterways.

Insects, mollusks, mammals, and birds have also been very successful invaders. Many common pest insects encountered in North America are alien invaders. These include the Argentine ant, the fire ant, the European earwig, and the gypsy moth. In turn, North and South American insects have been introduced throughout the world. For example, the Colorado beetle is an agricultural pest that was accidentally introduced to France in the early twentieth century and has now spread as far east as Russia. At the present time, the most infamous of these insect invaders in North America is the Africanized honey bee.

The European honey bee *(Apis mellifera lingustica)* was introduced to North and South America with the first European colonists. In addition to providing honey, these bees are extremely important for pollinating orchard trees and some other crops. The African subspecies of honey bee *(Apis mellifera scutellata)* is very similar in form and habit to its European counterpart. However, the African subspecies is extremely aggressive in protecting its hives. Swarms of the bees will pursue animals or people that disturb their hives. Sometimes, the intruders receive a lethal number of stings. Beekeeping for honey production and pollination is much more difficult with African bees than with the European subspecies. The African subspecies was brought to Brazil for breeding experiments to develop a hybrid that would exhibit the African subspecies' ability to survive tropical conditions and the European bee's more docile behavior. In 1956–1957, African bees escaped and established breeding populations in Brazil. The African bee can aggressively drive out European bees. However, in many cases, the African and European bees mate to produce hybrids. Unfortunately, the hybrid bees display the aggressive behavior of the African subspecies. Since hybrids are now so common, the aggressive strain of bees that is presently invading South and North America is referred to as the Africanized honey bee. The Africanized bees have been dispersing and colonizing South, Central, and North America at a rate of roughly 110 kilometers per year. They now occupy most temperate and tropical portions of South and Central America. The northward spread of the Africanized bee slowed when they reached the drier portions of northern Mexico in the late 1980s. However, by the mid-1990s they were established throughout the southwestern United States. By 1999 the Africanized bees had spread to the Pacific Coast of California and caused their first human fatalities in that state. It is thought that all wild populations of honey bee had become Africanized within about two years of their appearance in southern California. Today, Africanized bees are found in southern tier states from Florida westward to California. In Texas, the increasing dominance of Africanized bees corresponds with a linear increase in human deaths attributable to bee stings from 1992 to 2007.

Alien species of mollusks have also become major pests in North America. The large brown garden snail *(Helix aspersa)* that commonly destroys agricultural and ornamental plants in North America was introduced from Europe. Polynesians introduced at least two species of land snails to Hawaii. In recent years, the giant African land snail *(Achatina fulica)* has been accidentally introduced to a number of Pacific islands. This large mollusk causes damage to native flora and competes with species of native snails.

One of the most destructive alien mollusks in western Europe and North America is the zebra mussel *(Dreisena polymorpha)*. This small freshwater bivalve is native to the Caspian Sea of central Asia. A single female can release 5 million eggs annually, and the larvae of the species are free-swimming and disperse readily in fresh water. The building of shipping canals in the 1700s allowed the zebra mussel to invade eastern Europe. By the 1830s, the zebra mussel had spread as far west as England. The salt water of the Atlantic Ocean produced a barrier to stop the further spread of the mussel until around 1986–1988 when it appears possible that a freight ship that had taken on freshwater ballast in Europe released that ballast in Lake Erie. In that ballast were the larvae of the zebra mussel. The zebra mussel has now spread throughout the Great Lakes systems, the Mississippi River system, and the Hudson River.

Zebra mussels are a particularly damaging invader to both natural ecosystems and human activities. In North America, where they have few effective predators or parasites, they reach densities of up to 400,000 individual adults per square meter on any firm substrate from rocks to boat hulls to water pipes (Fig. 8.14). Zebra mussels exclude other mollusks and plants from these areas and have

FIGURE 8.14 Zebra mussels *(Dreissena polymorpha)* clog the intake pipes to a power station on Lake Erie. Following an accidental introduction into the lower Great Lakes in the late 1980s, zebra mussels have rapidly spread throughout much of eastern North America, causing economic damage and the loss of benthic habitat along many lakes and rivers (*Source*: Peter Yates / Science Source).

caused declines in native shellfish populations. It has been suggested that 140 different species of native North American shellfish might face serious threat, including extinction, due to zebra mussels. Water intake and outlet pipes quickly become plugged by the mussels. Zebra mussels are extremely effective filter feeders and remove large amounts of phytoplankton from the water, which, in turn, causes changes in the food chain and permits greater amounts of sunlight to penetrate lake and river waters. By decreasing the amount of plankton available to invertebrates and fish, it is thought that zebra mussels can limit the amount of fish that lakes can support. Zebra mussels can also decrease the amount of oxygen and calcium in waters. It has been suggested that the zebra mussel is fundamentally altering freshwater foodwebs and that few other human actions have had the impact on freshwater ecosystems that introduction of the zebra mussel has. The economic costs of zebra mussels are also appreciable. It has been estimated that costs to power plants and municipal drinking water infrastructure between 1989 and 2004 were $267 million. As destructive as the zebra mussel is, it is even more sobering to realize that it is only one of more than 140 alien species of fish, invertebrates, and plants to invade the Great Lakes system in the past 200 years. In any case, at present there appears to be no viable way to eradicate the invasive zebra mussel from North American waters.

Several mammal and bird species have become notorious invaders. The European rabbit (*Orryctalagus cuniculus*) was brought to the temperate regions of southern Australia in the 1860s. By 1900, it occupied the southeastern quarter of the continent and expanded almost to the northern coastline by 1980. Rabbits have proven extremely destructive to native Australian herbs and shrubs. The black rat (*Rattus rattus*), which infests almost all of temperate North America, is a native of Eurasia that has been introduced around the world. Interestingly, extremely stringent control measures at its borders, combined with a very cold climate, have kept the black rat from establishing in the province of Alberta, Canada. The Polynesian rat (*Rattus exulans*) was introduced to Hawaii by Polynesian settlers and poses a major threat to native mollusks, insects, and some birds. The native birds of Hawaii face an additional challenge in the form of large populations of the alien bird species myna (*Acridotheres tristis*), Japanese white-eye (*Zosterops japonica*), and red-billed leiothrix (*Leiothrix lutea*), which compete for food and nesting sites. In Europe, large flocks of Canada geese (*Branta canadensis*) have become nuisances in parks and ornamental waterways and drive out native waterfowl from other habitats.

Amphibians and reptiles have also been introduced to many regions by human activity and have proven to be destructive invaders. The cane toad (*Rhinella marina*, formerly *Bufo marinus*) is native to South and Central America and was brought to Australia in 1935 to control insect pests, but the voracious adults and tadpoles consume native insects, reptiles, birds, and even small mammals at an alarming rate. The speed at which the cane toad has spread across the Australian landscape has varied and has been most rapid in low-elevation regions with high temperatures and road networks providing for dispersal and habitat connectivity. Today, the cane toad can be found across a broad swath of the northeastern quarter of the continent of Australia. The species is still advancing its range, and invasion modeling suggests further spread across northern Australia and then down along the west coast. Australian scientists have concluded it is impossible at present to provide any broadscale control against this invasive species.

Predatory reptiles are often nonselective in prey and can both store and apportion a large part of their assimilated energy to growth and reproduction. This allows them to take advantage of many prey species and to persist in high numbers. They are a major threat to native prey species not adapted to such a predator. One of the most insidious reptile invasions occurred on the Pacific island of Guam. Beginning in the 1960s, scientists noticed significant declines in the native bird population of the island. During the 1980s, six of the eleven native bird species of Guam's forests went extinct. Surviving individuals of other species were captured and taken to breeding facilities to be kept alive in captivity. Scientists puzzled over the cause of the population declines and extinctions. After much study, it was determined that the culprit was a bird-eating snake called *Boiga irregularis* that was native to the Solomon Islands. It appears that the snake had been accidentally introduced to the island sometime during the 1940s. The snake approaches bird nests at night and eats adult birds, chicks, and eggs with proficiency. Due to a variety of prey and a lack of serious predators, the snake population grew to unnatural densities, and the native birds of Guam have paid the ultimate price for the introduction of this alien species. About 83% of the native bird species have now been driven to extinction on Guam.

FIGURE 8.15 Female Burmese python *(Python molurus bivittatus),* an invasive reptile species, captured by the Big Cypress National Preserve of Florida Scout Snake Program during the 2019–2020 breeding season (*Source:* Kirsten Hines / National Park Service).

The destructive impacts of invasive snake species are not limited to islands such as Guam. The Burmese python *(Python molurus bivittatus)* is a highly disruptive invasive species in the Everglades National Park and adjacent areas of Florida (Fig. 8.15). This predatory snake can reach lengths over 5 meters and weights of up to 74 kilograms. The Burmese python was likely introduced as escaped or released pets and has seen dramatic increases in population size since 2000. The pythons consume a wide variety of animals, ranging from birds to large mammals and even alligators. Aggressive removal and eradication campaigns in Florida have seen over 300 pythons removed per year from the Everglades National Park. Despite these efforts, python populations have increased in the state. Road surveys conducted in the park between 2003 and 2011 recorded decreases in rabbit, raccoon, opossum, and bobcat sightings by 87% or more, dependent upon the species. The pythons are known to consume at least 25 different species of bird, including the threatened wood stork *(Mycteria americana).* The Burmese python continues to be a growing threat to native ecosystems and species in Florida. However, it is only one of a number of such threats. Florida has the greatest number of introduced reptiles and amphibians of any U.S. state.

Destructive invading species need not be large organisms. Fungi, microscopic parasites, and bacteria have also been introduced to new regions by human activity with disastrous results. The North American deciduous forest has experienced catastrophic declines in elm trees *(Ulmus* spp.) and chestnut trees *(Castanea dentata)* due to introduced pathogens. Dutch elm disease is caused by the fungus *Ophiostoma* spp., which is native to Asia. The fungus is transported between trees by beetles. The disease was first detected in New York City in 1931. At that time, elm trees were one of the most widely planted trees in eastern North American cities. Within 20 years, the pathogen was considered an epidemic, and despite much effort by scientists and the government, millions of elm trees were lost. European elms have faced similar ravaging due to the introduction of the pathogen there.

A few disease-resistant strains of elm have been identified and bred and provide some hope for the future of the once ubiquitous urban tree. The fungal pathogen *Cryphonectria parasitica* is native to Asia. Following its accidental introduction in the early twentieth century, the fungus has caused the near total eradication of the native North American chestnut. The ecological impacts and changes in forest composition are profound because chestnut was once one of the most common trees in the eastern deciduous forest. The fungus is also causing considerable mortality to the European chestnut *Castanea sativa*.

The success of invading species often reflects several common factors. First, these species have been introduced to environments that provide physical conditions that are well suited for the survival of the invaders. Florida's warm climate and generally moist conditions provide such an environment for many reptiles and amphibians, for example. Second, the invaders generally have short generation times and produce large numbers of offspring. Third, the invading plants or animals have readily dispersed propagules. Fourth, the invading species have no serious natural predators, parasites, or pathogens in the new environment. Fifth, native competitors, if they exist, are less efficient than the invaders. Sixth, in the case of predatory invaders, prey species are abundant and lack effective defenses against the invader. In many cases, human activity, such as habitat disturbance or the hunting of native competitors or predator species, aids in the success of the invaders.

Can we determine which species have the potential to be invasive if a lot of information is available on the autecology and synecology of the invader and the plant and animal communities being invaded? In most cases, such information is in very short supply. A study of invasive plants in New Brunswick, Canada, produced an interesting result in terms of predicting the potential success of an invader. In that study, it was found that the size of the range of the invading plant in its native habitat was the best predictor of the species' potential to invade new territory. It appears that species that can occupy large geographic ranges in their native areas are often likely to have the right combination of reproductive traits, competitive ability, and niche breadth to successfully invade large areas when they are introduced by humans. However, organisms with small natural ranges are not always poor invaders. For example, the Monterey pine *(Pinus radiata)* has a tiny native range in small portions of coastal California but is successfully grown as a forestry species in many parts of the world. In New Zealand, the Monterey pine has spread from forestry plantations and invaded native vegetation. In general, it has proven very difficult to predict which organisms will be aggressive invaders and which ecosystems will be most prone to invasion.

For better, but mainly for worse, we have dramatically changed the biosphere by introducing thousands of alien species to new islands, continents, lakes, rivers, and seas. Some environments have been radically altered by these introductions; others have suffered relatively little disruption. However, it can be concluded that there are few, if any, large ecosystems left that do not host alien species introduced by humans in the past few centuries. Aside from the high Arctic, there are probably no national parks or other nature preserves in North America that do not contain significant numbers of alien species of plants and animals. The number and geographic scope of long-distance plant, animal, and microbe introductions since the start of the Columbian Exchange, and the associated effects of invasive species, appear to be unprecedented in the history of the earth. The ultimate ecological and human impacts remain to be determined.

KEY WORDS AND TERMS

Anemochores	Dispersal—active and passive	Jump dispersal
Anemohydrochores	Dispersal kernel	Logistic population growth
Anthropochore	Exponential population growth	Propagule
Barrier	Filter	Samara seeds
Biocontrol	Great American Biotic Interchange	Seasonal migration
Biologging	Harmonization	Stepping stones
Colonization	Hydrochore	Supertramp
Columbian Exchange	Invasion	Sweepstakes routes
Corridor	Invasive species	Waif dispersal events
Diffusion	Irruption	Zoochore

REFERENCES AND FURTHER READING

Anderson, R. C. (2009). Do dragonflies migrate across the western Indian Ocean? *Journal of Tropical Ecology* **25**, 347–358.

Aukema, J. E., Leung, B., Kovacs, K., Chivers, C., Britton, K. O., Englin, J., Frankel, S. J., Haight, R. G., Holmes, T. P., Liebhold, A. M., and McCullough, D. G. (2011). Economic impacts of non-native forest insects in the continental United States. *PLoS One* **6**, e24587.

Begon, M., Harper, J. L., and Townsend, C. R. (1996). *Ecology: Individuals, Populations and Communities*, 3rd edition. Blackwell Science, Oxford.

Bennett, K. D. (1988). Holocene geographic spread and population expansion of *Fagus grandifolia* in Ontario, Canada. *Journal of Ecology* **76**, 547–557.

Billings, W. D. (1990). The mountain forests of North America and their environments. In *Plant Biology of the Basin and Range* (ed. C. B. Osmund, L. F. Pitelka, and G. M. Hidy), pp. 47–86 Springer-Verlag, New York.

Blumler, M. A. (1995). Invasion and transformation of California's valley grassland: a mediterranean analogue ecosystem. In *Human Impact and Adaptation: Ecological Relations in Historical Times* (ed. R. Butlin and N. Roberts), pp. 308–332. Blackwell Science, Oxford.

Bradbury, I. R., Laurel, B., Snelgrove, P. V., Bentzen, P., and Campana, S. E. (2008). Global patterns in marine dispersal estimates: the influence of geography, taxonomic category and life history. *Proceedings of the Royal Society B: Biological Sciences* **275**, 1803–1809.

Bright, C. (1998). *Life out of Bounds: Bioinvasions in a Borderless World*. W. W. Norton, New York.

Brower, L. P., and Malcolm, S. B. (1991). Animal migrations: endangered phenomena. *American Zoologist* **31**, 265–276.

Brown, J. H., and Lomolino, M. V. (1998). *Biogeography*, 2nd edition. Sinauer Associates, Sunderland, MA.

Brzeziński, M., Romanowski, J., Żmihorski, M., and Karpowicz, K. (2010). Muskrat *(Ondatra zibethicus)* decline after the expansion of American mink *(Neovison vison)* in Poland. *European Journal of Wildlife Research* **56**, 341–348.

Buehlmann, C., Mangan, M., and Graham, P. (2020). Multimodal interactions in insect navigation. *Animal Cognition* **23**, 1129–1141.

Bukowski, E. (2019). Using the commons to understand the Dutch elm disease epidemic in Syracuse, NY. *Geographical Review* **109**, 180–198.

Bush, M. B., and Whittaker, R. J. (1991). Krakatau: colonization patterns and hierarchies. *Journal of Biogeography* **18**, 341–356.

Carlton, J. T. (2008). The zebra mussel *Dreissena polymorpha* found in North America in 1986 and 1987. *Journal of Great Lakes Research* **34**, 770–773.

Carrillo, J. D., Faurby, S., Silvestro, D., Zizka, A., Jaramillo, C., Bacon, C. D., and Antonelli, A. (2020). Disproportionate extinction of South American mammals drove the asymmetry of the Great American Biotic Interchange. *Proceedings of the National Academy of Sciences* **117**, 26281–26287.

Center, T. D., Purcell, M. F., Pratt, P. D., Rayamajhi, M. B., Tipping, P. W., Wright, S. A., and Dray, F. A. (2012). Biological control of *Melaleuca quinquenervia*: an Everglades invader. *Biocontrol* **57**, 151–165.

Chala, D., Zimmermann, N. E., Brochmann, C., and Bakkestuen, V. (2017). Migration corridors for alpine plants among the "sky islands" of eastern Africa: do they, or did they exist? *Alpine Botany* **127**, 133–144.

Chambers, J. C., Bradley, B. A., Brown, C. S., D'Antonio, C., Germino, M. J., Grace, J. B., Hardegree, S. P., Miller, R. F., and Pyke, D. A. (2014). Resilience to stress and disturbance, and resistance to *Bromus tectorum* L. invasion in cold desert shrublands of western North America. *Ecosystems* **17**, 360–375.

Chapman, B. B., Hulthen, K., Wellenreuther, M., Hansson, L., Nilsson, J., and Bronmark, C. (2014). Patterns of animal migration. In *Animal Movement Across Scales* (ed. L. Hansson and S. Akesson), pp. 11–35. Oxford University Press, Oxford.

Chapman, J. W., Reynolds, D. R., and Wilson, K. (2015). Long-range seasonal migration in insects: mechanisms, evolutionary drivers and ecological consequences. *Ecology Letters* **18**, 287–302.

Cheddadi, R., Birks, H. J. B., Tarroso, P., Liepelt, S., Gömöry, D., Dullinger, S., Meier, E. S., Hülber, K., Maiorano, L., and Laborde, H. (2014). Revisiting tree-migration rates: Abies alba (Mill.), a case study. *Vegetation History and Archaeobotany* **23**, 113–122.

Clark, J. S. (1998). Why trees migrate so fast: confronting theory with dispersal biology and the paleorecord. *American Naturalist* **152**, 204–224.

Clobert, J., Baguette, M., Benton, T. G., and Bullock, J. M. (2012). *Dispersal Ecology and Evolution*. Oxford University Press, Oxford.

Coates, P. S., Ricca, M. A., Prochazka, B. G., Brooks, M. L., Doherty, K. E., Kroger, T., Blomberg, E. J., Hagen, C. A., and Casazza, M. L. (2016). Wildfire, climate, and invasive grass interactions negatively impact an indicator species by reshaping sagebrush ecosystems. *Proceedings of the National Academy of Sciences* **113**, 12745–12750.

Cody, S., Richardson, J. E., Rull, V., Ellis, C., and Pennington, R. T. (2010). The great American biotic interchange revisited. *Ecography* **33**, 326–332.

Cousens, R., Dytham, C., and Law, R. (2008). *Dispersal in Plants: A Population Perspective*. Oxford University Press, Oxford.

Cremer, K. W. (1965). Dissemination of seed from *Eucalyptus regnans*. *Australian Forestry*, 33–37.

Crombie, A. C. (1945). On competition between different species of graminivorous insects. *Proceedings of the Royal Society of London, Series B* **132**, 362–395.

Crosby, A. W. (2003). *The Columbian Exchange: Biological and Cultural Consequences of 1492,* 30th anniversary edition. Praeger, Westport, CT.

Cuthbert, R. N., Diagne, C., Haubrock, P. J., Turbelin, A. J., and Courchamp, F. (2021). Are the "100 of the world's worst" invasive species also the costliest? *Biological Invasions*. doi.org/10.1007/s10530-021-02568-7.

D'Antonio, C. M., and Vitousek, P. M. (1992). Biological invasions by exotic grasses, the grass/fire cycle, and global change. *Annual Review of Ecology and Systematics* **23**, 63–87.

Davis, M. B. (1981). Quaternary history and the stability of forest communities. In *Forest Succession: Concepts and Application* (ed. D. C. West, H. H. Shugart, and D. B. Botkin), pp. 132–153 Springer-Verlag, New York.

Davis, M. B., Woods, K. D., Webb, S. L., and Futyma, R. P. (1986). Dispersal versus climate: expansion of *Fagus* and *Tsuga* into the upper Great Lakes region. *Vegetatio* **67**, 93–103.

Delcourt, P. A., and Delcourt, H. R. (1987). *Long-Term Forest Dynamics of the Temperate Zone: A Case Study of Late-Quaternary Forests in Eastern North America.* Springer-Verlag, New York.

Diamond, J. (1974). Colonization of exploded volcanic islands by birds: the 507 supertramp strategy. *Science* **184**, 803–806.

Doody, J. S., McHenry, C., Letnic, M., Everitt, C., Sawyer, G., and Clulow, S. (2018). Forecasting the spatiotemporal pattern of the cane toad invasion into north-western Australia. *Wildlife Research* **45**, 718–725.

Dorcas, M. E., Willson, J. D., Reed, R. N., Snow, R. W., Rochford, M. R., Miller, M. A., Meshaka, W. E., Andreadis, P. T., Mazzotti, F. J., Romagosa, C. M., and Hart, K. M. (2012). Severe mammal declines coincide with proliferation of invasive Burmese pythons in Everglades National Park. *Proceedings of the National Academy of Sciences* **109**, 2418–2422.

Dove, C. J., Snow, R. W., Rochford, M. R., and Mazzotti, F. J. (2011). Birds consumed by the invasive Burmese python *(Python molurus bivittatus)* in Everglades National Park, Florida, USA. *Wilson Journal of Ornithology* **123**, 126–131.

Drake, J. A., Mooney, H. A., di Castri, F., Groves, H., Kruger, F. J., Rejmanek, M., and Williamson, M. (1989). *Biological Invasions: A Global Perspective.* Wiley, New York.

Dray, F. A., Bennett, B. C., and Center, T. D. (2006). Invasion history of *Melaleuca quinquenervia* (Cav.) ST Blake in Florida. *Castanea* **71**, 210–225.

Ebenhard, T. (1988). Introduced birds and mammals and their ecological effects. *Swedish Wildlife Research* **13**, 1–107.

Edney-Browne, E., Brockerhoff, E. G., and Ward, D. (2018). Establishment patterns of non-native insects in New Zealand. *Biological Invasions* **20**, 1657–1669.

Elliott, K. J., and Swank, W. T. (2008). Long-term changes in forest composition and diversity following early logging (1919–1923) and the decline of American chestnut *(Castanea dentata)*. *Plant Ecology* **197**, 155–172.

Elton, C. S. (1958). *The Ecology of Invasions by Animals and Plants.* Methuen, London.

Fantle-Lepczyk, J. E., Haubrock, P. J., Kramer, A. M., Cuthbert, R. N., Turbelin, A. J., Crystal-Ornelas, R., Diagne, C., and Courchamp, F. (2022). Economic costs of biological invasions in the United States. *Science of the Total Environment* **806**, 151318.

Feurdean, A., Bhagwat, S. A., Willis, K. J., Birks, H. J. B., Lischke, H., and Hickler, T. (2013). Tree migration-rates: narrowing the gap between inferred post-glacial rates and projected rates. *PLoS One* **8**, e71797.

Froyd, C. A. (2005). Fossil stomata reveal early pine presence in Scotland: implications for postglacial colonization analyses. *Ecology* **86**, 579–586.

Goodwin, B. J., McAllister, A. J., and Fahrig, L. (1999). Predicting invasiveness of plant species based on biological information. *Conservation Biology* **13**, 422–426.

Gorman, M. L. (1979). *Island Ecology.* Chapman and Hall, London.

Greene, D. F., and Johnson, E. A. (1989). A model of wind dispersal of winged or plumed seeds. *Ecology* **70**, 339–347.

Greene, D. F., and Johnson, E. A. (1997). Secondary dispersal of tree seeds on snow. *Journal of Ecology* **85**, 329–340.

Grime, J. P., and Jeffrey, D. W. (1965). Seedling establishment in vertical gradients of sunlight. *Journal of Ecology* **53**, 621–642.

Grimm, W. C. (1967). *Familiar Trees of America.* Harper and Row, New York.

Gruenwald, N. J. (2012). Novel insights into the emergence of pathogens: the case of chestnut blight. *Molecular Ecology* **21**, 3896–3897.

Harper, J. L. (1977). *Population Biology of Plants.* Academic Press, New York.

Havel, J. E., Kovalenko, K. E., Thomaz, S. M., Amalfitano, S., and Kats, L. B. (2015). Aquatic invasive species: challenges for the future. *Hydrobiologia,* **750**, 147–170.

Hays, G. C., Mortimer, J. A., Ierodiaconou, D., and Esteban, N. (2014). Use of long-distance migration patterns of an endangered species to inform conservation planning for the world's largest marine protected area. *Conservation Biology* **28**, 1636–1644.

Hernández-Camacho, J. I., Alvarez-Leon, R., and Renjifo-Rey, J. M. (2006). Pelagic sea snake *Pelamis platurus* (Linnaeus, 1766) (Reptilia: Serpentes: Hydrophiidae) is found on the Caribbean coast of Colombia. *Memoria de la Fundación La Salle de Ciencias Naturales* **164**, 143–152.

Heywood, V. H. (1995). *Global Biodiversity,* Assessment. Cambridge University Press, Cambridge.

Higgins, S. I., and Richardson, D. M. (1996). A review of models of alien plant spread. *Ecological Modelling* **87**, 249–265.

Hoffmann, J. H., Moran, V. C., Zimmermann, H. G., and Impson, F. A. (2020). Biocontrol of a prickly pear cactus in South Africa: reinterpreting the analogous, renowned case in Australia. *Journal of Applied Ecology* **57**, 2475–2484.

Holland, R. A. (2010). Differential effects of magnetic pulses on the orientation of naturally migrating birds. *Journal of the Royal Society Interface* **7**, 1617–1625.

Hosking, J. R. (2012). *Opuntia* spp. In *Biological Control of Weeds in Australia* (ed. M. Julien, R. McFadyen, and J. Cullen), pp. 431–436. CSIRO, Collingwood, Australia.

Huggett, R. J. (1998). *Fundamentals of Biogeography.* Routledge, London.

Hulme, P. E. (2020). Plant invasions in New Zealand: global lessons in prevention, eradication and control. *Biological Invasions* **22**, 1539–1562.

Huntley, B., and Webb III, T. (1989). Migration: species' response to climatic variations caused by changes in the earth's orbit. *Journal of Biogeography* **16**, 5–19.

James, M. C., Sherrill-Mix, S. A., and Myers, R. A. (2007). Population characteristics and seasonal migrations of leatherback sea turtles at high latitudes. *Marine Ecology Progress Series* **337**, 245–254.

Johnson, J. C., Halpin, R., Stevens, D., Vannan, A., Lam, J., and Bratsch, K. (2015). Individual variation in ballooning dispersal by black widow spiderlings: the effects of family and social rearing. *Current Zoology* **61**, 520–528.

Jordano, P. (2017). What is long-distance dispersal? And a taxonomy of dispersal events. *Journal of Ecology* **105**, 75–84.

Kearney, M., Phillips, B. L., Tracy, C. R., Christian, K. A., Betts, G., and Porter, W. P. (2008). Modelling species distributions without using species distributions: the cane toad in Australia under current and future climates. *Ecography* **31**, 423–434.

Keller, R. P., Geist, J., Jeschke, J. M., and Kühn, I. (2011). Invasive species in Europe: ecology, status, and policy. *Environmental Sciences Europe* **23**(1), 1–17.

King, S. R., Schoenecker, K. A., and Manier, D. J. (2019). Potential spread of cheatgrass and other invasive species by feral horses in Western Colorado. *Rangeland Ecology & Management* **72**, 706–710.

Klein, I., Oppelt, N., and Kuenzer, C. (2021). Application of remote sensing data for locust research and management—a review. *Insects* **12**, 233.

Krebs, C. J. (1994). *Ecology*, 4th edition. Harper-Collins College Publishers, New York.

Krysko, K. L., Burgess, J. P., Rochford, M. R., Gillette, C. R., Cueva, D., Enge, K. M., Somma, L. A., Stabile, J. L., Smith, D. C., Wasilewski, J. A., and Kieckhefer III, G. N. (2011). Verified non-indigenous amphibians and reptiles in Florida from 1863 through 2010: outlining the invasion process and identifying invasion pathways and stages. *Zootaxa* **3028**, 1–64.

Lasky, J. R., Jetz, W., and Keitt, T. H. (2011). Conservation biogeography of the US–Mexico border: a transcontinental risk assessment of barriers to animal dispersal. *Diversity and Distributions* **17**, 673–687.

Liechti, F., Witvliet, W., Weber, R., and Bächler, E. (2013). First evidence of a 200-day non-stop flight in a bird. *Nature Communications* **4**, 2554.

Lillywhite, H. B., Sheehy III, C. M., Heatwole, H., Brischoux, F., and Steadman, D. W. (2018). Why are there no sea snakes in the Atlantic? *BioScience* **68**, 15–24.

Linck, E., Schaack, S. and Dumbacher, J. P. (2016). Genetic differentiation within a widespread "supertramp" taxon: molecular phylogenetics of the Louisiade White-eye (*Zosterops griseotinctus*). *Molecular Phylogenetics and Evolution* **94**, 113–121.

Lindgren, C., Pearce, C., and Allison, K. (2010). The biology of invasive alien plants in Canada. 11. *Tamarix ramosissima* Ledeb., T. chinensis Lour. and hybrids. *Canadian Journal of Plant Science* **90**, 111–124.

Long, J., and Siu, H. (2018). Refugees from dust and shrinking land: tracking the dust bowl migrants. *Journal of Economic History* **78**, 1001–1033.

MacDonald, G. M. (1993). Reconstructing plant invasions using fossil pollen analysis. *Advances in Ecological Research* **24**, 67–110.

MacDonald, G. M., and Cwynar, L. C. (1985). A fossil pollen based reconstruction of the late Quaternary history of lodgepole pine (*Pinus contorta* spp. *latifolia*) in the western interior of Canada. *Canadian Journal of Forestry Research* **15**, 1039–1044.

Mack, R. N. (1981). Invasion of *Bromus tectorum* L. into western North America: an ecological chronical. *Agroecosystems* **7**, 145–165.

Marcotrigiano, M. (2017). Elms and Dutch elm disease: a quick overview. In *Proceedings of the American Elm Restoration Workshop 2016; 2016 October 25–27* (ed. C. C. Pinchot, K. S. Knight, L. M. Haugen, C. E. Flower, and J. M. Slavicek).

Lewis Center, OH. Gen. Tech. Rep. NRS-P-174. Newtown Square, PA: U.S. Department of Agriculture, Forest Service, Northern Research Station: 2–5.

Matthysen, E. (2012). Multicausality of dispersal: a review. In *Dispersal Ecology and Evolution* (ed. M. Baguette, T. G. Benton, and J. M. Bullock), pp. 3–18. Oxford University Press, Oxford.

McGlone, C. M., Springer, J. D., and Covington, W. W. (2009). Cheatgrass encroachment on a ponderosa pine forest ecological restoration project in northern Arizona. *Ecological Restoration* **27**, 37–46.

McLachlan, J. S., Clark, J. S., and Manos, P. S. (2005). Molecular indicators of tree migration capacity under rapid climate change. *Ecology* **86**, 2088–2098.

Morain, S. A. (1984). *Systematic and Regional Biogeography*. Van Nostrand Reinhold, New York.

Mouritsen, H. (2018). Long-distance navigation and magnetoreception in migratory animals. *Nature* **558**, 50–59.

Mundinger, P. C., and Hope, S. (1982). Expansion of the winter range of the house finch. *American Birds* **36**, 347–353.

O'Meara, S., Larsen, D., and Owens, C. (2010). Methods to control saltcedar and Russian olive. *Saltcedar and Russian Olive Control Demonstration Act Science Assessment, U.S. Geological Survey Scientific Investigations Report 2009–5247*, 65–102.

Ongore, C. O., Aura, C. M., Ogari, Z., Njiru, J. M., and Nyamweya, C. S. (2018). Spatial-temporal dynamics of water hyacinth, Eichhornia crassipes (Mart.) and other macrophytes and their impact on fisheries in Lake Victoria, Kenya. *Journal of Great Lakes Research* **44**, 1273–1280.

Nathan, R. (2006). Long-distance dispersal of plants. *Science* **313**, 786–788.

Nathan, R. (2008). An emerging movement ecology paradigm. *Proceedings of the National Academies of Science* **105**, 19050–19051.

Pankiw, T. (2009). Reducing honey bee defensive responses and social wasp colonization with methyl anthranilate. *Journal of Medical Entomology* **46**, 782–788.

Pearce, C. M., and Smith, D. G. (2007). Invasive saltcedar (*Tamarix*): its spread from the American Southwest to the Northern Great Plains. *Physical Geography* **28**, 507–530.

Pearson, R. G. (2006). Climate change and the migration capacity of species. *Trends in Ecology & Evolution* **21**, 111–113.

Piersma, T., and Davidson, N. C. (1992). The migrations and annual cycles of five subspecies of knots in perspective. *Wader Study Group Bulletin* **64** (supplement), 187–197.

Pinchot, C. C., Flower, C. E., Knight, K. S., Marks, C., Minocha, R., Lesser, D., Woeste, K., Schaberg, P. G., Baldwin, B., Delatte, D. M., and Fox, T. D. (2017). *Development of New Dutch Elm Disease-Tolerant Selections for Restoration of the American Elm in Urban and Forested Landscapes*. Gen. Tech. Rep. PNW-GTR-963, pp. 53–63. U.S. Department of Agriculture, Forest Service, Pacific Northwest Research Station, Portland, Oregon.

Pollock, H. S., Savidge, J. A., Kastner, M., Seibert, T. F., and Jones, T. M. (2019). Pervasive impacts of invasive brown treesnakes drive low fledgling survival in endangered Micronesian Starlings (*Aplonis opaca*) on Guam. *The Condor* **121**, duz014.

Qian, H., and Ricklefs, R. E. (2006). The role of exotic species in homogenizing the North American flora. *Ecology Letters* **9**, 1293–1298.

Quammen, D. (1996). *Song of the Dodo: Island Biogeography in an Age of Extinctions*. Touchstone, New York.

Rapoport, E. H. (1982). *Areography: Geographical Strategies of Species*. Pergamon Press, Oxford.

Rappole, J. H. (1995). *The Ecology of Migrant Birds: A Neotropical Perspective*. Smithsonian Institution Press, Washington, DC.

Rayamajhi, M. B., Pratt, P. D., Center, T. D., and Van, T. K. (2010). Insects and a pathogen suppress Melaleuca quinquenervia cut-stump regrowth in Florida. *Biological Control* **53**, 1–8.

Reppert, S. M., and de Roode, J. C. (2018). Demystifying monarch butterfly migration. *Current Biology* **28**, R1009–R1022.

Ricciardi, A. (2007). Are modern biological invasions an unprecedented form of global change? *Conservation Biology* **21**, 329–336.

Ricciardi, A. (2013). Invasive species. In *Ecological Systems* (ed. R. Leemans), pp. 161–178. Springer, New York.

Ritter, M., and Yost, J. (2009). Diversity, reproduction, and potential for invasiveness of Eucalyptus in California. *Madroño* **56**, 155–167.

Roberts, L. (1990). Zebra mussel invasion threatens U.S. waters. *Science* **249**, 1370–1372.

Rockwood, L. L., and Witt, J. W. (2015). *Introduction to Population Ecology*. Wiley, Chichester, England.

Rutz, C., and Graeme, C. H. (2009). New frontiers in biologging science. *Biology Letters* **5**, 289–292.

Sánchez-Zapata, J. A., Donázar, J. A., Delgado, A., Forero, M. G., Ceballos, O., and Hiraldo, F. (2007). Desert locust outbreaks in the Sahel: resource competition, predation and ecological effects of pest control. *Journal of Applied Ecology* **44**, 323–329.

Sauer, J. D. (1988). *Plant Migration: The Dynamics of Geographic Patterning in Sea Plants*. University of California Press, Berkeley.

Sauer, J. R., Niven, D. K., Hines, J. E., Ziolkowski Jr., D. J., Pardieck, K. L., Fallon, J. E., and Link, W. A. (2017). *The North American Breeding Bird Survey, Results and Analysis 1966–2015*. Version 2.07. U.S. Geological Survey Patuxent Wildlife Research Center, Laurel, Maryland. https://www.mbr-pwrc.usgs.gov/bbs/specl15.html.

Savidge, J. A. (1987). Extinction of an island forest avifauna by an introduced snake. *Ecology* **68**, 660–668.

Scheltema, R. S. (1986). Long-distance dispersal by planktonic larvae of shoal-water benthic invertebrates among central Pacific islands. *Bulletin of Marine Science* **39**, 241–256.

Schneider, S. S., G. DeGrandi-Hoffman, and D. R. Smith. (2004). The African honey bee: factors contributing to a successful biological invasion. *Annual Review of Entomology* **49**, 351–376.

Schupp, E. W., Jordano, P., and Gómez, J. M. (2010). Seed dispersal effectiveness revisited: a conceptual review. *New Phytologist* **188**, 333–353.

Shigesada, N., and Kawasaki, K. (1997). *Biological Invasions; Theory and Practice*. Oxford University Press, Oxford.

Silvertown, J. W. (1982). *Introduction to Plant Population Ecology*. Longman, New York.

Silvertown, J. W., and Doust, J. L. (1993). *Introduction to Plant Population Biology*. Blackwell Scientific Publications, Oxford.

Simberloff, D. (2010). Invasive species. In *Conservation Biology for All* (ed. N. S. Sodhi and P. R. Ehrlich), pp. 131–152. Oxford University Press, New York.

Simberloff, D. (2013). *Invasive Species: What Everyone Needs to Know*. Oxford University Press, Oxford.

Simpson, G. G. (1940). Mammals and land bridges. *Journal of the Washington Academy of Science* **30**, 137–163.

Snell, R. S., and Cowling, S. A. (2015). Consideration of dispersal processes and northern refugia can improve our understanding of past plant migration rates in North America. *Journal of Biogeography* **42**, 1677–1688.

Strayer, D. L. (2009). Twenty years of zebra mussels: lessons from the mollusk that made headlines. *Frontiers in Ecology and the Environment* **7**, 135–141.

True-Meadows, S., Haug, E. J., and Richardson, R. J. (2016). Monoecious hydrilla—a review of the literature. *Journal of Aquatic Plant Management* **54**, 1–11.

Turbelin, A. J., Malamud, B. D. and Francis, R. A. (2017). Mapping the global state of invasive alien species: patterns of invasion and policy responses. *Global Ecology and Biogeography* **26**, 78–92.

Udvardy, M. D. F. (1969). *Dynamic Zoography*. Van Nostrand Reinhold, New York.

Urban, M. C., Phillips, B. L., Skelly, D. K., and Shine, R. (2008). A toad more traveled: the heterogeneous invasion dynamics of cane toads in Australia. *American Naturalist* **171**, E134–E148.

Van Kleunen, M., Dawson, W., Essl, F., Pergl, J., Winter, M., Weber, E., Kreft, H., Weigelt, P., Kartesz, J., Nishino, M., and Antonova, L. A. (2015). Global exchange and accumulation of non-native plants. *Nature* **525**, 100–103.

Vitousek, P. M., D'Antonio, C. M., Loope, L. L., and Westbrooks, R. (1996). Biological invasions as a global environmental challenge. *American Scientist* **84**, 468–478.

Webb, S. D. (1997). The Great American Faunal Interchange. In *Central America* (ed. A. G. Coates), pp. 97–122. Yale University Press, New Haven, CT.

Webb, S. D. (2006). The Great American Biotic Interchange: patterns and processes 1. *Annals of the Missouri Botanical Garden* **93**(2), 245–257.

Whittaker, R. J., and Bush, M. B. (1993). Dispersal and establishment of tropical forest assemblages, Krakatoa, Indonesia. In *Primary Succession on Land* (ed. J. Miles and D. W. H. Walton), pp. 147–160 Blackwell Scientific Publications, Oxford.

Williams, A. E., Duthie, H. C., and Hecky, R. E. (2005). Water hyacinth in Lake Victoria: why did it vanish so quickly and will it return? *Aquatic Botany* **81**, 300–314.

Wilmers, C. C., Nickel, B., Bryce, C. M., Smith, J. A., Wheat, R. E., and Yovovich, V. (2015). The golden age of bio-logging: how animal-borne sensors are advancing the frontiers of ecology. *Ecology* **96**, 1741–1753.

Wolf, K. M., and DiTomaso, J. (2016). Management of blue gum eucalyptus in California requires region-specific consideration. *California Agriculture* **70**, 39–47.

Woodroffe, C. D. (1987). Pacific island mangroves: distribution and environmental setting. *Pacific Science* **41**, 166–185.

Young, J., and Clements, C. D. (2009). *Cheatgrass: Fire and Forage on the Range*. University of Nevada Press, Reno.

Zacharin, R. F. (1978). *Emigrant Eucalypts: Gum Trees as Exotics*. Melbourne University Press, Carlton, Australia.

CHAPTER 9

EVOLUTION, SPECIATION, AND EXTINCTION

It is impossible to consider the great variety of plant and animal species that exist today and not ask the question, how did this diversity of life arise? We have seen that plants and animals possess extraordinary anatomical and physiological features that allow them to exist in many different environments. How do such adaptations develop? Even before the time of Aristotle, people contemplated these fundamental questions. Indeed, the origin of plant and animal species remains among the most vigorously debated issues in theology, philosophy, and science.

In addition to the plants and animals we see today, we are aware from geological and historical records that there are many species that lived in the past but no longer exist today. Everyone is familiar with the dinosaurs, but our information on these extinct animals comes to us only from fossils. Although modern birds descend from dinosaurs, the last traditional dinosaur species disappeared from the face of the earth some 65 million years ago. Not a single dinosaur exists today, and yet for over 100 million years (from the Triassic to the end of the Cretaceous), huge numbers of dinosaurs lived on the land, in the air, and in the sea. Why did all the hundreds of species of dinosaurs cease to exist some 65 million years ago?

What causes species to appear, adapt to certain environments, and disappear is a fundamental question for biogeographers. In this chapter we will consider how geography and environmental change may influence the evolutionary development of new species. We will then look at the role of geography and environmental change in the elimination of species through extinction. Finally, we will consider the positive role that the extinction of some species plays in promoting the development and success of other new species. Today, human actions are rapidly changing the environment in ways that can influence both the evolution and extinction of living species. Biogeographers must have a good grasp of the forces of evolution and extinction not just to understand the past, but to anticipate and plan for the future.

EVOLUTION AND SPECIATION

In order to study the way in which geography and environmental change have contributed to the great diversity of species, we need to consider two related processes—evolution and speciation. The term **evolution** refers to genetically controlled changes in physiology, anatomy, and behavior that occur to a species over time. Evolutionary change within an individual species or population is called microevolution, whereas evolutionary change that produces new taxonomic units such as species, genera, and families is termed macroevolution. **Speciation** refers to the development of two or more genetically differentiable species from a single common ancestor species. Speciation results from evolutionary change, but not all evolutionary change leads to the development of two or more species from a common ancestor. The different species that arise from the same ancestor are called a **clade.** The terms *cladogenesis* and *speciation* can be used synonymously. When speciation occurs, continued evolutionary divergence can lead to greater differences between ancestral and descendant species and the development of further new species, new genera, families, and higher taxonomic orders. The processes and products of speciation are at the foundation of macroevolution. Before we consider the processes of evolution and speciation in more detail, we should review some aspects of genetics and heredity as they relate to sexually reproducing species.

Some Basic Genetics

Let's focus our consideration of genetics on organisms like animals and plants that possess eukaryotic cells, which have a nucleus bound within a membrane. Let's also mainly consider sexually reproducing organisms. The basic physiology, anatomy, and behavior of species are controlled by chemical structures called **genes,** which are found in living cells and are passed from parents to offspring. The genes of plants and animals consist of molecules of **DNA** (deoxyribonucleic acid). The DNA segments are made up of various combinations of sugars and phosphates that are joined together by nitrogenous compounds consisting of base pairs of adenine, thymine, cytosine, and guanine bound by hydrogen. The DNA structure is referred to as the double helix and resembles a twisted ladder, with the sugars and phosphates forming the backbones of the sides and the nitrogenous compounds forming the rungs. The English scientist Francis Crick and his American colleague James Watson first described the basic structure of this crucial DNA molecule at Cambridge University in the late 1940s. The description of the DNA structure stands as one of the most important scientific discoveries of all time. It has been discovered that some genes are protein-coding genes that are transcribed into ribonucleic acid (RNA) and then create proteins. However, the vast majority of genes are noncoding genes and the RNA produced by them does not create proteins. Most research has focused on protein-coding genes as they are responsible for observable traits. Noncoding genes were initially thought to be "junk"—genetic remnants of no biological value—but subsequent work has shown that some can perform important genetic regulatory functions such as providing binding sites for proteins.

Within the cells of an organism, the genes are arranged along paired, threadlike structures called **chromosomes.** The point at which a gene is located on a chromosome is called a **locus** (the plural is *loci*). Different species have different numbers of chromosomes, different numbers of genes, and different types of genes located at the loci of the chromosomes. Humans have a total of 46 chromosomes arranged in 23 pairs. In total, the DNA in those chromosomes contains over 3 billion base pairs. The American black bear *(Ursus americanus)* has 37 chromosome pairs, while the common fruit fly *(Drosophila melanogaster)* has only 4 pairs. Plants and animals typically have the same number of chromosome pairs as each of their parents. Organisms and cells that have complete sets of chromosome pairs are said to be diploid. Certain reproductive cells, such as sperm from animals, contain only one-half of each chromosome pair. These cells are described as being haploid. During sexual reproduction, a haploid male cell fuses with a haploid female cell, and the two sets of chromosomes and associated genes combine. This results in the formation of a diploid embryo. Thus, each sexually reproducing plant or animal receives half of its genes and chromosomes from its male parent and half from its female parent.

Many thousands of different gene forms can be arranged along individual chromosomes. Differences in the specific chemical form of gene at a locus cause differences such as hair coloring, skin coloring, or eye coloring in humans. Such genetically controlled variation in appearance within a population is referred to as **polymorphism.** The different gene forms that exist for a given locus are called **alleles.** When a locus has different alleles associated with it, it is said to be **heterozygous.** The degree of heterozygosity within a species or population is an important measure of genetic diversity.

The nature of genes and chromosomes and the way in which sexually reproducing organisms pass along their genes to their progeny have important implications. First, the genes can best be thought of as particles of information that themselves do not change or blend during reproduction. The genes remain stable from generation to generation unless there is some mutation that arises. Second, if the chromosomes and loci of the male and female do not match perfectly, sexual reproduction generally cannot occur, blocking interbreeding between different species. Third, because chromosome numbers and the genes at each loci must match, when species reproduce, they produce offspring that are genetically similar to the parents. Therefore, the general physiology, anatomy, and behavior of the offspring will be like those of the parent. Fourth, since each offspring is a product of the recombination of chromosomes and genes from the male and female parent, the alleles found at the loci of the offspring will not exactly match either of the parents. Thus, the offspring can differ from the parents in traits for which there are different alleles, allowing for individual variability in traits, such as eye color, within the same species. Gender is also controlled by chromosomes and associated genes. Human males possess a Y chromosome and an X chromosome, while females possess two X chromosomes. Studies of the DNA of the Y chromosome allows the tracing of male genetics. The DNA found in the mitochondria,

TABLE 9.1 **Comparison of Human Genome with Other Species**

Species	Estimated Protein-Coding Genes	Total Chromosomes (2n)
Humans *(Homo sapiens)*	20,000–25,000	46
Domestic mouse *(Mus musculus)*	30,000	40
Chicken *(Gallus gallus)*	20,000–23,000	78
Fruit fly *(Drosophila melanogaster)*	14,000	8
Roundworm *(Caenorhabditis elegans)*	18,000	12
Rice *(Oryza sativa)*	51,000	24

Source: Data from Pray, 2008; Capanna and Castiglia, 2004; Trask, 2002; Sanford and Perry, 2001; Offner, 1996; Ahn et al., 1993; Borgaonkar, 1969; Makino, 1951.

which are organelles located within the cytoplasm of cells, is passed down maternally to both females and males and allows the tracing of maternal-line genetics specifically.

The complete range of genes present in a species is called the **genome.** Remarkable progress has been made in the development of chemical, instrumental, and analytic methods in molecular biology over the past six decades, allowing geneticists to develop refined estimates of genomes for over 10,000 different species, including humans. The science of **genomics,** which is the branch of molecular biology concerned with the measurement, traits, evolution, and mapping of species' genomes has accelerated at a staggering rate. This in turn has revolutionized some aspects of biogeography, as well as areas of ecology and conservation. It is estimated that the human genome consists of somewhere around 20,000–25,000 different protein-coding genes. The wide variety in human appearance—skin color, hair color, eye color, height, and so on is one visible product of this genetic diversity. Do the number of genes in a species' genome correlate with biological complexity, size, intelligence, and other traits? Specifically, we might ask does the number of protein-coding genes set humans apart? What is remarkable is that the humble roundworm *(Caenorhabditis elegans)* has almost as many protein-coding genes as humans. The rice plant *(Oryza sativa)* has more than twice as many such genes (Table 9.1). In fact, most plant species have more protein-coding genes than do humans. The largest eukaryotic genome yet discovered is the monocot plant *Paris japonica,* which is 50 times larger than the human genome. There does not appear to be any correlation between the size of a species' genome and the size or complexity of the species compared to others. In addition, for plants, it is thought that species with larger genomes grow more slowly and have greater difficulty facing environmental challenges such as pollution or climate change because of the relatively long time it takes to replicate their large genomes.

Differences in the physiology, anatomy, or behavior of different species or individuals of the same species are called **phenotypic variations.** These observable variations are often the result of genetic differences referred to as **genotypic variations.** For example, all white spruce trees have thin evergreen needles, whereas willows have broader deciduous leaves. These phenotypic differences between spruces and willows are genotypic and are inherited by the two plant species from their parents. Not all phenotypic differences are genotypic. The environmental conditions to which an organism is exposed can also influence the growth or form of the individual plant or animal. For example, the seed of white spruce trees *(Picea glauca)* planted in relatively warm forested sites can grow into a tree that is 30 meters tall. If the same seed is planted at a cold and unprotected site near the treeline, the spruce will grow as a very low shrub with sparse foliage. The difference between the tall spruce of the forest and the stunted spruce of the treeline has nothing to do with genetic differences; it is caused by differences in the environment. Thus, not all phenotypic differences are genetic.

Because both genetics and environmental conditions can influence the form or behavior of an organism, it is sometimes difficult to determine to what degree two individuals are genetically distinct. Two individuals may appear to be different simply because of the environmental conditions to which they have been exposed. This problem is particularly important for biogeographers. We frequently find that plants and animals of the same species living in different geographic regions display slightly different features or behavioral patterns. One way to assess if there is a genetic control behind such phenotypic differences in plant species is to conduct transplant experiments. In such experiments, the seeds from one region are transplanted into a different region and the development of the transplanted seedlings is observed. If the transplanted seedlings develop features that are typical of plants in their native habitat, but distinctive from the plants found in the transplant area, we can assume that the phenotypic differences we observe in plants from the different regions are indeed genetically controlled.

A classic transplant experiment was conducted on yarrow plants *(Achillea millefolium;* formerly classified as *Achillea lanulosa)* from California (Fig. 9.1). The yarrow plants grow from near sea level to alpine tundra at elevations of over 3600 meters. The yarrows growing at low elevations typically develop flower stalks that are up to 70 centimeters in height, while those found at the highest elevations grow stalks that are less than 20 centimeters tall. When transplanted to low elevation, the seeds from the alpine plants produced individuals that still grew to only 20 centimeters. Thus, the difference in size of the yarrows was experimentally shown to be a genetically controlled feature. Subsequent work on yarrows in the Sierra Nevada and western North America, including direct DNA analysis, support the presence of genotypic variation in the species related to variations in environmental conditions. It is likely that that the low stature of the alpine plants allows them to survive at high elevations because it keeps the flowering heads protected and closer to the warm ground surface. It has also been shown that the plants from high-elevation populations have greater rates of carbon assimilation through photosynthesis than low-elevation populations. Genetically and phenotypically distinct members of the same species that occur in different regions are referred to as *geographic races*. Those found in differing environmental settings are referred to as *ecological races,* or **ecotypes**. A continuous geographic gradient in a genetically controlled trait, such as the height of the yarrow plants, is called a **cline.**

Today, increasingly sophisticated biomolecular techniques are used to examine the exact genetic differences that occur across clines of plant and animal species. It is often found that within the geographic range of any species there are significant differences in the genetic composition of different geographic populations. These may reflect environmental differences, such as the ecotypic

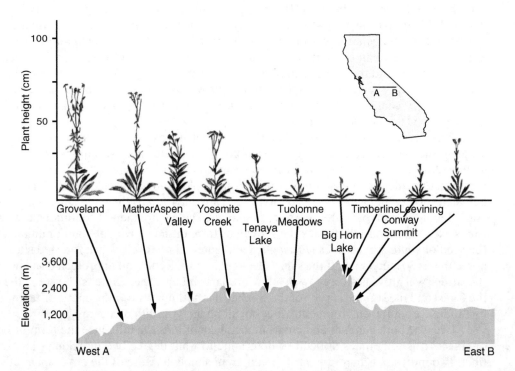

FIGURE 9.1   A cline in yarrow *(Achillea millefolium;* formerly classified as *Achillea lanulosa)* races found at different elevations in California. The high-elevation varieties are shorter than the low-elevation ones. When plants from both high and low elevations are grown together near sea level, the differences in height persist. This shows that the height differences are genotypic and the plants found at different altitudes represent genetically distinct races (adapted from Clausen et al., 1948; Krebs, 1994).

example of the yarrows, or the geographic history of the species. For example, lodgepole pine *(Pinus contorta* subsp. *latifolia)* trees in the Rocky Mountains of the United States have significantly higher numbers of different alleles than populations found in the Yukon Territories of Canada. This may reflect the loss of some alleles during the northward dispersal of the species after the end of the last glacial maximum. In the eastern United States, the Gulf of Mexico populations of widespread species of marine and shoreline animals such as the ribbed mussel *(Geukensia demissa),* black sea bass *(Centropristis stiata),* tiger beetle *(Cicindela dorsalis),* seaside sparrow *(Ammodramus maritimus),* and many other species are genetically distinct from populations of the same species that live along the Atlantic coastline. It is thought that these Atlantic-Gulf distinctions arise because the marine and shoreline populations in the two regions have long been separated from each other by the Miocene development and continued presence of the Florida Peninsula. Differences between the DNA sequences found in different species or separated populations of the same species are now widely used to estimate the timing of divergence or geographic separation of the two species or populations. These DNA-based approaches to reconstructing past divergences are referred to as **molecular clocks** (Technical Box 9.1). Molecular clocks have revolutionized the field of historical biogeography, which sought to infer the evolutionary histories of species from the geographic distributions of modern plants and animals or from fossil finds.

Many times, as in the case of the California yarrow plants, genetic differences within species cause the populations found in different geographic regions to have significant differences in morphology or physiology. For example, populations of the Douglas fir species *Pseudotsuga menziesii* can be found growing from central British Columbia southward along the Rocky Mountains to the mountains of northern Mexico and along the coast to northern California. Transplant experiments and DNA analysis show that there are a number of different phenotypes and genotypes within this large geographic range. One manifestation of this geographic diversity in genotypes is that trees from

TECHNICAL BOX 9.1

Molecular Clocks

In 1962, the Austrian-French biologist Émile Zuckerkandl and the American biochemist Linus Pauling noticed that differences in the amino acids in hemoglobin varied in proportion to the time since fossil records indicated different species had diverged from one another. They concluded that the rate of change in proteins was relatively constant over time and that the study of these differences provided a universal molecular clock that could be used to calculate the time in the past when species had diverged from one another. A few years later, the Japanese biologist Motoo Kimura suggested that since most genetic mutations are neutral and have no effect on the organism, they will appear and disappear more or less randomly at a constant rate. The longer two populations have been separated, the greater the difference should be in these neutral genes. The rate at which the new genetic mutations become fixed in a population is referred to as the substitution rate. If this rate in the appearance and disappearance of neutral genes is constant across all species, then one universal molecular clock could be employed to reconstruct the evolutionary divergences of all species. Today, molecular clocks are constructed for species using the mutation rates for DNA, RNA, or protein amino acid sequences. The mitochondrial DNA (mtDNA) in animals is neutral and evolves at a relatively fast rate and therefore is often used in molecular clock studies. However, research has shown that different species have different rates at which novel neutral genes appear and disappear. Ideally, each species' molecular clock should be calibrated against dated fossil or geological evidence of dates of divergence. One such study used 74 calibrations of the cytochrome b genes in a variety of birds. They calibrated the molecular clocks for each species based on the dates of fossils and geologic events such as the formations of mountain ranges and islands. These geologic events would have separated populations of the birds and started genetic diverges. The researchers estimated that the genes in the birds evolved at an average rate of approximately 1% per 1 million years and any two bird species diverged from each other at a rate of about 2% per 1 million years. This general "2% rule" is fairly widespread among birds, but does not hold for all bird species. Some may experience rates five times higher than this. Molecular clocks cannot be assumed to be constant across species. Molecular clocks are an important and still developing research tool in reconstructing evolutionary histories, but can be prone to considerable uncertainties.

coastal California are killed by temperatures of less than −16° C, while northern and high-elevation populations are much more frost-resistant. Recent research based on total genomic DNA analysis has shown that there are at least 12 different genes that display a significant association with cold hardiness in coastal Douglas fir trees.

The discussion of differences in the heights of yarrows or frost resistance in Douglas firs raises an interesting question. Can offspring of an organism inherit phenotypic traits that are acquired due to environmental conditions acting upon the organism? The answer is no. We now know that such acquired traits do not affect the alleles and chromosomes of the organism and are not passed down to offspring. If, for example, a bull elephant loses a tusk in a fight, the elephant's offspring will still develop two tusks. However, as late as the mid-twentieth century, there were scientists who argued that acquired traits could indeed be inherited.

We can see that genetics and heredity act to maintain the physiological, anatomical, and behavioral attributes of species from one generation to the next. How then do new genes and genetically controlled traits appear in species, and how do new species arise? Genes and chromosomes are subject to random mutations that create new alleles and chromosomal structures. Such mutations can give rise to new traits. Point mutations occur when individual genes experience changes in their chemical structure. Mutations also arise when chromosomes break into two or more smaller chromosomes and alter the chromosomal number of the organism. For example, some populations of the intertidal snail *Thais lappillus* have 26 chromosomes, while others have 36. In some cases, the broken chromosome fragments fuse in new or inverted configurations, and the gene loci are rearranged along the chromosome. In plants, it is relatively common for some offspring to have twice the chromosome number of either parent. The process that causes this is referred to as whole-genome duplication. The resulting plants are called **polyploids** and can be instantly reproductively isolated from the parent plants. This is extremely important in the evolution of new plant species. It has been estimated that some 30% to 50% of flowering plant species and up to 95% of fern species have experienced polyploidy events in their evolution. Finally, during recombination, some chromosome chains may not line up exactly, and some genes may be duplicated or deleted. All of these genetic mutations can give rise to new phenotypic traits. In most cases, however, major genetic mutations are lethal or do not allow successful reproduction by the mutant organism. However, some mutations may be benign or beneficial to the organism. In some cases, the mutant organism cannot mate with nonmutant members of the same species but can mate with siblings that share the mutation or reproduce asexually. In this case, instantaneous speciation occurs.

The genetic composition of a population can change over time as new genes arise via mutation and other genes are lost through chance processes. Such random changes are referred to as **genetic drift.** Over time, genetic drift can lead to the development of new species, particularly if the population is small and geographically isolated from other members of the same species. However, environmental conditions may also favor one genotype over another and lead to the progressive loss or increase in certain alleles. This process is not random and lies at the heart of the natural selection process and modern evolutionary theory, which are discussed below.

Historical Development of Evolutionary Theory

We have seen how genetically controlled traits are passed from parent to offspring. We have also seen that new traits can arise in species through genetic mutations. However, the capacity for the random generation of new traits or instantaneous speciation does not explain how species evolve specific adaptations that allow them to survive in new environments. The problem of how species evolve has baffled scientists from ancient times. Let's briefly consider the history of this scientific inquiry.

The idea that species could be transformed into new species was introduced in Greece at least a century before Aristotle (384–322 BCE). This idea, however, was at odds with Aristotle's view that species and genera were immutable, and the great philosopher argued against it. Aristotle's belief in the stability of species continued to be dominant until the time of the European Renaissance. Francis Bacon suggested in 1631 that there were some cases in which species appeared to be transmuted into other species and that this phenomenon warranted investigation. In 1766, the French scientist Comte de Buffon suggested that the number of animal families or genera of the earth represented the original number of animal species produced by divine creation. Under the influence of the environment, over a long period of time, these original species had subdivided into a number of degenerate species that now inhabit the planet. During the same period, Linnaeus suggested that new species could arise through interspecific mating. When two different species mate, the process is called **hybridization,** and the resulting progeny are called hybrids. Linnaeus saw hybridization as a means of both changing the characteristics of existing species and producing new species. A different force for creating new species was proposed by the French biologist Jean Baptiste Lamarck (1744–1829). In the early 1800s, Lamarck postulated that life first arose on earth as simple primitive organisms that developed in the primordial mud. He thought the impact of the environment on individual organisms resulted in their acquiring adaptive traits that they could then pass on to their offspring. For example, a giraffe feeding on savanna trees would stretch its neck to reach higher branches. As a result, the progeny of that giraffe would be born with a long neck. With each generation the necks of the giraffes would become longer. According to Lamarck, organisms that were exposed to different environmental conditions developed new adaptations and passed these down to their offspring. Thus, new species evolved to take advantage of different environments. At the root of Lamarck's theory was the hypothesis that acquired traits are passed down to the organism's progeny. We now know this is not possible.

The biological use of the term *evolution* appears to have begun at the time of Lamarck, although he never referred to his ideas using this term. Other nineteenth-century scientists, such as the Scottish geologist Charles Lyell, did use the word *evolution* in reference to Lamarck's ideas. Thus, the concept of evolutionary change and speciation was introduced into the language of biology and biogeography. The stage was now set for the development of the current theory of evolution. Our current ideas about evolution owe much to the work of two remarkable English naturalists and consummate biogeographers, Charles Darwin (Fig. 9.2) and Alfred Russel Wallace (Fig. 9.3).

Charles Darwin (1809–1882) came from a relatively affluent family and was educated in medicine at Edinburgh University and divinity at Cambridge University. By all accounts, he was not a particularly dedicated or gifted student in either subject. In 1831, he set out upon a 5-year voyage on the HMS *Beagle*. The British naval ship's mission was to chart the coast of South America and set up stations to aid navigation in the South Pacific. Darwin was an unpaid naturalist whose role was to

FIGURE 9.2 Charles Darwin (1809–1882). Darwin was only in his 20s when he embarked on the HMS *Beagle* for his scientific voyage around the world (*Source:* Henry Maull and John Fox / Wikipedia Commons / Public Domain).

FIGURE 9.3   Alfred Russel Wallace (1823–1913). Wallace began his scientific career as a freelance collector of exotic biological specimens and began his travels to places such as South America and the Malaysian Archipelago while in his 20s (*Source:* Sims / wellcomecollection).

make geological and biological observations, collect specimens, and provide appropriate company and intellectual stimulation for the upper-class captain of the ship. The young Charles Darwin turned out to be a very adventurous and energetic member of the expedition. Before and during the voyage, he read works by the biogeographer Alexander von Humboldt and the geologist Charles Lyell. Darwin accepted Lyell's ideas that the earth was ancient and prone to environmental change. Young Darwin was also familiar with incipient ideas of evolution through the writings of Lamarck and those of Charles Darwin's own grandfather, the physician-philosopher-biologist Erasmus Darwin. It is clear that during the voyage, the question of where and how different species arose was much on Darwin's mind.

During his explorations, Darwin observed several phenomena that would influence his thinking about how different species arose. He found many fossil remains and noted that some of these closely, but not exactly, resembled modern animal species. He noticed that in many cases bird or animal species would be common over a large area, but when one moved from that area to a different geographic region, the species were quickly replaced by other species. He visited islands such as the Galapagos and noted that some species of animals there resembled those on the mainland, while others were quite distinct. He also observed that the species of birds and tortoises on some of the Galapagos Islands appeared to represent different varieties from those found on other islands in the archipelago. He would later learn from experts in Britain that the different varieties of tortoises, finches, and mockingbirds he collected were actually different species.

When he returned to England, Darwin established his scientific reputation by publishing the geological observations he had made during the voyage. He also worked on the huge body of zoological and botanical information he had acquired on the expedition. In secret, he also began to use his data to try to answer the riddle of how species evolved. In this effort, he kept extensive notes of his thoughts and observations. In 1838, he read the writings of Thomas Malthus (1766–1834) and was greatly impressed by them. Malthus was an English clergyman and economist who set forth his views on human population growth and poverty in a treatise published in 1789. Malthus proposed that human population was capable of expanding in an exponential manner, while resources, such as agricultural produce, could increase only in a linear manner. Thus, over time the human population would grow beyond the ability of the resources to support it. This would lead to increasingly fierce competition, such as fights between people, the hoarding of resources by certain segments of society, or wars between nations. The result of such competition was that the weaker would be destroyed and the stronger would persist. Darwin later wrote that the ideas of Malthus at last provided him with "a theory by which to work" on his own ideas of natural evolution.

What was the concept that Darwin formulated? Darwin observed that all species have natural variability in physiology, anatomy, and behavior. In humans, for example, some people have black hair and some have blond hair. Some birds have slightly longer beaks than other individuals of the same species, while some horses have longer legs than other horses, and so on. He also observed that populations of plants and animals can grow rapidly to great sizes, but this growth is limited by intensive intraspecific and interspecific competition. If an individual possesses a trait that gives it an advantage in competition—say, for example, a longer beak that aids in hunting worms—that individual will be more likely to have sufficient energy to survive and produce healthy offspring. The trait that gives the individual a reproductive advantage over its competitors is in turn inherited by its offspring. These offspring are more numerous and more likely to survive than the offspring of competitors. Alternatively, individual birds of the same species that have slightly shorter beaks would be at a disadvantage in capturing prey and less likely to survive and breed. Over time, the long-beaked variety of bird will increase due to its success in feeding and reproduction, while the short-beaked variety will disappear. Darwin referred to this process as **natural selection.** Traits that provide an advantage in reproduction are selected for, whereas disadvantageous traits are selected against.

Natural selection provides a means for evolutionary change within a species. How does natural selection produce new species? As we will see, this is still a vexing problem. Darwin thought that new species could develop when individuals moved into new habitat and when natural selection within that new environment acted to promote certain traits and depress others. Darwin also considered the establishment of geographically separated populations by the dispersal of propagules to islands or distant continents to be an important force in allowing the development of new species.

Modern genetics had not been developed by Darwin's time, and he did not know about genes and how they pass along traits from parent to offspring. Despite his lack of knowledge of genetics, Darwin did not believe in Lamarck's ideas about acquired traits being inherited. It is amazing in some ways that Darwin developed his theory without knowing how new inheritable traits could arise as genetic mutations or how the recombination of genes during reproduction allows variability in offspring. If we wish to consider natural selection in light of modern genetics, we might say that Darwinian evolution works to select for genes that impart traits that lead to reproductive success. These genes are preserved in the population, while genes associated with traits that limit reproductive success are winnowed out.

Darwin first developed his theory in 1844 and 1845. At that point, he shared his ideas with a few close colleagues, such as the noted botanist Joseph Hooker. Darwin's ideas, that species could change over time and give rise to new species, were in some ways heretical to the teaching of the Church of England and the Christian religious beliefs prevalent at that time. He knew his theory would generate controversy, and he was very hesitant to publish it. Darwin enjoyed independent wealth, so he could continue to spend time refining his ideas and collecting evidence to support his theory of natural selection. Then, on June 18, 1858, he received an alarming letter from Malaysia that changed his plans.

The letter that caused Darwin so much distress came from a young English naturalist named Alfred Russel Wallace (1823–1913). Rather than being an enemy, Wallace was a great admirer of Darwin's work in geology and biology. Before we consider the contents of Wallace's letter, it is interesting to compare him to Darwin. Unlike Darwin, Wallace came from a relatively poor background and never attended university. Indeed, he had only 6 years of formal schooling and acquired his knowledge of biology largely though self-training. For a number of years, Wallace worked as a land surveyor in Great Britain. In 1848, he embarked on a voyage to the Amazon Basin with the entomologist Henry Bates to work as a self-employed collector of natural history specimens. Examples of exotic insects, birds, and flowers were very much in demand in Victorian England and could be sold at a profit. Wallace spent 4 years in the Amazon and wrote two books on his travels. In 1854, he departed on a collecting trip to Malaysia that was partially funded by the Royal Geographical Society. He spent the next 8 years collecting specimens and writing scientific articles. The conditions were incredibly difficult, and his challenges included severe bouts of malaria.

In his letter to Darwin, Wallace outlined a theory on how species evolve due to the pressure of competition, predation, and other environmental factors that favor some varieties and are deleterious to others. Wallace backed up his arguments with examples from the species of animals and insects he encountered on the various islands of Southeast Asia. He was particularly interested in the effect that the geographic separation of islands had on the eventual development of new species from common ancestors. Darwin saw at once that Wallace's ideas were extremely similar to his own theory of natural selection. How incredible that a young, poor, and relatively uneducated adventurer, working and living in the wilds of Malaysia, should so quickly develop the same theory of evolution as Darwin. The older scientist was the product of the best British education and had the luxury of spending over 20 years at home to work on his ideas. Of course, Darwin could have put Wallace's letter aside and quickly published his own theory. Thus, he would bring his life's work to fruition and secure his place in science by being the first to publish on the concept of natural selection. However, many have argued that such an action would not have been in keeping with the character of the man. Darwin confided his dilemma to his close friends Charles Lyell and Hooker. They helped arrange for a joint reading of separate papers by Darwin and Wallace at the Linnaean Society of London on July 1, 1858. The papers were then published as a single article entitled "On the tendency of species to form varieties; and on the perpetuation of varieties and species by means of natural selection." In 1859, Darwin expanded his ideas in a book entitled *On the Origin of Species by Means of Natural Selection, or the Preservation of Favoured Races in the Struggle for Life*. This pivotal volume went through several revisions in Darwin's time and remains in print today. However, debate has also continued to this day regarding the timing of the correspondences between Wallace and Darwin and the ultimate role Wallace's ideas played in Darwin's formulations regarding evolution.

Darwin and Wallace's concepts of evolution caused great controversy in both scientific and civil society. Bishops debated botanists over the origins of the human species. Famous geologists such as Louis Agassiz took exception to the theory, while others such as Charles Lyell were supporters. As he had feared, Darwin was the object of some personal attacks. The tempo of these attacks increased in 1871 when Darwin published his views that humans likely evolved from primates in the book *The Descent of Man, and Selection in Relation to Sex*. However, by the end of his life, the majority of the scientific community had embraced Darwin's views on evolution by natural selection. He was accorded many great honors, and when he died in 1882, a petition to the British Parliament accorded him burial in Westminster Abbey. This is the resting place of some of England's most famous figures. Wallace also had a long and distinguished scientific career. In fact, Wallace lived for 91 years and was granted many awards, including an honorary doctorate from Oxford University. He was, and remains, in the shadow of Darwin, however. It is telling that he was accorded a memorial medallion at Westminster Abbey but not burial there.

Even though over 150 years have passed since Darwin and Wallace's publication, and a huge body of research has been conducted during that time, an incredible amount of controversy continues to surround evolution. Much scientific debate has centered on five key questions: (1) Can new species develop if the founding populations are not somehow first separated geographically? (2) Do evolutionary changes occur gradually or rapidly? (3) Is some general direction toward increasing physiological or morphological complexity present in evolutionary history? (4) Does natural selection produce species that are perfectly adapted to their environments? (5) Has evolution produced a continuous increase in species diversity over geologic time? All of these questions are highly relevant for biogeographers. Let's tackle them individually and briefly review the current state of debate.

Isolation and Speciation

Darwin's insights and writings on evolution via natural selection are generally clear and detailed. However, on the question of how natural selection leads to the splitting of ancestral species into two or more new descendant species, Darwin was uncertain, and his writings reflect this doubt. He clearly thought that evolutionary changes caused by natural selection resulted ultimately in speciation, but he did not deeply delve into how such splits might occur. We should not be too hard on Darwin, for the question of how evolution results in speciation is still hotly debated today. It should be noted again that Darwin had no access to the modern science of genetics.

Here is the problem we face when we try to explain how evolutionary changes lead to speciation. In order for new biological species to be formed, there must be some barrier to reproduction between the newly formed species and the members of the original species. Evolutionary scientists refer to this barrier as reproductive isolation. Without isolation, the evolutionary changes that are developed within a population will be passed along throughout the entire species population, and although new traits may arise and spread in this manner, new species are not created. For speciation to occur, new genotypic traits must develop, and the individuals and populations that possess those traits must become reproductively isolated from populations of the original species. Once isolation occurs, the two species can undergo further differences in their genetic composition as further natural selection or genetic drift occurs.

Reproductive isolation is often classified as prezygotic isolation or postzygotic isolation. A zygote is the fertilized egg created during sexual reproduction. Prezygotic barriers preclude the formation of the zygote, while postzygotic barriers occur after a zygote is formed. Examples of postzygotic isolation effects are organisms that die or cannot reproduce after the zygote is formed. Mules, which are the offspring of horses and donkeys, cannot reproduce and are an example of postzygotic isolation. In some cases, mutations can cause instant postzygotic reproductive isolation and speciation because the mutant individuals can reproduce, but are genetically blocked from mating with members of the original nonmutant population. As noted earlier, many plant species have arisen because polyploidy occurred and produced new individuals with double the chromosomal number of the parent species. The mismatch in chromosome number between the original species and the polyploid offspring produces instant reproductive isolation. Let's explore the factors that produce prezygotic isolation by keeping individuals from interacting to produce zygotes.

How can reproductive isolation occur when there is no mutationally derived genetic barrier to reproduction to start with? The most obvious answer is geographic separation of the populations. Geographic barriers, including distance, preclude gene flow between populations and prevent new traits that evolve in one population from being introduced to other populations. In time, geographically separated populations may develop genetic traits that preclude interbreeding. Darwin himself was led to the conclusion that geographic separation was an important means for speciation to occur. The formation of new species by geographic isolation is called **allopatric speciation.**

Many phenomena can cause populations of species to be isolated geographically. A chance dispersal event may cause a population to become established on a distant island or continent. The honeycreeper species of Hawaii occur nowhere else in the world. Analysis of the mitochondrial DNA from 19 living or extinct Hawaiian honeycreeper taxa suggest that the Hawaiian honeycreeper species evolved from a single ancestral species of the Eurasian rosefinch genus *(Carpodacus)*. It has been calculated based on genetics and molecular clock analysis that an ancestral flock of rosefinches arrived on the islands about 7 million years ago. There are over 50 known extinct and living species of honeycreepers from the Hawaiian Islands. Large-scale geographic separation can explain why Hawaiian honeycreepers first speciated from continental rosefinches. In addition, there is evidence from molecular clock studies that some of the subsequent evolution of multiple species of honeycreepers is associated with the formation of new islands in the Hawaiian chain. The Hawaiian descendants of those original rosefinches experienced prezygotic isolation as they colonized the newly formed islands and stopped interacting with birds on other islands. The dispersal of the rosefinch to the remote islands produced isolation, and they evolved into different species and genera that are genetically and phenotypically distinct. In other cases, geologic events such the separating of landmasses by plate tectonics or climate change can cause isolation. For example, the 13 different species of beech *(Fagus)* trees found in Europe, Asia, and North America belong to the same genera and share a distant common ancestor. The genus was widely distributed across all three continents of the northern hemisphere in the Miocene. During late Miocene and post-Miocene times, as climate cooled and the increasing size of the North Atlantic Ocean and Arctic Ocean geographically fragmented the genus, allopatric speciation occurred between North American and Eurasian populations. The North American species *Fagus grandifolia* was present in its modern range of eastern North America and upland Mexico by 5 million to 2 million years ago and widely separated from the Eurasian species of the genus. It is distinctive genetically and phenotypically compared to species currently found in Eurasia.

Is geographic isolation required for speciation? This is perhaps one of the most contentious questions in evolutionary biology. The development of new species within the same geographic area is called

sympatric speciation. Lake Victoria in Africa supports over 300 different species of cichlids that likely arose from a common ancestor. The presence of multiple species of cichlid fish from common ancestral species in smaller isolated lakes, such as volcanic crater lakes in Africa and Lake Apoyo in Nicaragua have been held up as examples of speciation occurring in the absence of geographic isolation. The occurrence of such multiple speciation events on small islands or in isolated small waterbodies would seem to require that new species arise in the absence of large geographic barriers. In some instances, cichlid species may have evolved allopatrically after environmental changes caused lake levels to drop and populations to become isolated in remanent pools. However, it seems unlikely that all of the species of plants and animals we find could have evolved and speciated allopatrically due to large-scale geographic isolation. An analysis of mammal species in North America found that only about 45% of species are separated by geographical barriers from closely related species of the same genus.

The challenge of sympatric speciation is that individuals in the same geographic region have no physical barriers to cause reproductive isolation. New traits that do arise through mutations can be passed throughout the population. It is, therefore, difficult for genetically distinct and reproductively isolated new species to arise. Yet evidence exists that sympatric speciation does occur. For example, detailed study of the DNA structure and environmental history of cichlid fish populations in crater lakes suggests that a number of species must have evolved sympatrically within the lakes. Evolutionary ecologists are finding that the reproductive isolation required for sympatric speciation can occur between individuals of the same species within the same geographic region in a number of different ways.

Differences in life cycle timing, particularly the timing of reproduction, can cause reproductive isolation. For example, it has been shown that certain races of grasses that are genetically adapted to withstand toxic soil conditions flower at a different time than other races of the same species that grow nearby on normal soils. The difference in the timing of flowering prevents the species adapted to the toxic soils from interbreeding with the more numerous individuals that are adapted to growth on nontoxic soils. Similarly, many interfertile species of tropical plants, particularly orchids, remain reproductively isolated due to differences in their flowering season. Differences in flowering times prevent cross-fertilization and maintain separate species. Since the time of flowering is often controlled by day length or temperature, it is easy to see how populations of plants distributed along long latitudinal gradients or across large elevational gradients will flower at different times of the year and thus be reproductively isolated from other populations. One could imagine how California yarrow plants at low elevations would flower and be reproductively receptive to pollination early in the spring, while high-elevation plants would not flower until later in the summer. Temporal patterns can also reproductively isolate animal populations. One reason why two closely related species of North American flies *(Drosophila pseudoobscura* and *D. persimilis)* do not interbreed is because *D. pseudoobscura* is sexually active in the evening and *D. persimilis* is sexually active at night. A similar situation occurs for the orange- and white-colored races of sulfur butterflies *(Colias eurytheme)*. The orange-colored butterflies are active in the midday, while the white ones are active in the morning and late afternoon.

Mate choice in animals can be a complex process, and small differences in coloring or behavior may cause individuals to mate with one individual and not others. Barriers to breeding that are caused by behavior, particularly mate choice, are referred to as ethological isolation and sexual selection. There are many interfertile species and races of cichlid fish in the lakes of Africa. It has been observed that small differences in male coloring or courtship behavior are critical for female mate choice, and this keeps the cichlid races and species reproductively isolated and allows for sympatric evolution. Somewhat similar mate choice behavior appears responsible for sympatric speciation among stickleback fish *(Gasterosteus)* in lakes in British Columbia. The males of many tropical bird species have very distinctive plumage, while the females have plumage that is somewhat similar to that of other species. It is thought that mate choice by females is, in part, responsible for the reproductive isolation that has led to such high rates of speciation among these birds. In groups such as the birds of paradise found in New Guinea, the males must possess the proper plumage and engage in specific elaborate courtship behaviors in order to entice the female to mate. In North America, the mallard duck *(Anas platyrhynchos)* and pintail duck *(Anas acuta)* can physically interbreed but do not because the females do not recognize the courtship behaviors of males of the other species. Sexual selection has been argued by some to be one of the most important forces producing sympatric speciation, However, it remains uncertain what ultimately produces changing in the timing of reproduction or courtship displays and behaviors and if these differences originally arose allopatrically.

Finally, the difference between what we define as allopatric speciation and sympatric speciation may be largely a problem of scale. We might consider a large field of shrubs to be one geographic unit. However, for stenophagous insects that spend their entire life on one individual plant, such a field is really an ocean of isolated islands. Relatively little gene flow may occur between the insect populations on different plants. For example, at least 27 different species of weevils have evolved on the 6-kilometer-square island of Rapa in the South Pacific. This tiny island provides these small insects with a checkerboard of different habitat types and geographic locales for adaptation and speciation. For many species, even a fairly small area presents a patchwork of many different habitats. By restricting their activities to certain habitat types, animal populations become isolated from other populations. Genetic drift and natural selection for traits that impart reproductive success in the specific habitat occupied by the population lead to genetic differentiation and eventually to speciation. Habitat differences within a given area can also promote reproductive isolation by influencing mortality rates among offspring. For example, two closely related species of iris live in the Mississippi Delta region. *Iris fulva* grows on drier river banks, while *Iris giganticaerulea* grows in damp marshlands. The species are interfertile and can produce hybrids. However, these offspring are not successful in growing on either the dry or very wet sites and usually perish. Thus, genetic mixing between the two species is restricted by the inability of the hybrids to survive. This form of speciation, which is caused by the evolutionary divergence of populations that occupy different habitat in the same geographic area, is called **disruptive selection,** or **parapatric speciation.** Disruptive selection, in which individuals with extreme opposite values for genotypic traits are strongly selected for and individuals with intermediate values are selected against is likely a strong factor in cases of apparent sympatric speciation through natural selection (Fig. 9.4).

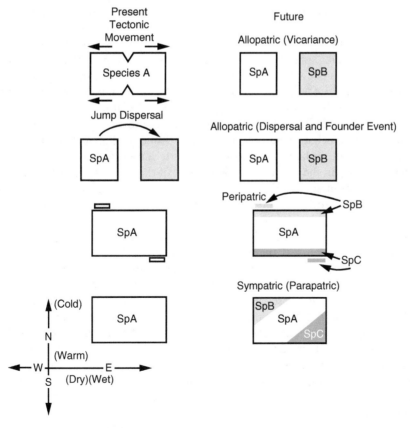

FIGURE 9.4 Different possible modes of geographic speciation. The rectangles represent land masses.

The Temporal Pattern of Evolution

Darwin clearly thought that evolution and speciation via natural selection occurred mainly as a slow and gradual process. New traits arise by mutations, and traits that infer greater reproductive success are selected for and eventually become dominant over many generations. Evolution in this manner is referred to as **phyletic gradualism.** If evolution proceeds in this manner, species should go through a number of transitional forms as natural selection works from generation to generation to favor certain adaptations. The earlier transitional forms should be preserved as fossils. The fossil history of horses over the past 55 million years provides evidence of such gradual evolution in the form of transitional fossils. Over time, the size and shape of the horses' long bones, skulls, and teeth changed as the group developed from a very small ancestor and speciated into the clade that includes horses, zebras, and donkeys today. However, even in Darwin's time, critics of natural selection pointed out that such transitional forms were rare in the fossil records of evolving species. Instead, the fossil record often shows that species persist for long periods of time with little outward change. Some paleontologists use the term *stasis* to refer to these long periods of time in which fossil lineages seem to undergo little change. The arrival of new species appears to occur very rapidly, often with no evidence of transitional forms. The fossil records of shellfish, for example, often include long periods of stasis and rapid speciation events. Interestingly, such periods of stasis often occur at the same time for many different species living within the same region. These episodes are called coordinated stasis. Darwin and many other evolutionists explained the lack of transitional forms by stating that the fossil record is often imperfect, and transitional forms might be missing because either they had not been fossilized due to geologic conditions or their fossils had not yet been discovered by paleontologists. However, by the middle of the twentieth century, after much paleontological research had been conducted, it became clear that many speciation events appear to have occurred very rapidly and there is little evidence for transitional forms existing in certain cases.

In the 1940s, the great biogeographer George Gaylord Simpson introduced the concept of quantum evolution to explain the apparent rapid appearance of new species. He proposed that many evolutionary changes arise from very small populations that lie at the periphery of the main ranges of the species or are geographically isolated. Simpson argued that evolutionary changes were more likely to occur in such small populations, but due to the small numbers of individuals involved, fossil evidence of such changes would be rare. Eventually, traits developed in such small populations could, if they were particularly beneficial, spread rapidly throughout the larger population. The fossil record would capture the arrival of these new forms, but not their development in small, isolated populations.

Simpson's ideas were developed and refined by the noted paleontologists Steven J. Gould and Neals Eldredge. Gould is famous for his popular writings on evolution, natural history, and the history of science. In the 1970s, Eldredge and Gould proposed that most new species arise from small peripheral or isolated populations. Such small populations are prone to relatively rapid speciation. In addition, many such peripheral and isolated habitats present different environmental conditions and selective pressures than are found in the main geographic range of the species. Due to natural selection related to their distinctive environments, and genetic drift, the peripheral populations will evolve traits that are different from those possessed by the main ancestral population of the species. The evolutionary changes that occur in such small populations are not likely to be well represented in the fossil record.

Eldredge and Gould (1977) postulated that environmental change also plays a role in the timing of large-scale speciation events. If the environment changed and produced conditions in the main range of the species that were more typical of those experienced by the peripheral races or species, the ancestral species would be replaced by the peripheral population. They felt that the rapid speciation events apparent in the fossil records of many species actually represented periods of environmental change that allowed peripheral races and species to replace the main ancestral species. For example, the rapid speciation events for shellfish in Lake Turkana, Africa, occur at the same time for many different taxa. Paleoenvironmental analysis shows that these speciation events are associated with environmental changes caused by shifts in lake levels. Between the speciation events there are long periods of coordinated stasis. Eldredge and Gould's concept of evolution is referred to as **punctuated equilibria.** They suggest that species persist for long periods of stasis when little evolutionary change occurs. These relatively stable species are then replaced by new species when the stability of the environment is punctuated by environmental changes. Some recent genetics research suggests that at a molecular level punctuated bursts of genomic instability may occur and multiple new mutations could be acquired simultaneously, preceded and followed by long periods of genetic stability.

So, which mode of evolution is correct, phyletic gradualism or punctuated equilibrium? When Simpson developed the concept of quantum evolution, he believed that both quantum evolution and gradualism operated in driving evolution. This is probably correct. Punctuated equilibrium seems particularly common for organisms such as shellfish, which occupy lakes possessing large central areas of similar habitat but also contain peripheral areas of shallow water or small bays that offer different environmental conditions. It is in these small peripheral environments that the small populations undergo evolutionary divergence. The punctuated equilibrium concept is attractive because it conforms to our understanding of how small populations can evolve differently and diverge from larger ancestral populations. In addition, the punctuated equilibrium concept explains why many fossil species seem to persist for so long and then are replaced without evidence of transitional forms. However, fossil evidence from some lineages, such as the horse, includes transitional forms that allow us to trace the gradual evolution of certain traits. Perhaps the conflict between phyletic gradualism and punctuated equilibrium is also a matter of scale. When we look at the fossils related to the formation of hooves in horses, we can see the transition that occurred over tens of millions of years between the four separate toes that the early genus *Hyracotherium* possessed, the three toes of intermediate taxa such as *Mesohippus*, and the single hoof possessed by modern horses *(Equus)*. However, we have little evidence of the transitional forms with reduced fourth toes that might have existed between four-toed *Hyracotherium* and its three-toed descendants. In addition, the fossil record shows that the transitions between four-toed, three-toed, and hooved horses occurred rapidly. Punctuated equilibrium provides insights on the pattern and mechanism of evolution and speciation that might occur at the scale of individual speciation events. Phyletic gradualism, identified by distinctive transitional forms, is perhaps most clearly apparent when we step back and look at the long histories of evolution encompassing many ancestor and descendant species.

It might be asked, how fast, in absolute terms such as years or generations, can evolutionary change and speciation occur due to natural selection? The actual number of years it takes for a species to evolve and speciate appears to be highly variable and dependent on both the genetics of the organism and the severity of selection pressure. Studies of fruit flies *(Drosophila subobscura)* that were accidentally introduced to California show that a genetically controlled increase in wing size developed in less than 20 years. Similarly, rapid changes have been seen in increasing wing sizes in Amazon bird populations that have become isolated in forest fragments. The fragmentation of tropical forest in Brazil and subsequent extinction of large seed-dispersing bird species have resulted in similarly rapid evolutionary reductions in the size of seeds produced by palm trees in the region. Today, the rapid changes in climate being produced by increasing greenhouse gases are being manifested in rapid genotypic and phenotypic evolutionary changes in a number of different plant and animal species. The evidence that is accumulating indicates that evolutionary changes driven by strong selection pressures can occur very rapidly.

Direction in Evolution

Is there a general direction in evolutionary development? We might ask if evolution leads to larger or more complex organisms. Many nineteenth- and twentieth-century scientists studying evolution felt that there was a general direction in evolution that leads to organisms that are more complex in form and function. Darwin was strongly influenced by the concept that there was a progression in evolutionary development from simple to more complex forms. He used terms such as *lower* and *higher* forms of life to rank the evolutionary development of plants and animals. Darwin considered humans to be the highest form of life and single-celled organisms, such as bacteria, to be the lowest forms. He saw evolution as the progressive development from lower to higher life forms.

Certainly, taking the entire sweep of the evolution of life over more than 500 million years, we see an initial development from simple single-celled organisms to more complex plants and animals. Most modern evolutionists do not subscribe to the idea that there is a general pattern of evolutionary development from simple to complex within subsequent clades. In fact, it is often quite difficult to define what is meant by the terms *simple* and *complex*. If you look at the flower of a grass plant, you will see a very tiny and simple arrangement of reproductive organs. The grass flower lacks petals, bright colors, and nectar. In contrast, the flowers of the magnolia are large, have showy petals, and

FIGURE 9.5   Example of evolutionary reduction in complexity. Simple grass flowers (Poaceae) versus a magnolia flower (Magnoliaceae). The magnolia flower appears similar to fossils of some of the earliest flowering plants, while grasses are a more recently evolved plant.

possess specialized organs for producing sweet-smelling nectar (Fig. 9.5). One might think that the simple form of the grass represents a primitive plant and the magnolia a more recently evolved complex one. From an evolutionary perspective, however, the grasses evolved fairly recently, in the late Cretaceous (90 million to 65 million years ago). In contrast, modern magnolias are similar to reconstructions of ancient flowering plants that developed their form of flower in the early Cretaceous (140 million to 130 million years ago), many millions of years before the grasses. The simplified flowers of the grasses are evolutionary reductions that evolved from seemingly more complex forms. There are many cases from the fossil records of land and marine plants and animals where morphologically complex species have given way to reduced forms over time.

One interesting general pattern that occurs in the evolutionary history of many species is a trend toward larger size as the lineage evolves. Horses, for example, started as a small dog-sized mammal and experienced a progressive increase in size as they evolved into the modern genus *Equus*. This general pattern of increasing size is referred to as **Cope's Rule.** A number of factors have been proposed to explain the increase in body size. These include things such as benefits of large size for competition, predation, escaping predation, ecological specialization, mate attraction, and fecundity. In a massive investigation of Cope's Rule using a big-data approach, researchers examined data for 17,208 genera of marine animals representing the past 542 million years. They found that body mass had increased by five orders of magnitude over that time and this pattern could not be the result of random chance. However, Cope's Rule is not universal across the animal kingdom. For example, studies of insects such as dragonflies and damselflies show no evidence of increasing body size over evolutionary history.

Speciation through evolutionary development is driven by natural selection to specific environmental conditions and by chance factors, such as mutations and genetic drift. Natural selection works to promote reproductive success above all else. Greater complexity in form or function does not necessarily translate to greater reproductive success. Thus, one cannot assume a general direction in evolution from complex to simple.

Perfection in Evolution

Can evolution and speciation be considered to produce perfect optimal adaptations to environmental conditions? The degree of morphological and physiological change that can occur within a species is governed by genetic constraints that limit the degree of change that can occur in the physiology and morphology of a species through evolution. In a classic discourse, Stephen Jay Gould used the example of the panda's thumb to illustrate this principle. Pandas must strip large amounts of leaves from bamboo plants in order to sustain themselves. The paw of the panda appears to have a set of digits and a thumb, which allows the animals to grab the bamboo and strip the leaves. However, pandas have evolved from carnivorous, clawed ancestors that did not possess a thumblike articulated digit. Upon closer inspection, the thumb of the panda is actually an elongated wrist bone that serves as an imperfect replacement for a real thumb provided by a flexible digit.

Some evolutionary scientists have also pointed out that the rate of adaptation through natural selection may be slower than the rate of environmental change that a species encounters. For example, many species persist for over 1 million years. In contrast, the Milankovitch cycles of climatic change occur on time scales of 100,000 years and less. These climatic cycles appear to have been important not just in the Quaternary, but earlier in geologic time as well. Consequently, a typical species must survive environmental changes that occur on 20,000-, 40,000-, and 100,000-year cycles. If a species is highly adapted to a particular climatic or environmental condition, it may not be able to survive the subsequent change in environment brought on by the Milankovitch cycles. The continuous disruptions caused by climatic changes may restrict the ability of species to accumulate very fine adaptations to particular environmental conditions through natural selection. Some evolutionary scientists have concluded that for the Quaternary period in particular, species are often adapted to climatic change. For example, during the last glacial maximum, the present location of the boreal forest biome in North America was largely covered by ice. Boreal forest plants and animals lived in southern areas such as the central plains and east coast where seasonal differences in hours of daylight are less extreme than in the modern boreal zone and where the climate was different than the modern boreal climate. Plant and animal species that have survived through the cycles of glaciation and nonglaciation must have enough phenotypic and genotypic plasticity to survive this wide range of environmental conditions. Certain traits found in these species may not be optimal for survival in the modern boreal zone but may have been important for survival during glacial periods. Thus, we must be careful in assuming that all species are optimally adapted to their present environments or that traits we observe today represent adaptations to modern environmental conditions.

Finally, there are often trade-offs between optimal efficiency in a given trait and the plethora of environmental factors that species must contend with to survive. As an example, a leopard might be able to run faster over open ground if it possessed the same curved, blunt, and semi-retractable claws as the much quicker cheetah. However, leopards are often found in a wide range of settings including forested and woodland habitat where their sharp protractile claws allow them to better grip bark and climb trees to their advantage, whereas cheetahs are relatively poor climbers and more typically inhabit grasslands and savannas where their high ground speed is an advantage.

Increasing Global Species Diversity

We might ask if there is a continuing growth trend in the total number of plant and animal species on earth over geologic time. It is obvious that there are more species today than there were when life first began in the Precambrian. In searching for long-term trends in species diversity we have to be careful of the bias offered by the fossil record. We have many more fossils and a much better understanding of the taxonomy of recent plants and animals that lived over the past few million years than we have for the plants and animals that existed in the more distant geologic past. Some paleontologists believe that the apparent increase in the number of species, genera, or families over the past 200 million years for some groups of organisms are more an artifact of the fossil record than a true representation of increasing diversity over recent geologic time.

One group of organisms with a reasonably good fossil record and evidence of long-term increasing species diversity are the tetrapods (Fig. 9.6). This group includes all of the land-dwelling, four-limbed vertebrate animals. This would include everything from frogs, to birds, to giraffes, to humans. Today, there are over 30,000 different species of tetrapods distributed across over 300 families. The first evidence of such land-dwelling animals appears in the Middle Devonian, about 398 to 385 million years ago. Since that time there appears to have been an exponential increase in the number of tetrapod species over time that cannot be explained solely as an artifact of an incomplete early fossil record. The increase in tetrapod diversity could reflect evolutionary driven adaptations that allowed new species to be capable of living in new habitat types and perhaps also avoid competition with other species. A similar exponential increase in species diversity has been suggested for land

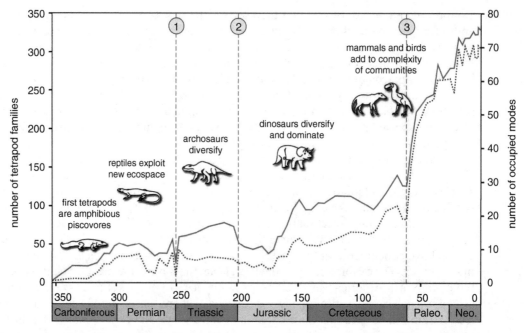

FIGURE 9.6   Increasing diversity of tetrapod families over the past 350 million years (*Source:* Sahney et al., 2010 / The Royal Society).

organisms in general, encompassing plants and invertebrates. In fact, it is likely that that there are more species present on earth today than in any other geological age. The increase in tetrapod families and other groups of organisms is not smooth however. It is clear that there were periods of time in the geologic past that experienced declining numbers of species and families, followed by recoveries. It is also notable that the rate of the evolution of new families has slowed in the past 25 million years. Just as is the case with populations of organisms occupying finite habitat, it seems unlikely that species diversity can increase in an exponential manner indefinitely.

GEOGRAPHY AND EVOLUTION: FOUNDER EFFECTS, BOTTLENECKS, VICARIANCE EVENTS, ADAPTIVE RADIATION, AND EVOLUTIONARY CONVERGENCE

Now that we have reviewed some basics, let's examine a little more closely how geography may impact on processes and patterns of evolution and speciation. It is thought that rates of evolutionary change and speciation are enhanced when very small populations are geographically separated from their parent population. In such cases, the smaller population often has a different genetic composition than is average for the larger populations. For example, imagine a small group of tourists that become stranded on a remote island. The group might only contain individuals with genes for blond hair and blue eyes. This is a subsample of the entire wide range of human genes for hair and eye color. Although the stranded tourists may reproduce and increase in population size, their offspring will have less genetic diversity than is evident in the entire human population. The populations that develop following such founder events appear to be particularly prone to speciation because they are already genetically different from the ancestral population and new alleles that may develop via mutations remain within the isolated populations. In addition, the isolated population is cut off from alleles that may develop in the ancestral population. The idea that such events promote rapid allopatric speciation in geographically isolated populations (Fig. 9.4) is called the **founder principle.**

Consider another scenario in which half of the stranded tourists sail away and half remain. Some of the genetic diversity that had existed on the island will have been lost because of the emigration of the people who sailed away. The decreased genetic diversity that results from a decreased population size and associated decrease in genetic diversity is called a **bottleneck.** Genetic bottlenecks generally occur when environmental factors, such as a rapid change in climate, cause the population of a species to be greatly reduced in size. During such catastrophic events, it is likely that certain genes will be lost from the species because all individuals carrying those genes perish. The small population that survives a bottleneck may produce offspring and increase in numbers. However, the post-bottleneck population does not contain the range of genetic diversity found in the original ancestral population. Over time, mutations lead to the introduction of new alleles and the development of new traits. Some of these traits may be advantageous and are selected for by natural selection. Others become common as a result of genetic drift. In any case, because the population is small, these alleles spread rapidly among it. If the population is isolated from the original ancestral population, these traits do not spread back to the original population.

Because of the founder effect, speciation is often more likely to occur at the edges of species range than in the central portions of their geographic distributions. Small isolated populations at the edges of species geographic distributions often have lower genetic diversity than is found in more central populations. Events that cause such peripheral populations to become geographically isolated from the main population accelerate genetic divergence and speciation through the founder effect. This geographic pattern of speciation is sometimes referred to as **peripatric speciation** (Fig. 9.4). A classic example of divergence and speciation due to the founder effect acting on geographically peripheral populations is provided by the distribution of different races and species of kingfisher birds *(Tanysiptera)* in New Guinea. The outer coastline and islands of New Guinea contain at least two different species and eight morphologically distinctive races of kingfishers. The main species, *Tanysiptera galatea,* is found at the center of the main island and on the largest of the surrounding islands. It is likely that the species of kingfishers on the small islands and coastal areas have diverged from *Tanysiptera galatea* into new races and one new species. The divergence is likely due to a combination of genetic drift and natural selection acting on small peripheral populations of birds. Rising sea levels at the close of the Pleistocene may have contributed to the geographic isolation and peripatric speciation of the outer island kingfisher populations by creating and isolating islands.

Many unique species of birds, invertebrates, and plants found on remote islands of the Pacific Ocean are the result of founder effects and population bottlenecks that occurred following long-distance dispersal and colonization. The colonizing populations of animals and plants arrived from the mainland or other islands and underwent evolution and allopatric speciation. The evolutionary development of an isolated species can be caused by genetic drift or natural selection. Through natural selection, a species on an isolated island might develop traits that are related to increased survival and reproduction on the island. Many species of island birds, such as the kiwi *(Apteryx australis)* in New Zealand, are flightless. The continental relatives of these birds are often good fliers. Hawaii, for example, had at least eight species of flightless geese that were closely related to the migratory Canada goose. Darwin himself noted the high proportion of flightless birds on islands. Flightlessness is also seen in many species of island insects. In addition, the seeds of a number of island plant species have reduced capacity for wind dispersal when compared with close relatives growing on the mainland. The inability to fly might be selectively favored on an island because it keeps individuals from getting blown out to sea. In many cases, large predators are absent on islands, so being able to fly is not as crucial to escape predation as it might be on the mainland.

Unfortunately, flightless bird species are extremely vulnerable when humans and alien predators arrive on islands. The arrival of human hunters and the introduction of predators such as cats, rats, and dogs to Hawaii contributed to a precipitous decline in the population of Hawaiian geese. All of the flightless geese species went extinct after the arrival of Polynesians about 1500 years ago. The single remaining species of Hawaiian goose, the nene *(Branta sandvicensis),* has shorter wings than its mainland relatives but can fly moderate distances (Fig. 9.7). Even with moderate flying abilities, the nene has not been excluded from the risk of extinction. By 1951, only 30 nene remained on the islands. The species survives today due to protection and propagation using captive populations.

In some cases, large populations become geographically separated by geologic events such as the splitting of continents. Such changes, which divide the ranges of species into geographically isolated distributions, are called **vicariance events** (Fig. 9.4). Once the populations are split by geographic barriers, allopatric speciation can occur. The divergence of North American and Eurasian deciduous trees into separate species is the result of such a vicariance event caused by the formation of the Atlantic Ocean and Arctic Ocean. Large-scale climatic changes can also produce vicariance events, which lead to the geographic isolation of species and the evolution of new species. During

FIGURE 9.7   The last surviving species of Hawaiian goose, the nene *(Branta sandvicensis)*. Like a number of species of bird and insect species that have evolved on remote islands, all the other species of Hawaiian goose have lost their ability to fly. Those species are now extinct. The nene can still fly moderate distances. *Source:* National Park Service / Public domain / https://www.nps.gov/havo/learn/nature/nene.htm

glacial periods, the Laurentide Ice Complex split the North American Boreal biome into western and eastern portions. Genetic analysis of jack pine *(Pinus banksiana),* common to eastern North America, and lodgepole pine *(Pinus contorta),* common to western North America, suggests that these two species developed from a common ancestral population. The range for the northern pines was split apart by advances of the Laurentide Ice Complex, and for long periods of time the ancestors of the two species were geographically isolated. During glacial periods, the ancestors of jack pine likely grew in the eastern United States, while the ancestors of lodgepole pine were restricted to the Rocky Mountains westward. The periods of separation caused by continental glaciation would have promoted allopatric divergence into the two modern species (Fig. 9.8).

The large populations created by continental-scale vicariance events contain large amounts of genetic diversity relative to the total diversity of the ancestral population. This contrasts with the lower amounts of genetic diversity contained in small founder event populations. In theory, evolutionary divergence and speciation will occur more slowly within these large vicariance populations and more quickly in small founding event populations. There is some evidence that species differentiation in large vicariance event populations does indeed occur relatively slowly. For example, the general morphology, favored climatic conditions, and growth habitat typical of the European beech tree *(Fagus sylvatica)* and the North American beech species *(Fagus grandifolia)* are extremely similar. A very close similarity in the niches of the beeches persists, despite the fact that the vicariance speciation event that led to the creation of these two species was the separation of North America from Eurasia some 30 million years ago. Many Eurasian species of plants, such as sycamores *(Platanus),* not only have similar niches with North American relatives, but reproduce easily with their North American representatives despite a separation of some 30 million years. Jack and lodgepole pine also hybridize when they grow together today (Fig. 9.8).

In the case of animals, a similar close correspondence in the niches of vicariant species was shown in an examination of 37 different pairs of upland butterfly, bird, and mammal species in southern Mexico. These species pairs were created when the formation of the Isthmus of Tehuantepec Lowlands split their ranges into two separate areas. Allopatric speciation then led to the evolution of separate species north and south of the lowlands. This vicariance event occurred some 10 million to 2.4 million years ago. Yet a careful analysis of the species niches shows that evolutionary divergence between the species to the north and south of the lowlands has not been very great.

Founder events and vicariance events can explain the evolutionary divergence of one species into two. However, in many cases, speciation produces more than one new species. The many species of

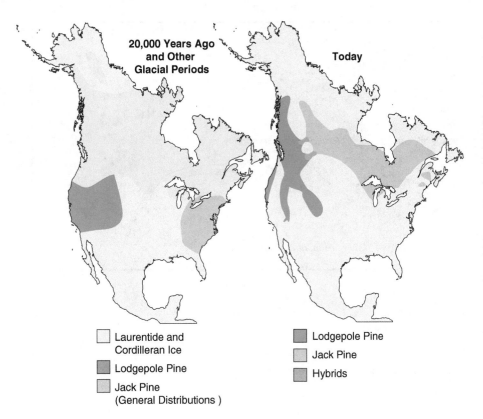

Laurentide and
Cordilleran Ice

Lodgepole Pine

Jack Pine
(General Distributions)

Lodgepole Pine

Jack Pine

Hybrids

FIGURE 9.8 The modern distribution of two closely related tree species, lodgepole pine *(Pinus contorta)* and jack pine *(Pinus banksiana)*. It is thought that divergence and the development of these two species occurred allopatrically and was influenced by the Pleistocene glaciations split of the ancestral pine distribution into separate western and eastern populations.

honeycreepers found on the Hawaiian Islands or cichlid fish species found in African lakes developed from the arrival of a single founding species at isolated islands or waterbodies. The development of many species from a single species often results from a process called **adaptive radiation.** During adaptive radiation, new species evolve from a common ancestor to fill all of the available niches in the colonized region. The niche space that is occupied by the adaptive radiation of a species is sometimes referred to as the adaptive zone. For example, the different honeycreepers on the Hawaiian Islands have a wide range of beaks that allow different species to take advantage of different foods. Some species of honeycreepers have short, thick beaks adapted for eating fruit and seeds (Fig. 9.9). Others have beaks that are adapted for prying bark apart to find insects. Finally, some species possess long and slender beaks with which they obtain nectar from flowers. Cichlid fish in African lakes display a similar wide variety of morphological adaptations to food and habitat that have developed through adaptive radiation (Fig. 9.9). By evolving traits that allow individuals to take advantage of different foods and habitats, and thus decrease niche overlap, competition is decreased between the species and carrying capacity is increased. Among other forms, some cichlids have large mouths adapted for eating fish, others have small mouths adapted for consuming zooplankton, and still others possess sharp, pointed mouths adapted to picking invertebrates off of aquatic vegetation and the lake bottom.

In comparison with the apparently slow rate of species divergence seen in the large vicariance events, evolutionary divergence and adaptive radiation by small founder populations often occur rapidly. The African lakes were extremely desiccated by dry climatic conditions between approximately 18,000 and 15,000 years ago, leading to large extinctions of fish species. The cichlid fish in Lake Victoria have evolved into hundreds of new species with extremely different niches in the period from 15,000 years ago. Experiments have been conducted using anole lizards of the species *Anolis sagrei* to try to assess how quickly evolutionary divergence might occur for small founding populations. A small number of anoles were experimentally introduced to several islands that have different vegetation covers. The islands were revisited, and the lizards were examined over a number of years.

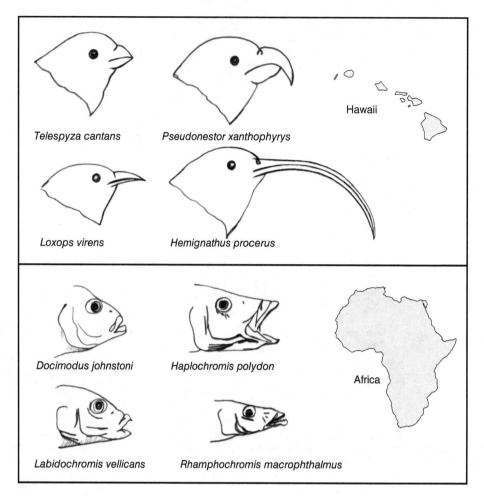

FIGURE 9.9   Two classic examples of adaptive radiation: honeycreepers from Hawaii (Raikow, 1977 / University of Hawaii) and cichlid fish from Africa (Fryer and Iles, 1972 / Oliver and Boyd).

The anoles displayed significant evolutionary divergence in shape and size after a period of only 10 to 14 years. Similarly, in the 1990s, Lake Victoria became highly eutrophic and water oxygen levels declined. Within less than two decades, the head shape and gill size of populations of the cichlid species *Haplochromis pyrrhocephalus* were observed to have diverged significantly from their ancestral population to a form that allowed for better survival in the oxygen-depleted waters.

The process of adaptive radiation is fundamental to the development of the great variety of plant and animal species that exist today and existed in the past. Since natural selection is a response to environmental conditions, we might expect that similar morphological and physiological traits will be selected for in different species that evolve in different regions that have similar environmental conditions. The development of similar morphological or physiological traits in unrelated species living in geographically separated regions is called **convergent evolution.** The classic examples of convergent evolution are the marsupial fauna of Australia and Tasmania, and the placental mammal fauna of North America and Eurasia. The continent of Australia has a host of environments, including deserts, savanna, tropical rainforests, temperate forests, and grasslands. Australia has been separated from all other continents for some 50 million years, and its native mammal fauna has been dominated by marsupial mammals (mammals that carry their young in sacks called marsupium where the young feed and continue to mature). Aside from some marsupial species in the Americas, such as the possum *(Dedelphis virginiana),* the mammal fauna of the rest of the world is comprised of placentals (mammals such as humans that do not carry young in external sacks). When Europeans first arrived in Australia, they found an amazing variety of marsupial species that were close counterparts in morphology and behavior to placental mammals found in similar environments outside of Australia. These included a marsupial "cat," a "wolf," a "mole," and even a marsupial version of a kangaroo rat (Fig. 9.10). These marsupials evolved

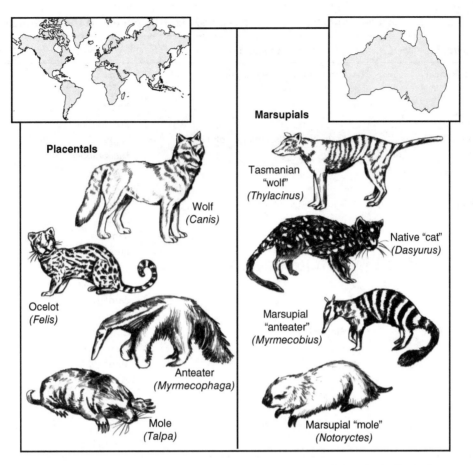

FIGURE 9.10   An example of convergent evolution: mammals and ecologically similar marsupials (*Source:* Baker and Garland, 1982 / Pearson).

via adaptive radiation to fill the many niches offered by the Australian environment. Similar traits were selected for in Australia as on the other continents, and convergent evolution produced similarities between the species of marsupials in Australia and their placental counterparts that lived elsewhere.

Convergent evolution also occurs in plants. The succulent euphorb plants that evolved in the deserts of Africa and Eurasia share gross similarities in form to the cacti that evolved in the Americas. The convergent evolution of plants is, of course, why the tropical forests in Africa, Southeast Asia, and South America look similar but are largely comprised of very different species of plants from continent to continent. The prevalence of convergent evolution in plants is crucial in producing the close linkage between general vegetation structure and climate that is apparent in the world's biomes.

One interesting aspect of adaptive radiation and convergent evolution is the very high degree of similarity that develops in the descendant species of common ancestors that evolve in geographically separate areas. The term **parallel evolution** is used to describe these instances where geographically isolated populations derived from the same ancestor evolve into morphologically and physiologically similar descendant species. For example, the islands of Cuba, Hispaniola, Jamaica, and Puerto Rico each support a number of distinctive species of anole lizards. Although the species that are present differ from island to island, all of these species appear to have evolved from the same common ancestor. On each island, species have evolved that have remarkably similar morphological adaptations and utilize similar habitats. For example, each island has a species with a large body and large toe pads that lives in tree crowns. Three of the islands also have a species with very slender bodies and long tails that lives on the ground. In fact, each of the islands possesses its own set of unique anole lizard species that are adapted for life either in tree canopies, upper trunks, twigs, midtrunks, lower trunks, or the ground (Table 9.2). It appears that on each island, competition caused the anoles to evolve into specialists that take advantage of the different habitats available in the forests of the islands. The close similarity between the canopy-dwelling species or lower trunk dwelling species on

TABLE 9.2   **Parallel Evolution in Anole *(Anolis)* Lizard Species in the Caribbean**

Habitat	Anole Morphology	Island	Species
Tree crown	Large body, large toe pads	Cuba	*A. equestris*
		Hispaniola	*A. ricodii*
		Jamaica	*A. garmani*
		Puerto Rico	*A. cuvieri*
Upper trunk	Large toe pads, change color	Cuba	*A. porcatus*
		Hispaniola	*A. chlorcynus*
		Jamaica	*A. grahami*
		Puerto Rico	*A. evermanni*
Twigs	Short body, slender legs	Cuba	*A. angusticeps*
		Hispaniola	*A. insolitus*
		Jamaica	*A. valenciennii*
		Puerto Rico	*A. occultus*
Midtrunk	Long forelimbs, flat body	Cuba	*A. loysiana*
		Hispaniola	*A. distichus*
		Jamaica	None
		Puerto Rico	None
Lower trunk	Stocky body, long hind limbs	Cuba	*A. sagrei*
		Hispaniola	*A. cybotes*
		Jamaica	*A. lineatopusi*
		Puerto Rico	*A. gundlachi*
Ground/bush	Slender body, long tail	Cuba	*A. alutaceus*
		Hispaniola	*A. olssani*
		Jamaica	None
		Puerto Rico	*A. pulchellus*

Source: Losos and de Queiroz, 1998.

the different islands reflects the fact that all of these geographically separate species evolved from a common colonizing ancestral species and share a common set of ancestral genes. Similar evidence of parallel evolution is seen in a number of other organisms, including stickleback fish from isolated lakes in British Columbia and fruit flies found on different Hawaiian Islands.

A final pattern of evolutionary development that is tightly tied by geography occurs when two unrelated species evolve traits that are tied to their interactions. This pattern is called **coevolution.** Mutualistic associations between plant and animal species are a result of coevolution and can develop only when the species are in close proximity. One example of coevolution in physical traits is provided by nectar-eating honeycreepers and the flowers that they feed from, and thus pollinate. The length and curvature of the bird's beaks and the floral tubes that they preferentially feed from are very similar. The shapes of the beaks and the flower tubes are different from those of mainland finches and close relatives of the Hawaiian flowers; this suggests coevolution between the two species on the islands.

EXTINCTION

Extinction is the loss of all individuals in the population of a given species, genus, family, or order. **Local extinction** occurs when a species or higher taxonomic order disappears in one or more geographic areas but persists in other regions. The history of the buffalo in North America provides an example of local extinction. In the late eighteenth century, the American buffalo *(Bison bison)* could be found across a continuous range that extended south from the northern plains of Canada to northern Mexico and east from the Rocky Mountains to New York (Fig. 9.11). The population size was likely in the millions. By 1888, factors such as hunting pressure from European settlers coupled with land-use changes had led to the local extinction of buffalo from all but a few tiny isolated areas of the interior plains. In addition, there is some evidence that Euro-American settlers saw extirpation of the buffalo

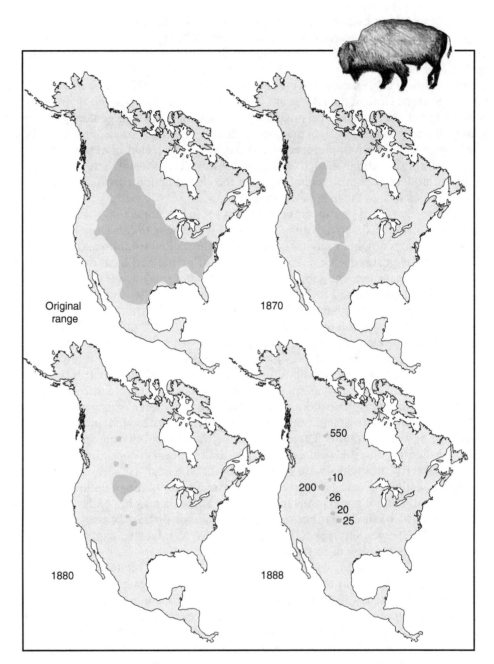

FIGURE 9.11 The widespread local extinction of the North American buffalo *(Bison bison)* throughout the interior of North America (*Source:* Grant 1991 / Columbia University Press).

as a way to destroy Native American populations, who were often highly dependent upon the animals for food, clothing, and other resources. By 1888, the total number of buffalo was likely less than 1000 animals. Captive breeding and protection of the remaining buffalo habitat have led to a resurgence in their numbers over the past hundred years. Today, there are about 20,000 wild buffalo in free-roaming herds on preserves in Canada, the United States, and Mexico. The species is still considered as near threatened in its wild state.

Global extinction refers to the loss of a species, or higher taxonomic order, over its entire range. Once global extinction occurs, the unique genome that was that species is lost forever. Global extinction for a species or genus is similar to death for an individual organism; it is final. The mammoth (*Mammuthus* spp.) provides an example of global extinction. Mammoths were relatives of the elephant and were extremely abundant during the Pleistocene. Prior to 12,000 years ago, species of mammoth were found across a huge range, extending from western Europe to northern Asia and across the unglaciated portions of North America. By about 10,000 years ago, all species of mammoth became extinct throughout almost all of their range on all three continents. The last mammoths probably lived on Wrangle Island in northeastern Siberia until about 4000 years ago. The global extinction of the mammoth has long been a topic of interest. It is likely that a combination of climatic and environmental change at the close of the last ice age and hunting by humans caused their extinction. Today there is ongoing consideration of the possibility of recreating this lost genus by recovering DNA from mammoth tissue that is frozen in permafrost. In theory, cloning procedures could be used to produce a new mammoth from well-preserved DNA. However, the technical hurdles, as well as the ecological and ethical considerations, to such efforts are substantial. Although there is great interest in the application of modern genomic technologies for the de-extinction of species, for the present, globally extinct species such as the mammoth are lost forever.

The extinction of one species can have a profound impact on the entire ecosystem in which it lives. For example, the loss of an important prey species can cause further extinctions because of the loss of food for higher predators. Such events are called **trophic cascades.** The loss of the mammoths may have caused such a trophic cascade of global and local extinction in North America about 10,000 years ago. The sabertooth cat (*Smilodon* spp.) was a very large and relatively common carnivore that existed in North America during the Pleistocene and likely fed upon mammoths and other large mammals. The sabertooth cat went extinct at about the same period as the mammoths. The extinction of the sabertooth cats was global, and we only know of these predators from their fossil remains. Trophic cascades can also have an impact on scavenger species. The California condor *(Gymnogyps californianus)* exists today only as reintroduced populations in highly protected mountain areas in California, Arizona, and Mexico. It is the largest bird in North America and can have a wing span of 3 meters. Fossil evidence shows that at the end of the Pleistocene, the California condor had a much more extensive geographic range, living as far east as New York State. It is thought that this large scavenger subsisted on the carcasses of large mammals such as the mammoth. The extinction of the mammoth and other large North American mammals likely led to the local extinction of the California condor over much of its range. When Europeans arrived, condors were found along the Pacific coast and parts of the Southwest. With European settlement and land-use changes, condor populations underwent a dramatic decline. The last small natural populations in southern California may have survived in part by feeding on the carcasses of whales and other sea mammals and on cattle carcasses following European colonization. The species survives today only because of intensive captive breeding-reintroduction programs and protection of its remaining habitat.

If death is the fate of all individuals, is global extinction the fate of all species? From the fossil record, we can deduce that over geologic time far more species have evolved and gone extinct than exist on the earth today. However, it is not always the case that species that go extinct in the fossil record actually represent evolutionary lineages and genomes that have completely died out. In many cases, species disappear from the fossil record because they evolved into new species. This is called phyletic extinction, and many of the genes of the original species persist in the descendant species. For example, 55 million years ago, there was a small herbivore mammal called *Hyracotherium* that browsed on shrubs and small trees. The members of this genus stood only 25 to 50 centimeters tall. From the fossil record, we can see that *Hyracotherium* went extinct before 40 million years ago. However, further analysis of the fossil record tells us that *Hyracotherium* is actually the ancestor of modern horses and their relatives in the genus *Equus*. Although apparently extinct, the *Hyracotherium* lineage evolved through various stages to become the modern genus *Equus*. In this case, many of the genes possessed by *Hyracotherium* have been preserved and passed down to its descendant species. When a species disappears without any descendant species, this event is called a true extinction. Unlike the modern horses, there are no descendant species from Pleistocene mammoths or sabertooth cats. These lineages are the victims of true extinctions.

When we further examine the fossil record for extinctions (both true and phyletic), we see that for commonly found organisms, such as marine invertebrates, the average rate of extinction over the past 500 million years has been something on the order of two complete families each million years. The majority of mammal genera have only lasted 10 million years or less. The majority of primate mammal genera, the group that includes humans, seem to persist for only 5 million

FIGURE 9.12   A living fossil, the horseshoe crab *(Limulus)*. These relatives of spiders have existed in forms similar to their modern one for over 400 million years. *Source:* National Park Service / Public domain / https://www.nps.gov/articles/ncbn-species-spotlight-atlantic-horseshoe-crab.htm

years or less. In general, individual species persist for fairly short periods, generally well less than 1 million to 2 million years for complex animals. There are some very interesting plants and animals that appear to have unusually long persistence, one of the most notable being horseshoe crabs (Order Xiphosura). These relatives of the spider have escaped extinction and have persisted with very little evolutionary change for the past 445 million years (Fig. 9.12). Another interesting example of a long-lived species is the coelacanth fish *(Latimeria chalumnae),* which appears to have persisted relatively unchanged in terms of appearance for about 70 million years. Indeed, the coelacanths were thought to have gone extinct about 70 million years ago in the Late Cretaceous. That was until a fishing crew caught a living one in 1938 off the coast of South Africa. When the ship docked, Marjorie Courtenay-Latimer, a 24-year-old museum curator, went to examine the catch for new specimens. She astutely recognized a large blue fish as being extremely unusual and worth further study. She ended up taking the fish back to the museum in a taxicab. Thanks to her efforts, and the work of her colleague J. L. B. Smith at Rhodes University, it was eventually proven to be the first living coelacanth ever identified. In her honor, the newly discovered living genus was named *Latimeria*. A number of other coelacanths have been caught since that time and two living species of *Latimeria* are now recognized. Ancient taxa that persist to the present day, such as the horseshoe crab and the coelacanth fish, are often called living fossils. The ginkgo tree *(Ginkgo biloba)* is an example of a living fossil plant. Fossil discoveries indicate that ginkgophytes similar to the modern species developed in the Permian over 200 million years ago and have remained essentially unchanged over the past 140 million years.

The fossil record of extinctions over time shows two interesting features. First, for plants or animals, there is continuous extinction of taxa throughout geologic time. For example, one estimate concluded that over the past 400 million years, plant families have gone extinct at an average base-level rate of about 1 every 4 million years. However, calculating such rates are difficult and contentious. These long-term base-level rates of extinction are sometimes called the **background extinction rate** or the *normal extinction rate*. In addition, there are clearly time periods where extinction rates become very high compared to the background extinction rate. Attention has been especially focused on what are called the **Big Five Mass Extinctions** (Table 9.3). The Big Five are (1) the Late Ordovician, 444 million years ago, when as much as 85% of the earth's species were lost; (2) the Late Devonian, 375 million years ago, when the loss of species may have reached 75%; (3) the Late Permian, 250 million years ago, when an astounding 90% or more of the earth's species may have been lost; (4) the Late Triassic, 200 million years ago, when it is estimated that around 70% of species were lost; and (5) the Late Cretaceous, 65 million years ago, when about 76% of the earth's species were lost, including the dinosaurs. Each of these mass extinction events marked the end of the geologic time period with which they are associated. The events were

TABLE 9.3 **The Big Five Extinction Events**

Geologic Time	Age, in Millions of Years	Possible Causes	Percentage of Species That Went Extinct
Late Ordovician	444	Climate cooling, glaciation, sea-level change, ocean and atmosphere chemistry changes	86%
Late Devonian	375	Ocean and atmosphere chemistry changes, possible role for vulcanism	75%
Late Permian	250	Siberian Traps vulcanism, climate warming, ocean and atmosphere chemistry changes	96%
Late Triassic	200	Atlantic Province vulcanism, climate warming, ocean and atmosphere chemistry changes	80%
Late Cretaceous	65	Asteroid impact near Yucatán, Mexico	76%

Source: Adapted from Barnosky et al., 2011.

sometimes multiphased and long in duration. For the most part, major extinction events such as the Big Five have been initially identified in the geologic record of marine life. They often have their strongest expression in the widespread extinction of marine species. This may reflect the low diversity of land organisms during early geologic history and the greater representation of marine sedimentary rocks and associated marine fossils in the geologic record. There is little evidence of Late Devonian terrestrial mass extinction, and land plant diversity actually increased during this time. Some of the best evidence for simultaneous marine and terrestrial mass extinctions comes from the Late Cretaceous at the time of the loss of the dinosaurs. It has been estimated that over 90% of mammal species and over 80% of mammal genera were also lost. The Late Cretaceous extinction of plant species may have exceeded 50% in some places.

For the most part, major extinction events such as the Big Five have been initially identified in the geologic record of marine life. They often have their strongest expression in the widespread extinction of marine species. This may reflect the low diversity of land organisms during early geologic history and the greater representation of marine sedimentary rocks and associated marine fossils in the geologic record. There is little evidence of Late Devonian terrestrial mass extinction, and land plant diversity actually increased during this time. Some of the best evidence for simultaneous marine and terrestrial mass extinctions comes from the Late Cretaceous at the time of the loss of the dinosaurs. It has been estimated that over 90% of mammal species and over 80% of mammal genera were also lost. The extinction levels for plant species may have exceeded 50% in some places.

It is believed that global mass extinctions coincide with catastrophic events such as large-scale vulcanism, rapid climatic change and glaciation, or the impact of large asteroids. There is compelling evidence that the Late Cretaceous extinctions, including the loss of the dinosaurs, appear to have been triggered by the impact of an asteroid in the vicinity of the Yucatán Peninsula of Mexico (Technical Box 9.2). Causes of some of the pre-Cretaceous extinction events remain the subject of intense research and some debate. However, it has been argued that large-scale vulcanism is plausibly at the root of many mass extinction events. Very large scale volcanic activity could induce mass extinctions through processes such as acidification of oceans, fresh waters, and soils; release of toxic metals; acid rain; damage to the ozone layer; and increases in ultraviolet radiation, darkening the atmosphere and producing cooling and decreased photosynthesis. Many have argued that today the world has entered the Sixth Mass Extinction. This event is being triggered not by volcanic activity or an asteroid, but by the actions of human beings. The Sixth Mass Extinction will be examined in greater detail in subsequent chapters of this book.

Let's turn back to considering background or normal extinctions. Biogeographers have long pondered what attributes cause some species to be prone to extinction while others are able to persist for many millions of years. We might first ask, what causes normal extinctions? In some cases, normal extinctions may relate to chance events. For example, a small population of a species may evolve on an isolated island and may not have the dispersal mechanisms needed to migrate away if the island is destroyed. In many cases, species may simply not have the phenotypic

Extinction of the Dinosaurs

Finding the cause of the late Cretaceous extinctions of the dinosaurs is one of the great detective stories in the history of geology, paleontology, and biogeography. Much of the credit for the evidence and arguments supporting the accepted theory can go to the father-and-son team of Luis and Walter Alvarez. The geologic boundary between the Cretaceous Period (the time during which the last dinosaurs lived) and the Paleogene is the time of interest. This important point in geological time is now referred to as the K-Pg Boundary, but was formerly known as the KT Boundary. You may see reference to the KT Boundary in older publications. The K-Pg Boundary is dated at about 65 million years ago. In many geologic sections, the sedimentary and fossil assemblage change at the boundary is relatively sharp and suggests a very rapid transition between the time of the dinosaurs and the subsequent time of the mammals. Although the span of time encompassed by the K-Pg Boundary is difficult to date precisely, the apparent sharpness of the boundary in many sections led some paleontologists to suggest that some catastrophic event had led to the extinction of the dinosaurs and many other organisms at the end of the Cretaceous. Luis Alvarez thought that by studying the concentration of a rare element called iridium he might be able to affix a firm age to the time that elapsed over the K-Pg Boundary and determine whether the extinction of the dinosaurs really was a rapid catastrophic event. Iridium found at the surface of the earth largely comes from small extraterrestrial particles and meteors and is deposited at a relatively constant rate. Alvarez reasoned that if the K-Pg Boundary occurred very quickly, it would have low concentrations of iridium. If the boundary encompassed a large amount of time, it should have higher amounts of iridium. Much to Alvarez's surprise, the samples of the K-Pg Boundary that he studied contained almost 100 times more iridium than he had estimated. He concluded that the only explanation for such high levels of iridium was the impact of a meteor or large comet on the surface of the earth during the time of the K-Pg Boundary. Evidence of high concentrations of iridium has subsequently been found in many other sites that possess sediments from the time of the K-Pg Boundary. In addition, geological and geophysical evidence shows that a very large asteroid impacted the earth in the area of the modern Yucatán Peninsula of Mexico at the time of the K-Pg Boundary. This huge geologic feature is called the Chicxulub Crater.

How would a large comet or meteor in Mexico cause global extinctions of terrestrial and marine life? The initial crater was likely some 40 kilometers deep and several hundred kilometers across. The heat and shock waves would have killed everything within several thousand kilometers of the impact. Hot volcanic ejecta would have rained down on earth, creating more fires over much of the planet. Huge tsunami waves would have inundated coastal areas and caused more deaths of plants and animals. Geological and paleontological evidence of such destruction by tsunamis has been found at sites along the Gulf of Mexico. Dust from the impact and fine charcoal from resulting fires would have darkened the atmosphere and caused global temperatures to plummet for several months. This would have been lethal to many plants and animals. The darkening sky would have decreased photosynthesis and the production of vitamin D in animals. Precipitation would have been extremely acidic due to increased levels of sulfur from the impact zone and the formation of nitric acid caused by the heating of the atmosphere. The acidic precipitation would have killed both terrestrial and aquatic organisms. Once the fine dust had settled, increased amounts of carbon dioxide and water vapor in the air caused by the impact would have created extreme greenhouse warming that would kill other organisms.

Perhaps what is surprising about the K-Pg Boundary is not that the dinosaurs went extinct, but rather how any other organisms such as amphibians, reptiles, birds, and mammals were able to survive. The case of the survival of birds is particularly intriguing. It is now recognized that birds are direct descendants of earlier dinosaurs and the closest relatives to the extinct ones that perished at the K-Pg Boundary. In fact, many paleontologists draw attention to this by referring to non-avian dinosaurs when writing about the extinct dinosaurs and referring to birds as avian-dinosaurs. The reason for the dinosaurs' particular sensitivity to the K-Pg Event is a topic of great scientific interest and continued research.

Could such a catastrophe occur again and cause our extinction? It is likely that other large comets or meteors will eventually hit the earth. In 1994, a relatively large comet impacted Jupiter and was observed by astronomers on earth. Perhaps such an event will not occur on earth for many millions of years. Maybe we will develop defenses against such an incident. What is more immediately troubling is the fact that a nuclear war could cause some of the same atmospheric events and global cooling as the impact of the Chicxulub asteroid 65 million years ago. Not only would any surviving humans face a prolonged period of intense cooling, called a nuclear winter, but they would also face dangerous levels of radiation. This human-caused catastrophe is hopefully more preventable than the impact of a comet or meteor.

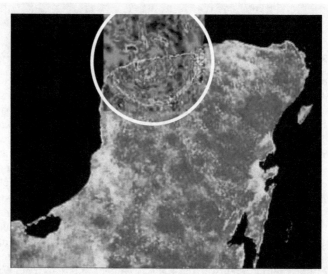

Satellite-derived gravity anomaly map showing the imprint of the Chicxulub Crater on the northern coast of the Yucatán Peninsula of Mexico (courtesy of NASA, https://www.nasa.gov/feature/asteroid-day-and-impact-craters).

plasticity and genetic diversity to adapt to long-term changes in their physical environment caused by an abiotic factor such as a climatic and environmental change. This type of abiotic extinction force has been referred to as the **Court Jester Hypothesis** of extinction. Biotic interactions can also be a factor in generating extinctions. The potential role of interspecific interactions, such as competition, in extinction has been elegantly summarized in what is called the **Red Queen Hypothesis** developed by the American biologist Leigh Van Valen in 1973. The name of the hypothesis comes from a line in Lewis Carroll's classic book *Through the Looking Glass*. In the book the Red Queen says, "It takes all the running you can do to stay in the same place." The point is that all of a species' competitors are continually evolving and becoming more competitive. If a species cannot evolve quickly enough to keep pace with the evolution of competing species, or prey or predators, it will go extinct because of interspecific interactions. There is debate over which factor, the abiotic changes highlighted by the Court Jester Hypothesis or the interspecific interactions highlighted by the Red Queen Hypothesis, is generally more important in driving background extinction rates. It is the case that both factors are at work and may operate preferentially at different geographic scales or time periods. Some scientists have argued that species interactions may be more important in driving extinctions of individual species at local scales over short time spans, while abiotic forces such as continental drift and climate change can exert more influence on taxonomically broader extinction events on continental scales over long time periods. However, for individual species, both biotic and abiotic extinction forces can operate at many different spatial and temporal scales.

However, evolutionary change and adaptation do not always protect a species against extinction. In some cases, species may become highly adapted for life in a very isolated geographic area, such as an island or a remote lake. If the environment changes, such species may have lost the ability to adapt to new conditions. Such overspecialization in an isolated site is referred to as an evolutionary trap. In addition, the loss of genetic diversity during evolution can fix species into modes of evolutionary development that become lethal. Such evolutionary paths are called blind alleys. This phenomenon is particularly likely for highly specialized species or those that develop after strong genetic bottlenecks. The loss of genetic diversity over time will be most problematic when the species has to contend with pressures brought on by environmental change.

A number of specific factors related to individual organisms and populations have been identified as important controls on the likelihood of extinction. Complexity, in terms of behavior or form, appears to be negatively correlated with the lifespan of species. Simple species, such as marine bivalves, mollusks, and plankton, have average durations of 10 million years before extinction. In contrast, mammal species rarely persist for more than 3 million years. It is thought that species that employ complex behavioral, physiological, or morphological adaptations in order to survive are less likely to persist if the environment in which they have evolved changes. Large animals appear more prone to extinction than small ones. The reason for this difference is that large animals require greater resources to support them and exist in a more coarse-grained environment. Should the environment change, a smaller animal may be able to take advantage of favorable microsites and survive. The larger animals cannot take advantage of such microsites or cannot obtain enough resources from them to survive. Predators at the top of the food chain are more sensitive to decreasing resources due to chance events and to trophic cascades than are primary consumers. Thus, animals at the higher trophic levels are more prone to extinction. Specialist species with narrow niches also appear more prone to extinction than generalist species. Species with narrow niche breadth are less likely to survive environmental changes.

Population size appears very important for species persistence. Species that have very small population sizes are prone to extinction due to chance events such as severe drought, whereas if a species has a large population size, there is a greater probability that some individuals will survive. Studies of the local extinction of tropical birds in Amazon forest patches over several decades indicate that populations with 50 individuals or more are about five times less likely to go extinct than those with 5 or less. Species that have rapid generation times and high birth/death ratios can produce rapid population regrowth after chance events and are more likely to evade extinction. Mathematical models have been used to explore the sensitivity of extinction rates to population size and birth rates/death rates. Such models show that species with very high birth rates relative to death rates can persist at relatively low population sizes for long periods of time. Species with low birth rates relative to death rates are prone to rapid extinction even when population sizes are high (Fig. 9.13).

Geography also appears to be one of the most important, or the most important, variable in determining the probable longevity of a species. In general, species that have large geographic distributions are less likely to go extinct than species with small ranges. This is because a chance event, such as a particularly severe winter or drought, is not likely to severely impact a large region as it is a small region. Thus, even though such chance events may cause local extinctions, they will not cause global extinctions. Studies of the local extinction of tropical birds on forest patches in the Amazon show that species living on 250-hectare patches have local extinction rates that are almost half the rate for species living on 24-hectare patches. Large geographic distributions also appear to be negatively correlated with extinctions during catastrophic events. For example, widely distributed marine invertebrates showed lower rates of extinction

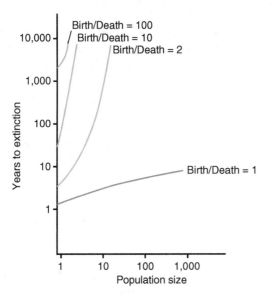

FIGURE 9.13 Mathematical modeling results suggest how the duration of a species prior to extinction is influenced by both population size and the birth rate/death ratio (*Source:* MacArthur and Wilson, 1967 / Princeton University Press).

during extinction events than invertebrates with small geographic distributions. Finally, species with high dispersal capability may be more likely to escape extinction than those with restricted abilities to disperse. Good dispersers can change their geographic distributions quickly during periods of environmental change. Species with poor dispersal capabilities may not be able to change their distributions quickly enough to escape extinction. Interestingly, dispersal capability does not always ensure survival. It has been found that marine bivalve species that released well-dispersed planktonic larvae had just as high an extinction rate as bivalves that did not have planktonic larvae.

Genetics plays an important role in extinction probability and is often influenced by population size and geographic range. Small populations restricted to small geographic areas are often characterized by inbreeding depression and low genetic diversity. Inbreeding depression results in reduced physiological vigor and reproduction capacity. Low genetic diversity produces a situation in which species do not have the capacity for the phenotypic and genotypic plasticity to adapt to changing biotic and abiotic conditions. These two genetic factors can promote extinction in small populations occupying small ranges.

THE RELATIONSHIP BETWEEN EXTINCTION, EVOLUTION, AND DIVERSITY

Although the extinction of a species is often compared to death, and thus has a very negative connotation, extinction does play a very positive role in the overall evolution of life. Extinction can serve to accelerate the tempo of evolution and speciation. Phyletic extinction, after all, represents the replacement of an ancestral species by its own descendant. If natural selection has been a strong force in the development of the new species, it can be assumed that it is better adapted to the environment in which it lives. When true extinction occurs, the removal of one species provides habitat and resources for competing species. The vacant niche space that is created by the extinction of one species can lead to evolutionary adaptation by other species to fill that niche. The end result can be increased diversity of species, genera, and families. Darwin saw these positive aspects of extinction when he formulated his ideas on evolution. Thus, both the Court Jester Hypothesis and the Red Queen Hypothesis are important concepts in understanding not just extinctions, but the drivers of evolution and increasing biological diversity.

In recent years, evolutionists have looked at the relationship between extinction and evolution in more detail. It has been found that genera and families that typically have high rates of species extinction also tend to have high speciation rates. In his classic investigations, American paleontologist Steven Stanley documented that on a broad scale the speciation rates for animals and higher plants are negatively correlated with the average length of time that species persist. Thus, species of marine bivalves often persist for 10 million years or more. The speciation rate for these bivalves is about one-quarter the rate for mammal species, which typically persist for less than 2 million years. It has been suggested that species that are found at low population sizes, with fragmented distributions, are the most likely to speciate allopatrically via founder effects. However, such small, isolated populations are also the most likely to go extinct.

Catastrophic extinctions play a very crucial role in generating speciation. When a catastrophic extinction event occurs, the loss of so many species creates an environment in which many potential niches are unfilled. The surviving species then typically experience high rates of adaptive radiation as they evolve and speciate to occupy the now vacant niches. A comparison of extinctions to rates by animals over the past 540 million years shows this relationship between mass extinction and adaptive radiation and speciation (Fig. 9.14). Catastrophic extinction events such as those at the end of the Permian or the end of the Cretaceous are followed by pulses in the evolution of new species, genera, and families.

As we close this chapter, we might reflect on the fact that the great diversity of mammals to which we belong occurred as adaptive radiation following the demise of the dinosaurs in the late Cretaceous (see Technical Box 9.2). Even though the Late Cretaceous extinction caused many mammal species to go extinct, within 10 million years the species diversity of mammals had increased to its modern level. Without the catastrophic extinction event that destroyed the dinosaurs and the subsequent adaptive radiation of mammals which it triggered, it is probable that humans would not have evolved, and you and I would not exist today.

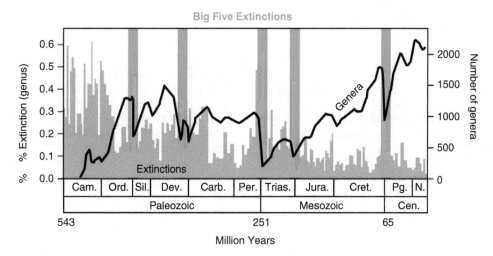

FIGURE 9.14   The relationship between mass extinctions of genera and recovery of marine via new genera (*Source:* Hull, P., 2015 / with permission of Elsevier).

KEY WORDS AND TERMS

Adaptive radiation
Allele
Allopatric speciation
Background extinction rate
Big Five Mass Extinctions
Bottleneck
Chromosome
Clade
Cline
Coevolution
Convergent evolution
Cope's Rule
Court Jester Hypothesis
Disruptive selection
DNA

Ecotypes
Evolution
Extinction
Founder principle
Gene
Genetic drift
Genome
Genomics
Genotypic variation
Global extinction
Heterozygous
Hybridization
Local extinction
Locus
Molecular clock

Natural selection
Parallel evolution
Parapatric speciation
Peripatric speciation
Phenotypic variation
Phyletic gradualism
Polymorphism
Polyploid
Punctuated equilibria
Red Queen Hypothesis
Speciation
Sympatric speciation
Trophic cascade
Vicariance event

REFERENCES AND FURTHER READING

Ahn, S., Anderson, J. A., Sorrells, M. E., and Tanksley, S. L. (1993). Homoeologous relationships of rice, wheat and maize chromosomes. *Molecular and General Genetics* **241**, 483–490.

Allio, R., Donega, S., Galtier, N., and Nabholz, B. (2017). Large variation in the ratio of mitochondrial to nuclear mutation rate across animals: implications for genetic diversity and the use of mitochondrial DNA as a molecular marker. *Molecular Biology and Evolution* **34**, 2762–2772.

Alvarez, L. W., Alvarez, W., Asaro, F., and Michel, H. V. (1980). Extraterrestrial cause for the Cretaceous-Tertiary extinction. *Science* **208**, 1095–1108.

Alvarez, W. (1997). *T. rex and the Crater of Doom*. Vintage Books, New York.

Arnegard, M. E., and Kondrashov, A. S. (2004). Sympatric speciation by sexual selection alone is unlikely. *Evolution* **58**, 222–237.

Baker, J. J. W., and Garland, E. A. (1981). *Study of Biology*. Addison-Wesley, Reading, MA.

Barluenga, M., Stölting, K. N., Salzburger, W., Muschick, M., and Meyer, A. (2006). Sympatric speciation in Nicaraguan crater lake cichlid fish. *Nature* **439**, 719–723.

Barnosky, A. D. (2001). Distinguishing the effects of the Red Queen and Court Jester on Miocene mammal evolution in the northern Rocky Mountains. *Journal of Vertebrate Paleontology* **21**, 172–185.

Barnosky, A. D., Matzke, N., Tomiya, S., Wogan, G. O., Swartz, B., Quental, T. B., Marshall, C., McGuire, J. L., Lindsey, E. L., Maguire, K. C., and Mersey, B. (2011). Has the Earth's sixth mass extinction already arrived? *Nature* **471**, 51–57.

Barraclough, T. G. (2019). *The Evolutionary Biology of Species*. Oxford University Press, Oxford.

Barton, N. H. (1989). Founder effect speciation. In *Speciation and Its Consequences* (ed. D. Otte and J. A. Endler), pp. 229–256. Sinauer Associates, Sunderland, MA.

Bennett, K. D. (1997). *Evolution and Ecology: The Pace of Life*. Cambridge University Press, Cambridge.

Benton, M. J. (2009). The Red Queen and the court jester: species diversity and the role of biotic and abiotic factors through time. *Science* **323**, 728–732.

Benton, M. J. (2016). Origins of biodiversity. *PLoS Biology* **14**, e2000724.

Bi, K. E., Linderoth, T., Singhal, S., Vanderpool, D., Patton, J. L., Nielsen, R., Moritz, C., and Good, J. M. (2019). Temporal genomic contrasts reveal rapid evolutionary responses in an alpine mammal during recent climate change. *PLoS Genetics* **15**, e1008119.

Boero, F., Belmonte, G., Bussotti, S., Fanelli, G., Fraschetti, S., Giangrande, A., Gravili, C., Guidetti, P., Pati, P., Piraino, S., Rubino, F., Saracino, O. D., Schmich, J., Terlizzi, A., and Geraci. S. (2004). From biodiversity and ecosystem functioning to the roots of ecological complexity. *Ecological Complexity* **1**, 101–109.

Bolnick, D. I., and Fitzpatrick, B. M. (2007). Sympatric speciation: models and empirical evidence. *Annual Review of Ecology, Evolution, and Systematics* **38**, 459–487.

Bond, D. P., and Grasby, S. E. (2017). On the causes of mass extinctions. *Palaeogeography, Palaeoclimatology, Palaeoecology* **478**, 3–29.

Borgaonkar, D. S. (1969). Observations on the chromosomes of one chicken *(Gallus domesticus)*. *Poultry Science* **48**, 331–333.

Brackman, A. C. (1980). *A Delicate Arrangement: The Strange Case of Charles Darwin and Alfred Russel Wallace*. Times Books, New York.

Brett, C. E., Ivany, L. C., and Schopf, K. M. (1996). Coordinated stasis: an overview. *Palaeogeography, Palaeoclimatology, Palaeoecology* **127**, 1–20.

Brown, E. J. (1995). *Charles Darwin: A Biography*. Knopf, New York.

Brown, J. H., and Lomolino, M. V. (1998). *Biogeography*, 2nd edition. Sinauer Associates, Sunderland, MA.

Brusatte, S. (2015). What killed the dinosaurs. *Scientific American* **313**, 54–59.

Brusatte, S. L., Butler, R. J., Barrett, P. M., Carrano, M. T., Evans, D. C., Lloyd, G. T., Mannion, P. D., Norell, M. A., Peppe, D. J., Upchurch, P., and Williamson, T. E. (2015). The extinction of the dinosaurs. *Biological Reviews* **90**, 628–642.

Capanna, E., and Castiglia, R. (2004). Chromosomes and speciation in *Mus musculus domesticus*. *Cytogenetic and Genome Research* **105**, 375–384.

Ceballos, G., García, A., and Ehrlich, P. R. (2010). The sixth extinction crisis: loss of animal populations and species. *Journal of Cosmology* **8**, 1821–1831.

Chamberlain, C. P., Waldbauer, J. R., Fox-Dobbs, K., Newsome, S. D., Koch, P. L., Smith, D. R., Church, M. E., Chamberlain, S. D., Sorenson, K. J., and Risebrough, R. (2005). Pleistocene to recent dietary shifts in California condors. *Proceedings of the National Academy of Sciences* **102**, 16707–16711.

Clausen, J., Keck, D. D., and Hiesey, W. M. (1948). *Experimental Studies on the Nature of Species. III. Environmental Response of Climatic Races of Achillea*. Carnegie Institute Publication No. 581, Washington, DC.

Cockburn, A. (1991). *An Introduction to Evolutionary Ecology*. Blackwell Scientific Publications, London.

Cody, M. L., and Overton, J. M. (1996). Short-term evolution of reduced dispersal in island plant populations. *Journal of Ecology* **84**, 53–62.

Condamine, F. L., Rolland, J., Höhna, S., Sperling, F. A., and Sanmartín, I. (2018). Testing the role of the Red Queen and Court Jester as drivers of the macroevolution of Apollo butterflies. *Systematic Biology* **67**, 940–964.

Costa, J. T. (2014). *Wallace, Darwin, and the Origin of Species*. Harvard University Press, Cambridge, MA.

Courtillot, V., and Gaudemer, Y. (1996). Effects of mass extinctions on biodiversity. *Nature* **381**, 146–148.

Coyne, J. A. (2010). *Why Evolution Is True*. Oxford University Press, Oxford.

Cwynar, L. C., and MacDonald, G. M. (1987). Geographical variation in lodgepole pine in relation to population history. *American Naturalist* **129**, 463–469.

Darwin, C. (1859). *On the Origin of Species by Means of Natural Selection, or the Preservation of Favoured Races in the Struggle for Life*. Murray, London. (Republished by Harvard University Press)

Darwin, C. (1871). *The Descent of Man and Selection in Relation to Sex*. Murray, London. (Republished by Random House)

Darwin, C. (1969). *The Autobiography of Charles Darwin*. W. W. Norton, New York.

D'Elia, J., Haig, S. M., Mullins, T. D., and Miller, M. P. (2016). Ancient DNA reveals substantial genetic diversity in the California Condor *(Gymnogyps californianus)* prior to a population bottleneck. *The Condor: Ornithological Applications* **118**, 703–714.

Denk, T., and Grimm, G. W. (2009). The biogeographic history of beech trees. *Review of Palaeobotany and Palynology* **158**, 83–100.

Devillers, C., and Chaline, J. (1993). *Evolution: An Evolving Theory*. Springer-Verlag, Berlin.

Diamond, J. M. (1984). Normal extinctions of isolated populations. In *Extinctions* (ed. M. N. Nitecki), pp. 191–246. University of Chicago Press, Chicago.

Drumm, D. T., and Kreiser, B. (2012). Population genetic structure and phylogeography of Mesokalliapseudes macsweenyi (Crustacea: Tanaidacea) in the northwestern Atlantic and Gulf of Mexico. *Journal of Experimental Marine Biology and Ecology* **412**, 58–65.

Eckert, A. J., Bower, A. D., Wegrzyn, J. L., Pande, B., Jermstad, K. D., Krutovsky, K. V., St. Clair, J. B., and Neale, D. B. (2009). Association genetics of coastal Douglas fir *(Pseudotsuga menziesii var. menziesii*, Pinaceae). I. Cold-hardiness related traits. *Genetics* **182**, 1289–1302.

Eldredge, N. (1989). *Macro-Evolutionary Dynamics: Species, Niches & Adaptive Peaks*. McGraw-Hill, New York.

Elewa, A. M., and Abdelhady, A. A. (2020). Past, present, and future mass extinctions. *Journal of African Earth Sciences* **162**, 103678.

Elmer, K. R., Reggio, C., Wirth, T., Verheyen, E., Salzburger, W., and Meyer, A. (2009). Pleistocene desiccation in East Africa bottlenecked but did not extirpate the adaptive radiation of Lake Victoria haplochromine cichlid fishes. *Proceedings of the National Academy of Sciences* **106**, 13404–13409.

Erickson, J. (1995). *A History of Life on Earth: Understanding Our Planet's Past*. Facts on File, New York.

Erwin, D. H. (1993). *The Great Paleozoic Crisis: Life and Death in the Permian*. Columbia University Press, New York

Erwin, D. H. (1996). The mother of mass extinctions. *Scientific American*, 72–78.

Fazekas, A. J., and Yeh, F. C. (2006). Postglacial colonization and population genetic relationships in the *Pinus contorta* complex. *Botany* **84**, 223–234.

Feduccia, A. (2014). Avian extinction at the end of the Cretaceous: assessing the magnitude and subsequent explosive radiation. *Cretaceous Research* **50**, 1–15.

Foster, C. S. (2016). The evolutionary history of flowering plants. *Journal and Proceedings of the Royal Society of New South Wales* **149**, 65–82.

Frankham, R. (2005). Genetics and extinction. *Biological Conservation* **126**, 131–140.

Fraser, D. R. (2019). Why did the dinosaurs become extinct? Could cholecalciferol (vitamin D3) deficiency be the answer? *Journal of Nutritional Science* **8**, 1–5.

Fryer, G., and IIes, T. D. (1972). *The Cichlid Fish of the Great Lakes of Africa: Their Biology and Evolution*. Oliver and Boyd, Edinburgh.

Futuyma, D. J. (1998). *Evolutionary Biology*, 3rd edition. Sinauer Associates, Sunderland, MA.

Futuyma, D. J., and Kirkpatrick, M. (2017). *Evolution*, 4th edition. Sinauer Associates, Sunderland, MA.

Galetti, M., Guevara, R., Côrtes, M. C., Fadini, R., Von Matter, S., Leite, A. B., Labecca, F., Ribeiro, T., Carvalho, C. S., Collevatti, R. G., and Pires, M. M. (2013). Functional extinction of birds drives rapid evolutionary changes in seed size. *Science* **340**, 1086–1090.

Goldschmidt, T. (1996). *Darwin's Dreampond*. MIT Press, Cambridge, MA.

Gould, S. J. (1989). *Wonderful Life: The Burgess Shale and the Nature of History*. W. W. Norton, New York and London.

Gould, S. J., and Eldredge, N. (1977). Punctuated equilibrium: the tempo and mode of evolution reconsidered. *Paleobiology* **3**, 115–151.

Grant, P. R., and Grant, B. R. (1989). Sypatric speciation and Darwin's finches. In *Speciation and Its Consequences* (ed. D. Otte and J. A. Endler), pp. 433–457. Sinauer Associates, Sunderland, MA.

Grant, V. (1991). *The Evolutionary Process*. Columbia University Press, New York.

Gurevitch, J. (1992). Differences in photosynthetic rate in populations of *Achillea lanulosa* from two altitudes. *Functional Ecology* **6**, 568–574.

Hamann, E., Pauli, C. S., Joly-Lopez, Z., Groen, S. C., Rest, J. S., Kane, N. C., Purugganan, M. D., and Franks, S. J. (2021). Rapid evolutionary changes in gene expression in response to climate fluctuations. *Molecular Ecology* **30**, 193–206.

Harries, P. J. (1993). Dynamics of survival following the Cenomanian-Turonian (Upper Cretaceous) mass extinction event. *Cretaceous Research* **14**, 563–583.

Harries, P. J. (1995). Recovery from mass extinction. *Palaios* **10**, 289–290.

Heasley, L. R., Sampaio, N., and Argueso, J. L. (2021). Systemic and rapid restructuring of the genome: a new perspective on punctuated equilibrium. *Current Genetics* **67**, 57–63.

Heim, N. A., Knope, M. L., Schaal, E. K., Wang, S. C., and Payne, J. L. (2015). Cope's rule in the evolution of marine animals. *Science* **347**, 867–870.

Hernández-Hernández, T., Miller, E. C., Román-Palacios, C., and Wiens, J. J. (2021). Speciation across the tree of life. *Biological Reviews* **96**, 1205–1242.

Ho, S. (2008) The molecular clock and estimating species divergence. *Nature Education* **1**, 168.

Ho, S. Y., and Duchêne, S. (2014). Molecular-clock methods for estimating evolutionary rates and timescales. *Molecular Ecology* **23**, 5947–5965.

Hone, D. W., and Benton, M. J. (2005). The evolution of large size: how does Cope's Rule work? *Trends in Ecology & Evolution* **20**, 4–6.

Hossfeld, U., and Olsson, L. (2013). The prominent absence of Alfred Russel Wallace at the Darwin anniversaries in Germany in 1909, 1959 and 2009. *Theory in Biosciences* **132**, 251–257.

Hull, P. (2015). Life in the aftermath of mass extinctions. *Current Biology* **25**, R941–R952.

Huntley, B., Bartlein, P. J., and Prentice, I. C. (1989). Climatic control of the distribution and abundance of beech (Fagus L.) in Europe and North America. *Journal of Biogeography*, **16**, 551–560.

Isenberg, A. C. (2020). *The Destruction of the Bison: An Environmental History, 1750–1920*, 20th anniversary edition. Cambridge University Press, Cambridge.

Jablonski, D. (2005). Mass extinctions and macroevolution. *Paleobiology* **31**, 192–210.

Jin, Y. G., Wang, Y., Wang, W., Shang, Q. H., Cao, C. Q., and Erwin, D. H. (2000). Pattern of marine mass extinction near the Permian-Triassic boundary in South China. *Science* **289**, 432–436.

Keller, E. F., and Lloyd, E. A. (1992). *Keywords in Evolutionary Biology*, p. 414. Harvard University Press, Cambridge, MA.

Kellman, M. C. (1975). *Plant Geography*. St. Martin's Press, New York.

Kerr, R. A. (1996). New mammal data challenge evolutionary pulse theory. *Science* **273**, 431–432.

Kerr, R. A. (1997). Climate-evolution link weakens. *Science* **276**, 1968.

Kerr, R. A. (1997). Cores document ancient catastrophe. *Science* **275**, 1265.

Kerr, R. A. (1999). A smoking gun for an ancient methane discharge. *Science* **286**, 1465.

Khibnik, A. I., and Kondrashov, A. S. (1997). Three mechanisms of Red Queen dynamics. *Proceedings of the Royal Society of London B* **264**, 1049–1056.

Kiessling, W., and Aberhan, M. (2007). Geographical distribution and extinction risk: lessons from Triassic-Jurassic marine benthic organisms. *Journal of Biogeography* **34**, 1473–1489.

Kin, A., and Błażejowski, B. (2014). The horseshoe crab of the genus *Limulus*: living fossil or stabilomorph? *PLoS One* **9**, e108036.

Knoll, A. H., Bambach, R. K., Canfield, D. E., and Grotzinger, J. P. (1996). Comparative Earth history and late Permian mass extinction. *Science* **272**, 452–457.

Kolbert, E. (2014). *The Sixth Extinction: An Unnatural History*. Bloomsbury Publishing, London.

Krebs, C. J. (1994). *Ecology*, 4th edition. Harper-Collins College Publishers, New York.

Kumar, S. (2005). Molecular clocks: four decades of evolution. *Nature Reviews. Genetics* **6**, 654–662.

Kutschera, U. (2003). A comparative analysis of the Darwin-Wallace papers and the development of the concept of natural selection. *Theory in Biosciences* **122**, 343–359.

Kyriazis, C. C., Wayne, R. K., and Lohmueller, K. E. (2021). Strongly deleterious mutations are a primary determinant of extinction risk due to inbreeding depression. *Evolution Letters* **5**(1), 33–47.

Lees, D. R., and Edwards, D. (1993). Evolutionary patterns and processes. In *Linnean Society Symposium Series*, vol. 14, p. 325. Academic Press, San Diego, CA.

Lehman, J., and Miikkulainen, R. (2015). Extinction events can accelerate evolution. *PloS One* **10**, e0132886.

Lerner, H. R., Meyer, M., James, H. F., Hofreiter, M., and Fleischer, R. C. (2011). Multilocus resolution of phylogeny and timescale in the extant adaptive radiation of Hawaiian honeycreepers. *Current Biology* **21**, 1838–1844.

Levins, R. (1968). *Evolution in Changing Environments*. Princeton University Press, Princeton, NJ.

Lieberman, B. S., and Eldredge, N. (2014). What is punctuated equilibrium? What is macroevolution? A response to Pennell et al. *Trends in Ecology & Evolution* **29**, 185–186.

Lloyd, D., Wimpenny, J., and Venables, A. (2010). Alfred Russel Wallace deserves better. *Journal of Biosciences* **35**, 339–349.

Longrich, N. R., Scriberas, J., and Wills, M. A. (2016). Severe extinction and rapid recovery of mammals across the Cretaceous-Palaeogene boundary, and the effects of rarity on patterns of extinction and recovery. *Journal of Evolutionary Biology*, **29**, 1495–1512.

Losos, J. B., Arnold, S. J., Bejerano, G., Brodie III, E. D., Hibbett, D., Hoekstra, H. E., Mindell, D. P., Monteiro, A., Moritz, C., Orr, H. A., and Petrov, D. A. (2013). Evolutionary biology for the 21st century. *PLoS Biology* **11**, e1001466.

Losos, J. B., and de Queiroz, K. (1997). Darwin's lizards. *Natural History* **106**, 34–38.

Losos, J. B., Warheit, K. I., and Schoener, T. W. (1997). Adaptive differentiation following experimental island colonization in *Anolis* lizards. *Nature* **387**, 70–73.

Lucas, S. G. (2021). Nonmarine mass extinctions. *Paleontological Research* **25**, 329–344.

MacArthur, R. H., and Wilson, E. O. (1967). *The Theory of Island Biogeography*. Princeton University Press, Princeton, NJ.

MacDonald, G. M., Beilman, D. W., Kuzmin, Y. V., Orlova, L. A., Kremenetski, K. V., Shapiro, B., Wayne, R. K., and Van Valkenburgh, B. (2012). Pattern of extinction of the woolly mammoth in Beringia. *Nature Communications* **3**, 1–8.

MacPhee, R. D. E. (1999). Extinctions in Near Time. In *Advances in Vertebrate Paleobiology* (ed. R. D. E. MacPhee and H. Seus), p. 394. Kluwer Academic/Plenum Publishers, New York.

Makino, S. (1951). *An Atlas of the Chromosome Numbers in Animals*, 2d edition. Iowa State College Press, Ames.

Martin, C. H., Cutler, J. S., Friel, J. P., Dening Touokong, C., Coop, G., and Wainwright, P. C. (2015). Complex histories of repeated gene flow in Cameroon crater lake cichlids cast doubt on one of the clearest examples of sympatric speciation. *Evolution* **69**, 1406–1422.

Mayr, E. (1942). *Systematics and the Origin of Species*. Columbia University Press, New York. (Republished by Dover Publications)

Mayr, E. (1963). *Animal Species and Evolution*. Harvard University Press, Cambridge, MA.

McElwain, J. C., Beerling, D. J., and Woodward, F. I. (1999). Fossil plants and global warming at the Triassic-Jurassic boundary. *Science* **288**, 1386–1390.

Minkoff, E. C. (1983). *Evolutionary Biology*. Addison-Wesley, Reading, MA.

Moghe, G. D., and Shiu, S. H. (2014). The causes and molecular consequences of polyploidy in flowering plants. *Annals of the New York Academy of Sciences* **1320**, 16–34.

Newell, N. D. (1967). Revolutions in the history of life. *Geological Society of America Special Paper* **89**, 63–91.

Nitecki, M. H. (1984). *Extinctions*. University of Chicago Press, Chicago.

Noor, E., and Milo, R. (2012). Efficiency in evolutionary trade-offs. *Science* **336**, 1114–1115.

Novak, B. J. (2018). De-extinction. *Genes* **9**, 548.

Offner, S. (1996). A plain English map of the chromosomes of the fruit fly *Drosophila melanogaster*. *American Biology Teacher* **58**, 462–469.

Payne, J. L., and Finnegan, S. (2007). The effect of geographic range on extinction risk during background and mass extinction. *Proceedings of the National Academy of Sciences* **104**, 10506–10511.

Pejchar, L., Medrano, L., Niemiec, R. M., Barfield, J. P., Davidson, A., and Hartway, C. (2021). Challenges and opportunities for cross-jurisdictional bison conservation in North America. *Biological Conservation* **256**, 109029.

Pellicer, J., Fay, M. F., and Leitch, I. J. (2010). The largest eukaryotic genome of them all? *Botanical Journal of the Linnean Society* **164**, 10–15.

Pennell, M. W., Harmon, L. J., and Uyeda, J. C. (2014). Is there room for punctuated equilibrium in macroevolution? *Trends in Ecology & Evolution* **29**(1), 23–32.

Pennisi, E. (2010). Biggest genome ever. *ScienceShot*. doi. org/10.1126/article.29806.

Peterson, A. T., Soberon, J., and Sanchez-Cordero, V. (1999). Conservatism of ecological niches in evolutionary time. *Science* **285**, 1265–1267.

Prakash, N., and Kumar, M. (2004). Occurrence of Ginkgo Linn. in early cretaceous deposits of South Rewa Basin, Madhya Pradesh. *Current Science* **87**, 1512–1515.

Pray, L. (2008) Eukaryotic genome complexity. *Nature Education* **1**, 96.

Quammen, D. (2007). *The Reluctant Mr. Darwin: An Intimate Portrait of Charles Darwin and the Making of His Theory of Evolution*. W. W. Norton, New York.

Raia, P., Carotenuto, F., Passaro, F., Fulgione, D., and Fortelius, M. (2012). Ecological specialization in fossil mammals explains Cope's rule. *American Naturalist* **179**, 328–337.

Raikow, R. J. (1976). The origin and evolution of the Hawaiian honeycreepers (*Drepaniidae*). *Living Bird* **15**, 95–117.

Ramsey, J., Robertson, A., and Husband, B. (2008). Rapid adaptive divergence in New World *Achillea*, an autopolyploid complex of ecological races. *Evolution: International Journal of Organic Evolution* **62**, 639–653.

Ramsey, J., and Schemske, D. W. (2002). Neopolyploidy in flowering plants. *Annual Review of Ecology and Systematics* **33**, 589–639.

Raup, D. M. (1986). Biological extinction in earth history. *Science* **231**, 1528–1533.

Raup, D. M. (1988). Diversity crises in the geological past. In *Biodiversity* (ed. E. O. Wilson and F. M. Peter), pp. 51–57. National Academy Press, Washington, DC.

Raup, D. M. (1991). *Extinction: Bad Genes or Bad Luck?* W. W. Norton, New York.

Reznick, D. N., Shaw, F. H., Rodd, F. H., and Shaw, R. G. (1997). Evaluation of the rate of evolution in natural populations of guppies *(Poecilia reticulata)*. *Science* **275**, 1934–1936.

Richards, R. J. (1992). *The Meaning of Evolution*. University of Chicago Press, Chicago.

Ridley, M. (1987). *The Darwin Reader*. W. W. Norton, New York.

Ripley, S. D., and Beehler, B. M. (1990). Patterns of speciation in Indian birds. *Journal of Biogeography* **17**, 639–648.

Ross, R. M., and Allmon, W. D. (1990). *Causes of Evolution: A Paleontological Perspective*, p. 479. University of Chicago Press, Chicago.

Rudkin, D. M., Young, G. A., and Nowlan, G. S. (2008). The oldest horseshoe crab: a new xiphosurid from Late Ordovician Konservat-Lagerstätten deposits, Manitoba, Canada. *Palaeontology* **51**, 1–9.

Ruse, M. (2009). The history of evolutionary thought. In *Evolution: The First Four Billion Years* (ed. M. Ruse and J. Travis), pp. 1–48. Belknap Press, Harvard University, Cambridge, MA.

Sahney, S., Benton, M. J., and Ferry, P. A. (2010). Links between global taxonomic diversity, ecological diversity and the expansion of vertebrates on land. *Biology Letters* **6**, 544–547.

Sanford, C., and Perry, M. D. (2001). Asymmetrically distributed oligonucleotide repeats in the *Caenorhabditis elegans* genome sequence that map to regions important for meiotic chromosome segregation. *Nucleic Acids Research* **29**, 2920–2926.

Sauer, J. D. (1990). Allopatric speciation: deduced but not detected. *Journal of Biogeography* **17**, 1–3.

Sauquet, H., Von Balthazar, M., Magallón, S., Doyle, J. A., Endress, P. K., Bailes, E. J., Barroso de Morais, E., Bull-Hereñu, K., Carrive, L., Chartier, M., and Chomicki, G. (2017). The ancestral flower of angiosperms and its early diversification. *Nature Communications* **8**, 16047.

Schlaepfer, M. A., Runge, M. C., and Sherman, P. W. (2002). Ecological and evolutionary traps. *Trends in Ecology & Evolution* **17**, 474–480.

Shapiro, B. (2015). Mammoth 2.0: will genome engineering resurrect extinct species? *Genome Biology* **16**, 1–3.

Shaw, J. H. (1995). How many bison originally populated western rangelands? *Rangelands Archives* **17**, 148–150.

Simpson, G. G. (1944). *Tempo and Mode in Evolution*. Columbia University Press, New York.

Simpson, G. G. (1953). *Major Features in Evolution*. Columbia University Press, New York.

Simpson, G. G. (1965). *The Geography of Evolution*. Chilton Book Co., Philadelphia.

Smith, C. H. (2013). A further look at the 1858 Wallace-Darwin mail delivery question. *Biological Journal of the Linnean Society* **108**, 715–718.

Smith, C. H. (2014). Wallace, Darwin and Ternate 1858. Notes and Records. *The Royal Society Journal of the History of Science* **68**, 165–170.

Solounias, N., Danowitz, M., Stachtiaris, E., Khurana, A., Araim, M., Sayegh, M., and Natale, J. (2018). The evolution and anatomy of the horse manus with an emphasis on digit reduction. *Royal Society Open Science* **5**, 171782.

Soltis, P. S., Marchant, D. B., Van de Peer, Y., and Soltis, D. E. (2015). Polyploidy and genome evolution in plants. *Current Opinion in Genetics and Development* **35**, 119–125.

Stanley, S. (1987). *Extinction*. Scientific American Books, New York.

Stanley, S. M. (1979). *Macroevolution: Pattern and Process*. W. H. Freeman, San Francisco.

Stanley, S. M. (1990). The general correlation between rate of speciation and rate of extinction: fortuitous causal linkages. In *Causes of Evolution: A Paleontological Perspective* (ed. R. M. Ross and W. D. Allmon), pp. 103–127. University of Chicago Press, Chicago.

Steadman, D. W., and Miller, N. G. (1987). California condor associated with spruce-jack pine woodland in the late Pleistocene of New York. *Quaternary Research* **28**, 415–426.

Stott, R. (2012). *Darwin's Ghosts: The Secret History of Evolution*. Random House, New York.

Strömberg, C. A. (2011). Evolution of grasses and grassland ecosystems. *Annual Review of Earth and Planetary Sciences* **39**, 517–544.

Terborgh, J., and Winter, B. (1980). Some causes of extinction. In *Conservation Biology. An Evolutionary-Ecological Perspective* (ed. M. S. Soule and B. A. Wilcox), pp. 119–133. Sinauer Associates, Sunderland, MA.

Thomas, A. (2014). *Introducing Genetics: From Mendel to Molecules*. Garland Science, New York.

Thomson, K. S. (1992). *Living Fossil: The Story of the Coelacanth*. W. W. Norton, New York.

Trask, B. J. (2002). Human cytogenetics: 46 chromosomes, 46 years and counting. *Nature Reviews Genetics* **3**, 769–778.

Vajda, V., and Bercovici, A. (2014). The global vegetation pattern across the Cretaceous-Paleogene mass extinction interval: A template for other extinction events. *Global and Planetary Change* **122**, 29–49.

van Noort, S., and Compton, S. G. (1996). Convergent evolution of Agaonine and Sycoecine *(Agaonidae, Chalcidoidea)* head shape in response to the constraints of host fig morphology. *Journal of Biogeography* **23**, 415–424.

Van Straalen, N. I., and Roelofs, D. (2006). *Introduction to Ecological Genetics*. Oxford University Press, Oxford.

Van Valen, L. (1973). A new evolutionary law. *Evolutionary Theory* **1**, 1–33.

Voje, K. L., Holen, Ø. H., Liow L. H., and Stenseth, N. C. (2015). The role of biotic forces in driving macroevolution: beyond the Red Queen. *Proceedings of the Royal Society B: Biological Sciences* **282**, 2015.0186.

Wallace, A. R. (1858). On the tendency of varieties to depart indefinitely from the original type. *Journal of the Proceedings of the Linnean Society (Zoology)* **3**, 45–62.

Waller, J. T., and Svensson, E. I. (2017). Body size evolution in an old insect order: no evidence for Cope's Rule in spite of fitness benefits of large size. *Evolution* **71**, 2178–2193.

Weir, J. T., and Schluter, D. (2008). Calibrating the avian molecular clock. *Molecular Ecology* **17**, 2321–2328.

Williamson, P. G. (1981). Paleontological documentation of speciation in Cenozoic molluscs from Turkana Basin. *Nature* **293**, 437–443.

Witte, F., Welten, M., Heemskerk, M., Van der Stap, I., Ham, L., Rutjes, H., and Wanink, J. (2008). Major morphological changes in a Lake Victoria cichlid fish within two decades. *Biological Journal of the Linnean Society* **94**, 41–52.

Xu, X., Zhou, Z., Dudley, R., Mackem, S., Chuong, C. M., Erickson, G. M., and Varricchio, D. J. (2014). An integrative approach to understanding bird origins. *Science* **346**. doi.org/10.1126/science.1253293.

REALMS, REGIONS, KINGDOMS, AND PROVINCES: THE BIOGEOGRAPHIC SUBDIVISIONS OF THE EARTH

Whether you climb to the alpine treeline in the mountains of northern Europe, northern Asia, or North America, you will encounter trees that belong to the same genera: pine *(Pinus)*, spruce *(Picea)*, fir *(Abies)*, or larch *(Larix)*. The coniferous trees found in the Alps are closely related to those found in the Rocky Mountains of Canada or the Ural Mountains of Russia. In contrast, if you were to venture to the mountains in New Zealand, Australia, or the southern tip of South America, you would find the alpine treeline occupied by species such as southern beech *(Nothofagus)*. You would not find a single native species of pine, spruce, fir, or larch. Similarly, not a single species of southern beech is native to North America, Europe, or Asia. Descending from the high elevations and walking across the savanna regions of Africa, you will encounter vast herds of gazelles *(Gazella)* but not a single native species of kangaroo *(Macropus)*. The savanna regions of Australia support several different species of kangaroos but not a single native species of gazelle. These differences in the families, genera, and species of plants and animals that dominate the fauna and flora of different parts of the world have fascinated biogeographers since the time of Buffon, de Candolle, and Von Humboldt in the eighteenth and nineteenth centuries. Why are the families, genera, and species that comprise the floras and faunas of some geographic areas of the world very similar to one another and very different from those found in other areas?

Why groups of closely related plants or animals are found together in particular regions and are absent from others has long been a central question in biogeography. The explanations of such patterns generally lie in both the geologic history of continents and islands, and the evolutionary history of the plants and animals that occupy them. In this chapter we will consider how biogeographers subdivide the earth into biogeographic regions on the basis of the geographic distributions of plant and animal families, genera, and species. We will examine why certain related species occur together in one region and are absent from other areas. We will see how earth history, evolution, and extinction are all key forces behind the development of modern biogeographic regions. Finally, we will briefly review the major biogeographic regions of the planet to consider their unique fauna, flora, and histories.

DEFINING BIOGEOGRAPHIC REALMS, REGIONS, KINGDOMS, AND PROVINCES

When we map out the global geographic distributions of mammal families, we discover some very interesting facts. Excluding bats and marine mammals, there are about 100 known families of mammals. Of these families, most are restricted to one or two continents, and no family is native to all of the continents of the world. We find broadly similar results when we study the global geographic distributions of birds, freshwater fish, or plant genera. When a species, genera, or family is restricted to one or a few geographic regions, it is referred to as being **endemic.** The term *endemic* is very relative. Lodgepole pine *(Pinus contorta)* is endemic to western North America, but within this region it grows from Alaska to southern California and from the Pacific Coast to the Rocky Mountains. In contrast, some species are narrow endemics that occur in only a very small area. For example, Santa Catalina mahagony *(Cercocarpus traskiae)* has a global natural distribution consisting of about a half dozen individuals

Biogeography: Introduction to Space, Time, and Life, Second Edition. Glen M. MacDonald.
© 2025 John Wiley & Sons, Inc. Published 2025 by John Wiley & Sons, Inc.
Companion website: www.wiley.com/go/macdonald/biogeography2e

that grow in Wild Boar Gully on Santa Catalina Island off the coast of southern California. Widely distributed families, genera, or species are said to be **cosmopolitan.** Owing in part to human transport, several mammal families are indeed truly cosmopolitan and are found today on all continents except for Antarctica. These include the Muridae (rats and mice), the Leporidae (hares and rabbits), and the Canidae (wolves and dogs). The degree of endemism in plants and animals increases with taxonomic level. Individual species are often far more endemic than the genera or families to which they belong. For example, the mountain mahogany genus *(Cercocarpus)* is widely distributed in the western United States and southward into Mexico. Its family (Rosaceae) has native species growing on every continent except Antarctica. The reasons for the generally wider distributions of genera and families compared to species will be discussed later in this chapter. In general, endemism is the common state for land plants and animals at the species level, while true cosmopolitan species are rare.

Recognizing that many families, genera, or species of plants and animals are endemic, biogeographers have sought to divide the earth into geographic regions that have similar species, genera, or families within each region. These geographic areas with similar fauna and flora have been variously called biogeographic realms, regions, kingdoms, and provinces, depending on the scientist writing, or the organisms studied, or the spatial scale of the unit. How can one objectively determine whether the assemblages of species, genera, or families present in two different areas are similar or distinct? Biogeographers often use mathematical indices to provide quantitative estimates of the floral or faunal similarities between regions. These indices are referred to as **coefficients of similarity.** Two of the most commonly applied coefficients of similarity are the Jaccard and the Simpson. The Jaccard coefficient has the following form:

$$\frac{C}{N_1 + N_2 - C}$$

where C is the number of families, genera, or species found in both regions; N_1 is the number of families, genera, or species found in region 1; and N_2 is the number of families, genera, or species found in region 2. The Simpson has a simpler form:

$$\frac{C}{N_1}$$

For both equations, a value of 1 indicates that the flora or fauna of both regions are identical. Biogeographers would say that the flora or fauna of regions are in **biogeographic harmony.** A value of 0 indicates that there is no similarity in the flora and fauna of the regions. By multiplying the coefficients by 100, we can transform them into percentages of similarity that range in value from 0% for regions that do not share any of the same families, genera, or species to 100% for areas that are completely harmonized.

When comparisons of biogeographic similarity are carried out, we find that the higher the taxonomic order being examined, the higher the biogeographic similarity. For example, if we compare the common coniferous trees near the Arctic treeline of Canada with those of northern Russia, we will find that at the level of family or genus, the biogeographic similarity coefficient is 1. However, if we compare the two regions at the level of species, the coefficient is 0. The reason for this difference is that both regions are occupied by the same families and genera of coniferous trees, but those genera are represented by the distinctively different species in Canada and Russia (Table 10.1).

Several reasons can be cited as to why higher taxonomic levels are more cosmopolitan. First, families and genera often include many different species. These species might have different niches, which allows the genera or family to occupy a wide variety of environments, although each individual species might be much more restricted in its distribution. Second, better dispersal abilities or chance events might lead to long-distance dispersal and colonization by one species of a genera or family but not by others. The range expansion experienced by the individual species creates a larger geographic range for the genus and family. Third, because evolution works directly at the level of the species, new species evolve and diverge much more quickly than do entire genera and families. Therefore, new species may evolve and exist as narrow endemics, while the family and genus to which they belong remain relatively cosmopolitan in distribution. Fourth, extinction due to either normal or catastrophic events may eliminate individual species of a genus or family, but other species of the family or genus might escape due to differences in their niches or geographic distributions. Thus, an individual species may go extinct in a region, but the family or genus might persist through the survival of other member species.

If we examine the global geographic distributions of plant and animal species and the coefficients of similarity between different continents, we notice some interesting patterns. Each continent

TABLE 10.1   Biogeographic Similarity Between Treeline Conifers in Canada and Northern Russia

Country	Family	Genus	Species
Canada	Pinaceae	*Larix*	*laricina*
	Pinaceae	*Picea*	*glauca*
	Pinaceae	*Picea*	*mariana*
	Pinaceae	*Pinus*	*banksiana*
Russia	Pinaceae	*Larix*	*sibirica*
	Pinaceae	*Picea*	*obovata*
	Pinaceae	*Pinus*	*sylvestris*
Simpson coefficient	1.0	1.0	0.0

possesses families of mammals or plants that are not found on other continents. Thirteen different families of native mammals occupy the North American continent at present. Interestingly, none of those 13 families is native to Africa. However, five of the North American families are found in northern Eurasia. As a result, the Simpson index of similarity is 0% between North America and Africa and 38% between North America and northern Eurasia. If we consider the continental distribution of flowering plants, we find that much greater similarity exists between the floras of the different continents compared to the mammal faunas. For example, the Simpson index of similarity between the flowering plant families of North America and northern Eurasia is 47%. The similarity between the plant families of Australia and those of South America, Africa, and southern Asia ranges from 69% to 63%. Neither the mammal families nor the flowering plant families are completely harmonized between any of the continents. Each continent contains families of mammals and plants that are unique. We see a similar pattern when we examine birds, reptiles, or freshwater fish.

Our brief examination of the worldwide distributions of mammals and flowering plants raises three interesting questions. First, why do different continents contain not just individual families, but groups of families that are not present on other continents? Second, why does Australia have no mammal families in common with other continents, while Eurasia and North America share so many families with each other and other regions? Third, why are the flowering plant families more similar from continent to continent while the mammal families are so different? Such questions have long intrigued biogeographers.

The tendency for different regions to possess unique species, genera, or families is called **biogeographic provincialism.** The modern scientific study of biogeographic provincialism can be traced back to the work of nineteenth-century European phytogeographers such as F. Schouw and A. de Candolle, while the current concept of subdividing the earth into biogeographic regions based on endemism in fauna or flora owes much to Philip Sclater. Sclater was a British ornithologist responsible for the scientific discovery and description of over 1000 different bird species from all over the world. Sclater was a friend of Charles Darwin and very active in the Royal Geographic Society of London. He was very well informed regarding the biological and geographical discoveries of his time. Based on his data regarding the global geographic distributions of birds, plants, and animals, Sclater argued that previously proposed biogeographic subdivisions of the earth were based on latitude, longitude, and to some degree climate and did not take into account the provincialism observed in the geographic distributions of flora and fauna. He observed that, even though places such as tropical Africa and South America have similar climates, they are occupied by very different species of plants and animals. Sclater proposed that the earth should instead be subdivided on the basis of the geographic distributions of related species, genera, or families of organisms rather than on geographic position, environmental conditions, or vegetation structure. Sclater's approach is, of course, quite different from the biome approach for subdividing the biosphere (see Chapter 6). He believed that areas where a species is found today likely represent the geographic area in which the species first developed. Therefore, Sclater reasoned that divisions of the earth based on species distributions would provide insights into biological history. Sclater developed his subdivision using passerine birds (perching birds such as sparrows), for this was the group he knew best. He presented his results in 1857 before the Linnaean Society of London. Interestingly, Sclater's presentation occurred one year before the reading of the work by Darwin and Wallace on evolution to the same society. The system proposed by Sclater was called the *Schema Avium Distributionis Geographicae.* This hierarchical scheme divided the earth into two super regions: the New World (North and South America) and the Old World. The land-masses were then divided into six biogeographic regions based on the similarity in the bird species within each region and the dissimilarity from the bird species found in other regions.

Sclater's concept of biogeographic regions was further developed by the great evolutionist and biogeographer Alfred Wallace. In 1876, Wallace presented a subdivision of the earth into biogeographic regions based on the distributions of vertebrates, with particular reference to nonflying mammals. By focusing on nonflying mammals, the impact of long-distance dispersal between different regions is decreased and the boundaries between regions should be clearer. The regions recognized by Wallace correspond relatively closely to those proposed by Sclater, although some of the names differed. The Sclater-Wallace system of faunal regions still remains widely used today and forms the basis of more recent systems. The Sclater-Wallace system divides the major landmasses of the earth into six biogeographic faunal regions (Fig. 10.1). These six faunal regions as typically recognized today are the Nearctic, Palearctic, Neotropical, Ethiopian or African, Oriental, and Australian.

The general success of the Sclater-Wallace system can be assessed by looking at the relationship between modern terrestrial mammal distributions and the proposed faunal regions. Of the 90 most common mammal families, 11 have relatively cosmopolitan distributions and are native to all continents except Australia and Antarctica. These families are the Soricids (shrews), Sciurids (squirrels

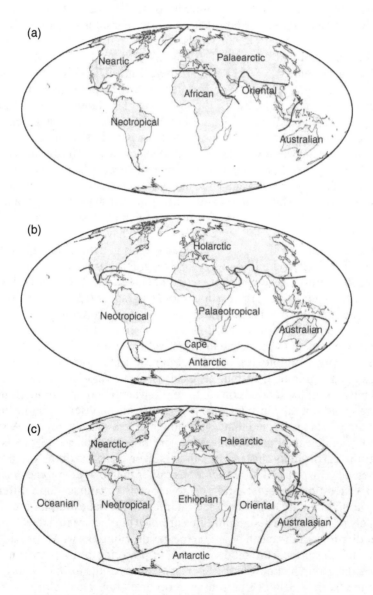

FIGURE 10.1   A generalized representation of the world's faunal regions (largely based on mammalian fauna), floral regions (largely based on vascular plants), and Pielou's (1979) combined floral and faunal biogeographic regions.

and chipmunks), Cricetids (hamsters and lemmings), Cervids (deer), Ursids (bears), Felids (cats), Mustelids (weasels and badgers), Bovids (cows), Murids (rats), Canids (dogs), and Leporids (rabbits). These families are called the 11 wandering families because of their relatively cosmopolitan distributions. Fifty-one (57%) of the 90 mammal families are endemic to only one faunal region. Most families are restricted to one or two of the recognized regions, and this supports the general applicability of the scheme. In addition, one can examine the degree of endemicity within each region to see how unique each region's mammal fauna is. The degree of edemicity ranges from 91% for the Australian region to 3% for the Palearctic. Therefore, the Australian region is the most clearly differentiable on the basis of mammal fauna, while the Palearctic is hard to differentiate from the adjacent Oriental, African, and Nearctic regions.

A number of efforts have been undertaken to subdivide the earth into biogeographic regions based on plant distributions. These are sometimes referred to as floral kingdoms. One of the earliest attempts was by Engler, who distinguished four floral realms in the nineteenth century. A scheme developed by the Soviet botanist A. Takhtajan in the twentieth century subdivides the earth into six floral regions (Fig. 10.1) on the basis of flowering plant distributions. The widely recognized floral regions include the Holarctic, Neotropical, Paleotropical, Cape, Australian, and Antarctic. It is sometimes difficult to define clear biogeographic regions on the basis of plant families. Only about 18% of the world's plant families are endemic to one floral region. In some cases, such as the Holarctic, the floral regions encompass more than one faunal region. In contrast, the Cape floral region of Africa is a tiny area of endemic plants compared to the African faunal region. Despite its small size, the unusual flora of the Cape of Good Hope area of South Africa contains over 500 different species of plants, many of which are narrow endemics found only in the Cape region.

In theory, biogeographic regions should follow a hierarchy with **biogeographic realms,** which incorporate several continents and large landmasses, being the largest; followed by the **biogeographic regions,** which subdivide the earth at the continental level; followed by **biogeographic provinces,** which subdivide smaller areas. Provinces are sometimes further divided into subdivisions. Some researchers use the term **biogeographic kingdoms** for larger regions. To simplify our discussion, we are broadly employing the term *biogeographic region* here. It must be kept in mind that biogeographic realms, regions, and provinces are based on the evolutionary and genetic relationships

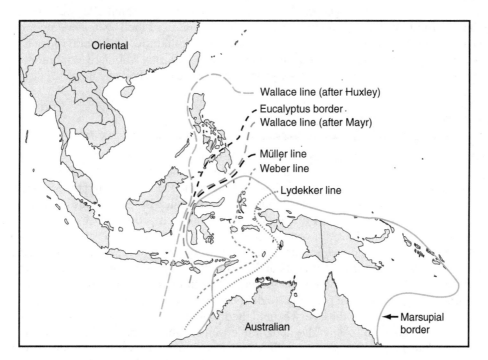

FIGURE 10.2 The Wallacea region of southeastern Asia and Australia and several of the biogeographic lines that have been proposed to separate the Australasian and Oriental regions. The northwestern range limits of the Australian groups, the marsupials and eucalyptus, are also shown (adapted from Zacharin, 1978).

between plant and animal species within each region and are thus very different from biomes, which are based on the similarity of vegetation structure and climate within each biome.

The division of regions into provinces relies on the careful identification and mapping of endemic species within each region. For example, the Nearctic region, which consists of North America, can be divided into broad floral provinces that represent differing areas such as the Arctic, the boreal-temperate forests and grasslands, the western mountains and deserts, and the Mexican-Caribbean region. California, because of its high degree of plant endemism, is generally recognized as a separate subprovince, or province. Over 2600 vascular plant species (about 42% of the vascular flora) are endemic to the California province. Within California, a number of biogeographic subdivisions are recognizable on the basis of high numbers of local endemic plant species. Mexico also has a high number of reginal and local endemic plant and animal species and can be subdivided into 14 distinct biogeographic provinces.

Despite the long history of study, the floral and faunal regions of the earth are still the subject of research and debate. In the late twentieth century, the Canadian biogeographer and ecologist E. C. Pielou proposed a scheme to divide the world into a number of biogeographic regions based on the combination of both faunal and floral distributions (Fig. 10.1). The general regionalization proposed by Pielou is similar to that of Wallace for his faunal regions. Pielou adds a separate Antarctic region. Pielou and Wallace also differ regarding the locations of specific regional boundaries and the treatment of oceanic islands. More recently, the British biogeographer Barry Cox reevaluated the global floral and faunal regions and concluded that the Cape and Antarctic floral regions should be deleted as separate entities from the adjacent regions. He also suggested modifying the names of some of the other biogeographic regions. More extensive and precise databases of the geographic distributions of plants and animals, coupled with sophisticated statistical and modeling techniques, are also refining the demarcation of biogeographic regions. As an example, one such study of North American mammals conducted by the Mexican research team of Tania Escalante, Juan Morrone, and Gerardo Rodríguez-Tapiawas used an initial data set consisting of 245,818 geographical observations for 710 species. Although the broad biogeographic regions of Nearctic and Neotropical faunal regions were supported by the study, the Mexican transitional boundary between the two regions and the smaller subdivisions within the Nearctic region appear more complex than previously believed. Another study has incorporated data on the geographic distributions and phylogenies of over 21,000 species, including amphibians and birds in addition to mammals, to provide an updated and more inclusive scheme for the world's zoogeographic regions. What is notable is that after well over a century of data collection and research, the major biogeographic regions described by biogeographers such as Wallace have retained much of their validity even as detailed boundaries and subdivisions have been revised. The biogeographic regions of the earth and their boundaries will likely continue to be a subject of research and refinement as databases grow and analytic tools become more sophisticated.

DETERMINING THE BOUNDARIES BETWEEN REGIONS

As we have seen, Australia has very different mammal fauna and somewhat distinctive flowering plant flora when compared to mainland Asia. Despite the obvious differences in the fauna and flora of Australia and Thailand, it is not so easy to determine the exact geographic boundary between the Australian and Oriental regions. Such boundaries are called **biogeographic lines.** The most famous of these boundaries is **Wallace's Line,** which Wallace himself proposed to separate his Australian and Oriental regions (Fig. 10.2). In general, plants and animals to the west of Wallace's Line are dominated by Oriental species, while Australian species dominate to the east. However, the exact location of the boundary between Australasia and the Oriental region is hard to fix because many groups of animals and plants have different geographic limits in the region. Consequently, many different biogeographic lines have been proposed to delineate the Australian and Oriental regions (Fig. 10.2). For example, marsupials are the overwhelmingly dominant mammal in Australia and Tasmania but are not found in mainland Asia. The geographic range limits of marsupials extend northwestward to Sulawesi but do not extend to the Philippines or the islands west of Timor. However, some lines would include some of these smaller nonmarsupial islands within the Australian region, while other proposed lines would not. Eucalyptus trees are a dominant element of Australian vegetation, comprising 95% of the forest trees on that continent. Eucalyptus are found only as far northwest as New Guinea, Sulawesi, and the southern Philippines. A number of smaller islands east of Wallace's Line that do not contain native eucalyptus are still grouped within the Australian region, while the Philippines are not. Some more recent lines would also exclude Sulawesi and the Philippines; others would not.

The region of Wallace's Line is a narrow archipelago of islands separated by relatively deep water. The channels between the islands provide barriers to the dispersal of land plants and animals and contribute to a relatively clear boundary between the Oriental and Australian biogeographic regions. As we have seen, even here the exact location of the line between the biogeographic regions is not so clear-cut. It is even more difficult to differentiate biogeographic lines when they cross continuous land areas. One example of this is the line separating the Palearctic and Ethiopian regions. At least 18 different biogeographic lines have been proposed for this region. The locations of these lines extend from the Mediterranean Coast southward to the southern edge of the Sahara Desert. In the case of the Palearctic and Ethiopian regions, the Sahara provides a barrier to dispersal by some plant and animal species but not to others. A similar difficulty is encountered in delineating the boundary between the Nearctic and Neotropical regions. Based on the northern range limits of South American mammal families and the southern range limits of North American mammal families, it would be possible to suggest a number of different biogeographic lines to separate the regions. The lines could be drawn as far north as central Mexico and as far south as Panama.

Although large barriers, such as the Atlantic and Pacific oceans, clearly delineate regions such as the Neotropical from the Ethiopian, determining sharp boundaries and distinct biogeographic lines between regions that are closely adjacent or share continental areas is much more difficult. Such less distinct boundary areas are referred to as biogeographic transition zones. Some authorities consider the area between the Malaysian Peninsula and Australia to be a transition zone between the Australian and Oriental regions. They call this transition zone Wallacea in honor of Wallace. The Nearctic and the Neotropical, the Palearctic and the Oriental, and the Ethiopian and Palearctic regions might all be considered to be separated by transition zones.

It is not surprising that the exact boundaries between biogeographic regions are not sharply definable. Different species from each region will have different environmental requirements, dispersal and colonization abilities, and histories. All of these factors contribute to the blurring of the boundaries between realms, regions, and provinces. After well over 100 years of study, Wallace's Line still remains a topic of research and debate. Instead of asking why boundaries are not distinct, it is perhaps better to ask why biogeographic regions exist in the first place. We will tackle this question in the next section.

FACTORS BEHIND THE BIOGEOGRAPHIC REGIONS

As we have seen, the question as to why regions of the earth possess groups of plants and animals that differ, often dramatically, from the families, genera, and species found in other regions has long intrigued biogeographers. Prior to an understanding of evolution and plate tectonics, a complete scientific answer to this question was impossible. We now know that the three most important factors behind the existence of clearly definable faunal and floral regions are: (1) the present locations of biogeographic barriers, (2) the history of the continental plates, and (3) the evolutionary history of modern animal and plant families. Let's look at each of these factors in turn.

As we discussed earlier, the boundaries between the present biogeographic regions are largely formed by physical barriers to the dispersal of terrestrial plants and animals. The sea, both in the form of large bodies, such as the Pacific and Atlantic oceans, or smaller bodies of water, such as the channels between the western and eastern islands of the Malaysian Archipelago, is a major barrier to terrestrial organisms. Deserts and mountains also form environmental barriers that serve as boundaries between biogeographic regions. The hot and extremely arid conditions of the Sahara Desert and Arabian Desert restrict exchanges of fauna and flora between the Palearctic and Ethiopian regions. The cold conditions of the highlands and mountains of the Tibetan Plateau and the Himalayas help isolate the Oriental region from the Palearctic. Some boundaries are formed by physical and biological barriers. Although the Isthmus of Panama provides a landbridge between the Nearctic and

Neotropical regions, the tropical climate of the southern end and competition from the associated tropical vegetation and animal communities preclude the northward or southward migration of most temperate, desert, or tundra species. Similarly, the arid desert conditions of northern Mexico and the adjacent United States produce a barrier that also blocks the passage of most temperate, tropical, or tundra organisms.

Of course, the physical and biological conditions of the boundaries between the biogeographic regions have been subject to changes. The Bering Strait serves as the modern boundary between the Nearctic and Palearctic regions. Here, Alaska and Siberia are divided by less than 100 kilometers of ocean. During glacial periods, the eustatic sea level dropped enough so that the Palearctic and Nearctic regions were joined by a large landbridge known as Beringia. However, the Arctic climate and treeless tundra environment of the region was far too harsh to allow the dispersal of many species of plants and animals. Drops in sea level also produced land connections between the mainland of Asia or Australia and many of the islands in the Malaysian Archipelago. However, the main channels between islands such as Borneo and Sulawesi in the Malay Archipelago are deeper than 120 meters and remained inundated by the sea even during glacial periods. As a result, the ocean served as a barrier between the Australasian and Oriental regions even during glacial periods.

As we saw in Chapter 7, the geographic positions of the continents have changed over time. Thus, some continents, such as Europe and Asia, have been joined as a continuous landmass for at least 50 million years. Others such as South and North America have become linked by a landbridge only in the past few million years. Australia has been isolated from all other continents, including Antarctica, for over 50 million years. It is not surprising, therefore, that Europe and northern Asia form one modern biogeographic region, the Palearctic. It is also not surprising that the Australasian region is the most distinct of all the biogeographic regions.

The biogeographic regions we are considering here are based on the modern distributions of mammal and flowering plant species. The evolutionary history of these groups explains why the mammal fauna of the different biogeographic regions is more distinctive than the plant flora, why marsupials dominate the native mammal fauna of Australia, and why no southern beech trees are native to the northern hemisphere. Let's briefly consider the evolutionary history of the mammals and flowering plants as it is known today.

Evolution of the Mammals

It may be surprising to learn that the first primitive mammals appeared at about the same time as the first dinosaurs. Fossils suggest that the first true mammals, which are the stem *taxa* of the mammalian phylogenetic tree, appeared and underwent initial evolutionary divergence sometime during the Late Triassic to Early Jurassic, between 208 million and 178 million years ago. The first mammals possibly appeared in Laurasia. Finding and deciphering the earliest fossil evidence of mammals remains a matter of great scientific interest and debate. Both dinosaurs and mammals evolved from reptiles. Mammals likely descended from a group of reptiles called the therapsids, which appeared in the Permian about 270 million years ago. Hair, which typifies mammals, likely evolved from scales. One of the first known mammals was a very small shrewlike creature called Megazostrodon. These earliest mammals likely laid eggs. The only surviving egg-laying mammals are the monotremes, which include the duckbill platypus (*Ornithorhynchus anatinus*) and echidnas (*Tachyglossus* and *Zaglossus*) found in Australia, Tasmania, and New Guinea. Fossilized teeth are by far the most common remains found of the earliest mammals. These teeth indicate that some early mammal species fed on plants, others fed on insects, and still others were omnivores. Although they experienced appreciable evolutionary divergence during the Jurassic, many mammal stem lineages remained small in size, and their populations remained small. Mammal diversification appears to have slowed in the later Jurassic and subsequent Cretaceous. During this time, the dinosaurs grew in physical size, abundance, and range of adaptations and came to be the dominant large animals on land, sea, and air. This would all change with the close of the Cretaceous about 65 million years ago.

The end of the Cretaceous Period represents one of the greatest catastrophic extinction events that the earth has experienced (see Chapter 9). The dinosaurs, which had been so dominant, disappeared along with many other terrestrial and marine animals and plants. It seems indisputable that this event was caused by the impact of a large meteor or comet near the present-day Yucatan Peninsula of Mexico. Fires, tidal waves, acidic precipitation, changes in the ozone layer, and later climatic cooling caused by the extraterrestrial strike destroyed the dinosaurs directly and through trophic cascade events. Some mammals, however, survived this catastrophe, even though many species apparently did go extinct at this time. The small size of the mammals, along with their warm-blooded physiology and wide range of foods, likely contributed to their survival.

If the Mesozoic Era was the age of the dinosaurs, the subsequent Cenozoic Era has been the age of the mammals. Today, the crown of the mammalian phylogenetic tree contains over 6400 extant species of mammals grouped into 27 orders. When we look at the fossil evidence for the initial evolutionary development of modern-day mammalian orders, it seems that many originated from a period of explosive speciation following the close of the Cretaceous. Rare and exquisite fossil evidence of this explosion of mammalian diversity in the Paleogene is found in the Messel Quarry of Germany (Technical Box 10.1). The classical explanation for this evolutionary pattern is that strong adaptive

TECHNICAL BOX 10.1

The Messel Fossil Record and the Rise of the Mammals

Fossils recovered from an abandoned coal quarry near Frankfurt, Germany, present one of the most spectacular glimpses into the early diversification of the mammals and development of the modern mammalian fauna. About 47 million years ago, during the Eocene Epoch of the Paleogene Period, the Messel site was a lake and highly organic sediments collected in its basin. Over geologic time, these sediments became the brown coal that was mined from the site in the nineteenth and twentieth centuries. Included within the coal-bearing sediments were the abundant remains of plants, insects, fish, amphibians, reptiles, birds, and mammals. It has been speculated that episodic mass mortality events possibly caused by blooms of cyanobacteria that poisoned the water, or the release of volcanic gases, may have killed the animals in and near the lake. Low levels of oxygen in the deep waters and rapid burial by fine sediment helped preserve the remains. Although studies of early mammal fossils generally rely on individual teeth or disarticulated bones, at Messel, complete articulated skeletal remains have been found for all of the mammal species represented there. Aside from skeletal fossils, evidence of soft tissue and even stomach contents have been preserved in some instances. At the time the fossils were deposited, the Messel region had a somewhat tropical climate, and the diversity of plants and animals preserved is particularly high.

Over 30 genera and 40 species of mammals, representing 13 different orders, have been identified among the Messel fossils. These provide important insights into the diversity of both extinct and modern extant mammalian clades in the Early Cenozoic Period. Some of the important mammal genera from Messel (and their likely modern relatives) include the following:

Hedgehogs

Pholidocercus

Macrocranion

Pangolins

Eomanis

Eurotamandua

Rodents

Ailuravus

Masillamys

Horses

Propalaeotherium

Tapirs and Rhinoceroses

Hyrachyus

Carnivores

Paroodectes

Bats

Palaeochiropteryx

Primates

Darwinius

Europolemur

Godinotia

Metatherian (early marsupial)

Peradectes

In 2009, the remarkably complete skeleton of an early primate, *Darwinius masillae,* was described from Messel. The find included an outline of soft tissue. It is the most complete fossil of an early primate ever recovered. The juvenile female appears to have been an early form of lemur. Some speculated that *Darwinius masillae* could represent an extremely early ancestor to humans and named the fossil Ida. However, this interpretation is not widely accepted. Ida's species is related to human's primate ancestors, but is not likely one.

The Messel fossils were largely brought to the attention of the world by amateur paleontologists. Their efforts contributed much to our knowledge of early mammal species. Today, Messel is a United Nations World Heritage Site and continues to provide important scientific insights into the evolutionary and environmental history of life in the Eocene.

Cast of a bat fossil *(Paleochiropteryx tupainodon)* from the Messel beds.

radiation occurred as the mammals evolved to fill the many vacant niches caused by the catastrophic extinction of the dinosaurs and other animals. However, there is evidence of initial divergence occurring in major extant mammal types commencing in the Late Jurassic and Early Cretaceous. At that time, Laurasia and Gondwanaland were in the process of breaking into the modern continents. In addition, North America was split by a shallow sea (called an epeiric sea) that formed when the sea level rose and flooded the central part of the continental plate. Allopatric speciation caused by the breakup of Laurasia and Gondwanaland and the development of epeiric seas would have contributed to this early divergence. Even if this is the case, the sparsity and small size of mammal fossils from the Cretaceous indicate that mammals remained small in number and body size until the burst of adaptive radiation that followed the Late Cretaceous extinction of the dinosaurs.

One of the most important events in the history of mammal evolution is the development of the divergent marsupial and placental mammals. These two modern groups and their fossil forms are known as the *metatherian mammals (marsupials)* and the *eutharian mammals (placentals)*. The term *Theria* is used to refer collectively to Eutheria and Metatheria mammal clades. The egg-laying monotremes, such as the platypus, belong to the *Australosphenida clade*. Extant monotremes are extremely reduced in population size, diversity, and geographic distribution than the modern marsupial and placental mammals. The earliest origins of the Eutheria, Metatheria, and Australosphenida mammalian clades remains a topic of intense research and debate. Molecular DNA-based estimates do not always agree with fossil evidence, and the discovery of new fossils and the reinterpretation of existing samples are not uncommon. An example of the current state of research is provided by the work of the Chinese-American team of Shundong Bi, Xiaoting Zheng, Xiaoli Wang, Natalie Cignetti, Shiling Yang, and John Wible. In a 2018 paper, they concluded that molecular clock estimates of the split between placental and marsupial mammals suggest that the divergence took place in the Jurassic period. Fossil evidence supported this and indicated the presence of the placental mammal genus *Juramaia* in northeastern China as early as 160 million years ago (Late Jurassic). It had been thought that the earliest marsupial fossils were from the genus *Sinodelphys*, found in Early Cretaceous deposits in eastern Asia. These fossils were estimated to be 125 million years old. However, Bi and colleagues reclassified the genus as being eutherian, and concluded that the earliest marsupials appear to have evolved in North America around 100 million years ago in the Cretaceous. The origins and development of the extant mammalian crown species and extinct mammalian stem clades will undoubtedly remain a topic of keen scientific research and further revision.

Today, marsupials are absent from Europe, represented by only one species in North America north of Mexico (the opossum *Didelphis virginiana*), and they form a relatively small fauna in South America. They remain dominant in the mammal fauna only in Australia. What happened to this once far more widespread group of mammals? The restricted distribution of the marsupials today is tied to continental drift and the rise of the placental mammals. An epeiric sea separated Asia from Europe during the Late Cretaceous and kept the marsupials from colonizing Asia. India, Africa, and New Zealand were also separated from the other continents by oceans, and marsupials never colonized these regions. By 50 million years ago, placental mammals had spread to Europe, North America, Africa, India, and South America (Fig. 10.3). Mammal orders such as the Primates (which includes

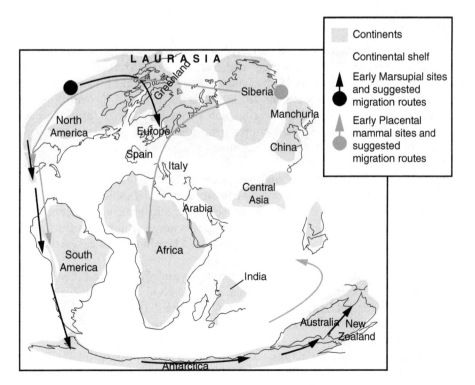

FIGURE 10.3 The Late Cretaceous and Paleogene global dispersal of marsupial and placental mammals (*Source:* Osborne and Tarling, 1996 / Henry Holt and Company).

monkeys and apes), Carnivores (which includes dog and cats), and the Lagomorpha (which includes rabbits and hares) were present by this time. Placental mammals were not, however, able to cross over to Australia. At this time, Australia was separated from the other continents, largely leaving the continent to the marsupials and monotremes.

Elsewhere, the arrival and diversification of the placentals likely caused the decline of the marsupials due to competition. The gestation of the young within the uterus is thought to be more efficient than gestation in the external sack found on marsupials, and this may have provided a reproductive advantage to placentals. In Europe, competition from placentals led to the complete extinction of the marsupials, while in North and South America the marsupials were greatly reduced in importance and diversity. The early separation of Australia from the other continents protected the marsupials there from competition with any mammals except bats, and later rats.

During the Paleogene and Neogene, adaptive radiation allowed marsupial mammals to fill the many environments offered by Australia. Elsewhere, adaptive radiation and allopatric speciation caused by events such as the separation of Europe and North America promoted speciation among the placental mammals. Orders such as the Chiroptera (bats), Rodentia (rodents), Odontoceti (includes toothed marine species such as dolphins and sperm whales), Mysticeti (includes filter-feeding marine species such as gray whales), Proboscidea (elephants), Perissodactyla (includes horses and rhinoceroses), and Artiodactyla (includes cattle, pigs, sheep, antelopes, and giraffes) all evolved between about 50 million and 35 million years ago. Convergent evolution (see Chapter 9) by marsupials in Australia produced species that were often strikingly similar in form and behavior to the placental mammals dominant elsewhere.

The fact that the continents were roughly in their modern locations during the Paleogene and Neogene meant that the many terrestrial mammals that evolved in Africa would have considerable distances and barriers to overcome in order to migrate to North America and vice versa. The isolation of Australia meant that almost no exchange of mammals could occur. Thus, the differences between the mammal species found in the modern faunal regions is strongly influenced by the relatively late speciation of the mammals and the breakup of Laurasia and Gondwanaland into separate landmasses by this time.

Evolution of the Flowering Plants

Although much mystery still remains concerning the evolution and spread of the mammals, the history of the flowering plants (angiosperms—which are sometimes referred to as a class [Angiospermae] or a division [Magnoliophyta]) could be considered even cloudier. Today, the crown of the angiosperm phylogenetic tree contains over 295,000 species arranged in over 60 orders. The time and place of the first appearance of the flowering plants have long been a topic of great interest. Physical evidence of the earliest angiosperms comes from fossilized leaves, stems, fruits, pollen, and, very rarely, flowers. In addition, there has been much study of modern plant morphology and genetic structure in order to determine which living species in the crown of the phylogenetic tree might be most closely related to the ancient stem ancestors of the angiosperms. Numerous molecular clock studies have been conducted on individual flowering plant families and angiosperms in general. Despite intensive efforts for over 200 years, scientists still have not reached consensus on which plant was the ancestor to the angiosperms, when and where the angiosperms first evolved, and what the earliest angiosperms looked like. Indeed, Charles Darwin himself called the origin of the flowering plants an "abominable mystery."

Most agree that the cone-bearing seed plants (gymnosperms) and the flowering plants share a common ancestor in the distant geologic past. There is good fossil evidence that early angiosperms, including a number resembling modern magnolias *(Magnolia)*, were present in the Early Cretaceous. Angiosperms became increasingly abundant during this geologic period. Between roughly 100 million and 65 million years ago, angiosperms increased from less than 1% of the flora to well over 50%. Many of the modern plant families appeared in the Middle to Late Cretaceous. During the Paleogene, angiosperms increased to comprise 90% or more of the earth's total flora. The diversification and spread of the flowering plants in the Cretaceous and Paleogene coincided with a strong decline in the diversity and abundance of gymnosperms. It has been suggested that the primary driver of the decline in the previously dominant gymnosperms was competition from the flowering plants. When and where did these successful plants first originate and spread from?

Fossil evidence, including pollen grains, which are particularly robust fossils and can be widely distributed and preserved, provide unequivocal evidence of angiosperms in the Early Cretaceous, some 140 million to 120 million years ago. Fossils of branches, leaves, and inflorescences of the aquatic flowering plant *Archaefructus liaoningensis* have been described from 124-million-year-old deposits in China. Similar fossils for the aquatic plant *Montsechia vidalii* have been found in deposits in Spain dating back 130 million years. However, fossil pollen grains present evidence of angiosperm origins dating back as far as the Jurassic and the Late Triassic over 200 million years ago. Some of the earliest fossil pollen evidence of angiosperms comes from the rift zone between the current South American and African continents. Recently, a flower bud from the fossil plant *Florigerminis jurassica* has been described from 164-million-year-old deposits in China. This, along with the pollen records, appears to push the earliest fossil evidence for flowering plants firmly into the Jurassic.

Molecular clock evidence has been compiled from massive data sets of DNA from living plant species from throughout the world. A very comprehensive study that examined the genetics of 2351 species of living angiosperms concluded that the origin of the angiosperms dates to the Late Triassic, over 200 million years ago. The study found evidence of major diversifications leading to the evolution of modern plant families occurring in the subsequent Jurassic and Cretaceous. The researchers also concluded that the orders Amborellaceae, Nymphaeales, and Austrobaileyales represent the earliest extant clades. Living Nymphaeales are aquatic plants with widespread distribution, while Amborellaceae and Austrobaileyales are terrestrial woody plants. The Amborellaceae is represented by only one species, *Amborella trichopoda*, which is endemic to New Caledonia in the South Pacific. The researchers acknowledged the paucity of fossil evidence of angiosperms during the Jurassic and referred to this as "the puzzle of the Jurassic gap."

The geographic location for the origin of the earliest angiosperms and the geography of their initial spread remains debated. As we have seen in the preceding discussions, early evidence of angiosperms has been found in Europe, China, Africa/South America, and Australasia (Fig. 10.4). Genetics evidence suggests that tropical clades diverged and diversified earlier than those associated with temperate regions. Although the timing and place of initial angiosperm evolution remain intensely debated, it seems from the global dominance of the angiosperms that they underwent significant diversification and spread in the Jurassic and Cretaceous, while the land connections related to the super continent of Pangaea were still significant.

As noted earlier, the expansion of the angiosperms led to the displacement and extinction of many gymnosperms. It is thought that angiosperms possess several competitive advantages over other plants such as gymnosperms and ferns. For example, angiosperms have more efficient vascular systems and are better able to withstand drought. Angiosperms have been shown to have photosynthetic capacity. The ovaries and pollen grains of angiosperms are often better protected from desiccation than are the cones and pollen of gymnosperms or the sporangia and spores of ferns and mosses. Finally, the coevolution of pollen and fruit-dispersing animals with the flowering plants produced additional and more efficient mechanisms for pollen and seed dispersal than exists for gymnosperms or ferns, which generally rely on the wind.

Gymnosperms were largely displaced by flowering plants in the tropics. However, gymnosperms such as pine and spruce, particularly in high elevation and high latitudes in the northern hemisphere, continued to be dominant. The retention of needles during the winter is an advantage for conifers in regions where summers are short. Evergreens are able to begin photosynthesis quickly once temperatures warm up. In addition, the loss of viable seeds due to the desiccation of gymnosperm cones is not such a problem in these cooler regions. Finally, the large distance between the high latitudes of the northern and southern hemisphere and the ocean barrier between the continents of Laurasia and those of Gondwanaland may have restricted the northward dispersal of cold-adapted angiosperms such as southern beech and lessened the competitive pressure on the northern gymnosperms. The modern-day restriction of southern beech to the high latitudes and high elevations of the southern hemisphere, and the restriction of pines to the northern hemisphere (Fig. 10.5), is directly related to the Late Cretaceous evolution and distributions of these two genera. *Nothofagus* evolved in the southern hemisphere around 70 million to 90 million years ago. The southern beeches

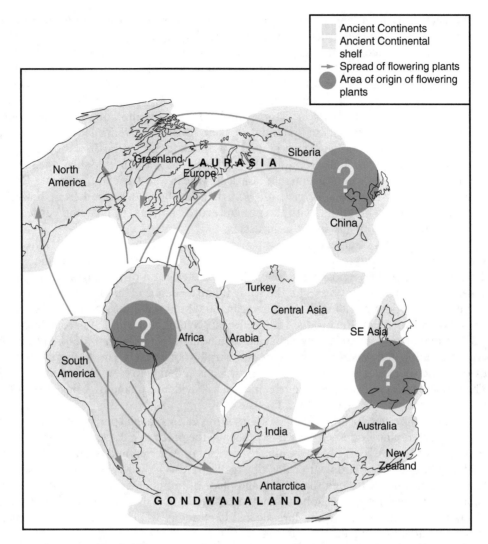

FIGURE 10.4 The Late Jurassic to Cretaceous global dispersal of angiosperms (*Source:* Osborne and Tarling, 1996 / Henry Holt and Company).

were then able to disperse to the Gondwanaland related landmasses of Australia, New Zealand, New Guinea, and Antarctica at this time (Fig. 10.4). Pines evolved about 100 million to 130 million years ago in the northern hemisphere. Subsequently, pines continued to evolve and disperse across the Laurasian-related landmasses of Europe, northern Asia, and North America. By the close of the Cretaceous, pines were present in the mid- to higher latitudes of the northern hemisphere, and southern beech was present in the high latitudes of the southern hemisphere.

The history of southern beech and pine provides an interesting contrast to the general success of the angiosperms. Over the past 65 million years, the range of pine in the northern hemisphere has become relatively large, and many species have evolved. Today, there are over 100 pine species in existence. Pines can be found growing from the Arctic treeline of Eurasia and North America to the tropics of southeastern Asia and Central America. Southern beech is now represented by a handful of species, and its range is restricted to cool sites that occupy only small portions of the southern hemisphere.

Although many details of angiosperm evolution remain to be resolved, it is clear that diversification and spread occurred earlier for flowering plants than for mammals. The relatively early evolution, diversification, and spread of the angiosperms during the Cretaceous, compared with the latter evolution and spread of the mammals, help to explain why the differences between the floral regions are not as great as the differences between the mammal regions.

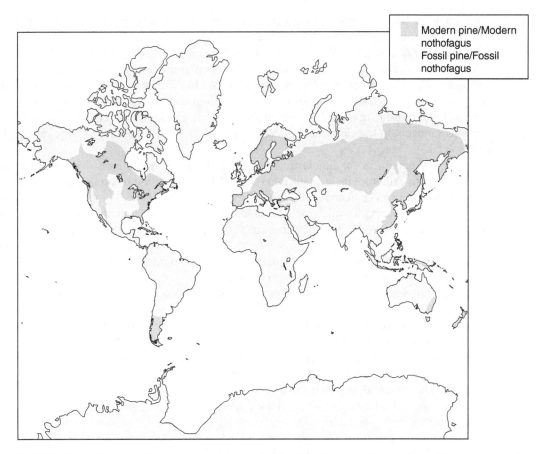

FIGURE 10.5 The Late Cretaceous–Paleogene distribution of the southern beeches *(Nothofagus)* and pines *(Pinus)* along with the modern distributions of the two genera.

THE BIOGEOGRAPHIC REGIONS

In the following consideration of a simplified approach to biogeographic regions, we will focus on some of the better-known endemic fauna and flora of five major regions and discuss some of the highlights of their biogeographic history. We will be focusing mainly on terrestrial biogeographic regions as defined by mammals and flora (Fig. 10.1). We briefly consider marine regions at the close of the chapter. As we will see, the timing of when different continental plates either diverged or converged because of plate tectonics is critical in explaining the modern distributions of many mammal and plant families. Although Pielou recognizes distinctive Antarctic and Oceanic faunal-floral regions, we will not include these in our discussion, for the Antarctic possesses no terrestrial mammals, and at the level of family, the flora and fauna of the Oceanic region can be considered an amalgamation of elements that dispersed from continental regions.

Nearctic and Palearctic Regions—The Holarctic

The close similarities in the fauna and flora of the Nearctic and Palearctic regions have led some biogeographers to treat these two areas as one super region called the Holarctic. In view of the close relationships between their flora and fauna, and the historical linkages during the Cenozoic, we will discuss the two regions together here. The Nearctic region includes all of North America, Greenland, and most of Mexico. The Palearctic includes Europe, northern Africa, and northern Asia. The mammal fauna of the Nearctic and Palearctic is smaller than that of tropical regions. The Nearctic supports 13 families of terrestrial mammals, while the Palearctic supports 18 families. The mammal fauna of the Nearctic shares many families and genera with the Palearctic. Twenty-one species of native mammals are found in both regions. The Nearctic fauna is overwhelmingly dominated by placentals. The one living marsupial genus is the opossum. In contrast, there are over 100 different genera of placental mammals in the Nearctic. Well-known species that are now endemic to the Nearctic include North American buffalo *(Bison bison),* muskox *(Ovibos moschatus),* pronghorn antelope *(Antilocapra americana),* bighorn sheep *(Ovis canadensis),* North American elk *(Cervus candensis),* prairie dog *(Cynomys* spp.), grizzly bear *(Ursus horribilis),* and coyote *(Canis latrans).* During the Pleistocene, the number of large mammal species in the Nearctic was much larger than it is today. A catastrophic extinction event at the close of the Pleistocene led to a major decline in large mammal diversity (see Chapter 12).

Many genera of mammals that are found in the Palearctic also occur in the Nearctic. Some better-known examples include buffalo *(Bison bonansus,* called bison in Eurasia), moose *(Alces alces,* called elk in Eurasia), bear *(Ursus* spp.), caribou *(Rangifer tarandus,* similar to reindeer in Eurasia), and lynx *(Lynx* spp.). In addition, the Palearctic contains genera that do not occur in the modern native fauna of the Nearctic. These include well-known mammals such as monkeys, pandas *(Ailuropoda),* tigers *(Panthera),* horses *(Equus),* and camels *(Camelus).* However, it should be noted that during the Pleistocene, horses and camels were also found in the Nearctic.

The angiosperm flora of the Nearctic shares many of its families and genera with the Palearctic. The Nearctic is richer, possessing over 90 different native angiosperm families compared to around 70 in the Palearctic. In both regions, species of maple, oak, ash, elm, hazel, birch, basswood, poplar, and beech dominate the deciduous forests. The two regions also share many conifer genera. The Nearctic and Palearctic have species of pine, spruce, fir, and larch dominating boreal vegetation. The Arctic shrubs and herbs of the Nearctic and the Palearctic are quite similar, even at the level of species. Species such as Arctic willow *(Salix arctica)* and avens *(Dryas integrifolia)* are common to both regions. The desert flora of the two regions is, however, quite different in composition. Cacti, for example, are found only in the Nearctic and Paleotropical regions. A number of desert genera are shared between the Nearctic and Neotropical regions. Some well-known endemic trees of the Nearctic region include the tulip *(Lirodendron tulipifera),* coastal redwood *(Sequoia sempervirons),* and Douglas fir *(Pseudotsuga menziesii).* The Palearctic also has many notable endemic trees such as the olive *(Olea europaea),* pistachio *(Pistacia vera),* and Old World cedars *(Cedrus* spp.) of the European and the Mediterranean region; the dawn redwood *(Metasequia glyptostroboides)* of China; and the red cedar *(Cryptomeria japonica)* of Japan.

The close relationship between the mammals and plants of the Nearctic and Palearctic can be explained by the similar longitudinal positions of the two regions, and thus their similar range of climate and environments. In addition, faunal and flora exchanges have been facilitated by the close connection between Europe, Asia, and North America during much of the past 65 million years. Finally, the environments of both regions were strongly impacted by the climatic and geographic changes associated with Late Cenozoic cooling and resulting Quaternary glaciations. In the past, the two regions have been closely connected in both the North Atlantic and the North Pacific regions. At approximately 60 million to 50 million years ago, during the Paleogene, 34 of the 60 mammal genera known from Europe also occurred in North America. Early primates, ancestral to Old World and New World monkeys and other primates, likely evolved in North America at this time and dispersed to Europe and Africa. A similar close correspondence is evident in European and North American land plants during this period. Exchange of fauna and flora from Europe and North America was enabled by a land linkage in the area of northeastern Canada, Greenland, and northern Europe.

As the North Atlantic grew, the barrier between North America and Europe increased over time. At the same time, the European and Asian plates fused, allowing dispersal of land plants and animals across Eurasia. Well before 14 million years ago, Eurasia was one large landmass extending from the Atlantic to the Pacific. By this time, northeastern Asia and northwestern North America were almost joined in the area of Beringia. Sea-level changes due to glaciation during the later Cenozoic produced intermittent land connections between the two continents. The drops in sea level appear to have allowed important exchanges to take place between Palearctic and Nearctic mammals. The dispersal of horses is a particularly good example. The ancestor of the grazing horses was the genus *Hipparion.* It first evolved in North America about 13 million years ago. A large decline in sea level at about 11.5 million years ago allowed the *Hipparion* horses to migrate across the Bering land-bridge and enter Eurasia. The drop in sea level was caused by the establishment and growth of the

Antarctic ice cap. The development of that ice cap caused sea level to drop and allowed dispersal between Asia and North America but, of course, led to the extinction of southern beech and almost all other terrestrial organisms in Antarctica. Within a few hundred thousand years, horses had spread across Asia to Europe and Africa. The spread of horses and other grazing mammals was aided by the first development of grasslands due to the evolution of grasses (Poaceae).

The current distributions of many of the mammal genera found in the Nearctic and Palearctic can be traced to later dispersal across Beringia. Bears likely originated in Eurasia in the Neogene and spread to North America in a number of different dispersal events. Other members of the order Carnivora, which includes dogs and cats, likely also first evolved in Eurasia and dispersed to North America. Wolves and cats experienced several waves of migration both eastward and westward across the landbridge during the Neogene and Quaternary. Buffalo, moose, caribou, and the pronghorn antelope also have their origins in Eurasia and arrived in North America via the Beringian route. Camels, interestingly enough, evolved in North America some 50 million years ago and dispersed to Eurasia via Beringia in the Quaternary.

Climatic cooling during the Quaternary, together with the associated continental glaciations, had a dramatic impact on the mammalian fauna of the Nearctic and Palearctic. Cold-adapted mammals such as the muskox, reindeer, and polar bear evolved in the Palearctic during the last 1 million years and dispersed across the northernmost portions of the two regions. The onset of cooler conditions and the disruptions caused by the shifts between glacial and nonglacial conditions also produced significant extinctions of mammals and plants during the Quaternary. The Neogene floras of Europe and eastern North America were relatively similar in diversity. During the Quaternary, the European flora, particularly in terms of tree species, has been reduced significantly compared to that of North America. The relatively low mammal diversity of Europe has also been attributed to the impact of climatic cooling and the disruption of environments by glacial cycles during the Quaternary.

Neotropical Region

The Neotropical region includes South and Central America and the adjacent islands. It supports an interesting assortment of mammal types and has a higher diversity of mammals than the Nearctic, consisting of some 23 families. Well-known endemic mammal genera of the Neotropical region include llamas *(Lama),* guinea pigs *(Cavia),* and chinchillas *(Chinchilla).* The Neotropical region also shares many families and genera with the Nearctic, including rabbits, shrews, squirrels, cats, foxes, raccoons, mountain lions (pumas), and deer. In addition, the mammal fauna of the Neotropical region has some interesting similarities with other regions. Llamas belong to the family Camelidae and are related to the Palearctic camels. The Neotropical region also supports two families of native monkeys that are more primitive but distantly related to those found in the Palearctic, Oriental, and Ethiopian regions. One interesting feature of the Neotropical region is the diversity of endemic rodents found there. This includes 12 different endemic families and close to 200 endemic species. Finally, the Neotropical region contains the highest diversity of marsupial mammals next to Australia. The marsupials contribute three families and over 80 different species to the mammal fauna of the Neotropical region.

The flora of the Neotropical region is notable for its variety. The region possesses over 130 different families of flowering plants, making it the most diverse of all floral regions. Over 50 families of angiosperms are endemic to the Neotropical region. These range from plants of the alpine tundra to the rainforest and the desert. A number of other angiosperm families are shared with the adjacent Nearctic region. Some, such as the cacti, have relatively continuous distributions across the Isthmus of Panama. A number of shared plant genera and species show an interesting pattern of disjunction in their distributions between the Neotropical and the Nearctic. A surprisingly large number of desert species, including the creosote bush *(Larrea tridentata),* which is very common in California, are found both in the North American desert and in the desert regions of South America. A large number of Neotropical plant families and genera are also found in the Australasian or Ethiopian regions. Notable examples besides southern beeches include southern hemisphere coniferous tree genera, *Podocarpus* and *Araucaria,* and many genera and species of the family Proteaceae. The genera *Podocarpus* and *Weinmannia* are found in Australian, Ethiopian, and Neotropical regions. In addition, several genera of cushion plants found in the mountain tundra of New Zealand and Tasmania also occur in Chile. These include plants of the genera *Donatia, Phyllanchne, Drapetes,* and *Gaimardia.*

The high diversity of the Neotropical fauna and flora, and their linkages with other biogeographic regions, reflect both the environmental variety and the geologic history of the region. The Neotropical region includes environments ranging from alpine tundra to temperate forest, tropical rainforest, dry tropical forest, savanna, grasslands, and desert. The range of potential niches is therefore greater than in the Nearctic or Palearctic regions. The relative importance of the marsupials in the mammal fauna is explained by the former physical connection between South America and the other continents of Gondwanaland. During the early mammal and angiosperm radiations of the Cretaceous, South America was joined to Australia by Antarctica. Africa had already begun to move northwestward and separate from Antarctica by the Cretaceous, about 95 million years ago. By about 55 million years ago, South America was well separated from Antarctica but had not yet moved far enough northward to be linked to North America. Its mammal fauna, dominated by marsupials, and its flora, containing many Gondwana taxa, evolved in isolation for almost 40 million years. This history explains the linkages between the Neotropical, Australasian, and Ethiopian fauna and flora.

The presence of fauna and flora typical of the Nearctic region is explained by the development of a landbridge between the two regions, which occurred with the formation of the Isthmus of Panama about 3 million years ago (see Chapter 8). Some migration across the islands that lay in the area of the present isthmus likely began earlier than this. It is thought that the primate ancestors of the Neotropical monkeys and the ancestors of the diverse Neotropical rodents arrived via these biogeographic stepping stones. The establishment of a landbridge around 3 million years ago allowed the northward migration of the marsupial opossum and the southward migration of placental deer and cats. This so-called Great American Interchange (see Chapter 8) led not only to the introduction of species to the Neotropical region, but it also caused the widespread extinction of many marsupial species in the Neotropical region.

Ethiopian (African/Paleotropical) Region

The Ethiopian region consists of sub-Saharan Africa and adjacent portions of the Arabian Peninsula. The island of Madagascar has a different geologic history than Africa and is generally excluded from the Ethiopian region. The Ethiopian region possesses the most diverse mammal fauna of all biogeographic regions (30 families) and has an extremely diverse angiosperm flora (117 families). There are many well-known endemic mammal genera in the Ethiopian region, including zebras *(Equus burchelli)*, giraffes *(Giraffa camelopardalis)*, gazelles *(Gazella subgutturosa)*, wildebeests *(Connochaetes taurinus)*, African elephants *(Loxodonta africana)*, baboons *(Papio* spp.), chimps *(Pan troglodytes)*, and gorillas *(Gorilla gorilla)*. Although the Ethiopian region shares a number of mammal families with the adjacent Palearctic and Oriental regions, it shares few land mammal families with the Neotropical and none with the Australian region. In contrast, the Ethiopian region shares a number of angiosperm plant families with the Neotropical and Australian regions.

An association between the fauna of the Ethiopian region and that of the Palearctic can be attributed to the geologic history of the two regions. The first mammal fossils found in Africa date from about the Eocene, some 55 million to 40 million years ago. However, despite the lack of direct evidence, it is likely that some early placental mammals were present in Africa in the Late Cretaceous. The early separation of Africa from Australia, South America, and Antarctica prevented the dispersal of the southern marsupial fauna to the Ethiopian region. By the Early Miocene, about 20 million years ago, the African continent had moved northward far enough to collide with Eurasia in the vicinity of Arabia and Turkey. This produced a landbridge between the Ethiopian and Palearctic regions. The landbridge facilitated the southward migration of animals such as the ancestors of the primates, camels, and zebras, which originated in the Nearctic region, and carnivores such as the ancestors of lions and jackals, which originated in the Palearctic region. The connection between Africa and Eurasia also allowed the northward migration of elephants and their relatives, the mammoths, to the Palearctic and Oriental regions. The order Proboscidea, to which elephants belong, likely evolved in Africa in the Late Eocene, some 40 million years ago. The mammoths, which eventually colonized most of the Palearctic and Nearctic, migrated northward from Africa at about 2 million years ago. At the same time, the genus *Equus,* which includes modern horses and zebras, migrated westward from the Nearctic and colonized the Palearctic and Ethiopian regions.

The association between the Ethiopian angiosperm flora and that of the Neotropical and Australian regions reflects the fact that during the radiation of the angiosperms in the Late Cretaceous, Africa was still close enough to these southern continents to exchange flora. This explains the presence of many common plant genera in the Ethiopian, Australasian, and Neotropical regions today. However, the relatively early separation of Africa likely precluded the dispersal of others such as southern beech to the Ethiopian region.

Oriental Region

The Oriental region includes the Indian subcontinent and adjacent portions of southern Asia. Although it is small in area compared to the regions we have discussed thus far, it has a very high diversity of both mammals (20 families) and angiosperms (108 families). Some well-known endemics include the gibbon family *(Hylobatidae)*, Orangutans *(Pongo pygmaneus)*, and the flying lemur family *(Cynocephalidae)*. A number of well-known species from the Oriental region have relatives in the Ethiopian region. These include the Indian elephant *(Elephas maximus)* and the Indian rhinoceros *(Rhinoceros unicornis)*, which have relatives in Africa. The Bengal tiger of India *(Panthera tigris tigris)* has a close relative in the Siberian tiger *(Panthera tigris altaica)*, which is found in the southeastern portion of the Palearctic. The tropical angiosperms of the Oriental region have many relatives in both the Ethiopian and Australian regions, while the montane vegetation contains Palearctic-Nearctic elements, including pines. A well-known endemic tree of the Oriental region is teak *(Tectona grandis)*. In addition, the tropical forests of the region are distinctive in being characterized by a large number of trees of the family Diptocarpaceae.

The mammalian fauna of the Oriental region has close ties to the Ethiopian and Palearctic regions and none with the adjacent Australian region. In contrast, the flora has closer ties with the Australian, Ethiopian, and Neotropical regions. The explanation for these relationships lies in the rather complex geologic history of the region. During the Early Cretaceous, India and Madagascar were part of Gondwanaland and were connected to the southern continents. By the Late Cretaceous, some 90 million years ago, the Indian landmass had separated from the rest of the southern continents and had begun a northward movement. By this time, the Indian flora contained many elements of the Australian and African angiosperm floras. When India eventually collided with the Asian continent, these plants were able to expand to other portions of the Oriental region.

The transport of flora to Asia by the Indian subcontinent is evident in the similarities between the Oriental, Neotropical, and Australasian floras. However, there is no evidence that a mammal fauna of either southern marsupials or placentals existed in India prior to its contact with Asia. India came into contact with Asia by the Middle Miocene, some 14 million years ago. The transport of southern species to Asia by the northward-moving Indian subcontinent represents a special type of dispersal mechanism referred to as an **ark.** The name is a poetic homage to the story of Noah's ark from the Judeo-Christian bible. Some of the earliest mammal fossils from India date from the Miocene and include fauna from both the Palearctic and Ethiopian regions.

By the Middle Miocene, Australia and New Guinea had moved northward and were in close proximity to the Malaysian Archipelago. However, Australia was never directly connected to the Oriental region via a landbridge, and this has prevented a natural exchange of mammal fauna. It did not preclude the migration of many other plants and animals, however. Species of the plant family Proteacea dispersed and colonized southeastern Asia from Australasia. Angiosperm families that first evolved in Asia, such as the Myrtaceae, dispersed southward to Australasia. At this time, bats and birds also dispersed across the stepping stone route from the Oriental region to Australasia.

Many of the Oriental mammals and angiosperms evolved in Southeast Asia. The Southeast Asian region has been in the lower latitudes for over 100 million years, resulting in the evolution of a flora and fauna distinctive from the temperate, boreal, and Arctic fauna and flora of the adjacent Palearctic. However, much interchange did take place between the Palearctic and Southeast Asian portions of the Oriental regions. The primate ancestors of the Asian monkeys, gibbons, and orangutans migrated from the Palearctic during the Cenozoic. Fossils of the Diptocarpaceae, a family of trees that may have first originated in Gondwanaland, and expanded to the Oriental region, have been found in Paleogene deposits from Alaska.

Australian Region

The Australian region includes Australia, Tasmania, New Guinea, and in some regional subdivisions New Zealand. Despite its relatively small size, it is the most distinct of all the biogeographic regions. Its native mammal fauna is dominated by marsupials. Native placental mammals not introduced by humans appear limited to bats and rodents. Other placental mammals, such as the dingo dog, and perhaps the rat, were likely brought to Australia by humans. Unique and well-known endemic mammals include kangaroos *(Macropus)*, wallabees *(Wallabia),* and koala bears *(Phascolarctos cinereus)*. In addition to marsupials, the region contains the only surviving egg-laying monotremes, including the duckbill platypus *(Ornithorhynchus anatinus)* and the echidnas *(Tachyglossus* and *Zaglossus)*. Although Australia and Tasmania have a rich marsupial fauna, the islands of New Zealand, because of their early separation from Gondwanaland, had no nonflying mammals until the arrival of humans. The mammal diversity of the Australasian region in terms of families is relatively small. The angiosperm diversity is slightly lower than that of any other continental region (about 90 families) and shows affinities with the Oriental, Ethiopian, and Neotropical regions.

In contrast to the mammals, in which all the marsupials and monotremes are endemic, only 18 of the angiosperm families are endemic to Australasia. Two very dominant endemic genera are *Eucalyptus* and *Melaleuca*. The eucalyptus is the quintessential Australian tree. It occurs in Australia, New Guinea, Tasmania, and smaller adjacent islands. There are many different species that grow from alpine tundra to dry savanna to tropical rainforests. Eucalyptus accounts for about 95% of the tree biomass in Australia. The genus, however, does not occur in New Zealand. Fossil evidence suggests that the family of plants to which the eucalyptus belongs, the Myrtaceae, evolved in the Oriental region during the Late Cretaceous and likely migrated to Australia following the sundering of Australia's connection to the Antarctic and New Zealand during the Paleogene. The many modern species of Australian eucalyptus are the result of adaptive radiation by this Asian angiosperm family. Although the eucalyptus has not been able to disperse from the Australasian region to the Oriental region via the stepping stone route of the Malaysian Archipelago, other plants and animals have been able to disperse. We have already mentioned the possible northward movement of the Proteaceae family. The ancestors of the widespread bird group the Corvini, which includes crows, ravens, and magpies, first evolved in Australasia and then spread throughout the world after migrating across the Malaysian Archipelago in the Miocene. These birds, which have their origins in Australasia, can now be found as far away as the Arctic treeline of the Nearactic.

Marine Biogeographic Regions

Although oceans cover over 70% of the earth's surface, attempts to subdivide the marine environment into biogeographic regions similar to those for land areas have been limited and not widely applied. It has been suggested that an absence of complete physical barriers between many ocean basins and resulting lack of clear biogeographic boundaries makes such efforts questionable. However, an analysis of the geographic distribution of 65,000 species of marine animals and plants suggests that the world's oceans can be subdivided into 30 distinctive biogeographic realms based on the degree of endemism of the benthic and pelagic fauna and flora (Fig. 10.6). Factors such as land barriers, temperature, salinity, light, and environmental heterogeneity all influence marine life. In addition to latitude and longitude, water depth and location relative to coastlines are important in delineating the geography of the marine realms. Continental shelves comprise 18 of the realms and 12 of the realms are deep sea offshore locations. On average, about 42% of the species in each realm were endemic to that realm alone. The highest degree of endemicity (> 50%) was found in Realms 2 (Black Sea 84%), 14 (Red Sea 74%), 25 (Coastal Chile 68%), 1 (Inner Baltic Sea 63%), 10 (South-East Pacific 59%), 23 (Tropical East Atlantic 57%), and 15 (Tasman Sea—New Zealand 57%). The lowest endemicity (17%) was found in Realm 30 (Southern Ocean). The degree of endemicity also varied by organism type. Benthic arthropod and mollusk species exhibited the highest degree of endemicity. Planktonic microorganisms, which can be dispersed wide distances by currents, and wide-ranging large fish, turtles, birds, and mammals displayed lower average rates of endemicity.

Subdivision of the marine environment into biogeographic realms assumes that evolution has played an important role in the development of traits that allow some species to exist in one environment and not another. There is much evidence that allopatric processes due to isolation by distance or environmental barriers in the marine environment has contributed to speciation. However, the

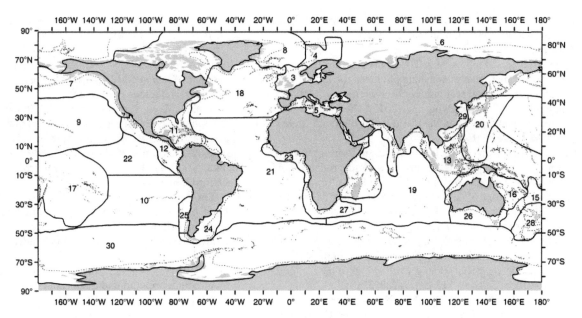

FIGURE 10.6   Global marine biogeographic realms: 1. Inner Baltic Sea; 2. Black Sea; 3. Northeast Atlantic; 4. Norwegian Sea; 5. Mediterranean Sea; 6. Arctic Seas; 7–9. North Pacific and North American Boreal; 10. Southeast Pacific; 11. Caribbean and Gulf of Mexico; 12. Gulf of California; 13. Indo Pacific and Indian Ocean; 14. Gulf of Aqaba, Aden, Suez, Red Sea; 15. Tasmin Sea; 16–22. Open Atlantic, Pacific and Indian Oceans; 23. Gulf of Guinea; 24–25. Southern South America; 26–28. Southern South Africa, South Australia, New Zealand; 29. Northwest Pacific; 30. Southern Ocean (*Source:* Costello et al., 2017 / Springer Nature / CC BY 4.0.).

marine realms presented here do not explicitly take into account the phylogenies and evolutionary histories of the species that define each realm. It is notable that these realms have been defined on the basis of species endemicity, and their definitions are much weaker when attempts are made to use higher taxonomic levels such as genus or family endemicity to define them. Undoubtedly, as data sets pertaining to the physical environment, species, and genetics of the earth's marine ecosystems grow, our understanding of the biogeography of the oceans will increase dramatically. The marine realm represents an exciting frontier area for biogeography.

KEY WORDS AND TERMS

Ark	Biogeographic province	Coefficient of similarity
Biogeographic harmony	Biogeographic provincialism	Cosmopolitan
Biogeographic kingdom	Biogeographic realm	Endemic
Biogeographic line	Biogeographic region	Wallace's Line

REFERENCES AND FURTHER READING

Ali, J. R., and Heaney, L. R. (2021). Wallace's line, Wallacea, and associated divides and areas: history of a tortuous tangle of ideas and labels. *Biological Reviews* **96**(3), 922–942.

Archibold, O. W. (1995). *Ecology of World Vegetation*. Chapman and Hall, New York.

Axelrod, D. I. (1952). A theory of angiosperm evolution. *Evolution* **6**, 29–60.

Bi, S., Wang, Y., Guan, J., Sheng, X. and Meng, J. (2014). Three new Jurassic euharamiyidan species reinforce early divergence of mammals. *Nature* **514**, 579–584.

Briggs, J. C. (1995). *Global Biogeography*. Elsevier, Amsterdam.

Brown, J. H., and Gibson, A. C. (1983). *Biogeography*. C. V. Mosby, St. Louis.

Brown, J. H., and Lomolino, M. V. (1998). *Biogeography*, 2nd edition. Sinauer Associates, Sunderland, MA.

Burge, D. O., Thorne, J. H., Harrison, S. P., O'Brien, B. C., Rebman, J. P., Shevock, J. R., Alverson, E. R., Hardison, L. K., Rodríguez, J. D., Junak, S. A., and Oberbauer, T. A. (2016). Plant diversity and endemism in the California Floristic Province. *Madroño* **63**, 3–206.

Burgin, C. J., Colella, J. P., Kahn, P. L., and Upham, N. S. (2018). How many species of mammals are there? *Journal of Mammalogy* **99**, 1–14.

Celik, M. A., and Phillips, M. J. (2020). Conflict resolution for Mesozoic mammals: reconciling phylogenetic incongruence among anatomical regions. *Frontiers in Genetics* **11**, 0651.

Christenhusz, M. J., and Byng, J. W. (2016). The number of known plants species in the world and its annual increase. *Phytotaxa* **261**, 201–217.

Close, R. A., Friedman, M., Lloyd, G. T., and Benson, R. B. (2015). Evidence for a mid-Jurassic adaptive radiation in mammals. *Current Biology* **25**, 2137–2142.

Coiro, M., Doyle, J. A., and Hilton, J. (2019). How deep is the conflict between molecular and fossil evidence on the age of angiosperms? *New Phytologist* **223**, 83–99.

Condamine, F. L., Silvestro, D., Koppelhus, E. B., and Antonelli, A. (2020). The rise of angiosperms pushed conifers to decline during global cooling. *Proceedings of the National Academy of Sciences* **117**, 28867–28875.

Cook, L. G., and Crisp, M. D. (2005). Not so ancient: the extant crown group of *Nothofagus* represents a post-Gondwanan radiation. *Proceedings of the Royal Society B: Biological Sciences* **272**, 2535–2544.

Costello, M. J., Tsai, P., Wong, P. S., Cheung, A. K. L., Basher, Z., and Chaudhary, C. (2017). Marine biogeographic realms and species endemicity. *Nature Communications* **8**, 1057.

Cox, B. (2001). The biogeographic regions reconsidered. *Journal of Biogeography* **28**, 511–523.

Cox, C. B., and Moore, P. D. (1993). *Biogeography: An Ecological and Evolutionary Approach*, 5th edition. Blackwell Scientific Publications, Oxford.

Crepet, W. L. (1998). The abominable mystery. *Science* **282**, 1653–1654.

Cui, D. F., Hou, Y., Yin, P., and Wang, X. (2022). A Jurassic flower bud from China. *Geological Society, London, Special Publications*, 521.

Cvetković, T., Hinsinger, D. D., Thomas, D. C., Wieringa, J. J., Velautham, E., and Strijk, J. S. (2022). Phylogenomics and a revised tribal classification of subfamily Dipterocarpoideae (Dipterocarpaceae). *Taxon* **71**, 85–102.

Darlington Jr., P. J. (1957). *Zoogeography: The Geographical Distribution of Animals*. Wiley, New York.

Darlington Jr., P. J. (1965). *Biogeography of the Southern End of the World*. Harvard University Press, Cambridge, MA.

Doyle, J. A. (2012). Molecular and fossil evidence on the origin of angiosperms. *Annual Review of Earth and Planetary Sciences* **40**, 301–326.

Erickson, J. (1995). *A History of Life on Earth: Understanding Our Planet's Past*. Facts on File, New York.

Escalante, T., Morrone, J. J., and Rodríguez-Tapia, G. (2013). Biogeographic regions of North American mammals based on endemism. *Biological Journal of the Linnean Society* **110**, 485–499.

Ficetola, G. F., Mazel, F., and Thuiller, W. (2017). Global determinants of zoogeographical boundaries. *Nature Ecology & Evolution* **1**, 0089.

Franzen, J. L., Gingerich, P. D., Habersetzer, J., Hurum, J. H., von Koenigswald, W., and Smith, B. H. (2009). Complete primate skeleton from the middle Eocene of Messel in Germany: morphology and paleobiology. *PLoS One* **4**, e5723.

Friis, E. M., Chaloner, W. G., and Crane, P. R. (1987). *The Origins of Angiosperms and Their Biological Consequences*. Cambridge University Press, Cambridge.

Frisch, W., Meschede, M., and Blakey, R. C. (2011). *Plate Tectonics: Continental Drift and Mountain Building*. Springer, Heidelberg, Germany.

Godthelp, H., Wroe, S., and Archer, M. (1999). A new marsupial from the early Eocene Tingamarra local fauna of Murgan, Southeastern Queensland: a prototypical Australian marsupial? *Journal of Mammal Evolution* **6**, 289–313.

Hallam, A. (1994). *An Outline of Phanerozoic Biogeography*. Oxford University Press, Oxford.

Hassanin, A., Veron, G., Ropiquet, A., Jansen van Vuuren, B., Lécu, A., Goodman, S. M., Haider, J., and Nguyen, T. T. (2021). Evolutionary history of Carnivora (Mammalia, Laurasiatheria) inferred from mitochondrial genomes. *PloS One* **16**, e0240770.

Hedges, S. B., Parker, P. H., Sibley, C. G., and Kumar, S. (1996). Continental breakup and the ordinal diversification of birds and mammals. *Nature* **381**, 226–229.

Holt, B. G., Lessard, J. P., Borregaard, M. K., Fritz, S. A., Araújo, M. B., Dimitrov, D., Fabre, P. H., Graham, C. H., Graves, G. R., Jønsson, K. A., and Nogués-Bravo, D. (2013). An update of Wallace's zoogeographic regions of the world. *Science* **339**, 74–78.

Janis, C. M. (1993). Tertiary mammal evolution in the context of changing climates, vegetation and tectonic events. *Annual Review of Ecology and Systematics* **24**, 467–500.

Joyce, W. G., Micklich, N., Schaal, S. F., and Scheyer, T. M. (2012). Caught in the act: the first record of copulating fossil vertebrates. *Biology Letters* **8**, 846–848.

Kemp A. (2005). *The Origin and Evolution of Mammals*. Oxford University Press, Oxford.

Kielan-Jaworowska, Z., Cifelli, R.L., and Luo, Z. X. (2004). *Mammals from the Age of Dinosaurs: Origins, Evolution, and Structure*. Columbia University Press, New York.

Koblmüller, S., Vilà, C., Lorente-Galdos, B., Dabad, M., Ramirez, O., Marques-Bonet, T., Wayne, R. K., and Leonard, J. A. (2016). Whole mitochondrial genomes illuminate ancient intercontinental dispersals of grey wolves (*Canis lupus*). *Journal of Biogeography* **43**, 1728–1738.

Koenigswald, W. V., Braun, A., and Pfeiffer, T. (2004). Cyanobacteria and seasonal death: a new taphonomic model for the Eocene Messel lake. *Paläontologische Zeitschrift* **78**, 417–424.

Kreft, H., and Jetz, W. (2010). A framework for delineating biogeographical regions based on species distributions. *Journal of Biogeography* **37**, 2029–2053.

Kremenetski, C., Liu, K., and MacDonald, G. M. (1998). The late Quaternary dynamics of pines in northern Asia. In *Ecology and Biogeography of Pinus* (ed. D. M. Richardson), pp. 95–106. Cambridge University Press, Cambridge.

Kurtén, B. (1969). Continental drift and evolution. *Scientific American* **220**, 54–65.

Lambeck, K., Rouby, H., Purcell, A., Sun, Y., and Sambridge, M. (2014). Sea level and global ice volumes from the Last Glacial Maximum to the Holocene. *Proceedings of the National Academy of Sciences* **111**, 15296–15303.

Li, H. T., Yi, T. S., Gao, L. M., Ma, P. F., Zhang, T., Yang, J. B., Gitzendanner, M. A., Fritsch, P. W., Cai, J., Luo, Y., and

Wang, H. (2019). Origin of angiosperms and the puzzle of the Jurassic gap. *Nature Plants* **5**, 461–470.

Lusk, C. H., Wright, I., and Reich, P. B. (2003). Photosynthetic differences contribute to competitive advantage of evergreen angiosperm trees over evergreen conifers in productive habitats. *New Phytologist* **160**, 329–336.

MacDonald, G. M., Cwynar, L. D., and Whitlock, C. (1998). The late Quaternary dynamics of pines in northern North America. In *Ecology and Biogeography of Pinus* (ed. D. M. Richards), pp. 122–136. Cambridge University Press, Cambridge.

Meredith, R. W., Janečka, J. E., Gatesy, J., Ryder, O. A., Fisher, C. A., Teeling, E. C., Goodbla, A., Eizirik, E., Simão, T. L., Stadler, T., and Rabosky, D. L. (2011). Impacts of the Cretaceous Terrestrial Revolution and KPg extinction on mammal diversification. *Science* **334**, 521–524.

Morain, S. A. (1984). *Systematic and Regional Biogeography*. Van Nostrand Reinhold, New York.

Morrone, J. J., Escalante, T., and Rodriguez-Tapia, G. (2017). Mexican biogeographic provinces: map and shapefiles. *Zootaxa* **4**, 277–279.

Muller, P. (1974). *Aspects of Zoogeography*. Dr. W. Junk Publishers, The Hague.

Newham, E., Gill, P. G., Brewer, P., Benton, M. J., Fernandez, V., Gostling, N. J., Haberthür, D., Jernvall, J., Kankaanpää, T., Kallonen, A., and Navarro, C. (2020). Reptile-like physiology in Early Jurassic stem-mammals. *Nature Communications* **11**, 1–13.

O'Brien, S. J., Menotti-Raymond, M., Murphy, W. J., Nash, W. G., Wienberg, J., Stanyon, R., Copeland, N. G., Jenkins, N. A., Womack, J. E., and Graves, J. A. M. (1999). The promise of comparative generics in mammals. *Science* **286**, 458–481.

O'Leary, M. A., Bloch, J. I., Flynn, J. J., Gaudin, T. J., Giallombardo, A., Giannini, N. P., Goldberg, S. L., Kraatz, B. P., Luo, Z. X., Meng, J., and Ni, X. (2013). The placental mammal ancestor and the post–K-Pg radiation of placentals. *Science* **339**, 662–667.

Osborne, R., and Tarling, D. (1996). *The Historical Atlas of the Earth: A Visual Exploration of the Earth's Physical Past*, p. 192. Henry Holt and Co., New York.

Pearson, L. C. (1995). *The Diversity and Evolution of Plants*. CRC Press, Boca Raton, FL.

Pickrell, J. (2019). How the earliest mammals thrived alongside dinosaurs. *Nature* **574**, 468–473.

Pielou, E. C. (1979). *Biogeography*. Wiley, New York.

Ramirez-Barahona, S., Sauquet, H., and Magallon, S. (2020). The delayed and geographically heterogeneous diversification of flowering plant families. *Nature Ecology & Evolution* **4**, 1232–1238.

Richardson, D. M., and Rundel, P. W. (1998). Ecology and biogeography of *Pinus*: an introduction. In *Ecology and Biogeography of Pinus* (ed. D. M. Richardson), pp. 3–46. Cambridge University Press, Cambridge.

Rieseberg, L. H., and Gerber, D. (1995). Hybridization in the Catalina Island mountain mahogany (*Cercocarpus traskiae*): RAPD evidence. *Conservation Biology* **9**, 199–203.

Roberts, N. (2014). *The Holocene*, 3rd edition. Wiley Blackwell, Oxford.

Romano, M., and Cifelli, R. L. (2015). 100 years of continental drift. *Science* **350**, 915–916.

Romer, A. S. (1966). *Vertebrate Paleontology*, 3rd edition. University of Chicago Press, Chicago.

Rose, K. D. (2012). The importance of Messel for interpreting Eocene Holarctic mammalian faunas. *Palaeobiodiversity and Palaeoenvironments* **92**, 631–647.

Rougier, G. W., Wible, J. R., and Novacek, M. J. (1998). Implications of Deltatheridium specimens for early marsupial history. *Nature* **396**, 459–463.

Rueda, M., Rodriguez, M. A., and Hawkins, B. A. (2013). Identifying global zoogeographical regions: lessons from Wallace. *Journal of Biogeography* **40**, 2215–2225.

Savage, R. J. G. (1986). *Mammal Evolution: An Illustrated Guide*. Facts on File, New York.

Scotese, C. R. (2021). An atlas of Phanerozoic paleogeographic maps: the seas come in and the seas go out. *Annual Review of Earth and Planetary Sciences* **49**, 679–728.

Simpson, G. G. (1940). Mammals and land bridges. *Journal of the Washington Academy of Science* **30**, 137–163.

Simpson, G. G. (1953). *Major Features in Evolution*. Columbia University Press, New York.

Simpson, G. G. (1961). Historical zoogeography of Australian mammals. *Evolution* **15**, 413–446.

Simpson, G. G. (1965). *The Geography of Evolution*. Chilton Book Co., Philadelphia.

Soltis, D. E., Bell, C. D., Kim, S., and Soltis, P. S. (2008). Origin and early evolution of angiosperms. *Annals of the New York Academy of Sciences* **1133**, 3–25.

Sun, G., Dilcher, D. L., Zheng, S., and Zhou, Z. (1998). In search of the first flower: a Jurassic angiosperm, *Archaefructus*, from Northeast China. *Science* **282**, 1692–1695.

Switek, B. J. (2010). Ancestor or adapiform? Darwinius and the search for our early primate ancestors. *Evolution: Education and Outreach* **3**, 468–476.

Thorne, R. F. (1992). Classification and geography of the flowering plants. *Botanical Review* **58**, 225–327.

Vilhena, D. A., and Antonelli, A. (2015). A network approach for identifying and delimiting biogeographical regions. *Nature Communications* **6**, 1–9.

Wang, B., and Wang, X. R. (2014). Mitochondrial DNA capture and divergence in Pinus provide new insights into the evolution of the genus. *Molecular Phylogenetics and Evolution* **80**, 20–30.

Wang, X., and Zheng, X. T. (2012). Reconsiderations on two characters of early angiosperm *Archaefructus*. *Palaeoworld* **21**, 193–201.

Webb, S. D. (1997). The Great American faunal interchange. In *Central America* (ed. A. G. Coates), pp. 97–122. Yale University Press, New Haven, CT.

White, A. E., Dey, K. K., Stephens, M., and Price, T. D. (2021). Dispersal syndromes drive the formation of biogeographical regions, illustrated by the case of Wallace's Line. *Global Ecology and Biogeography* **30**, 685–696.

Zacharin, R. F. (1978). *Emigrant Eucalypts: Gum Trees as Exotics*. Melbourne University Press, Carlton, Australia.

Zavada, M. S. (2007). The identification of fossil angiosperm pollen and its bearing on the time and place of the origin of angiosperms. *Plant Systematics and Evolution* **263**, 117–134.

CHAPTER 11

BIOGEOGRAPHY AND HUMAN EVOLUTION

Where does the human race come from? How has the environment shaped us? How have we shaped the environment? These are basic questions that we all ponder. People seek answers for such universal questions through many different avenues. After the publication of his book on the origin of species in 1859, Charles Darwin directly tackled the problem of human origins in his two-volume treatise, *On the Descent of Man, and Selection in Relation to Sex*. In this work, published in 1871, Darwin made the then startling proposition that humans were related by a common ancestor to the great apes, such as chimpanzees and gorillas. Darwin argued that, like other species, we have evolved to our present physical form and degree of intelligence through natural selection. The work and its tenets were, and remain, roundly condemned by some. Others consider Darwin's work a seminal stroke of genius and courage. Since the time of Darwin, scientific research into the evolution of humans has been a major focus of work by anthropologists, biologists, geologists, and paleontologists. This is also the case for biogeographers, who have long been interested in how geography and environmental change have influenced human evolution and the early spread of humans around the globe. Indeed, the history of our own species remains one of the most fascinating and important of all biogeographic stories. Because of the intensity of interest and research, it is also a story that continues to develop as scientists make new discoveries and we learn more about our ancestral origins.

In this chapter, we will review the current state of knowledge regarding the history and geography of human evolution. We will start by looking at our relationship to other closely related species of mammals, the primates. We will then trace the evolutionary and geographic history of the primates and human ancestors. We will also consider the role that environmental change may have played in the human evolutionary process. We will then examine the geographic dispersal of humans throughout the world and the startling rise in human population size.

THE PRIMATE LINKAGE

Humans *(Homo sapiens*, sometimes referred to as *Homo sapiens sapiens)* are considered by biologists to belong to the order of placental mammals called the **Primates.** The Primate Order includes animals such as lemurs, tarsiers, monkeys, apes, and humans. The Primate Order is divided by primatologists into two suborders and then three infraorders. Humans belong to the suborder Haplorrhini, which also includes tarsiers (although that assignment is debated), monkeys, and apes. These species are referred to as anthropoids due to their similarity to humans. The other suborder, Strepsirhini, includes lemurs, lorises, and indriids, which are considered a somewhat primitive group of primates. These species are referred to as prosimians. Within the Haplorrhini, humans belong to an infraorder called the Catarrhini, which includes the Old World monkeys and the apes. The New World monkeys of South and Central America appear to be more primitive than the Catarrhini and are grouped in the infraorder Platyrrhini. Humans, the small apes (gibbons), and the great apes (orangutans, chimps, gorillas) are grouped together in a superfamily called the Hominoidea. Humans and the great apes

Biogeography: Introduction to Space, Time, and Life, Second Edition. Glen M. MacDonald.
© 2025 John Wiley & Sons, Inc. Published 2025 by John Wiley & Sons, Inc.
Companion website: www.wiley.com/go/macdonald/biogeography2e

TABLE 11.1   Taxonomic Classification of Modern Humans

Order	Primates
Suborder	Haplorrhini
Infraorder	Catarrhini
Superfamily	Hominoidae
Family	Hominidae
Subfamily	Homininae
Genus	*Homo*
Species	*Homo sapiens*

fall within the Family Hominidae in recognition of the similarity of these other primates to humans. We and the great African apes are typically grouped as members of the Primate subfamily Homininae, and humans are the only living members of the genus *Homo*. The orangutans of Asia are generally classified as belonging to the subfamily Ponginae. The extinct ancestors of modern humans are referred to as Hominini. Given this taxonomic classification of humans (Table 11.1), what is the basis for claiming that humans are related to other primates?

The genetic relationship between humans and other primates is readily apparent through DNA analysis. Humans share well over 90% of their genome with our closest relatives, the chimps *(Pan troglodytes)*, the bonobos *(Pan paniscus)*, and gorillas *(Gorilla gorilla gorilla)*. Jared Diamond has referred to humans as the "third chimpanzee." The orangutans are also genetically close relatives of humans. In fact, geneticists have concluded that a large proportion of the protein-coding genes found in humans are also found in most primates. A close relationship to primates is also supported by similarities in the blood and protein chemistry of humans and primates. Humans share many physical characteristics with other primates. The primates are pentadactyl, having five fingers and toes. The hands of the primates, and in most cases the feet, are prehensile, which means they can grasp objects. Other primate species sometimes use hand gestures to communicate. Human feet are perhaps the least prehensile of the primates, but most of us can still grasp a pencil with our toes. The primate clavicle and scapula (shoulder blade) allow for a great range of movement and a high degree of strength in the arms. Many primates can brachiate, which means they can use the arms to move from branch to branch with the body suspended beneath. Brachiating is what human children do when they swing hand over hand along monkey bars. Most nonprimate mammals cannot brachiate. The primates have a reduced sense of smell compared to many other mammals. Thus, the snout on most primates, particularly monkeys and apes, is small and relatively flat. Most primates also have stereoscopic vision because the eyes are placed on the face in such a way that both eyes can focus on the same object at once. Most primates also have color vision and possess a unique central pit in the retina called the fovea centralis, which allows sharp focusing on one object while retaining visual perception of surrounding features. Of the primates, only lemurs do not possess color vision, stereoscopy, and a fovea centralis. The primate cranium is generally high and vaulted. The primate vertebral column (the spine) attaches lower on the primate skull than is the case for other mammals. This attachment allows for great mobility of the primate head. The above features of the primate skull help produce the similarity we see when we compare the heads and faces of humans, apes, and monkeys. Primate jaws contain teeth for grinding food (molars), puncturing (canine teeth), and scraping (incisors). This wide range of tooth types suggests an adaptation to a generalist diet. Human dentition is quite similar to chimp and ape dentition, although our canines are greatly reduced.

Relative to other terrestrial mammals, the brains of primates are large compared to their body mass and they are physiologically complex. Since the great cognitive power of humans is often invoked to set us apart from other animals, the evolution of primate brain size has drawn much scientific interest. There have been posited a number of explanations for the relatively large brains of primates, including unique genetic mutation, environmental demands requiring high levels of cognition, and the cognitive demands of complex social interactions.

The shared primate features of prehensile hands and feet, shoulder and arm traits, and color stereoscopic vision are potentially explained as adaptations to life in trees. Most primates are arboreal, meaning they live mainly in trees. Others, such as chimps, spend some time on the ground but climb trees to escape predators or secure food. Even modern humans are reasonably good climbers compared to many other mammals. Being able to judge the distance between branches using forward-looking stereoscopic vision, seeing color differences that might help identify food or predators within dense forest canopies, being able to climb using prehensile hands and feet, and moving swiftly by brachiating are all advantageous to arboreal animals. Cretaceous and Paleogene diversification of angiosperms, with colored fruits and flowers, may have contributed to the value of color vision. Primates give birth to relatively small numbers of offspring. Lemurs produce two or three offspring at each birth, but monkeys and apes generally give birth to only one offspring at a time. Again, this is possibly an adaptation to arboreal existence. Primate mothers must care for their offspring for several months in the case of lemurs and up to four years in the case of chimps. In addition to feeding and protecting them, the mothers must carry offspring with them along branches. Broods of more than one offspring would be difficult to accommodate.

That humans are relatively closely related to primates is indisputable given the overwhelming genetic, physiological, morphological, and behavioral similarities. What remains of intense interest and debate is when, where, and how our shared common human ancestors diverged from the other primates and developed into the modern human species. Molecular clock analysis (see Technical Box 9.1) is an approach that has been commonly used to explore humans' shared ancestral roots. The degree of difference between the DNA of different species tells us the time that has elapsed since species divergence occurred (Technical Box 11.1). Some molecular clock estimates suggest that the human lineage diverged from the ancestors of chimps as early as 12 million years ago and from

TECHNICAL BOX 11.1

Molecular Clocks and Our Common Ancestors

The common individual from whom the later species and individuals descended is referred to as the most recent common ancestor (MRCA) or the last common ancestor (LCA). Determining the age of the MRCA between humans and other primates, and between living humans, has long been of great interest. In the 1960s, Allan Wilson and Vincent Sarich of the University of California at Berkeley analyzed the genetically controlled differences in blood proteins from humans and other primates and concluded that divergence between ape and human ancestors occurred 5 million to 6 million years ago. Subsequent molecular clock research at Berkeley focused on mitochondrial DNA (mtDNA). Mitochondrial DNA is found in small energy-producing structures in the cytoplasm of cells. Unlike nuclear DNA, which controls inherited traits such as eye color, mtDNA is inherited only from the maternal parent. Thus, the mtDNA of animals should be an exact copy of the female parent. However, it is known that mtDNA has a high rate of spontaneous mutation. This rate of mutation is about ten times faster than nuclear DNA. Thus, over time, the mtDNA of a lineage will diverge through mutation from the original maternal pattern. Breeding patterns or natural selection does not influence the divergence rates of mtDNA. These properties make mtDNA an ideal molecular clock. The longer that two lineages have been separated, the greater the divergence in their mtDNA. In the 1980s, Rebecca Cann, Mark Stoneking, and Allan Wilson analyzed the mtDNA differences in 147 women of different races and origins, including Africans, Asians, Caucasians, New Guineans, and Aboriginal Australians. They concluded that mtDNA evidence pointed to a common African ancestor for all living humans and that this ancestral woman lived about 200,000 years ago. The terms *African Eve* or *Mitochondrial Eve* have sometimes been applied to this hypothetical ancestor. The research drew strong attention and much debate with some researchers studying molecular clocks, paleontology, and archaeology debating the precision of the determinations and some of the assumptions of this pioneering analysis.

Geneticists have also developed molecular estimates for male common ancestry through analysis of the Y chromosome. As it happens, the human Y chromosome contains a large amount of nonrecombinant DNA that is passed directly from father to son in an unaltered state except in the case of mutations. The number of mutational changes in nonrecombinant DNA increases over generations, providing a basis for a molecular clock for male lineages. The application of equivalent analysis to Y chromosome and MtDNA has led some researchers to conclude that the most recent male common ancestor (sometimes called the *Y chromosome Adam*) lived in Africa between 120,000 and 156,000 years ago.

In general, based on the DNA evidence, we may all share common female and male ancestors who likely lived in Africa between 200,000 and 100,000 years ago. This is, however, an area of continued keen scientific research and debate.

gorillas around 15 million years ago, whereas others place the divergence more recently at perhaps 5 million and 10 million years ago. The search for fossils and other physical evidence of the ancestors of modern primates and modern humans is one of the most intriguing adventures in modern science. Tracing the fossil evidence of primate and human evolution also helps us see how geography and environment have been such crucial factors in the development of modern species. In the next section we will embark on an exploration of the paleontological and anthropological evidence from our ancient past. It should be borne in mind that new evidence and analyses continue to arise and will further inform this fascinating story beyond what can be reported now. Be on the lookout for future breakthroughs!

EARLY PRIMATES

Our knowledge of early primate evolution comes largely from the fossil record. The most abundant fossils are often jaws and teeth; these remains are fairly robust compared to other parts of the skeleton. Analysis of the shape and wear of fossil teeth also allows for inferences about the type of food the animal depended on. The fossil record of the early primates is sparser than we might hope because these animals were arboreal and fossilization is rare in forested settings. Organisms living in environments such as lakes or swamps have a much better chance of being buried by sediment and preserved as fossils after death. Fossil evidence for the first undisputed primates extends back about 56 million years, while molecular clock evidence suggests an origin older than 70 million years. From the evidence at hand, it seems possible that primates arose during the Late Cretaceous about 70 million years ago. An extinct suborder called the Plesiadapiformes is considered a proto-primate or archaic primate that shared features with the later primates. They appear in the North American fossil record in the very early Paleogene shortly after 65 million years ago. This notably follows the extinction of the nonavian dinosaurs and the resulting burst of mammalian evolution and speciation. The Plesiadapiformes were small animals, the size of squirrels and house cats. They were tree-dwelling and had skeletal features that are similar to those of some modern primates, but their dentition was more rodentlike, except for their molars, which were like those of the primates. The Plesiadapiformes hunted small insects and ate plant material. They appear to have had somewhat larger brains relative to their body mass than other mammals with similar niches, such as tree shrews. The Plesiadapiformes became very abundant in North America and Europe. These two continents were still joined in the North Atlantic sector at this time. The climate of the earth was generally warm and moist, and this resulted in an extensive and continuous forest habitat. Despite their abundance in the Paleocene, by about 37 million years ago, the Plesiadapiformes had gone extinct.

By the Eocene, about 55 million to 38 million years ago, two groups of true primates had become abundant in the northern hemisphere. One group resembled the modern tarsiers, and the other group resembled the modern lorises and lemurs. These two fossil groups are called the Omomyids and the Adapids, respectively, and are likely ancestral to modern tarsiers of the Haplorrhini primate suborder and lemurs of the Strepsirhini primate suborder. The fossils of *Darwinius masillae* found in the Eocene-aged Messel fossil beds described in Chapter 9 are of an Adapiforme primate. During this period, primates were present in Africa, Eurasia, and North America. As a result of plate tectonics, the African continent had moved far enough northward so that primates could disperse between Eurasia and Africa. A continuation of generally warm and moist conditions promoted extensive forest cover that suited the early primates. Toward the end of the Eocene, there was a great decrease in Strepsirhini primates. The global decline in Strepsirhini primates was likely due to increasing competition from newly evolving rodents and primates of the Catarrhini infraorder, which are ancestors of modern monkeys, apes, and humans. The early lemurs went extinct in North America. The Adapiforme primates also went extinct in North America. No later primates are known from that continent from 26 million years ago until the arrival of humans in the Late Quaternary. Primates continued to live in Central and South America. Platyrrhini monkeys exist in South and Central America today, and fossil deposits show that they have been there since the Late Eocene to Oligocene, some 36 million years. It is possible that they dispersed from Africa to South America, perhaps via rafting on floating plant debris. The early arrival and evolutionary history of the South and Central American Platyrrhini primates remain a topic of research and debate.

Some of the best fossil evidence of the early Catarrhine primates, including anthropoid primates, comes from the Fayum Depression in Egypt. These Late Eocene and Oligocene deposits date from about 37 million to 29 million years and are extremely rich in primate fossils. Included in the fossil fauna is a small, monkey-like primate of the genus *Parapithecus* that appears ancestral to the Old World monkeys. The name *Parapithecus* can be translated from its Greek roots as meaning "close to ape." In addition, the deposits contain fossils of the apelike primate of the genus *Aegyptopithecus* ("Egyptian ape"). Many believe that *Aegyptopithecus* is an ancestor of the modern apes and humans. Paleontologists have also found some earlier primates that are potential ancestors of *Aegyptopithecus*, modern apes, and humans. These include the small primates of the genus *Catopithecus*, which are found in the early portion of the Fayum deposits. In addition, fossils have been recovered of an Early Eocene primate that lived in northwestern Africa some 50 million years ago and possessed teeth that are very much like those of modern apes. Finally, teeth and jaw fragments of a potential early ancestor to anthropoid primates have been found in deposits in Myanmar (Burma). One such Asian primate is called *Amphipithecus*. Some paleontologists have suggested that the early ancestors of the great apes and humans may have first evolved in Eurasia. During the Eocene, Africa, Europe, and Asia had become closely connected and were largely forested, so a widespread distribution of primates in Africa and Eurasia is not surprising. The North Atlantic corridor between Europe and North America had been sundered at this point however. Cooling temperatures, particularly at high latitudes during the Late Oligocene and Eocene, may have kept these newly evolved primates from migrating to the New World via Beringia.

What was *Aegyptopithecus* like? These small primates were tree dwellers, with a long tail and prehensile hands and feet. They were quadrupeds, walking using four legs. In terms of size, they were about as big or slightly larger than a modern house cat. The skulls of *Aegyptopithecus* had protected bony eye sockets that are typical of modern apes and humans. Its snout was reduced, the visual cortex region of the brain was enlarged, and the olfactory lobes (area of the brain related to the sense of smell) were reduced. Finally, the dentition of *Aegyptopithecus* was similar to that of apes. It was clearly much closer to modern apes than it was to lorises and tarsiers. *Aegyptopithecus* appears to have dwelt in tree canopies and were fruit-eaters. It has been speculated that they lived in social groups.

The Late Oligocene and the first half of the Miocene, from about 23 million to 14 million years ago, saw a great increase in apelike primate diversity and geographic range. Extensive forest cover still existed across Africa and most of Eurasia at this time, and it is thought that these apes were quadrupedal (using all four limbs to walk) tree dwellers. However, there is evidence that some were capable of upright stance. Catarrhini apes were present from Africa to western Europe to southeastern Asia. Many different genera have been described from the fossil record. Some, such as *Pliopithecus*, found in Europe were relatively small and similar to modern gibbons. Others, such as *Gigantopithecus* from Asia, were considerably larger than modern gorillas. With such an abundance, it is difficult to trace which Miocene primate might be ancestral to humans. Identifying the Miocene primates ancestral to humans and the Miocene primates that are simply relatives of our primate ancestors remains controversial.

The abundance and diversity of the Miocene apes seem to have been much higher than that of apes today. What happened to this array of different species? Despite their large numbers, large geographic range, and great diversity, the apelike primates suffered geographic range constrictions, declines, and extinctions in the Late Miocene and Pliocene. Among the reasons may have been changing landscapes, increasing competition from monkeys in shrinking forested areas, increasing pressure from new predators, and other changes in the environment. In the temperate and tropical regions where the primates were concentrated, it appears that forest cover had begun to decline and large areas of savanna were developing. Some primate species survived these changes through evolutionary adaptations to more open ground environments. The ancestors of modern humans were one such apelike primate. In contrast to many other primates, ancestral members of the human lineage would experience evolutionary radiation and disperse across much of the earth during the subsequent Pliocene and Pleistocene. To consider a widely accepted candidate for a direct ancestor to modern humans, we will move forward in time to about 4 million years ago.

THE HOMINIDS: *AUSTRALOPITHECUS*

The term **hominid** refers to humans and extinct bipedal primates that are believed to be directly ancestral or closely related to humans. A number of Miocene-Pliocene primates have been suggested to be early hominids, such *Sahelanthropus, Orrorin,* and *Ardipithecus.* The first widely accepted hominids and potential ancestors of modern humans are of the genus *Australopithecus,* which appeared in Africa about 4 million years ago. As we shall see, bipedalism is a critical factor in defining the hominids. Before discussing the Australopithecines (also referred to as the Australopiths) in detail, let's consider the events that may have led up to the development of the bipedal primates and eventually the hominids.

Charles Darwin identified bipedalism as a central feature of human evolution. Of all the modern primates, humans are the only species that is truly bipedal. Fossil skeletal evidence indicates that some Miocene primates could stand on two legs. Monkeys and apes can walk on two legs and do so quite rapidly for short distances. In general, however, chimps prefer to move along the ground using both their arms and legs for propulsion. In gorillas and chimps, this takes the form of knuckle walking. Humans do not knuckle walk. Our bodies are clearly adapted for movement on two feet. Alone among all other primates, and indeed all other mammals, humans are fully bipedal. Although we take human bipedalism for granted as our natural condition, it is not so easy to explain how this characteristic developed. Primates clearly have their origins in the trees. As discussed, many facets of human morphology and physiology can be attributed to adaptations to arboreal life. In trees, quadrupedal movement using prehensile feet and hands is clearly more efficient (not to mention safer!) than bipedalism. Like humans, some other modern primates, such as chimps, gorillas, and baboons, are essentially ground-dwelling animals, yet they have remained largely quadrupedal. So the questions we must answer are, what made our primate ancestors, which were initially creatures of the forest, climb down from the trees and take up life on the ground, and why did they become bipedal? Let's consider the development and implications of bipedalism.

The movement of some ape species away from a strictly arboreal existence likely began by the Miocene. Bipedalism likely arose from arboreal-dwelling primates that possessed long backs and were capable of upright stance. The gibbon genus *Hylobates* found today in Asia is perhaps a modern analogue. The name *Hylobates* means "forest walker" in Greek. An important factor that may have contributed to the adaptive value of bipedalism was environmental change. From about 11 million to 8 million years ago in the Miocene, a trend toward cooler, drier, and more variable climatic conditions developed into the subsequent Pliocene and Pleistocene. This led to a fragmentation of the once continuous African and southern Eurasian forest cover. This trend of climatic and vegetation change is coincident with the diversification of grasses and associated browsing fauna and predators. Deforestation in Africa saw formerly extensive forests replaced in large regions by deserts, thorn scrub, savanna, and woodland during the Late Miocene, Pliocene, and Pleistocene (Fig. 11.1). Not only was the world of the Old World primates becoming more arid, it was also becoming more variable in

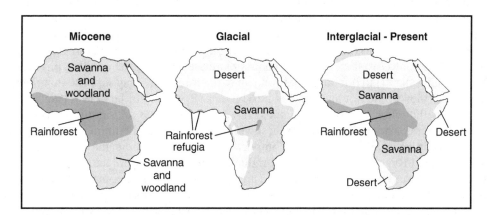

FIGURE 11.1 Modern distribution of African rainforest, possible Pleistocene glacial distribution, and possible distribution during the Miocene (adapted from Foley, 1987).

climate and vegetation distribution. The Milankovitch cycles of insolation variability (see Chapter 7) were becoming increasingly important forces in the climate system during the Late Tertiary. Increasing concentrations of fine terrestrial dust in deep-sea cores off the coast of Africa provide direct evidence of the drying of the continent, loss of forest cover, and increasing variability of climate.

The fragmentation in forests, the decrease in total forest cover, and variable climate and shifting vegetation cover posed a challenge to the arboreal primates and led to many extinctions. Some primate groups, however, were able to adapt to the new environments. Initially, these species may have favored diverse habitats found at forest margin sites where they could take advantage of savanna and grassland resources, yet also feed and seek protection in the trees. *Australopithecus* fossils are found with fossil fauna and flora indicative of diverse habitats of forest, woodland, and savanna. Analysis of tooth wear and the carbon isotopes of tooth enamel of *Australopithecus* fossils suggest a diverse plant diet with forest and savanna indicators. During the Pliocene and Early Pleistocene, African climate was becoming increasingly drier and more variable, and savanna was expanding at the expense of forest. So changes in climate and vegetation may have prompted our ancestors to leave the trees and take advantage of diverse habitats.

On the open ground of woodlands, the savanna, or the forest's edge, bipedalism is a more energy-efficient form of movement over long distances than is quadrupedalism. More ground can be crossed with less energy, and this would be important in moving between widely scattered patches of trees. A bipedal stance allows the individual to pick up and carry food efficiently. It is difficult to transport much material if you rely on both your legs and forelimbs to walk. Being able to transport food off of the savanna to the safety of trees to eat would be a distinct advantage on the predator-filled plains of Pliocene and Pleistocene Africa. The fossil evidence tells us that Australopithecines were good climbers, and we can assume they used trees for protection and for foraging for food. Bipedal movement also makes it easier to transport food and infants from site to site. Chimps do carry objects over short distances, but their infants typically must cling to their mothers when being transported, and falling is a significant factor in chimp infant mortality. Bipedalism places the head and arms of the individual relatively high above the landscape. This allows for better visibility when searching for food and water on the savanna. Being tall presents other significant advantages. It facilitates the foraging of leaves and fruits from higher bushes and small trees. It assists in the spotting of predators and identification of safe places to run to, such as climbable trees.

In the case of the Australopithecines, they were generalists in terms of diet and had teeth, like ours, that allowed for biting, scraping, and chewing a wide variety of foodstuffs. Unlike baboons, which retain very large and dangerous canine teeth, the Australopithecine canines were, like ours, very much reduced. Therefore, unlike the modern baboons, they had little natural defensive weapons against predators on the savanna. What a bipedal individual can do to defend itself, however, is use its free hands for throwing objects and swinging sticks. From the use of objects, rather than teeth, for defense, we can see how the use of tools for other tasks would be a logical development. Once hominids came to depend on and develop tools, evolution would favor dexterity of the hands and increasing mental capability required to devise, construct, and apply tools. Consequently, bipedalism was essential for the further evolutionary cognitive development that resulted in modern humans. Even if forest fragmentation and savanna development did not initiate the movement to bipedalism, movement on two feet allowed human ancestors to move onto and take advantage of more open and diverse habitats such as the savanna. As a parting thought on the movement of human ancestors from the trees and onto two feet, we might contemplate the words of the eminent physical anthropologist William Howells: "Had our ancestors gone on in the forests, we may suppose that nothing would have happened: they would not have turned into bipeds; we would not be here" (1993, p. 73).

Now that we have briefly examined the potential causes and implications of bipedalism, let's return to our consideration of the earliest widely acknowledged hominid, *Australopithecus*. The first person to discover and recognize the importance of Australopithecine fossils was a young medical doctor named Raymond Dart. He was originally from Australia and had studied medicine in Britain. In 1923, Dart went to South Africa to take up a professorship in anatomy at Witwatersrand University. At that time, the mining of limestone from ancient cave deposits was producing a number of interesting mammal and reptile fossils. In 1924, Dart was given a piece of stone with a skull embedded inside. After much laborious picking and cleaning, using among other things a pair of knitting needles, Dart recovered the majority of an immature primate's skull, including a natural limestone cast of the brain.

The skull was apelike but possessed molars that were similar to those of humans. The brain, though small, had a number of features that were more human than ape. Intriguingly, the manner in which the spine attached to the skull suggested a bipedal stance. Dart concluded that the fossil belonged to an intermediary species between apes and humans. In 1925, he published his findings, naming the newly discovered primate *Australopithecus africanus* (southern African ape).

Dart's discovery was not received with much enthusiasm or interest. Indeed, it took Dart and his colleague, a Scottish medical doctor named Robert Broom, many years of hard work excavating South African limestone caves to make their case that *Australopithecus africanus* was bipedal and potentially ancestral to modern humans. By 1950, Dart and Broom had found enough skulls indicative of an upright position and a bipedal pelvis that anthropologists accepted their conclusions. *Australopithecus africanus* was firmly placed with the hominids. The work from South Africa indicated that *Australopithecus africanus* was bipedal and was intermediate in skull and brain features between modern apes and humans. It is now recognized that with the discovery of Australopithecus, Raymond Dart revolutionized thinking about human evolution and our origin. However, exactly how old the Australopithecine fossils were was uncertain.

In 1959, another chapter in the Australopithecine saga opened. In that year, the renowned Kenyan-British paleontologists Louis and Mary Leakey began digging in the Olduvai Gorge region of east Africa. The gorge contains deposits from lake, shoreline, and terrestrial environments that were laid down over the past 2 million years. The area is extremely arid today and bisected by gullies and valleys, providing for many good geologic exposures. In addition, volcanic deposits occur in many parts of the Olduvai rock record, and this allows for the dating of the fossil finds using modern geochronological techniques. It is one of the richest areas in the world for early hominid fossils. For many years, the Leakeys, and more recently their son Richard, worked the deposits at Olduvai and other regions of east Africa discovering not only further Australopithecines but other hominids as well. Since that time, others have found Australopithecines in a number of sites in Tanzania and Kenya and as far north as the Afar region of Ethiopia. In 1974, a 40% complete skeleton of an *Australopithecus* female was found by the American physical anthropologist Donald Johanson and his team in Ethiopia (Fig. 11.2). This remarkable 3.2-million-year-old fossil was named "Lucy," after the Beatles song "Lucy in the Sky with Diamonds," which the anthropologists were listening to during the excavation. Lucy is revered in Africa and some people in Ethiopia refer to her as "Dinkinesh" (you are marvelous) and those from the Afar region where she was found call her "Heelomali" (she is special). Lucy has been the subject of intense scientific research since her discovery. She was bipedal, stood about 104 to 106 centimeters tall, and was between 12 and 18 years in age. Forensic anthropology studies have suggested she may have died due to a fall from a tree, a landslide, or an animal attack. In 1975, Johanson and his colleagues found the remains of a group of 17 Australopithecines that he concluded had died together in some sort of natural disaster such as a flood or slope failure. However, the questions of if the group died at one time and in one place and what may have caused their demise remain debated. The fossils included the remains of nine adults and eight adolescents and children. This group was called "The First Family." Perhaps the most amazing find has been a set of bipedal footprints that were most likely made by two *Australopithecus* walking side by side. The tracks are at least 3.65 million years old and were discovered by Mary Leakey's team in 1976 at Laetoli in Tanzania. The footprints were left by hominids and other animals, crossing a landscape that had recently been covered by fine volcanic ash.

Based on the anatomy and dating of the African finds, we know that Australopithecines first appeared around 4 million years ago and disappeared 1.4 million years ago (Fig. 11.3). The Ethiopian paleontologist Yohannes Haile-Selassie has made key contributions in describing and dating early *Australopithecus* from his native country. Some of these finds date from 3.8 million to 3.4 million years ago. The work of Haile-Selassie has also shone important light on the species diversity of Australopithecines in the Middle Pliocene. How many species of *Australopithecus* there were remains a matter of debate. It is accepted that between 4 million and 2 million years ago a number of distinctive species of *Australopithecus* existed in Africa (including potentially *A. garhi, A. africanus, A. sediba, A. afarensis, A. anamensis, A. bahrelghazali,* and *A. deyiremeda*). Some authorities conclude that one of the earliest species, *Australopithecus afarensis,* persisted until about 3 million years ago when at least two distinctive clades arose. One is represented by *Australopithecus africanus* and descendants (referred to as the Gracile Australopithecines), which were somewhat lightly built and

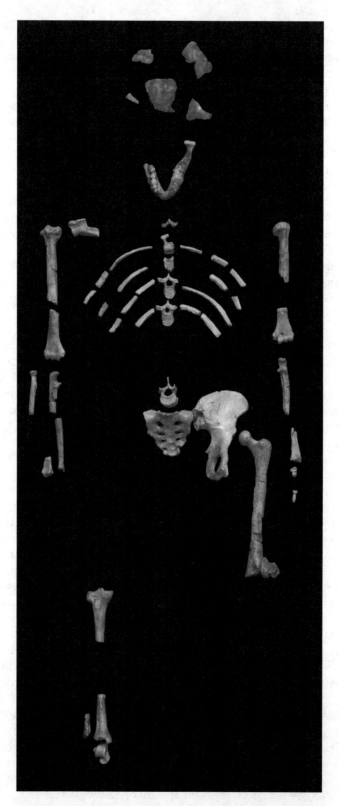

FIGURE 11.2 Lucy—the fossil skeleton of a female *Australopithecus afarensis* from the Afar region of Ethiopia. Lucy is one of the most complete early hominid skeletons ever found. She walked upright and stood about 110 to 120 centimeters (about 4 feet) tall. She was perhaps 12 to 18 years old at the time of death (*Source:* 120 / Wikimedia Commans).

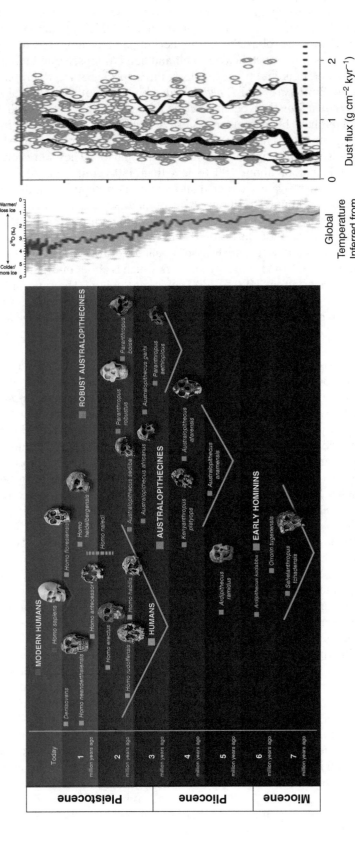

FIGURE 11.3 A history of the hominids. Authorities can differ on number and designation of the species of *Australopithecus, Paranthropus,* and *Homo.* The progressive cooling and increasing variability of the planet's climate and the development of glacial interglacial cycles during the transition from the Miocene to the Pleistocene as evident in cores from the world's oceans is shown. African aridification and forest decline during this time is evidenced by increasing dust deposition in the Atlantic Ocean. Some scientists believe that there is a linkage between global cooling and drying, increasing climatic variability, expansion of the African savanna, and hominid evolution (*Source:* The Trustees of the Natural History Museum / https://www.nhm.ac.uk/discover/the-origin-of-our-species.html / last accessed December 27, 2023; Zachos et al., 2001; Andersson, 2009; Polissar et al., 2019).

had dentition consistent with an omnivore diet that included soft vegetable matter. This group persisted until about 2 million years ago. The other lineage is referred to as the robust Australopithecines. They lived from about 2.5 million years ago until approximately 1 million years ago. Originally included in the genus *Australopithecus,* the robust Australopithecines are now more commonly given their own genus — *Paranthropus.* They were much heavier built and had jaw muscles and teeth suggesting a diet of coarser vegetable matter, nuts, and tubers. At least three species of *Paranthropus* are known from fossil remains *(P. boisei, P. robustus,* and *P. aethiopicus).* Finally, another potential Australopithecine species, *Kenyanthropus platyops,* has been described from a single fossil skull found in Kenya. Australopithecine remains have been found in eastern Africa from South Africa to Ethiopia (Fig. 11.4). It is likely that they were restricted in geographic distribution to the continent of Africa.

What were the Australopithecines like in form and behavior? They were relatively small creatures, with an upright height of about 100 to less than 150 centimeters and a weight of 30 to 40 kilograms. Males were significantly larger than females. For comparison, the average height for mature and healthy modern humans is about 160 centimeters for females and 170 centimeters for males. Average weight for modern humans is around 60 kilograms and also varies by gender. The lower bodies, including their feet, of Australopithecines looked more like those of modern humans than apes. However, their skulls were much more apelike. Australopithecines had relatively small brains, ranging from about 380 to 485 cubic centimeters, which is about one-third the size of a modern human brain, which averages about 1100 to 1300 cubic centimeters. The ratio of brain size to body mass was also much less than that of modern humans. The general shapes of their brains indicate that they might not have had the same full capacity for abstract thinking and language as modern humans. However, brain casts and details from the interior of the cranium show the incipient development of anatomical features associated with both abstract thought and speech. Although difficult to prove conclusively, it has been suggested that Australopithecines lived in social groups. Such groups are typical of chimps and baboons, as well as humans. There is no concrete evidence of the use of fire by Australopithecines. However, recent finds suggest that *Australopithecus* and *Paranthropus* were creating and using primitive stone tools such as hammerstones and sharp stone flakes by 3.3 million to 2.8 million years ago. The primitive tools have been classified as **Oldowan Tradition** tools. The appearance of these tools marks the start of the archaeological period called the **Lower Paleolithic** (Old Stone Age). The Oldowan tool assemblage typically includes large round cobbles that were likely used as hammers, cobbles with sharpened edges that were used as hand axes, and sharp stone flakes that were used as blades. There is some evidence that *Paranthropus* used stone tools in the butchery of hippopotami. As simple as they were, the Oldowan tools would have greatly increased the food resources for hominids, particularly in terms of high-protein meat. The increase in high-protein food decreased the foraging time needed to find enough food to sustain the individual or group. This would have allowed more time for tool making and social interaction and may have contributed to the further evolution of intelligence and communication. Paleoenvironmental reconstructions of habitat, dental wear analysis, and biochemical analysis of Australopithecine fossils indicates that they made use of a wide variety of habitats and had a varied diet that included fruits and nuts typical of the forest, but also grasses typical of the savanna. Australopithecine skeletons suggest that they were better climbers than modern humans and likely used forest trees for food sources and to escape predators.

Are the Australopithecines ancestral to modern humans? This topic has been debated for many years. Some scientists, such as Louis Leakey, have argued that they are not. These scientists believe that *Australopithecus* is related to our genus but had too many apelike features at a relatively late date to serve as an ancestor. In this case, the Australopithecines are just one of many mammal genera that have gone extinct in the past and left no living lineage of descendants. Many other scientists believe that the gracile line of *Australopithecus africanus* is a very likely ancestor to the genus *Homo.* If we assume for a moment that the generalist gracile Australopithecines evolved into the genus *Homo* and thus persisted in this manner, what happened to the robust line? All agree that the robust Australopithecines of the genus *Paranthropus* are unlikely to have been ancestral to any modern primates or humans. Their potential greater dependence on a low-protein diet, perhaps consisting largely of coarse vegetable matter, nuts, and tubers, may have made them unable to persist through increasingly pronounced environmental shifts that typified the later portion of the Pleistocene. In addition, the robust Australopithecines may also have been driven to extinction by competition and predation by our ancestors, the more generalist hominid cousins of the genus *Homo.*

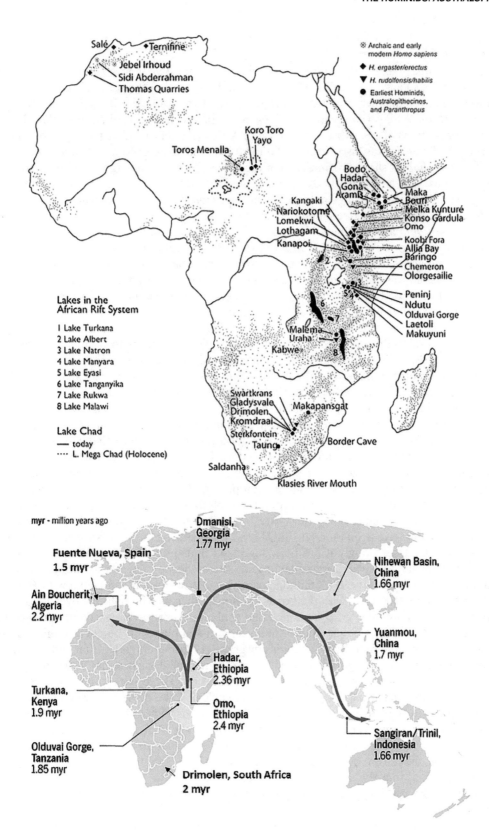

FIGURE 11.4 The distribution of important *Australopithecus, Paranthropus,* and *Homo* fossil and tool sites in Africa (from Schrenk et al., 2015). Early ages for *Homo* in Africa and spread of the genus to Eurasia by *Homo erectus* (modified from Gibbon, 2016).

THE HOMINIDS: EARLY *HOMO*

Some scientists believe that the earliest evidence of our own genus, *Homo,* comes from east Africa and suggests an age of about 2.5 million to 2.4 million years at the start of the Pleistocene. The first of these fossil remains were found by the Leakeys at the Olduvai Gorge in 1960 and were formally named *Homo habilis* (handy man) in 1964. Since that time, fossils of *Homo habilis* have been found in other portions of east Africa and southern Africa (Fig. 11.4). Subsequently, another early hominid, *Homo rudolfensis*, was discovered in the Lake Turkana basin of Kenya. Not all paleontologists agree that *Homo habilis* should be included in the genus *Homo*. Some argue that its relatively small brain size, and its arms and upper body more adapted to climbing than later members of the genus *Homo,* suggest it should be classified as a species of *Australopithecus*. Here we will assume *Homo habilis* and *Homo rudolfensis* are indeed members of our genus.

Homo habilis was a small hominid with a skeleton and size somewhat similar to the gracile Australopithecines. The big difference between the Australopithecines and *Homo habilis* is the skull. In the new genus, the brain increased to over 600 cubic centimeters. To accommodate this larger brain, the skull was higher and the face more vertical in appearance. The cheek bones, however, were still relatively broad compared to those of modern humans. *Homo rudolfensis* had a larger skull than *Homo habilis* with a brain size of about 750 cubic centimeters. In many ways, the anatomical features of *Homo habilis* appear to make it a good candidate for an evolutionary intermediary between the gracile Australopithecines and the genus *Homo*. There appears to be much variability within the *Homo habilis* fossil skulls, and some scientists have suggested this might represent more than one species. It is notable that during the period 2.5 million to 1.5 million years ago, other hominid species, such as *Homo rudolfensis*, *Australopithecus sediba*, and *Paranthropus robustus*, were also living in Africa. Evidence from the Drimolen cave site in South Africa suggests that *Australopithecus*, *Paranthropus*, and *Homo* were all living in the same region about 2 million years ago.

Aside from the larger brain, there is good physical evidence that *Homo habilis* represents a major step toward the development of modern human intelligence. Associated with some of the *Homo habilis* sites is evidence for the manufacture and use of specialized tools of the Oldowan Tradition. Such tools found at Hadar in Ethiopia are about 2.5 million years old, and many sites have revealed tools that are 2 million years old. The tools that we know of are made of stone. It is likely that *Homo habilis* also made tools from wood, bones, and sinew, but these have not been preserved. The stone tools that have been found do not change much during the 1-million-year period of *Homo habilis*'s existence. Interestingly, it has been suggested that some features of a few Oldowan-aged stone objects from the Olduvai Gorge may provide the evidence of symbolic thought and decorative art among hominids as early as 1.8 million years ago. Oldowan tools associated with *Homo habilis* have been found in association with the bones of animals such as wild pigs and even elephants. The bones of these mammals show clear evidence that tools were used to break them open for marrow or to scrape meat off bones. Some sites contain large assemblages of butchered animals and discarded tools, suggesting the butchering activities by groups of hominids rather than individuals. Group activity in hunting, scavenging, butchering, and protecting meat caches may have promoted the further development of social interaction and communication. There is no doubt that *Homo habilis* used meat as an important part of its diet. What is unclear is the degree to which meat was obtained by hunting versus the scavenging of kills made by predators such as lions. Chimps in the wild have been known to hunt other mammals, such as monkeys, so it is likely that *Homo habilis* both hunted and scavenged meat. Without tools, humans are poorly equipped in terms of teeth or claws to hunt, skin, and remove meat from other animals. There is no evidence that *Homo habilis* employed a controlled use of fire, and thus the meat they consumed was eaten raw. Although they likely lived and worked together in groups, it appears unlikely that *Homo habilis* had the physiological or mental capacity for modern speech.

For the next step in our study of human evolution, we will leave Africa and travel to Trinil, Java, in the year 1891. In that year, Eugène Dubois, a young medical officer in the Dutch Army, made a startling discovery. Dubois had studied anatomy in Holland and had become intrigued with Darwin's ideas on the evolution of humans. He was also familiar with the works of Alfred Russel Wallace, who as early as 1855 had suggested that humans and apes might share a common ancestor, and the fossils of that ancestor might be found in Southeast Asia. Dubois concluded that Southeast Asia was indeed a likely spot to find fossil evidence of early human ancestry. He was so convinced of his conclusions that he joined the Dutch Army in order to get stationed in the colony of the Dutch East Indies, which today forms modern Indonesia. At first, Dubois had no official support for his paleontological endeavors, but after publishing several speculative articles, he was able to secure full authorization, along with two engineers and a team of convict laborers.

For two years, Dubois searched the island of Sumatra in vain for early hominid fossils. He then moved his expedition to Java, where he discovered a skull cap and a femur that appeared to share features of both humans and apes. In 1894, he published his account and suggested that this was indeed an early hominid. Dubois called the fossil species *Pithecanthropus erectus* (upright apeman). The fossils that Dubois recovered were popularly referred to as "Java Man." Both scientists and the public greeted Dubois' conclusions with considerable skepticism. We should remember that his findings were announced about 30 years before Dart's early hominid discoveries in South Africa and almost 70 years before the Leakeys' work in east Africa. Dubois was shattered by the poor reception of his discovery, and for a long period of time he simply kept the fossils hidden beneath the floorboards of his home in Holland. We now know that Dubois was correct; his fossils were indeed from an early hominid that we classify today as *Homo erectus*. We also know that *Homo erectus* is younger than *Homo habilis* and older than *Homo sapiens*.

As the twentieth century progressed, more *Homo erectus* fossils were found in Java and in China. One of the most famous of these was from Zhoukoudian near Beijing. The Zhoukoudian fossils, which came to be known as "Peking Man" (Peking is an older spelling for the capital city of China, Beijing), were unfortunately lost during World War II. During the last 50 years, fossils of *Homo erectus* have also been found in several parts of Africa and Eurasia (Fig. 11.4). There is much temporal and geographic variability in details of the skulls and other bones of *Homo erectus*. A number of *Homo erectus* bones, including several skulls, have been described from approximately 1.8-million-year-old deposits at Dmanisi, Georgia. They indicate that these hominids were of much smaller stature and smaller in brain size than later *Homo erectus*. As with other early hominids, there is debate on the taxonomy of these finds, and some paleontologists recognize different species or subspecies of *Homo* during this early period. Some authorities classify the earliest *Homo erectus*–type fossils from Africa as a separate species, *Homo ergaster*. We will generally stick with a conservative interpretation, and all of these finds will be referred to as *Homo erectus*. The age of *Homo erectus* fossils spans the period of about 2 million to less than 100,000 years ago. The appearance of *Homo erectus* coincides with increasingly cold, dry, and variable climatic conditions associated with the Pleistocene and its glacial-interglacial cycles. This has led some paleontologists to suggest a link between climatic change and the evolution of *Homo erectus*. They suggest that in many cases *Homo erectus* may have been one of a suite of predators hunting large mammals in relatively open landscapes. The dates from the different localities show that the oldest *Homo erectus* fossils come from Africa and then the country of Georgia. The earliest finds from eastern Asia date slightly later, around 1.7 million to 1.66 million years ago. *Homo erectus* appears to have also dispersed into southern Europe about 1.35 million years ago. The youngest fossils of *Homo erectus* come from southeastern Asia where finds in Indonesia have been dated to 100,000 and perhaps as young as 50,000 years ago. This suggests that *Homo erectus* persisted in Indonesia for over 1.6 million years. In any case, with *Homo erectus*, we have the first concrete evidence of hominids leaving Africa and spreading through Asia and Europe. This was an incredible step and sets the geographic stage for many important events in the evolution of humans.

The later *Homo erectus* have been typified as being relatively tall and strong individuals, some possibly reaching heights of over 180 centimeters. However, the fossil remains display variability and many *Homo erectus* were likely shorter than modern humans. One of the most complete skeletons is that of an adolescent boy found by Richard Leakey and Alan Walker at Nariokotome in Kenya. It seems that the boy had died in shallow water and his body had been quickly covered by sediment; this preserved his skeleton from scavengers. According to some estimates, had he reached adulthood, the so-called Nariokotome Boy could have stood close to 180 centimeters tall and weighed about 70 kilograms. Aside from the skulls, the skeletons of *Homo erectus* are very similar to those of modern humans and appear well adapted to long and relatively fast movement across open country. There is much less difference in the sizes of male and female *Homo erectus* than is the case with earlier hominids. Their skulls were slightly wider than those of modern humans, and they had a very prominent bony ridge above their eyes. They also had strongly sloping foreheads and receding chins. A relatively wide nasal opening suggests a somewhat modern nose shape and nasal passages. These features have been interpreted as an adaptation to conserve body moisture during breathing. Both the stature and nasal configuration of *Homo erectus* have been interpreted as evolutionary adaptations to life on the hot, dry, open savanna. The brain capacity of later *Homo erectus* ranged from 900 to 1200 cubic centimeters and was therefore larger than that of earlier hominids but slightly smaller than modern humans. The brain shape and throat of *Homo erectus* may have allowed something similar to modern speech, but this remains controversial among physical anthropologists.

The geographic expansion of *Homo erectus* out of Africa and into Eurasia is an extraordinary chapter in human history. Hominids now ranged from the Atlantic Coast in Europe to the shores of the Indian Ocean in Africa and to the Pacific Coast in southeastern Asia. No other Quaternary primate and few other terrestrial mammals aside from hominids have had such a huge range. It is likely that the evolution of *Homo erectus* and its geographic expansion were generated in part by environmental change. As the Pleistocene proceeded, conditions in Africa and adjacent portions of Europe and Asia became generally drier, while the variations in climate associated with Milankovitch cycles became more pronounced. Increasing areas of savanna and open landscapes may have promoted the adaptations to dry conditions and mobility seen in *Homo erectus*. In addition, the movement of *Homo erectus* and eventual expansion beyond Africa may have been promoted by the adjustments in foraging areas in response to changes in climate and vegetation. It is reasonable to think that *Homo erectus* populations shifted their ranges as the distributions of vegetation and animal resources shifted in response to climatic change.

The geographic expansion of *Homo erectus* also appears to have been aided by increasing cultural sophistication. The Lower Paleolithic stone tools most often associated with *Homo erectus* are called the **Acheulean Tradition** and include much more refined and varied choppers and scrapers, as well as finely developed hand axes and cleavers. Tool-making materials were sometimes imported from appreciable distances. Importantly in terms of hominid cognitive development, there is evidence of art and symbolism in the form of collections of red ochre, decorative carved patterns, and even small human figurines associated with Acheulean finds. The Acheulean Tradition extends in time from about 1.7 million years ago to around 130,000 years ago. It should be noted that the stone tools found at the 1.8-million-year-old *Homo erectus* site at Dmanisi in Georgia seem to be of the more primitive Oldowan assemblage. It appears that the Acheulean Tradition technologies arose in Africa and then eventually spread to Eurasia, but did not accompany the first wave of humans that spread from Africa. How the expanding geographic ranges and habitats used by *Homo erectus* may have influenced the development of tool making can only be speculated upon. Similar levels of workmanship in stone, but not exactly similar Acheulean tool kits, are found from Africa to China. Asian artefacts of this age are sometimes considered a different tradition. It has been shown that the Acheulean technology was used in processing meat. The Acheulean tool kit and analysis of the dentition of *Homo erectus* suggest an increased proportion of meat in diet compared to earlier hominids.

Did *Homo erectus* employ the controlled use of fire? Evidence of fire and even hearths have been discovered at some *Homo erectus* sites. An early association between evidence of fire and the presence of *Homo erectus* about 1 million years ago is found at Wonderwerk Cave in South Africa. The archaeological record suggests that fire was being collected and maintained by *Homo erectus* by 400,000 years ago. There is strong evidence from features such as hearths that fire was widely used for purposes such as warmth, cooking, and processing materials by at least 200,000 years ago. In addition to direct evidence, it has been argued that populations of *Homo erectus* could not have survived in relatively cool regions of Asia and Europe if they did not possess fire. Fire not only provided warmth and protection against predators, but it also greatly enhanced food resources. Cooking can release additional nutrients and proteins in meats and vegetables. It can make food easier to process and eat. Cooking can also kill bacterial contamination in meats and detoxify certain plants that would be poisonous if eaten uncooked. The adaptation of cooked foods and the increase in available nutrition attributable to this may have been one of the most important developments in human evolution. The availability of cooked foods, coupled with a higher proportion of protein from meats, would impact many aspects of hominid evolution, ranging from jaw and tooth structure to digestive physiology to cognitive ability to social structure. The role of changing diets on human evolution is an area of ongoing and intense research.

In addition to stone tools and fire, it is likely that *Homo erectus* fashioned tools from wood, bamboo, and bone, and they were probably able to cross water on rafts. Populations in cool areas may have used plant matter and animal skins for clothing and to construct simple shelters. Paleontological and pathological evidence suggests that *Homo erectus* lived, hunted, and/or scavenged in groups. Remarkably, *Homo erectus* footprints have been preserved in the shoreline sediments of an ancient lake in Kenya and provide evidence that groups of males traveled together beside the lake 1.5 million years ago. This suggests not only group activity, but potentially a sex-based social structure related to tasks such as food gathering. As geographically expansive and technologically advanced as they were compared with earlier hominids, we have yet to find evidence for complex ritualistic behavior such as formal burial practices for the dead.

THE HOMINIDS: *HOMO SAPIENS* AND RECENT COUSINS

All humans alive today are members of the species *Homo sapiens*. A subspecies name is sometimes included in the form of *Homo sapiens sapiens* to differentiate anatomically modern humans (sometimes abbreviated as AMH) from older fossils of *Homo sapiens*. These older fossils possess features such as a more prominent brow line and a thicker, less globular skull than is found in present-day humans. These early forms are generally referred to as archaic *Homo sapiens* when they are classified as *Homo sapiens*. In many cases, however, what formerly might have been classified as an archaic *Homo sapiens* is now designated as another species of *Homo*. For example, *Homo neanderthalensis* (Neanderthals) has in the past been referred to as a subspecies—*Homo sapiens neanderthalensis*—rather than recognized as its own species. We will follow the convention of classifying anatomically modern humans as *Homo sapiens* and archaic forms as distinct species, such as *Homo neanderthalensis*.

It is likely that *Homo sapiens* evolved from descendants of *Homo erectus*, but the history of this remains uncertain. The species *Homo heidelbergensis* and *Homo rhodesiensis* have been suggested as the most recent links between the earlier *Homo erectus* and later *Homo sapiens*. Some fossils formerly classified as Archaic *Homo sapiens* fossils are now often grouped into *Homo heidelbergensis*. Other potential intermediate species include *Homo ergaster* and *Homo antecessor*. However, all of these hominid relations are murky over the Middle Pleistocene period from about 1 million to 300,000 years ago. The name *Homo heidelbergensis* comes from a fossil jaw found in 1907 at Mauer, near Heidelberg in Germany. The deposits there are about 600,000 years old. The specimen for *Homo rhodesiensis* was found in cave deposits from Zambia in 1921. It is estimated that these deposits are about 300,000 years old. It has been argued that *Homo rhodesiensis* is simply a variety of *Homo heidelbergensis*. Alternatively, it has been suggested that both the taxon names *Homo heidelbergensis* and *Homo rhodesiensis* should be abandoned and the African and eastern Mediterranean fossils be assigned to a new species called *Homo bodoensis*. It is further argued that *Homo bodoensis* is the ancestral species to *Homo sapiens* and *Homo heidelbergensis* from western and northern Europe represents the ancestor of *Homo neanderthalensis*. This proposal has been contested however. What is certain is that there will continue to be lively debates about the identity and evolution of our ancestral species during the Middle Pleistocene.

Let's turn our attention back to modern humans of the species *Homo sapiens*. In 1967, Richard Leakey found the skull of an early *Homo sapiens* near the Omo River in Kenya. This skull dates to about 130,000 years ago and at that time provided the earliest evidence of essentially modern humans. As more evidence has accrued, some of the earliest dates for *Homo sapiens* along with some of their stone tools now come from Morocco. These dates have a range of about 350,000 to 300,000 years. *Homo sapiens* fossils of over 230,000 years in age have been reported from deposits in eastern Africa. What sets *Homo sapiens* apart from earlier species of *Homo*? Anatomically modern humans/*Homo sapiens* have a larger and rounded braincase, lack of a brow-ridge, a short jaw, a prominent chin, and a narrow pelvis compared to other species in the genus *Homo*. With *Homo sapiens* we also find increasingly sophisticated tool assemblages.

Formerly there was much debate regarding the geographic location where *Homo sapiens* first evolved. The differing theories can be summarized as follows. There is the **Out of Africa Model,** whereby *Homo sapiens* evolved first in Africa and then spread through Europe and Asia. They either displaced and caused the extinction of most other species of *Homo* or absorbed them through interbreeding. *Homo sapiens,* with their larger cranial capacity and demonstrably superior technical capabilities, was likely a superior competitor than earlier species of *Homo.* It has also been proposed that *Homo sapiens* may have evolved from an earlier species of *Homo* in southern Asia and then spread to Australia, northern Asia, Europe, and Africa. This is called the **Out of Asia Model.** Finally, some researchers suggest that there may have been two separate Asian origins for the evolution of *Homo sapiens.* Modern northern Asians may have originated from populations of *Homo erectus* in China, and southern Asians and Australians may have originated from populations living in Java and adjacent islands. The scenario in which *Homo sapiens* evolved separately at two or more geographic locations is called the **Multiregional Model.** In this case, it is possible that parallel evolution by hominids would have produced similar morphological and physiological traits in geographically separated populations. These hominids were still closely related enough to interbreed when they did come into contact. Today, it is widely agreed on the basis of fossils, archaeological evidence, and genetics that modern humans of the species *Homo sapiens* first evolved in Africa and subsequently dispersed to Eurasia and the rest of the world.

When modern humans arose in the form of the species *Homo sapiens,* they were not the only hominid species present on earth. We know that between 300,000 years ago and about 30,000 years ago several other hominid species have been identified that appear to overlap in age with *Homo sapiens.* These include *Homo erectus, Homo heidelbergensis/Homo rhodesiensis/Homo bodoensis, Homo neanderthalensis, Homo floresiensis, Homo luzonensis,* and the Denisovans. The Denisovans were discovered during excavations of Denisova Cave in the Altai Mountains of Siberia. They are only known from fossil fragments and DNA recovered from that site and a few sites in Asia. They lack a formal species designation at this time.

The history of the Neanderthals is an extremely interesting story. The first Neanderthal fossils that were recognized as representing extinct hominids were found in 1856 in the Neander Valley of Germany. The quarrying of cave deposits for limestone unearthed a rather unusual hominid skull cap. The skull cap was quite thick, had a pronounced sloping of the forehead, and displayed a very thick, prominent brow. The nineteenth-century scientists did not know what to make of this odd skull cap. It was suggested at first that it might have been the remains of a lost Cossack soldier from the Napoleonic Wars. Alternatively, it was also proposed that the skull cap might have belonged to an infirm Dutchman. An English anthropologist, William King, suggested that the skull might be from an extinct human and classified the fossil as a new species, *Homo neanderthalensis.* Over the next 50 years, Neanderthal remains continued to surface in European excavations or were recognized in existing museum collections. By the early twentieth century, it was widely accepted that Neanderthals were indeed an extinct hominid closely related to modern humans. We now know that most Neanderthal fossils date from approximately 200,000 years ago to about 30,000 years ago. The numerous Neanderthal fossil sites provide a fairly clear picture of the geographic distribution of these hominids. They appear to have lived in Europe into central Asia and the Near East. It now seems quite clear that Neanderthals did not live in far eastern or southeastern Asia, nor did they live in sub-Saharan Africa. It is probable that they evolved from *Homo heidelbergensis* in Europe or western Asia.

Because of their sloping skull, prominent brow, recessed chin, and heavy build, it was supposed that Neanderthals represented a primitive ancestor to modern humans. However, both the conclusions that Neanderthals were evolutionarily primitive or were a primitive ancestral ancestor to modern humans are wrong. The stocky build, thick bones, large nose, and heavy jaws and dentition of the Neanderthals were genotypic features and can be seen in the skeletons of Neanderthal children. It is possible that the Neanderthals evolved some features as adaptations to life as hunters in cold environments of Europe and central Asia. The stocky build would have maintained body warmth during the cold winters. The large muscles and thick bones would have been a benefit to hunters who went after game such as deer, mammoth, and woolly rhinoceros. The maximum population size and widest geographic extent of Neanderthals appear to have occurred during the period 100,000 to 40,000 years ago, which coincides with generally cool, but not glacial, conditions (see Chapter 7). Neanderthals have been referred to by some as a highly cold-adapted hominid in terms of physical traits. However, the potential cold adaptations of Neanderthals remains a topic of lively debate, and it is notable that they did not persist into the time of the last glacial maximum about 20,000 years ago.

The Neanderthal tools from this period are classified as **Middle Paleolithic** and are typified by the **Mousterian Tradition,** which may have evolved directly from the older Acheulean Tradition. The

Mousterian tools were more finely made and more diverse, including borers, simple saws, and points that were attached to wooden poles to form spears. Neanderthal spears have been found embedded in the fossil remains of large mammals and provide evidence of the hunting skills of the Neanderthals. Neanderthals used fire and certainly fashioned clothing and shelter from animal hides. They lived in groups and cared for infirm members. They also had ritualistic practices that are evident in formal burials that have been found in some cave sites. This includes the placing of flowers in the grave or the arrangement of animal skulls around the grave. Carved bone and even abstract cave art have been found at some Neanderthal sites and provide further evidence of symbolic abstract thought. Many anthropologists believe that Neanderthals had the capacity for speech, but perhaps not the full range used by modern humans. Impressively, the world's first known musical instrument is a small flute made from bone and found at a 60,000- to 50,000-year-old Neanderthal site in Slovenia. Rather than being evolutionarily primitive, the Neanderthals were intelligent and possessed specialized adaptations to live by hunting and gathering in a cool climate region.

What was the relationship of the Neanderthals to us? Fully modern ancestral humans of the species *Homo sapiens* likely evolved in Africa around the time that Neanderthals were present in Europe and the Near East. Thus, modern humans evolved separately from Neanderthals and not as their evolutionary descendants. Molecular clock analysis estimates are variable and suggest that *Homo neananderthalensis* and *Homo sapiens* shared a common ancestor and diverged sometime between around 800,000 years ago and 300,000 years ago. Neanderthals and modern humans appear to share over 98% of their DNA. Although we share common ancestry, our similarity in behavior and physical appearance to Neanderthals has long been a matter of conjecture. Early anthropological reconstructions portrayed Neanderthals as stooped, brutish creatures with hanging arms and apelike faces. Later reconstructions led to the suggestion that if we were to meet a Neanderthal on a crowded subway we would not notice much difference between him and other passengers. The eminent physiologist, ecologist, and biogeographer Jared Diamond argued differently. He points out that although Neanderthals were no slouching brutes, the huge barrel chest, massive arms, giant bulbous nose, deeply receding chin, prominent brow ridge, and sloping forehead would appear jarring to modern humans riding on a subway, no matter how crowded.

What became of the Neanderthals has long been a topic of debate. A short answer is that as a recognizable fossil hominid, they went extinct 40,000 to 30,000 years ago, replaced in their range in Eurasia by *Homo sapiens*. Why might this have happened? Modern humans have been referred to as possessing a generalist/specialist niche, whereby our physical traits, coupled with increased cognitive abilities and associated enhanced capacity for language, social interactions, and technological developments, allowed *Homo sapiens* to adapt to and occupy a more diverse range of environments than other hominids. Our skulls, dentition, and physique suggest a more generalist set of adaptations to a warmer climatic environment than the Neanderthals. The generalist species *Homo sapiens* may have had a much more energy-efficient build for all but the coldest conditions. Modern humans were potentially better able to travel long distances for foraging or to escape inhospitable climatic conditions than Neanderthals. In addition, it is possible that *Homo sapiens* used a wider variety of food sources than Neanderthals. Finally, the speed of technical innovation shown by *Homo sapiens* in tool development and tool diversity indicates a more rapid ability to shift and develop tools and survival techniques to cope with changing environments.

Did modern humans actually drive Neanderthals extinct? There is evidence that Neanderthals and humans lived in the same geographic regions during the end of the Neanderthal period. The rapid decline in Neanderthals between about 40,000 and 30,000 years ago coincides with the expansion of anatomically modern *Homo sapiens* across Europe and adjacent portions of Asia. There is no clear evidence that either group fought with or hunted the other. At first, we might think the idea of hominids hunting other hominids preposterous, but we know that both Neanderthals and modern humans practiced cannibalism, so it is not outrageous to think that the two species might have hunted one another also. If they did not generally interact directly, perhaps *Homo sapiens* were simply better hunters and foragers, and it has been suggested that resource competition by modern humans led to the decline of the Neanderthals. It is possible that Neanderthals could not adopt a more generalist diet or evolve new tools and strategies to cope with dwindling game. Finally, it has been proposed the Neanderthals were less capable of adapting to the climatic changes of this time than were modern humans.

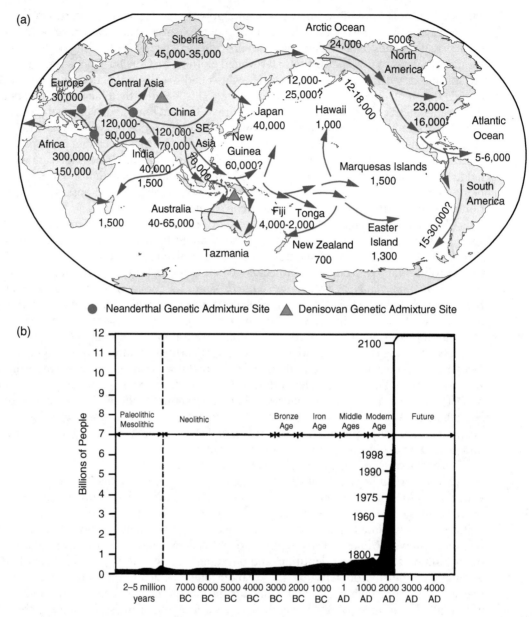

FIGURE 11.5 (*a*) The general geographic spread of modern humans (*Homo sapiens*) and sites of with evidence of overlap/genetic admixture with Neanderthals and Denisovans. (based on information from Achilli et al., 2013; Armitage et al., 2011; Becerra-Valdivia and Highman, 2020; Bennett et al., 2021; Bourgeon et al., 2017; Bunbury et al., 2021; Dominguez and Netea, 2019; Fleontov et al., 2019; Gamble, 1984; Highman et al., 2011; Kinaston et al., 2013; Laghlin and Harper, 1988; Mosquera et al., 2013; Oppenheimer, 2012; Pitulko et al., 2016; Prates, 2020; Sheppard, 2011; Veeramah and Hammer, 2014; Valentin et al., 2016; Walter et al; 2017) (*b*) Hominid population growth over the past 5 million years (McFalls, 1991).

Although the exact causes, or combination of causes, may never be known, by about 30,000 years ago the Neanderthals disappear from the fossil record. By around the same time, *Homo floresiensis* and *Homo luzonensis* from southeastern Asia and the Denisovans in central Asia were also extinct. However, much less is known about these more recently discovered hominid cousins and their histories than we know about the Neanderthals. Regardless, by sometime between 30,000 and 14,000 years ago, our species, *Homo sapiens*, was alone as the last hominid standing. This is, however, not the end of the Neanderthal and Denisovan stories and their legacy. Anatomically, modern humans existed in the same areas as Neanderthals and Denisovans in Europe and portions of Asia between 50,000 and 30,000 years ago (Fig. 11.5). DNA analysis of Neanderthals, Denisovans, and modern humans indicates that all three species interbred and had offspring during this period. Some of those children lived to pass their genes down to us. Today,

Neanderthal DNA makes up about 2% of the genome of non-African human populations. Denisovan DNA is found in modern Asian and Indigenous American populations, reaching 5% in Melanesians. Many of us alive today can count Neanderthal and/or Denisovan ancestors as part of our own family tree.

THE GEOGRAPHIC EXPANSION OF MODERN HUMANS

The geographic expansion and population growth of *Homo sapiens* is one of the most dramatic and significant events in the recent biogeographic history of the earth (Fig. 11.5). At 200,000 years ago, it is possible that *Homo sapiens* were present only in Africa. By 120,000 to 70,000 years ago, they were also present across portions of southern Asia. The early *Homo sapiens* were adapted to life in relatively warm conditions on the open countryside and utilized a variety of Middle Paleolithic tools. They created art, practiced ritual burials, and displayed other indications of sophisticated culture. There is evidence that anatomically modern humans colonized Europe and northern Siberia by 45,000 to 35,000 years ago. The late spread of humans to northern regions of Europe and Asia may indicate that a relatively high degree of technical and cultural development was required before people could settle such very cold regions. In Europe, the earliest *Homo sapiens* were widespread by about 40,000 years ago. The earliest anatomically modern humans in Europe were formerly called Cro-Magnons after a cave near Les Eyzies, France, where some of their remains were found. These particular modern humans were often tall and slender people with large craniums. The average size of the early modern humans in Europe may have been slightly larger than that of most subsequent humans until the twentieth century. This is because they apparently obtained a highly nutritious diet from hunting and gathering compared to later agrarian peoples.

The preserved stone, bone, and wooden artifacts of the early European and western Asian *Homo sapiens* display much greater variety and greater finesse than the Mousterian Tradition artifacts of the Neanderthals. We classify these early modern human artifacts and the culture they reflect as **Upper Paleolithic.** Some important new tools included chisels and wooden spear throwers that increased the distance a projectile could be launched. We also see the clear development of representational art. Wood, bone, and stone carvings of animals were made, as were Venus figurines, which are carvings and baked clay figures of women with greatly enhanced hips and breasts. Venus figures are found from western Europe to the Near East to Siberia and indicate the sharing of artistic and symbolic ideas over very great distances even in this early period. We also find evidence of highly abstract thought in the form of representational art associated with *Homo sapiens*. Cave paintings of pigs have been described from the Leang Tedongnge Cave on the island of Sulawesi in Indonesia. These artworks date from about 45,000 to 32,000 years ago. The cave paintings, attributed to anatomically modern humans, are considered the earliest well-dated representational cave art in the world. Figurative cave paintings from Namibia depict animals such as zebra and rhinoceros and date to 30,000 years ago. Similarly, cave paintings from locations like Chauvet Cave and nearby locales in southeastern France depict animals such as horses, aurochs, and mammoths. The age of the artwork is debated, but possibly lies between 37,000 and 28,000 years ago. The walls of at least 200 caves in France and Spain contain beautiful, somewhat abstracted paintings of animals. Early paintings and drawings also appear on rocks at a number of sites in Australia. Across Europe, Asia, Australia, and Africa, we see greater geographic variability in tools and other artifacts than was the case with Neanderthals and earlier species of hominids. People were devising new tools, art forms, and other innovations to suit the resources and requirements of their environments.

At the time that *Homo sapiens* were expanding throughout Europe and continental Asia, they were also colonizing Australia. Although *Homo erectus* may have been able to cross small areas of water, the larger water passages between southeastern Asia and New Guinea–Australia kept these earlier hominids from reaching Australia. There is firm evidence of human presence in Australia about 40,000 years ago, and some evidence suggests human presence as early as 65,000 years ago.

The timing and arrival routes of the first humans in North and South America remain highly controversial. The drop in sea level that occurred during the last glacial maximum lowered the Bering Sea enough to produce a landbridge between Siberia and Alaska. There is evidence that some humans crossed the Bering landbridge from Siberia and entered the ice-free portions of Alaska and the Yukon Territory of Canada by 24,000 years ago. During this period, the region of Beringia supported a number of large mammals, including mammoth and bison, and would have provided good hunting grounds for humans. Genetic evidence from modern human populations suggests a linkage between North American and South American native peoples and people living in Siberia. As the Cordilleran and Laurentide ice sheets melted between 15,000 and 13,000 years ago, a corridor of tundra, grassland, and shrubby habitat opened between Alaska and the Great Plains. This corridor would have served as a route for the rapid southward migration of humans and animals into the heart of North America. There is also some evidence that prehistoric people arriving in North and South America may have used several different migration routes. There is evidence that people may have moved south from Beringia along the Pacific coastline of North America. In this case, human migration into central North America would not have had to wait until the continental ice sheets had retreated. By about 16,000 to 13,000 years ago, people were present throughout much of North and South America. The initial peopling of the New World via Siberia is well supported by evidence from anatomical and genetic analysis that shows an affinity between native peoples of North and South America and people in northeastern Asia. However, geneticists debate if the evidence from the analysis of the DNA of modern populations of American Indigenous peoples supports early or late migration hypotheses or multiple waves of migration.

Prehistoric Asian migration to North America continued into more recent time. The ancestors of the modern Navajo and Apache of the Southwest and the Na-Dene of northwestern Canada likely migrated from northeastern Asia in the last 7000 to 4000 years. Their languages are related to one another, and all show linkages to northern Asian languages. Genetics suggests an affinity with modern eastern Siberian populations. The Aleuts and Inuits of northern Alaska, northern Canada, and Greenland are genetically related to the Chukchis and other native peoples in northern Siberia. They also share genetic affinities to the Na-Dene. The first humans in the High Arctic of the New World, including the ancestors of the modern Inuit, likely migrated from Asia and expanded across the Arctic in two waves between about 5000 and 1000 years ago.

There are other portions of the inhabited world that were settled relatively recently by humans. These are islands that required relatively long voyages to reach them. Jamaica and the Lesser Antilles of the Caribbean region were settled about 8000 to 5000 years ago. Madagascar, off the coast of Africa, was settled about 1500 years ago. Some islands, including several in the Indian Ocean and the Commander Islands off the coast of Kamtchatka, Russia, appear to have first been reached by humans only in the last few centuries.

Perhaps the most impressive story of human expansion to remote islands involves the peopling of Hawaii and the other islands of the Pacific. The Pacific Islands appear to have been settled by waves of migrants originating in southeastern Asia. The first humans arrived on the western Melanesian Islands near New Guinea about 45,000 years ago. These early peoples were able to cross large expanses of water but were not able to spread into the remote islands of the eastern Pacific. One of the most important later groups to expand in Melanesia were the people of the Lapita Culture who spread into Melanesia at around 3500 years ago and reached islands as far eastward as Fiji. The Lapita were a relatively sophisticated people who in addition to being excellent sailors, practiced agriculture and produced ceramic pottery. The Lapita peoples may have included both Polynesians and Melanesians. Subsequent human migrants did not spread beyond the area of Fiji until the Polynesian expansion phase occurring between about 2100 and 700 years ago. Genetic studies indicate that the Polynesians are most closely related to peoples living on the islands of Southeast Asia, particularly the Philippines. The Polynesians were clearly exceptional sailors who colonized islands as far north as Hawaii and eventually as far east as Easter Island. To the south, Polynesians most likely reached New Zealand about 800 to 700 years ago. In addition to their initial colonization voyages, Polynesian peoples in Hawaii maintained marine trade linkages with Tahiti for a number of years following their arrival on Hawaii.

The geographic spread of modern humans has been followed by growth in population size. Over most of the time during which the genus *Homo* has existed, the population level of hominids has remained relatively low. Until about 10,000 years ago, the entire hominid population of the planet at any one time was likely a few million individuals. During the more recent period, human population growth was promoted in part by the colonization of new parts of the world by people. Some anthropologists have suggested that from the appearance of *Homo sapiens* onward there has been a positive feedback linkage between human population size and cultural and technical innovation. The hunting and gathering practices of early *Homo sapiens* appear to have required mobility and large ranges for movement. There is some evidence that as *Homo sapiens* populations grew in the **Mesolithic** (Middle

Stone Age) and subsequent **Neolithic** (New Stone Age) (see Chapter 12), the movement of people and the range available for foraging became more restricted due to competition with other bands of humans. This may have spurred on changes in social organization and technologies that allowed larger numbers of people to be supported in smaller areas. These successes in turn created greater population pressure, fueling new innovations in social organization and technology.

During the period 10,000 years ago to about 200 years ago, there was a more marked and steady increase in the world population of humans (Fig. 11.5). Only a small portion of this growth can be attributed to the colonization of uninhabited regions. The increasing size of population was largely promoted by cultural innovations such as the development of agriculture and the development of pottery and then tools made of metals such as bronze and iron. These innovations increased the carrying capacity of the land for humans. The paleontologist, Niles Eldredge, has suggested that the development of agriculture for humans can be likened to a major increase in the niche of a species. The initial development of agriculture and the associated evolution of human society has been referred to as the **Agricultural Revolution,** or the *Neolithic Revolution* because it represented a break from the earlier hunting and gathering economies of the Mesolithic. More people could be supported in a given area, and this was associated with increasing social and technological sophistication.

In the last 200 years, the size of the human population has grown remarkably, from about half a billion to 8 billion people. The recent explosive growth in human population can be attributed to increasing rates of industrial, technical, medical, and scientific development and the rapid transfer of such new developments throughout the world. These developments have served to increase agricultural output and decrease human mortality rates in many parts of the world. The rapid development of new technologies and the shift of the socioeconomic system away from an agrarian base toward a techno-industrial base that has occurred since about 1800 is called the **Industrial Revolution.** In addition to leading to increased human population size, the Industrial Revolution has seen a shift of the world's human population to the cities from rural areas. The remarkable geographic expansion and population growth of *Homo sapiens* have had a large impact on the environment and other species with which we share the earth. In the next chapter we will examine the development and spread of agriculture and the role of our species as an agent of both evolution and extinction.

KEY WORDS AND TERMS

Acheulean Tradition
Agricultural Revolution
Bipedal
Hominid
Industrial Revolution
Lower Paleolithic

Mesolithic
Middle Paleolithic
Mousterian Tradition
Multiregional Model
Neolithic
Oldowan Tradition

Out of Africa Model
Out of Asia Model
Primate
Upper Paleolithic

REFERENCES AND FURTHER READING

Abbate, E., Albianelli, A., Azzaroli, A., Benvenuti, M., Tesfamariam, B., Bruni, P., Cipriani, N., Clarke, R. J., Ficcarelli, G., Macchiarelli, R., Napoleone, G., Papini, M., Rook, L, Sagri, M., Tecle, R. M., Torre, D., and Villa, I. (1998). A one-million-year-old *Homo* cranium from the Danakil (Afar) depression of Eritrea. *Nature* **393**, 458–460.

Achilli, A., Perego, U. A., Lancioni, H., Olivieri, A., Gandini, F., Hooshiar Kashani, B., Battaglia, V., Grugni, V., Angerhofer, N., Rogers, M. P., and Herrera, R. J. (2013). Reconciling migration models to the Americas with the variation of North American native mitogenomes. *Proceedings of the National Academy of Sciences* **110**, 14308–14313.

Agustí, J., and Rubio-Campillo, X. (2017). Were Neanderthals responsible for their own extinction? *Quaternary International* **431**, 232–237.

Aiello, L. C. (2010). Five years of *Homo floresiensis. American Journal of Physical Anthropology* **142**, 167–179.

Andersson, C. (2009). Neogene climates. In *Encyclopedia of Paleoclimatology and Ancient Environments. Encyclopedia of Earth Sciences Series* (ed. V. Gornitz), pp. 609–612). Springer, Dordrecht, Germany.

Argue, D., Groves, C. P., Lee, M. S., and Jungers, W. L. (2017). The affinities of *Homo floresiensis* based on phylogenetic analyses of cranial, dental, and postcranial characters. *Journal of Human Evolution* **107**, 107–133.

Armitage, S. J., Jasim, S. A., Marks, A. E., Parker, A. G., Usik, V. I., and Uerpmann, H. P. (2011). The southern route "out of Africa": evidence for an early expansion of modern humans into Arabia. *Science* **331**, 453–456.

Bae, C. J. (2013). Archaic *Homo sapiens*. *Nature Education Knowledge* **4**, 4.

Bae, C. J., Douka, K., and Petraglia, M. D. (2017). On the origin of modern humans: Asian perspectives. *Science* **358**, 9067.

Becerra-Valdivia, L., and Higham, T. (2020). The timing and effect of the earliest human arrivals in North America. *Nature* **584**, 93–97.

Begun, D. R. (2006). Planet of the apes. *Scientific American Special Editions* **16**, 4–13.

Begun, D. R. (2010). Miocene hominids and the origins of the African apes and humans. *Annual Review of Anthropology* **39**, 67–84.

Behrensmeyer, A. K. (2008). Paleoenvironmental context of the Pliocene AL 333 "first family" hominin locality, Hadar Formation, Ethiopia. *Geological Society of America Special Papers* **446**, 203–214.

Behrensmeyer, A. K., and Reed, K. E. (2013). Reconstructing the habitats of Australopithecus: paleoenvironments, site taphonomy, and faunas. In *The Paleobiology of Australopithecus. Vertebrate Paleobiology and Paleoanthropology* (ed. K. Reed, J. Fleagle, and R. Leakey), pp. 41–60. Springer, Dordrecht, Germany.

Bello, S. M. (2021). Boning up on Neanderthal art. *Nature Ecology & Evolution* **5**, 1201–1202.

Benefit, B. R., and McCrossin, M. L. (2015). A window into ape evolution. *Science* **350**, 515–516.

Bennett, M. R., Bustos, D., Pigati, J. S., Springer, K. B., Urban, T. M., Holliday, V. T., Reynolds, S. C., Budka, M., Honke, J. S., Hudson, A. M., and Fenerty, B. (2021). Evidence of humans in North America during the last glacial maximum. *Science* **373**, 1528–1531.

Bergström, A., Stringer, C., Hajdinjak, M., Scerri, E. M., and Skoglund, P. (2021). Origins of modern human ancestry. *Nature* **590**, 229–237.

Böhme, M., Spassov, N., Fuss, J., Tröscher, A., Deane, A. S., Prieto, J., Kirscher, U., Lechner, T., and Begun, D. R. (2019). A new Miocene ape and locomotion in the ancestor of great apes and humans. *Nature* **575**, 489–493.

Bond, M., Tejedor, M. F., Campbell Jr., K. E., Chornogubsky, L., Novo, N., and Goin, F. (2015). Eocene primates of South America and the African origins of New World monkeys. *Nature* **520**, 538–541.

Bourgeon, L., Burke, A., and Higham, T. (2017). Earliest human presence in North America dated to the last glacial maximum: new radiocarbon dates from Bluefish Caves, Canada. *PloS One* **12**, e0169486.

Brown, P., Sutikna, T., Morwood, M. J., Soejono, R. P., Jatmiko, Saptomo, E. W., Due, R.A. (2004). A new small-bodied hominin from the Late Pleistocene of Flores, Indonesia. *Nature* **431**, 1055–1061.

Brumm, A., Oktaviana, A. A., Burhan, B., Hakim, B., Lebe, R., Zhao, J. X., Sulistyarto, P. H., Ririmasse, M., Adhityatama, S., Sumantri, I., and Aubert, M. (2021). Oldest cave art found in Sulawesi. *Science Advances* **7**, eabd4648.

Buck, L. T., and Stringer, C. B. (2014). *Homo heidelbergensis*. *Current Biology* **24**, R214–R215.

Bunbury, M. M., Petchey, F., and Bickler, S. H. (2022). A new chronology for the Maāori settlement of Aotearoa (NZ) and the potential role of climate change in demographic developments. *Proceedings of the National Academy of Sciences* **119**, e2207609119.

Callaway, E. (2017). Oldest *Homo sapiens* fossil claim rewrites our species' history. *Nature* **546**, 289–293.

Cann, R. L., Stoneking, M., and Wilson, A. C. (1987). Mitochondrial DNA and human evolution. *Nature* **325**, 31–36.

Carotenuto, F., Tsikaridze, N., Rook, L., Lordkipanidze, D., Longo, L., Condemi, S., and Raia, P. (2016). Venturing out safely: the biogeography of Homo erectus dispersal out of Africa. *Journal of Human Evolution* **95**, 1–12.

Cartmill, E. A., Beilock, S., and Goldin-Meadow, S. (2012). A word in the hand: action, gesture and mental representation in humans and non-human primates. *Philosophical Transactions of the Royal Society B: Biological Sciences* **367**, 129–143.

Carvalho, S., Biro, D., Cunha, E., Hockings, K., McGrew, W. C., Richmond, B. G., and Matsuzawa, T. (2012). Chimpanzee carrying behaviour and the origins of human bipedality. *Current Biology* **22**, R180–R181.

Charlier, P., Coppens, Y., Augias, A., Deo, S., Froesch, P., and Huynh-Charlier, I. (2018). Mudslide and/or animal attack are more plausible causes and circumstances of death for AL 288 ("Lucy"): a forensic anthropology analysis. *Medico-Legal Journal* **86**, 139–142.

Chazan, M. (2017). Toward a long prehistory of fire. *Current Anthropology* **58**, S351–S359.

Ciochon, R. L., and Bettis Iii, E. A. (2009). Asian Homo erectus converges in time. *Nature* **458**, 153–154.

Clarke, R. J., Pickering, T. R., Heaton, J. L., and Kuman, K. (2021). The earliest South African Hominids. *Annual Review of Anthropology* **50**, 125–143.

Clemens, S. C., and Tiedemann., R. (1997). Eccentricity forcing of Pliocene-Early Pleistocene climate revealed in a marine oxygen-isotope record. *Nature* **385**, 801–807.

Clottes, J. (1999). Archaeology: new light on the oldest art. *Science* **283**, 920–922.

Cohen, A. S., Du, A., Rowan, J., Yost, C. L., Billingsley, A. L., Campisano, C. J., Brown, E. T., Deino, A. L., Feibel, C. S., Grant, K., and Kingston, J. D. (2022). Plio-Pleistocene environmental variability in Africa and its implications for mammalian evolution. *Proceedings of the National Academy of Sciences* **119**, e2107393119.

Conde-Valverde, M., Martínez, I., Quam, R. M., Rosa, M., Velez, A. D., Lorenzo, C., Jarabo, P., Bermúdez de Castro, J. M., Carbonell, E., and Arsuaga, J. L. (2021). Neanderthals and Homo sapiens had similar auditory and speech capacities. *Nature Ecology & Evolution* **5**, 609–615.

Crocker, A. J., Naafs, B. D. A., Westerhold, T., James, R. H., Cooper, M. J., Röhl, U., Pancost, R. D., Xuan, C., Osborne, C. P., Beerling, D. J., and Wilson, P. A. (2022). Astronomically controlled aridity in the Sahara since at least 11 million years ago. *Nature Geoscience* **15**, 671–676.

Culotta, E. (1999). Neanderthals were cannibals, bones show. *Science* **286**, 18–19.

Dannemann, M., and Kelso, J. (2017). The contribution of Neanderthals to phenotypic variation in modern humans. *American Journal of Human Genetics* **101**, 578–589.

Darwin, C. (1871). *The Descent of Man and Selection in Relation to Sex*. Murray, London. (Republished by Random House)

Darwin, C. (1969). *The Autobiography of Charles Darwin*. W. W. Norton, New York.

Defleur, A., White, T., Valensi, P., Slimak, L., and Cregut-Bonnoure, E. (1999). Neanderthal cannibalism at Moula-Guercy, Ardeche, France. *Science* **286**, 128–131.

De la Torre, I. (2011). The origins of stone tool technology in Africa: a historical perspective. *Philosophical Transactions of the Royal Society B: Biological Sciences* **366**, 1028–1037.

De la Torre, I. (2016). The origins of the Acheulean: past and present perspectives on a major transition in human evolution. *Philosophical Transactions of the Royal Society B: Biological Sciences* **371**, 20150245.

Delson, E., and Harvati, K. (2006). Return of the last Neanderthal. *Nature* **443**, 762–763.

Delson, E., and Stringer, C. (2022). The naming of *Homo bodoensis* by Roksandic and colleagues does not resolve issues surrounding Middle Pleistocene human evolution. *Evolutionary Anthropology: Issues, News, and Reviews* **31**, 233–236.

De Menocal, P. B. (1995). Plio-Pleistocene African climate. *Science* **270**, 53–59.

De Menocal, P. B. (2004). African climate change and faunal evolution during the Pliocene-Pleistocene. *Earth and Planetary Science Letters* **220**, 3–24.

Derevianko, O. P., Shunkov, M. V., and Kozlikin, M. B. (2020). Who were the Denisovans? *Archaeology, Ethnology & Anthropology of Eurasia* **48**, 3–32.

Détroit, F., Mijares, A. S., Corny, J., Daver, G., Zanolli, C., Dizon, E., Robles, E., Grün, R., and Piper, P. J. (2019). A new species of *Homo* from the Late Pleistocene of the Philippines. *Nature* **568**, 181–186.

Diamond, J. (1992). *The Third Chimpanzee: The Evolution and Future of the Human Animal*. Harper Perennial, New York.

Disotell, T. R. (2012). Archaic human genomics. *American Journal of Physical Anthropology* **149**, 24–39.

Dominguez-Andres, J. and Netea, M.G. (2019). Impact of historic migrations and evolutionary processes on human immunity. *Trends in Immunology* **40**, 1105–1119.

Dunbar, R. I., and Shultz, S. (2017). Why are there so many explanations for primate brain evolution? *Philosophical Transactions of the Royal Society B: Biological Sciences* **372**, 20160244.

Dupont, L. M., Rommerskirchen, F., Mollenhauer, G., and Schefuß, E. (2013). Miocene to Pliocene changes in South African hydrology and vegetation in relation to the expansion of C4 plants. *Earth and Planetary Science Letters* **375**, 408–417.

Fagan, B. (2011). *Cro-Magnon: How the Ice Age Gave Birth to the First Modern Humans*. Bloomsbury Publishing, New York.

Fagan, B. M. (1998). *People of the Earth: An Introduction to World Prehistory*, 9th edition. Longman, NY.

Falk, D. (2019). How Australopithecus provided insight into human evolution. *Nature* **575**, 41–42.

Fisher, S. E., and Ridley, M. (2013). Culture, genes, and the human revolution. *Science* **340**, 929–930.

Fleagle, J. G. (2013). *Primate Adaptation and Evolution*. Academic Press, San Diego.

Flegontov, P., Altınışık, N. E., Changmai, P., Rohland, N., Mallick, S., Adamski, N., Bolnick, D. A., Broomandkhoshbacht, N., Candilio, F., Culleton, B. J., and Flegontova, O. (2019). Palaeo-Eskimo genetic ancestry and the peopling of Chukotka and North America. *Nature* **570**, 236–240.

Foley, R. (1987). *Another Unique Species*. Longman, London.

Fox, R. C., and Scott, C. S. (2011). A new, early Puercan (earliest Paleocene) species of Purgatorius (Plesiadapiformes, Primates) from Saskatchewan, Canada. *Journal of Paleontology* **85**, 537–548.

Fu, Q., Hajdinjak, M., Moldovan, O. T., Constantin, S., Mallick, S., Skoglund, P., Patterson, N., Rohland, N., Lazaridis, I., Nickel, B., and Viola, B. (2015). An early modern human from Romania with a recent Neanderthal ancestor. *Nature* **524**, 216–219.

Galway-Witham, J., and Stringer, C. (2018). How did *Homo sapiens* evolve? *Science* **360**, 1296–1298.

Gamble, C. (1994). *Timewalkers: The Prehistory of Global Colonization*. Harvard University Press, Cambridge, MA.

Gaur, R. (2015). Louis Leakey and Mary Leakey: the first family of palaeoanthropology. *Resonance* **20**, 667–679.

Gibbons, A. (2011). A new view of the birth of Homo sapiens. *Science* **331**, 392–394.

Gibbons, A. (2011). Who was Homo habilis—and was it really Homo? *Science* **332**, 1370.

Gibbons, A. (2016). Meet the frail, small-brained people who first trekked out of Africa. *Science* **22**. doi.org/10.1126/science.aal0416.

Gibbons, A. (2023). Should an also-ran in human evolution get more respect? *Science* **379**, 522–523.

Gilpin, W., Feldman, M. W., and Aoki, K. (2016). An ecocultural model predicts Neanderthal extinction through competition with modern humans. *Proceedings of the National Academy of Sciences* **113**, 2134–2139.

Gingerich, P. D. (2012). Primates in the Eocene. *Palaeobiodiversity and Palaeoenvironments* **92**, 649–663.

Gitschier, J. (2010). All About Mitochondrial Eve: An Interview with Rebecca Cann. *PLoS Genetics* **6**, e1000959.

Gowlett, J. A. (2016). The discovery of fire by humans: a long and convoluted process. Philosophical *Transactions of the Royal Society B: Biological Sciences* **371**, 2015.0164.

Grehan, J. R., and Schwartz, J. H. (2009). Evolution of the second orangutan: phylogeny and biogeography of hominid origins. *Journal of Biogeography* **36**, 1823–1844.

Grehan, J. R., and Schwartz, J. H. (2011). Evolution of human-ape relationships remains open for investigation. *Journal of Biogeography* **38**, 2397–2404.

Groucutt, H. S., Petraglia, M. D., Bailey, G., Scerri, E. M., Parton, A., Clark-Balzan, L., Jennings, R. P., Lewis, L., Blinkhorn, J., Drake, N. A., and Breeze, P. S. (2015). Rethinking the dispersal of Homo sapiens out of Africa. *Evolutionary Anthropology: Issues, News, and Reviews* **24**, 149–164.

Grün, R., Pike, A., McDermott, F., Eggins, S., Mortimer, G., Aubert, M., Kinsley, L., Joannes-Boyau, R., Rumsey, M., Denys, C., and Brink, J. (2020). Dating the skull from Broken Hill, Zambia, and its position in human evolution. *Nature* **580**, 372–375.

Haile-Selassie, Y., Gibert, L., Melillo, S. M., Ryan, T. M., Alene, M., Deino, A., Levin, N. E., Scott, G., and Saylor, B. Z. (2015). New species from Ethiopia further expands Middle Pliocene hominin diversity. *Nature* **521**, 483–488.

Haile-Selassie, Y., Latimer, B. M., Alene, M., Deino, A. L., Gibert, L., Melillo, S. M., Saylor, B. Z., Scott, G. R., and Lovejoy, C. O. (2010). An early Australopithecus afarensis postcranium from Woranso-mille, Ethiopia. *Proceedings of the National Academy of Sciences* **107**, 12121–12126.

Haile-Selassie, Y., Melillo, S. M., Vazzana, A., Benazzi, S., and Ryan, T. M. (2019). A 3.8-million-year-old hominin cranium from Woranso-Mille, Ethiopia. *Nature* **573**, 214–219.

Hammond, A. S., Mavuso, S. S., Biernat, M., Braun, D. R., Jinnah, Z., Kuo, S., Melaku, S., Wemanya, S. N., Ndiema, E. K., Patterson, D. B., and Uno, K. T. (2021). New hominin remains and revised context from the earliest Homo erectus locality in East Turkana, Kenya. *Nature Communications* **12**, 1939.

Harrod, J. B. (2014). Palaeoart at two million years ago? A review of the evidence. *Arts* **3**, 135–155.

Haviland, W. A. (1989). *Anthropology*, 5th edition. Holt, Rinehart and Winston, New York.

Herries, A. I., Martin, J. M., Leece, A. B., Adams, J. W., Boschian, G., Joannes-Boyau, R., Edwards, T. R., Mallett, T., Massey, J., Murszewski, A., and Neubauer, S. (2020). Contemporaneity of Australopithecus, Paranthropus, and early Homo erectus in South Africa. *Science* **368**, eaaw7293.

Higham, T., Compton, T., Stringer, C., Jacobi, R., Shapiro, B., Trinkaus, E., Chandler, B., Gröning, F., Collins, C., Hillson, S., and O'Higgins, P. (2011). The earliest evidence for anatomically modern humans in northwestern Europe. *Nature* **479**, 521–524.

Holden, C. (1999). Were Spaniards among the first Americans? *Science* **286**, 1467–1468.

Holliday, T. W. (2012). Body size, body shape, and the circumscription of the genus Homo. *Current Anthropology* **53**, S330–S345.

Howells, W. (1993). *Getting Here.* Compass Press, Washington, DC.

Hublin, J. (1999). The quest for Adam. *Archaeology* **52**, 26–32.

Hublin, J. J., Ben-Ncer, A., Bailey, S. E., Freidline, S. E., Neubauer, S., Skinner, M. M., Bergmann, I., Le Cabec, A., Benazzi, S., Harvati, K., and Gunz, P. (2017). New fossils from Jebel Irhoud, Morocco and the pan-African origin of Homo sapiens. *Nature* **546**, 289–292.

Hudjashov, G., Endicott, P., Post, H., Nagle, N., Ho, S. Y., Lawson, D. J., Reidla, M., Karmin, M., Rootsi, S., Metspalu, E., and Saag, L. (2018). Investigating the origins of eastern Polynesians using genome-wide data from the Leeward Society Isles. *Scientific Reports* **8**, 823.

Janis, C. M. (1993). Tertiary mammal evolution in the context of changing climates, vegetation and tectonic events. *Annual Review of Ecology and Systematics*, 467–500.

Jensen-Seaman, M. I., and Hooper-Boyd, K. A. (2013). Molecular clocks: determining the age of the human-chimpanzee divergence. *Encyclopedia of Life Sciences (ELS)*. doi.org/10.1002/9780470015902.a0020813.pub2.

Johanson, D., Johanson, L., and Edgar, B. (1994). *Ancestors: In Search of Human Origins.* Villard Books, New York.

Johanson, D. C. (2004). Lucy, thirty years later: an expanded view of *Australopithecus afarensis*. *Journal of Anthropological Research* **60**, 465–486.

Josenhans, H. F., Daryl, R. P., and Southon, J. (1997). Early humans and rapidly changing Holocene sea levels in the Queen Charlotte Islands–Hecate Strait, British Columbia, Canada. *Science* **277**, 71–74.

Jouve, G., Pettitt, P., and Bahn, P. (2020). Chauvet Cave's art remains undated. *L'Anthropologie* **124**, 102765.

Kay, R. F. (2015). New World monkey origins. *Science* **347**, 1068–1069.

Kinaston, R. L., Walter, R. K., Jacomb, C., Brooks, E., Tayles, N., Halcrow, S. E., Stirling, C., Reid, M., Gray, A. R., Spinks, J., and Shaw, B. (2013). The first New Zealanders: patterns of diet and mobility revealed through isotope analysis. *PloS One* **8**, e64580.

Laghlin, W. S., and Harper, A. B. (1988). Peopling of the continents: Australia and America. In *Biological Aspects of Human Migration, Cambridge Studies in Biological Anthropology* (ed. C. G. N. Mascie-Taylor and G. W. Lasker), pp. 14–40. Cambridge University Press, Cambridge.

Langdon, J. H. (2016). Case Study 19. Leaving Africa: Mitochondrial Eve. In *The Science of Human Evolution*, pp.151–158. Springer, Cham, Switzerland.

Leakey, M. (1999). Perspectives on the past. *Archaeology* **52**, 32–36.

Leakey, M., and Walker, A. (1997). Early Hominid fossils from Africa. *Scientific American*, 74–79.

Leder, D., Hermann, R., Hüls, M., Russo, G., Hoelzmann, P., Nielbock, R., Böhner, U., Lehmann, J., Meier, M., Schwalb, A., and Tröller-Reimer, A. (2021). A 51,000-year-old engraved bone reveals Neanderthals' capacity for symbolic behaviour. *Nature Ecology & Evolution* **5**, 1273–1282.

Lehtonen, S., Sääksjärvi, I., Ruokolainen, K., and Tuomisto, H. (2011) Who is the closest extant cousin of humans? Total-evidence approach to hominid phylogenetics via simultaneous optimization. *Journal of Biogeography* **38**, 805–808.

Lepre, C. J., and Quinn, R. L. (2022). Aridification and orbital forcing of eastern African climate during the Plio-Pleistocene. *Global and Planetary Change* **208**, 103684.

Lepre, C. J., Roche, H., Kent, D. V., Harmand, S., Quinn, R. L., Brugal, J. P., Texier, P. J., Lenoble, A., and Feibel, C. S. (2011). An earlier origin for the Acheulian. *Nature* **477**, 82–85.

Liddy, H. M., Feakins, S. J., and Tierney, J. E. (2016). Cooling and drying in northeast Africa across the Pliocene. *Earth and Planetary Science Letters* **449**, 430–438.

López-Torres, S., Selig, K. R., Prufrock, K. A., Lin, D., and Silcox, M. T. (2018). Dental topographic analysis of paromomyid (Plesiadapiformes, Primates) cheek teeth: more than 15 million years of changing surfaces and shifting ecologies. *Historical Biology* **30**, 76–88.

Lordkipanidze, D., Ponce de León, M. S., Margvelashvili, A., Rak, Y., Rightmire, G. P., Vekua, A., and Zollikofer, C. P. (2013). A complete skull from Dmanisi, Georgia, and the evolutionary biology of early Homo. *Science* **342**, 326–331.

MacDonald, G. M., Beilman, D. W., Kuzmin, Y. V., Orlova, L. A., Kremenetski, K. V., Shapiro, B., Wayne, R. K., and Van Valkenburgh, B. (2012). Pattern of extinction of the woolly mammoth in Beringia. *Nature Communications* **3**(1), 893.

MacDonald, G. M., and McLeod, T. K. (1996). The Holocene closing of the "Ice-Free Corridor." *Quaternary International* **32**, 87–95.

Maier, P. A., Runfeldt, G., Estes, R. J., and Vilar, M. G. (2022). African mitochondrial haplogroup L7: a 100,000-year-old maternal human lineage discovered through reassessment and new sequencing. *Scientific Reports* **12**, 1–14.

Maslin, M. A., Brierley, C. M., Milner, A. M., Shultz, S., Trauth, M. H., and Wilson, K. E. (2014). East African climate pulses and early human evolution. *Quaternary Science Reviews* **101**, 1–17.

MCabe, S. C. J. (2016). Aegyptopithecus. *The International Encyclopedia of Primatology*, pp. 1–2. Wiley-Blackwell, New York.

McFalls Jr., J. A. (1991). Population: a lively introduction. *Population Bulletin* **46**, 1–47.

McPherron, S. P., Alemseged, Z., Marean, C. W., Wynn, J. G., Reed, D., Geraads, D., Bobe, R., and Béarat, H. A. (2010). Evidence for stone-tool-assisted consumption of animal tissues before 3.39 million years ago at Dikika, Ethiopia. *Nature* **466**, 857–860.

Melchionna, M., Di Febbraro, M., Carotenuto, F., Rook, L., Mondanaro, A., Castiglione, S., Serio, C., Vero, V. A., Tesone, G., Piccolo, M., and Diniz-Filho, J. A. F. (2018). Fragmentation of Neanderthals' pre-extinction distribution by climate change. *Palaeogeography, Palaeoclimatology, Palaeoecology* **496**, 146–154.

Miller, E. R., Gunnell, G. F., Seiffert, E. R., Sallam, H., and Schwartz, G. T. (2018). Patterns of dental emergence in early anthropoid primates from the Fayum Depression, Egypt. *Historical Biology* **30**, 157–165.

Moorjani, P., Amorim, C. E. G., Arndt, P. F., and Przeworski, M. (2016). Variation in the molecular clock of primates. *Proceedings of the National Academy of Sciences* **113**, 10607–10612.

Mosquera, M., Ollé, A., and Rodríguez, X. P. (2013). From Atapuerca to Europe: tracing the earliest peopling of Europe. *Quaternary International* **295**, 130–137.

Noonan, J. P., Coop, G., Kudaravalli, S., Smith, D., Krause, J., Alessi, J., Chen, F., Platt, D., Paabo, S., Pritchard, J. K., and Rubin, E. M. (2006). Sequencing and analysis of Neanderthal genomic DNA. *Science* **314**, 1113–1118.

Ocobock, C., Lacy, S., and Niclou, A. (2021). Between a rock and a cold place: Neanderthal biocultural cold adaptations. *Evolutionary Anthropology: Issues, News, and Reviews* **30**, 262–279.

Oppenheimer, S. (2012). Out of Africa, the peopling of continents and islands: tracing uniparental gene trees across the map. *Philosophical Transactions of the Royal Society B: Biological Sciences* **367**, 770–784.

O'Rourke, D. H., and Raff, J. A. (2010). The human genetic history of the Americas: the final frontier. *Current Biology* **20**, R202–R207.

Patrick, V. K. (2010). Peopling of the Pacific: a holistic anthropological perspective. *Annual Review of Anthropology* **39**, 131–148.

Petr, M., Hajdinjak, M., Fu, Q., Essel, E., Rougier, H., Crevecoeur, I., Semal, P., Golovanova, L. V., Doronichev, V. B., Lalueza-Fox, C., and de la Rasilla, M. (2020). The evolutionary history of Neanderthal and Denisovan Y chromosomes. *Science* **369**, 1653–1656.

Pettitt, P., and Bahn, P. (2015). An alternative chronology for the art of Chauvet cave. *Antiquity* **89**, 542–553.

Pitulko, V. V., Tikhonov, A. N., Pavlova, E. Y., Nikolskiy, P. A., Kuper, K. E., and Polozov, R. N. (2016). Early human presence in the Arctic: evidence from 45,000-year-old mammoth remains. *Science* **351**, 260–263.

Plummer, T. W., Oliver, J. S., Finestone, E. M., Ditchfield, P. W., Bishop, L. C., Blumenthal, S. A., Lemorini, C., Caricola, I., Bailey, S. E., Herries, A. I., and Parkinson, J. A. (2023). Expanded geographic distribution and dietary strategies of the earliest Oldowan hominins and Paranthropus. *Science* **379**, 561–566.

Polissar, P. J., Rose, C., Uno, K. T., Phelps, S. R., and de Menocal, P. (2019). Synchronous rise of African C4 ecosystems 10 million years ago in the absence of aridification. *Nature Geoscience* **12**, 657–660.

Pollard, K. S. (2009). What makes us human? *Scientific American* **300**, 44–49.

Posth, C., Wißing, C., Kitagawa, K., Pagani, L., van Holstein, L., Racimo, F., Wehrberger, K., Conard, N. J., Kind, C. J., Bocherens, H., and Krause, J. (2017). Deeply divergent archaic mitochondrial genome provides lower time boundary for African gene flow into Neanderthals. *Nature Communications* **8**(1), 16046.

Potts, R. (1996). Evolution and climate variability. *Science* **273**, 922–923.

Prates, L., Politis, G. G., and Perez, S. I. (2020). Rapid radiation of humans in South America after the last glacial maximum: a radiocarbon-based study. *PloS One* **15**, e0236023.

Profico, A., Di Vincenzo, F., Gagliardi, L., Piperno, M., and Manzi, G. (2016). Filling the gap. Human cranial remains from Gombore II (Melka Kunture, Ethiopia; ca. 850 ka) and the origin of *Homo heidelbergensis*. *Journal of Anthropological Sciences* **94**, 1–24.

Prüfer, K., Munch, K., Hellmann, I., Akagi, K., Miller, J. R., Walenz, B., Koren, S., Sutton, G., Kodira, C., Winer, R., and Knight, J. R. (2012). The bonobo genome compared with the chimpanzee and human genomes. *Nature* **486**, 527–531.

Qin, P., and Stoneking, M. (2015). Denisovan ancestry in East Eurasian and native American populations. *Molecular Biology and Evolution* **32**, 2665–2674.

Raghavan, M., DeGiorgio, M., Albrechtsen, A., Moltke, I., Skoglund, P., Korneliussen, T. S., Grønnow, B., Appelt, M., Gulløv, H. C., Friesen, T. M., and Fitzhugh, W. (2014). The genetic prehistory of the New World Arctic. *Science* **345**, 1255832.

Reynolds, S. C., Bailey, G. N., and King, G. C. (2011). Landscapes and their relation to hominin habitats: case studies from *Australopithecus* sites in eastern and southern Africa. *Journal of Human Evolution* **60**, 281–298.

Richter, D., Grün, R., Joannes-Boyau, R., Steele, T. E., Amani, F., Rué, M., Fernandes, P., Raynal, J.-P., Geraads, D., Ben-Ncer, A., Hublin, J. J., and McPherron, S. P. (2017). The age of the hominin fossils from Jebel Irhoud, Morocco, and the origins of the Middle Stone Age. *Nature* **546**, 293–296.

Rifkin, R. F., Henshilwood, C. S., and Haaland, M. M. (2015). Pleistocene figurative art mobilier from Apollo 11 cave, Karas region, Southern Namibia. *South African Archaeological Bulletin* **70**, 113–123.

Rifkin, R. F., Prinsloo, L. C., Dayet, L., Haaland, M. M., Henshilwood, C. S., Diz, E. L., Moyo, S., Vogelsang, R., and Kambombo, F. (2016). Characterising pigments on

[30,000]-year-old portable art from Apollo 11 Cave, Karas Region, southern Namibia. *Journal of Archaeological Science: Reports* **5**, 336–347.

Rizal, Y., Westaway, K. E., Zaim, Y., van den Bergh, G. D., Bettis III, E. A., Morwood, M. J., Huffman, O. F., Grün, R., Joannes-Boyau, R., Bailey, R. M., and Westaway, M. C. (2020). Last appearance of Homo erectus at Ngandong, Java, 117,000–108,000 years ago. *Nature* **577**, 381–385.

Roach, N. T., Hatala, K. G., Ostrofsky, K. R., Villmoare, B., Reeves, J. S., Du, A., Braun, D. R., Harris, J. W., Behrensmeyer, A. K., and Richmond, B. G. (2016). Pleistocene footprints show intensive use of lake margin habitats by *Homo erectus* groups. *Scientific Reports* **6**, 1–9.

Roberts, D. L., Jarić, I., Lycett, S. J., Flicker, D., and Key, A. (2023). *Homo floresiensis* and *Homo luzonensis* are not temporally exceptional relative to Homo erectus. *Journal of Quaternary Science*. doi.org/10.1002/jqs.3498.

Roberts, P., and Stewart, B. A. (2018). Defining the "generalist specialist" niche for Pleistocene Homo sapiens. *Nature Human Behaviour* **2**, 542–550.

Roberts, R., Walsh, G., Murray, A., Olley, J., Jones, R., Morwood, M., Tuniz, C., Lawson, E., Macphail, M., Bowdery, D., and Naumann, I. (1997). Luminescence dating of rock art and past environments using mud-wasp nests in northern Australia. *Nature* **387**, 696–699.

Rodríguez-Vidal, J., d'Errico, F., Pacheco, F. G., Blasco, R., Rosell, J., Jennings, R. P., Queffelec, A., Finlayson, G., Fa, D. A., Gutierrez Lopez, J. M., and Carrión, J. S. (2014). A rock engraving made by Neanderthals in Gibraltar. *Proceedings of the National Academy of Sciences* **111**, 13301–13306.

Rogers, J., and Gibbs, R. A. (2014). Comparative primate genomics: emerging patterns of genome content and dynamics. *Nature Reviews Genetics* **15**, 347–359.

Roksandic, M., Radović, P., Wu, X. J., and Bae, C. J. (2022). *Homo bodoensis* and why it matters. *Evolutionary Anthropology: Issues, News, and Reviews* **31**, 240–244.

Roksandic, M., Radović, P., Wu, X. J., and Bae, C. J. (2022). Resolving the "muddle in the middle": the case for *Homo bodoensis* sp. nov. *Evolutionary Anthropology: Issues, News, and Reviews* **31**, 20–29.

Rosen, K. H., Jones, C. E., and DeSilva, J. M. (2022). Bipedal locomotion in zoo apes: Revisiting the hylobatian model for bipedal origins. *Evolutionary Human Sciences* **4**, E12.

Sadier, B., Delannoy, J. J., Benedetti, L., Bourlès, D. L., Jaillet, S., Geneste, J. M., Lebatard, A. E., and Arnold, M. (2012). Further constraints on the Chauvet cave artwork elaboration. *Proceedings of the National Academy of Sciences* **109**, 8002–8006.

Samuels, J. X., Albright, L. B., and Fremd, T. J. (2015). The last fossil primate in North America, new material of the enigmatic Ekgmowechashala from the Arikareean of Oregon. *American Journal of Physical Anthropology* **158**, 43–54.

Savage, R. J. G. (1986). *Mammal Evolution: An Illustrated Guide*. Facts on File, New York.

Scally, A., and Durbin, R. (2012). Revising the human mutation rate: implications for understanding human evolution. *Nature Reviews Genetics* **13**, 745–753.

Scally, A., Dutheil, J. Y., Hillier, L. W., Jordan, G. E., Goodhead, I., Herrero, J., Hobolth, A., Lappalainen, T., Mailund, T., Marques-Bonet, T., and McCarthy, S. (2012). Insights into hominid evolution from the gorilla genome sequence. *Nature* **483**, 169–175.

Senut, B., Pickford, M., Gommery, D., and Ségalen, L. (2018). Palaeoenvironments and the origin of hominid bipedalism. *Historical Biology* **30**, 284–296.

Schrein, C. M. (2015) Lucy: A marvelous specimen. *Nature Education Knowledge* **6**, 2.

Schrenk, F., Kullmer, O., and Bromage, T. (2015). The earliest putative Homo fossils. In *Handbook of Paleoanthropology* (ed. W. Henke and I. Tattersall), pp. 2145–2165. Springer, Berlin, Heidelberg, Germany.

Sheppard, P. J. (2011). Lapita colonization across the Near/Remote Oceania boundary. *Current Anthropology* **52**, 799–840.

Shipton C. (2020). The unity of Acheulean culture. In *Culture History and Convergent Evolution* (ed. H. S. Groucutt), pp. 13–27. Springer International Publishing, New York.

Slicox, M. T. (2014). Primate origins and the Plesiadapiforms. *Nature Education Knowledge* **5**, 1.

Slon, V., Mafessoni, F., Vernot, B., de Filippo, C., Grote, S., Viola, B., Hajdinjak, M., Peyrégne, S., Nagel, S., Brown, S., and Douka, K. (2018). The genome of the offspring of a Neanderthal mother and a Denisovan father. *Nature* **561**, 113–116.

Sponheimer, M., and Lee-Thorp, J. A. (1999). Isotopic evidence for the diet of an early hominid, *Australopithecus*. *Science* **283**, 368–370.

Stein, P. L., and Rowe, B. M. (1974). *Physical Anthropology*. McGraw-Hill, New York.

Steiper, M. E., and Seiffert, E. R. (2012). Evidence for a convergent slowdown in primate molecular rates and its implications for the timing of early primate evolution. *Proceedings of the National Academy of Sciences* **109**, 6006–6011.

Strait, D. S. (2010). The evolutionary history of the Australopiths. *Evolution: Education and Outreach* **3**, 341–352.

Stringer, C., and Galway-Witham, J. (2017). On the origin of our species. *Nature* **546**, 212–214.

Su, D. F., and Haile-Selassie, Y. (2022). Mosaic habitats at Woranso-Mille (Ethiopia) during the Pliocene and implications for *Australopithecus* paleoecology and taxonomic diversity. *Journal of Human Evolution* **163**, 103076.

Sussman, R. W., Tab Rasmussen, D., and Raven, P. H. (2013). Rethinking primate origins again. *American Journal of Primatology* **75**, 95–106.

Tankersley, K. B. (1999). A matter of superior spearpoints. *Archaeology* **52**, 60–63.

Tattersall, I. (1997). Out of Africa again . . . and again? *Scientific American*, 60–67.

Tattersall, I. (1999). Rethinking human evolution. *Archaeology* **52**, 22–26.

Tattersall, I. (2009). Human origins: out of Africa. *Proceedings of the National Academy of Sciences* **106**, 16018–16021.

Thorpe, S. K., McClymont, J. M., and Crompton, R. H. (2014). The arboreal origins of human bipedalism. *Antiquity* **88**, 906–914.

Timmermann, A. (2020). Quantifying the potential causes of Neanderthal extinction: abrupt climate change versus competition and interbreeding. *Quaternary Science Reviews* **238**, 106331.

Townsend, C., Ferraro, J. V., Habecker, H., and Flinn, M. V. (2023). Human cooperation and evolutionary transitions in individuality. *Philosophical Transactions of the Royal Society B* **378**, 20210414.

Turk, M., Turk, I., and Otte, M. (2020). The Neanderthal musical instrument from Divje Babe I Cave (Slovenia): a critical review of the discussion. *Applied Sciences* **10**, 1226.

Ungar, P. S., and Sponheimer, M. (2011). The diets of early hominins. *Science* **334**, 190–193.

Urciuoli, A., and Alba, D. M. (2023). Systematics of Miocene apes: state of the art of a neverending controversy. *Journal of Human Evolution* **175**, 103309.

Valentin, F., Détroit, F., Spriggs, M. J., and Bedford, S. (2016). Early Lapita skeletons from Vanuatu show Polynesian craniofacial shape: implications for Remote Oceanic settlement and Lapita origins. *Proceedings of the National Academy of Sciences* **113**, 292–297.

Veeramah, K. and Hammer., M. F. (2014). The impact of whole-genome sequencing on the reconstruction of human population history. *Nature Reviews Genetics* **15**, 149–162.

Vernot, B., Tucci, S., Kelso, J., Schraiber, J. G., Wolf, A. B., Gittelman, R. M., Dannemann, M., Grote, S., McCoy, R. C., Norton, H., and Scheinfeldt, L. B. (2016). Excavating Neandertal and Denisovan DNA from the genomes of Melanesian individuals. *Science* **352**, 235–239.

Vidal, C. M., Lane, C. S., Asrat, A., Barfod, D. N., Mark, D. F., Tomlinson, E. L., Tadesse, A. Z., Yirgu, G., Deino, A., Hutchison, W., and Mounier, A. (2022). Age of the oldest known Homo sapiens from eastern Africa. *Nature* **601**, 579–583.

Vila, C., Savolainen, P., Maldonado, J. E., Amorim, I. R., Rice, J. E., Honeycutt, R. L., Crandall, K. A., Lundébergy, J., and Wayne, R. K. (1997). Multiple and ancient origins of the domestic dog. *Science* **276**, 1687–1689.

Walker, A., and Shipman, P. (1996). *The Wisdom of Bones*. Weidenfeld and Nicolson, London.

Walter, R., Buckley, H., Jacomb, C., and Matisoo-Smith, E. (2017). Mass migration and the Polynesian settlement of New Zealand. *Journal of World Prehistory* **30**, 351–376.

Waters, M. R. (2019). Late Pleistocene exploration and settlement of the Americas by modern humans. *Science* **365**, 5447.

Wayne, R. K. (1996). Conservation genetics in the Canidae. In *Conservation Genetics: Case Histories from Nature* (ed. J. C. Avise and J. L. Hamrick), pp. 75–118. Springer, New York.

Wayne, R. K., Leonard, J. A., and Cooper, A. (1999). Full of sound and fury: the recent history of ancient DNA. *Annual Review of Ecology and Systematics* **30**, 457–477.

Wilmshurst, J. M., Hunt, T. L., Lipo, C. P., and Anderson, A. J. (2011). High-precision radiocarbon dating shows recent and rapid initial human colonization of East Polynesia. *Proceedings of the National Academy of Sciences* **108**, 1815–1820.

Will, M., Pablos, A., and Stock, J. T. (2017). Long-term patterns of body mass and stature evolution within the hominin lineage. *Royal Society Open Science* **4**, 171339.

Wilson, A. K., and Sarich, V. M. (1969). A molecular time scale for human evolution. *Proceedings of the National Academy of Science* **63**, 1089–1093.

Wolpoff, M. H., Hawks, J., Frayer, D. W., and Hunley, K. (2001). Modern human ancestry at the peripheries: a test of the replacement theory. *Science* **291**, 293–297.

Wood, B. (2020). Birth of Homo erectus. *Evolutionary Anthropology: Issues, News, and Reviews* **29**, 293–298.

Wong, K. (2014). The 1 percent difference. *Scientific American* **311**, 100.

Wrangham, R. (2017). Control of fire in the Paleolithic: evaluating the cooking hypothesis. *Current Anthropology* **58**, S303–S313.

Wroe, S., William, Parr, W. C. H., Ledogar, J. A., Bourke, J., Evans, S. P., Fiorenza, L., Benazzi, S., Hublin, J. J., Stringer, C., Kullmer, O., Curry, M., Rae, T. C., and Yokley, T. R. (2018). Computer simulations show that Neanderthal facial morphology represents adaptation to cold and high energy demands, but not heavy biting. *Proceedings of the Royal Society B: Biological Sciences* **285**, 85.

Wuethrich, B. (1999). Proto-Polynesians quickly settled Pacific. *Science* **286**, 2054–2056.

Zachos, J., Pagani, M., Sloan, L., Thomas, E., and Billups, K. (2001). Trends, rhythms, and aberrations in global climate 65 Ma to present. *Science* **292**, 686–693.

Zhang, Y., and Harrison, T. (2017). *Gigantopithecus blacki*: a giant ape from the Pleistocene of Asia revisited. *American Journal of Physical Anthropology* **162**, 153–177.

Zimmer, C. (1999). Kenyan skeleton shakes ape family tree. *Science* **285**, 1335–1337.

Zin-Maung-Maung-Thein, Thaung-Htike, Soe, A. N., Sein, C., Maung-Maung, and Takai, M. (2017). Chapter 9 review of the investigation of primate fossils in Myanmar. *Geological Society, London, Memoirs* **48**, 185–206.

CHAPTER 12

HUMANS AS A FORCE IN EVOLUTION AND EXTINCTION

In Chapter 11, we saw how the history of humans is itself a fascinating biogeographic story in which evolutionary forces, extinction, dispersal, geography, and environment all played a role. The end result of this saga was the rise of modern humans to a global population size of 8 billion individuals and a geographic distribution that includes every continent, with forays to the depths of the ocean and the nearer reaches of outer space. What impact have the numerous and widespread members of the human species had on the rest of the biosphere? We could easily focus an entire textbook on this important question. In this chapter we will have to restrict our scope to looking at the role of humans in two key areas. First, we will examine how humans have influenced the evolutionary development of plant and animal species, particularly through domestication. The domestication of plants and animals and the development of agriculture are central to the growth of complex human societies, technological advancements, and increases in human population size that have occurred over the past 10,000 years. Second, we will consider the sobering evidence of how humans have become a major force in the extinction of plant and animal species around the world either directly by hunting or by causing landscape alterations and environmental change.

HUMANS AS AN EVOLUTIONARY FORCE

Humans can influence the evolutionary development of species in a number of different ways. Changes in the landscape caused by agricultural or urban land clearance can lead to strong selective pressure for traits that allow wild species to live in open fields or utilize crop and pasture plants for food. The removal of predators by humans can cause the remaining species to evolve adaptations that decrease defenses against predation but increase foraging efficiency or reproductive rates. Hunting pressure by humans can cause compensatory evolutionary developments in prey species. For example, there is evidence that hunting by Neanderthals and modern humans led to a decrease in the body size of the spur tortoise *(Testuda gracea)* in the Mediterranean region. Small body size may have been a beneficial adaptation because smaller individuals are less visible in the landscape and less attractive to hunters as they provide lower nutritional value than large prey. In modern times, a correlation has been observed between trophy hunting by humans and a trend toward decreased body size and horn length in bighorn sheep *(Ovis canadensis)* in Alberta, Canada. In Zambia, a 300% increase in the population proportion of tuskless female elephants *(Loxodonta africana)* was found to coincide with increased hunting by illegal ivory traders. Simply put, female elephants that lacked tusks were of no interest to the ivory hunters and were more likely to survive, reproduce, and pass on the genes responsible for a lack of tusk development.

In recent times, the introduction of pesticides and pollutants into the environment has exerted selective pressure and led to evolutionary change in some organisms. One widespread and economically significant example of how human-introduced selective pressure influenced the genotypic and phenotypic characteristics of other species comes from the evolutionary response of insects and mites to pesticides. When the use of insecticide chemicals such as pyrethroids became widespread in the 1950s, there were very few instances of natural resistance by insects. By the late 1980s, many

Biogeography: Introduction to Space, Time, and Life, Second Edition. Glen M. MacDonald.
© 2025 John Wiley & Sons, Inc. Published 2025 by John Wiley & Sons, Inc.
Companion website: www.wiley.com/go/macdonald/biogeography2e

populations of insect and mite species were almost completely resistant to common pesticides. Over 400 species of insects were reported to have developed resistance to one or more classes of pesticides in this 30-year period. Mosquitos can transmit a number of diseases to people, including malaria, and are one of the many groups of insects displaying pyrethroid resistance. Individuals that possessed genes producing resistance to pesticides were selected for because they were able to survive and reproduce, while individuals without such genes were killed by the pesticides.

Animal and Plant Domestication

The most important examples of human impact on the evolution of other species come from the domestication of plants and animals. In addition, the process of domestication has been of critical importance to the geographic spread and population increase of humans over the past 12,000 years. The importance of plant and animal domestication and the rise of agriculture to humans cannot be overstated. Domesticated plants and animals are the basis of agriculture, and without agriculture the complex, technically innovative societies and large human populations that exist today could not have come into being. Agriculture promoted people becoming sedentary (living for a prolonged period in one place), establishing permanent villages and towns, and developing stratified societies that included specialized and dedicated segments such as farmers, artisans, soldiers, religious leaders, educators, and governors. Because of the crucial importance of domestication and agriculture, some biogeographers, including the influential figure Carl O. Sauer of the University of California at Berkeley, have made the patterns and processes of plant and animal domestication, and resulting landscape change, a central focus of their work. Understanding the origins of domesticated plants and animals is of more than strictly academic interest. The ancestral species of crop plants may often still be found in areas where early domestication occurred and may provide important crop strains and genetic resources for breeding and agriculture today. It has been shown, for example, that careful use of native varieties of potatoes *(Solanum tuberosum)* can aid in maintaining sustainable agricultural yields in Andean farming regions. Most important crop plants experience significant declines in yield when exposed to even moderate salinity. Scientists are studying the physiology of the ancestral halophytic sea beet *(Beta vulgaris maritima)* to see how salt tolerance might be improved in the cultivated beet varieties. Let's explore some of the fascinating biogeographic research into domestication and the development of agriculture.

Domestication refers to the complex evolutionary process in which the use of a plant or animal species by humans selects for physiological, morphological, or behavioral traits in the domesticated species that differentiates it from its wild ancestors. Domestication is the basic process by which plant and animal species come to depend on humans for survival, while in turn providing humans with practical benefits such as food and other resources. Many domesticated species, such as corn *(Zea mays*—also called *maize)*, provide food. Others, such as garden roses *(Rosa* spp.), are used for nonpractical purposes such as providing color in gardens. During active domestication, humans selectively breed plants or animals to propagate and enhance desired traits. These traits can include docility and trainability; increased meat, milk, or wool production in animals; or increased fruit size and production in plants. In many cases, the resulting domesticated species are very markedly different in morphology and behavior from their wild progenitors. For example, the wild ancestor of domesticated corn is possibly one of the wild species of the grass genus *Zea* that grow in Mexico and Central America. Collectively, they are referred to as *teosintes*. Some of the first archaeological evidence of corn domestication comes from ears and kernels preserved in caves in the Tehuacán Valley of Mexico. Through a process of natural mutation and then domestication that extended from about 8000 to 500 years ago, people in the highlands of southern Mexico produced increasingly large eared and large seeded domestic corn from small-eared and small-seeded beginnings (Fig. 12.1). Domesticated corn provides several advantages over teosinte for humans. Domesticated corn has larger and more nutritious soft kernels and far more kernels per ear. It has been estimated that a single kernel of modern corn contains as much nutrition as an entire ear of the earliest domesticated variety. In addition, the kernels of domesticated corn do not fall off the ear easily. This makes it simple and efficient to harvest corn. However, the adherence of the kernels to the ears means that domesticated corn cannot disperse its own seeds. Corn seed dispersal must be done by humans. As a result, domesticated corn is entirely dependent on humans for its survival as a species. If humans did not plant and tend domesticated corn, it would go extinct almost immediately. On the flip side, domesticated corn is now grown

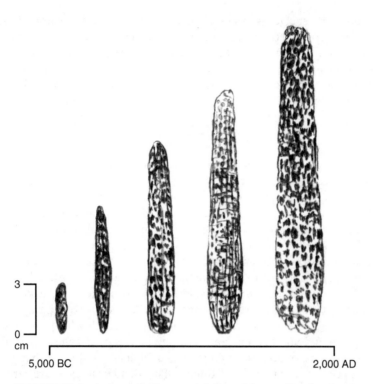

FIGURE 12.1   A generalized representation of the gradual development of modern corn (maize) from teosinte.

in huge quantities throughout the world, making it far more abundant and widespread than ancestral teosinte. In this light, domestication is viewed by some as a symbiotic relationship between humans and domesticated plant and animal species. Many domesticated plants and animals would not survive without human intervention. Although domestication and agriculture are tightly intertwined, the word *agriculture* has a different specific meaning than domestication. **Agriculture** is the cultivation of domesticated plants and animals for human use. It is thought that over the past 12,000 years, some 1000–2500 domesticated plant species have been derived from about 120–160 plant families. These along with some 25 domesticated animal species form the basis of modern agriculture that nourishes us, along with providing other things such as the beauty of gardens and companionship from pets.

Charles Darwin was much interested in the selective pressures that humans apply to species during domestication. He saw that the domestication process was much like natural selection in nature, except that during domestication, human actions were the most important selection force. Darwin recognized that humans applied both conscious selection when they purposefully bred or removed certain individuals to achieve desired genotypically controlled traits and unconscious selection when maintenance or removal of certain genotypes was done inadvertently. We can consider both conscious and unconscious selection practiced by humans to be forms of **artificial selection,** in contrast to natural selection where the selective pressures are generated by the natural physical and biological environment.

Homo sapiens are the only hominids conclusively shown to practice domestication and agriculture. Earlier hominids, and early *Homo sapiens,* were **hunters and gatherers** (also called hunter-gatherers) who relied on naturally occurring vegetation, fruits, nuts, carrion, and game for subsistence. Hunters and gatherers generally do not establish permanent settlements such as villages. They are typically nomadic and move their camps in response to the movement of game and changes in the season. However, it should be noted that in places such as coastal British Columbia and the Pacific Northwest of the United States, socially complex fishing-hunting-gathering communities did establish sedentary and semi-sedentary settlements to take advantage of marine resources. Subsistence by hunting and gathering is still practiced by people in some remote areas, such as portions of Australasia, Africa, and South America.

The earliest domesticated species was probably the dog *(Canis familiaris)*. The domestic dog appears to be a direct descendant of the grey wolf *(Canis lupus)*. The modern dog, encompassing all of the different breeds we have today, was created by domestication by humans. We can speculate how domestication of the dog might have occurred. Wolves are relatively intelligent and social animals that hunt and live in socially ordered packs. They have excellent memories. Wolves have a wide-ranging diet and can tolerate inbreeding. They were predisposed by these traits to domestication. However, the exact means by which the wild wolf became domesticated remains a topic of speculation. It is likely that humans and wolves competed for the same game and perhaps for protection in caves. Because of this close proximity, it is easy to imagine how wolf cubs could have been acquired and raised by humans. Wolf cubs that were particularly vicious would perhaps have been abandoned or killed by their human captors. Other cubs that were not inclined to bond with the human group would have wandered away from the humans. Eventually, selection for traits such as loyalty and non-aggressive behavior toward humans would be dominant in the captive wolf population. The domesticating wolves would be dependent upon humans for food, shelter, and protection. The bond between the species would strengthen. After thousands of years, domestication has produced what many consider a new species, the domestic dog. Fossil evidence from *Homo sapiens* sites in Europe suggests that the domestication of the wolf and evolution of the modern dog were taking place in Eurasia and North America by 15,000 to 11,000 years ago. There is some fossil and genetic evidence of domestication commencing as early as 30,000 to 40,000 years ago. It is probable that domestication of the dog occurred at multiple sites in the Late Pleistocene and Early Holocene. The specific history and geography of this important event remains to be resolved. Regardless of the initial date and place of dog domestication, all of the amazing breeds of dogs, from the tiny chihuahuas to huge Saint Bernards, were developed through artificial selection and resulting evolutionary change caused by humans.

Some particularly good fossil and genetic evidence of the rate and timing of early animal domestication comes from the history of goats *(Capra)* and sheep *(Ovis)* in southwestern Asia. The ancestors of the modern domestic goats and sheep lived in the Zagros Mountains of Iran and Iraq and in southeastern Anatolia. They were hunted by Neanderthals and modern humans starting at least 40,000 years ago. Modern goats and sheep are smaller in size and have different horns and coats than their wild ancestors. The coats of wild sheep contain far less wool than is the case for domesticated sheep. The remains of sheep and goats found at hominid sites that are older than about 10,500 years ago show no evidence of domestication. In addition, it is clear that hunters were killing and eating female and male goats and sheep of all ages. After about 10,000 years ago, there is a shift toward a greater abundance of pre-adult male goat and sheep skeletons at hominid sites. It is possible that at this time humans realized that by taking pre-adult rams and sparing females and sexually mature rams for reproduction, the populations of sheep and goats would remain abundant for hunting. Many parts of the world have similar hunting laws today, allowing, for example, the hunting of male deer but not female ones. Shortly following the shift to selective hunting of young males, the development of domesticated sheep and goat species is evidenced by decreases in the size of goat and sheep skeletons and changes in horn morphology. It seems clear that shortly after 10,500 years ago people were keeping herds of these animals. By 9000 years ago, goat and sheep were being introduced and raised by people at sites that were outside the natural ranges of wild goats and sheep.

Thus, it may be that human interaction with goats, sheep, and other domesticated herding animals first took the form of hunting, then selective hunting, and finally herding. In some locations, such as the Zagros Mountains, goat and sheep hunters may have begun the process of domestication by driving small wild herds into areas closer to human habitation or where the herds could be supported by water and vegetation but also be hunted easily. True domestication, caused by artificial selection and resulting genetic changes, could ensue only if natural wild populations were not allowed to breed with the herded ones. Natural populations may have been kept away from human herds either by fencing or eradication. It is likely that, at first, inadvertent actions by people kept the wild populations away from herded animals. Humans may have observed how some of the herded individuals had desirable traits, such as long coats, high milk production, or large amounts of meat, and selectively bred animals in the herds that had those traits.

Over the past 10,000 years, many animals have been domesticated in different parts of the world, and we often have archaeological information on the early period of domestication (Fig. 12.2). For example, pigs *(Sus scrofa)* and taurine cattle *(Bos taurus)* were domesticated in southwestern

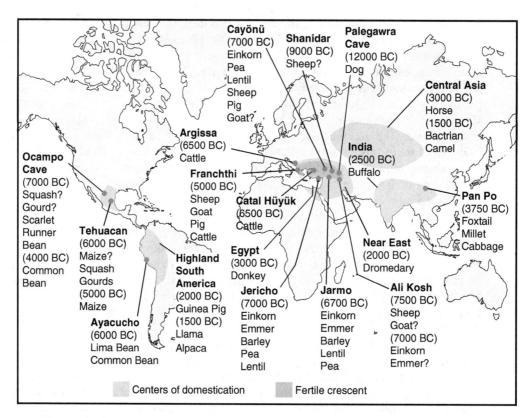

FIGURE 12.2 Some important areas and minimum ages for domestication of selected animals and plants around the world (*Source:* Simmons, 1989 / U.S. Department of Energy / Public Domain).

Asia at about the same time as sheep and goats. Over time, the archaeological record shows that the domesticated pigs became smaller in overall size and dentition than their wild ancestors. Taurine cattle developed smaller horns. Indicine cattle *(Bos indicus)* were first domesticated in the Indus Valley region about 8000 years ago. Both domesticated species of cattle descended from ancestral wild aurochs *(Bos primigenius)*. The keeping of horses for food may have begun on the Eurasian steppes 7000 years ago. The horse *(Equus equus)* and bactrian camel *(Camelus bactrianus)* were domesticated in central Asia by between 5000 and 3500 years ago. The domestication of these two animals revolutionized human transportation because they could be ridden. Water buffalo *(Bubalus bubalis)* were domesticated in India and adjacent southeastern Asia about 4500 years ago. Llama *(Lama glama)* and alpaca *(Vicugna pacos)* were domesticated in South America about 4500 and 3500 years ago. In areas such as central Eurasia, portions of the Near East, northern Eurasia, and parts of Africa, human cultures developed that were organized around herding domesticated animals such as sheep, horses, cattle, camels, and reindeer. These people supplemented the meat and milk obtained from their herds with hunting and gathering. The human herders migrated and moved along with their herds during seasonal changes in grazing areas. This form of subsistence is called **pastoral nomadism** and is still practiced in some areas of the world today.

Some of the earliest evidence for the domestication of plants comes from southwestern Asia and dates from about the same period as the domestication of goats and sheep. Indeed, a variety of important food plants, including wheat *(Triticum)*, rye *(Secale cereale)*, barley *(Hordeum vulgare)*, lentils *(Lens culinaris)*, chickpeas *(Cicer arietinum)*, and peas *(Pisum sativum)*, were all domesticated in the Near East about 11,000 to 9000 years ago. In southwestern Asia, the archaeological sites with evidence of early plant domestication occur along an inverted crescent that extends from the mouth of the Tigris and Euphrates rivers, north to eastern Turkey, and then south to the coastal regions of Lebanon and Israel (Fig. 12.2). This region of early plant domestication and agriculture is called the **Fertile Crescent.** The transition here from hunting and gathering to agricultural societies is often called the **Neolithic Revolution** due to the profound changes it wrought for humans and the

FIGURE 12.3 Archaeological excavation through the ancient settlement layers and remains of structures at Tell Umm el-Marra in northwestern Syria. The earliest structures are about 4700 years old.

environment. Today, many of the wild ancestors of domesticated plants such as wheat, barley, lentils, peas, and chickpeas are found in the vicinity of the Fertile Crescent. There is evidence of wild cereal gathering by hunters and gatherers in this region dating back to at least 19,000 years ago. Many of the archaeological sites with evidence of early domestication, agriculture, and sedentary human populations occur as low mounds, locally called tells (Fig. 12.3). The tells are formed by the remains of human habitation. In most cases, these remains are dominated by the melted and broken mud bricks and building stones of structures such as houses and village walls. Many tells have been occupied and abandoned many times and have layers of artifacts dating from many different periods. In some instances, the earliest buildings and artifacts date from the very dawn of domestication. Studying the plant and animal remains from the tells shows how the process of domestication and development of agriculture went hand in hand with the development of permanent villages. The biblical city of Jericho is represented by such a tell that dates back to 10,000 years ago and provides evidence for the early domestication of wheat. Perhaps the best-known settlement of this age is Catal Hüyük located in eastern Turkey (Fig. 12.2). By about 9500 years ago, this settlement may have supported some 10,000 inhabitants based on early agriculture combined with hunting and gathering.

We can speculate how hunting-and-gathering peoples living in the Fertile Crescent and other parts of the world initially domesticated plants. These people likely used the wild ancestors of domesticated plants as food sources for a very long period prior to domestication. Flint sickle blades from the Fertile Crescent date back to 14,000 years ago and suggest the focused harvesting of the wild ancestors of wheat and other cereals. Many of the wild progenitors of domesticated plants favor disturbed environments such as burned areas or areas with exposed mineral soils. These plants may have seeded into land near campsites that were disturbed by human activity. Humans would have been able to observe the ecology of these plants at close quarters. They could have observed on what types of soils the food plants grew best and which varieties of food plants provided the most seeds or fruit. The domestication of plants likely began as selective cultivation of wild food plants, perhaps by transporting seeds to new campsites or changing local environmental conditions to suit favored plants. Hunting-and-gathering people who lived in more recent times have been shown to practice such cultivation. For example, Australian aborigines were known to have replanted small wild yams after harvesting the larger ones. They also used burning and flooding to alter environmental conditions to benefit the growth of certain edible plants. Early human cultivators would also have noticed that some individuals of the same species produced larger seeds and fruits or more abundant seeds and fruits. They may have selectively planted seeds from those individuals.

As in the case of animals, restricting the flow of wild genes into the plant population that is being domesticated is very important. However, it is much more difficult to restrict the flow of pollen than it is to control contact between cultivated and wild animals. Pollen is transported by the wind or insects and can easily move from wild to domesticated plants growing near each other. Interestingly, many of the ancestors of important early domesticated plants, such as wheat, typically self-pollinate more frequently than they are fertilized by other individuals. Therefore, it is less likely that pollen from a nonfavored wild stock of plants would introduce genes into plant populations that humans were cultivating.

By at least 14,400 years ago, hunter-gatherers in northeastern Jordan were using wild grain seeds to make flatbread. Seeds from archaeological sites in the Fertile Crescent show that between 11,000 and about 8000 years ago, cultivation of wild plants led to domesticated varieties of grains such as wheat and barley that had larger seeds, more nutritious seeds, larger seed production, and in many cases, seeds that more strongly adhered to the plant. An important example of such evolutionary change due to domestication is provided by the early history of wheat. Because of its nutritious seeds, high yield, wide potential food uses, and good environmental tolerance, wheat is the most important domesticated grain in the world today. About 20% of the calories consumed by the entire human population are derived from wheat. Archaeological evidence and data from the modern genetics of domesticated and wild wheat provide information on the history of this important crop plant. A number of wild species of the wheat genus are found in southwestern Asia today. One of the earliest cultivated species was probably wild einkorn wheat *(Triticum boeoticum)*, which through artificial selection gave rise to domesticated einkorn *(T. monococcum)*. Domesticated einkorn has larger seeds that are more strongly attached to the rachis (central stem) of the plant compared to wild einkorn. A similar history is apparent for domesticated emmer wheat *(T. dicoccum)*, which arose from the wild species *T. dicoccoides*. Some taxonomists now combine both einkorn species together as *T. monococcum*. Archaeological evidence indicates that both wild einkorn and wild emmer wheat were being cultivated in the Fertile Crescent by around 11,000 to 10,000 years ago. In addition to cultivation within the natural distributions of these species, wheat had also been introduced and cultivated in areas outside its native range on the higher floodplains of the Euphrates River. Modern bread wheat *(T. aestivum)* appears to be a hybrid produced by cross-fertilization between emmer wheat and a wild goatgrass species *(Aegilops tauschii)*. In addition to being a hybrid, modern bread wheat is also a hexaploid, having a chromosomal count of 42. Emmer wheat, which is a tetraploid (chomosomal count of 28), is probably a polyploid progeny of einkorn wheat, which has a chromosome count of 14. Polyploid plants often have larger seeds than normal diploid varieties. Aside from having larger seeds, bread wheat also has seeds that lack the plumed bracts that cover the seeds of other wheat. The naked seeds of bread wheat are much easier to process for food. Events such as the mutation that caused polyploidy in the wheats and cross-fertilization between emmer and *A. tauschii* were fortuitous accidents that improved the food value of wheat. By selecting these hybrids and polyploids for cultivation, humans capitalized on these chance genetic events. As is the case with corn, should humans cease to cultivate it, bread wheat would not be able to disperse adequate numbers of seeds for the species to survive.

Although many important domestic species come from southwestern Asia, people around the world domesticated plants in the Holocene (Fig. 12.2). As we have already seen, corn domestication occurred in Central America and spread to North and South America. Lima beans, common beans, potato, and squash also come from Central and South American areas. Sorghum, finger millet, and some yams *(Dioscorea)* come from Africa. Domesticated Asian rice *(Oryza sativa)* is from southern Asia. Foxtail millett *(Setaria italica)* and cabbage *(Brassica oleracea)* come from China.

Questions of the Origin and Spread of Agriculture

Two questions regarding domestication and agriculture have long fascinated biogeographers and archaeologists. First, why did hunting-and-gathering people turn to agriculture? Second, what was the geographic pattern of agricultural origin and spread? Surprisingly, the impetus for people to abandon a hunting-and-gathering lifestyle and take up agriculture is not that clear. Experiments have shown that in the natural environment of southwestern Asia, a hunter-gatherer can obtain up to 50 kilocalories of food energy for every 1 kilocalorie of energy expended on work. In contrast, because of energy needed for field preparation, sowing seeds, and protecting crops, low-technology subsistence farming nets only about 17 kilocalories of energy for every 1 kilocalorie expended by the farmer. From a per capita energy perspective, there seems no reason to shift from hunting and gathering to subsistence agriculture. In addition, it has also been shown that subsistence farming leaves less free time for recreation and other activities than does hunting and gathering in relatively resource-rich environments. Finally, of the fruits and seeds available to hunter-gatherers, grains such as wheat are not the most nutritious, flavorful, or easiest to process. From the individual's viewpoint, hunting and gathering is a relatively attractive mode of life compared to subsistence agriculture. Two factors may help explain how and why hunters and gatherers turned to agriculture. The first is the external pressure caused by environmental change, and the second is the intrinsic dynamics of human populations and societies, particularly cognitive changes and pressure from population growth.

The premise that changes in human culture are driven mainly by changes in environment is often referred to as **environmental determinism.** In addition, environmental determinists believe that differences in present cultures and economies can largely be explained by geographic differences in environment. Such theories were particularly popular with geographers, anthropologists, and biologists in the nineteenth and early twentieth centuries, but largely discredited today. In the mid-twentieth century, some of the most divisive debates within academic geography erupted over the validity of environmental determinism. In the early twentieth century, the American geologist Raphael Pumpelly and the Australian archaeologist V. Gordon Childe advanced the **Oasis Theory** to explain how environmental change led to the domestication of plants and animals and the development of agriculture in southwestern Asia and adjacent Africa. They supposed that the climate of southwestern Asia and northern Africa had been cool and wet during Pleistocene glaciations. The Oasis Theory proposed that during the last glacial maximum, about 20,000 years ago, the climate of the region was moist, vegetation was lush, and game was plentiful. As the global climate warmed at the close of the Pleistocene, southwestern Asia and northern Africa became warmer and drier. Under these conditions, productive vegetation, game animals, and humans all became restricted to sites near water, such as river valleys, lake shores, and oases. By being forced into such close contact, humans were able to observe the ecology of plants and animals that they relied upon. Eventually, this led to the selection of species and individuals for cultivation. As attractive as Pumpelly and Childe's original hypothesis sounds, it has a fatal flaw. Since the 1960s, fossil pollen records have been recovered from the sediments of a number of lakes in southwestern Asia. Similar records have been obtained from dried-up lakes in Egypt. These records show that the climate of the last glacial maximum was actually quite dry. Vegetation in southwestern Asia and adjacent Africa was sparser and more desert-like during the Late Pleistocene than it was 11,000 to 8000 years ago when initial domestication and the development of agriculture occurred. Archaeological evidence of Neolithic wheat and barley agriculture during the Early Holocene can be found at sites that now lie in the barren western desert of Egypt (Fig. 12.4). In the Early Holocene, these regions were moister than they were during the last glacial maximum and moister than today up until about 5000 to 4000 years ago. This is exactly the opposite of what Childe had envisioned when he constructed the Oasis Theory. The original hypothesis has been discredited.

The possibility that climatic change may have prompted plant and animal domestication and the development of agriculture in southwestern Asia has more recently been reexamined. Evidence from a number of different sources suggests that a general pattern of warming and increasing moisture during the Pleistocene-Holocene transition was interrupted by a rapid and pronounced episode of cooler conditions around 12,900 to 11,700 years ago. This episode is referred to as the **Younger Dryas.** The Younger Dryas event was centered on the North Atlantic region but affected the climate of much of the world. The cause of the Younger Dryas remains widely debated. It has been suggested by some researchers that decreases in plant and game resources during this period of rapid climatic change may have caused extreme stress to hunters and gatherers in the Fertile Crescent region. The movement toward domestication of plants and animals and the development of agriculture may have been a response to a rapid decline in natural resources during the Younger Dryas. However, this theory regarding abrupt climatic changes associated with the Younger Dryas triggering the development of agriculture in the Fertile Crescent region is not universally accepted and remains vigorously

FIGURE 12.4 The author holds a portion of a Neolithic grindstone used to process grains such as wheat and barley. The artifact was found in the desert of western Egypt and provides evidence of moist conditions in this now very arid region and its suitability for agriculture in the Early Holocene.

researched. Nor does it explain the development of agriculture in other parts of the world where the transition from hunting and gathering occurred later in the Holocene.

When we consider the world at large, two important facts argue against environmental determinism as the sole or universal factor that led to the domestication of plants and animals and subsequent development of agriculture. First, the domestication of plants and animals occurred at different times in different parts of the world. It is often difficult to find any convincing evidence for significant environmental changes during initial domestication in these other regions. For example, the domestication of corn in Central America did not begin until about 8000 years ago, and there is no evidence of major environmental changes coincident with this process. Second, domestication and the shift to agriculture was not an instantaneous event. Evidence from the Fertile Crescent shows that it took centuries to millennia for the inhabitants to shift from dependence on wild plants, including wild einkorn wheat, and wild animals such as gazelle *(Gazella subgutturosa)* to goats, sheep, and domesticated grains. During the next 2000 to 3000 years, additional plants were domesticated in the region. In Mexico and Central America, it took perhaps 5000 years from the early start of the domestication of corn until agricultural crops contributed as much as hunting or gathering to the food used by humans. It appears that once domestication and the development of agriculture is set in motion, there is a cultural dynamic that propels it, regardless of environmental change. However, the timeframe between the start of domestication and the development of full agricultural societies can be long. There are also some cases where groups regress back to more foraging-based ways of life.

The isolated hunter-gatherer in an environment rich with food plants and game has little obvious inducement to develop and adopt an agricultural lifestyle. Agriculture does, however, over time provide several advantages when considered in the context of an increasing population of humans that are not isolated from other humans. First, although the effort required to obtain energy via subsistence farming is greater, a cultivated landscape provides much more energy per unit of land than does the harvesting of wild plants. More people can thereby be supported on a smaller area of land. Second, agricultural systems that permit the storage of grains or the keeping of herds allow people to use these reserves if environmental factors, such as drought or a particularly cold year, decrease the yields of crops. Most hunter-gatherers can only move on and try their luck elsewhere, or in extreme cases perish, during periods of difficult climatic conditions. Third, in most environments, hunting and gathering requires mobility of populations. They must be able to move to follow game or take advantage of seasonal changes in plant availability. It is difficult to make large movements with the very young, the aged, and the infirm. In addition, as human populations increased, such movements would bring different bands of people into contact and conflict over hunting-and-gathering grounds. It would be advantageous for a band to remain near and defend a particularly good fishing or hunting site, even if this meant that natural plant resources were less than optimal. In view of these considerations, it has been suggested that one crucial factor in prompting the development of agriculture was the growth and geographic expansion of *Homo sapiens* populations during the Holocene. Domesticated plants and animals allowed for a sedentary life, where the danger of conflict during foraging trips would be reduced and the population size for defending resources in a small area would be maximized. With permanent habitation and larger populations, the stratified societies that developed would increase in power to defend and eventually dominate other groups. Permanent settlements of large populations also meant that people were now tied to agriculture in order to sustain society. A dynamic feedback was set in place that continued to drive domestication and the intensification of agricultural production.

The spread of agriculture has long been a topic of intense interest to biogeographers. In the first half of the twentieth century, a number of researchers speculated that domestication and agriculture developed in the Fertile Crescent and then spread from there throughout the world. Alternatively, Carl O. Sauer proposed that domestication and agriculture first developed in southeastern Asia and then spread from there. To support such claims, some scientists have pointed to the antiquity of agriculture in the Fertile Crescent relative to other parts of the world. Purported evidence for a southeastern Asian origin for agriculture came from the fact that the bottle gourd *(Langenaria siceraria)* was an early domesticant in south Asia and was also found domesticated in South America. The problem with such theories that envision one, and only one, center of the origination of domestication and agriculture is that there is no convincing archaeological or historical evidence to support the movement of people, crops, or technologies from the Fertile Crescent or Asia over great distances to areas such as South America in the Early Holocene. In the case of bottle gourd, recent genetics research suggests that gourds from Africa were carried to South America by ocean currents and subsequently domesticated at several different sites in the New World. The initiation of corn domestication in Mexico or potato domestication in South America began about 8000 years ago and appears to be local innovations that took advantage of existing plants. Similarly, the domestication of finger millet in Africa or rice in Asia also appears to have been a local innovation. There is no reason to suspect that people in many parts of the world were not all capable of independently domesticating plants and animals and developing agriculture. In fact, it appears that domestication of plants and animals and the development of early agrarian societies developed independently at different sites within the Fertile Crescent region.

Despite the widespread ability of humans to domesticate plants and animals and to develop agriculture, both archaeological and genetic evidence from crop plants shows that certain areas of the world were **centers of domestication** for the early development of agriculture. These were areas where wild ancestral species of domesticated plant and animal species lived. Other areas either developed agriculture after contact with people and crops from one of these centers, or they never developed it at all. Some notable centers of initial plant domestication and agricultural development include southwestern Asia; a broad band of central Africa, India, and eastern Asia; Central America; and South America (Fig. 12.2).

The geographic diffusion of agriculture from centers of domestication has long been of interest to biogeographers and anthropologists. The geographic diffusion of domesticated plants and animals in Europe, southern Asia, and pre-Columbian North America has been studied with particular intensity. Crops such as wheat and barley moved from the core of the Fertile Crescent to adjacent areas of southwestern Asia rapidly (Fig. 12.5). By 8000 years ago, these crops and agricultural practices had spread to Greece. Between 7500 and 6500 years ago, grains originating in the Fertile Crescent were grown across much of the European continent from southern Spain to Poland and Ukraine. By 6000 to 5500 years ago, the growing of grain had spread to the climatic limits of such crops in Scandinavia and Britain. The rate at which agriculture spread across Europe has been estimated to average about 1 kilometer per year, with faster movement along coastlines. Domesticated Asian rice is the

Spread of Neolithic Agriculture in Europe

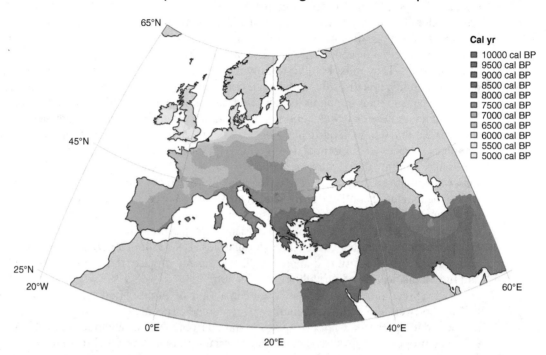

Spread of Neolithic Agriculture in Southeast Asia

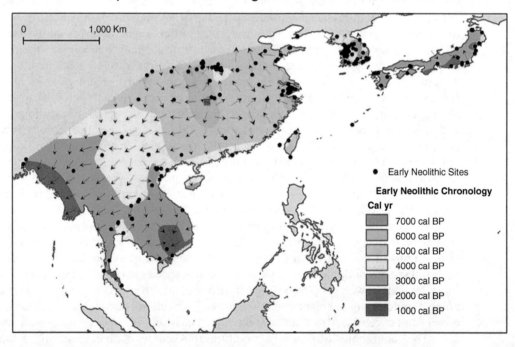

FIGURE 12.5 The initial spread of Neolithic agriculture in Europe and in Southeast Asia (modified from Vander Linden and Silva, 2018; Cobo et al., 2019).

foundation of Neolithic agricultural development in eastern and southeastern Asia. The domestication of rice may have occurred at several places by 8000 years ago with the earliest evidence coming from China's Yangtze River region. Rice cultivation subsequently spread to regions such as India, Thailand, and Japan by 3000 years ago (Fig. 12.5). The spread of rice cultivation is estimated to have occurred at a rate of 0.72–0.92 kilometers per year. In North America, the cultivation of corn spread from

Mesoamerica to the southwestern United States by about 4000 years ago and reached its northeastern climatic limits of corn cultivation in southern Ontario about 1300 to 1000 years ago. The overall rate of northward spread of maize cultivation in North America is somewhere in the range of .9 to 1 kilometer per year, similar to the rates of agricultural spread in Europe and Asia. Corn cultivation also spread southward into western South America and reached Chile between 4000 and 2000 years ago. As domesticated species are introduced to new climatic regions and zones, further selection by humans for traits compatible with the new environments can drive additional evolution of the species. For example, corn grown in the northern United States would have to be adapted to a shorter growing season than corn grown in lowland sites in Mexico.

The diffusion of domesticated plants and animals, as well as agricultural practices, in Europe, North America, and other parts of the world was likely facilitated by many factors, including the movement of people and the transfer of materials and ideas from one group of people to another. Detailed genetic studies of Neolithic human remains and modern populations indicate that the westward and northward migration of agrarian peoples from the vicinity of the Fertile Crescent contributed to the spread of agriculture across Europe. This produced admixtures of genes from Mesolithic European hunter-gatherers and Neolithic farmers from western Asia, which is still apparent today in the genetics of modern European human populations. The latter arrival of Yamnaya culture pastoralists from the western Eurasian steppes in the Bronze Age (about 5000 years ago) brought domesticated horses and dairy cattle and added another genetic element to the mix found in modern European human populations. The Indo-European languages that dominate Europe today were also introduced by these migrating people from the steppes. In contrast, current evidence from the study of the diffusion of rice cultivation in Asia suggests a stronger role for the transmission of rice cultivars and practices between peoples rather than population movement.

The cultivation of wheat and other grains, along with domesticated animals such as sheep, cattle, and pigs, formed the foundation for the rise of cities and civilizations in the Fertile Crescent. The diffusion of these crops and animals across Europe was followed by the diffusion of technical practices such as pottery making and metal smelting. Similarly, rice cultivation and the domestication of animals such as water and swamp buffalo, chickens, and pigs in Asia lay at the foundation of the rich and complex civilizations that subsequently developed. In the Americas, the cultivation of corn, along with beans and squash, satisfied a similar function, serving as the foundation for the Mesoamerican civilizations, such as the Maya and Aztecs, and the Inca in South America. **Mesoamerica** is an archaeological term that refers to southern Mexico and northern central America. The pre-Columbian Pueblo builders, or Anasazi people, of the southwestern United States built considerable towns and village complexes, such as those found at Mesa Verde in Colorado, Wupatki National Monument in Arizona, and Chaco Canyon in New Mexico, on the basis of the same three crops. Not only were these crops introduced from Mesoamerica, but the presence of ceremonial ball courts in pre-Columbian sites throughout the Southwest and Mesoamerica shows that other aspects of culture and technology diffused along with agriculture.

The spread of both nomadic pastoralism, based on domesticated animals, and agriculture, based on crops and animals, occurred in many other regions of the world. In some cases, climate, soils, and available crops led to complex stratified societies based on a variety of crops and animals. By 2000 years ago, such societies were present in Mesoamerica and South America, much of northern Africa, Europe, and southern Asia. In other regions, such as southwestern and eastern North America, simpler agricultural societies, based on a smaller number of domesticated species and often having only semipermanent villages, had also developed. Nomadic pastoralism was dominant in grassland and desert regions of Eurasia and Africa. European expansion, following the fifteenth century, would lead to further expansion of agriculture to areas such as Australia and central and northwestern North America. It would also lead to the worldwide diffusion of crops such as wheat from the Near East, rice from Asia, and corn from the Americas. Today, for example, California is one of the largest producers of rice in the world, while China is a major producer of corn. The environmental transformation that the global spread of agriculture has wrought is massive in scope. Croplands and pasturelands now occupy about 40% of the earth's land surface. This transformation of huge areas of the planet's ecosystems has created widespread changes in climatic, hydrologic, geomorphologic, and geochemical processes and conditions that have had profound and often deleterious implications for plants and animals. On the other hand, the world's human population of 8 billion people depend upon the food provided by this incredible transformation. To keep pace with increasing global population, the earth's cropland area increased by 9% between 2003 and 2019. Much of this recent increase in cropland has occurred in Africa and South America.

Domestication originated and took hold in some areas and not in others primarily because the availability of wild plants and animals that can be domesticated varies from geographic region to geographic region. The Fertile Crescent contained many wild plant and animal species such as the ancestors of wheat, barley, rye, lentils, chickpeas, sheep, goats, and pigs, which were ideal for domestication. Both grains, which are high in carbohydrates, and animals, which could be domesticated and furnish meats high in proteins, were available. This abundance and variety of plants and animals that could be domesticated explain the rapid development of an agriculture-based society there. Southern Mexico and Central and South America also had a number of species suitable for domestication, such as corn and potatoes, but had fewer high-yield plants and no large animals aside from the llama and alpaca for domestication. These more limited resources may explain the relatively slow rate of transition from hunting and gathering to a more or less completely agricultural society in Central and South America. Finally, areas such as much of North America, northwestern Europe, Australia, and other regions simply did not possess the variety of plant and animal species that could be domesticated and form the basis for the transition to a fully agricultural society. In northwestern Europe, North America, or Australia, for example, there are no native species of plants that possess the similar combined properties of nutrition, crop yield, harvesting-processing ease, and range of growth conditions that have made wheat, Asian rice, and corn such important crops around the world. Although humans have shown a remarkable ability to shape the evolutionary development of many plants and animals, they have required the proper raw materials to accomplish this. Those raw materials are the plant and animal species that humans were able to domesticate because of their morphology, physiology, or behavior. Perhaps now, with the advent of biochemical techniques that allow the alteration of genetic codes and the splicing of new genes into organisms, we are shaping the physiology and morphology of domesticated species and have the potential to fashion new ones that even more specifically suit our perceived needs. By 2020, over 500 such genetic modifications had been made across 32 different crops. The wisdom of such molecular manipulations of the evolutionary process remains highly controversial however.

MODERN HUMANS AS A FORCE OF EXTINCTION

The growth and geographic spread of *Homo sapiens* populations has been a remarkable episode in the biogeographic history of the earth. However, this growth and expansion have not been without significant cost. Perhaps the most regrettable byproduct of human success has been the global extinction of a great many plant and animal species. Unfortunately, the rate at which our species has caused other species to go extinct appears directly tied to human population size. As human population has exploded over the past few centuries, so has the number of plants and animals that we have lost forever to extinction. Niles Eldredge has compared the magnitude of these recent extinctions to five earlier mass extinctions represented in the geologic record. These are the Late Ordovician, Late Devonian, Late Permian, Late Triassic, and Late Cretaceous extinction events (see Chapter 9). He coined the term **sixth extinction** to refer to the episode of extinctions over the past 10,000 years or so that have been caused by humans. All of the five earlier mass extinctions were caused by catastrophic geologic events or strikes by extraterrestrial objects such as meteors. Although it is arguable that the number of extinctions over the past 10,000 years does not yet equal the massive loss of species during these earlier geologic periods, neither have these earlier extinctions been caused by the actions of one species—that species being *Homo sapiens*. The term *sixth extinction* has become widely used to describe the recent loss of species attributable to human actions over the past 500 years.

Human beings have caused the extinction of other species through both direct and indirect processes. In a number of cases, direct predation by humans has led to extinction. In other cases, the removal or reduction of prey species has created trophic cascades that have led to the extinction of predators. Humans have also introduced alien species into environments, and these have driven other species to extinction through predation or competition. Land clearance and modification are cited as the leading causes of recent extinctions and have caused species of plants and animals to go extinct due to loss of natural habitat. In many cases, extinctions have been due to combinations of factors, such as aggressive hunting coupled with the destruction of habitat due to land clearance.

Extinctions caused by modern humans can be divided into two general categories. The first category is a series of **prehistoric extinctions** coinciding with the initial geographic expansion of human beings from Africa and southern Eurasia. The second category is a series of **historic extinctions** caused by the combination of European colonization and socioeconomic expansion with the rapid global increase in human population and transformation of the earth's landcover over the past 500 years. Let's look at both prehistoric and historic extinctions.

Prehistoric Extinctions

Prehistoric extinction events can be hard to attribute, but those thought to be caused by the initial geographic expansion of modern humans can be regarded as being similar to the introduction of a nonselective (or euryphagous) predator into a new environment. Much of the remarkable success of modern humans in terms of geographic range and population size can be attributed to our relatively wide generalized niche. Humans are omnivores adapted to take advantage of a number of different prey species and plants. The increasingly sophisticated tools and social organization of prehistoric modern humans made them a very efficient predator. If a favored prey species goes extinct, humans can shift reliance to other animal prey and plants. Prey species that coevolve in the same geographic region as a nonselective predator will evolve adaptive strategies to persist in the face of this predation. However, when nonselective predators colonize or are artificially introduced to an environment where prey species have not coevolved with them, the prey often are not adapted to escape or otherwise withstand predation by the new predator. The results can often be extinction of the prey species. Hominids lived and evolved in Africa for most of their history. Other animals here evolved with hominids as predators. In the past 60,000 years, modern humans spread to regions where animals had evolved without the presence of hominid predators. We have less evidence of prehistoric humans causing widespread extinction of animals in Africa or southern Asia where humans had a long presence. We do have evidence of large prehistoric extinction events in areas such as North America, South America, Australia, and the islands of the South Pacific where modern humans have spread in the last 60,000 years. Let's look at some of these examples for evidence of large-scale animal extinctions and the potential role of humans.

Fossil evidence from many sites shows that until the Early Holocene, about 12,000 years ago, North America contained a richness of large mammal species that is remarkable by modern standards. Large mammals are those species in which the adults are greater than 44 kilograms in weight. The large mammal fauna are often referred to as the **megafauna.** Among other things, the megafauna contained species of mammoth *(Mammuthus primigenius, M. jeffersonii,* and *M. columbi)* and mastodon *(Mammut americanum)* related to elephants (Fig. 12.6), horses *(Equus* spp.), camels

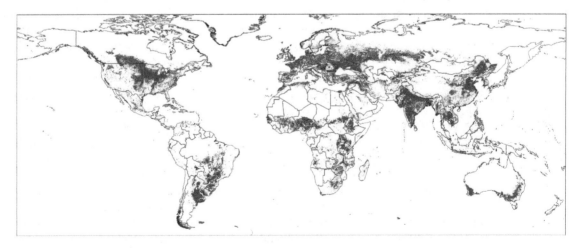

FIGURE 12.6 World extent of areas dominated by cropland (does not include pastureland and rangeland), 2000–2019 (*Source:* Potapov et al., 2022 / Springer Nature / CC by 4.0.).

FIGURE 12.7   The skeleton of a Jefferson mammoth *(Mammuthus jeffersonii)* at the American Museum of Natural History in New York. This species of mammoth was approximately the same size as living elephants and was present in midwestern and eastern North America during the end of the Pleistocene. Mammoths were the largest mammals in North America at that time. By around 12,000 years ago, mammoths in North America had become extinct in most regions (*Source:* Photo-Denis Finnin / American Museum of Natural History).

(Camelops hesternus), Shasta ground sloths *(Nothrotheriops shastensis),* sabertooth cats *(Smilodon fatalis),* and short-faced bear *(Arctodus simus).* The youngest radiocarbon dates from the remains of these extinct mammals suggest that all of the North American representatives of these large mammals, and other notable large mammal species, disappeared around 13,000 to 5000 years ago (Fig. 12.7). The last surviving species of the extinct megafauna were mammoths, isolated on some of the islands adjacent to Alaska and Siberia. In fact, some 38 genera of North American mammals went extinct with the close of the Pleistocene. When Europeans arrived, the largest mammals remaining in North America were species such as moose *(Alces alces)* and bison *(Bison bison)* in temperate and boreal environments, and muskox *(Ovibus moschatus)* in the tundra of the far north. Interestingly, small mammals did not appear to suffer the same high rate of extinction during the period of large mammal extinctions.

Several hypotheses have been posited to explain the megafaunal extinctions that occurred in North America at the close of the Pleistocene. One hypothesis attributes the extinctions to climatic change. According to this hypothesis, large mammals could not cope with the rapid changes in climate and vegetation at the end of the last glacial maximum and beginning of the Holocene. This was because larger mammals were more sensitive to climatic and environmental conditions and had larger resource requirements, larger individual feeding ranges, smaller populations, and lower reproductive rates than smaller species. A critique of this argument is that these same genera, and indeed many of the same species, had survived earlier glacial to interglacial transitions over the Pleistocene. What was so different about the end of the last glacial episode that made it lethal to the North American large mammals? One related hypothesis is that the rapid climatic and environmental changes associated with the Younger Dryas around 13,000 years ago may have placed exceptional stress on megafauna populations and contributed to their extinction (Fig. 12.8). Some scientists have suggested that one or more extraterrestrial objects hitting the surface of the earth or exploding in the atmosphere may have both triggered the Younger Dryas climatic event and directly induced megafaunal extinctions. The extinctions would have occurred through climatic and environmental changes, including the generation of widespread fires. This hypothesis has been sharply criticized by other researchers and remains fiercely debated.

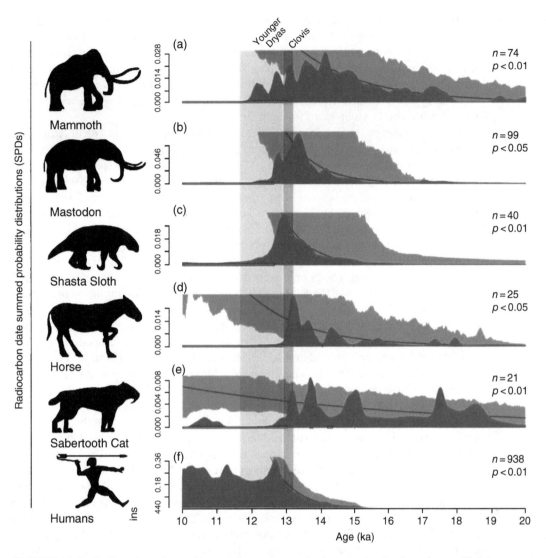

FIGURE 12.8 Radiocarbon date evidence of presence and subsequent extinction of some Pleistocene mega-fauna and the initial presence of humans in the conterminous United States and southern Canada during the Late Pleistocene and Early Holocene (*Source:* Broughton and Weitzel, 2018 / Springer Nature / CC by 4.0.).

Another hypothesis attributes the North American large mammal extinction directly to humans. This hypothesis was particularly well developed and articulated by the American geoscientist Paul S. Martin of the University of Arizona. As we have reviewed earlier, there is compelling evidence that humans entered Alaska and the Yukon Territory from Siberia about 15,000 years ago and spread southward as the Laurentide ice sheet melted about 13,000 years ago. Martin theorized that around 13,000 years ago, newly arrived immigrants brought with them a hunting-and-gathering culture from Siberia that included the hunting of large game such as horse, bison, and mammoth. When these hunters first arrived in the interior of North America, they would have found many different species of large mammals for hunting. These species would never have been preyed upon by human hunters and would not have developed physical or behavioral adaptations to human hunting pressure. Martin believes that the humans would have had great initial success in hunting and as a result experienced rapid population growth. Ease of hunting, coupled with increasing numbers of human hunters, would lead to local extinction of large mammal prey species. Local extinction of this type by a predator is called **overkill.** This in turn would cause humans to move generally southward in search of new prey. The geographic expansion of humans southward in search of new prey explains the apparent rapid spread of humans from Canada to the southern tip of South America 12,000 years ago. Behind the southward-moving front of the large and geographically expanding population of human hunters would be a sparser population of humans who would likely hunt the remaining large mammals. Individuals of the large mammals that escaped the first arrival of these human hunters would likely be hunted and potentially driven to extinction by the humans who remained behind the advancing front. The hypothesized expanding front of human hunters is sometimes called the kill front. The movement and actions of the human hunters under Martin's scenario are somewhat like a rapidly advancing army, and his hypothesis is referred to as the **Blitzkrieg Theory of Pleistocene Overkill.** (*Blitzkrieg* is the German word for lightning war.)

What evidence is there to substantiate the Blitzkrieg Theory in North America? Modern humans may have reached North America between 24,000 and 16,000 years ago, prior to the megafaunal extinction. The evidence of such early migrants and their tools remains sparse. However, later evidence of human presence and their technologies is much more abundant. The stone tools used by people in much of North America and portions of Mesoamerica from around 13,000 and 12,000 years ago are very similar. They are characterized by the presence of large points with fluted bases. The fluting appears to have been used to fasten the points to spear shafts. These very distinctive artifacts are called **Clovis points** after the town near the archaeological site in New Mexico where the points were first discovered. The first Clovis points were found in 1929 by a teenager named Ridgely Whiteman, who reported his find to the Smithsonian Institution. The people who used these points are often referred to as Paleoindians. The points and other features of the Paleoindian tool kit are somewhat similar to prehistoric tools from older archaeological sites in Siberia. The Siberian peoples associated with these tools appear to have hunted mammoths and can be linked genetically with modern North American native peoples. It is likely, however, that the distinctive fluting of the Clovis point technology developed in North America. In North America, there are a number of archaeological sites where Clovis points and tools are found in association with butchered mammoths. The site in New Mexico where the points were first discovered included the bones of butchered mammoths. All of these types of sites date from about 13,000 to 12,000 years ago, and many contain the remains of several mammoths. Clovis points are also found in association with the remains of other living and extinct mammals. This evidence shows that at the time of the Late Pleistocene extinctions, the Paleoindians in North America were generalist hunters and were capable of hunting very large mammals. Recent studies also suggest that habitat changes caused by human-set fires may have also contributed to the megafaunal extinctions in portions of North America.

The evidence that humans played a role in the extinction of many large mammals in North America appears very strong. There is also evidence of an association between mammoth extinction and human hunters in adjacent Siberia. However, it cannot be said that human hunting alone was responsible. It is likely that the North American biota was under stress during the Late Pleistocene as climate warmed and vegetation changed. The foraging conditions and ranges for the large mammals were shifting geographically and, in some cases, contracting. Rapid climatic shifts of the Younger Dryas would have exerted additional stresses on large mammals. The addition of humans, who were highly sophisticated and generalist predators, was potentially more than the already environmentally stressed large mammal populations could withstand in some areas. Vegetation alteration by the use of fire by humans may have also had an impact on the megafauna at this time. The removal of herbivores such as mammoths, mastodons, horses, and camels likely produced trophic cascades that led to the extinction of large predators such as sabertooth cats. Interestingly, the disappearance of the Clovis point technology, which appears centered on large game such as mammoths, may reflect the disappearance of the megafaunal prey species and a shift by humans to smaller game.

The magnitude of Late Pleistocene mammal extinctions is no less severe in Australia than in North America. Indeed, the Australian extinctions were broader in scope. All marsupials weighing more than 100 kilograms disappeared between about 60,000 and 15,000 years ago. Although the fossil evidence from Australia is not as extensive as that from North America, it has been estimated that over 50 species of mammals, including 29 megafaunal species, went extinct in Australia and New Guinea during the Late Pleistocene. During glacial periods, New Guinea, Australia, and Tasmania formed one landmass that is referred to as **Sahul.** The Australian mammalian fauna was dominated by marsupials, the largest being the *Diprotodon optatum,* which was the size of a hippopotamus. All of the largest marsupials, those weighing more than 100 kilograms, went extinct. Reptiles and birds also went extinct during this period. These include a 7-meter-long carnivorous monitor lizard (*Megalania prisca*) and a flightless ostrich-sized bird (*Genvornis newtoni*). Most of the extinct Australian megafauna had disappeared by about 46,000 years ago.

What was the potential role of humans in the Australian Late Pleistocene extinctions? There is good evidence that humans first arrived in Australia about 60,000 to 43,000 years ago. This time period corresponds with the last known ages for the large flightless bird *Genyornis newtoni* and a number of other marsupial species. Hunting by humans was likely a factor in these extinctions. Fossil pollen and charcoal in lake sediments also show evidence of increased fires and losses of forest and productive shrubland during the period 60,000 to about 20,000 years ago. Humans may have been responsible for setting frequent fires, which transformed the interior of the continent to more sclerophyllous shrubland or desertlike conditions. Thus, humans may have killed off species both by hunting and by creating fires that changed the vegetation. Detailed morphological and chemical analysis of the bones and eggs of extinct Australian vertebrates suggests that many were browsers who lived by eating the leaves of shrubs and trees. The replacement of forest and shrubland by sparser vegetation and desert would have destroyed their habitat. The loss of habitat explains why so many small as well as large animals went extinct during the Late Pleistocene in Australia. Trophic cascades would have led to the extinction of large predators such as *Megalania prisca*. Natural climatic change likely also played a role in the latter prehistoric faunal extinctions in Australia. During the last glacial maximum 20,000 years ago, Australia became very dry and experienced increased desertification. Some authorities argue these climatic and vegetative changes were the most important drivers of mammalian extinction.

In both North America and Australia, it might be argued that natural climatic change, in addition to human activity, contributed to prehistoric extinctions. The exact balance between these forces may remain difficult to resolve and engender further research and debate. In the case of the Pacific Islands, we have clear evidence of the power of prehistoric humans to cause widespread extinction in the absence of any other environmental changes. The eastward expansion of Polynesians between 2100 and 700 years ago brought humans to many uninhabited islands such as Hawaii and New Zealand. Although these remote islands did not possess mammal faunas aside from bats, they did have distinctive bird faunas (avifauna). New Zealand, for example, had a number of species of large flightless birds called moas (Aves, Dinornithiformes). These birds ranged in size from 20 kilograms to a staggering 250 kilograms. Archaeological evidence shows that when humans first arrived on the islands of New Zealand about 700 years ago, they slaughtered and ate moas in huge numbers. It is estimated that all of the moa species were driven to extinction within about 150 to 200 years following Polynesian settlement. Following the extinction of the moas, there is archaeological evidence that humans turned their focus to other prey, including now extinct species of penguin and petrels which nested along the coast. At least 76 species of birds went extinct in New Zealand following Polynesian colonization. Similar avifaunal extinctions are evident on other Pacific Islands that were formerly uninhabited by humans. Hawaii was discovered and occupied by humans about 1000 years ago. Fossil evidence from Hawaii shows that at least 35 species, and perhaps over 56 species, of birds went extinct following the arrival of the Polynesians. In some cases, these birds were hunted principally to supply feathers for ornamentation of clothing and jewelry. Flightless birds were particularly vulnerable to extinction. Hawaii had a number of flightless birds, including eight species of geese and an ibis species that went extinct following the arrival of the Polynesians. On average, about 50% of bird species on Pacific islands went extinct following Polynesian arrival and later European settlement.

Aside from direct hunting, several other factors contributed to the extinction of island biota. The Polynesians brought agriculture with them as they expanded across the Pacific. They transported and introduced domesticated animals, including dogs and pigs. In addition, they inadvertently introduced rats. All these mammals preyed upon bird, reptile, and invertebrate faunas and contributed to Pacific Island extinctions. Pigs and rats also preyed upon plants and their seeds, and rats were likely an important contributing factor to the extinction of island plant species such as the native palms on Rapa Nui (Easter Island). The Polynesians also brought a number of crop species and edible plants, including coconut *(Cocos nucifera)*, taro *(Colocasia esculenta)*, yams *(Discorea)*, breadfruit *(Artocarpus communis)*, sweet potato *(Ipomea batatas)*, and other plants, together with many weedy species. In order to grow taro, the Polynesians cleared large areas of lowland vegetation. On wetter sites, the Polynesians practiced slash-and-burn agriculture. Finally, they cut down additional forest at low to mid-elevations for building materials and firewood. Following these clearances, much of the original island vegetation was replaced by plants introduced by the Polynesians. Today on the Hawaiian Islands, much of the vegetation below 760 meters in elevation is dominated by plants introduced by the Polynesians and later by the Europeans. It has been estimated that about 130 species, or 10%, of Hawaii's native plants were driven to extinction. Undoubtedly, this landscape alteration on Hawaii significantly contributed to the extinction of birds and other animals, such as snails and insects, by destroying habitat. The Polynesians dramatically changed the landscape on other islands. When the Europeans first arrived and settled in New Zealand between 1642 and 1800, they found the lowland landscape largely dominated by grasslands and extensive areas of bracken ferns *(Pteridium aquilinum)*. Evidence from fossil soils, wood, and pollen shows that, prior to the arrival of the Polynesians, these areas had been dominated by forests of trees such as southern beech *(Nothofagus)*. Charcoal found in soils shows that the forest clearance was caused by burning. Some of the forest was likely cleared to facilitate the cultivation of sweet potato. However, there was also much clearing beyond the agricultural limits of this crop. Other areas were likely burned to ease the hunting of moas and to encourage the growth of bracken, which was used as food by the Polynesians.

Although we have focused on North America, Australia, and the Pacific Islands, the geographic expansion of *Homo sapiens* in the Late Pleistocene and Holocene is associated with increased extinction rates in many other parts of the world. Such areas include northern Eurasia, islands in the Mediterranean Sea, and the island of Madagascar off Africa. Humans arrived in Madagascar about 2000 years ago. All animals with a body mass larger than 12 kilograms were driven to extinction shortly thereafter. It would be wrong to think that the Indigenous Americans, Australians, or Polynesian islanders are in any way worse than other groups of humans in terms of their intentional impact on the environment and resulting extinctions. The aim of these and other peoples was not to produce massive extinctions of other species; it was simply to survive.

Historic Extinctions

The past 500 years have witnessed a wave of extinctions of plants and animals that goes far beyond what would be considered a normal background rate of species loss and is greater in magnitude and geographic scope than any of the prehistoric extinction events associated with humans (Fig. 12.8). In the case of mammals, a conservative estimate is that at least 85 species have gone extinct since 1500. Many of the mammal species lost were rodents. Some 3% of the newly extinct mammals belonged to our own order, the Primates. One analysis of historical records concluded that at least 51 mammal species have disappeared in the time period from 1750 to the present. The others went extinct sometime between 1500 and 1750. Over 70% of those mammalian extinctions occurred on islands, many in the Caribbean, where at least 29 mammal species have gone extinct since 1500. The Caribbean losses include a species of monkey *(Xenothrix mcgregori)* that once lived on Jamaica. Many of the continental mammals that went extinct during the historic period were from Australia. At least 28 Australian land mammal species have gone extinct since the start of European settler colonialism in 1788. The historic mammal extinctions in Australia range from small marsupials to species as large as kangaroos. In contrast, North America has experienced the extinction of a number of mammalian subspecies, but limited mammal species extinction since 1500. One famous subspecies of mammal that has gone extinct in North America is the California grizzly bear, which was a variety of the grizzly bear *(Ursus arctos californicus)*. Hunting and intentional poisoning by Spanish, Mexican, and American settlers was the primary cause of the demise of this distinctive variety of grizzly. One of the last California grizzly bears died in the San Francisco Zoo in the early twentieth century. For birds, the rate of extinction over the past 500 years has been even more severe than for mammals. The evidence suggests that 159 to well over 200 species of birds have gone extinct in the historic period.

Species that have evolved on and are endemic to islands have been particularly vulnerable to extinction since 1500. One recent study concluded that 95% of the mammal and bird species that have gone extinct since 1500 were endemic to islands and the relatively small and isolated continent of Australia. The rate of extinction for continental mammal and bird species since 1500 may have been up to 7.4 and 5.9 times higher than expected natural extinction rates, respectively. For island mammal and bird species, the rate of extinction may have been up to 702 and 844 times higher than expected natural extinction rates.

Freshwater fish have also suffered greatly increased rates of extinction in the historic period. Over the past 100 years, somewhere between 80 to over 170 species of fish have gone extinct. Many of the extinct species were those that lived in lakes, rather than streams and rivers. A number of those species have been cichlid fish from Lake Victoria in Africa. Lake fish often cannot escape from overfishing or environmental changes caused by humans. Historic extinctions have also been high for many other groups of organisms. Over the past 500 years, 299 to 638 species of mollusks have been lost, and 63 to over 190 species of insects have disappeared. Island mollusks have been particularly susceptible to extinction. Of the flowering plants, various studies suggest that somewhere between 124 and 713 species have gone extinct. Again, a high proportion of these extinctions have occurred on islands. Unfortunately, the known historic extinctions such as presented in Fig. 12.8 often only represent conservative minimum estimates of the actual number of species lost. Undoubtedly, hundreds of species went extinct before they were known to scientists. This is particularly true for insects, mollusks, and smaller plants. For example, a recent analysis concluded that it is likely some 3,000–5,100 mollusk species have gone extinct since 1500. Another study concluded that it is possible that 250,000 to 500,000 insect species have gone extinct in the past 200 years, the vast majority unknown to science. We will likely never know how many species have gone extinct and what most of these lost organisms were like. What we can be certain of is that thousands of species of plants and animals have gone extinct over the past 500 years.

The extinctions of plants and animals observed over the past 500 years have occurred in conjunction with the unprecedented growth of the world's human population. Our numbers have grown from about 450 million people to almost 8 billion. We are products and beneficiaries of this growth, but this growth has also increased many environmental pressures conducive to increased rates of extinction, including habitat alteration, increased hunting and fishing, pollution, and introduced alien species of predators, competitors, and pathogens. As the human population has grown since 1500, so too has the tempo of extinctions. One estimate suggests that over the past 100 years the rate of extinction for vertebrate species has surged to 100 times what would be expected under natural conditions. Habitat alteration is perhaps one of the most important forces driving historic extinctions. As human populations have grown, they have altered terrestrial landscapes for agriculture (see Fig. 12.9), urban development, and resource extraction such as mining and logging. Such alterations of the landscape destroy the habitat of organisms. Human alteration of the environment also impacts heavily on freshwater and coastal ecosystems. In the case of extinct fish, about 70% of the species lost over the past 100 years were victims to habitat alterations such as the damming of rivers, changes in lake water levels, additional sedimentation due to human land use, changes in river channels and lake bottoms, and disruption of aquatic vegetation.

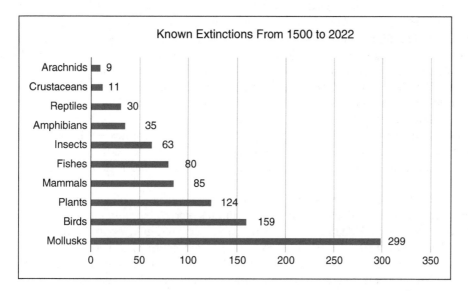

FIGURE 12.9 Conservative account of historical species extinctions based on numbers of known extinctions from the IUCN Red List (IUCN, 2022). Actual numbers of extinctions, particularly for plants and smaller organisms, are much higher.

We also have good historical evidence that aggressive hunting played a central role in the historical extinctions of a number of animals such as the great auk *(Pinguinus impennis)* and Steller's seacow *(Hydrodarmalis gigas)*. The great auk was a flightless seabird of the North Atlantic that was very much like the penguins of the southern hemisphere. It lived in great numbers on rocky coastlines from Scotland to Newfoundland, Canada. During the 1800s, European sealers, fishermen, and sailors began to intensively hunt the great auk for food and to provide fish bait. Hunters were able to herd the birds straight from the shore to their deaths in the holds of ships. In addition, egg collectors in Europe paid high prices for great auk eggs; the taking of eggs for collectors may have been a major factor in the final demise of the birds. By 1844, the great auk had gone extinct. Steller's seacow, which was similar to manatees found in Florida, was first discovered in the Bering Sea in 1741 by George W. Steller, the naturalist on a Russian expedition led by Vitus Bering. The seacow became a prime prey for Russian sealers, and by 1768 it was driven to extinction. In addition to hunting animals for food, sport, and collecting, people have hunted predatory animals to protect their domesticated livestock. A large predatory marsupial called the Tasmanian Tiger *(Thalacinus cynocephalus)* was a direct victim of eradication efforts. European colonists, perceiving the Tasmanian Tiger to be a major threat to sheep herds, launched a vigorous hunting campaign against the animals and by the 1930s it had disappeared.

Pollution has been an especially important factor in the extinction of many fish and other aquatic species. Many rivers and lakes have received large amounts of pollution either through the direct dumping of sewage and waste or through inadvertent spills and the runoff of polluted waters from urban and agricultural areas. Pollution has caused or played a major role in the historic extinction of three species of chub *(Evarra)* from the Mexico City region. It has also been blamed for the extinction of fish species such as *Chondrostoma scodrensis* in Lake Scadar, Montenegro, and *Brycon acuminatus* and *Leptolebias marmoratus* in Brazil.

As European exploration and colonization led to frequent travel and trade between continents and islands, alien species of plants and animals were introduced either by accident or design. In Chapter 4 we discussed how the accidental introduction of mosquitoes carrying avian malaria led to the extinction of many bird species in Hawaii. Organisms that are purposely introduced to new areas can have similar disastrous results. It is often not possible to foresee the extinctions that may result from such introductions. The French introduced the African land snail *(Achatina fulica)* to the islands of French Polynesia as a source of food. However, the African snails quickly became serious pests to native vegetation and crops. To remedy this situation, the predatory snail *Euglandia rosea* was introduced to the islands to combat the African land snail. Unfortunately, this nonselective predator began to prey upon native island snails. Predation by *Euglandia rosea* caused the extinction of many native Pacific Island snail species.

In many cases, historic extinctions result from a combination of factors, as is illustrated by the dodo bird *(Raphus cucullatus)*. The dodo was first discovered on Muritius Island by European sailors in 1507. The flightless birds were easy prey, and so the island became a stop for providing meat for sailing ships. Europeans also cleared the native forests, which likely deprived the dodos of food. In addition, cats, rats, and pigs introduced by the Europeans probably preyed on the dodo nestlings and eggs. By the 1660s to 1680s, the dodo was extinct. During the same period, a similar fate befell the dodo's nearest relatives, *Raphus solitaires* and *Pezophaps solitaria,* flightless birds that lived on nearby islands in the Indian Ocean. This extinction event represents the loss of the entire Raphidae family to which the three species belonged. In more recent times, we have even clearer historical evidence of how a number of interacting factors produced the extinction of freshwater cichlids in Lake Victoria and the continental extinction of the North American passenger pigeon. Let's look at both of these examples.

It is unknown precisely how many species of cichlid fish *(Haplochromis)* have gone extinct in Lake Victoria in recent decades, but some estimates place the number at over 100. Many of these species were very specialized in terms of feeding and occurred in low population numbers in small areas of the lake. They were thus particularly prone to extinction. The widespread extinction of cichlids corresponds with a number of human-induced changes in Lake Victoria. During the twentieth century, the Nile perch *(Lates niloticus)* was introduced to the lake to develop a fishery that would support people living around the lake. The perch are predators that feed on the cichlids. After the introduction of perch, the cichlids also experienced increased fishing pressure. Cichlids were caught for use as bait or were inadvertently caught in perch nets. Cichlids were also killed as a byproduct of the application of fish poison to catch perch. By the late 1980s, an introduced aquatic plant, water hyacinth *(Eichhornia crassipes),* had spread throughout the lake and covered many shallows with dense floating foliage. The plant cover destroyed some cichlid habitat by causing decreased sunlight penetration and decreased water oxygen concentrations. Finally, land clearance for agriculture and increased human population density around the lake led to increased inputs of pesticides, fertilizer, and sediment, all of which contributed to fish mortality and the loss of cichlid species. Some of the changes caused by humans to Lake Victoria were instituted with the best intentions of aiding local peoples, but the end result has been a catastrophic disruption of the ecosystem and the extinction of many species. One ray of positive news is that in recent years some cichlid species thought extinct have experienced a resurgence.

In many cases, the animals that have gone extinct in historic times have been species that had small population sizes and were highly endemic. In Chapter 9, we saw how such species are generally more prone to extinction. The small population sizes, highly endemic distributions, and often limited dispersal capability of island species made them particularly susceptible to extinction due to human hunting, introduced predator and competitor species, and landscape change. Historic extinctions, however, have certainly not been restricted to species with small population sizes or endemic distributions. Passenger pigeons *(Ectopistes migratorius)* were beautiful birds that resembled mourning doves and turtle doves (Fig. 12.10). When the Europeans first arrived in North America, the most abundant bird in the continent was the passenger pigeon. It is estimated that 5 billion passenger pigeons inhabited the eastern half of North America. That is almost equal to the total number of all birds that inhabit the continent today! The passenger pigeons were strong fliers, and the geographic range of the birds extended from the Gulf of Mexico to southern Canada. They traveled in great flocks that numbered in the millions, and it is said that their passing would darken the sky. During the mid-1800s, two factors began to seriously impact the passenger pigeon. First, land clearance for agriculture intensified and expanded westward from the Atlantic Coast states to the edges of the Great Plains. This destroyed the forest habitat of the pigeons. The pigeons favored oaks, chestnuts, and beech as sources of food. The birds also required trees to nest. The cutting of a favored nesting tree could destroy over 100 nests. Second, the pigeons were hunted in great numbers. Random shots into the sky would bring down several birds when flocks were passing. It is estimated that one day of hunting using shotguns could yield up to 25,000 birds. The birds were also caught in nets when resting and got stuck in sticky lime spread out on the ground. Millions were killed each year. The expansion of the railroad in the 1850s led to increased hunting as the birds could be easily transported to markets. Some of these birds were eaten by humans, and others were used to feed pigs. Many birds were

FIGURE 12.10   Museum specimen of the extinct North American passenger pigeon *(Ectopistes migratorius).*

ground up and used for fertilizer. Pigeons were also killed simply to keep them from eating newly planted crops. To get some perspective on this wide-scale slaughter, it is worth noting that in 1869 one Michigan county sent 7.5 million birds to markets in the east. During the 1870s and 1880s, a drastic decline was observed in the number of passenger pigeons. By 1880, the entire state of Michigan shipped only about 500,000 birds eastward. Even though commercial hunting eventually ceased, the decline could not be halted. By the 1890s, the passenger pigeon had disappeared from most of its native range. On September 1, 1914, the last living passenger pigeon died in captivity at the Cincinnati Zoo. The most abundant bird on the North American continent was driven to annihilation in a little over a century. A monument to the passenger pigeon was subsequently erected in a park in Wisconsin. The words on the monument are a sad but fitting epitaph for far too many species of plants and animals: "This species became extinct through the avarice and thoughtlessness of man."

KEY WORDS AND TERMS

Agriculture
Artificial selection
Blitzkrieg Theory of Pleistocene Overkill
Centers of domestication
Clovis points
Domestication
Environmental determinism
Fertile Crescent
Historic extinctions
Hunters and gatherers
Megafauna
Mesoamerica
Neolithic Revolution
Oasis Theory
Overkill
Pastoral nomadism
Prehistoric extinctions
Sahul
Sixth extinction
Younger Dryas

REFERENCES AND FURTHER READING

Abbo, S., Lev-Yadun, S., and Gopher, A. (2010). Agricultural origins: centers and noncenters; a Near Eastern reappraisal. *Critical Reviews in Plant Science* **29**, 317–328.

Alberto, F. J., Boyer, F., Orozco-terWengel, P., Streeter, I., Servin, B., de Villemereuil, P., Benjelloun, B., Librado, P., Biscarini, F., Colli, L., and Barbato, M. (2018). Convergent genomic signatures of domestication in sheep and goats. *Nature Communications* **9**, 813.

Arranz-Otaegui, A., Gonzalez Carretero, L., Ramsey, M. N., Fuller, D. Q., and Richter, T. (2018). Archaeobotanical evidence reveals the origins of bread 14,400 years ago in northeastern Jordan. *Proceedings of the National Academy of Sciences* **115**, 7925–7930.

Baker, H. G. (1970). *Plants and Civilization*, 2nd edition. Wadsworth Publishing, Belmont, CA.

Baldini, J. U., Brown, R. J., and Mawdsley, N. (2018). Evaluating the link between the sulfur-rich Laacher See volcanic eruption and the Younger Dryas climate anomaly. *Climate of the Past* **14**, 969–990.

Balter, M. (1999). A long season puts Catalhoyuk in context. *Science* **286**, 890–891.

Balter, M. (2010) The tangled roots of agriculture. *Science* **327**, 404–406.

Becerra-Valdivia, L., and Higham, T. (2020). The timing and effect of the earliest human arrivals in North America. *Nature* **584**, 93–97.

Bell, M., and Walker, M. J. C. (1992). *Late Quaternary Environmental Change: Physical and Human Perspectives*. Wiley, New York.

Bellwood, P. (2023). *First Farmers: The Origins of Agricultural Societies*, 2nd edition. Wiley, Hoboken, NJ.

Blumler, M. A. (1992). Independent inventionism and recent genetic evidence on plant domestication. *Economic Botany* **46**, 93–111.

Blumler, M. A., and Byrne, R. (1991). The ecological genetics of domestication and the origins of agriculture. *Current Anthropology* **32**, 23–54.

Boyer, A. G. (2008). Extinction patterns in the avifauna of the Hawaiian islands. *Diversity and Distributions* **14**, 509–517.

Brace, S., Diekmann, Y., Booth, T. J., van Dorp, L., Faltyskova, Z., Rohland, N., Mallick, S., Olalde, I., Ferry, M., Michel, M., and Oppenheimer, J. (2019). Ancient genomes indicate population replacement in Early Neolithic Britain. *Nature Ecology & Evolution* **3**, 765–771.

Broughton, J. M., and Weitzel, E. M. (2018). Population reconstructions for humans and megafauna suggest mixed causes for North American Pleistocene extinctions. *Nature Communications* **9**, 5441.

Brown, J. H., and Lomolino, M. V. (1998). *Biogeography*, 2nd edition. Sinauer Associates, Sunderland, MA.

Carey, J. (2023). Unearthing the origins of agriculture. *Proceedings of the National Academy of Sciences* **120**, e2304407120.

Carlson, A. E. (2010). What caused the Younger Dryas cold event? *Geology* **38**, 383–384.

Caro, T., Rowe, Z., Berger, J., Wholey, P., and Dobson, A. (2022). An inconvenient misconception: climate change is not the principal driver of biodiversity loss. *Conservation Letters* **15**, e12868.

Ceballos, G., Ehrlich, P. R., Barnosky, A. D., García, A., Pringle, R. M., and Palmer, T. M. (2015). Accelerated modern human–induced species losses: entering the sixth mass extinction. *Science Advances* **1**, e1400253.

Charmet, G. (2011). Wheat domestication: lessons for the future. *Comptes rendus biologies* **334**, 212–220.

Cobo, J. M., Fort, J., and Isern, N. (2019). The spread of domesticated rice in eastern and southeastern Asia was mainly demic. *Journal of Archaeological Science* **101**, 123–130.

Cowie, R. H., Bouchet, P., and Fontaine, B. (2022). The Sixth Mass Extinction: fact, fiction or speculation? *Biological Reviews* **97**, 640–663.

Cowie, R. H., Regnier, C., Fontaine, B., and Bouchet, P. (2017). Measuring the sixth extinction: what do mollusks tell us? *Nautilus* **131**, 3–41.

Cuddihy, L. W., and Stone, C. P. (1990). *Alteration of Native Hawaiian Vegetation*. University of Hawaii, Honolulu.

Diamond, J. (2007). Easter island revisited. *Science* **317**, 1692–1694.

Diamond, J. M. (1984). Normal extinctions of isolated populations. In *Extinctions* (ed. M. N. Nitecki), pp. 191–246. University of Chicago Press, Chicago.

Diamond, J. M. (1992). *The Third Chimpanzee: The Evolution and Future of the Human Animal*. Harper Perennial, New York.

Diamond, J. M. (1997). *Guns, Germs and Steel: The Fates of Human Societies*. W. W. Norton, New York.

Diamond, J. M. (2000). Archaeology: blitzkrieg against the moas. *Science* **287**, 2170–2171.

Diamond, J. M. (2002). Evolution, consequences and future of plant and animal domestication. *Nature* **418**, 700–707.

Douché, C., and Willcox, G. (2023). Identification and exploitation of wild rye (*Secale* spp.) during the Early Neolithic in the Middle Euphrates Valley. *Vegetation History and Archaeobotany*, 1–15.

Duncan, R. P., and Blackburn, T. M. (2007). Causes of extinction in island birds. *Animal Conservation* **10**, 149–150.

Eldredge, N. (1999). Cretaceous meteor showers, the human "niche" and the sixth extinction. In *Extinctions in Near Time* (ed. R. D. E. MacPhee), pp. 1–15. Kluwer Academic/Plenum, New York.

Ellis, E. C., Kaplan, J. O., Fuller, D. Q., Vavrus, S., Klein Goldewijk, K., and Verburg, P. H. (2013). Used planet: A global history. *Proceedings of the National Academy of Sciences* **110**, 7978–7985.

Emery-Wetherell, M. M., McHorse, B. K., and Davis, E. B. (2017). Spatially explicit analysis sheds new light on the Pleistocene megafaunal extinction in North America. *Paleobiology* **43**, 642–655.

Fearn, M. L., and Lui, K. B. (1995). Maize pollen of 3500 B.P. from southern Alabama. *American Antiquity* **60**, 109–117.

Firestone, R. B., West, A., Kennett, J. P., Becker, L., Bunch, T. E., Revay, Z. S., Schultz, P. H., Belgya, T., Kennett, D. J., Erlandson, J. M., and Dickenson, O. J. (2007). Evidence for an extraterrestrial impact 12,900 years ago that contributed to the megafaunal extinctions and the Younger Dryas cooling. *Proceedings of the National Academy of Sciences* **104**, 16016–16021.

Flannery, T. F. (1999). Paleontology: debating extinction. *Science* **283**, 182–183.

Flannery, T. F., and Roberts, R. G. (1999). Late Quaternary extinctions in Australasia. In *Extinctions in Near Time Causes, Contexts and Consequences* (ed. R. D. E. MacPhee), pp. 239–269. Kluwer Academic/Plenum, New York.

Foley, J. A., DeFries, R., Asner, G. P., Barford, C., Bonan, G., Carpenter, S. R., Chapin, F. S., Coe, M. T., Daily, G. C., Gibbs, H. K., and Helkowski, J. H. (2005). Global consequences of land use. *Science* **309**, 570–574.

Gamble, C. (1994). *Timewalkers: The Prehistory of Global Colonization*. Harvard University Press, Cambridge, MA.

Georghiou, G. P. (1986). The magnitude of the resistance problem. In *Pesticide Resistance: Strategies and Tactics for Management* (ed. Committee on Strategies for the Management of Pesticide Resistant Pet Populations), pp. 14–43. National Academy Press, Washington, DC.

Gill, J. L., Williams, J. W., Jackson, S. T., Lininger, K. B., and Robinson, G. S. (2009). Pleistocene megafaunal collapse, novel plant communities, and enhanced fire regimes in North America. *Science* **326**, 1100–1103.

Goldschmidt, T. (1996). *Darwin's Dreampond*. MIT Press, Cambridge, MA.

Graham, R. W., Belmecheri, S., Choy, K., Culleton, B. J., Davies, L. J., Froese, D., Heintzman, P. D., Hritz, C., Kapp, J. D., Newsom, L. A., and Rawcliffe, R. (2016). Timing and causes of mid-Holocene mammoth extinction on St. Paul Island, Alaska. *Proceedings of the National Academy of Sciences* **113**, 9310–9314.

Gross, M. (2013). The paradoxical evolution of agriculture. *Current Biology* **23**, R667–R670.

Gross, M. (2022). Shopping with hunter-gatherers. *Current Biology* **32**, R596–R599.

Guiry, E. J., Orchard, T. J., Royle, T. C., Cheung, C., and Yang, D. Y. (2020). Dietary plasticity and the extinction of the passenger pigeon *(Ectopistes migratorius)*. *Quaternary Science Reviews* **233**, 106225.

Haak, W., Lazaridis, I., Patterson, N., Rohland, N., Mallick, S., Llamas, B., Brandt, G., Nordenfelt, S., Harney, E., Stewardson, K., and Fu, Q. (2015). Massive migration from the steppe was a source for Indo-European languages in Europe. *Nature* **522**, 207–211.

Hancock, J. F. (2005). Contributions of domesticated plant studies to our understanding of plant evolution. *Annals of Botany* **96**, 953–963.

Harrison, I. J., and Stiassny, M. L. J. (1999). The quiet crisis. In *Extinctions in Near Time* (ed. R. D. E. MacPhee), pp. 271–331. Kluwer Academic/Plenum, New York.

Hartman, G., Bar-Yosef, O., Brittingham, A., Grosman, L., and Munro, N.D. (2016). Hunted gazelles evidence cooling, but not drying, during the Younger Dryas in the southern Levant. *Proceedings of the National Academy of Sciences* **113**, 3997–4002.

Haviland, W. A. (1989). *Anthropology*, 5th edition. Holt, Rinehart and Winston, New York.

Heiser Jr., C. B. (1973). *Seed to Civilization: The Story of Man's Food*. W. H. Freeman, San Francisco.

Heywood, V. H. (1995). *Global Biodiversity*, Assessment. Cambridge University Press, Cambridge.

Hocknull, S. A., Lewis, R., Arnold, L. J., Pietsch, T., Joannes-Boyau, R., Price, G. J., Moss, P., Wood, R., Dosseto, A., Louys, J., and Olley, J. (2020). Extinction of eastern Sahul megafauna coincides with sustained environmental deterioration. *Nature Communications* **11**, 2250.

Holdaway, R. N., and Jacomb, C. (2000). Rapid extinction of the moas (Aves: Dinornithiformes): model, test and implications. *Science* **287**, 2250–2254.

Holdaway, R. N., Worthy, T. H., and Tennyson, A. J. (2001). A working list of breeding bird species of the New Zealand region at first human contact. *New Zealand Journal of Zoology* **28**, 119–187.

Holdaway, S., Phillipps, R., Emmitt, J., and Wendrich, W. (2016). The Fayum revisited: reconsidering the role of the Neolithic package, Fayum north shore, Egypt. *Quaternary International* **410**, 173–180.

Huan, X., Lu, H., Jiang, L., Zuo, X., He, K., and Zhang, J. (2021). Spatial and temporal pattern of rice domestication during the Early Holocene in the lower Yangtze region, China. *The Holocene* **31**, 1366–1375.

Hufford, M. B., Bilinski, P., Pyhäjärvi, T., and Ross-Ibarra, J. (2012). Teosinte as a model system for population and ecological genomics. *Trends in Genetics* **28**, 606–615.

Hunt, T. L., and Lipo, C. P. (2009). Revisiting Rapa Nui (Easter Island) "Ecocide" 1. *Pacific Science* **63**, 601–616.

IPBES (2019). Global assessment report on biodiversity and ecosystem services of the Intergovernmental Science-Policy Platform on Biodiversity and Ecosystem Services (ed. E. S. Brondizio, J. Settele, S. Díaz, and H. T. Ngo). IPBES Secretariat, Bonn, Germany.

Isern, N., Zilhão, J., Fort, J., and Ammerman, A.J. (2017). Modeling the role of voyaging in the coastal spread of the Early Neolithic in the West Mediterranean. *Proceedings of the National Academy of Sciences* **114**, 897–902.

IUCN (2022). *The IUCN Red List of Threatened Species. Version 2022-2*. https://www.iucnredlist.org.

Jackson, A. (2014). Added credence for a late Dodo extinction date. *Historical Biology* **26**, 699–701.

Jeffries, M. J. (1997). *Biodiversity and Conservation*. Routledge, London.

Jennings, T. A., and Waters, M. R. (2014). Pre-Clovis lithic technology at the Debra L. Friedkin site, Texas: comparisons to Clovis through site-level behavior, technological trait-list, and cladistic analyses. *American Antiquity* **79**, 25–44.

Johnson, C. N. (2006). *Australia's Mammal Extinctions: A 50,000 Year History*. Cambridge University Press, Melbourne.

Katzenberg, M. A., John Staller, R. T., and Benz, B. (2006). Prehistoric maize in southern Ontario. In *Histories of Maize: Multidisciplinary Approaches to the Prehistory, Linguistics, Biogeography, Domestication, and Evolution of Maize* (ed. John Staller, Robert Tykot, and Bruce Benz), pp. 263–273. Routledge, New York.

Kavar, T., and Dovč, P. (2008). Domestication of the horse: genetic relationships between domestic and wild horses. *Livestock Science* **116**, 1–14.

Keigwin, L. D., Klotsko, S., Zhao, N., Reilly, B., Giosan, L., and Driscoll, N. W. (2018). Deglacial floods in the Beaufort Sea preceded Younger Dryas cooling. *Nature Geoscience* **11**, 599–604.

Kennedy, L. M., and Horn, S. L. (1997). Prehistoric maize cultivation at the La Selva Biological Station, Costa Rica. *Biotropica* **29**, 368–370.

Kirch, P. V. (1982). The impact of prehistoric Polynesians on the Hawaiian ecosystem. *Pacific Science* **36**, 1–14.

Kirch, P. V. (2011). When did the Polynesians settle Hawaii? A review of 150 years of scholarly inquiry and a tentative answer. *Hawaiian Archaeology* **12**, 3–26.

Kistler, L., Montenegro, Á., Smith, B. D., Gifford, J. A., Green, R. E., Newsom, L. A., and Shapiro, B. (2014). Transoceanic drift and the domestication of African bottle gourds in the Americas. *Proceedings of the National Academy of Sciences* **111**(8), 2937–2941.

Klümper, W., and Qaim, M. (2014). A meta-analysis of the impacts of genetically modified crops. *PloS One* **9**, e111629.

Kolbert, E. (2014). *The Sixth Extinction: An Unnatural History*. Bloomsbury, London.

Kumar, K., Gambhir, G., Dass, A., Tripathi, A. K., Singh, A., Jha, A. K., Yadava, P., Choudhary, M., and Rakshit, S. (2020). Genetically modified crops: current status and future prospects. *Planta* **251**, 1–27.

Lamb, H. F., Damblon, F., and Maxted, R. W. (1991). Human impact on the vegetation of the Middle Atlas, Morocco, during the last 5000 years. *Journal of Biogeography* **18**, 519–532.

Larson, G., and Fuller, D. Q. (2014). The evolution of animal domestication. *Annual Review of Ecology, Evolution, and Systematics* **45**, 115–136.

Larson, G., Piperno, D. R., Allaby, R. G., Purugganan, M. D., Andersson, L., Arroyo-Kalin, M., Barton, L., Climer Vigueira, C., Denham, T., Dobney, K., and Doust, A. N. (2014). Current perspectives and the future of domestication studies. *Proceedings of the National Academy of Sciences* **111**, 6139–6146.

Lazaridis, I., Patterson, N., Mittnik, A., Renaud, G., Mallick, S., Kirsanow, K., Sudmant, P. H., Schraiber, J. G., Castellano, S., Lipson, M., and Berger, B. (2014). Ancient human genomes suggest three ancestral populations for present-day Europeans. *Nature* **513**, 409–413.

Lev-Yadun, S., Gopher, A., and Abbo, S. (2000). The cradle of agriculture. *Science* **288**, 1602–1603.

Librado, P., Khan, N., Fages, A., Kusliy, M. A., Suchan, T., Tonasso-Calvière, L., Schiavinato, S., Alioglu, D., Fromentier, A., Perdereau, A., and Aury, J. M. (2021). The origins and spread of domestic horses from the Western Eurasian steppes. *Nature* **598**, 634–640.

Liu, N., Xu, Q., Zhu, F., and Zhang, L. E. E. (2006). Pyrethroid resistance in mosquitoes. *Insect Science* **13**, 159–166.

Loehle, C., and Eschenbach, W. (2012). Historical bird and terrestrial mammal extinction rates and causes. *Diversity and Distributions* **18**, 84–91.

Louys, J., Braje, T. J., Chang, C. H., Cosgrove, R., Fitzpatrick, S. M., Fujita, M., Hawkins, S., Ingicco, T., Kawamura, A., MacPhee, R. D., and McDowell, M. C. (2021). No evidence for widespread island extinctions after Pleistocene hominin arrival. *Proceedings of the National Academy of Sciences* **118**, e2023005118.

MacDonald, G. M., Beilman, D. W., Kuzmin, Y. V., Orlova, L. A., Kremenetski, K. V., Shapiro, B., Wayne, R. K., and Van Valkenburgh, B. (2012). Pattern of extinction of the woolly mammoth in Beringia. *Nature Communications* **3**, 893.

MacPhee, R. D. E. (1999). *Extinctions in Near Time*. Kluwer Academic/Plenum, New York.

MacPhee, R. D. E., and Flemming, C. (1999). *Requiem aeternam*. In *Extinctions in Near Time* (ed. R. D. E. MacPhee), pp. 333–371. Kluwer Academic/Plenum, New York.

Maeda, O., Lucas, L., Silva, F., Tanno, K. I., and Fuller, D. Q. (2016). Narrowing the harvest: increasing sickle investment and the rise of domesticated cereal agriculture in the Fertile Crescent. *Quaternary Science Reviews* **145**, 226–237.

Marnetto, D., Pankratov, V., Mondal, M., Montinaro, F., Pärna, K., Vallini, L., Molinaro, L., Saag, L., Loog, L., Montagnese, S., and Costa, R. (2022). Ancestral genomic contributions to complex traits in contemporary Europeans. *Current Biology* **32**, 1412–1419.

Martin, P. S. (1984). Catastrophic extinctions and Late Pleistocene blitzkrieg: two radiocarbon tests. In *Extinctions* (ed. M. H. Nitecki), pp. 153–189. University of Chicago Press, Chicago.

Martin, P. S., and Steadman, D. W. (1999). Prehistoric extinctions on islands and continents. In *Extinctions in Near Time* (ed. R. D. E. MacPhee), pp. 17–55. Kluwer Academic/Plenum, New York.

Martinez-Soriano, J. P., and Avina-Padilla, K. (2009). Ustilago and the accidental domestication of maize. *Nature Proceedings*, pp. 1–1. doi.org/10.1038/npre.2009.3240.1.

Matthews, T. J., Wayman, J. P., Cardoso, P., Sayol, F., Hume, J. P., Ulrich, W., Tobias, J. A., Soares, F. C., Thébaud, C., Martin, T. E., and Triantis, K. A. (2022). Threatened and extinct island endemic birds of the world: distribution, threats and functional diversity. *Journal of Biogeography* **49**, 1920–1940.

Matson, R. G. (2016). *The origins of southwestern agriculture*. University of Arizona Press, Tucson.

McGee, M. D., Borstein, S. R., Neches, R. Y., Buescher, H. H., Seehausen, O., and Wainwright, P. C. (2015). A pharyngeal jaw evolutionary innovation facilitated extinction in Lake Victoria cichlids. *Science* **350**, 1077–1079.

Mech, L. D., and Janssens, L. A. (2022). An assessment of current wolf Canis lupus domestication hypotheses based on wolf ecology and behaviour. *Mammal Review* **52**, 304–314.

Meltzer, D. J. (2020). Overkill, glacial history, and the extinction of North America's Ice Age megafauna. *Proceedings of the National Academy of Sciences* **117**, 28555–28563.

Meltzer, D. J., and Mead, J. I. (1982). The timing of Late Pleistocene mammalian extinctions in North America. *Quaternary Research* **19**, 130–135.

Merrill, W. L., Hard, R. J., Mabry, J. B., Fritz, G. J., Adams, K. R., Roney, J. R., and MacWilliams, A. C. (2009). The diffusion of maize to the southwestern United States and its impact. *Proceedings of the National Academy of Sciences* **106**, 21019–21026.

Miller, G. H., Fogel, M. L., Magee, J. W., Gagan, M. K., Clarke, S. J., and Johnson, B. J. (2005). Ecosystem collapse in Pleistocene Australia and a human role in megafaunal extinction. *Science* **309**, 287– 290.

Miller, G. H., Magee, J. W., Johnson, B. J., Fogel, M. L., Spooner, N. A., McCulloch, M. T., and Ayliffe, L. K. (1999). Pleistocene extinction of *Genyornis newtoni*: human impact on Australian megafauan. *Science* **283**, 205–209.

Miller, G., Magee, J., Smith, M., Spooner, N., Baynes, A., Lehman, S., Fogel, M., Johnston, H., Williams, D., Clark, P., and Florian, C. (2016). Human predation contributed to the extinction of the Australian megafaunal bird Genyornis newtoni~ 47 ka. *Nature Communications* **7**, 10496.

Ming, L., Siren, D., Hasi, S., Jambl, T., and Ji, R. (2022). Review of genetic diversity in Bactrian camel (Camelus bactrianus). *Animal Frontiers* **12**, 20–29.

Moore, A. M., Kennett, J. P., Napier, W. M., Bunch, T. E., Weaver, J. C., LeCompte, M., Adedeji, A. V., Hackley, P., Kletetschka, G., Hermes, R. E., and Wittke, J. H. (2020). Evidence of cosmic impact at Abu Hureyra, Syria at the younger Dryas Onset (~12.8 ka): High-temperature melting at > 2200 C. *Scientific Reports* **10**, 4185.

Njiru, M., Mkumbo, O. C., and Van der Knaap, M. (2010). Some possible factors leading to decline in fish species in Lake Victoria. *Aquatic Ecosystem Health & Management* **13**, 3–10.

O'Brien, M. J., and Buchanan, B. (2017). Cultural learning and the Clovis colonization of North America. *Evolutionary Anthropology: Issues, News, and Reviews* **26**, 270–284.

O'Keefe, F. R., Dunn, R. E., Weitze, E. M., Waters, M. R., Martinez, L., Binder, W. J., Southon, J. R., Cohen, J. E., Meachen, J. A., DeSantis, L. R. J., Kirby, M. E., Ghezzo, E., Coltrain, J. B., Fuller, B. T., Farrell, A. B., Takeuchi, G. T., MacDonald, G., Davis, E. B., and Lindsey, E. L. (2023). Pre-Younger Dryas megafaunal extirpation and anthropogenic state shift in southern California. *Science* **381**. doi.org/10.1126/science.abo359.

Orton, D., Anvari, J., Gibson, C., Last, J., Bogaard, A., Rosenstock, E., and Biehl, P. F. (2018). A tale of two tells: dating the Çatalhöyük West Mound. *Antiquity* **92**, 620–639.

Ottoni, C., Girdland Flink, L., Evin, A., Geörg, C., De Cupere, B., Van Neer, W., Bartosiewicz, L., Linderholm, A., Barnett, R., Peters, J., and Decorte, R. (2013). Pig domestication and human-mediated dispersal in western Eurasia revealed through ancient DNA and geometric morphometrics. *Molecular Biology and Evolution* **30**, 824–832.

Outram, A. K., Stear, N. A., Bendrey, R., Olsen, S., Kasparov, A., Zaibert, V., Thorpe, N., and Evershed, R. P. (2009). The earliest horse harnessing and milking. *Science* **323**, 1332–1335.

Ovodov, N. D., Crockford, S. J., Kuzmin, Y. V., Higham, T. F., Hodgins, G. W., and van der Plicht, J. (2011). A 33,000-year-old incipient dog from the Altai Mountains of Siberia: evidence of the earliest domestication disrupted by the Last Glacial Maximum. *PloS One* **6**, e22821.

Pereira, H. M., Navarro, L. M., and Martins, I. S. (2012). Global biodiversity change: the bad, the good, and the unknown. *Annual Review of Environment and Resources* **37**, 25–50.

Perry, G. L., Wheeler, A. B., Wood, J. R., and Wilmshurst, J. M. (2014). A high-precision chronology for the rapid extinction of New Zealand moa (Aves, Dinornithiformes). *Quaternary Science Reviews* **105**, 126–135.

Pitt, D., Sevane, N., Nicolazzi, E. L., MacHugh, D. E., Park, S. D., Colli, L., Martinez, R., Bruford, M. W., and Orozco-terWengel, P. (2019). Domestication of cattle: two or three events? *Evolutionary Applications* **12**, 123–136.

Ponting, C. (1992). *A Green History of the World*. Penguin, Middlesex, United Kingdom.

Potapov, P., Turubanova, S., Hansen, M. C., Tyukavina, A., Zalles, V., Khan, A., Song, X. P., Pickens, A., Shen, Q., and Cortez, J. (2022). Global maps of cropland extent and change show accelerated cropland expansion in the twenty-first century. *Nature Food* **3**, 19–28.

Prentiss, A. M., Walsh, M. J., Foor, T. A., Bobolinski, K., Hampton, A., Ryan, E., and O'Brien, H. (2020). Malthusian cycles among semi-sedentary fisher-hunter-gatherers: The socio-economic and demographic history of housepit 54, Bridge River site, British Columbia. *Journal of Anthropological Archaeology* **59**, 101181.

Price, M. D., and Arbuckle, B. S. (2015). Early pig management in the Zagros flanks: reanalysis of the fauna from Neolithic Jarmo, northern Iraq. *International Journal of Osteoarchaeology* **25**, 441–453.

Purugganan, M. D. (2019). Evolutionary insights into the nature of plant domestication. *Current Biology* **29**, R705–R714.

Purugganan, M. D. (2022). What is domestication? *Trends in Ecology & Evolution* **37**, 663–671.

Purugganan, M. D., and Fuller, D. Q. (2009). The nature of selection during plant domestication. *Nature* **457**, 843–848.

Pyšek, P., Hulme, P. E., Simberloff, D., Bacher, S., Blackburn, T. M., Carlton, J. T., Dawson, W., Essl, F., Foxcroft, L. C., Genovesi, P., and Jeschke, J. M. (2020). Scientists' warning on invasive alien species. *Biological Reviews* **95**, 1511–1534.

Quammen, D. (1996). *Song of the Dodo: Island Biogeography in an Age of Extinctions*. Touchstone, New York.

Racimo, F., Woodbridge, J., Fyfe, R. M., Sikora, M., Sjögren, K. G., Kristiansen, K., and Vander Linden, M. (2020). The spatiotemporal spread of human migrations during the European Holocene. *Proceedings of the National Academy of Sciences* **117**, 8989–9000.

Rasmussen, M., Anzick, S. L., Waters, M. R., Skoglund, P., DeGiorgio, M., Stafford Jr., T. W., Rasmussen, S., Moltke, I., Albrechtsen, A., Doyle, S. M., and Poznik, G. D. (2014). The genome of a Late Pleistocene human from a Clovis burial site in western Montana. *Nature* **506**, 225–229.

Renssen, H., Mairesse, A., Goosse, H., Mathiot, P., Heiri, O., Roche, D. M., Nisancioglu, K. H., and Valdes, P. J. (2015). Multiple causes of the Younger Dryas cold period. *Nature Geoscience* **8**, 946–949.

Riehl, S. (2016). The role of the local environment in the slow pace of emerging agriculture in the Fertile Crescent. *Journal of Ethnobiology* **36**, 512–534.

Riehl, S., Zeidi, M., and Conard, N. J. (2013). Emergence of agriculture in the foothills of the Zagros Mountains of Iran. *Science* **341**(6141), 65–67.

Ripple, W. J., and Van Valkenburgh, B. (2010). Linking top-down forces to the Pleistocene megafaunal extinctions. *BioScience* **60**, 516–526.

Roberts, D. L., and Solow, A. R. (2003). When did the dodo become extinct? *Nature* **426**, 245–245.

Roberts, N. (1999). *The Holocene*. Blackwell, Oxford.

Rowley-Conwy, P. (2011). Westward Ho! The spread of agriculture from Central Europe to the Atlantic. *Current Anthropology* **52**, S431–S451.

Rozema, J., Cornelisse, D., Zhang, Y., Li, H., Bruning, B., Katschnig, D., Broekman, R., Ji, B., and van Bodegom, P. (2015) Comparing salt tolerance of beet cultivars and their halophytic ancestor: consequences of domestication and breeding programmes. *AoB Plants* **7**, pii: plu083. doi. org/10.1093/aobpla/plu083.

Rule, S., Brook, B. W., Haberle, S. G., Turney, C. S. M., Kershaw, A. P., Johnson, C. N. (2012). The aftermath of megafaunal extinction: ecosystem transformation in Pleistocene Australia. *Science* **335**, 1483–1486.

Saltré, F., Rodríguez-Rey, M., Brook, B. W., Johnson, C. N., Turney, C. S., Alroy, J., Cooper, A., Beeton, N., Bird, M. I., Fordham, D. A., and Gillespie, R. (2016). Climate change not to blame for Late Quaternary megafauna extinctions in Australia. *Nature Communications* **7**, 10511.

Sandom, C., Faurby, S., Sandel, B. and Svenning, J. C. (2014). Global Late Quaternary megafauna extinctions linked to humans, not climate change. *Proceedings of the Royal Society B: Biological Sciences* **281**, 20133254. doi. org/10.1098/rspb.2013.3254.

Sauer, C. D. (1952). *Agricultural Origins and Dispersals*. MIT Press, Cambridge.

Schorger, A. W. (1955). *The Passenger Pigeon, Its Natural History and Extinction*. University of Wisconsin, Madison.

Shiferaw, B., Smale, M., Braun, H. J., Duveiller, E., Reynolds, M., and Muricho, G. (2013). Crops that feed the world 10. Past successes and future challenges to the role played by wheat in global food security. *Food Security* **5**, 291–317.

Shirai, N. (2016). Establishing a Neolithic farming life in Egypt: a view from the lithic study at Fayum Neolithic sites. *Quaternary International* **412**, 22–35.

Silva, F., and Vander Linden, M. (2017). Amplitude of travelling front as inferred from 14C predicts levels of genetic admixture among European early farmers. *Scientific Reports* **7**, 1–9.

Simmons, I. G. (1989). *Changing the Face of the Earth*. Blackwell, Oxford.

Skoglund, P., Malmström, H., Raghavan, M., Storå, J., Hall, P., Willerslev, E., Gilbert, M., T. P., Götherström, A., and Jakobsson, M. (2012). Origins and genetic legacy of Neolithic farmers and hunter-gatherers in Europe. *Science* **336**, 466–469.

Smith, B. D. (2005). Reassessing Coxcatlan Cave and the early history of domesticated plants in Mesoamerica. *Proceedings of the National Academy of Sciences* **102**, 9438–9445.

Spengler III, R. N. (2020). Anthropogenic seed dispersal: rethinking the origins of plant domestication. *Trends in Plant Science* **25**, 340–348.

Steadman, D. W. (1993). Biogeography of Tongan birds before and after human impact. *Proceedings of the National Academy of Science* **90**, 818–822.

Steadman, D. W. (1995). Prehistoric extinctions of Pacific Island birds: biodiversity meets zooarchaeology. *Science* **267**, 1123–1131.

Stein, P. L., and Rowe, B. M. (1974). *Physical Anthropology*. McGraw-Hill, New York.

Stewart, M., Carleton, W. C., and Groucutt, H. S. (2021). Climate change, not human population growth, correlates with Late Quaternary megafauna declines in North America. *Nature Communications* **12**, 965.

Stiner, M. C., Munro, N. D., Surovell, T. A., Tchernov, E., and Bar-Yosef, O. (1999). Paleolithic population growth pulses evidenced by small animal exploitation. *Science* **283**, 190–194.

Stone, C. P., and Scott, J. M. (1985). *Hawaii's Terrestrial Ecosystems: Preservation and Management.* University of Hawaii Press, Honolulu.

Stuart, A. J. (2015). Late Quaternary megafaunal extinctions on the continents: a short review. *Geological Journal* **50**, 338–363.

Sullivan, A. P., Bird, D. W., and Perry, G. H. (2017). Human behaviour as a long-term ecological driver of non-human evolution. *Nature Ecology & Evolution* **1**, 0065.

Surovell, T. A., Holliday, V. T., Gingerich, J. A., Ketron, C., Haynes Jr., C. V., Hilman, I., Wagner, D. P., Johnson, E., and Claeys, P. (2009). An independent evaluation of the Younger Dryas extraterrestrial impact hypothesis. *Proceedings of the National Academy of Sciences* **106**, 18155–18158.

Swarts, K., Gutaker, R. M., Benz, B., Blake, M., Bukowski, R., Holland, J., Kruse-Peeples, M., Lepak, N., Prim, L., Romay, M. C., and Ross-Ibarra, J. (2017). Genomic estimation of complex traits reveals ancient maize adaptation to temperate North America. *Science* **357**, 512–515.

Sweatman, M. B. (2021). The Younger Dryas impact hypothesis: review of the impact evidence. *Earth-Science Reviews* **218**, 103677.

Tankersley, K. B. (1999). A matter of superior spearpoints. *Archaeology* **52**, 60–63.

Tanno, K. I., and Willcox, G. (2006). How fast was wild wheat domesticated? *Science* **311**, 1886–1886.

Thalmann, O., and Perri, A. R. (2018). Paleogenomic inferences of dog domestication. In *Paleogenomics. Population Genomics* (ed. C. Lindqvist and O. Rajora), pp. 273–306 Springer, Cham, Switzerland.

Tierney, J. E., Pausata, F. S., and deMenocal, P. B. (2017). Rainfall regimes of the Green Sahara. *Science Advances* **3**, e1601503.

Turvey, S. T., Kennerley, R. J., Nuñez-Miño, J. M., and Young, R. P. (2017). The last survivors: current status and conservation of the non-volant land mammals of the insular Caribbean. *Journal of Mammalogy* **98**, 918–936.

Vallebueno-Estrada, M., Rodríguez-Arévalo, I., Rougon-Cardoso, A., Martínez González, J., García Cook, A., Montiel, R., and Vielle-Calzada, J. P. (2016). The earliest maize from San Marcos Tehuacán is a partial domesticate with genomic evidence of inbreeding. *Proceedings of the National Academy of Sciences* **113**, 14151–14156.

Vander Linden, M., and Silva, F. (2018) Comparing and modeling the spread of early farming across Europe. *PAGES Magazine* **26**, 28–29.

Van Hoesel, A., Hoek, W. Z., Pennock, G. M., and Drury, M. R. (2014). The Younger Dryas impact hypothesis: a critical review. *Quaternary Science Reviews* **83**, 95–114.

Vigne, J. D. (2011). The origins of animal domestication and husbandry: a major change in the history of humanity and the biosphere. *Comptes rendus biologies* **334**, 171–181.

Vila, C., Savolainen, P., Maldonado, J. E., Amorim, I. R., Rice, J. E., Honeycutt, R. L., Crandall, K. A., Lundeberg, J., and Wayne, R. K. (1997). Multiple and ancient origins of the domestic dog. *Science* **276**, 1687–1689.

Warner, R. E. (1968). The role of introduced diseases in the extinction of the endemic Hawaiian avifauna. *Condor* **70**, 101–120.

Waters, M. R., 2019. Late Pleistocene exploration and settlement of the Americas by modern humans. *Science* **365**(6449). eaat5447. doi.org/10.1126/science.aat5447

Waters, M. R., Stafford Jr., T. W., and Carlson, D. L. (2020). The age of Clovis—13,050 to 12,750 cal yr BP. *Science Advances* **6**, eaaz0455. doi.org/10.1126/sciadv.aaz0455.

Wayne, R. K., Leonard, J. A., and Cooper, A. (1999). Full of sound and fury: the recent history of ancient DNA. *Annual Review of Ecology and Systematics* **30**, 457–477.

Webb, S. (2008). Megafauna demography and Late Quaternary climatic change in Australia: a predisposition to extinction. *Boreas* **37**, 329–345.

Weiss, E., and Zohary, D. (2011). The Neolithic Southwest Asian founder crops: their biology and archaeobotany. *Current Anthropology* **52**, S237–S254.

Wilkin, S., Ventresca Miller, A., Fernandes, R., Spengler, R., Taylor, W. T. T., Brown, D. R., Reich, D., Kennett, D. J., Culleton, B. J., Kunz, L., and Fortes, C. (2021). Dairying enabled early bronze age Yamnaya steppe expansions. *Nature* **598**, 629–633.

Willcox, G. (2013). The roots of cultivation in southwestern Asia. *Science* **341**, 39–40.

Woinarski, J. C., Burbidge, A. A., and Harrison, P. L. (2015). Ongoing unraveling of a continental fauna: decline and extinction of Australian mammals since European settlement. *Proceedings of the National Academy of Sciences* **112**, 4531–4540.

World Conservation Monitoring Center. (1992). *Global Biodiversity: Status of the Earth's Living Resources.* Chapman and Hall, London.

Wroe, S., and Field, J. (2006). A review of the evidence for a human role in the extinction of Australian megafauna and an alternative interpretation. *Quaternary Science Reviews* **25**, 2692–2703.

Wuethrich, B. (1999). Proto-Polynesians quickly settled Pacific. *Science* **286**, 2054–2056.

Yacobaccio, H. D. (2021). The domestication of South American camelids: a review. *Animal Frontiers* **11**, 43–51.

Zeder, M. A. (2012). The domestication of animals. *Journal of Anthropological Research* **68**, 161–190.

Zeder, M. A. (2015). Core questions in domestication research. *Proceedings of the National Academy of Sciences* **112**, 3191–3198.

Zeder, M. A., and Hesse, B. (2000). The initial domestication of goats *(Capra hircus)* in the Zagros Mountains. *Science* **287**, 2254–2257.

Zhang, Y., Colli, L., and Barker, J. S. F. (2020). Asian water buffalo: domestication, history and genetics. *Animal Genetics* **51**, 177–191.

Zimmerer, K. S. (1991). The regional biogeography of native potato cultivars in highland Peru. *Journal of Biogeography* **18**, 165–178.

Zohary, D., and Hopf, M. (1994). *Domestication of Plants in the Old World*, 2nd edition. Clarendon Press, Oxford.

Zörb, C., Geilfus, C. M., and Dietz, K. J. (2019). Salinity and crop yield. *Plant Biology* **21**, 31–38.

Zuo, X., Lu, H., Jiang, L., Zhang, J., Yang, X., Huan, X., He, K., Wang, C., and Wu, N. (2017). Dating rice remains through phytolith carbon-14 study reveals domestication at the beginning of the Holocene. *Proceedings of the National Academy of Sciences* **114**, 6486–6491.

PART III

THEORY AND PRACTICE

CHAPTER 13

DESCRIPTION AND INTERPRETATION OF BIOGEOGRAPHIC DISTRIBUTIONS

What can the size and shape of the geographic distributions of plants and animals tell us about their ecology and evolutionary history? For example, does examination of the size and shape of the geographic distributions of bird species allow us to draw conclusions regarding underlying general physiological relationships between birds and the environment? Is it possible that by analyzing maps of the modern distribution of a plant species we can untangle the long history of evolution and dispersal that a species and its ancestors have experienced? For over 150 years, biogeographers have pondered and debated these questions. The ecological analysis of geographic distributions remains of keen interest and continues to evolve as a subdiscipline within biogeography and ecology. In the early 1980s, the Argentine biogeographer Eduardo Rapoport applied the word **areography** to the ecological analysis of modern plant and animal distributions, while the American biogeographer James Brown employed the term **macroecology** to describe such work. Many historical biogeographers have made their life's work out of the careful analysis of modern geographic distributions and the use of these distributions to infer evolutionary history. Today, molecular genetics analysis of living and fossil organisms plays an increasingly dominant role in reconstructing evolutionary relationships and biogeographic histories. Such work can be referred to as **phylogenetic** biogeography and is of great importance.

In this chapter we will explore some of the ecological and evolutionary interpretations that have been made from the modern geographic distributions of plants and animals. We will see that many modern distributions are the product of fascinating biogeographic histories. In our analysis of geographic distributions, we will use the term **range** to refer to the area permanently occupied by a plant or animal taxon. This term can also be applied to the browsing, foraging, or hunting area of an individual animal, but we will not use the word in that sense in this chapter. Any meaningful analysis of the size and shape of a species range requires reliable mapping of the geographic distributions of plant and animal species. We will first consider how geographic range maps are generated. We will then look at some general relationships that have been observed between the physiology and morphology of species and the sizes and shapes of their geographic ranges. We will follow this by outlining some of the frequently encountered general distributional patterns that biogeographers have come to recognize. We will then look at how these general patterns have been interpreted in terms of species ecology and evolutionary history. We will conclude by examining the ways in which the analysis of modern distributions has been used to infer the evolutionary history of plants and animals—and how modern genetic analysis has advanced such efforts.

MAPPING BIOGEOGRAPHIC DISTRIBUTIONS

In this book we have examined many maps portraying the distributions of different plant and animal species. We have assumed that these range maps are valid representations of the spatial distributions of these organisms, and we have based some important conclusions on the maps. In the past two decades, technological developments in areas such as computing, geographic information sciences (GIS), remote sensing, ecological modeling, the application of citizen scientist approaches, and ecological

Biogeography: Introduction to Space, Time, and Life, Second Edition. Glen M. MacDonald.
© 2025 John Wiley & Sons, Inc. Published 2025 by John Wiley & Sons, Inc.
Companion website: www.wiley.com/go/macdonald/biogeography2e

DNA analysis have significantly advanced the field of species distribution mapping. Let's now consider how such biogeographic maps are made and the uncertainties that they may contain.

Biogeographic maps depicting the distributions of plants and animals are still often derived from spatial networks of scientific observations that record the presence or absence of species. In collecting such data, plant and animal species are identified in the field or collected and taken back to the laboratory for identification. Sometimes feathers, tissue, or blood collected in the field are analyzed for DNA and compared to DNA reference libraries to conclusively identify the species or subspecies present. The use of this **DNA barcoding** approach can be extremely effective. One analysis of over 600 different North American bird species demonstrated that 94% of the species possessed distinctive genetic barcode gene clusters. Recently, attention has been drawn to the analysis of **environmental DNA (eDNA)** from water or soil samples to detect the presence of plant and animal species in current or past times. Field observations and identifications that do not require collecting or harming organisms are best because they are the least disruptive to the environment. Data on species presence, absence, and abundance may be collected during targeted research programs or can be harvested from museum specimen collections and historical documentary sources. The latitude, longitude, elevation, and environmental setting of each collection site are carefully recorded. In some cases, these data are used to produce dot maps that simply show the sites where the species has been found and recorded (Fig. 13.1). The more sites visited, the more accurate the maps will be.

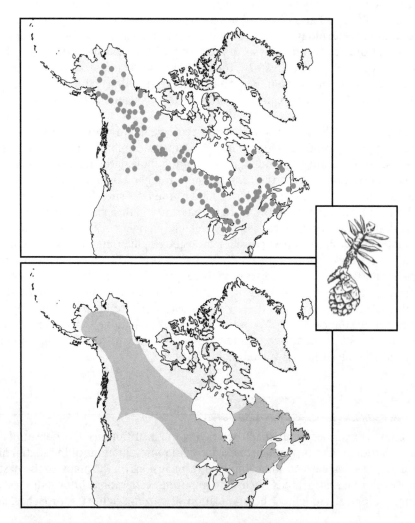

FIGURE 13.1   A dot map based on recorded collection sites for black spruce *(Picea mariana)* in North America (*Source:* Porsild and Cody, 1980 / National Museums of Canada) and a choropleth map based on the same data. Detailed surveys based on field observations and air photo analysis show that the density of spruce decreases dramatically near the northern treeline. The actual northern boundary is relatively diffuse and composed of scattered disjunct individuals and stands.

Citizen scientist approaches, which collect georeferenced observations from nonexperts about plant and animal species they encounter, are rapidly expanding such biogeographic map databases. Citizen scientist observations have a long history in ecology and biogeography. The Christmas Bird Count of the National Audubon Society provides an example that dates back to 1900. Smartphone technology and associated networking have revolutionized the collection of species distribution data by citizen scientists. iNaturalist, eBird, and Pl@ntNet are examples of mobile phone apps that can be used to help identify species and collect georeferenced data on species occurrence taking advantage of a phone's GPS capacity. Research projects can take advantage of such **crowdsourcing** approaches to greatly expand the geographic coverage of observational networks of plant and animal species occurrences. There are now an abundance of projects that collect and analyze input from citizen scientists in regards to plant and animal species. The United States provides a useful catalog of such crowdsourcing projects that citizen scientists can contribute to at https://www.citizenscience.gov/catalog/. Although there were initially strong doubts about the scientific integrity of citizen scientist observations, more and more studies are confirming the general veracity of such data. Today, there are millions of citizen scientists contributing crowdsourced ecological and biogeographic data around the world. Anyone reading this book can become a citizen scientist and make important contributions to biogeography!

Throughout this book we have presented maps that use shading or boundary lines to represent the geographic ranges of species. These types of maps, in which the area occupied by an entity (such as a plant or animal species) is delineated by continuous shading or circumscribed by a line, are called **choropleth maps.** The data from dot maps are extrapolated to make choropleth maps. The extrapolations are done using knowledge of the environmental requirements of the species and the environmental conditions of the area being mapped. Information from air photos or other remote sensing imagery is sometimes used to supplement field observations. The development and application of computer models to produce range estimate maps for plants and animals is an area of great growth and application for biogeography, ecology, and conservation (Technical Box 13.1). Choropleth maps are popular because they rapidly convey a general impression of the geographic areas in which the species is found and the areas in which it is absent (Fig. 13.1). However, information from biogeographic choropleth maps must be used with some caution.

TECHNICAL BOX 13.1

Janet Franklin—pioneer in species distribution modeling (with permission of J. Franklin).

Species Distribution Models

The use of computer modeling to infer geographic distributions of species based on observed occurrences coupled with physiological, ecological, and environmental data is an important tool for deriving dot and choropleth maps of species distributions. These **species distribution models** (SDMs) have become one of the most commonly used tools in biogeography and ecology. Geographers, due to their interest and capability in mapping and the integration of biological and physical environmental data, have been particularly active in this field. The American geographer Janet Franklin is considered one of the leading pioneers in the field of SDMs. Her 2010 book *Mapping Species Distributions: Spatial inference and Prediction* is a highly cited standard text on the topic. Among Janet Franklin's honors was election to the National Academy of Sciences in 2014 in recognition of her research on "describing spatial patterns and inferring underlying pattern-generating processes via simulation modeling."

Two general approaches are typically used to produce SDMs. Correlative SDMs are based on observed empirical relations between a species geographic distribution and the distribution of environmental variables such as climatic conditions. Correlative models, particularly those linking distribution to climate, are the most widely used today. Mechanistic SDMs use information about the physiology of species and environmental variables relevant to the physiological functioning to model the environmental conditions under which a species might be expected to occur and then map where such conditions are met. One way of thinking about this is that the environmental data are used to determine a niche for the species, and the species is then predicted to occur where those niche conditions are met. One important strength of SDMs is that given the proper information on past or future environmental conditions, they can predict past and future species ranges. This makes SDMs particularly useful for studies of the impacts of climate change. The application of SDMs has grown in frequency and in precision with the dramatic growth of computing hardware, software, and databases over the past two decades. The large data sets provided by citizen scientists have also helped facilitate the growing application of SDMs. Such modeling of species ranges is not without uncertainties however. One assessment of over 100 different studies indicated that the SDMs performed best at predicting simple presence or absence and declined in ability to predict properties such as population size or genetic diversity. There are a number of different SDMs that are commonly used and they may produce differing estimates of species ranges. As observational data sets have grown and model algorithms have improved, SDMs have become increasingly better. SDMs will undoubtedly increase in precision and grow in use as an important tool in determining and analyzing the geographic and temporal distributions of plant and animal species.

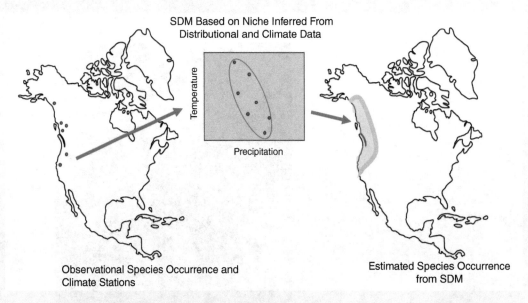

A simplified example of a correlative SDM used to infer the range of a hypothetical North American plant species based on observations of reported locations of the species and the climatic conditions at those locations.

Choropleth maps often include large areas in which the presence or absence of the species is extrapolated and not directly observed. In some cases, such extrapolations have been shown to be incorrect. For example, it has long been thought that bigcone Douglas fir *(Pseudotsuga macrocarpa)* grows both in the mountains of southern California and in adjacent highlands of the Baja Peninsula of Mexico. Because the general environment of the Mexican mountains is similar to that in southern California, the range of the bigcone Douglas fir was inferred to extend into Mexico. Biogeographic maps have been made that show the species growing in northern Mexico. Recent detailed botanical surveys found that there are actually no bigcone Douglas firs in Mexico. It can be concluded that the species is a very narrow endemic restricted to southern California. Incorrect extrapolations of the presence or absence of the species can produce erroneous maps and misleading interpretations of biogeographic distributions.

Simple choropleth maps present the illusion that the plant or animal species is distributed evenly and continuously throughout the shaded or delineated area. Such even and continuous distributions do not occur in nature. Within the mapped boundary of the species, there will be areas of high- and low-population density. Neither are plants and animals distributed continuously within their mapped geographic ranges. Instead, they occur as scattered individuals and groups (metapopulations) separated by small and large areas where they are absent. Such discontinuous distributions are referred to as being punctiform. Classic examples of how species occur in such a discontinuous manner come from the detailed mappings of the distribution of wild plants such as clematis *(Clematis fremontii)* in Missouri or wild animals such as pocket gophers *(Thomomys monticola)* in California. Thus, a choropleth map may present plants and animals as continuously distributed throughout their ranges (Fig. 13.2). However, when the distribution is examined in increasing detail, it is seen that the species typically occur as relatively small metapopulations and individuals that are separated by large and small areas of unoccupied land. In many cases, the distributions of species become extremely fragmented

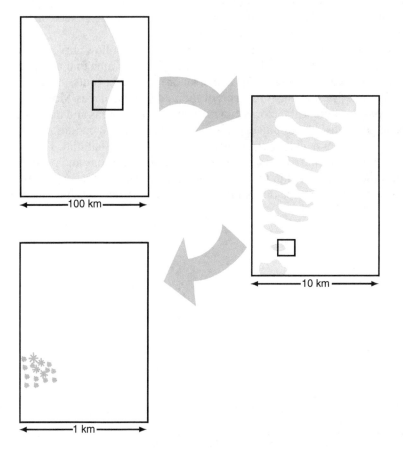

FIGURE 13.2 The typical geographic distribution of a plant species. The distribution appears continuous on a choropleth map, but when viewed at a smaller scale, we can see that the species actually occurs as scattered metapopulations; when viewed at an even smaller scale, the species is seen to occur as scattered individuals.

into isolated metapopulations or individuals in the vicinity of the species geographic limits. Choropleth maps do not convey this fragmentation and the diffuse nature of geographic range limits. In summary, biogeographic maps are useful tools, but are often somewhat crude approximations of the detailed spatial distributions of plants and animals.

BIOGEOGRAPHY OF RANGE SIZE AND RANGE SHAPE

We have seen in preceding chapters that some plants and animals, such as modern humans, have extremely large geographic ranges. Other taxa, such as the Devil's Hole pupfish *(Cyprindon diabolis),* which is restricted to one tiny desert spring in Nevada, or the Catalina mahogany *(Cercocarpus traskiae),* which is restricted to one small valley on Catalina Island off the coast of California, have very small ranges. We might ask, is there a typical or average range size for species? If we plot the range sizes of North American and Central American bird species (Fig. 13.3), we find that the majority of species have ranges of less than 5 million square kilometers. A few species have ranges greater than 10 million square kilometers. A range of 10 million square kilometers is about equal to the size of Canada. The same pattern is observed in plotting global patterns of range sizes for birds. For North and Central American mammal species, we see a similar pattern. Most mammal species have a range size of less than 2 million square kilometers. Most groups of organisms show a similar pattern: the majority of species have medium to small ranges, and only a few species have very large ranges. Regardless of which continent is studied, it has been found that most terrestrial animal species have ranges that are less than one-fifth to one-fourth the size of the continent they live on. In general, a lognormal curve having the shape of a backward *J* describes the relationship between species numbers and geographic range sizes for most groups of plants and animals (Fig. 13.3). This typical distribution, consisting of many species with small ranges and a few species with large ranges, is sometimes called the **hollow curve distribution.** As discussed below, the mean range size in such distributions will vary depending on the taxonomic group being considered.

What are the potential ecological controls that produce the hollow curve distribution of species ranges? This phenomenon is often linked to niche breadth. If we assume that the species being studied have stable geographic ranges, not limited by dispersal or historical events, the hollow curve suggests that most species are relative specialists in terms of their ecological niches. Most species are excluded by physical conditions or biological factors from many geographic areas and habitats. A relatively few species have broad generalist niches and can survive diverse physical conditions and compete effectively with other species over a wide range of geographic regions or habitats. The majority of species survive by being adapted to a narrower set of environmental conditions and perhaps being competitively superior to other species within the geographic region where those environmental conditions exist. The widespread occurrence of the hollow curve indicates that the mix of environmental and

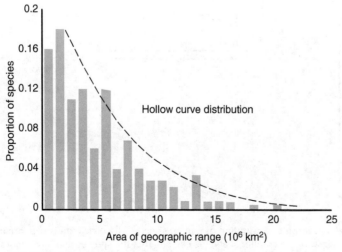

FIGURE 13.3 Size of the geographic ranges for North and Central American land birds (*Source:* Brown, 1995 / University of Chicago Press).

competitive pressures leading to adaptation for persistence in a select geographic area is common. However, the nature and overall strength of the relationship between niche breadth and geographic range is an area of continued research.

American ecologist and biogeographer James Brown observed that species with large geographic ranges generally occur in high densities throughout their range, while species with smaller ranges tend to occur at relatively low densities. Brown concluded that this is evidence of the key role that niche breadth plays in determining both geographic range and density. He suggested that generalist species can use a wide variety of habitat and ubiquitous resources, which allows these species to be both common on a local scale and widely distributed geographically. A contrasting explanation for the relationship of range size and population density comes from metapopulation theory. According to metapopulation theory, the species that develop large geographic ranges and high population densities are superior to other species in terms of dispersal capability and resistance to extinction. In addition, since metapopulations are by definition geographically separated from other populations of the same species, individual events such as catastrophic disturbances or outbreaks of disease will likely not affect all of the populations. It is these properties, particularly resistance to large-scale synchronous declines in population, that allow for large population sizes. However, since resistance to extinction is higher for generalist species than for specialists, the importance of large niche breadths in promoting large geographic ranges for metapopulation distributions cannot be ruled out. The relationship between range size, abundance, and niche size remains an area of active research.

We might ask, are there systematic differences in the range sizes of different types of organisms? When we compare different groups of organisms, we find that insects have generally smaller average range sizes than vertebrates. Amphibians and freshwater fish tend to have smaller range sizes than mammals and birds. Birds tend to have larger range sizes than mammals. North American birds have an average range size of over 3 million square kilometers. North American mammals have an average range size of around 1.5 million square kilometers. The relatively large range size of birds may relate to better dispersal capability, broader niches, and/or greater ability to take advantage of a wider variety of microhabitats. Not all flying organisms have range sizes that are particularly large, however. The average range sizes for North and Central American bat species are about the same as for nonflying mammal species. In general, marine species have larger range sizes than terrestrial species.

When we look in more detail at birds and mammals, we find that predators generally have larger ranges than herbivores. In Africa, large carnivores such as lions and hyenas have an average range size of about 9 million square kilometers. In contrast, large herbivores such as elephants and zebras have an average range size of about 4 million square kilometers. In North America, the average range size of predatory owls is about 6 million square kilometers, while for seed-eating sparrows it is about 3.5 million square kilometers. The relatively large size of predator ranges likely reflects the fact that the higher trophic levels that predators occupy requires them to utilize a wide range of different prey species and be able to hunt and survive in a wide range of environments in order to obtain adequate food.

Examination of the relationship between the body size of animals and their geographic ranges shows that both large and small species can possess large geographic ranges (Fig. 13.4). However, there are few very large bodied species that occupy very small ranges. Species with large body masses require greater amounts of food and larger browsing, hunting, and foraging areas than small-bodied species. In general, the carrying capacity of a given area is less for a large-bodied species than for a small one. If a large-bodied species has a small geographic range, it will have a low population size and be particularly prone to extinction. Thus, a large range size is important for the survival of large-bodied species. Following from this consideration, and our discussion of the differences in the range sizes of carnivores and herbivores, we can see why large carnivores in particular require very large range sizes to survive and small herbivores can persist with small ranges. It is not surprising that the average range size for predatory cat and canine species in North America is about 6 million square kilometers, while for herbivorous small rodents the average range size is less than 800,000 square kilometers.

Evolution and speciation events also play a role in determining the present range sizes for plant and animal species. Minor landscape features, such as streams or small mountain ranges, may be barriers for seed dispersal or the movement of small nonflying animals. The presence of such barriers restricts gene flow between populations and promotes allopatric speciation. These allopatric events create new species that initially have small geographic ranges. Such small landscape features are not significant barriers for larger animals and flying organisms. Biogeographic maps provide some evidence

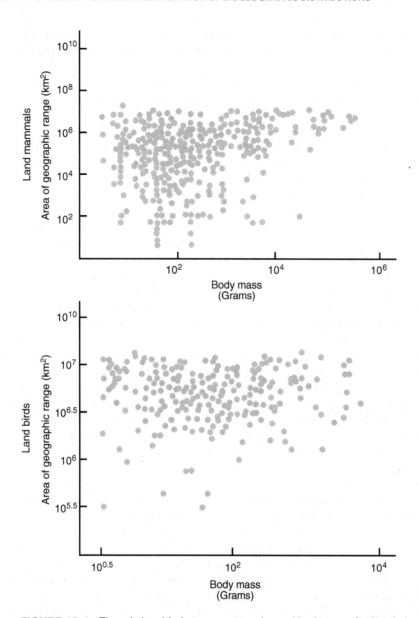

FIGURE 13.4 The relationship between range size and body mass for North American terrestrial mammal and bird species. Small- and medium-sized animals can have large or small ranges, but few large animals have small ranges (*Source:* Brown, 1995 / University of Chicago Press).

for the role of landscape diversity and physical barriers in the development of small geographic ranges through allopatric speciation. North and Central American mammal species with ranges of less than 10,000 square kilometers are generally small bodied and are most commonly found in mountainous regions, islands, and peninsulas. The rugged terrain of mountainous regions and the waterbodies surrounding islands and peninsulas produce barriers that promote allopatric speciation.

Some very interesting patterns emerge when we examine large-scale geographic variations in range sizes. The average range sizes for plants and animals that live at high latitudes or high elevations often tend to be larger than the range sizes of species that live at low latitudes and low elevations. There are many more mammal and bird species with very small ranges in southern Central America than in North America. The pattern of decreasing range size toward the equator, although not universal, is called **Rapoport's Rule** in recognition of the Argentine ecologist and biogeographer Eduardo Hugo Rapoport. Larger ranges at high latitudes and high elevations may in part relate to more generalized and broader niches. This is referred to as the climatic variability hypothesis. It has been argued that, in general, high latitude and high elevation environments are more variable in terms of seasonal and annual climates, generating broader niches in plants and animals. However, this is not accepted by all scientists and the applicability of Rapoport's Rule and potential reasons for larger range sizes at high latitudes and high elevations are still being vigorously debated. In addition, the relationship between latitude and range size appears greater for the northern hemisphere than the southern hemisphere.

The general shapes of geographic ranges reveal additional interesting patterns that can have ecological significance. If dispersal alone controls the shape of a species range, it has been hypothesized that ranges should most often be circular. This has been referred to as the neutral model. Alternatively, if climate is a major controller of a species range, it might produce elongation of the range longitudinally since at macroscales climate typically changes more significantly with latitude than with longitude. However, physical features such as north-south–oriented mountain ranges can be longitudinal dispersal barriers and also longitudinal corridors of similar climate. An analysis of the ranges of 11,582 different plant and animal species from around the world showed that neither the neutral model nor a macroclimatic model adequately explains all the variations in range shape found in the data. The shapes of plant and animal species with small ranges favor more circular (neutral) than elongated ranges. The greater tendency for ranges of widely distributed species to be elongated along one axis relative to the other can often be shown to be controlled by climate and landforms. In North America, species that have large geographic ranges typically possess ranges that are longer in longitude than they are in latitude. In contrast, animals with small ranges tend to have latitudinal axes that are long compared to their longitudinal axes (Fig. 13.5). It appears that animals with large ranges are typically distributed parallel to longitude along broad climatic zones. In contrast, rivers

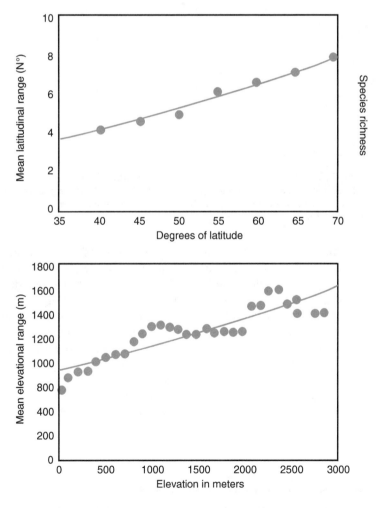

FIGURE 13.5   The relationship between the latitude and elevation at which a species is found and the degrees of latitude or meters of elevation that its range spans. The examples here are for tree species in North America and tree species in Costa Rica (*Source:* Brown, 1995 / University of Chicago Press).

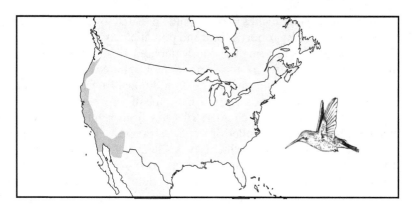

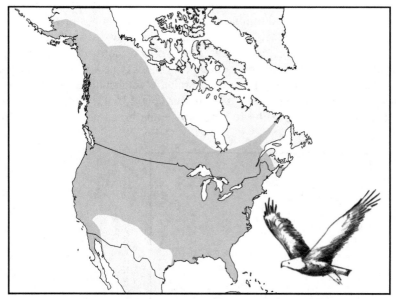

FIGURE 13.6   The latitudinally elongated geographic range of Anna's hummingbird *(Calypte anna),* which is a small-bodied nectar-eating species with a small geographic range. The longitudinally elongated range of the bald eagle *(Haliaeetus leucocephalus),* which is a large-bodied predator species possessing a large geographic range (*Source:* Robbins et al., 1983 / St. Martin's Press).

and mountain systems, which produce more local climatic regions and also present barriers to the distributions of species with small ranges, often run from north to south in North America. Examples of north-to-south mountain ranges include the Sierra Nevada, the Rockies, and the Appalachians. These landscape features and barriers appear most important in controlling the range shape for small-bodied species with small ranges. This pattern can be seen for many mammals and a number of bird species. For example, the total geographic range (winter and breeding ranges) of Anna's hummingbird *(Calypte anna)* is restricted to a narrow latitudinal strip of land west of the main coastal mountains that runs south to north from Baja California to Washington State. In contrast, the total natural range of the bald eagle *(Haliaeetus leucocephalus)* extends from the western coastline of Alaska to the Atlantic coastline of eastern North America (Fig. 13.6). The bald eagle is excluded from desert regions in the south and tundra regions to the north and occupies a broad longitudinal band of more temperate climatic conditions between these two extremes.

COMMON BIOGEOGRAPHIC DISTRIBUTIONAL PATTERNS

After examining many range maps for plants and animals, biogeographers have come to recognize a number of geographic distributional patterns that are repeated for different taxa. In many cases, the study of modern ecological and dispersal processes and the examination of geologic history and the fossil record allow us to determine the causes of such distributions. The distributions themselves can also provide evidence regarding the ecology and evolutionary history of the organisms. Of course, in making such inferences, we must be careful not to become circular in our reasoning. A rich vocabulary has been developed to describe different geographic distributional patterns and the factors that are likely to have created them. In this section, we will investigate some of the more important distributional patterns recognized by biogeographers.

Endemic and Cosmopolitan Distributions Revisited

In Chapter 10 we saw that biogeographers use the term *cosmopolitan* to describe plants and animals that possess very large geographic ranges distributed over most continents. Taxa with restricted geographic ranges are classified as endemics. Most of the time these terms are used in a rather loose fashion. Rapoport has fashioned more precise definitions that classify geographic distributions relative to both the number of biogeographic regions in which the species occurs and the absolute size of its geographic range. According to this system, *endemics* are species that are found in only one biogeographic region. Characteristic species are those found in two regions. Species that occur in three or four regions are classified as semicosmopolitan. Those species that are found in five or more regions are cosmopolitans. Species that have a large range (greater than 4,400,000 square kilometers) within the biogeographic region in which they are found are classified as macro-areal. Those with small ranges are classified as micro-areal (less than 1000 square kilometers). Meso-areal species have intermediate range sizes. For example, the skua (*Stercorarius skua*) is a seabird found in a small area of southern South America (Neotropical region), in Iceland, and in the Faeroe Islands (Palearctic region). This bird would be classified as a micro-areal characteristic species. Humans, who are found over large areas of all biogeographic regions, are classified as cosmopolitan macro-areal species. Macro-areal cosmopolitan species are also sometimes called **pandemics.** There are very few truly pandemic species.

Endemics can also be subclassified by their history. Species that have just recently evolved and speciated often have small population sizes and geographic ranges. These are referred to as neoendemics. It is assumed that they are endemic to the area in which they evolved. The higher elevations of the volcanic mountain Mauna Kea on the island of Hawaii support a beautiful subspecies of silversword plant (*Argyroxiphium sandwicense sandwicense*). The subspecies is found nowhere else on the Hawaiian Islands or anywhere else in the world for that matter. As the island of Hawaii is only about a million years old, it is likely that this plant is a neoendemic that evolved on the island. Today, only about 40 individuals of this highly endemic subspecies grow in the wild. Part of its small population size today is due to human disruption of its habitat through factors such as the introduction of alien herbivores including goats and pigs. Fortunately, several hundred of the endangered silversword have been planted elsewhere.

Some endemics once had larger geographic ranges that have now been reduced in size. The ranges of species may be reduced in size because of physical environmental factors such as climate change or because of increased competition or predation from newly arrived or recently evolved species. Endemics that had much larger geographic ranges in the past are called paleoendemics. The coastal redwood tree (*Sequoia sempervirens*) is a good example of a paleoendemic. Today, the species is only found along a very narrow strip of coastline in northern California and a small portion of adjacent Oregon. During the Paleogene, the ancestors of the redwoods were extremely widespread across North America and Eurasia. Climatic cooling and drying during the Neogene and the Quaternary caused the extinction of the redwoods over the vast majority of their geographic range. We will consider the history of this interesting species further a little later in this chapter.

Continuous Zonal Biogeographic Distributions

Many semicosmopolitan and cosmopolitan plant and animal taxa have geographic distributions that form broad longitudinal bands that follow longitudinal temperature belts. Taxa are restricted to these bands because of their physiological sensitivity to temperature or possibly in some cases because of their inability to compete with species that dominate other climatic zones. There are five such zonal distributional patterns. **Circumpolar** taxa have distributions that circle the Arctic or Antarctic regions and lie within the colder climates of the high latitudes. A good example of a circumpolar taxon is the arrow grass genus *(Scheurchzeria)*. Species belonging to the arrow grass genus can be found growing at high northern latitudes from northern Europe to northeastern Siberia and from Alaska to the Atlantic Coast of Canada. The terms *Boreal* (northern hemisphere) and *Austral* (southern hemisphere) are used to describe plants that are distributed in the cooler areas of the mid- to high latitudes. The spruce genus *(Picea)* is an excellent example of a boreal taxon. Species of spruce are typical of the northern forests of Alaska, Canada, and northern Eurasia. **Temperate** plants and animals are found in the intermediate zone typical of the middle latitudes. Maples *(Acer)* provide an example of a northern hemisphere temperate taxon. **Pantropical** organisms are found in the tropical zone. The palm family (Aracaceae) is a good example of a pantropical taxon.

Disjunct Distributions

The term **disjunct distribution** is used to refer to geographic ranges that are divided into two or more geographically separated locations. Such disjunct distributions and their causes have long been of interest to biogeographers and are of widely recognized importance today in addressing ecological, evolutionary, and conservation questions. One analysis of over 10,000 bird, amphibian, crocodilian, and mammal species found that about 11% displayed disjunct distributions with 500 kilometers or more separating the disjunct populations. Birds were the most commonly disjunct taxa (19.3% of species analyzed) and had the greatest average distance between disjunct populations (2000 kilometers). They were followed by mammals (6.4%) and then amphibians (2.8%). Disjunct species can be found from the tropics to the higher latitudes. At a continental scale, South America had the largest number of disjunctions.

Amphiregional disjunct distributions occur when macro- and meso-areal taxa have two widely separated geographic ranges. The root *amphi* comes from Greek and can be translated as meaning "on two sides." Plants and animals that are amphiregional have distributions that are divided into two distinct ranges that are usually separated by biogeographic barriers. Some common distributions of this kind include **amphitropical,** taxa which occur on either side of the tropics but not in the tropical zone itself. The Chenopodiaceae family of plants, which includes sugarbeets and spinach, has an amphitropical distribution. It is likely that amphitropical plants are excluded from equatorial regions by high temperatures or competition with tropical plants. North and South America have over 200 different species of temperate vascular plants that have such amphitropical disjunct distributions. One example is the woodland flower species *Osmorhiza chilensis,* which is found in southern South America and central portions of North America.

Amphioceanic disjunct taxa occur on opposite sides of oceans. Amphiatlantic and amphipacific taxa occur along the western and eastern coastal areas of the Atlantic and Pacific oceans, respectively. There are over 100 genera and higher taxa of flowering plants that have amphipacific distributions. In the South Pacific, the southern beech *(Nothofagus),* which is found in eastern Australia, New Zealand, and western South America, is an example of a southern amphipacific taxon. Many shore-dwelling plants and animals, as well as shallow-water marine species, have amphipacific and amphiatlantic distributions.

A number of terrestrial and marine species are found in cold polar regions, but not in the warmer mid- and low latitudes. Such organisms are referred to as being **bipolar.** Many members of the whale family Balaenidae and the freshwater fish family Petromyzonidae display bipolar distributions. Intolerance to warm temperatures and competitive exclusion probably keep such taxa from occupying warmer temperate and tropical areas.

Dispersal Disjunctions

Biogeographers often classify disjunct distributions on the basis of how the disjunction is thought to have occurred. Dispersal disjunctions take place when a jump dispersal event allows a population of a plant or animal species to become established a long distance from the main population. Dispersal disjunctions are particularly common for flying animals and for plants that possess seeds prone to long-distance dispersal. The least bittern *(Ixobrychus exilis),* with breeding populations throughout eastern North America and as disjunct breeding populations in California and Oregon, is probably a dispersal disjunction (Fig. 13.7). During the winter, the birds migrate to Mexico and Central America.

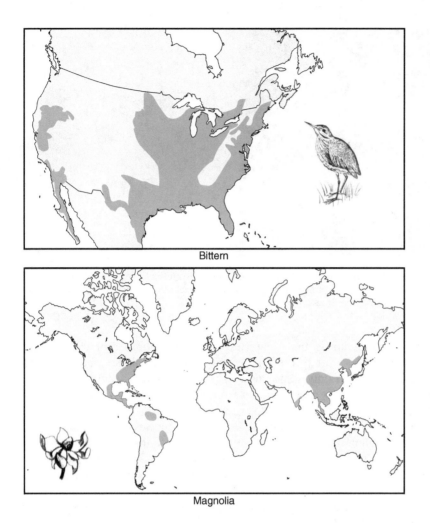

FIGURE 13.7 Two examples of disjunct distributions. The least bittern *(Ixobrychus exilis)* is considered an example of a geographic disjunct distribution caused by long-distance dispersal. The geographic distribution of the magnolia family (Magnoliaceae) is considered an example of disjunction caused by long-term climatic change (*Source:* Robbins et al., 1984; Heywood, 1978).

Naturally established populations of hydrochores such as coconut palms *(Cocos nucifera)* found on isolated tropical islands are also examples of dispersal disjunctions.

Climatic Disjunctions

A climatic disjunction occurs when climatic change makes a portion of a taxon's range uninhabitable and splits a once continuous geographic distribution into two or more separate parts. A classic example of a climatic disjunction is the magnolia family of trees (Fig. 13.8) and shrubs (Magnoliaceae). Fossil evidence shows that this ancient family of plants was distributed over much of the northern hemisphere during the Cretaceous. The general cooling and drying of the world's climate that occurred during the Cenozoic eliminated magnolia from most of its range. The family survives today in the Americas and in eastern Asia (Fig. 13.7). The alligator genus, which is represented today with one species in China *(Alligator sinensis)* and one in North America *(Alligator mississipiensis)* (see Chapter 1), is a similar example of climatic disjunction created by climatic cooling during the Cenozoic, coupled with the development of the Bering Sea separating Asia and North America. We can find many other plant genera and species with similar distributions and some animals. For example, some aphid, butterfly, and fish taxa present evidence of being North American–Asian climatic disjuncts. Like the magnolia, the disjunct distributions reflect the cooling and drying of global climate during the Cenozoic that caused the fragmentation of once continuous temperate forest that stretched across Eurasia and North America.

FIGURE 13.8   A sweetbay magnolia *(Magnolia virginiana)* from the Everglades of Florida. Magnolias represent one of the most ancient families of flowering plants (Magnoliaceae) and were once very widely distributed (*Source:* nickkurzenko / Adobe Stock).

Climatic changes during the Quaternary caused a number of climatic disjunctions within continents. For example, prior to the Quaternary, moles (Talpidae) were found throughout North America. Today, they occur only along the west coast and in the eastern deciduous and mixed-wood forest region. Cold conditions prevent moles from inhabiting the northern portion of North America and the Rocky Mountains. Dry conditions restrict moles from living in the western grasslands and deserts. A number of other small mammals, such as some species of chipmunks *(Eutamius),* which are found on isolated high mountains in the deserts of eastern California, Nevada, and Utah are climatic relicts. These species had larger continuous ranges in the Great Basin region during the last glacial maximum when the climate of the region was cooler and moister. Conifer forest extended to lower elevations in the Great Basin during that time. The onset of dry conditions and the replacement of low-elevation conifer woodlands by desert during the Holocene led to the present disjunct distributions of these small mammals.

Gorillas *(Gorilla)* have a particularly interesting climatically disjunct distribution in Africa. One population of the species *(Gorilla gorilla)* is found in lowland tropical rainforest along the west coast of Africa, and another disjunct population *(Gorilla beringei)* is found in upland tropical rainforest in the highlands of central Africa. The two populations are separated by savanna vegetation, which is inappropriate for the forest-dwelling gorillas. During the Early Holocene and other warm periods in the Pleistocene, the climate of northern portions of Africa was both warm and moister than it is today, and it is likely that continuous rainforest developed in central Africa linking the coastal and upland rainforest. The range of the gorilla species was probably continuous from the coast to the uplands during such periods.

Geologic Disjunctions

Geologic disjunctions occur when geologic processes such as plate tectonic movement split once continuous ranges into two or more separate parts. The distribution of the southern beech *(Nothofagus)* in Australia, New Zealand, and South America is an example of a geologic disjunction. Fossil evidence shows that this genus was widespread on the supercontinent of Gondwanaland (see Chapter 10). The breakup of Gondwanaland contributed to the present disjunct distribution of the genus. Many plant families in the southern hemisphere share a similar history and show a similar geographic pattern of disjunction. The large flightless birds of the superorder Ratites, including the ostriches (Struthionidae) of Africa, the emus (Dromaiidae) of Australia, the now extinct moas (Dinornithidae) of New Zealand, and the now extinct rheas (Rheidae) of South America, evolved from a common ancestral lineage that also experienced disjunctions during the breakup of Gondwanaland. As we shall see later in this chapter, some aspects of these disjunctions for *Nothofagus* and the ratites likely reflect dispersal events in addition to the breakup of Gondwanaland.

Evolutionary Disjunctions

Evolutionary disjunctions occur when two new species develop in different portions of the geographic range of a widespread common ancestor. The subsequent extinction of the common ancestral species leaves the two new species disjunct from each other. Many amphitropical species of plants may have arisen in this manner. Species of a number of different tree and shrub genera such as *Acacia* and mesquite *(Prosopis)* are found in both the deserts of southwestern North America and the deserts of Chile and Peru. The mesquite species *Prosopis juliflora* found in the southwestern United States and the closely similar species *Prosopis chilensis* found in Chile are likely evolutionary disjunct species.

Biogeographic Relicts

Biogeographic relicts are taxa that once had larger distributions but have become narrow endemics. Two types of biogeographic relicts are generally recognized. The first type are climatic relicts, which are species whose geographic ranges have been constricted due to recent climatic changes. Warming of the climate at the end of the Pleistocene created many such climatic relicts. These climatic relicts are sometimes referred to specifically as glacial relicts. One good example is the muskox *(Ovibos moschatus),* which is found today only in the Arctic regions of North America. Fossil evidence shows that muskox ranged across large portions of central North America and northern Eurasia during Pleistocene glacial periods. Similar fossil evidence shows that the alpine marmot *(Marmota marmota)* ranged across much of Europe during the last glacial maximum, but its natural historic range is restricted today to the Alps.

A number of plant species have glacial relict distributions. One classic example is the Norwegian mugwort *(Artemisia norvegica),* which ranged across central Europe during the last glacial maximum but is today restricted to small populations in the Scottish Highlands, the coastal mountains of Norway, and the Ural Mountains of Russia. In North America, several glacial relict species of pines and cypresses occur as endemics along the coast of California. These include the Torrey pine *(Pinus torreyana),* which grows today only as small populations near San Diego and on Santa Rosa Island, and the Monterey pine *(Pinus radiata),* which is found in small scattered locations along the central California coast as well as on Cedros Island and Guadalupe Island in Mexico. Fossil evidence, including pollen from marine sediments collected off the California coast, suggests that pines and cypresses formed extensive coastal forests and were widely distributed along the coastlines of California and adjacent Mexico during the Late Pleistocene. Cooler and moister conditions associated with the Late Pleistocene allowed these trees to occupy large ranges along the coast. Warmer and drier conditions became established during the Holocene, and this led to the restriction of the pines and cypresses to their present small and disjunct distributions. The pines may have been able to survive on islands and a few areas of the coast because these sites are typified by unusually large amounts of fog. The fog increases moisture availability and keeps temperatures and evaporation rates low during the summer months.

Evolutionary relicts are paleoendemics that are survivors of formerly more widespread and diverse evolutionary lineages. The tuatara *(Sphenodon punctatus)* is a lizard-like reptile that is found only in New Zealand. It is also the only native reptile of that country. The tuatara is the last surviving species of an ancient reptilian order called the Rhynsoles. Prior to the Cretaceous, this order had many species and was geographically widespread. The early separation of New Zealand from the rest of Gondwanaland protected the tuatara lineage from contact and competition with later evolving reptiles and mammals. The lack of reptile and mammal predators in New Zealand was an important factor in the survival of the tuatara there.

There are many evolutionary relicts among plant species. Cycads were an extremely diverse and geographically widespread group of plants prior to the Cretaceous. Fossil cycads are found from the equatorial regions to Siberia and from Greenland to Australia. Today, they survive as about 100 species that are restricted to tropical and subtropical regions. Only one native genus containing one species, *Zamia floridana,* remains in the whole of the United States. This cycad is relatively common in Florida. Similarly, the tulip tree *(Liriodendron tulipifera)* of the southeastern United States is an evolutionary relict of a once more diverse and widespread family that extended across Eurasia and North America in the Late Cretaceous and Paleogene. Today, only two species remain—the American tulip tree and a related species in southeastern Asia *(Liriodendron chinense).* As is the case with the magnolias, increasingly cool and dry climates that developed during the Neogene and Quaternary drove the tulip tree to extinction across most of its range. Thus, tulip trees and magnolias can be classified as both evolutionary relicts and climatic disjuncts.

The coastal redwood of California and its nearest relative, the dawn redwood of China (*Metasequoia glyptostroboides*), are interesting examples of disjunct paleoendemics that are evolutionary and climatic relicts. Unraveling the history of these two relicts is a very interesting chapter in biogeographic research. As we have seen, the California redwood is found only along a narrow strip of coastline extending from central California to southwestern Oregon. The dawn redwood occurs as a tiny population in the Szechuan Province of central China. It has long been known from fossil evidence that the ancestors of the California redwood were extremely widespread in the northern hemisphere during the Cretaceous and Paleogene. Fossils of redwoods have been found throughout North America, from the Arctic to the modern deserts of the Great Basin. Redwood fossils have also been found throughout Eurasia. In the early 1940s, the Japanese botanist Shigeru Miki described a new species of redwood from fossils found near Kobe, Japan. The arrangement of the needles and the cones of the fossil species was different from that of the living California redwoods and previously collected fossils. No living species of trees were like the new fossil redwood, so Miki created a new genus and species for the fossils—*Metasequoia glyptostroboides*. Coincidentally, during the 1940s, Tisang Wang, a Chinese forester, reported finding a small population of a previously unknown tree species in a remote area of the Szechuan Province of China. Subsequent comparison of the newly discovered tree, now called the dawn redwood, and fossil material showed that the living trees belonged to the fossil species *Metasequoia glyptostroboides*. The living dawn redwood turns a beautiful gold color and loses its needles during the winter. California redwoods are evergreen. By being deciduous, the dawn redwood can tolerate cold better than the California redwood. Fossils of dawn redwood have now been recognized from North America and Greenland in addition to Asia. The species dates to the Cretaceous and appears to have remained morphologically and genetically unchanged since that time. The dawn redwood truly is a living fossil! As was the case for the tulip trees and the magnolias, increasing cold and dry conditions during the Late Cenozoic led to the huge decrease in the range of the redwoods and their present status as evolutionary and climatic relicts. Today, dawn redwoods are grown in gardens throughout temperate regions. A species that was unknown only a few decades ago is regaining part of its former large range through human intervention.

BIOGEOGRAPHIC DISTRIBUTIONS AND THE RECONSTRUCTION OF EVOLUTIONARY HISTORY

Can the modern distribution of organisms be used to reconstruct the evolutionary history of plant and animal species? This question has been a central concern and source of contention for biogeographers for almost 200 years. Let's step back in time and look at the dawn of these debates. In 1843, two British naval ships, the HMS Erebus and the HMS Terror, returned to England from a voyage of exploration around Antarctica and the adjacent continents. Returning with the expedition was Joseph D. Hooker, a young English doctor and naturalist. He brought back with him a large collection of specimens and a vexing scientific puzzle. Hooker had noted that many of the same plant families and genera could be found in the widely separated floras of New Zealand, Tasmania, southern Australia, and southern South America. In addition, he had found fossil wood on the remote and treeless Kerguelen Islands off the coast of Antarctica. After deciding that long-distance dispersal could not have produced these distributional patterns, Hooker concluded that these plants were the surviving relicts of a once extensive forest that had grown on all of the continents and major islands of the southern hemisphere. He suggested that the southern land areas must have once been joined together and geologic events must have caused the breakup, fragmenting their flora and leading to the evolution of new species.

We now know that Hooker was correct. The geographic distributions of organisms and fossils he studied were plausibly a result of the breakup of the Gondwanaland supercontinent. As we saw in Chapter 9, geologic events that cause the breaking apart of formerly continuous geographic ranges are called vicariance events. In the twentieth century, arguments for the primacy of vicariance events in the evolution of new species was most forcefully championed by Leon Croizat. We will learn more about Croizat and his ideas later in this chapter.

Sadly, the HMS Erebus and the HMS Terror, along with all their crewmembers, would be lost exploring the Canadian Arctic in 1845. Hooker remained in England and went on to a very long life and distinguished career. As fate would have it, he co-chaired the 1858 session of the Linnaean Society during which Charles Darwin and Alfred Wallace's papers on natural selection were read. Ironically, both Wallace and Darwin disagreed with Hooker regarding the relative importance of the geologic breakup of landmasses versus long-distance dispersal in generating the biogeographic disjunctions seen in the southern hemisphere. Wallace and Darwin believed that dispersal was very effective over geologic time scales and could explain many of the disjunct distributions of plants and animals that are present today. In addition, Darwin believed that dispersal was fundamental to evolution and to the formation of new species. He formulated the center of origin model to link geographic distributions to evolutionary history. He postulated that evolution and speciation occur as species disperse and spread outward from central points of origin. Darwin suggested that speciation during dispersal is driven by the new environmental selection pressures encountered when species disperse into new areas. In the twentieth century, famous biogeographers and evolutionists such as George Gaylord Simpson and Ernst Mayr argued for the importance of dispersalist evolution in the creation of new species. After almost 200 years, the debate still continues between vicariance biogeographers and dispersalists—and can be extremely heated. With the contentious nature of this debate in mind, let's consider both vicariance and dispersalist approaches to reconstructing evolutionary history on the basis of modern distributions.

Centers of Origin and the Dispersalist Model

The search for the centers of the evolutionary origin of plant and animal species has long been a preoccupation of biogeographers. Many different criteria have been proposed to identify the geographic locations in which species have originated. In the 1940s, the biogeographer Stanley Cain listed 13 different rules that had been proposed to indicate the geographic origin for plant or animal species. One commonly applied criterion is to identify the area in which the greatest number of related species are found. For example, if we map out the geographic distribution of pheasant species (Phasianinae), we find that the greatest number of genera occur in western China in the Sino-Himalayas region. Thus, some have concluded that pheasants originated in western China and that continued evolution and speciation have led to the development of many species there, while dispersal and further evolutionary development by some of these species have led to the development of new pheasant species that occupy other areas of Asia and Africa. If, however, we map the world distribution of palm (Arecacea) genera, we see a pattern that is more difficult to interpret. In the case of palms, you see high species diversity in tropical regions of South America, Africa, and southeastern Asia. Did the palms originate in South America, Africa, or southeastern Asia? Without fossils or other evidence, it is impossible to say on the basis of modern geographic distribution exactly where the pheasants or palms originated. Even if there is only one geographic area with highest diversity, it does not unequivocally mean that is the center of origin for the taxa. For example, the highest species diversity for living members of the horse and zebra genus *(Equus)* occurs in Africa. However, we know from fossil evidence that the ancestors of horses first evolved in North America and later dispersed to Africa. Until the reintroduction of horses by the Spanish, there were no surviving species of horses in North America. As for pheasants and palms, recent genetic analysis coupled with SDMs suggests an African origin for pheasants around 33 million years ago. Early fossils of palms, dating back as far as about 90 million years ago, come from a number of locations including North America and Europe, far removed from the modern tropics. There are some cases where current centers of diversity may be centers of evolutionary origin. For example, the Coral Triangle region of Australasia has a very high diversity of marine species. Genetic analysis suggests it is indeed a center of origin for the evolution and diversification of dwarfgobies (Gobiidae). We can conclude that the current location of the highest diversity of species or genera does not necessarily mean that is the geographic center of origin for that taxon group.

Another common method for identifying the center of origin is to look for the most evolutionarily primitive or advanced members of the genus or family being studied. Some scientists have argued that primitive species, which are assumed to be close in form to the supposed ancestral species, are most likely not to have dispersed far from the center of origin. Two major proponents of this hypothesis were Willi Hennig and Lars Brundin. The primitive characteristics are called pleisio-morphs. Many flowering plants with apparently primitive characteristics are found in the western South Pacific, including portions of Fiji, New Caledonia, New Guinea, eastern Australia, and the Malay Archipelago. This has led some scientists to identify this region as the center of origin for flow-ering plants. Other scientists argue, however, that fossil evidence points to South America or China as the origin of flowering plants. In complete contrast, some biogeographers, such as P. J. Darlington, have suggested that the centers of origin will possess the most evolutionarily advanced species. The supposed advanced characteristics are called apomorphs. These biogeographers contend that high genetic diversity at the center of origin promotes evolution and speciation. The newly evolved forms outcompete primitive forms and relegate the latter to survival at the margins of the geographic range of the taxon. One problem in general is determining what constitutes a "primitive" or "advanced" morphological feature. Such classifications have been shown to differ from researcher to researcher. In the case of flowering plants, DNA analysis suggests that the genus *Amborella*, found in New Caledonia, appears to represent an early sister lineage to all flowering plants. Perhaps the flowering plants did originate in that region.

The preceding discussion shows that it is difficult to confidently determine the evolutionary center of the origin of a species on the basis of modern geographic distributions alone. In a careful review of the 13 different criteria used to identify centers of origin, Cain concluded that none of them was completely reliable. His conclusions have been echoed by many other biogeographers. What then do we do with Darwin's theory regarding the evolutionary importance of natural selection during dispersal from the center of origin? Biogeographers such as Hennig and Brundin suggest that Darwin was essentially correct and that there should be a gradation from primitive pleisomorphs at the center of origin to more advanced apomorphs at the geographic range boundaries of the taxon. Brundin called this the rule of deviation, while Hennig coined the term progression rule to describe this phe-nomenon. There is, however, little direct field evidence to support the widespread presence of clear geographic gradients of primitive to advanced forms in either plants or animals. The best evidence for the presence of such progressions comes from the distributions of insects on some island chains. For example, fruit flies *(Drosophila)* on the Hawaiian Islands show a progressive increase in supposed apomorphic traits as one goes from the oldest islands, such as the outer islands beyond Kauai, to the younger islands, such as Hawaii. Such progressions appear to be somewhat rare, however. In any case, as Croizat forcefully argued, Darwin never presented clear and convincing arguments of exactly how dispersal promotes evolution and speciation at the edges of the expanding ranges, nor did he explain why it is that new traits developed at the edges are not introduced throughout the entire range of the species. Finally, we should remember that the movement of continents and climatic changes over geologic time may be such that the modern geographic distribution of a taxon bears little to no resem-blance to its ancient geographic place of evolutionary origin and earlier distribution.

Cladistic Biogeography

One of Brundin's and Hennig's most important contributions was to help link the analysis of geo-graphic distributions directly to phylogenetic reconstructions. Phylogenetic reconstructions of specia-tion history and the evolutionary linkages between species are often presented as branching trees called taxon cladograms (Fig. 13.9). The formulation and study of such diagrams is called **cladistics.** Hennig pioneered the cladistic method in the 1950s. The endpoints on the taxon cladograms are modern taxa, such as the different species, genera, or families of plants or animals that exist today. The con-struction of the cladogram starts with measuring a set of physical traits or the genetics for a group of species and determining which traits or genes are common to all of the species being studied, which traits are shared by some, and which are unique to individual species. This is at the heart of the science of systematics. The use of DNA to develop cladograms is referred to as **molecular systematics.** It is assumed that the traits common to all species are derived from the common ancestor of the plants or animals. Species that share many traits are assumed to be more recently differentiated from each other and more closely related than species that share few common traits. The species that are similar to each other in terms of physical traits are placed close together on the cladogram. Thus, the

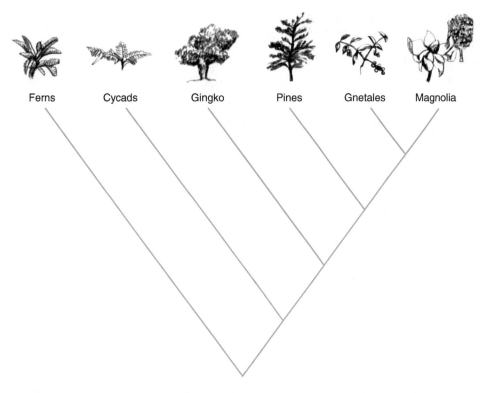

FIGURE 13.9 A simplified cladogram for magnolias (representing angiosperms) and ferns, cycads, gingkos, pines and gnetales. The ordering and splits in such cladograms are based on a number of traits related to reproductive structures, leaf shape, plant chemistry, and genetics.

proximity of the different taxa to each other on the cladogram infers their evolutionary relationship. The nodes for the branches of the cladogram indicate the relative timing of evolutionary divergence. The positioning of the nodes corresponds to the timing of the loss of shared characteristics or the appearance of new characteristics. The nodes represent points of evolutionary divergence between lineages.

We can examine a simplified representation of a cladogram for plants to see how taxon cladograms are actually constructed (Fig. 13.9). Hennig produced such a cladogram based on 29 morphological traits. The base of the cladogram groups together cycads, pines, gingko trees, gnetales, and magnolias and separates them from the ferns. All of these groups of plants share common traits such as possessing vascular cambium and seeds. Ferns are not included in this group because they do not possess seeds. The next group to be split off is the cycads because they do not have features such as ancillary branching and true pollen. Pines are split off because among other things pines possess needles rather than broad leaves and do not possess true flowers. The gingkos do have broad leaves that are roughly similar to those associated with magnolias and other flowering plants, but they are split off because they do not possess true flowers. At the top of the cladogram, magnolias are split from gnetales because the latter do not possess essential oils in their leaves, have a different method of producing pollen grains, and do not have flowers similar to magnolia and other angiosperms. From the cladogram, we would infer that magnolias are most similar in form and evolutionary history to gnetales and most distant from ferns.

The different endpoints (taxa) on the cladogram can also be thought of in terms of geographic areas. Older species that are similar to the ancestral plant or animal will be grouped together at one end of the cladogram. These species will have many shared primitive traits. According to the progression rule espoused by some biogeographers, these taxa will be found in the central portions of the geographic range of the taxon (Fig. 13.10). Younger species that have diverged recently will be found at the other end of the cladogram and will be found in geographic areas to which the taxa have most recently dispersed. Thus, the endpoints on the cladogram can represent not only different species but different geographic areas as well. In some cases, the geographic areas where species are found are

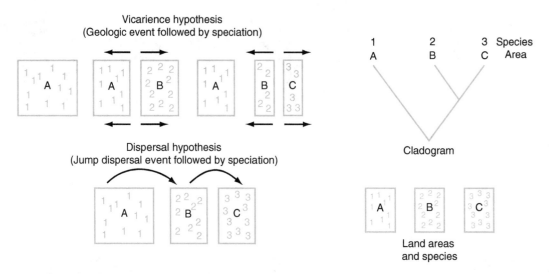

FIGURE 13.10   A three species and three area taxon-area cladogram with vicariance and dispersalist models of geographic speciation that would produce such a cladogram.

listed on the cladograms. Cladograms that list and categorize the geographic areas in which the taxa are found are called area cladograms. The linking of cladograms with biogeographic history is referred to as cladistic biogeography and is an important tool for deducing evolutionary history.

In the 1960s, Brundin produced cladograms for southern hemisphere midge (Chironomidae) species in which he replaced the scientific name for each species on the cladogram with the name of the geographic region in which the species was found. In doing so, he produced the first area cladogram. This research remains highly influential and highly cited. The now classic cladogram of Brundin tied the evolutionary linkages between different midge species to the geographic regions in which they lived. The taxon and area cladograms for the southern hemisphere midges strongly reflect the geologic and geographic history of Gondwanaland. Groups of midges such as the genera *Afrochlus* and *Archaeochlus* are found only in Africa and are morphologically distinctive from other groups because of Africa's early separation from the rest of Gondwanaland and the long period in which midges there have been isolated from other southern hemisphere midges. Genera such as *Parochlus* and *Podonomus* occur on New Zealand, South America, and Australia, but are represented by different species in each of the three areas. These land areas were joined together longer than was the case for Africa and the rest of Gondwanaland and thus share genera. However, sufficient time has elapsed since the separation of New Zealand, South America, and Australia for distinctive species to arise in each region. Molecular systematics research based on analysis of four distinct genes appears to support the basic outline of Brundin's findings, but also provides evidence of important dispersal events occurring after the breakup of Gondwanaland.

Panbiogeography and Vicariance Models

Deep dissatisfaction with the ambiguous nature of center of origin approaches and Darwin's model of dispersal and geographic evolution has led to a critical reevaluation of the role of geography in reconstructing evolutionary history. The most strident early voice in this reassessment was the Leon Croizat. While working in Venezuela in the 1950s, he developed a new method of linking geography and evolutionary history that is called **panbiogeography.** Croizat analyzed the distributions of hundreds of plants and animals and realized that in many cases the species of very different groups, such as plants and insects, shared similar geographic patterns of endemism and disjunction. Croizat noted that there were great differences in the dispersal capabilities of these different species. It did not seem likely, therefore, that these shared patterns of endemism and disjunction could have arisen by long-distance dispersal. Like Hooker, Croizat decided that the only logical explanation for such repeated patterns of disjunction was that these taxa had once had much larger distributions and that geologic events had split these once continuous ranges. Croizat argued that long-distance dispersal events were rare and of little importance to the development of new species. Rather, most speciation occurred following vicariance events that split the ranges of species into two or more separate and isolated geographic regions.

To support his ideas and provide evidence of the past vicariance events that led to the splitting of ranges, Croizat amassed data on hundreds of disjunct taxa and then drew lines connecting the ranges of related disjunct species. From this exercise, he showed that there were a number of frequently repeated lines connecting related disjunct taxa. He called these lines tracks. For example, he found that many genera of plants and animals had disjunct distributions along the Pacific coasts of North America and across the ocean along the Pacific coasts of Asia, so he drew lines along the coasts and across the Pacific that represented tracks connecting the distributions of the related taxa. When the same tracks appeared for many different taxa, Croizat called these lines generalized tracks. Croizat called the points where tracks intersected each other nodes. As a number of generalized tracks crossed the oceans, he concluded that they represented past land connections that had been cut by geologic changes (Fig. 13.11). Croizat found that a number of tracks joined taxa on both sides of the South Pacific Ocean, the Indian Ocean, and the South Atlantic Ocean. The connections that span oceans or other prominent features of the geographic landscape, such as mountain chains, are called baselines. Oddly, Croizat originally dismissed continental drift as a mechanism for severing connections between disjunct populations. In fact, he explicitly denounced Wegener's ideas regarding continental drift. Instead, Croizat argued that the continents had never shifted position, and he hypothesized the creation and destruction of all sorts of improbable landbridges across the oceans. The locations of these "lost" connections coincide with the biogeographic tracks identified by Croizat. He argued for now submerged landbridges across the Pacific, Indian, and Atlantic oceans that joined South America, Africa, and Australia in the past. He believed that the biogeographic evidence of past connection was far more reliable than the geologic evidence of the past geography of the globe. The current geologic evidence does not support many of his assertions of past landbridges.

Croizat argued that, by examining geographic distributions and constructing tracks, one could deduce both evolutionary history and geologic history. For example, fig trees *(Ficus)* are found in South and Central America, Africa, Asia, and Australia. Systematics studies based on fig morphology suggest that the Australasian and American species are more closely related to each other than they are to the African species. Croizat's tracks, however, suggest a closer relationship in flora and fauna and greater past geographic connections between Africa and Australasia than between America and

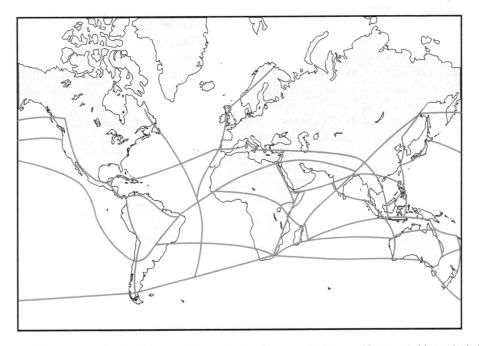

FIGURE 13.11   Some of the most important tracks suggested by panbiogeographic analysis. The tracks join the disjunct distributions of related plant and animal taxa (after a variety of sources including Rosen, 1988; Brown and Lomolino, 1998; Craw et al., 1999).

Australasia. Croizat therefore argued that the systematics studies were flawed and that the figs of Australasia were most closely related to those from Africa. With more than 800 species to contend with, the systematics of *Ficus* still remains to be fully resolved.

Croizat's ideas were startling to many mid-twentieth-century biogeographers. Those who disagreed with him found much ammunition to refute his claims. Croizat made a number of mistakes in taxonomy and geology. He dismissed the occurrence of any significant long-distance dispersal, even in cases where empirical evidence for jump dispersals was overwhelming. In addition, his proposed landbridges often had no basis in geologic fact. Today, the evidence for plate tectonics and continental drift is too overwhelming and clear to be overlooked. Finally, the arrogance and scorn evident in many of Croizat's writings did little to engender respect for or civil discourse on his ideas. Today, very few biogeographers strictly adhere to Croizat's original panbiogeographic concept and interpret both evolutionary and geologic history solely from the basis of his drawn tracks.

Although the classical panbiogeographic models and methods proposed by Croizat have relatively few adherents today, some of his arguments regarding geography and speciation have been much more influential. During the latter half of the twentieth century, a number of biogeographers built on some of Croizat's ideas to formulate a new approach to reconstructing evolutionary history from geographic distributions. Notable among these biogeographers are G. J. Nelson, N. Platnick, and D. E. Rosen from the United States and C. J. Humphries from Britain. The approach they pioneered was to combine Croizat's concepts on the primacy of vicariance events in causing speciation with Hennig's systematic organization of taxa on cladograms. This approach has been called **vicariance biogeography.** Following from this method, the branches in cladograms for plants and animals can be thought of as directly representing the different geographic vicariance events that have occurred during the history of the species. Thus, each node where the cladogram splits is taken to be a vicariance event (Fig. 13.10). The endpoints of the cladograms represent both modern species and the geographic areas in which the species developed following vicariance events. According to the vicariance approach, the cladogram presents both the history of evolution and the geologic and geographic history of the earth.

The early vicariance biogeographers followed Croizat's tenet that dispersal events were rare and unimportant in producing disjunction and allopatric speciation. They assumed that all disjunction and speciation arose from the splitting of once continuous ranges by geologic events. They believed that modern distributions were a reflection of these past events. In producing taxon and area cladograms, the vicariance biogeographers argued that only evidence of physical similarity between taxa or geographic locations should be considered. They felt that the fossil record could be misleading and would corrupt the production of cladograms if it was incorporated in their construction. More recently, some vicariance biogeographers have used fossil and geologic evidence as tests of their cladograms and have proposed phylogenetic and geographic histories. They have also increasingly recognized that climatic changes as well as geologic changes can create biogeographic barriers and produce vicariance events.

Heated debates about panbiogeography and vicariance biogeography have continued into the twenty-first century. Critics have denounced the dismissal of long-distance dispersal events, climatic change, fossil and genetic evidence, and the importance of natural selection incorporated in strict panbiogeography orthodoxy. Some scientists have referred to the approach as "misleading" and "moribund" and questioned if such studies should even be published in mainstream science journals. Panbiogeography supporters have argued that these attacks misrepresent the modern and more nuanced applications of panbiogeography.

BEYOND STRICT PANBIOGEOGRAPHY/VICARIANCE BIOGEOGRAPHY: DNA AND THE PHYLOGEOGRAPHIC REVOLUTION

The classical vicariance model of speciation and the interpretation of cladograms as records of vicariance events are embraced by panbiogeographers and were previously more widely accepted among biogeographers in general. Vicariance events are undoubtedly important in explaining many microevolutionary and macroevolutionary events. One weakness is that the interpretation of taxon and area cladograms can be seriously compromised if parallel evolution has occurred and produced morphologically similar members of a clade in separate areas. In the classic construction of clad-ograms, all groupings and splits are based on similarities and differences in physical traits. If the same physical trait arises in two different species that live in two separate geographic areas, it will lead to a grouping of the two species and will raise the possibility that the two geographic regions were once joined. Cladistics and vicariance biogeographic analysis provide information on the relative timing of speciation events but do not yield direct evidence of the actual number of years since two lineages diverged. Since the actual age of divergence cannot be determined by cladograms alone, biogeogra-phers often must speculate on the timing and causes of vicariance events that may have led to allopat-ric speciation and the divergence of evolutionary lineages. Often such inferences make use of accepted geologic histories and the fossil record. Finally, even if jump dispersal events are rare, they are likely to be important over time and are not accommodated within the framework of classical panbiogeography/ vicariance biogeography.

In recent years, the development of more refined molecular systematics analysis has led to the testing of the proposed evolutionary relationships that have arisen through traditional taxonomic and area cladograms. This direct genetic approach, to which the term *phylogeography* was applied in the mid-1980s, is based on molecular systematics and is becoming an increasingly important focus of bio-geographic research. Much of the current work in phylogeography is based on analysis of mitochon-drial DNA (mtDNA) (see Chapter 11), but that is not the only line of genetic evidence being employed. One example of the potential impact of the molecular genetics approach is provided by the examination of the relationship between turtles, lizards, crocodiles, and birds. Classic cladistic studies based on physical traits suggest that crocodiles are most closely related to birds and less closely related to turtles. Comparison of the mtDNA of modern crocodiles, turtles, lizards, and birds shows that the turtles are a sister group of both birds and crocodiles. The combination of genetic analysis with traditional taxon and area cladograms is an important means of advancing our confi-dence in interpretations of evolutionary history.

Phylogeography can be applied to explicitly geographic questions regarding the distribution and evolution of species. In particular, molecular clock analysis (see Chapter 11) has been used to determine the timing of presumed vicariance events that have split range limits and led to the allopat-ric development of new species or subspecies. Such studies often do provide evidence of important past vicariance events and associated speciation. Brett Riddle of the University of Nevada and his colleagues used mtDNA and the molecular clock approach to determine the likely timing of vicari-ance events that have led to the development of new species and subspecies of animals such as mice, lizards, and toads in the southwestern United States and adjacent Mexico. It had long been thought that geographically distinct species and subspecies of these organisms arose during the past 10,000 years or so when increasing aridity following the close of the last glacial period led to expansion of dry desert and fragmented the previously continuous geographic ranges of many mice, lizards, and other plant and animal species. However, detailed genetic analysis of these different geographic pop-ulations has revealed surprisingly large genetic differences between the different species and subspe-cies. The large genetic differences suggest that geographic separation of many of the different species and subspecies may have occurred millions of years in the past and may be due to environmental and geologic changes that occurred as early as the Paleogene and Neogene. Riddle refers to vicariance events that have occurred in the recent geologic past, say the past 10,000 to 20,000 years, as shallow history and those that occurred millions of years ago as deep history. On the basis of the newly acquired phylogeographic evidence, he has suggested that such deep history vicariance events are far more common and important than biogeographers had previously realized.

Molecular DNA approaches have been applied to some of the classic questions of systematics and the evolutionary histories of important plants and animals. We earlier discussed the thoughts of Hooker on disjunct southern hemisphere plants. The southern beeches *(Nothofagus)* were one of the taxa that drew his attention. As we have seen, the disjunct distribution of *Nothofagus* species in Australia, New Guinea, New Zealand, and South America has long drawn the interest of panbioge-ographers and has been explained by vicariance events linked to the breakup of Gondwanaland. Genetic sequencing of *Nothofagus* growing in Australia, New Zealand, and Chile and comparisons with fossil pollen records have produced evidence supporting the importance of vicariance events related to the breakup of Gondwanaland. However, the results of these studies also suggest that long-distance dispersal was likely an important process in the arrival of *Nothofagus* in New Zealand.

Divergence of southern beech from other members of the Fagales family occurred more than 84 million years ago. The radiation of present species lineages likely commenced 65 million to 40 million years ago. The results are consistent with some vicariance events associated with rifting of Australia, Antarctica, and South America. However, these events are too late to explain the appearance and speciation of *Nothofagus* in New Zealand, which separated from Gondwanaland about 80 million years ago.

One of the most interesting examples of the challenges to classical vicariance biogeography posed by molecular systematics is the history of the large flightless palaeognath birds of the superorder Ratites (Fig 13.12). As we discussed previously, the ratites include the ostriches (Struthionidae) of Africa, the extinct Madagascan elephant birds (Aepyornithidae), the emus (Dromaiidae) of Australia, the cassowaries (Casuariidae) of Australia and New Guinea, the now extinct moas (Dinornithidae) of New Zealand, and the now extinct rheas (Rheidae) of South America. The kiwi birds of New Zealand (Apterygidae) are also a ratite. It had long been posited that these widely separated species of large flightless birds of the southern hemisphere evolved from a common ancestral lineage, presumed to be large and flightless birds that evolved in Gondwanaland. The ratites then became disjunct during the breakup of Gondwanaland in the Cretaceous, and the modern taxa were the product of vicariance. However, the analysis of mitochondrial DNA and the development of molecular clock estimates of divergence from samples of extinct and living ratite species suggests that this model is incorrect, or at least incomplete. It appears from the DNA evidence that ancestral ratites may have dispersed via flight to different geographic areas and that parallel evolution led to the development of flightlessness and gigantism independently in different locations such as Africa, Madagascar, Australia, and New Zealand. Recent phylogenetic research indicates that the ancestor of the ratites (infraclass Palaeognathae) originated in Laurasia rather than Gondwana and likely dispersed southward through flight. The tinamou birds (Tinamidae) of Mesoamerica and South America are a sister group of the ratites. The tinamou can fly and share a common Palaeognathae ancestor with the ratites. Phylogenetics based on molecular systematics is revolutionizing our understanding of the ratites. Such studies for other taxa often show that past evolutionary histories include both vicariance and dispersal events.

Although uncertainties remain in things such as molecular clock estimates of divergence, the results of molecular systematics analyses, such as the studies of *Nothofagus* and the ratites, are

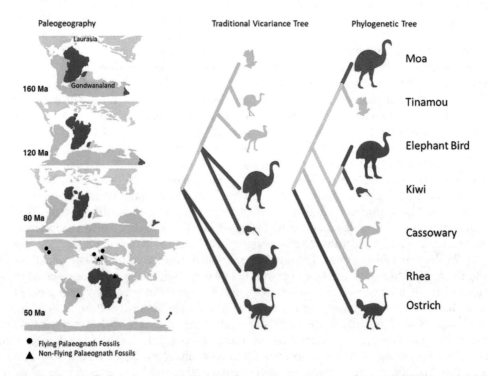

FIGURE 13.12   Paleogeographic history of continental movement from 160 million to 50 million years ago with fossil ancestral ratite sites (Palaeognath Fossils). Traditional vicariance biogeography cladistics tree for ratites and the contrasting tree based on phylogenetics (adapted from Mitchell et al., 2014).

making important progress. With advances in sequencing techniques and declining analytic costs, molecular systematics studies are increasingly common and will continue to be the focus of intense research. Such studies will further inform our understanding of species relationships and biogeographic histories in ways that traditional cladistics—based on morphology alone and overly simplified models of panbiogeography, vicariance biogeography, and dispersalist biogeography—cannot achieve. One important contribution of molecular systematics is that it can allow the direct testing of vicariance versus dispersalist hypotheses. As the Spanish biogeographer Isabel Sanmartín has stated, "This has revolutionized the discipline, allowing it to escape the dispersal versus vicariance dilemma and to address a wider range of evolutionary questions, including the role of ecological and historical factors in the construction of biomes or the existence of contrasting patterns of range evolution in animals and plants" (2012, p. 555).

KEY WORDS AND TERMS

Amphioceanic
Amphiregional
Amphitropical
Areography
Biogeographic relicts
Bipolar
Choropleth maps
Circumpolar
Citizen scientist

Cladistics
Crowdsourcing
Disjunct distribution
DNA barcoding
Environmental DNA (eDNA)
Hollow curve distribution
Macroecology
Molecular systematics
Panbiogeography

Pandemic
Pantropical
Phylogenetic biogeography
Range
Rapoport's Rule
Species distribution models
Temperate
Vicariance biogeography

REFERENCES AND FURTHER READING

Amano, T., Lamming, J. D., and Sutherland, W. J. (2016). Spatial gaps in global biodiversity information and the role of citizen science. *Bioscience* **66**, 393–400.

Ariza, M., Fouks, B., Mauvisseau, Q., Halvorsen, R., Alsos, I. G., and de Boer, H. J. (2023). Plant biodiversity assessment through soil eDNA reflects temporal and local diversity. *Methods in Ecology and Evolution* **14**, 415–430.

Avise, J. C., and Hamrick, J. L. (1996). *Conservation Genetics: Case Histories from Nature*. Chapman and Hall, New York.

Baselga, A., Lobo, J. M., Svenning, J. C., and Araujo, M. B. (2012). Global patterns in the shape of species geographical ranges reveal range determinants. *Journal of Biogeography* **39**, 760–771.

Bock, C. E., and Ricklefs, R. E. (1983). Range size and local abundance of some North American songbirds: a positive correlation. *American Naturalist* **122**, 295–299.

Brown, J. H. (1995). *Macroecology*. University of Chicago Press, Chicago.

Brown, J. H., and Lomolino, M. V. (1998). *Biogeography*, 2nd edition. Sinauer Associates, Sunderland, MA.

Bush, G. L. (1993). A reaffirmation of Santa Rosalina, or why there are so many kinds of small animals. In *Evolutionary Patterns and Processes* (ed. D. R. Lees and D. Edwards), pp. 229–249. Academic Press, London.

Cardillo, M., Dinnage, R., and McAlister, W. (2019). The relationship between environmental niche breadth and geographic range size across plant species. *Journal of Biogeography* **46**, 97–109.

Caudullo, G., Welk, E., and San-Miguel-Ayanz, J. (2017). Chorological maps for the main European woody species. *Data in Brief* **12**, 662–666.

Chen, J., Hao, Z., Guang, X., Zhao, C., Wang, P., Xue, L., Zhu, Q., Yang, L., Sheng, Y., Zhou, Y., and Xu, H. (2019). Liriodendron genome sheds light on angiosperm phylogeny and species-pair differentiation. *Nature Plants* **5**, 18–25.

Chiari, Y., Cahais, V., Galtier, N., and Delsuc, F. (2012). Phylogenomic analyses support the position of turtles as the sister group of birds and crocodiles (Archosauria). *Bmc Biology* **10**, 1–15.

Clement, W. L., Bruun-Lund, S., Cohen, A., Kjellberg, F., Weiblen, G. D., and Rønsted, N. (2020). Evolution and classification of figs (*Ficus,* Moraceae) and their close relatives (Castilleae) united by involucral bracts. *Botanical Journal of the Linnean Society* **193**, 316–339.

Cook, L. G., and Crisp, M. D. (2005). Not so ancient: the extant crown group of Nothofagus represents a post-Gondwanan radiation. *Proceedings of the Royal Society B: Biological Sciences* **272**, 2535–2544.

Cox, C. B., and Moore, P. D. (1993). *Biogeography: An Ecological and Evolutionary Approach*, 5th edition. Blackwell Scientific Publications, Oxford.

Cranston, P. S., Hardy, N. B., Morse, G. E., Puslednik, L., and McCluen, S. R. (2010). When molecules and morphology concur: the "Gondwanan" midges (Diptera: Chironomidae). *Systematic Entomology* **35**, 636–648.

Craw, R. C., Grehan, J. R., and Heads, M. J. (1999). *Panbiogeography*. Oxford University Press, New York.

Crisp, M. D., Trewick, S. A., and Cook, L. G. (2011). Hypothesis testing in biogeography. *Trends in Ecology & Evolution* **26**, 66–72.

Croizat, L. (1978). Deduction, induction, and biogeography. *Systematic Zoology* **27**, 265–287.

Croizat, L. (1981). Biogeography: past, present, and future. In *Vicariance Biogeography* (ed. G. Nelson and D. E. Rosen), pp. 501–523. Columbia University Press, New York.

Darlington Jr., P. J. (1957). *Zoogeography: The Geographical Distribution of Animals*. Wiley, New York.

Devictor, V., Whittaker, R. J., and Beltrame, C. (2010). Beyond scarcity: citizen science programmes as useful tools for conservation biogeography. *Diversity and Distributions* **16**, 354–362.

Elith, J., and Leathwick, J. R. (2009). Species distribution models: ecological explanation and prediction across space and time. *Annual Review of Ecology, Evolution, and Systematics* **40**, 677–697.

Feldman, M. J., Imbeau, L., Marchand, P., Mazerolle, M. J., Darveau, M., and Fenton, N. J. (2021). Trends and gaps in the use of citizen science derived data as input for species distribution models: a quantitative review. *PLoS One* **16**, e0234587.

Franklin, J. (1995). Predictive vegetation mapping: geographic modelling of biospatial patterns in relation to environmental gradients. *Progress in Physical Geography* **19**, 474–499.

Franklin, J. (2010). *Mapping Species Distributions: Spatial Inference and Prediction*. Cambridge University Press, Cambridge.

Franklin, J. (2023). Species distribution modelling supports the study of past, present and future biogeographies. *Journal of Biogeography* **50**, 1533–1545.

Gaston, K. J. (1991). How large is a species' geographic range? *OIKOS* **61**, 434–438.

Gaston, K. J. (1996). Species-range-size distributions: patterns, mechanisms and implications. *Trends in Ecology and Evolution* **11**, 197–201.

Glover, K. C., George, J., Heusser, L., and MacDonald, G. M. (2021). West Coast vegetation shifts as a response to climate change over the past 130,000 years: geographic patterns and process from pollen data. *Physical Geography* **42**, 542–560.

Graham, C. H., and Hijmans, R. J. (2006). A comparison of methods for mapping species ranges and species richness. *Global Ecology and Biogeography* **15**, 578–587.

Grehan, J. R. (1991). Panbiogeography 1981–91: development of an earth/life synthesis. *Progress in Physical Geography* **15**, 331–363.

Gura, T. (2013). Citizen science: amateur experts. *Nature* **496**, 259–261.

Hanski, I., and Gyllenburg, M. (1997). Uniting two general patterns in the distribution of species. *Science* **275**, 397–400.

Harley, M. M. (2006). A summary of fossil records for Arecaceae. *Botanical Journal of the Linnean Society* **151**, 39–67.

Harshman, J., Braun, E. L., Braun, M. J., Huddleston, C. J., Bowie, R. C., Chojnowski, J. L., Hackett, S. J., Han, K. L., Kimball, R. T., Marks, B. D., and Miglia, K. J. (2008). Phylogenomic evidence for multiple losses of flight in ratite birds. *Proceedings of the National Academy of Sciences* **105**, 13462–13467.

Heads, M. (2005). Dating nodes on molecular phylogenies: a critique of molecular biogeography. *Cladistics* **21**, 62–78.

Heads, M. (2015). Panbiogeography, its critics, and the case of the ratite birds. *Australian Systematic Botany* **27**, 241–256.

Hedges, S. B., and Poling, L. L. (1999). A molecular phylogeny of reptiles. *Science* **283**, 998–1001.

Heusser, L. E., and Siroko, F. (1997). Millennial pulsing of environmental change in southern California from the past 24 ky: a record of Indo-Pacific ENSO events? *Geology* **25**, 243–246.

Heywood, V. H. (1978). *Flowering Plants of the World*. Oxford University Press, Oxford.

Hubbell, S. P. (2001). The unified neutral theory of biodiversity and biogeography. *Princeton University Press Monographs in Population Biology* **32**, 392 pp. doi.org/10.1515/9781400837526.

Huggett, R. J. (1998). *Fundamentals of Biogeography*. Routledge, London.

Humphries, C. J., Ladiges, P. Y., Roos, M., and Zandee, M. (1998). Cladistic biogeography. In *Analytical Biogeography: An Integrated Approach to the Study of Animal and Plant Distributions* (ed. A. A. Myers and P. S. Giller), pp. 371–404. Chapman and Hall, London.

Humphries, C. J., and Parenti, L. R. (1986). *Cladistic Biogeography*. Clarendon Press, Oxford.

Hutchinson, G. E. (1959). Homage to Santa Rosalia, or why are there so many kinds of animals? *American Naturalist* **93**, 145–159.

Jaeger, J. R., Riddle, B. R., and Bradford, D. F. (2005). Cryptic Neogene vicariance and Quaternary dispersal of the red-spotted toad *(Bufo punctatus)*: insights on the evolution of North American warm desert biotas. *Molecular Ecology* **14**, 3033–3048.

Juvik, O. J., Nguyen, X. H. T., Andersen, H. L., and Fossen, T. (2016). Growing with dinosaurs: natural products from the Cretaceous relict *Metasequoia glyptostroboides* Hu & Cheng—a molecular reservoir from the ancient world with potential in modern medicine. *Phytochemistry Reviews* **15**, 161–195.

Kambach, S., Lenoir, J., Decocq, G., Welk, E., Seidler, G., Dullinger, S., Gégout, J. C., Guisan, A., Pauli, H., Svenning, J. C., and Vittoz, P. (2019). Of niches and distributions: range size increases with niche breadth both globally and regionally but regional estimates poorly relate to global estimates. *Ecography* **42**, 467–477.

Kearney, M., and Porter, W. (2009). Mechanistic niche modelling: combining physiological and spatial data to predict species' ranges. *Ecology Letters* **12**, 334–350.

Kearney, M. R., Wintle, B. A., and Porter, W. P. (2010). Correlative and mechanistic models of species distribution provide congruent forecasts under climate change. *Conservation Letters* **3**, 203–213.

Kerr, K. C., Stoeckle, M. Y., Dove, C. J., Weigt, L. A., Francis, C. M., and Hebert, P. D. (2007). Comprehensive DNA barcode coverage of North American birds. *Molecular Ecology Notes* **7**, 535–543.

Knapp, M., Stöckler, K., Havell, D., Delsuc, F., Sebastiani, F., and Lockhart, P. J. (2005). Relaxed molecular clock provides evidence for long-distance dispersal of *Nothofagus* (southern beech). *PLoS Biology* **3**, e14.

Köckemann, B., Buschmann, H., and Leuschner, C. (2009). The relationships between abundance, range size and niche breadth in Central European tree species. *Journal of Biogeography* **36**, 854–864.

Kosmala, M., Wiggins, A., Swanson, A., and Simmons, B. (2016). Assessing data quality in citizen science. *Frontiers in Ecology and the Environment* **14**, 551–560.

La Sorte, F. A., and Somveille, M. (2021). The island biogeography of the eBird citizen-science programme. *Journal of Biogeography* **48**, 628–638.

Lee-Yaw, J. A., McCune, J. L., Pironon, S., and Sheth, S. N. (2022) Species distribution models rarely predict the biology of real populations. *Ecography* **6**, e05877.

Ma, J. (2007). A worldwide survey of cultivated Metasequoia glyptostroboides Hu & Cheng (Taxodiaceae: Cupressaceae) from 1947 to 2007. *Bulletin of the Peabody Museum of Natural History* **48**, 235–253.

Matsunaga, K. K., and Smith, S. Y. (2021). Fossil palm reading: using fruits to reveal the deep roots of palm diversity. *American Journal of Botany* **108**, 472–494.

McGann, M. (2015). Late Quaternary pollen record from the central California continental margin. *Quaternary International* **387**, 46–57.

McGlone, M. S. (2016). Once more into the wilderness of panbiogeography: a reply to Heads (2014). *Australian Systematic Botany* **28**, 388–393.

Minnich, R. A. (1982). *Pseudotsuga macrocarpa* in Baja California? *Madrono* **29**, 22–31.

Mitchell, K. J., Llamas, B., Soubrier, J., Rawlence, N. J., Worthy, T. H., Wood, J., Lee, M. S., and Cooper, A. (2014). Ancient DNA reveals elephant birds and kiwi are sister taxa and clarifies ratite bird evolution. *Science* **344**, 898–900.

Morrone, J. J. (2015). Track analysis beyond panbiogeography. *Journal of Biogeography* **42**, 413–425.

Morrone, J. J. (2021). (Croizat's) dangerous ideas: practices, prejudices, and politics in contemporary biogeography. *History and Philosophy of the Life Sciences* **43**(2), 77.

Mourelle, C., and Ezcurra, E. (1997). Rapoport's rule: a comparative analysis between South and North American columnar cacti. *American Naturalist* **150**, 131–142.

Muller, P. (1974). *Aspects of Zoogeography*. Dr. W. Junk Publisher, The Hague.

Muller, P. (1980). *Biogeography*. Harper and Row, New York.

Myers, A. A., and Giller, P. S. (1988). *Analytical Biogeography*. Chapman and Hall, New York.

O'Brien, S. J., Menotti-Raymond, M., Murphy, W. J., Nash, W. G., Wienberg, J., Stanyon, R., Copeland, N. G., Jenkins, N. A., Womack, J. E., and Graves, J. A. M. (1999). The promise of comparative genomics in mammals. *Science* **286**, 458–481.

Orme, C. D. L., Davies, R. G., Olson, V. A., Thomas, G. H., Ding, T. S., Rasmussen, P. C., Ridgely, R. S., Stattersfield, A. J., Bennett, P. M., Owens, I. P. F., and Blackburn, T. M. (2006). Global patterns of geographic range size in birds. *PLoS Biology* **4**, e208.

Ornduff, R. (1974). *Introduction to California Plant Life*. University of California Press, Berkeley.

Owen, M. G., and Taylor, Z. P. (2023). Patterns of disjunction in western hemisphere birds, amphibians, crocodilians, and mammals. *Physical Geography* **44**, 782–797.

Patterson, C. (1981). Methods of paleobiogeography. In *Vicariance Biogeography* (ed. G. Nelson and D. E. Rosen), pp. 446–489. Columbia University Press, New York.

Pedrotti, F. (2013). *Plant and Vegetation Mapping*. Geobotany Studies. Springer, Berlin, Heidelberg.

Pena, C., Nylin, S., Freitas, A. V., and Wahlberg, N. (2010). Biogeographic history of the butterfly subtribe Euptychiina (Lepidoptera, Nymphalidae, Satyrinae). *Zoologica Scripta* **39**, 243–258.

Pielou, E. C. (1979). *Biogeography*. Wiley, New York.

Pintor, A. F., Schwarzkopf, L., and Krockenberger, A. K. (2015). Rapoport's Rule: do climatic variability gradients shape range extent? *Ecological Monographs* **85**, 643–659.

Polanco Fernández, A., Marques, V., Fopp, F., Juhel, J. B., Borrero-Pérez, G. H., Cheutin, M. C., Dejean, T., González Corredor, J. D., Acosta-Chaparro, A., Hocdé, R., and Eme, D. (2021). Comparing environmental DNA metabarcoding and underwater visual census to monitor tropical reef fishes. *Environmental DNA* **3**, 142–156.

Porsild, A. E., and Cody, W. J. (1980). *Vascular Plants of the Continental Northwest Territories, Canada*. National Museum of the Natural Sciences, Ottawa.

Rapoport, E. H. (1982). *Areography: Geographical Strategies of Species*. Pergamon Press, Oxford.

Ren, Z., Zhong, Y., Kurosu, U., Aoki, S., Ma, E., von Dohlen, C. D., and Wen, J. (2013). Historical biogeography of Eastern Asian–Eastern North American disjunct Melaphidina aphids (Hemiptera: Aphididae: Eriosomatinae) on Rhus hosts (Anacardiaceae). *Molecular Phylogenetics and Evolution* **69**, 1146–1158.

Richards, C. M., Falk, D. A., and Montalvo, A. M. (2016). Population and ecological genetics in restoration ecology. In *Foundations of Restoration Ecology* (ed. M. A. Palmer, J. B. Zedler, and D. A. Falk), pp. 123–152. Island Press, Washington, DC.

Riddle, B. R. (1995). Molecular biogeography in the pocket mice *(Perognathus* and *Chaetodipus)* and grasshopper mice *(Onychomys):* the Late Cenozoic development of a North America aridlands rodent guild. *Journal of Mammalogy* **76**, 283–301.

Riddle, B. R. (1996). The molecular phylogenetic bridge between deep and shallow history in continental biotas. *Trends in Ecology and Evolution* **11**, 207–211.

Riddle, B. R., Dawson, M. N., Hadly, E. A., Hafner, D. J., Hickerson, M. J., Mantooth, S. J., and Yoder, A. D. (2008). The role of molecular genetics in sculpting the future of integrative biogeography. *Progress in Physical Geography* **32**, 173–202.

Robbins, C. S., Bruun, B., and Zim, H. S. (1983). *A Guide to Field Identification: Birds of North America*. Golden Press, New York.

Ronquist, F., and Sanmartín, I. (2011). Phylogenetic methods in biogeography. *Annual Review of Ecology, Evolution, and Systematics* **42**, 441–464.

Rosen, B. R. (1988). Biogeographic patterns: a perceptual overview. In *Analytical Biogeography* (ed. A. A. Myers and P. S. Giller), pp. 23–55. Chapman and Hall, New York.

Ruggiero, A., and Werenkraut, V. (2007). One-dimensional analyses of Rapoport's rule reviewed through meta-analysis. *Global Ecology and Biogeography* **16**, 401–414.

Sanmartín, I. (2012). Historical biogeography: evolution in time and space. *Evolution: Education and Outreach* **5**, 555–568.

Santini, L., Benítez-López, A., Maiorano, L., Čengić, M., and Huijbregts, M. A. (2021). Assessing the reliability of species distribution projections in climate change research. *Diversity and Distributions* **27**, 1035–1050.

Shabani, F., Kumar, L., and Ahmadi, M. (2016). A comparison of absolute performance of different correlative and mechanistic species distribution models in an independent area. *Ecology and Evolution* **6**, 5973–5986.

Silvertown, J. (2009). A new dawn for citizen science. *Trends in Ecology & Evolution* **24**, 467–471.

Simpson, M. G., Johnson, L. A., Villaverde, T., and Guilliams, C. M. (2017). American amphitropical disjuncts: perspectives from vascular plant analyses and prospects for future research. *American Journal of Botany* **104**, 1600–1650.

Slatyer, R. A., Hirst, M., and Sexton, J. P. (2013). Niche breadth predicts geographical range size: a general ecological pattern. *Ecology Letters* **16**, 1104–1114.

Stevens, G. C. (1989). The latitudinal gradient in geographical range: how so many species coexist in the tropics. *American Naturalist* **133**, 240–256.

Stevens, G. C. (1992). The elevational gradient in altitudinal range: an extension of Rapoport's latitudinal rule to altitude. *American Naturalist* **140**, 893–911.

Sun, Y. H., Xie, C. X., Wang, W. M., Liu, S. Y., Treer, T., and Chang, M. M. (2007). The genetic variation and biogeography of catostomid fishes based on mitochondrial and nucleic DNA sequences. *Journal of Fish Biology* **70**, 291–309.

Suprayitno, N., Narakusumo, R. P., von Rintelen, T., Hendrich, L., and Balke, M. (2017). Taxonomy and biogeography without frontiers—WhatsApp, Facebook and smartphone digital photography let citizen scientists in more remote localities step out of the dark. *Biodiversity Data Journal* **5**, e19938.

Thieltges, D. W., Hof, C., Borregaard, M. K., Matthias Dehling, D., Brändle, M., Brandl, R., and Poulin, R. (2011). Range size patterns in European freshwater trematodes. *Ecography* **34**, 982–989.

Tocheri, M. W., Dommain, R., McFarlin, S. C., Burnett, S. E., Troy Case, D., Orr, C. M., Roach, N. T., Villmoare, B., Eriksen, A. B., Kalthoff, D. C., and Senck, S. (2016). The evolutionary origin and population history of the grauer gorilla. *American Journal of Physical Anthropology* **159**, 4–18.

Tornabene, L., Valdez, S., Erdmann, M., and Pezold, F. (2015). Support for a "Center of Origin" in the Coral Triangle: cryptic diversity, recent speciation, and local endemism in a diverse lineage of reef fishes (Gobiidae: Eviota). *Molecular Phylogenetics and Evolution* **82**, 200–210.

Udvardy, M. D. F. (1969). *Dynamic Zoography*. Van Nostrand Reinhold, New York.

Ushio, M., Murata, K., Sado, T., Nishiumi, I., Takeshita, M., Iwasaki, W., and Miya, M. (2018). Demonstration of the potential of environmental DNA as a tool for the detection of avian species. *Scientific Reports* **8**, 4493.

Van Den Ende, C., White, L. T., and van Welzen, P. C. (2017). The existence and break-up of the Antarctic land bridge as indicated by both amphi-Pacific distributions and tectonics. *Gondwana Research* **44**, 219–227.

Waters, J. M., Trewick, S. A., Paterson, A. M., Spencer, H. G., Kennedy, M., Craw, D., Burridge, C. P., and Wallis, G. P. (2013). Biogeography off the tracks. *Systematic Biology* **62**, 494–498.

Wen, J., Ree, R. H., Ickert-Bond, S. M., Nie, Z., and Funk, V. (2013). Biogeography: where do we go from here? *Taxon* **62**, 912–927.

White, E. P., Ernest, S. M., Kerkhoff, A. J., and Enquist, B. J. (2007). Relationships between body size and abundance in ecology. *Trends in Ecology & Evolution* **22**, 323–330.

Whitton, F. J., Purvis, A., Orme, C. D. L., and Olalla-Tárraga, M. Á. (2012). Understanding global patterns in amphibian geographic range size: does Rapoport rule? *Global Ecology and Biogeography* **21**, 179–190.

Williams, D. M., and Ebach, M. C. (2004) The reform of palaeontology and the rise of biogeography—25 years after "ontogeny, phylogeny, paleontology and the biogenetic law" (Nelson, 1978). *Journal of Biogeography* **31**, 685–712.

Xi, Z., Rest, J. S., and Davis, C. C. (2013). Phylogenomics and coalescent analyses resolve extant seed plant relationships. *PLoS One* **8**, e80870.

Yang, M. Q., Li, D. Z., Wen, J., and Yi, T. S. (2017). Phylogeny and biogeography of the amphi-Pacific genus *Aphananthe*. *PLoS One* **12**, e0171405.

Yonezawa, T., Segawa, T., Mori, H., Campos, P. F., Hongoh, Y., Endo, H., Akiyoshi, A., Kohno, N., Nishida, S., Wu, J., and Jin, H. (2017). Phylogenomics and morphology of extinct paleognaths reveal the origin and evolution of the ratites. *Current Biology* **27**, 68–77.

Yu, F., Groen, T. A., Wang, T., Skidmore, A. K., Huang, J., and Ma, K. (2017). Climatic niche breadth can explain variation in geographical range size of alpine and subalpine plants. *International Journal of Geographical Information Science* **31**, 190–212.

Zhang, M. (2011). A cladistic scenario of southern Pacific biogeographical history based on *Nothofagus* dispersal and vicariance analysis. *Journal of Arid Land* **3**, 104–113.

Zurell, D., Franklin, J., König, C., Bouchet, P. J., Dormann, C. F., Elith, J., Fandos, G., Feng, X., Guillera-Arroita, G., Guisan, A., and Lahoz-Monfort, J. J. (2020). A standard protocol for reporting species distribution models. *Ecography* **43**, 1261–1277.

CHAPTER 14

THE GEOGRAPHY OF BIOLOGICAL DIVERSITY

Imagine that you are making a biogeographic fieldtrip and traveling eastward from Alaska to Labrador, Canada. The route of your trip will transect the entire boreal biome of North America. Being a good biogeographer, you keep a list of the plants and animals encountered along the way. After several weeks, you have covered hundreds of kilometers, crossed into Canada, and have seen only a handful of different species of trees. By the time you finally arrive at the Atlantic coastline, you have traveled more than 6000 kilometers through the boreal forest and would be lucky to have encountered more than perhaps a dozen different species of trees. During the thousands of kilometers of your travels, had you been observant and lucky, you would have seen perhaps 100 different bird species and 30 species of mammals. If you had taken a similar fieldtrip across Costa Rica from the Pacific Coast to the Atlantic Coast, some of your route would have been through a portion of the tropical forest biome. Here you might have encountered more than 50 different tree species within a distance of a few hundred meters. As you traveled across Costa Rica, you could have found several hundred tree species, about 600 species of birds, and around 140 species of mammals, all in a distance of less than 200 kilometers from coast to coast! The obvious question that arises from these two fieldtrips is, why is there such a striking difference in the number of plant and animal species in the North American boreal forest biome compared to tropical Costa Rica?

Documenting and understanding such huge differences in the geographic distribution of biological diversity are central concerns for biogeographers. In this chapter we will begin by considering exactly what biological diversity is and how it can be measured. We will then examine some geographic patterns in the distribution of biological diversity on continents and the oceans. We will also examine some of the factors that have been postulated to control global patterns of biological diversity. As we conclude the chapter, we will look at the very interesting distribution of biological diversity on islands. We will see why the study of island biological diversity has long held a particular fascination for biogeographers and ecologists.

WHAT IS BIOLOGICAL DIVERSITY?

Biological diversity, sometimes called **biodiversity,** can be broadly defined as the number, variety, and variability of living organisms. Biodiversity is often simply defined as the number of different species in a given geographic area. The greater the number of species present, the greater the biodiversity. For example, Mexico has over 500 mammalian species and the province of British Columbia in Canada has 137 mammalian species. Further to the north, Alaska has 112 mammalian species. Thus, we could say that British Columbia has a lower mammalian biodiversity than Mexico, but a slightly higher diversity than Alaska. This simple measure of biodiversity, based solely on the number of different species in a given area, is also called **species richness.** In many large-scale biogeographic studies, biodiversity is synonymous with species richness.

When conducting comparative studies of biodiversity in different regions, it is important that we compare areas that are similar in size. In general, species richness increases with the size of the area

Biogeography: Introduction to Space, Time, and Life, Second Edition. Glen M. MacDonald.
© 2025 John Wiley & Sons, Inc. Published 2025 by John Wiley & Sons, Inc.
Companion website: www.wiley.com/go/macdonald/biogeography2e

sampled. If, for example, we conduct surveys of plant species richness, we might start by laying out a plot that is 1-by-1 meter in size. We could identify all of the different plant species in the plot and determine the species richness of that particular sampling site. In most instances, if we increase the size of the plot, we will discover additional species. Thus, if we increased our plot to 10-by-10 meters in size, we might encounter twice as many plant species. If we increase our sampling area to 100-by-100 meters, we might again double the number of species we discover. The larger the area that is sampled, the greater the number of species that we will find. The positive relationship between the size of the area surveyed and the number of species encountered is called the **species-area relationship (SAR).** In general, species-area relationships are not linear. This typical nonlinear relationship between sampling area and species richness is called the species-area curve. A common formula for the species-area curve is a power function:

$$S = cA^z$$

where S is species richness, c is a constant based on the number of species per unit of area in the geographic region in which the study is conducted, A is the size of the sampling plot analyzed for species richness, and z is a constant that represents the mathematical relationship between area and species richness.

Regions with large numbers of species will have large values of c. If we conducted our study of plants in Costa Rica, which has very high biodiversity, the value for c would be high. If we conducted a similar study in the boreal forest of Alaska, the value for c would be low. If we use a 1-by-1 meter plot, the value for A will be 1 square meter. If we use a 10-by-10 meter plot, the value for A will be 100 square meters. Finally, z can be thought of graphically as the slope of the function that describes the relationship between area and species richness. For study areas in which species richness rises quickly with increasing area, the value for z will be high. The species-area curve resulting from the equation above can be made into a straight line by a simple logarithmic transformation:

$$\log S = (\log c) + z(\log A)$$

However, the simple power function may not be appropriate for many species-area relationships, particularly when applied beyond local sites to larger continental scales. For example, recent empirical research and theory suggests that at such larger scales, the curves will steepen as the size of the sampling area approaches the size of the individual species' geographic ranges. This steepening of the species-area curve can be seen in the logarithmic plots of the curves for amphibians, birds, and mammals from Africa, Eurasia, North America, South America, and Australia (Fig. 14.1).

There are several reasons why species richness increases as the size of the sampling area increases. First, small sampling plots may not include all of the species present in an area because of random chance in the selection of where the sampling will be conducted. Additional species might be present in the vicinity but might be missed because of the small size and choice of location of the sample site. Second, small sampling sites contain only a small selection of very local microhabitats and may not provide the right environmental conditions for many species. An increase in the size of the sampling area will increase the variety of local habitats and the number of species that are represented. Greatly increasing the size of the sampling area will incorporate areas representing wider ranges of environmental conditions and will in turn increase the number of species that will be present. The problem with very large sample plots is that it is difficult to completely survey all of the species of plants or animals present. In most cases, sampling is done by establishing a number of small sample plots or sampling stations that are randomly or systematically distributed across the landscape. In this manner, it is hoped that both the habitat diversity and the species diversity of the study area will be represented and sampled in an efficient manner.

When used for comparative purposes, species richness values are often reported as the number of species per hectare or the number of species per square kilometer. However, in many cases, censuses of species richness are conducted for political units such as states or countries. The area of the different states or countries will vary, and this will affect the species richness that is reported. If we compare the species richness for different states, countries, or continents, it is important to remember the impact that differences in area will have on total species richness. Larger areas will have more species than smaller ones if all else is equal.

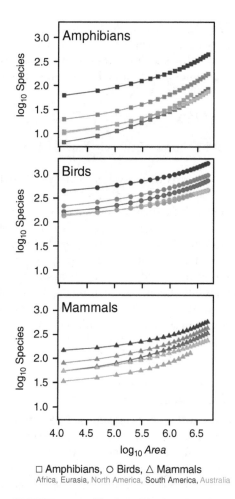

FIGURE 14.1   The logarithmic plots of the relationship between sampling area and amphibian, bird, and mammal species richness in Africa, Eurasia, North America, South America, and Australia. Note the increasing steepness of the curves at large area scales (*Source:* Storch et al., 2012 / Springer Nature).

Species richness is a simple and useful means of measuring biodiversity. Species richness, however, does not tell a complete story regarding biodiversity. Simple species richness does not, for example, tell us anything about the rarity of species or the evenness of species richness. What do we mean by the term *species evenness*? Consider that we might find that two areas have exactly the same species richness, say, 10 species of birds. In one of the two areas, we might discover that the individuals of the 10 bird species occur in roughly equal numbers. In the other area, we might find that 90% of the birds all belong to one species and that the 9 other species of birds are rare. The species richness in the first area is evenly distributed, while the species richness in the second area is unevenly distributed.

To capture information on both species richness and evenness, ecologists use a variety of mathematical indices of species diversity that include data for both richness and evenness. Sites that have many species, each occurring at about the same proportion, will generate high diversity values. Sites that have few species and are dominated by the individuals of one or two of those species will have low diversity values. These indices incorporate information on both the number of species present and the proportion of the flora and fauna that each species contributes. One commonly applied mathematical measure of diversity is the Shannon index *(H′)* which takes into account both species richness and evenness. The index has the following form:

$$H' = -\Sigma p_i \ln p_i$$

where p_i is the proportion of the *i*th species and ln is the natural logarithm. To use the index, one simply counts all the individuals of all the species in a study site, transforms these counts into proportions, and then applies the Shannon index formula. In natural systems, the value of H' has been found to range from 1.5 for systems with low species richness and evenness to 3.5 to species with high species richness and evenness. The Shannon index for species diversity incorporates both species richness and evenness. An estimate of species evenness alone may be obtained from the Shannon index value (H') using the following equation:

$$E = H' / \ln S$$

where E is species evenness and S is the species richness of the site (the total number of species identified or collected).

Although species evenness is an important property of ecosystems, rarely is it adequately measured for natural environments. It is very time consuming to sample and identify all individuals in a given area. In addition, rare species are extremely difficult to count adequately and are generally underrepresented in studies of species evenness. There is no general rule or accepted theory as to how even species distributions will be in natural environments. Some systems seem to have a large number of very rare species, while other systems are typified by an abundance of species that are neither particularly rare nor overwhelmingly dominant. A long-term study of butterfly assemblages in the North Saskatchewan River Valley of Alberta, Canada, found a negative relationship between species richness and evenness. It has been suggested that different ecological factors and processes may be important in determining species richness and evenness, rendering the relationships between these two measures of biodiversity inconsistent and site and habitat specific.

HOW MANY DIFFERENT SPECIES ARE THERE ON EARTH?

An obvious question we might ask is, how many different plant and animal species are there on earth? After more than 250 years of taxonomic research, the total number of species still remains a mystery. Let's leave viruses and bacteria aside and focus on eukaryotic organisms such as animals, plants, fungi, and algae. There are differing estimates, but according to the International Union for Conservation of Nature (IUCN) Red List, there are about 2.13 million species of plants and animals known to science. This tally ranges from algae to higher plants to insects, amphibians, reptiles, birds, and mammals (Fig. 14.2). About 50% of all known species are insects. Of these, Coleoptera (beetles)

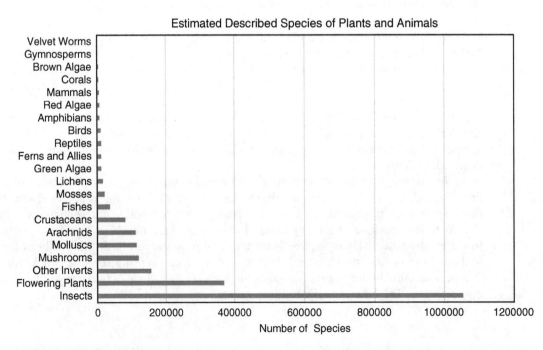

FIGURE 14.2   The distribution of described species among different plant and animal groups as of 2021 (adapted from IUCN Red List, 2021).

contribute 40% to 50% of the known species. Approximately 17% of the known species are flowering plants. Mammals, birds, reptiles, and amphibians together make up about 1.7% of the world's known species according to the IUCN estimates.

All scientists agree that there are many species that have not yet been discovered. Past estimates have placed the total number of eukaryotic species on earth from about 3 million to over 100 million. The eminent British ecologist Robert May referred to this large uncertainty as embarrassing. Some mathematically based extrapolations have provided estimates of about 5 million to 8.7 million for eukaryotic species. Despite the large differences between estimates, these studies point to the conclusion that there are many more unknown species on earth than those that have already been described by science. Of this great unknown, there are thought to be relatively fewer species of birds and mammals yet to be discovered compared to invertebrates. The rate of discovery of flowering plant, bird, and mammal species was relatively high between the eighteenth and early twentieth centuries and then somewhat tapered off. In contrast, the rate of discovery of new species of invertebrates such as insects, arachnids, and crustaceans has accelerated in the past few decades. About 15,000 to 18,000 new species are described by scientists each year. On average, over half are new invertebrates. Of these, most are insects with about 7000 new species described annually (Table 14.1). In looking at these aggregate global numbers, it is important to remember that there may be a bias in overrepresentation of terrestrial species to marine species. There are about 242,000 known marine species, and an average of about 2300 new marine species are described each year. As marine exploration progresses, including in the deep ocean, this bias will lessen.

The discovery of new species of birds and mammals sometimes occurs under somewhat amusing circumstances. For example, a new species of bird, the Udzungwaq partridge *(Xenoperdix udzungwensis),* was discovered by Danish biologists in Tanzania in 1991. In this case, the scientists were having dinner and found some unusual-looking bird legs in a pot of "chicken" stew that was being served at a field camp. Subsequent searching in the surrounding countryside turned up living members of this previously undescribed species of bird. Comparison of the Udzungwaq partridge to other species using genetics and fossil material shows it to be a paleoendemic species that is an evolutionary relict of a group of birds that once ranged from Southeast Asia to Africa. In 1994, a new species of tree kangaroo of the genus *Dendrolagus* was discovered by scientists working in New Guinea. This species came to light when a zoologist's dog noticed the animal and began to give chase.

In our discussion above, we have focused on biodiversity based on a classic taxonomy of plants and animals. However, it is important to remember that large and geographically dispersed populations of single species often include many different and distinct genotypes. Biologists and biogeographers are increasingly interested in studying genetic biodiversity as well as traditional taxonomic species biodiversity. By using genetic analysis, the geographic patterns of genetic variability within a single species can be recorded, mapped, and analyzed to study the relationship between genetic diversity, geography, the environment, and human impact. For example, analysis of grey wolf *(Canis lupus)* populations in Eurasia shows that the remaining small populations in Portugal, Sweden, Estonia,

TABLE 14.1 **Approximate Average Numbers of New Species Described per Year in Recent Decades**

Organisms	New Species Described per Year
Mammals	25
Birds	6
Reptiles	66
Amphibians	84
Fish	151
Mollusks	600
Insects	7000
Spiders	200
Vascular plants	2000

Source: Adapted from Rosenberg, 2014; Christenhusz and Byng, 2016; Stork, 2018; Jäger et al., 2021; Liu et al., 2022.

Italy, Israel, Iran, and China are all genetically distinctive from each other on the basis of their mtDNA. The Eurasian populations are also distinctive from North American populations of grey wolf. Although all of these populations are considered the same species, they display significant genetic biodiversity within the species. Interestingly, the genetic biodiversity within the Eurasian wolf populations is greater than the diversity within the North American populations. It is possible that the high degree of geographic diversity in the genetic composition of the Eurasian wolves is due in part to the fragmentation of their range by human activities.

Despite their genetic differences, the various wolves discussed above can still interbreed and produce reproductively viable offspring. In some cases though, organisms that are taxonomically indistinguishable possess enough genetic differentiation that they are essentially reproductively isolated. These **cryptic species** taxonomically appear to be one species, but are actually two or more different species based on the biological species definition. One example is the skipper butterfly *(Astraptes fulgerator),* which is common in Central America. The adults are almost taxonomically indistinguishable, but DNA analysis has revealed that the group consists of at least 10 cryptic species that are largely reproductively isolated. For many species, such genetic work is in its early stages. However, genetic analyses will be an ever more important means of measuring the true biodiversity of the earth. The DNA barcoding approach efficiently uses the information from short genetic sequences taken from one or a few specific gene regions of a species, rather than analyzing the entire genome. Increasingly, barcoding provides a widely applied means of rapidly assessing the genetic differentiation within taxonomically identified species, thus revealing cryptic species.

LATITUDINAL AND ALTITUDINAL DIVERSITY GRADIENTS

If you were exploring the equatorial rainforest of South America, you would encounter a host of beautiful swallowtail butterflies (Papilionidae). In fact, with careful work, you might discover as many as 80 different species. Moving northward to Mexico at about 30° north, you might find 30 or so different swallowtail species. By the time you reached the U.S.-Canada border between about 40° north and 50° north, you would encounter only 18 species; at 60° north latitude in Alaska, the species richness of swallowtails would drop to only 4. Your curiosity piqued by this latitudinal decline in species richness, you could conduct a similar transect southward in South America and find that between 40° south and 50° south, at the tip of the continent, you encounter only one species. Taking your explorations to Africa and Europe, and Australia and Asia, you would find the same relationship between the species richness of swallowtail butterflies and latitude. The highest species richness is found at the equator and declines toward the poles (Fig. 14.3). There is also a dip in species richness across the vast and barren expanse of the Sahara Desert in Africa.

The pattern of decreasing species richness toward the northern and southern polar regions is not limited to the swallowtail butterflies. It is widespread in insect groups. Decreasing species richness toward the north and south poles is also observed in birds, mammals, amphibians, and reptiles (Fig. 14.4). Similar to the butterflies, there is a dip in species richness for mammals, birds, amphibians, and reptiles across the Sahara Desert and Arabian Desert regions. The pattern of decreasing species richness with increasing north and south latitude is strong enough to trump the usual species-area relationships discussed earlier in this chapter. The small country of Costa Rica in central America (51,000 square kilometers) supports over 900 bird species, while the huge northern country of Canada (9.98 million square kilometers) supports about 460.

The pattern of increasing diversity toward the low latitudes holds up for plants and is apparent both north and south of the equator. The number of tree species on a typical plot of forest in Europe increases from 2 to 3 in far northern Scandinavia to around a dozen in northern continental Europe. Farther south, the equatorial rainforest of Africa contains hundreds of different tree species. A 50-hectare plot in Zaire can contain as many as 450 different tree species. Similar to animals, there is a decline in plant species richness over the region of the Sahara Desert and the Arabian Desert, with a pronounced increase in the Mediterranean region. South of the equator in South America, the number of orchid species increases from 15 at the southern tip of the continent to over 2500 in the tropical rainforest near the equator. Although vascular plants show an overall strong poleward decline in species richness, such a pattern is not as clear for mosses.

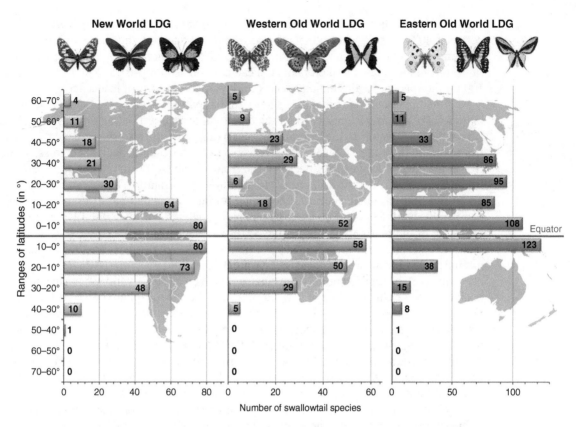

FIGURE 14.3 Swallowtail butterfly species richness in 10° latitudinal bands for North and South America and Europe, Africa and western Asia, and eastern Asia and Australia (*Source:* Codamine et al., 2012 / John Wiley & Sons).

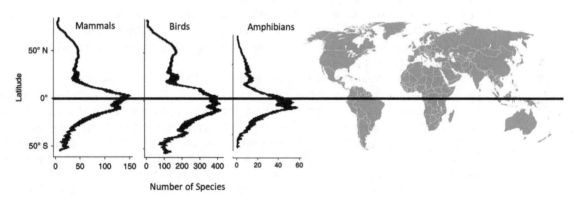

FIGURE 14.4 Global species richness for mammals, birds, and amphibians for 1° latitudinal bands (Saupe et al., 2019 / Springer Nature).

This pervasive pattern of decreasing species richness from the tropics to the polar regions has been noted for over 200 years and is revealed in hundreds of scientific studies on numerous different groups of organisms. It is referred to as the **latitudinal diversity gradient (LDG).** Genetics research on a number of organisms indicates that the genetic diversity within mammal, bird, reptile, amphibian, and fish species is also higher at the lower latitudes than at higher latitudes. Although globally widespread, the LDG appears to be strongest across the North and South American continental land mass. In addition, the overall species richness for some groups of terrestrial organisms has been found to be higher in the southern hemisphere than in the northern hemisphere. Given the pervasive nature of the latitudinal diversity gradient among different groups of organisms, it is not surprising that the aggregate effect is that the tropical rainforests of South and Central America, Africa, southeastern

Asia, and Australasia are the most biologically diverse regions in the world. It is estimated that between 40% and 75% of all plant and animal species on the earth live in tropical rainforests. The rainforests are particularly rich in plant, bird, amphibian, and insect species. A patch of tropical rainforest that is a few kilometers in area can contain as many as 1500 plant species, 400 bird species, 100 reptile species, 60 amphibian species, and many thousands of insect species.

A number of marine organisms show a similar pattern of increasing species diversity toward lower latitudes. The species richness of planktonic foraminifera (free-floating protists) increases by a factor of 8 from the polar waters to the tropics. Radiolaria (also protists) also show an increase in species richness from poles to the equator. Small planktonic invertebrates called copepods increase from about 20 species in the Arctic Ocean to 80 species in the tropical Pacific. There are about twice as many species of fish in the tropical Atlantic as there are in the far North Atlantic. The Bay of Bengal off India supports about five times as many species of bottom-dwelling worms and mollusks as do similar sized areas of comparable benthic habitat off Cape Cod in New England. However, the LDG is often not as clear and pervasive in the marine realm as in the terrestrial realm. The peak species richness for a number of benthic (bottom dwelling) marine taxa is found at 10° to 20° north rather than at the equator. For some marine groups, particularly pelagic (open water) taxa, there is a bimodal distribution with peak richness occurring both north and south of the equator between about 20° and 40° north and south. For fish, the peak in species richness is biased in favor of the northern hemisphere between about 15° and 30° north. For a number of planktonic groups, the peak in species richness is in temperate, rather than tropical, waters. The western tropical Pacific has strikingly high species richness across a variety of organisms, including mangroves, seagrasses, corals, and fish, than similar low latitudes in the eastern Pacific or the Atlantic. Coastal-dwelling fish taxa seem to have a stronger propensity for a maximum in species richness at the equator than these other groups. Work continues on assessing the taxonomic richness of the world's oceans and the geographic distributions of species richness.

Many freshwater organisms also have high species diversity in tropical regions compared to temperate and Arctic areas. The Great Lakes region supports about 170 species of fish. There are over 600 species of freshwater fish known from Central America and over 2900 known from the Amazon. All of Europe possesses a reported 546 species of freshwater fish, while the Congo Basin supports over 1500 species. Other research has shown that there are many times more species of aquatic insects in tropical streams in Central and South America as there are in the temperate streams of North America. The species richness of freshwater fish species, amphibians, and some insects such as dragonflies clearly reflects the LDG; however, there are other groups, particularly among insects, that do not adhere to this general pattern.

Although the pattern of increasing diversity from high to low latitudes is widely pervasive, in the terrestrial realm there are geographic exceptions. As illustrated earlier by the Sahara Desert, arid regions, even those near the equator, have very low species diversity of plants and animals despite their geographic location. Very high elevation regions in the tropics typically have low biodiversity compared to lowlands. Peninsulas have lower species diversity than adjacent large mainland areas. The species richness on peninsulas also generally declines toward the tip. The impact of the **peninsula effect** can be seen in decreased tree, mammal, and/or bird diversity in Florida and Baja California. In these areas, species diversity decreases southward toward the tip of the peninsula. Diversity also decreases toward the tip of the Yucatan Peninsula in Mexico. Islands have lower species diversity than adjacent mainland areas. We will discuss the biodiversity of islands in detail below.

There are some notable exceptions to the LDG. Some groups of organisms, such as whales and some bird species such as the sandpiper birds of the family Scolopacidae, achieve greatest species richness at high latitudes. Only 2 species of sandpipers are found along the Pacific Coast of southern California, and 14 species are found along the Bering Sea coastline of Alaska. Diving beetles (Colymbetinae) have 140 known species and display higher species richness in temperate regions compared to the tropics. Grasses (Poaceae) also show higher diversity in temperate regions. For many groups of marine organisms, such as benthic invertebrates in the southern hemisphere, there is no clear increase in species diversity toward the tropics. The examples above are not alone; other groups of plants and animals, such as pines, aphids, or the Ichneumonidae family of wasps, reach maximum species richness in the middle latitudes and decrease in diversity toward the Arctic and the equator. These groups are sometimes referred to as representing the inverse Latitudinal Diversity Gradient (iLDG). They are exceptions, however, to a more general pattern of decreasing diversity from the tropics to the higher latitude regions.

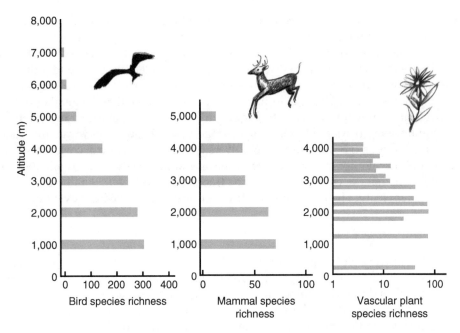

FIGURE 14.5 Declining species richness of birds, mammals, and vascular plants at higher elevations in the Himalayan Mountains (after *Source:* Begnn et al., 1996; Hunter and Yonzon, 1992; Whittaker, 1997).

On a global basis, there is a general pattern of decreasing species richness at very high elevations. Tundra ecosystems found at high elevations are poorer in bird, mammal, and plant species than lower elevation forests. For example, in the Himalayan Mountains of Nepal, there are over 300 bird species found at 500 meters of elevation and none above 8000 meters (Fig. 14.5). The same trend can be seen for Himalayan mammals and vascular plant species. However, high mountains in more arid and semi-arid regions often show a different geographical pattern of species richness. The lowest elevations typically have low species diversity. Species diversity for plants, birds, and nonflying mammals reaches a maximum in mid-elevation sites and then declines as high alpine environments and treeline are approached. The diversity of coniferous tree species on the eastern slopes of the Sierra Nevada Mountains in California displays this pattern. The greatest coniferous tree diversity occurs between 2000 and 3000 meters of elevation and decreases downslope toward the Great Basin Desert and upslope toward the alpine treeline. Mid-elevation peaks in species richness are not always restricted to arid regions and appear to be common for plant taxa. The patterns and controls of elevational diversity gradients are themselves a topic of intense interest and research.

Understanding what causes the large-scale latitudinal gradients in species diversity is an important topic for research in global biogeography. The global diversity gradients and the large regional differences in biodiversity are also important for understanding local species diversity. Analysis of species diversity for a number of different plant and animal taxa shows that the species richness of individual local sites is highly correlated with the broader species diversity of the large geographic region in which the sites are located. Large regions that have high levels of species richness can be expected to have high species richness in local habitats when compared to similar habitat in regions that have low overall levels of species richness.

CONTROLS ON GEOGRAPHIC GRADIENTS OF SPECIES DIVERSITY

Why does biodiversity differ geographically, and in particular, why is biodiversity higher in tropical regions than temperate and polar regions? This question has perplexed biogeographers since it was raised by Alfred Russel Wallace in 1878. Two broad categories of explanations have been proposed to account for geographic differences in biodiversity. Some biogeographers have believed that modern geographic differences in biodiversity may be explained by the past history of species evolution and extinction. Other biogeographers considered geographic differences in biodiversity to be the product of modern environmental conditions and biological processes. Explanations of the first category are referred to as **historical theories of biodiversity.** Explanations of the second category are referred to as **equilibrium theories of biodiversity.** The historical explanations assume that the modern patterns of biodiversity are not in true equilibrium with modern environmental conditions but instead reflect the impacts of past events. The equilibrium explanations assume that the number of species in a given area reflects the present physical and biological environment of that area. Let's examine some historical and equilibrium theories of biodiversity and the LDG. Be aware, however, that there is a plethora of over 30 theories on the generation of biodiversity and our review can cover only a few examples in limited detail. In addition, we will concentrate on the modern LDG. Examination of the geologic record indicates that the current LDG can be traced back to about 30 million years ago and has not been a persistent feature in the geographic distribution of species richness over the entire Phanerozoic.

Historical Theories

Theories that modern global gradients in biodiversity reflect past events in the evolution and extinction of species were developed largely to explain the high species richness of the tropics compared to the lower biodiversity of the temperate and polar regions. This general line of reasoning has a long history that can be traced back to Wallace. The classic historical theory focused on the fact that the glacial periods of the Pleistocene resulted in severe disruptions of the environment in polar and temperate regions. According to the theory, advancing ice and cold during glaciations caused widespread extinctions of plants and animals at the higher latitudes. It is surmised that rates of evolution are too slow to produce significant adaptive radiation and rebuild species richness in the intervals between glacial periods. So, as long as the earth alternates between glacial and nonglacial conditions, species diversity will remain low at the high latitudes.

The classic historical theory further holds that the tropics have not experienced significant environmental disruption during glacial periods. Thus, extinction rates have remained low compared to those of the high latitudes. Long periods of stable environment have allowed for the evolutionary development of a multitude of new species. This line of general reasoning, that long periods of environmental stability lead to high species richness, is called the **stability-time hypothesis.** If the earth does not experience another glacial period, it could be supposed that decreased extinction rates and the evolutionary development of new species will, over time, lead to higher and higher species diversity at the mid- and high latitudes. Eventually, the difference in biodiversity between the tropics and other parts of the world would disappear. There is evidence for higher speciation over time in the tropics, but the evidence for lower extinction rates is mixed.

Although the premise that long periods of environmental stability reduce extinction rates and allow for evolution and increased speciation may be correct, the evidence suggests that the tropical regions were not immune to environmental changes during the glacial and nonglacial periods of the Pleistocene. There is now much evidence that tropical land areas cooled by several degrees and became significantly drier than present during the last glacial maximum. Some tropical lakes dried up, areas of sand dunes and desert expanded, and forested areas decreased in extent. Records of conditions in the tropics during the last glacial maximum, about 20,000 years ago, come from the sediments and fossils in deep sea cores, fossil pollen from lake sediments, fossil soils, cave deposits, and other lines of evidence. Climate models have also been used to study this issue. Many of these records, though not all, suggest cooling and arid conditions and concomitant shrinkage and fragmentation of the tropical rainforest (Fig. 14.6). Decreased atmospheric CO_2, which increased vegetation sensitivity to moisture stress, coupled with increased fires may have been important contributing factors to rainforest shrinkage and fragmentation. Shorter term climatic variations during the late Pleistocene also may have exacerbated these vegetation changes. Fossil pollen records and modeling studies from some sites in Australia show a replacement of tropical rainforest by dry forest and savanna during the period 26,000 to 10,000 years ago. However, other studies suggest a resilient rainforest vegetation persisted in some regions of Amazonia and Australia despite the changes in temperature and precipitation during the last glacial maximum. The low latitudes have not been immune to climatic changes over the Quaternary. However, there remains much debate over the resilience of the biota to such climatic changes and the actual configuration of vegetation in the tropics during the glacial periods of the Pleistocene.

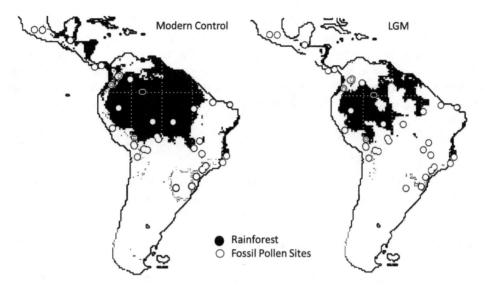

FIGURE 14.6 Possible scenario of shrinkage and fragmentation of tropical rainforest in South and Central America during the last glacial maximum (LGM) due to climatic change coupled with decreased atmospheric CO_2 and increased fires. Estimates based on climate model and fossil pollen data (*Source:* Sato et al., 2021).

Contrary to the classic stability-time hypothesis, some biogeographers have suggested that fragmentation of the tropical rainforest during glacial episodes may have actually promoted high diversity by allowing allopatric speciation to occur within the isolated forests that existed along rivers and in areas of moist habitat. In addition, genetic studies of modern rainforest species suggest that genetic divergence and speciation are most likely to occur along ecotones between rainforest and savanna. The relative area of such ecotonal sites would have increased as rainforest shrunk and fragmented during glacial periods. It is entirely plausible that allopatric speciation and ecotonal speciation in the fragmented tropical forests would lead to increased species diversity, and this is supported by simulation models. However, it does not explain why such high levels of speciation did not also occur in areas such as Siberia and central Europe where glacial ice was absent during glaciations and boreal and temperate forests were similarly fragmented, potentially leading to increases in the relative area of ecotonal habitat. As we have seen, the biodiversity of boreal and temperate forests is much lower than for rainforests. In addition, the empirical evidence for the large-scale shrinkage and fragmentation of rainforest in regions such as the Amazon and portions of Australia remains debated, with many suggesting a high degree of resilience to climatic changes during glacial periods. No doubt, research will long continue on this important topic.

Time certainly can contribute to differences in biodiversity. For example, Lake Baikal in Russia lies in a portion of Siberia that has never been glaciated. The lake began to form over 30 million years ago. Lake Baikal contains over 2500 species of plants and animals. Many of these species are endemic to Lake Baikal and evolved there after the formation of the lake. There are 57 species of fish in Lake Baikal and half are endemic. Great Bear Lake in Canada is a similarly sized, deep, and cold waterbody, but did not appear until about 10,000 years ago when glacial ice retreated from northern Canada. In contrast to Lake Baikal, Great Bear Lake has only 15 species of fish and none are endemic to the lake.

Looking more deeply back in time than the Pleistocene, biogeographers have sought to find relationships between long-term earth history and latitudinal diversity. Prior to the Quaternary, tropical conditions were much more widespread. Even the current Canadian Arctic areas had a warm climate and supported lush vegetation and alligators about 50 million years ago. Tropical conditions appear to have been much more typical of the earth over geologic time than are cooler temperate and polar conditions. Much of the current diversity in terrestrial and marine biota can be traced to clades that originated 16 million years ago or more and developed initially under tropical conditions. Climatic cooling over the past few million years could have led to greater diminishment of species richness at higher latitudes as members of these clades were unable to persist outside the tropics.

According to this theory, much of the modern LDG can be explained by climates, environments, and evolutionary events in the deeper geologic past. This explanation assumes that the niches for many species remained stable for long periods of time, producing extinctions as higher latitudes cooled, but allowing persistence in warmer tropical areas. This theory is referred to as the **tropical niche conservatism hypothesis (TCH).** Supporting evidence for the TCH has been found in a number of studies of plants and animals. However, acceptance is not universal. The TCH is an exciting area of current research.

Equilibrium Theories

A number of different equilibrium theories of biodiversity have been proposed to explain differences in species richness at small and large scales, including the LDG. In most cases, the processes invoked in these theories can be extended to smaller scales than the LDG.

Niche-Based Species Carrying Capacity To begin our study of equilibrium diversity theories, let's consider biodiversity within the context of environmental gradients and niche theory (see Chapters 3 and 4). In theory, areas with high numbers of species might reflect one of three conditions.

First, areas of high biodiversity could contain large gradients for resources and offer a wide range of habitat and niches for different species to utilize. For example, a large geographic area, such as the Amazon Basin, may extend across a wide variety of climatic and soil conditions and thus provide a wide range of habitat and potential niche space that can support a large number of different species. A small region, such as Costa Rica, may offer a wide range of environmental conditions because its diverse topography creates different environments in the mountains and lowlands. In theory, these regions can support a large number of species because each resource gradient is long and there is a range of habitat that can support different species with different ranges of tolerance and niches (Fig. 14.7). If such areas of habitat diversity are large, they can support bigger species populations and experience lower extinction rates and higher speciation rates as a result.

Second, the range of habitat in two regions may be similar and the resource gradient lengths may be equal, but the length of the distribution of individual species along the gradients may be short in one region because they have specialized niches. Thus, if we assume that competition between species precludes too much overlap along gradients and between niches, more species are accommodated along the resource gradient if each species is a specialist and has a narrow distribution. The development of such specialized niches might reflect the influence of intense interspecific competition on natural selection. High levels of interspecific competition in one area might produce a large number of specialist species with small niches, while lack of competition in another area might allow a few species to persist with very large niches. Accordingly, the high species diversity in the tropics might reflect a long period of intense competition producing species with narrow specialized niches.

Third, key resources may be more abundant in one area than in another. The gradients for the area of high species diversity would be typified by a high abundance of resources, while the gradients for the low-diversity area would be typified by a low abundance of resources. The greater abundance of resources could allow either a relaxation of competitive pressure, thereby allowing a greater number of generalist species to survive, or it could allow greater numbers of species with highly specialized niches to survive if interspecific competition was an important factor.

If species maintain their niches over long periods of time and successfully persist, then the niche relations above could help explain biodiversity at different scales in many different settings. For example, a recent global meta-analysis of termite species found evidence for narrower thermal niches in the tropics, and this helped explain higher species richness there. The premise that there is a carrying capacity for species richness determined by niche abundance has a long history, has many adherents, and permeates some of the theories presented in the sections that follow, but it also remains contested. A number of recent syntheses of empirical data and modeling results conclude that, on the whole, the data often do not support the premise that latitudinal differences in species richness is limited by niche space. For example, a global study of 308 different lizard species found no evidence that competition drove dietary niche size or that this explained the LDG in lizards.

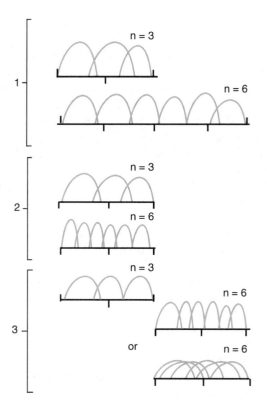

FIGURE 14.7 Hypothetical resource gradients and niche spaces for species in low- and high-diversity geographic areas. The first example represents a situation in which the high-diversity region has greater environmental variability and thus greater gradient length and available niche space. The second example represents a situation in which gradient lengths are equal, but the region with high biodiversity accommodates more species because the organisms living there possess more specialized niches and require less length along the resource gradient. The third example represents a situation in which resource abundance is higher in the area with high biodiversity and thus can accommodate more species either because the abundance of the resource allows the area to support very specialized species that exist using a small portion of the resource gradient and niche space, or they must compete less for the abundant resources and have greater gradient and niche overlap.

In contrast to the niche-based paradigm outlined above, the American ecologist Steve Hubbell proposed the **neutral theory of biodiversity and biogeography.** This theory posits that all trophically similar species in a community are ecologically identical and have little to no niche differentiation. For example, tropical trees on the same site often have the same general environmental requirements, and species diversity can be argued to be driven by random events of establishment, death, dispersal, speciation, and extinction. As you might imagine, the neutral theory has been the topic of much research and some pointed debate. Some empirical data combined with modeling appear to contradict the assumption of no niche differentiation among shared community species. One example is found in the study of Indo-Pacific coral reef communities. The variability in the communities studied was found to be much higher than predicted by the neutral model, and niche differences were thought to be a likely contributing factor. Another combination of experimentation and modeling was conducted using annual plant communities associated with serpentine soils in California. That study found clear evidence of niche differentiation among plant species and concluded that these differences in niche were important in maintaining the biodiversity in the plant community. Research and debate on the roles of the niche in biodiversity and the veracity of the neutral theory will continue. For our purposes in this chapter, we will not abandon differential niches as being of potential importance in contributing to biodiversity.

Environmental Heterogeneity/Habitat Diversity Greater environmental heterogeneity can generate increased habitat diversity. In many regions, species diversity is positively correlated with habitat diversity. This relationship seems particularly strong at larger spatial scales. A meta-study of 1148 different data points scattered throughout the world and encompassing plants and animals along with a variety of environmental measures of heterogeneity showed a positive relation between heterogeneity and species richness in 88% of the cases. Similarly, a meta-analysis of plant and animal species richness measured across 24 different island groups and isolated mountain tops found that in 83% of the cases studied, species richness increased with the increasing habitat diversity of the island.

Greater habitat diversity correlates positively with greater resource gradient length and greater available niche space. Habitat diversity can be generated by differences in the physical or biological environment. For example, areas with complex topography present a greater number of habitats and wider variability of habitat types than do areas with flat topography. Topography influences habitat diversity at both small and large scales. At the small scale, a field with gently rolling topography might include relatively dry hilltop soils, mesic hillslope soils, and moist soils in low-lying areas. Habitat would thus be available for plants and animals that required both dry soil conditions and mesic or wet conditions. The rolling topography of the field described above would present more habitat diversity and have a greater species diversity than an adjacent flat field that presented only mesic soil conditions. At the large scale, mountainous regions can include a wide range of environments, extending from hot and dry low-elevation deserts, cooler and moister mid-elevation regions supporting woodlands and forest, and cold high-elevation tundra. In addition, the diverse topography of mountainous regions can lead to the reproductive isolation of species populations and produce increased rates of allopatric speciation. In Central America, the mountainous and topographically diverse landscapes of Costa Rica support many more tree species than the relatively flat areas of adjacent southern Nicaragua. In the United States, mammal biodiversity is highest in the topographically diverse areas of the Pacific coastal mountains, Rocky Mountains, and Appalachian Mountains.

Although topographic diversity may explain local and regional differences in biodiversity, it does not provide an adequate explanation for the LDG. Both large sections of the Amazon rainforest and the boreal forest of central Canada are relatively flat areas, with topographic habitat diversity generated by low uplands, small-scale landscape features, and floodplains. As we have mentioned, the Amazon rainforest, however, has much greater biodiversity than the Canadian boreal forest. Similarly, tropical mountains have much higher biodiversity than the high-latitude mountains of western Canada and adjacent Alaska despite arguably similar topographic heterogeneity. In fact, the LDG for mountains is steeper in terms of poleward declines in species richness than is the case for lowlands. Nor can diversity in topography or ocean bathymetry conditions explain the increasing gradient of marine species richness from the polar areas to the equator.

Vegetation complexity is an important biological contributor to habitat diversity for both plants and animals. A highly diverse and stratified forest provides differences in plant composition and microclimate in each stratum and presents a wide range of habitats for birds, mammals, reptiles, amphibians, and insects. In addition, the different plant species that dominate different strata provide different food sources. Complex vegetation also produces a diversity of biochemical compounds that influence plants and animals. Such phytochemical complexity is recognized as an important influence upon plant and animal diversity. Consider a tropical rainforest with its typically complex stratified canopy. Some animals, such as many species of tropical butterflies, are adapted to life on top of the sunny and relatively dry upper canopy surface. Primates are often adapted to life within the upper to middle canopy, taking advantage of tree limbs for movement and living off fruits. Some amphibians live on the lower vegetation and moist cooler floor of the forest. It has been shown that many tropical beetle species are adapted to life in a very specific section of the vegetation strata and are absent from all other areas. This helps generate extremely high diversities of beetles in tropical rainforests. A single *Lubea seemannii* tree in the Panamanian rainforest may be host to over 400 species of canopy-dwelling beetles, with the lower limbs and trunks hosting an additional 200 species. Phytochemical complexity can be extreme in the tropics. For example, the genus *Piper* has been shown to be associated with 667 different metabolite chemical compounds. There is a direct positive effect between the phytochemical diversity and the diversity of associated herbivore insects. Differences in forest strata diversity, even in less complex forests, can have an impact on habitat diversity and species diversity. A classic study of bird species diversity in the deciduous forest of the eastern United States showed a strong positive correlation between plant species diversity and height of the foliage in forests, and the species richness of birds (Fig. 14.8). Similar relationships are evident in many different forest types and geographic regions. For example, one recent study has shown that average passerine bird species richness in eucalyptus forests and woodlands eastern Australian similarly increases with canopy height.

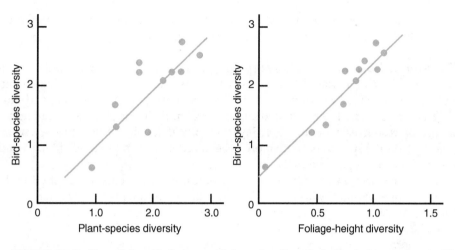

FIGURE 14.8   The relationship between bird species diversity in the deciduous forest of the eastern United States and plant species diversity and vegetation structure (*Source:* MacArthur and MacArthur, 1961 / The University of Chicago Press).

Does increasing vegetation complexity really explain the latitudinal gradients in species richness from the poles to the equator? Again, the answer is no. First of all, increased vegetation stratification and biological habitat diversity do not explain why biodiversity in tropical deciduous forest, savanna, and grasslands is higher than structurally similar vegetation in mid- and high latitudes. An area of African savanna can contain dozens of species of large mammal herbivores; a similar sized area of forest-tundra vegetation supports only two or three large mammal herbivores. Second, attributing the latitudinal gradient of species diversity to increasing vegetation structure does not explain what produces the gradient of increasing plant species diversity and vegetation complexity from poles to tropics in the first place. It is the higher diversity of plant species in the tropics that produces much of the complex vegetation stratification found in the rainforest.

Large Land Areas of the Tropics   It has been suggested that the large continuous land area of the tropical zone in South America, Africa, and Australasia is an important contributor to the development and maintenance of high species diversity. You may have the impression that the tropics are small relative to the large land areas of northern Eurasia and North America. However, the relative size of tropical areas is understated in the Mercator projection maps that are typically used to show the world. The Mercator projection "stretches" the area of the high latitudes and "shrinks" the area of the low latitudes in order to present the spherical earth on a two-dimensional map surface. If you look at the relative land area of the tropics on a globe or on alternative map projections that portray the relative size of the tropics and high latitudes correctly, you will see that the land area of the tropics is actually very large. For example, on Mercator projections, the island of Greenland, which lies close to the north pole in the vicinity of 70° to 80° north latitude, appears to be larger than Australia, which lies in the vicinity of 12° to 30° south latitude. In reality, Australia covers some 7,682,000 square kilometers and is over three times larger than Greenland, which covers only 2,175,000 square kilometers.

The presence of large areas of continuous habitat in the equatorial zone, such as the Amazon rainforest, provides high carrying capacities and allows species to have large populations and large geographic distributions. As we have discussed in Chapter 9, large population size decreases the risk of extinction. Large geographic distributions that support large population sizes lead to decreased risk of extinction. Thus, if distributed over large areas, species can survive with narrow ranges of tolerance along resource gradients and develop specialized niches that would lead to unsustainable small populations if the geographic area of the species distribution was small. The problem with this hypothesis is that it does not explain why the huge boreal biome has much lower species richness than smaller biomes such as the temperate deciduous forest, Mediterranean, or coastal rainforest biomes. In addition, it does not explain latitudinal gradients in marine biodiversity. The cold southern oceans around Antarctica are spatially large but have a much lower diversity of many marine groups than does the tropical Pacific or Atlantic. The large area of land in the tropical belt undoubtedly is a contributor to the LDG, but not the sole underlying explanation.

Environmental Stability   It has been suggested that environments in which climate is relatively stable on the short time scale (days, months, seasons, and years) promote higher biodiversity than unstable environments. Environments with low amounts of daily, seasonal, and annual variability allow species to become finely adapted and to develop the most efficient form and behavior to take advantage of resources without requiring tradeoffs that allow them to cope with variability in environment. As the species become more specialized and are more efficient in extracting resources, more species can be supported by given resources.

As we have seen in previous chapters, tropical environments are typified by low amounts of daily and seasonal variability in climate. Midlatitude environments can experience large variations in daily temperatures and huge variations in seasonal temperatures. The short- and long-term variability of the higher latitudes requires species to be more generalistic in terms of niche breadth. Because they are generalists, they cannot extract resources as efficiently, but they survive because they are adaptable to a wide range of conditions.

One attraction of the climatic stability hypothesis is that it explains both the decrease in species richness at high latitudes and the increasing size of species ranges that are found at high latitudes. The larger ranges found at high latitudes are considered to be a reflection of the more generalized niches of the species there. If we assume that competition between species is important in determining how many species can be supported in a given environment, we can postulate that since the niches of the higher latitude species are more general, not as many species can be supported there.

The stability theory has three drawbacks in explaining the geographic distribution of biodiversity. First, it is unclear why short-term climatic instability should be so limiting to speciation via natural selection. Why could species not evolve adaptations to climatic variability without sacrificing their ability to utilize resources? Second, many areas of the earth have very stable environmental conditions but experience low species diversities. For example, the deep ocean is extremely stable today in terms of temperature and salinity, but it has lower species diversity of fish and invertebrates than many shallow-water areas, which are far less environmentally stable. Large saline lakes in the desert, such as Mono Lake in California or the Great Salt Lake in Utah, have very stable daily or seasonal conditions compared to adjacent small alpine lakes in the Sierra Nevada and Rocky Mountains. Yet the saline lakes have much lower diversity of aquatic organisms. Third, latitudinal diversity gradients are seen in many groups of species for which there is no evidence for more general niche breadth or wider geographic distributions at high latitudes.

Disturbance   In 1978, the eminent ecologist J. H. Connell proposed a theory that links biodiversity to disturbance. The idea that Connell proposed is called the ***intermediate disturbance hypothesis.*** According to this hypothesis, if an ecosystem remains free of disruption by disturbance, the stable homogeneous environmental conditions will favor some species, but will lead to the extinction of species for which the stable habitat is not favorable or which are prone to competitive exclusion by species that are favored by the undisturbed environment. The loss of species produces decreased biodiversity. Disturbances cause physical and biological changes and promote spatial heterogeneity that can favor species that would not survive in a stable undisturbed system. On the other hand, if disturbance occurs too frequently and is too severe, it will lead to the extinction of disturbance-sensitive species that have long generation times or occur in low numbers and are prone to chance extinction. Thus, maximum levels of biodiversity can be expected from intermediate levels of disturbance that occur frequently enough and cause enough disruption and environmental heterogeneity for the survival of species that could not persist in an undisturbed system, but are not so frequent or severe as to cause the extinction of long-lived and disturbance-sensitive species.

Connell developed his hypothesis largely on the basis of the differences in disturbance frequency and biodiversity that he observed in the Great Barrier Reef of Australia. He initially applied the theory there and to rainforests. Boreal forest fires provide a good terrestrial example of how intermediate disturbance regimes can promote species diversity. If fires were excluded, many boreal sites would become dominated by long-lived conifer species such as white spruce *(Picea glauca)* or balsam fir *(Abies balsamifera)*. Other boreal plant species, including grasses, many herbs, shrubs such as willow, and deciduous trees such as aspen *(Populus tremuloides)* and paper birch *(Betula tremuloides)*, cannot survive in the shaded conditions under a spruce or fir canopy. Because of their size and longevity, spruce and fir are superior competitors for light and soil moisture. Given time, the shade intolerant species would likely go extinct, at least locally, if there were no disturbances to open the coniferous forest canopy. Fires, windthrow, and insects create small and large patches in the coniferous forest that allow the shade-intolerant species to persist by taking advantage of the temporary high-light intensity in these gaps. If, however, severe fires were to become extremely frequent, occurring at intervals of, say, 10 years (which is less time than it takes spruce or fir to reach sexual maturity), then spruce, fir, and other fire-sensitive long-lived species would go extinct.

The intermediate disturbance hypothesis predicts that biodiversity will be the greatest in systems in which the rate of displacement of species by competition is low and the rate of disturbance is also low. Thus, if it takes 50 years for a population of slow-growing trees to develop a dense canopy and root system that causes the local extinction of shade-intolerant grasses, herbs, and shrubs, and it takes 30 years for the spruce trees to reach reproductive age, then a disturbance frequency of one event every 40 years would be optimal for maintaining species diversity. Disturbances once every 40 years would give the spruce time to reach reproductive maturity but would not allow the trees to completely displace the grasses, herbs, and shrubs.

Some field studies and computer modeling experiments support the potential importance of intermediate disturbance regimes in generating biodiversity. However, a number of other studies and simulations appear to refute the concept. A large-scale study in Ghana found the theory was plausible in regards to dry tropical forest tree diversity, but not as applicable to rainforests as previously held. The hypothesis encounters several problems in explaining global gradients of species diversity. To explain global biodiversity patterns, it would have to be assumed that tropical ecosystems provide a combination of lower exclusion rates of species from undisturbed sites and lower disturbance rates than other ecosystems. It is difficult to develop evidence for or against this proposition. In many areas of the world, both on land and in the ocean, long-term disturbance rates and competitive exclusion rates are unknown. It is becoming increasingly clear that even tropical rainforest ecosystems are prone to disturbance from a number of sources, including fires, hurricanes, windthrow, flooding, landslides, and pathogens. Many tropical environments may have disturbance rates that are not much less than areas of the temperate deciduous forest or temperate rainforest. It is also unclear what biological or physical factors would ultimately cause tropical species to experience particularly low rates of exclusion relative to disturbance rates.

Competition   It has been suggested that the natural selection and evolution of species in the mid- to high latitudes are driven by adaptation to physical stresses related to climate, such as cold or aridity. In contrast, evolution in the warm and moist tropics is argued to be driven more by interspecific competition. The high degree of competition in the lower latitudes leads to species developing specialized niches that restrict their distribution in terms of habitat and along resource gradients. Species are competitively superior within narrow ranges of environmental conditions. Specialized niches and narrow habitat preferences decrease direct competition with other species. This increases the number of species that can survive locally, each species using a narrow range of microhabitats. For example, in the tropics, many frog species breed in ponds and other standing water similar to the breeding habitat of most temperate frogs. However, other species are adapted to breed in the small pockets of moisture in the forest floor or canopy, including the water that collects in the foliage or flowers of plants such as epiphytic bromeliads. The wide range of specialized breeding areas used by tropical frog species decreases competition for breeding space.

Some evidence of the possible impact of competition on the niche sizes of tropical animals comes from studies of anole lizards *(Anolis* spp.*)* in the Caribbean. The niche breadth of the anole species has been measured in terms of the size of prey they eat. Anoles that hunt prey of only a certain size are considered specialists, whereas anoles that eat prey of a variety of sizes are considered generalists. When a single lizard species occupies an island, it generally will hunt and eat a larger range of prey of different sizes. When several different species of anoles occur together, each species often has a relatively narrow range of prey sizes that it will hunt and eat. The narrow ranges of prey size decrease the overlap in prey species hunted by each lizard species. Thus, competition between species for prey is decreased.

The competition theory suffers serious drawbacks when applied to explain the LDG. First, most plant species, especially trees, are limited by a relatively few resources, such as light, water, heat, and soil nutrients. Many of them do not display narrow and nonoverlapping requirements for these resources, even in tropical settings. There is in fact little convincing evidence that competition and narrow niche specialization can explain tropical tree diversity. Second, it still must be explained why competition is such a potent force in the lower latitudes, but not in temperate and Arctic regions. If competition alone was responsible for biodiversity, why wouldn't similar decreases in niche breadths and high rates of biodiversity develop everywhere? Modeling studies suggest that strong competition can serve to steepen the LDG by excluding new species from expanding poleward into environments that were already occupied by better dispersing and more generalist species.

Predation It has been argued that the high numbers of predator and parasite species, particularly in the tropics, maintain high biodiversity by keeping prey populations low and decreasing the competitive exclusion of one prey species by another. The reduction in competition in turn allows for more prey species to evolve, and this leads to the coevolution of additional specialist and generalist predators. The shift toward greater biodiversity at one trophic level due to predation-competition effects can be transmitted to other trophic levels to influence the species richness of the entire ecosystem. In this regard, the role of predation in the promotion of species diversity has to be viewed in terms of the interaction between predation and competition, and the propagation of those effects beyond the immediate interacting species. The predation model has been applied in a very interesting manner for an explanation of tropical tree diversity. The American ecologist Daniel Janzen spent decades working in Costa Rica to understand its biodiversity. He noted that tropical forests contain high numbers of different tree species but that most tree species occur as isolated individuals. In general, stands dominated by one or two species do not occur in mature tropical lowland forests, which means that no species appears to be capable of effectively excluding others at a landscape scale. Janzen suggested that seed predation may explain the spatial isolation of rainforest trees from other members of the same species. This in turn explains the apparent lack of competitive exclusion of some tree species by others. He suggests that specialized seed predation is particularly common in tropical forests. Indeed, insects are extremely important seed predators in the tropics, and many are very specific regarding which species they will eat. Very high concentrations of these specific seed predators are common in the vicinity of their host trees. Thus, most seeds that are distributed near the parent tree are devoured. The loss of these seeds keeps the area close to the parent tree free of its own offspring. Seeds that are dispersed away from the parent tree may end up at sites away from other members of their species. Such areas will, of course, lack the specific predators that prey on the tree species. Seeds transported to these areas have a greater chance of survival and germination. The inability of trees to establish close to other members of the same species keeps the populations of tree species small and dispersed. This decreases competitive exclusion and allows a greater number of tree species to coexist.

Applying the predator-competition model of species diversity to explain large biodiversity gradients raises two issues. First, field and laboratory experiments show that large numbers of predators lead to increased prey species diversity in some cases, but to decreased prey diversity in others. For example, the species richness of unicellular organisms found in the small pools of water that form in the leaves of the pitcher plant *(Wyeomyia smithii)* decreases as the number of predaceous mosquito larvae living in the pools increases. Second, and most important, the theory does not explain why there should be much greater numbers of predators and parasites in the low latitudes in the first place.

Productivity It has been suggested that regions with high primary productivity will develop high biodiversity because the vegetation in such areas produces more energy that can support more species at the higher trophic levels. The high amounts of available energy allow for primary consumer species to have relatively specialized niches in terms of food. The greater the number of primary consumers, the greater the number of predators that can be supported by the environment. The large number of different herbivorous species promotes even greater diversity of the plants through the coevolution of specialized herbivores, pollinators, and seed dispersers. At a global scale, there is indeed a general correlation between primary productivity and latitudinal differences in biodiversity. The typical primary productivity of tropical rainforests is some 2200 grams per square meter per year, whereas temperate forests produce only about 1200 grams per square meter per year, and boreal forests produce only about 800 grams per meter per year. These differences in productivity can be directly correlated with the relative levels of plant and animal biodiversity in these biomes.

When examined at smaller scales, the relationship between primary productivity and biodiversity can break down. For example, estuaries, marshes, and swamps have some very high rates of primary productivity. They can demonstrate rates of gross primary productivity that are among the highest for aquatic systems and equal to some forests. Yet these systems have extremely low diversities of both plants and animals. Many estuary marshes are dominated by one or a few species of vascular plants such as pickleweeds *(Salicornia),* cordgrass *(Spartina),* or mangroves. Some regional studies of plant biodiversity in terrestrial settings also present evidence that contradicts the hypothesis that high biodiversity is a result of high productivity. It has been suggested there is a humped relationship, with diversity reaching a peak at intermediate productivity levels. For example, comparisons of plant species diversity and primary productivity in Australia and South Africa show that the highest plant biodiversity is correlated with intermediate levels of primary productivity. Areas with very high or very low primary productivity have lower plant biodiversity. However, it has been argued, based on analysis of global data, that such humped relationships do not extend to the scale of the LDG. This then begs the question, what ultimately produces higher productivity in the tropics to begin with?

Temperature and the Metabolic Theory of Biodiversity (MTB)   It is obvious that one of the clearest environmental gradients between the equator and the poles is temperature. It is therefore not surprising that the global or continental species richness of many terrestrial and marine organisms ranging from trees to marine phytoplankton shows a significant positive relationship with temperature. It is interesting to note that study of the LDG over the entire 500-million-year span of the Phanerozoic indicates that the gradient is best developed in geologic times when the earth is generally cool and there is a high contrast in temperature between equatorial and polar regions. How temperature might serve to drive biodiversity is a topic of great interest and much research. Warmer conditions, perhaps coupled with long growing seasons, might, for example, allow for greater primary productivity and provide more energy to support greater diversity in plants and associated animals. Freezing temperatures and prolonged winter conditions might decrease primary productivity and also be lethal to many species of plants and animals.

The **metabolic theory of biodiversity (MTB)** is an interesting eco-evolutionary theory that attempts to relate warmer temperatures to increased rates of metabolism and to increased evolutionary rates of speciation. Higher temperatures can cause increased metabolic rates, shorter generation times, and increased mutation rates in populations. This should promote speciation. In addition, biomolecules are most stable at around 20° C ambient temperature. At low temperatures, they become more unstable. Thus, mutations might have a greater probability of success and persistence in warmer conditions and be more deleterious to the organism in cold conditions. Analysis of planktonic foraminifera has revealed a positive relationship between temperature and speciation rates, which has been interpreted as resulting from the MTB. Similar conclusions have been drawn from a study of over 540,000 global observations of phytoplankton. A global analysis of tree diversity data from approximately 1.3 million forest plots also detected a positive relationship between temperature and species richness, which was considered consistent with the MTB. A wide variety of organisms have been analyzed in the study of the MTB and show a great deal of variability in the relationship between temperature and species richness. In addition, many studies also indicate that other environmental variables, such as moisture or the duration of days with temperature below freezing, have a significant effect on the temperature-species richness relationships. A recent global study of approximately 332,000 seed plants concluded that there was no convincing evidence of higher diversification rates in warm climates and no support for the general applicability of the MTB in explaining seed plant biodiversity at a global scale. The study suggested that other historical factors, such as embodied in the tropical niche conservatism hypothesis, and ecological factors, such as carrying capacity and habitat heterogeneity, were more plausible explanatory factors for the LDG. Debate and research on the MTB will likely continue.

A Synthesis

It appears that no single general rule can be applied to explain all the differences in biodiversity at large and small scales. Explaining the large-scale gradient between the high latitudes and the low latitudes is perhaps of greatest interest to biogeographers. Almost 30 years ago, the British biogeographer Kevin Gaston, who has done much analysis of this problem, commented as follows: "The search for a single explanation for latitudinal gradients in diversity may be attempting to answer two questions at once, how available energy is converted into individual organisms and how these individuals are distributed amongst species" (1996, p. 471). In 2023, following a further three decades of intense research, the Chilean ecologist Ricardo Segovia summed up the state of knowledge thus: "The geographic variation in species richness over broad spatial scales is probably the most intriguing pattern of biodiversity. While correlations between species richness and environmental variables have been widely documented . . . , general law or even a single, coherent explanation for these correlations is still elusive" (2023, p. 1). The large difference in biodiversity between the high and low latitudes likely reflects a number of factors.

First, history, such as embodied in the tropical niche conservatism hypothesis, cannot be excluded as a contributing factor to the high biodiversity in the tropics and the relatively low biodiversity found at high latitudes. Although global glacial periods did result in cooler and drier conditions in the tropics, the magnitude of these changes was small compared to the large changes in climate and environment that occurred in the mid- and high latitudes. During glacial periods, the geographic ranges of many high-latitude plant and animal species were displaced by thousands of kilometers. The fossil record shows that in both Europe and North America, many species of temperate plants went extinct during the transition from the warm and moist conditions of the early to mid-Cenozoic to the colder and more variable conditions of the Pleistocene. We also know that the species composition of mid- and high-latitude plant and animal communities was markedly different during glacial and nonglacial periods. The reshuffling of community composition could work against the creation of new species by coevolution or competitive interactions. Finally, the tropical communities, particularly rainforests, were probably fragmented during glacial periods but were not dislocated in terms of latitude as was the case for Arctic and temperate communities. Environmental factors such as day length or the degree of contrast between winter and summer climates did not change as much for tropical species as it did for high-latitude species that experienced large latitudinal shifts in their ranges during glacial and nonglacial periods. Although tropical communities were fragmented, their species composition may not have changed as much as that of temperate and Arctic communities. The fragmentation of the rainforests and other tropical systems during glacials was also likely of importance as it promoted allopatric speciation.

History alone, however, cannot explain all the differences in species diversity between high and low latitudes. It is also likely that much of the LDG is related to area, habitat heterogeneity, climate, and energy/productivity. The tropical land areas are large and offer a high degree of heterogeneity in factors such as topography and vegetation canopies. At a global scale, a very clear relationship exists between climate, primary productivity, and species diversity. For example, areas with high annual evapotranspiration rates are typified by high species diversity of plants and animals (Fig. 14.9). High evapotranspiration rates occur in regions with high amounts of solar energy, heat, and moisture. Light, heat, and moisture are key resources for plants and animals. High amounts of light, heat, and moisture undoubtedly lead to higher primary productivity, and at a global scale, this provides energy and carrying capacity to support a wider range of species. The low-latitude areas of highest biodiversity are also areas where seasonal differences in climate are relatively small. The ability of plants to photosynthesize and produce energy throughout the year means that vegetative food and fruit are available at all times. Many tropical bird, mammal, and insect species, particularly obligate fruit eaters, cannot survive in temperate or Arctic settings because they do not have access to adequate or appropriate food in winter. Warm climates throughout the year also allow reptiles, amphibians, and insects to remain active in summer and winter. Many tropical predatory birds rely on the consumption of reptiles or insects during summer and winter and would not survive in areas with cold winter climates. As noted above in the

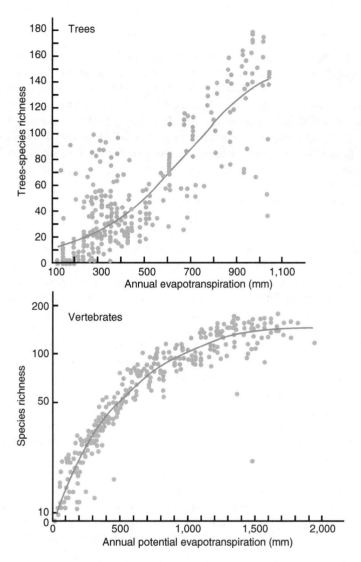

FIGURE 14.9 The relationship between energy and moisture as measured by evapotranspiration and the species diversity of trees and vertebrates in North America (Currie, 1991).

discussion of the metabolic theory of biodiversity, warm temperatures may have also contributed to greater rates of mutation and speciation, although this remains controversial.

There is undoubtedly a positive feedback relationship between the ability of the tropics to support high biodiversity because of available light, heat, and water, and low seasonal variability in energy, and the evolutionary development of additional new species. The wide diversity of plant species and the resulting complex canopy structure that develops in the tropical rainforests produce a wide range of habitat types for animals and smaller plants. As we discussed earlier, increasing niche specialization in response to competition can be a strong contributor to biodiversity in areas such as the tropical rainforest where resources are abundant. Abundant prey in high-productivity areas allows the evolution of prey-specific predators that can exert pressure on the population size of their prey species, keeping that species from becoming dominant and competitively excluding other species. In summary, we can say that the high biodiversity of the tropics relative to higher latitudes has defied one single simple explanation and is more realistically the product of many historical and equilibrium factors.

ISLAND BIOGEOGRAPHY

Islands have always held a special attraction for biogeographers. The fascination with islands extends back to the foundations of the discipline. Darwin's ideas regarding evolution were very strongly influenced by the biological diversity of the finch species that he observed on different islands of the Galapagos Archipelago. So, too, was Wallace influenced by the similarities and differences in bird species and other organisms that he observed on the islands of the Malaysian Archipelago. Islands have been very important in many subsequent studies of the role of geography in evolution. Examples that we have already discussed in this book include the work of Ernst Mayr on kingfisher species on New Guinea and surrounding islands and the many studies of fruit flies, honeycreepers, and plant species on the Hawaiian Islands.

Many biogeographers have focused on intriguing biological attributes and phenomena that are characteristic of island organisms and ecosystems. In 1880, Wallace published a classic book on this topic called *Island Life*. In 1965, Sherwin Carlquist published an updated synthesis of island biology that was also entitled *Island Life*. From these writings, we can identify a suite of characteristics that are often found in insular species of plants and animals and distinguish them from mainland ancestors and relatives. Such features include the loss of dispersibility, the development of gigantism or dwarfism, the loss of antipredator defensive features and behaviors, the development of woody shrubs and trees from species that occur as soft-stemmed herbs on continents, and the development of highly specialized niches.

The study of species richness on islands has had a long and important history in biogeography. In the nineteenth century, the distinguished botanist Joseph Hooker analyzed Darwin's plant collections from the Galapagos Islands and attempted to decipher the cause of interisland differences in plant species richness. In the early twentieth century, O. Arrhenius studied the biota of islands in the Baltic Sea and produced the first paper on the species-area curve. This paper precipitated heated debate on the meaning of species-area patterns with the noted American ecologist H. A. Gleason (see Chapter 6). The analysis and interpretation of patterns of biodiversity on islands became a topic of much interest among ecologists and biogeographers from that point onward. In 1967, Robert MacArthur and Edward O. Wilson published a seminal book in the field of biogeography. Their book, entitled *The Theory of Island Biogeography,* provides a theoretical construct linking species-area relationships on islands to dispersal and extinction processes. In the decades since its publication, MacArthur and Wilson's theory of island biogeography has remained the subject of intense interest, research, and vigorous debate. Before we examine the theory of island biogeography, let's take a look at the generalized patterns in island biodiversity that influenced MacArthur and Wilson and underlie the theory.

Geographic Patterns of Island Biodiversity

Study after study has shown that the species richness of islands is strongly correlated with island size. Larger islands support more species of plants and animals than do smaller islands. This pattern is seen on local to global scales. Analysis of the Galapagos Islands clearly shows a strong correlation between island size and plant species richness. The larger islands in the archipelago support over 300 species of land plants. The smaller islands support a handful of species. The big island of Hawaii is over 10,000 square kilometers in area and supports over 30 native bird species today. Islands in the Hawaiian Archipelago that are less than 400 square kilometers in area support fewer than 10 species of native birds. A classic analysis of biodiversity on the Caribbean islands showed that large islands such as Cuba and Hispaniola support over 50 species of reptiles and amphibians, while small islands such as Saba and Redonda support less than 10 species. Intermediate-sized islands, such as Jamaica and Puerto Rico, support an intermediate number of species (Fig. 14.10). A global analysis of the species richness of both native and introduced plant and bird species on oceanic islands has revealed similarly strong relationships between island size and species richness (Fig. 14.11). This suggests the general relationship between island size and species richness is robust at a range of spatial scales, across islands in different environments, and for long established and recently established populations.

The pattern of increasing species richness on the Caribbean and oceanic islands has also been found on the smaller islands located in lakes. For example, the number of vertebrate species found on islands in Lake Michigan is larger on big islands than on small islands. In addition to true islands,

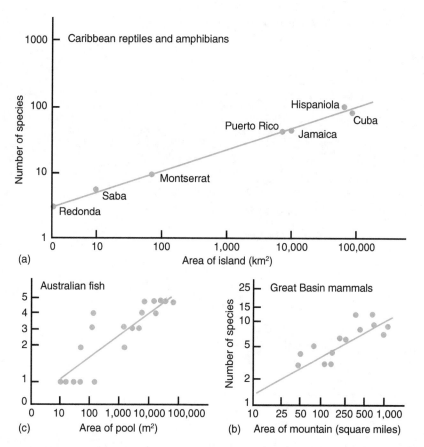

FIGURE 14.10 The relationship between island area and species richness for reptiles and amphibians in the Caribbean (a). The relationship between fish species richness and spring-pond size in the Australian desert (b), and between boreal mammal species richness and montane habitat area on isolated mountains in the Great Basin of North America (c) (MacArthur and Wilson, 1967; Kodric-Brown and Brown, 1993; and Brown, 1971).

many other forms of fragmented habitat can be thought of as islands. One example is lakes and ponds that are not connected to other waterbodies by rivers and streams. These isolated waterbodies are islands of aquatic habitat surrounded by land. The analysis of desert spring-ponds in Australia shows that the species richness of fish generally increases in proportion to the size of the ponds (Fig. 14.10). Isolated mountain tops surrounded by warmer or drier lowlands are another form of island. For example, a number of bird and mammal species typical of the boreal biome can be found on high isolated mountains in the Great Basin region of the United States. Lowland areas in the Great Basin are too hot and dry for these mammal species to survive. Thus, these isolated mountains can be likened to islands of boreal habitat in a sea of desert. In fact, in the southwestern United States, such ranges are referred to as sky islands. Analysis of the bird and mammal species richness shows that the number of boreal species found on Great Basin mountains is highly correlated with the area of alpine habitat provided by the individual mountains and ranges (Fig. 14.10).

Larger islands generally have greater numbers of species than smaller islands for all of the same reasons that larger sampling plots on continents have more species than small plots. We discussed these reasons in some detail in the preceding section. At this point, we might then ask two questions. First, are the species-area curves for islands essentially the same for different island groups, or do they differ from one set of islands to another? Second, are species-area curves from islands similar to the species-area curves obtained from continents? Note that all of the graphs presented here use log scales to linearize the graphed relationship between island area and species richness. We know that this relationship is nonlinear and we can use the species-area curve approach discussed earlier in this chapter and compare the relationships between island area and species richness with different island groups and with continental areas using the species-area curve equation described previously

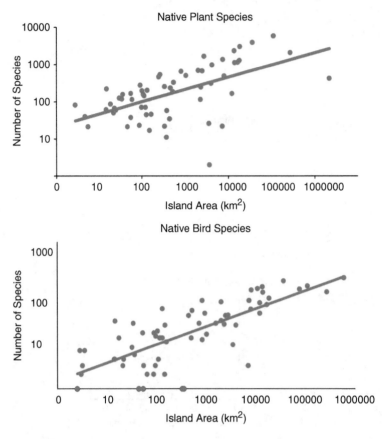

FIGURE 14.11 The relationship between island area and species richness for native plant species and native bird species from a global analysis of oceanic islands (*Source:* Blackburn et al., 2016 / John Wiley & Sons).

($S = cA^z$). If we look at the relationship between island area and land plant and animal species richness on the Galapagos or the Caribbean islands, the species-area curve equations are as follows:

For Galapagos plants:

$$S = 28.6A^{0.32}$$

For Caribbean amphibians and reptiles:

$$S = 3.3A^{0.30}$$

The values for the constant c in the equations is larger for the Galapagos plants because there are more plant species in the Galapagos than there are reptile and amphibian species in the Caribbean. Interestingly, the values of z for both the Galapagos plants and Caribbean reptile and amphibian fauna are roughly the same (0.32 and 0.30, respectively). This implies that the relative rate of increase in species related to increasing island area is the same for both systems. In fact, a wide range of island studies have produced values of z that average around 0.30. In contrast, the species-area equation for land birds in North America is as follows:

$$S = 40.0A^{0.17}$$

The high value for c relates to the large number of bird species on the continent. The value for z obtained from the species-area equation for North American birds is about half the value obtained for island systems. For species-area equations derived for different-sized areas on continents or other large landmasses, the value of z tends to range from 0.15 to 0.25. This means that the slope between the number of species and area increases more quickly for islands than for sites of different sizes located on continents.

Why should species-area curves be different for islands and continents? The likely reason is that much of the biota that we find on different-sized plots on continents may be supported by the

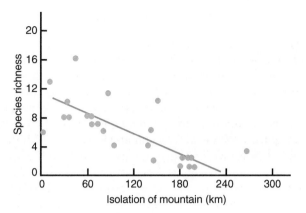

FIGURE 14.12 The impact of increasing isolation on the species richness of boreal mammals found on mountains in the southwestern United States. Isolated mountains have fewer species than those located close together (Lomolino et al., 1989).

resources of the plot and resources from the surrounding environment outside the plot. Birds that nest in a continental plot area, for example, may forage for food both within the plot area and in adjacent regions. In addition, a plot on the continent may support a small population of plants and animals that are prone to local extinction, but immigration from populations from just outside the plot boundaries keeps the local population from going extinct. For island organisms, the resources on the island may be all that is available to support them, and thus fewer individuals and fewer species can be supported on an island surrounded by inhospitable ocean than on a plot of land surrounded by a continent of similar habitat. Finally, immigration to islands is often difficult, and small island populations are more prone to local extinction than small populations on continental plots.

Differences in island size appear to explain a considerable amount of the variation in species diversity between different islands. However, island area is not the only important control on species richness. When we examine islands that are close to continents, we find that they tend to have higher species richness than islands that are located far away from continents. For example, islands in the South Pacific that are located less than 800 kilometers from the large land areas of New Guinea and Australia have up to 10 times as many terrestrial and freshwater bird species as islands that are located more than 3200 kilometers from New Guinea and Australia. In the case of the South Pacific, both island size and island isolation appear to be important determinants of species richness. Low species richness on isolated islands has been observed in many other studies. For example, the species richness of birds in isolated patches of montane forest in Venezuela, Colombia, and Ecuador decreases with distance from the main chain of the Andes Mountains. In a similar manner, the species richness of boreal mammals on mountains in the southwestern United States is highest on mountains that are close to other ranges (Fig. 14.12).

Before we proceed, we might ask how the species on islands are related to the species on other nearby islands or continents. In the case of both true islands and island-like areas of isolated habitat, it has been found that the species living on the islands are often subsets of the larger species pool found on the closest mainland. In the case of groups of islands, the species of plants and animals found on smaller islands are generally subsets of the species found on the larger islands. It is rare that the flora or fauna found on a small island located near a continent or a large island would contain an entirely different group of species than the mainland or larger island. This phenomenon of similarity between island and continental species is referred to as the **nested pattern of insular communities** and is very widespread. Not all species on islands, however, are subsets of the species found on the mainland or larger islands. There are many island species that have evolved on islands and are endemics. In general, the number of island endemics increases with the degree of isolation of the island. The islands of Hawaii are extremely isolated from other islands and the mainland and have many endemic plants and animals. The Channel Islands of California lie close to North America and have only a few endemic species.

The Equilibrium Theory of Island Biogeography: Historical Roots

Why precisely do small isolated islands possess fewer species than large islands that are near continents? The **equilibrium theory of island biogeography** was first advanced by MacArthur and Wilson in a paper in 1963 to explain the near-consistent relationships between species richness and island size and island isolation. It is an exceptional theory in terms of its conceptual simplicity, scope of applicability, and potential implications for species conservation. It is the most well known theoretical concept to arise from biogeography in the twentieth century. For over 60 years, the theory has remained at the heart of much research and vigorous debate. MacArthur and Wilson's 1967 book has been cited in over 26,000 subsequent studies. Let's start by looking at the people behind the idea.

In 1953, a young mathematician named Robert MacArthur arrived at Yale University to begin a Ph.D. degree under the supervision of the noted ecologist G. E. Hutchinson. Today, Hutchinson is remembered primarily for his work on lake ecology. However, he had broad interests in all areas of ecology and evolution, including the study of biodiversity. For his part, MacArthur had an avid interest in birds, and this is one factor that brought the young mathematician to do a Ph.D. in ecology. Under Hutchinson's supervision, MacArthur applied his mathematical expertise to questions of niche partitioning and community structure for warblers. He established himself as a leader in ecology by applying the rigor and logic of mathematics to study the community organization and geographic distribution of plants and animals. By 1965, MacArthur was a faculty member at Princeton University. During this time, he began to interact with a young biologist named Edward O. Wilson who was a specialist in the study of ants. Wilson had been fascinated by ants since his youth. By the 1960s, Wilson had become a faculty member at Harvard. He had a more traditional biology education than MacArthur and spent long seasons in the tropics collecting ants. Wilson saw the value of mathematics in interpreting ecological data and had even taken undergraduate classes in the early 1960s to learn calculus. Both MacArthur and Wilson felt biogeography was a science that was rich with information on species distributions but poor in rigorous theory to explain these distributions. In 1962, the two men turned their attention to the application of mathematics to try and explain species-area relationships. By 1963, they had published some of their ideas in a paper entitled "An Equilibrium Theory of Insular Zoogeography." In 1967, their ideas were expanded in the book *The Theory of Island Biogeography*. The theory they laid out in their book combined extinction, dispersal, and geography to explain the patterns of biodiversity observed on islands.

The theory of island biogeography generated immediate interest among biogeographers and ecologists. Ten years after publication of the book, at least 121 studies were published that cited the theory of island biogeography. Twenty years after the book's publication, there were over 2000 such published citations. Today, the theory is cited in hundreds of papers each year. MacArthur and Wilson quickly became major figures in ecology and biogeography. After the book was published, they pursued individual research projects. Wilson remained a towering presence in the study of biodiversity. Much of his attention focused on the use of science to aid in the conservation of endangered species. Some of his positions, such as the use of biological and ecological principles to study society, have been controversial, but he became a distinguished figure with many awards to his credit. What of Robert MacArthur? He continued his ecological work, including a detailed field study of birds, in Panama. In 1972, he published a new and important book, *Geographical Ecology: Patterns and Distributions of Species*. By this point, however, MacArthur was battling against time. As he worked on the 1972 book, he knew he was terminally ill with cancer. Robert MacArthur passed away shortly after finishing his second book. He was only in his early 40s. We will never know what new ideas he might have developed and the role he would have played in the development of biogeography had he lived longer (Fig. 14.13).

So what is the equilibrium theory of island biogeography? The theory is predicated on the assumption that the species richness of most islands is in a state of equilibrium with processes of immigration and extinction. Immigration through dispersal and colonization events brings new species to islands. Extinction removes species. New species continuously arrive at the island by dispersal, but the addition of these species is counterbalanced by the continuous process of extinction that removes some of the new arrivals and some of the older species. Extinction of species is in turn counterbalanced by further new arrivals, some of which are successful in colonizing the island. Assuming

FIGURE 14.13 Robert H. MacArthur (1930–1972) who along with Edward O. Wilson was the co-originator of the equilibrium theory of island biogeography. Although he died tragically young, the ecological and biogeographic concepts he proposed during his short career continue to stimulate much research by biogeographers and ecologists (Courtesy of Princeton University Libraries).

that no environmental change occurs, the species richness of the island will vary only slightly around a stable mean. However, since new species are arriving and older species are going extinct in a continual process, the species composition of the island will be continuously changing. The changing species composition is called **species turnover.**

The best way to consider the theory of island biogeography is to view it graphically. In a graphical format we can present the number of species along the horizontal axis and the rate of extinction and colonization along the vertical axis (Fig. 14.14). The number of species on an island can vary from 0 to P, where P equals the number of species on the mainland from which the island is colonized. The graph presents two curves—one for the rate of immigration by colonizing species and the other for the rate of extinction of species on the island. The rate of immigration declines as the number of species on the island increases. This is because if there are greater numbers of species on the island, there will be fewer resources and less habitat available for colonization. In other words, the greater the number of species, the less unoccupied niche space there is. When many species are already present, new colonizing species face stiff competition and cannot successfully immigrate. The rate of extinction increases with the number of species that are on the island because the more species there are, the greater the chance of random extinction events occurring to some species. If more species are competing for a limited amount of resources, the population sizes of the species will be small due to resource limitations and species with small population sizes will be more prone to extinction. The equilibrium number of species that the island can support is the point on our graph where the rates of immigration and extinction intersect. This is designated as S on the graph. At this point, the successful establishment of new immigrants is exactly balanced by the extinction of other species.

The theory of island biogeography states that the equilibrium number of species on an island is determined by rates of immigration and extinction. What happens then if we perturb these rates? Suppose that a storm from the mainland blows dozens of new bird species to an island. Following this event, the number of species on the island increases. Because there are now greater numbers of species on the island, the rate of extinction will increase and the rate of successful immigration will decrease. The species numbers will then decline because there will be more extinctions than immigrations. Over time the number of species will move back down to the original equilibrium size S. Similarly,

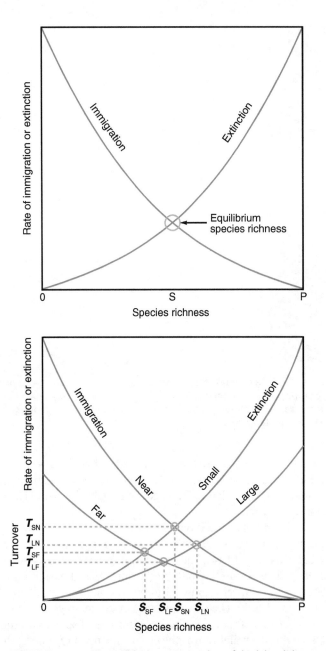

FIGURE 14.14   A graphical representation of the island theory of biogeography (Blackburn et al., 2016 / Harper & Row). **S** equals the equilibrium species number on islands of different sizes and degrees of isolation (**LN** = Large Near, **LF** = Large Far, **SN** = Small Near, **SF** = Small Far). **S′** and **S″** are nonequilibrium species numbers following an extreme extinction event (**S′**) or colonization event (**S″**). **T** is the turnover rate on different islands depending on island size and isolation. **P** is the total number of species available from the mainland and is equal to the species richness of the mainland.

imagine if some catastrophe causes massive and sudden extinctions on the island. Then the number of species will decrease. Now that the number of species is low, competition pressure is relaxed, the rate of immigration will increase, and the rate of extinction will decrease. Over time, the number of species will increase back to the equilibrium point *S*.

According to MacArthur and Wilson, the size of the island strongly affects rates of extinction. Islands that are small have fewer resources and less habitat diversity. Therefore, small islands cannot support either large populations or large numbers of species. The species that are on these islands must live in close proximity, share limited habitat, and experience stiff intraspecific and interspecific competition for resources. The intraspecific competition keeps population sizes low. The small size of the islands and the small size of the plant and animal populations mean the species are prone to random extinction events due to events such as fires, hurricanes, or flooding. The small resource base and small populations also mean that extinction due to competitive exclusion is more likely on small islands.

How does the theory help explain why small isolated islands have fewer species than large islands close to the mainland? Again, the best way to consider this question is to look at the graph (Fig. 14.14). The proximity of islands to the mainland largely influences rates of immigration. Islands that are near the mainland are more likely to receive colonists by jump dispersal than are islands that are far away from the mainland. The model predicts that the equilibrium number of species is controlled by island size and island isolation as follows. Islands that are small *(S)* and far *(F)* from the mainland will experience the highest rates of extinction and lowest rates of immigration and thus will have the lowest number of species (S_{SF}). Islands that are large *(L)* and near *(N)* the mainland will experience the lowest rates of extinction and the highest rates of immigration and will have the highest number of species (S_{LN}). Islands that are large and far or small and near the mainland will have intermediate numbers of species $(S_{LF}$ and $S_{SN})$. Thus, the relationship between species numbers and island geography can be summarized as follows:

$$S_{LN} > S_{LF} \sim S_{SN} > S_{SF}$$

According to island biogeography theory, equilibrium species numbers are determined by the impact of island size and isolation on rates of extinction and immigration. If a small island near the mainland has higher rates of extinction and immigration than a large island that is far away from the mainland, then the smaller island must experience higher rates of species turnover *(T)*. Because species are arriving and going extinct at a relatively high rate on the small island close to the mainland, the species composition must be changing relatively rapidly (T_{SN}). The inverse is true for a large distant island. It will have a low rate of turnover (T_{LF}) because extinctions will be rarer and immigration events fewer. Thus, according to the theory of island biogeography, turnover rates are related to island size and island isolation in the following manner:

$$T_{SN} > T_{LN} \sim T_{SF} > T_{LF}$$

The theory of island biogeography is especially powerful because it not only offers a simple explanation for why species diversity varies according to island size and isolation, but it also provides a set of hypotheses on species richness and turnover that can be used to test the theory. The first hypothesis is that species richness and island size and isolation are related. A second test of the island biogeography theory is presented by the prediction that species turnover is a continuous process on islands and is related to island size and isolation.

As we have noted, many studies of islands and island-like habitat have shown that small isolated islands have lower species richness than large islands located close to the mainland. Indeed, it was this observation that first led MacArthur and Wilson to develop the theory.

A more crucial test of the island biogeography theory is presented by the prediction that species turnover is a continuous process on islands and is related to island size and isolation in the following manner:

$$T_{SN} > T_{LN} \sim T_{SF} > T_{LF}$$

MacArthur and Wilson's ideas on continuous species turnover on islands were much influenced by evidence from the islands of Krakatoa. The islands are located in the Malaysian Archipelago and were created when a huge volcanic eruption destroyed the original island of Krakatoa in 1883. These islands—Sertung, Anak Krakatoa, Panjang, and Rakata—were almost devoid of any life immediately after the eruption. The islands were quickly colonized by plants and animals from the large islands of Sumatra and Java, which lie about 40 kilometers away from the Krakatoa group. Smaller islands that lie between Sumatra and the Krakatoa group also provided colonists or acted as stepping stones for colonizing species. By 1935, tropical rainforest had been established on the Krakatoa Islands. Censuses of bird populations carried out in 1908, 1991–1921, and 1932–1934 showed that bird species richness increased quickly until 1921 and then stabilized somewhat (Table 14.2). Although bird species richness did not increase significantly after 1921, there were turnovers in the bird species present on the islands. After 1921, some new colonists became established, and other species went extinct. The data

TABLE 14.2   Changes in Freshwater and Terrestrial Bird Species on Two Islands of the Krakatoa Group (Rakata and Sertung) Following the 1883 Eruption

Year	No. of Species Found	No. of New Species	No. of Extinctions
1908	14	14	[Unknown]
1919–1921	60	48	2
1932–1934	64	11	7

Source: MacArthur and Wilson, 1967 / PRINCETON UNIVERSITY PRESS.

from Krakatoa regarding seemingly continuous species turnover on the islands were central to MacArthur and Wilson's theory of island biogeography.

Some of the most interesting field studies in biogeography have been conducted in attempts to test the hypothetical relationship between species turnover and island size and isolation presented in the theory of island biogeography. These tests can be broadly divided into two types: (1) the examination of plant and animal census data from unmanipulated environments and (2) the direct experimental manipulation of island ecosystems to observe if the predictions made by island biogeography theory hold true.

Some of the earliest attempts to test the island biogeography predictions regarding turnover rates were conducted by Jared Diamond and his colleagues. Their analyses were based on bird census data from the Channel Islands of California, small islands near New Guinea, and the Farnes Islands of Britain. In the case of the Channel Islands, land bird census data from 1917 were compared with census data from 1968. The census data suggest that a number of species disappeared from the islands during the twentieth century, while other new species colonized. Consistent with island biogeography theory, the total number of bird species on the different islands did not change greatly between 1919 and 1968. However, the exact species of birds that were present on each island had changed significantly. On some islands, 62% of the land bird species found in the 1960s were different from the species found in 1917. In general, large islands had lower rates of turnover than smaller islands. For example, the largest island, Santa Cruz, had a turnover rate of 17%, while the tiny island of Santa Barbara had a 62% turnover rate. Interestingly, turnover rates did not appear to be strongly influenced by distance from the North American continent. However, all of the islands are relatively close to the mainland, and for birds the different distances involved may not be great enough to be a significant barrier to dispersal. More detailed analysis of year-by-year species counts on the Channel Islands and the Farnes Islands also provided evidence of continuous turnover of species. The annual species turnover rates ranged from 3 species to over 20 species per year. Similar evidence of continuous species turnover was reported from other island studies, including an examination of birds on the island of Mona near Puerto Rico.

A classic experimental test of island biogeography theory was conducted in the 1960s by Edward Wilson and his student Daniel Simberloff. They conducted the experiment by manipulating the environment of small islands in the Florida Keys. The islands are located in shallow waters and are created by the growth of salt-tolerant mangrove trees *(Rhizophora mangle)*. The four islands they studied are all very small in size, ranging from 75 to 250 square meters in area and are located about 0.002 to 1.2 kilometers from the mainland. Since all the islands are small, size is not an important variable in the experiment, but distance from the mainland is. The islands support terrestrial arthropod faunas that are comprised of 10 to 40 species of insects and arachnids. The mainland species pool is over 1000 species of arthropods. Simberloff and Wilson used the insecticide methyl bromide to exterminate all the arthropods in the islands. They then conducted careful repeated censuses of the islands' arthropod fauna for over two years. The census allowed them to document the return of the arthropods to the islands and to determine whether arthropod species diversity was consistent with the patterns predicted by island biogeography theory. Consistent with island biogeography theory, they found that species turnover was a relatively constant process on all of the islands. They also found that the lowest rates of immigration and turnover occurred on the most distant island. The highest species richness and the highest rates of species turnover and immigration were found on the island closest to the mainland.

After less than a year, the species richness of the islands had returned to its pre-experiment levels. This result suggests that island biogeographic processes of immigration and extinction had returned the islands to their equilibrium state of species richness. However, only about 19% to 40% of the arthropod species on the islands before the experimental defaunation were present in the new fauna of the islands. Why would the pre-experiment and post-experiment species composition be different, and is this finding consistent with island biogeography theory? Since all of the islands are small, they can only support small populations of arthropods. Such small populations are prone to random extinction events. Thus, there is a large degree of chance determining exactly which species will be present on one of the small islands at a given time. So the difference between the pre-experiment and post-experiment species composition is not inconsistent with island biogeography theory.

The Theory of Island Biogeography Today

In the decades since its introduction, the theory of island biogeography has spawned a huge amount of research and been applied in many areas. The theory has achieved particular attention in regard to its potential application for the conservation of endangered species. This intensive research on island biogeography has raised many questions about the assumptions of the theory and about attempts to test it. At present, the theory is neither universally accepted nor universally rejected. Parts of it have been critically questioned and modifications proposed. Although experimental work on the scale of the manipulations of mangrove island biota described above remains exceptional, much progress has been made in the collection of important empirical data, including genetics data, and in modeling and theory. The research prompted by MacArthur and Wilson's theory has certainly led to a much more extensive and deeper understanding of the factors that affect species richness on islands and elsewhere. Let's look at some areas where its initial positive tests of the theory have been criticized and then examine some of the additional insights that study of island biogeography theory has helped to generate. As we shall see, some of these findings suggest that the original model can be improved.

Classic Tests Revisited At the heart of island biogeography is the premise that differences in immigration and extinction rates lead to a continuous process of species turnover that maintains species diversity at an equilibrium size. The early studies of bird species turnover of islands off California and England conducted by Diamond and his colleagues appeared to support this contention. However, other scientists have pointed out that environmental changes caused by humans may have been largely responsible for the differences between the bird species found in early census studies and the species found more recently on the Californian and English islands. For example, some of the birds that disappeared from the Californian islands between the different census periods were predators such as the bald eagle *(Haliaeetus leucocephlus),* osprey *(Pandion haliaetus),* and peregrine falcon *(Falco peregrinus).* These species were likely the victims of pesticide poisoning. Some of the new immigrants were species such as starlings *(Sturnus vulgaris)* and house sparrows *(Passer domesicus)* that were recently introduced to North America by humans. The most recent analysis of the bird fauna of the Channel Islands has shown that most of the species turnover has been among species of birds from the continental North America, which are nonbreeding visitors or species that maintain low, unstable breeding populations on the islands. Endemic island species and subspecies, such as the island scrub jay *(Aphelocoma insularis),* have had relatively stable populations and have shown no evidence of turnover. Some changes in bird species composition may have been caused by changing land cover on islands. For example, the bird fauna on the island of Mona near Puerto Rico was likely impacted by vegetation changes caused by goat herds that were introduced there. Finally, studies of several island systems, such as research on the avifauna of islands off Mexico in the Sea of Cortez and in the Pacific Ocean, have been unable to detect clear evidence of species turnover as predicted by the MacArthur and Wilson model.

The experimental defaunation of the Florida mangroves by Simberloff and Wilson yielded results consistent with island biogeography theory. However, the turnover of species on these small islands was so rapid that reliable curves for immigration and extinction could simply not be calculated and compared with the hypothetical immigration and extinction curves of MacArthur and Wilson. In addition, subsequent work has shown that many of the arthropods collected on the islands represent highly transitory species. Individuals of these species can move rapidly between sites even on the mainland. The appearance and disappearance of these species may not represent true immigration and extinction events. Later work by Simberloff led him to conclude that the rates of extinction suggested by his original study were overestimates of the true rates.

Small Island Effect   Island biogeography theory assumes that species richness will decrease with decreasing island size in a continuous manner. The smallest island should have a smaller number of species than the next largest island and so on. However, MacArthur and Wilson themselves suggested that this relationship might not hold on very small islands. Indeed, it has been shown in several cases that if only the smallest islands of island groups are studied, there is often no relationship between island size and species richness. The reason for this may be that these islands are so small that none of them can support large enough population sizes to produce significantly decreased extinction rates. It may also be the case that at these small sizes differences in habitat diversity related to island area are nil.

Rescue Effect   The MacArthur-Wilson model assumes that the distance of an island from the mainland affects species diversity by controlling immigration rates. Immigration rates impact on extinction rates by influencing competitive exclusion. According to the model, islands that are near the mainland should have higher rates of immigration, prompting higher rates of extinction and thus higher rates of species turnover. However, more recent studies have shown that the proximity of an island to the mainland also may serve to decrease extinction rates. In some studies, it has been shown that, contrary to theory, islands that are near the mainland have significantly lower turnover rates than distant islands. Examples of such studies include the species richness of arthropods on isolated patches of thistles, the distributions of small mammals on isolated piles of rocky habitat, and land birds on small islands in Lake Gatun, Panama. The Lake Gatun study shows that islands near the shore of the lake have very low turnover rates, islands an intermediate distance from the shore have the highest turnover rates, and islands located far from shore again have low turnover rates. The decreased turnover rates observed in the near islands have been attributed to a phenomenon called the **rescue effect.** The rescue effect occurs when small populations of a species are rescued from the brink of extinction by the arrival of new immigrants of the same species. Isolated islands are less likely to receive frequent immigrants, so small populations on these islands are more likely to go extinct. The rescue effect has been shown to be important in maintaining some populations of continental bird species on the California Channel Islands. MacArthur and Wilson did not anticipate the rescue effect when they produced their model.

Target Area Effect   The theory of island biogeography assumes that immigration rates are controlled solely by the distance of the island from the mainland. The more isolated an island is, the less likely it is that organisms will be able to disperse to it from the mainland. However, it has been suggested that island size will also influence rates of immigration. Larger islands provide a bigger target for dispersing organisms. The positive relationship between island size and immigration rates is called the **target area effect.** For actively dispersing animals that travel between the mainland and islands by flying, swimming, or otherwise crossing barriers, larger islands are more likely to be seen and colonized than smaller islands. For example, an analysis of the travel of land mammals across snow-covered ice in the Saint Lawrence River showed that these animals were more likely to head toward the larger islands than smaller islands as they traversed the ice. Similar results have been seen in the dispersal of shrews to small islands in Finland. The target area effect also influences immigration by passive dispersers such as plants. Islands with large areas are simply more likely to intercept waterborne and airborne seeds because they offer more surface area as a target. Studies along the northeastern coast of Australia show that the species richness of the seaborne seeds is positively correlated with the beach area of islands. Thus, the relatively high species richness observed on large islands may reflect higher rates of immigration in addition to lower extinction rates.

Biological Factors and the Nonequilibrium Effect   In order to be as general as possible, the theory of island biogeography treats all species as being essentially equal in terms of immigration and extinction. However, the many important differences between different groups of organisms and even within clades and guilds will substantially impact on immigration and extinction rates. For example, the decrease in species richness observed as one moves away from the mainland is much more pronounced for terrestrial mammals, amphibians, and reptiles than it is for birds and bats. This is of course because flying animals are more likely to cross water barriers and arrive at distant islands. However, by floating on logs and other flood debris, many species of terrestrial animals have been carried to various relatively distant islands. Thus, although it will take long periods of time, these animals do have the potential to reach distant islands. Therefore, it is possible that many distant islands may have achieved equilibrium species richness for flying and wind-dispersed organisms, but may not yet have reached equilibrium species richness for less efficiently dispersed plants and animals. This is one of many biological factors that can cause a nonequilibrium effect in which the species richness is not in equilibrium with the processes of immigration and extinction as envisioned by island biogeography theory.

The theory of island biogeography does not take into account the impact that plants and animals might have in facilitating immigration by other species. For example, many insects can feed on only one or two species of plants. Thus, the insect species diversity of an island will be highly dependent on the plant species diversity. Birds can be relatively specific about the vegetation they require for nesting or the fruit and insects they eat. Thus, bird fauna richness can also be dependent on certain plant species being present in order for successful immigration and establishment to occur. The inability of some species to successfully immigrate to an island may make it impossible for other species to immigrate and thus promote nonequilibrium conditions.

The theory of island biogeography predicts that periods of extinction should be followed by a rapid increase in immigration because of the relaxation of competitive exclusion. However, it is also possible that the extinction of some species may make it impossible for other immigrants to become established. For example, if a hurricane were to cause the extinction of most large fruit trees on a small tropical island, the successful immigration of insects, birds, or bats that depend on those species of fruit trees for food would be precluded until a new population of fruit trees became established. If such disturbances were frequent enough, there might never be enough time between them for the island biota to ever reach equilibrium.

The English biogeographer Robert J. Whittaker (not to be confused with the American ecologist Robert H. Whittaker) and his colleagues have conducted considerable analysis of the biota of the Krakatoa island group to study how dispersal and biological interactions have affected species richness there. Their data extend forward in time from the 1931 records from Krakatoa that MacArthur and Wilson considered when they formulated the theory of island biogeography. The more recent work on Krakatoa shows that the species richness of plants that disperse their seeds in sea water leveled off before 1920 and has remained relatively stable. The species richness of wind-dispersed plants continues to increase, but the rate of increase slowed considerably after the 1920s. In contrast, the species richness of animal-dispersed plants has continued to increase at a rapid rate. Bird species richness on the Krakatoa islands is just slightly below that found on other nearby islands that did not suffer from the catastrophic volcanic eruption. However, the species richness of nonflying mammals on the Krakatoa islands remains far below that of nearby similar islands. Thus, after more than 100 years, the Krakatoa island group may be at or near equilibrium species composition for some well-dispersing groups of plants and animals, but far from equilibrium species richness for poorly dispersing groups such as nonflying mammals. Some animals may be absent from the island because of poor dispersal capabilities, whereas others may be absent because their host plant species have not yet become established on the islands. How long will it take for islands of Krakatoa to reach equilibrium species richness for all groups of plants and animals? Whittaker and his colleagues predict that this will take thousands of years.

Other studies support the contention that the development of true equilibrium in species richness on islands can take very long periods time and perhaps never be achieved. A survey and comparison of the bat and nonflying mammal fauna of many ancient islands located in the Pacific far from mainland areas suggest that even after tens of thousands of years or more the species richness of nonflying mammals remains far below what would be predicted as equilibrium by the theory of island biogeography. The dispersal of nonflying animals is such a slow process that islands may never have a true equilibrium fauna. Given the slow rate at which islands reach equilibrium and the relatively high frequencies of disturbances such as fires or hurricanes that can cause extinctions, many islands probably never reach a state of steady turnover rates and equilibrium species numbers as predicated by the theory of island biogeography.

A final biological factor we might consider in the case of nonequilibrium island biotas is the impact of humans. As we have seen in Chapters 8 and 12, the arrival of humans on remote islands has generated pronounced peaks in extinctions of native fauna and introductions of new species. In some cases, the original species richness in terms of native plants and animals may never be truly known. All modern island data sets pertaining to species richness are to a greater or less extent influenced by the impact of humans.

Historical Factors and the Nonequilibrium Effect   A nonequilibrium state between species richness, island size, and island isolation can be caused by a number of historical factors. The recent geologic/geomorphic history of the island can be extremely important in this regard. Many modern islands that are located close to continents were actually joined to the mainland during the last glacial maximum about 20,000 years ago. This is because sea level was about 120 meters lower at that time. Such islands are called landbridge islands. Islands that have never been connected to the mainland are called oceanic islands. Tasmania was connected to Australia during the last glacial maximum and is a landbridge island. The Hawaiian Islands have never been a part of the mainland and are oceanic islands. When landbridge islands were connected to the mainland, there was no dispersal barrier to terrestrial organisms, and they likely possessed flora and fauna similar to that of the mainland. With the rise of sea level between 15,000 and 6000 years ago, the landbridge islands became separated from the mainland. When they became islands, their populations of plants and animals became cut off from the larger mainland populations. Since their small populations were now restricted to a limited area of land, extinction rates would have increased and caused a decline in species richness. The decline in species richness that occurs when a region, such as a landbridge island, is cut off from a larger landmass or area of similar habitat is called **species relaxation.**

How quickly does species relaxation bring island biodiversity into equilibrium with island size? The answer varies from island to island. For example, there is good evidence that some of the species richness in boreal mammals seen on isolated mountains in the desert areas of the Great Basin and the southwestern United States is a reflection of species richness that developed during the last glacial maximum. It is also thought that the bird species richness on islands off New Guinea may have developed when the islands were connected to New Guinea during the last glacial period. In the cases of both the Great Basin and New Guinea, species richness may still be declining very slowly due to the process of species relaxation that began when these mountain environments and islands became isolated at the end of the Pleistocene. In the 10,000 years since the close of the Pleistocene, these regions may not yet have reached equilibrium species richness. In contrast to these potentially slow rates of species relaxation, we can also find evidence of extremely fast declines in biodiversity on newly created islands. For example, a number of very small islands were created when the Chagras River was dammed, and Lake Gatun was created during the building of the Panama Canal. The canal was completed in 1914, and the largest of the newly created islands, Barro Colorado, had lost 27% of the species of land birds that existed in the area of the island prior to the creation of Lake Gatun. The island may now be close to equilibrium species richness for birds. It is likely that rates of species relaxation vary for different groups of organisms and on different groups of islands. In general, rate of species relaxation might be expected to be fastest on small islands because they support small, and thus extinction-prone, populations.

When applied to long-time scales, the theory of island biogeography weakens because it excludes the impact of evolution and speciation in generating biodiversity on islands. We know from much evidence such as the diversity of endemic honeycreepers on Hawaii or the many endemic species of Darwin's finches on the Galapagos Islands that island speciation can be an important process in generating island biodiversity. It might be argued that the theory of island biogeography was designed to address time scales of years to millennia, while evolutionary processes occur over periods of millions of years. However, not all speciation events are so slow. Some differences in the modern biodiversity of islands and island-like habitats may result from differences in rates of evolution and not from equilibrium processes of island biogeography. The hundreds of endemic species of cichlids found in Lake Victoria in Africa is far greater than the total fish diversity of comparable size lakes in the temperate zones. As Lake Victoria was completely dry during part of the last glacial maximum, much of this diversity arose from speciation events that occurred over the past 10,000 years or so.

We still have much to learn about contemporary rates of evolution on islands, particularly those in the tropics, and how island species richness might relate to speciation rather than processes of immigration and extinction. We might speculate, however, that speciation rates will be highest on relatively large and isolated islands such as those of Hawaii. Large islands offer a wider variety of habitats and available niche space that promote adaptive radiation. Larger islands can also support larger populations, which increases the probability of new traits arising through chance mutations and also increases the probability of species escaping extinction. Isolated islands are not as likely to receive immigrant species that will be capable of filling all available niches. This leaves niche space available for filling through adaptive radiation. The large number of endemic species found on the large remote oceanic islands of Hawaii supports this contention.

Detailed analysis of species turnover rates on islands such as Jamaica, the California Channel Islands, and the Revillagigedo Islands of Mexico shows that the endemic species that have evolved on these islands have very low rates of turnover or extinction unless they are disturbed by human land-use changes and the introduction of alien species. In contrast, species that have recently dispersed from the mainland areas to an island exhibit relatively high rates of extinction and turnover. We should recognize that island species dynamics reflect two very different groups of species. One group can be called **continental taxa** because they represent species that have their main populations and centers of origin on continents. The other group can be called **insular taxa** because they have their evolutionary origins on the island. The continental taxa experience high turnover rates because they are often unable to maintain viable population sizes on islands inasmuch as they are not well adapted to conditions on the island, unable to compete with the endemic island taxa, or unable to persist at the small population sizes dictated by the limited carrying capacity of the island. In contrast, unless the island environment is altered physically or biologically, the insular taxa are more resistant to extinction because they are well adapted to the island and to maintaining themselves, even if population sizes are low. It has been suggested that once an endemic island biota has evolved, it occupies the available niche space, and this leads to the resistance of successful colonization by plants and animals dispersing from the continent. The apparent high turnover rates and seeming inability of island biota to resist colonization by new species today are most likely a result of the large changes humans have caused to island environments.

Islands can also change in size and topographic diversity as geologic and geomorphic forces act upon them over time. Some volcanic islands may exist above sea level for only a few million years before they erode to below sea level; others may last tens of millions of years. In either case, the islands can undergo profound changes in size and topography as they age. Robert J. Whittaker and colleagues have tackled the complexities related to the equilibrium theory of island biogeography: long time periods, the geomorphological dynamics of island formation and history, and evolutionary speciation. From their effort, they developed a **general dynamic model (GDM) of island biogeography** (Fig. 14.15). The model integrates the impacts of changing island habitat diversity (and most likely size too) on carrying capacity, speciation and extinction rates, and species richness. They postulate that the species

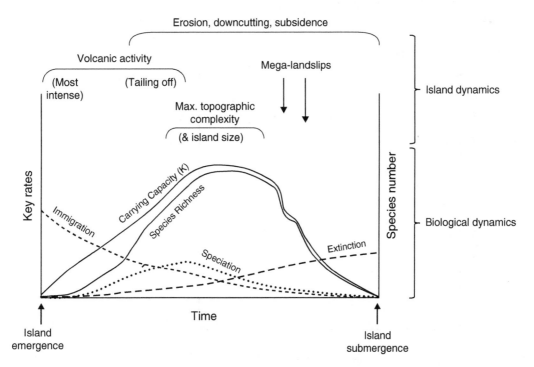

FIGURE 14.15 The relationship between the geologic/geomorphologic dynamics of volcanic oceanic island formation, history, and eventual submergence and the biological factors contributing to changes in species richness as the island evolves (*Source:* Whittaker et al., 2008 / John Wiley & Sons).

richness on the islands is a function of rates of immigration, speciation, and extinction. The model is based on the premise that oceanic islands are relatively short-lived on geologic time scales and have a humped-shaped trend in carrying capacity related to topographic complexity and changing island size. In their model, species richness is closely associated with carrying capacity. Carrying capacity is also linked to rates of speciation and extinction. Some predictions from the model have been tested using observational data for plants, insects, and other invertebrates from islands of different sizes, topographies, and geologic ages, including the Canaries, Galápagos, Marquesas, Azores, and Hawaiian Islands. The GDM has attracted much interest from biogeographers and will undoubtedly be a driver of further research into the detailed geologic histories and biota of oceanic islands.

It is difficult to conclude that the flora and fauna of remote islands are in a true state of equilibrium between size, isolation, and species richness of plants and animals. Leaving the equilibrium questions aside, it has also been demonstrated that because of factors such as the rescue effect and large target effect, we cannot accept the assumption that immigration rates are controlled solely by relative isolation or that extinction rates are controlled solely by island size. The GDM has focused attention on the important role that long-term geologic history of islands can play in the biodiversity of oceanic islands.

We might try to incorporate the rescue effect, the large target effect, and evolutionary speciation into a slightly revised version of island biogeography theory. The revisions are best considered in their graphical form (Fig. 14.16). In considering the rescue effect, island biogeography theory might take into account both island size and island isolation when predicting relative extinction rates. Islands located near the mainland experience lower rates of extinction by virtue of the rescue effect. Remember the original theory postulated that extinction rates were controlled solely by island size. In a revised version of the theory, it could be postulated that large near islands experience the lowest extinction rates and small isolated islands suffer the highest rates of extinction. Islands that are small and near the mainland, or large and faraway, experience intermediate rates of extinction. To convey the relationship between island isolation and extinction rates, we need to revise the familiar island biogeography graph to include three curves to represent extinction rates on small far islands, small near and large far islands, and large near islands.

The impact of island size on immigration rates due to the large target effect can also be incorporated with a slight revision to the theory. In this case, we can postulate that immigration rates will be highest on large near islands and lowest on small far islands. Large near islands are close to the source of immigrants and present a large target. Small far islands are far from the dispersal center of immigrants

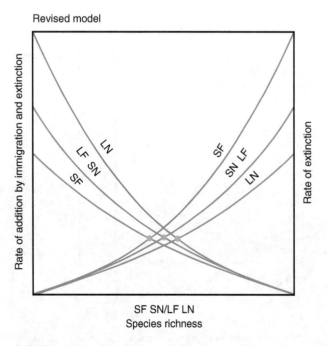

FIGURE 14.16 A revised view of the theory of island biogeography by the author that takes into account the rescue effect on near islands, the target effect of large islands, and the fact that the addition of species to island biota occurs both through immigration and evolution.

and present a small target. Islands that are large and far, or small and near, will have intermediate rates of immigration because either they are not close to dispersal sources but offer a large target area, or they are close to dispersal centers but provide only a small target for dispersing organisms. The inclusion of three immigration curves, for large near islands, large far and small near islands, and small far islands expands the island biogeography theory to incorporate the impact of target size on immigration rates.

What impact does evolution have on species richness? Evolution might be thought of as a factor that, combined with immigration, produces the rate of species additions to islands. On short time scales of years to millennia, species additions will mainly reflect immigration. Over longer time scales, the evolution of endemic island species will be an increasingly more important factor in determining island species richness. Taking both immigration and evolution together, it is plausible to speculate that the large near islands will have high rates of immigration because they are near a dispersal source and moderately high rates of speciation because they offer a range of habitat and niche space for adaptive radiation. In addition, large islands support large enough population sizes to generate a reasonable chance of introducing new genotypes through mutation. Small and isolated islands will have lower rates of immigration and low rates of speciation because they offer little niche space for adaptive radiation. Because small islands can support only small populations, the chance occurrence of new genotypes by mutation decreases, and the chance that endemic species will go extinct due to chance events increases. Therefore, we might predict that the rate of species additions due to immigration and evolution will be highest on large near islands, lowest on small far islands, and intermediate on large far and small near islands. Therefore, by replacing the term *immigration* with the term *species additions*, which recognizes that the rate of addition of species on islands reflects both immigration and evolution, we can incorporate the influence of evolution into our model along with the influences of the rescue effect and island target area (see Fig. 14.14).

The revised model of island biogeography predicts that equilibrium species richness should be highest on large near islands, intermediate on large far and small near islands, and smallest on small far islands ($S_{LN} > S_{LF} \sim S_{SN} > S_{SF}$). This is very consistent with the data on island species richness we have reviewed in this chapter. This revised model predicts no systematic differences in the turnover rates for the different island types. Is this reasonable at this time? Most field studies that originally presented evidence of species turnover rates consistent with the original theory of island biogeography have been shown to be flawed and biased by the impact of recent human activity. It can be argued that the differences in turnover rates predicted by the original theory of island biogeography might simply be an artifact that was produced because MacArthur and Wilson's original model does not recognize the impacts of rescue effect, target size, and evolution. Since the revised model incorporates evolutionary change, it implies that rates of species additions and extinctions are best measured over periods of thousands to millions of years rather than over years and decades. This is more consistent with the concepts embodied in the GDM. Once an endemic biota is established, true turnover rates on undisturbed island systems might be very slow and be detectable only over very long periods of time. Similarly, movement toward any sort of true equilibrium in species richness by colonization and adaptive radiation will be slow and perhaps will never actually be achieved in the life of an individual island.

The model presented here is only one of a number of approaches for refining the theory of island biogeography. Many attempts have been made, and many more will be made in the future. Much more work will be done on collecting field data on the plants and animals that live on islands to test these ideas—and to develop new insights. Perhaps this continued research is the best compliment we can pay to the memory of Robert MacArthur.

KEY WORDS AND TERMS

Biodiversity
Continental taxa
Cryptic species
Equilibrium theories of biodiversity
Equilibrium theory of island biogeography
General dynamic model (GDM) of island biogeography
Historical theories of biodiversity

Insular taxa
Intermediate disturbance hypothesis
Latitudinal diversity gradient (LDG)
Metabolic theory of biodiversity (MTB)
Nested pattern of insular communities
Neutral theory of biodiversity and biogeography
Peninsula effect
Rescue effect

Species area relationship (SAR)
Species relaxation
Species richness
Species turnover
Stability-time hypothesis
Target area effect
Tropical niche conservatism hypothesis (TCH)

REFERENCES AND FURTHER READING

Adams, R. I., and Hadly, E. A. (2013). Genetic diversity within vertebrate species is greater at lower latitudes. *Evolutionary Ecology* **27**, 133–143.

Adler, P. B., Seabloom, E. W., Borer, E. T., Hillebrand, H., Hautier, Y., Hector, A., Harpole, W. S., O'Halloran, L. R., Grace, J. B., Anderson, T. M., and Bakker, J. D. (2011). Productivity is a poor predictor of plant species richness. *Science* **333**, 1750–1753.

Allen, A. P., and Gillooly, J. F. (2006). Assessing latitudinal gradients in speciation rates and biodiversity at the global scale. *Ecology Letters* **9**, 947–954.

Allen, A. P., and White, E. P. (2003). Effects of range size on species-area relationships. *Evolutionary Ecology Research* **5**, 493–499.

Alongi, D. M. (2020). Carbon balance in salt marsh and mangrove ecosystems: a global synthesis. *Journal of Marine Science and Engineering* **8**(10), 767.

Arrhenius, O. (1921). Species and area. *Journal of Ecology* **9**, 95–99.

Begon, M., Harper, J. L., and Townsend, C. R. (1996). *Ecology: Individuals, Populations and Communities*, 3rd edition. Blackwell Science, Oxford.

Blackburn, T. M., Delean, S., Pyšek, P., and Cassey, P. (2016). On the island biogeography of aliens: a global analysis of the richness of plant and bird species on oceanic islands. *Global Ecology and Biogeography* **25**, 859–868.

Blackburn, T. M., and Gaston, K. J. (1997). The relationship between geographic area and the latitudinal gradient in species richness in New World birds. *Evolutionary Ecology* **11**, 195–204.

Boltovskoy, D., and Correa, N. (2017). Planktonic equatorial diversity troughs: fact or artifact? Latitudinal diversity gradients in Radiolaria. *Ecology* **98**, 112–124.

Borregaard, M. K., Amorim, I. R., Borges, P. A., Cabral, J. S., Fernández-Palacios, J. M., Field, R., Heaney, L. R., Kreft, H., Matthews, T. J., Olesen, J. M., and Price, J. (2017). Oceanic island biogeography through the lens of the general dynamic model: assessment and prospect. *Biological Reviews* **92**, 830–853.

Borregaard, M. K., Matthews, T. J., and Whittaker, R. J. (2016) The general dynamic model: towards a unified theory of island biogeography? *Global Ecology and Biogeography* **25**, 805–816.

Bouchet, P., Decock, W., Lonneville, B., Vanhoorne, B., and Vandepitte, L. (2023). Marine biodiversity discovery: the metrics of new species descriptions. *Frontiers in Marine Science* **10**, 929989.

Brown, J. H. (1995). *Macroecology*. University of Chicago Press, Chicago.

Brown, J. H. (2014). Why are there so many species in the tropics? *Journal of Biogeography* **41**, 8–22.

Brown, J. H., and Kodric-Brown, A. (1993). Highly structured fish communities in Australian deep springs. *Ecology* **74**, 1847–1855.

Brown, J. H., and Lomolino, M. V. (1998). *Biogeography*, 2nd edition. Sinauer Associates, Sunderland, MA.

Burgin, C. J., Colella, J. P., Kahn, P. L., and Upham, N. S. (2018). How many species of mammals are there? *Journal of Mammalogy* **99**, 1–14.

Bush, G. L. (1993). A reaffirmation of Santa Rosalina, or why there are so many kinds of small animals. In *Evolutionary Patterns and Processes* (ed. D. R. Lees and D. Edwards), pp. 229–249. Academic Press, London.

Bush, M. B. (2017). The resilience of Amazonian forests. *Nature* **541**, 167–168.

Bush, M. B., and Whittaker, R. J. (1991). Krakatau: colonization patterns and hierarchies. *Journal of Biogeography* **18**, 341–356.

Cai, L., Kreft, H., Taylor, A., Denelle, P., Schrader, J., Essl, F., van Kleunen, M., Pergl, J., Pyšek, P., Stein, A., and Winter, M. (2023). Global models and predictions of plant diversity based on advanced machine learning techniques. *New Phytologist* **237**, 1432–1445.

Caley, M. J., and Schluter, D. (1997). The relationship between local and regional diversity. *Ecology* **78**, 70–80.

Carlquist, S. (1965). *Island Life*. Natural History Press, Garden City, NY.

Carlquist, S. (1974). *Island Biology*. Columbia University Press, New York.

Case, T. J., and Cody, M. L. (1983). *Island Biogeography in the Sea of Cortez*. University of California Press, Berkeley.

Cerezer, F. O., de Azevedo, R. A., Nascimento, M. A. S., Franklin, E., de Morais, J. W., and de Sales Dambros, C. (2020). Latitudinal gradient of termite diversity indicates higher diversification and narrower thermal niches in the tropics. *Global Ecology and Biogeography* **29**, 1967–1977.

Chaudhary, C., Saeedi, H., and Costello, M. J. (2016). Bimodality of latitudinal gradients in marine species richness. *Trends in Ecology & Evolution* **31**, 670–676.

Chesson, P., and Kuang, J. J. (2008). The interaction between predation and competition. *Nature* **456**, 235–238.

Chesson, P. L., and Case, T. J. (1986). Overview: nonequilibrium community theories: chance, variability, history, and coexistence. In *Community Ecology* (ed. J. Diamond and T. J. Case), pp. 229–239. Harper and Row, New York.

Christenhusz, M. J., and Byng, J. W. (2016). The number of known plant species in the world and its annual increase. *Phytotaxa* **261**, 201–217.

Clark, J. S. (2012). The coherence problem with the unified neutral theory of biodiversity. *Trends in Ecology & Evolution* **27**, 198–202.

Colinvaux, P. A. Ice-age Amazonia revisited. *Nature* **340**, 188–189.

Colinvaux, P. A., DeOliveira, P. E., Moreno, J. E., Miller, M. C., and Bush, M. B. (1996). A long pollen record from lowland Amazonia; forest end cooling in glacial times. *Science* **274**, 86–88.

Condamine, F. L., Sperling, F. A., Wahlberg, N., Rasplus, J. Y., and Kergoat, G. J. (2012). What causes latitudinal gradients in species diversity? Evolutionary processes and ecological constraints on swallowtail biodiversity. *Ecology Letters* **15**, 267–277.

Connell, J. H. (1978). Diversity in tropical rain forests and coral reefs. *Science* **199**, 1301–1310.

Connell, J. H. (1980). Diversity and the coevolution of competitors, or the ghost of competition past. *Oikos* **35**, 131–138.

Cook, R. E. (1969). Variation in species density of North American birds. *Systematic Zoology* **18**, 63–84.

Cooper, M. (2022). The inverse latitudinal gradient in species richness of forest millipedes: Pachybolidae Cook, 1897. *Journal of Entomological and Zoological Studies* **10**, 5–8.

Costello, M. J., May, R. M., and Stork, N. E. (2013). Can we name Earth's species before they go extinct? *Science* **339**, 413–416.

Cox, C. B., and Moore, P. D. (1993). *Biogeography: An Ecological and Evolutionary Approach*, 5th edition. Blackwell Scientific Publications, Oxford.

Cronin, T. M., and Raymo, M. E. (1997). Orbital forcing of deep-sea benthic species diversity. *Nature* **385**, 624–627.

Crowe, T. M., Bowie, R. C., Bloomer, P., Mandiwana, T. G., Hedderson, T. A., Randi, E., Pereira, S. L., and Wakeling, J. (2006). Phylogenetics, biogeography and classification of, and character evolution in, gamebirds (Aves: Galliformes): effects of character exclusion, data partitioning and missing data. *Cladistics* **22**, 495–532.

Currie, D. J. (1991). Energy and large-scale patterns of animal- and plant-species richness. *American Naturalist* **137**, 27–49.

Currie, D. J., and Paquin, V. (1987). Large-scale biogeographical patterns of species richness of trees. *Nature* **329**, 326–327.

Curtis, J. R., Robinson, W. D., Rompré, G., Moore, R. P., and McCune, B. (2021). Erosion of tropical bird diversity over a century is influenced by abundance, diet and subtle climatic tolerances. *Scientific Reports* **11**, 10045.

deMenocal, P. B. (1995). Plio-Pleistocene African climate. *Science* **270**, 53–59.

DeSalle, R., and Goldstein, P. (2019). Review and interpretation of trends in DNA barcoding. *Frontiers in Ecology and Evolution* **7**, 302.

Diamond, J. M. (1971). Comparison of faunal equilibrium turnover rates on a tropical island and a temperate island. *Proceedings of the National Academy of Sciences* **68**, 2742–2745.

Diamond, J. M. (1975). Assembly of species communities. In *Ecology and Evolution of Communities* (ed. M. L. Cody and J. M. Diamond), pp. 342–444. Belknap Press of Harvard University Press, Cambridge, MA.

Diamond, J. M., and May, R. M. (1976). Island biogeography and the design of natural preserves. In *Theoretical Ecology: Principles and Applications* (ed. R. M. May), pp. 163–186. W. B. Saunders Co., Philadelphia.

Dinesen, L., Jensen, F. P., Sonne, J., Levinsky, I., and Mulungu, E. (2023). Status and habitat description of the globally threatened Udzungwa Forest Partridge *Xenoperdix udzungwensis* thirty years after discovery. *Bird Conservation International* **33**, e1.

Dobson, A. P. (1996). *Conservation and Biodiversity*. Scientific American Library, New York.

Dornelas, M., Connolly, S. R., and Hughes, T. P. (2006). Coral reef diversity refutes the neutral theory of biodiversity. *Nature* **440**, 80–82.

Duchêne, D. A., and Cardillo, M. (2015). Phylogenetic patterns in the geographic distributions of birds support the tropical conservatism hypothesis. *Global Ecology and Biogeography* **24**, 1261–1268.

Eberle, J. J., Fricke, H. C., Humphrey, J. D., Hackett, L., Newbrey, M. G., and Hutchison, J. H. (2010). Seasonal variability in Arctic temperatures during early Eocene time. *Earth and Planetary Science Letters* **296**, 481–486.

Economo, E. P., Narula, N., Friedman, N. R., Weiser, M. D., and Guénard, B. (2018). Macroecology and macroevolution of the latitudinal diversity gradient in ants. *Nature Communications* **9**, 1778.

Ellerton, D., Shulmeister, J., Woodward, C., and Moss, P. (2017). Last Glacial Maximum and Last Glacial-Interglacial Transition pollen record from northern NSW, Australia: evidence for a humid late Last Glacial Maximum and dry deglaciation in parts of eastern Australia. *Journal of Quaternary Science* **32**, 717–728.

Elmberg, J., Nummi, P., Poysa, H., and Sjoberg, K. (1994). Relationships between species number, lake size and resource diversity in assemblages of breeding waterfowl. *Journal of Biogeography* **21**, 75–84.

Erwin, T. L. (1997). Biodiversity at its utmost: tropical forest beetles. In *Biodiversity II* (ed. M. L. W. Reaka-Kudla, D. E. Wilson, and E. O. Wilson), pp. 27–40. Joseph Henry Press, Washington, DC.

Fenton, I. S., Pearson, P. N., Dunkley Jones, T., and Purvis, A. (2016). Environmental predictors of diversity in recent planktonic foraminifera as recorded in marine sediments. *PloS One* **11**, e0165522.

Field, R., Hawkins, B. A., Cornell, H. V., Currie, D. J., Diniz-Filho, J. A. F., Guégan, J. F., Kaufman, D. M., Kerr, J. T., Mittelbach, G. G., Oberdorff, T., and O'Brien, E. M. (2009). Spatial species-richness gradients across scales: a meta-analysis. *Journal of Biogeography* **36**, 132–147.

Fox, J. W. (2013). The intermediate disturbance hypothesis should be abandoned. *Trends in Ecology & Evolution* **28**, 86–92.

Francisquini, M. I., Lorente, F. L., Pessenda, L. C. R., Junior, A. A. B., Mayle, F. E., Cohen, M. C. L., França, M. C., Bendassolli, J. A., Giannini, P. C. F., Schiavo, J., and Macario, K. (2020). Cold and humid Atlantic rainforest during the last glacial maximum, northern Espírito Santo state, southeastern Brazil. *Quaternary Science Reviews* **244**, 106489.

Frank, K., Krell, F. T., Slade, E. M., Raine, E. H., Chiew, L. Y., Schmitt, T., Vairappan, C. S., Walter, P., and Blüthgen, N. (2018). Global dung webs: high trophic generalism of dung beetles along the latitudinal diversity gradient. *Ecology Letters* **21**, 1229–1236.

Freyhof, J., and Kottelat, M. (2007). *Handbook of European Freshwater Fishes*. Spring, Berlin.

Gainsbury, A., and Meiri, S. (2017). The latitudinal diversity gradient and interspecific competition: no global relationship between lizard dietary niche breadth and species richness. *Global Ecology and Biogeography* **26**, 563–572.

Gaston, K. J. (1996). Biodiversity—congruence. *Progress in Physical Geography* **20**, 105–112.

Gaston, K. J. (1996). Biodiversity—latitudinal gradients. *Progress in Physical Geography* **20**, 466–476.

Gaston, K. J. (1996). Species-range-size distributions: patterns, mechanisms and implications. *Trends in Ecology and Evolution* **11**, 197–201.

Gaston, K. J., and Spicer, J. I. (2004). *Biodiversity: An Introduction*, 2nd edition. Blackwell, Oxford.

Gillman, L. N., and Wright, S. D. (2014). Species richness and evolutionary speed: the influence of temperature, water and area. *Journal of Biogeography* **41**, 39–51.

Gillman, L. N., Wright, S. D., Cusens, J., McBride, P. D., Malhi, Y., and Whittaker, R. J. (2015). Latitude, productivity and species richness. *Global Ecology and Biogeography* **24**, 107–117.

Gleason, H. A. (1922). On the relationship between species and area. *Ecology* **3**, 158–162.

Goldschmidt, T. (1996). *Darwin's Dreampond*. MIT Press, Cambridge.

Gorman, M. L. (1979). *Island Ecology*. Chapman and Hall, London.

Grassle, J. F. (1991). Deep-sea benthic biodiversity. *Bioscience* **41**, 464–469.

Gray, K. W., and Rabeling, C. (2023). Global biogeography of ant social parasites: exploring patterns and mechanisms of an inverse latitudinal diversity gradient. *Journal of Biogeography* **50**, 316–329.

Guilhaumon, F., Gimenez, O., Gaston, K. J., and Mouillot, D. (2008). Taxonomic and regional uncertainty in species-area relationships and the identification of richness hotspots. *Proceedings of the National Academy of Sciences* **105**, 5458–5463.

Guo, Q., Kelt, D. A., Sun, Z., Liu, H., Hu, L., Ren, H., and Wen, J. (2013). Global variation in elevational diversity patterns. *Scientific Reports* **3**, 3007.

Häggi, C., Chiessi, C. M., Merkel, U., Mulitza, S., Prange, M., Schulz, M., and Schefuß, E. (2017). Response of the Amazon rainforest to late Pleistocene climate variability. *Earth and Planetary Science Letters* **479**, 50–59.

Harpole, W. (2010). Neutral theory of species diversity. *Nature Education Knowledge* **3**, 60.

He, F., Legendre, P., and LaFrankie, J. V. (1996). Spatial pattern of diversity in a tropical rain forest in Malaysia. *Journal of Biogeography* **23**, 57–74.

Heaney, L. R. (2007). Is a new paradigm emerging for oceanic island biogeography? *Journal of Biogeography* **34**, 753–757.

Hebert, P. D., Penton, E. H., Burns, J. M., Janzen, D. H., and Hallwachs, W. (2004). Ten species in one: DNA barcoding reveals cryptic species in the neotropical skipper butterfly *Astraptes fulgerator*. *Proceedings of the National Academy of Sciences* **101**, 14812–14817.

Heino, J. (2011). A macroecological perspective of diversity patterns in the freshwater realm. *Freshwater Biology* **56**, 1703–1722.

Helmus, M. R., Mahler, D. L., and Losos, J. B. (2014). Island biogeography of the Anthropocene. *Nature* **513**, 543–546.

Henriques-Silva, R., Kubisch, A., and Peres-Neto, P. R. (2019). Latitudinal-diversity gradients can be shaped by biotic processes: new insights from an eco-evolutionary model. *Ecography* **42**, 259–271.

Heywood, V. H. (1995). *Global Biodiversity Assessment*. Cambridge University Press, Cambridge.

Higgs, N. D., and Attrill, M. J. (2015). Biases in biodiversity: wide-ranging species are discovered first in the deep sea. *Frontiers in Marine Science* **2**, 61.

Hillebrand, H. (2004). On the generality of the latitudinal diversity gradient. *American Naturalist* **163**, 192–211.

Hortal, J., Triantis, K. A., Meiri, S., Thébault, E., and Sfenthourakis, S. (2009). Island species richness increases with habitat diversity. *American Naturalist* **174**, E205–E217.

Hubbell, S. (2001). *The Unified Neutral Theory of Biodiversity and Biogeography*. Princeton University Press, Princeton, NJ.

Hunter, M. L., and Yonzon, P. (1992). Altitudinal distributions of birds, mammals, people, forests and parks in Nepal. *Conservation Biology* **7**, 420–423.

Huntley, B. (1993). Species-richness in north-temperate zone forests. *Journal of Biogeography* **20**, 163–180.

Huston, M. A. (1994). *Biological Diversity: The Coexistence of Species on Changing Landscapes*. Cambridge University Press, Cambridge.

Hutchinson, G. E. (1959). Homage to Santa Rosalia, or why are there so many kinds of animals? *American Naturalist* **93**, 145–159.

IUCN (2021). *The IUCN Red List of Threatened Species*. https:// nc.iucnredlist.org/redlist/content/attachment_ files/2021-3_RL_Stats_Table_1a_v2.pdf.

Jäger, P., Arnedo, M. A., Azevedo, G. H. F., Baehr, B., Bonaldo, A. B., Haddad, C. R., Harms, D., Hormiga, G., Labarque, F. M., Muster, C., Ramirez, M., JavierIcon, S., and Adalberto J. (2021). Twenty years, eight legs, one concept: describing spider biodiversity in Zootaxa (Arachnida: Araneae). *Zootaxa* **4979**, 131–146.

Jansson, R., Rodríguez-Castañeda, G., and Harding, L. E. (2013). What can multiple phylogenies say about the latitudinal diversity gradient? A new look at the tropical conservatism, out of the tropics, and diversification rate hypotheses. *Evolution* **67**, 1741–1755.

Janzen, D. H. (1970). Herbivores and the number of tree species in tropical forests. *American Naturalist* **104**, 501–528.

Jeffries, M. J. (1997). *Biodiversity and Conservation*. Routledge, London.

Jenkins, C. N., Pimm, S. L., and Joppa, L. N. (2013). Global patterns of terrestrial vertebrate diversity and conservation. *Proceedings of the National Academy of Sciences* **110**, E2602–E2610.

Jones, H. L., and Diamond, J. M. (1976). Short-time-base studies of turnover in breeding bird populations on the California Channel Islands. *Condor* **78**, 526–549.

Jost, L. (2006). Entropy and diversity. *Oikos* **113**, 363–375.

Jullien, M., and Thiollay, J. (1996). Effects of rainforest disturbance and fragmentation: comparative changes of the raptor community along natural and human-made gradients in French Guiana. *Journal of Biogeography* **23**, 7–25.

Kato, M. (2000). *The Biology of Biodiversity*. Springer-Verlag, Hong Kong.

Kellman, M., and Tackaberry, R. (1993). Disturbance and tree species coexistence in tropical riparian forest fragments. *Global Ecology and Biogeography Letters* **3**, 1–9.

Kerkhoff, A. J., Moriarty, P. E., and Weiser, M. D. (2014). The latitudinal species richness gradient in New World woody angiosperms is consistent with the tropical conservatism hypothesis. *Proceedings of the National Academy of Sciences* **111**, 8125–8130.

Kershaw, A. P. (1978). Record of last interglacial glacial cycle from north-eastern Queensland. *Nature* **272**, 159–161.

Kershaw, A. P. (1981). Quaternary vegetation and environments. In *Ecological Biogeography of Australia* (ed. A. Keast), pp. 81–101. Junk, The Hague.

Kier, G., Kreft, H., Lee, T. M., Jetz, W., Ibisch, P. L., Nowicki, C., Mutke, J., and Barthlott, W. (2009). A global assessment of endemism and species richness across island and mainland regions. *Proceedings of the National Academy of Sciences* **106**, 9322–9327.

Kier, G., Mutke, J., Dinerstein, E., Ricketts, T. H., Küper, W., Kreft, H., and Barthlott, W. (2005). Global patterns of plant diversity and floristic knowledge. *Journal of Biogeography* **32**, 1107–1116.

Kinlock, N. L., Prowant, L., Herstoff, E. M., Foley, C. M., Akin-Fajiye, M., Bender, N., Umarani, M., Ryu, H. Y., Şen, B., and Gurevitch, J. (2018). Explaining global variation in the latitudinal diversity gradient: meta-analysis confirms known patterns and uncovers new ones. *Global Ecology and Biogeography* **27**, 125–141.

Krebs, C. J. (1994). *Ecology*, 4th edition. Harper-Collins College Publishers, New York.

Kreft, H., and Jetz, W. (2007). Global patterns and determinants of vascular plant diversity. *Proceedings of the National Academy of Sciences* **104**, 5925–5930.

Kress, W. J., García-Robledo, C., Uriarte, M., and Erickson, D. L. (2015). DNA barcodes for ecology, evolution, and conservation. *Trends in Ecology & Evolution* **30**, 25–35.

Lack, D. (1976). *Island Biology Illustrated by the Land Birds of Jamaica*. University of California Press, Berkeley.

Laiolo, P., Pato, J., and Obeso, J. R. (2018). Ecological and evolutionary drivers of the elevational gradient of diversity. *Ecology Letters* **21**, 1022–1032.

Lawlor, T. E. (1986). Comparative biogeography of mammals on islands. *Biological Journal of the Linnean Society* **28**, 99–125.

Levine, J. M., and HilleRisLambers, J. (2009). The importance of niches for the maintenance of species diversity. *Nature* **461**, 254–257.

Liang, J., Gamarra, J. G., Picard, N., Zhou, M., Pijanowski, B., Jacobs, D. F., Reich, P. B., Crowther, T. W., Nabuurs, G. J., De-Miguel, S., and Fang, J. (2022). Co-limitation towards lower latitudes shapes global forest diversity gradients. *Nature Ecology & Evolution* **6**, 1423–1437.

Lima Jr., W. J. S., Cohen, M. C. L., Rossetti, D. D. F., and França, M. C. (2018). Late Pleistocene glacial forest elements of Brazilian Amazonia. *Palaeogeography, Palaeoclimatology, Palaeoecology* **490**, 617–628.

Lin, H. Y., Corkrey, R., Kaschner, K., Garilao, C., and Costello, M. J. (2021). Latitudinal diversity gradients for five taxonomic levels of marine fish in depth zones. *Ecological Research* **36**, 266–280.

Liu, J., Slik, F., Zheng, S., and Lindenmayer, D. B. (2022). Undescribed species have higher extinction risk than known species. *Conservation Letters* **15**, e12876.

Lomolino, M. V. (1996). Investigating causality of nestedness of insular communities: selective immigrations or extinctions? *Journal of Biogeography* **23**, 699–703.

Lomolino, M. V., and Davis, R. (1997). Biogeographic scale and biodiversity of mountain forest mammals of western North America. *Global Ecology and Biogeography Letters* **6**, 57–76.

Lomolino, M. V., Riddle, B. R., and Whittaker, R. J. (2017). *Biogeography: Biological Diversity Across Space and Time*, 5th edition. Sinauer Associates, Sunderland, MA.

Ma, M. (2005). Species richness vs evenness: independent relationship and different responses to edaphic factors. *Oikos* **111**, 192–198.

MacArthur, R. H. (1972). *Geographical Ecology*. Harper and Row, New York.

MacArthur, R. H., and MacArthur, J. W. (1961). On bird species diversity. *Ecology* **42**, 594–598.

MacArthur, R. H., and Wilson, E. O. (1963). An equilibrium theory of insular zoogeography. *Evolution* **17**, 373–387.

MacArthur, R. H., and Wilson, E. O. (1967). *The Theory of Island Biogeography*. Princeton University Press, Princeton, NJ.

MacDonald, Z. G., Nielsen, S. E., and Acorn, J. H. (2017). Negative relationships between species richness and evenness render common diversity indices inadequate for assessing long-term trends in butterfly diversity. *Biodiversity and Conservation* **26**, 617–629.

Mannion, P. D., Upchurch, P., Benson, R. B., and Goswami, A. (2014). The latitudinal biodiversity gradient through deep time. *Trends in Ecology & Evolution* **29**, 42–50.

Marin, J., and Hedges, S. B. (2016). Time best explains global variation in species richness of amphibians, birds and mammals. *Journal of Biogeography* **43**, 1069–1079.

Maslin, M. A., Ettwein, V. J., Boot, C. S., Bendle, J., and Pancost, R. D. (2012). Amazon Fan biomarker evidence against the Pleistocene rainforest refuge hypothesis? *Journal of Quaterny Science* **27**, 451–460.

Maslin, M. A. B., and Stephen J. (2000). Reconstruction of the Amazon Basin: effective moisture availability over the past 14,000 years. *Science* **290**, 2285–2287.

May, R. M. (1988). How many species are there on earth? *Science* **358**, 278–279.

May, R. M. (1990). Taxonomy as destiny. *Nature* **347**, 129–130.

May, R. M. (2010). Tropical arthropod species, more or less? *Science* **329**, 41–42.

Mayle, F. E., Burbridge, R., and Killeen, T. J. (2000). Millennial-scale dynamics of southern Amazonian rain forests. *Science* **290**, 2291–2294.

Mayr, E. (1942). *Systematics and the Origin of Species*. Columbia University Press, New York. (Republished by Dover Editions)

McCain, C. M. (2005). Elevational gradients in diversity of small mammals. *Ecology* **86**, 366–372.

McCain, C. M. (2009). Global analysis of bird elevational diversity. *Global Ecology and Biogeography* **18**, 346–360.

McCain, C. M. (2010). Global analysis of reptile elevational diversity. *Global Ecology and Biogeography* **19**, 541–553.

Meave, J., and Kellman, M. (1994). Maintenance of rain forest diversity in riparian forests of tropical savannas: implications for species conservation during Pleistocene drought. *Journal of Biogeography* **21**, 121–135.

Miller, E. C., and Román-Palacios, C. (2021). Evolutionary time best explains the latitudinal diversity gradient of living freshwater fish diversity. *Global Ecology and Biogeography* **30**, 749–763.

Mittelbach, G. G., Schemske, D. W., Cornell, H. V., Allen, A. P., Brown, J. M., Bush, M. B., Harrison, S. P., Hurlbert, A. H., Knowlton, N., Lessios, H. A., and McCain, C. M. (2007). Evolution and the latitudinal diversity gradient: speciation, extinction and biogeography. *Ecology Letters* **10**, 315–331.

Möls, T., Vellak, K., Vellak, A., and Ingerpuu, N. (2013). Global gradients in moss and vascular plant diversity. *Biodiversity and Conservation* **22**, 1537–1551.

Mooney, H. A., Cushman, J. H., Medina, E., Sala, O. E., and Schulze, E. (1996). *Functional Roles of Biodiversity: A Global Perspective.* Wiley, Chichester, England.

Mora, C., Tittensor, D. P., Adl, S., Simpson, A. G., and Worm, B. (2011). How many species are there on Earth and in the ocean? *PLoS Biology* **9**, e1001127.

Morinière, J., Van Dam, M. H., Hawlitschek, O., Bergsten, J., Michat, M. C., Hendrich, L., Ribera, I., Toussaint, E. F., and Balke, M. (2016). Phylogenetic niche conservatism explains an inverse latitudinal diversity gradient in freshwater arthropods. *Scientific Reports* **6**, 26340.

Mourelle, C., and Ezcurra, E. (1997). Rapoport's rule: A comparative analysis between South and North American columnar cacti. *American Naturalist* **150**, 131–142.

Muir, A. M., Leonard, D. M., and Krueger, C. C. (2013). Past, present and future of fishery management on one of the world's last remaining pristine great lakes: Great Bear Lake, Northwest Territories, Canada. *Reviews in Fish Biology and Fisheries* **23**, 293–315.

O'Dwyer, J. P., and Green, J. L. (2010). Field theory for biogeography: a spatially explicit model for predicting patterns of biodiversity. *Ecology Letters* **13**, 87–95.

Petherick, L. M., Moss, P. T., and McGowan, H. A. (2017). An extended last glacial maximum in subtropical Australia. *Quaternary International* **432**, 1–12.

Pianka, E. R. (1966). Latitudinal gradients in species diversity: a review of concepts. *American Naturalist* **100**, 33–46.

Pielou, E. C. (1975). *Ecological Diversity.* Wiley, New York.

Powell, M. G., Beresford, V. P., and Colaianne, B. A. (2012). The latitudinal position of peak marine diversity in living and fossil biotas. *Journal of Biogeography* **39**, 1687–1694.

Preston, F. W. (1960). Time and space and the variation in species. *Ecology* **41**, 611–627.

Puurtinen, M., Elo, M., Jalasvuori, M., Kahilainen, A., Ketola, T., Kotiaho, J. S., Mönkkönen, M., and Pentikäinen, O. T. (2016). Temperature-dependent mutational robustness can explain faster molecular evolution at warm temperatures, affecting speciation rate and global patterns of species diversity. *Ecography* **39**, 1025–1033.

Quammen, D. (1996). *Song of the Dodo: Island Biogeography in an Age of Extinctions.* Touchstone, New York.

Quintero, I., and Jetz, W. (2018). Global elevational diversity and diversification of birds. *Nature* **555**, 246–250.

Rabosky, D. L., Chang, J., Title, P. O., Cowman, P. F., Sallan, L., Friedman, M., Kaschner, K., Garilao, C., Near, T. J., Coll, M., and Alfaro, M. E. (2018). An inverse latitudinal gradient in speciation rate for marine fishes. *Nature* **559**, 392–395.

Randlkofer, B., Obermaier, E., Hilker, M., and Meiners, T. (2010). Vegetation complexity—the influence of plant species diversity and plant structures on plant chemical complexity and arthropods. *Basic and Applied Ecology* **11**, 383–395.

Reaka-Kudla, M. L., Wilson, D. E., and Wilson, E. O. (1997). *Biodiversity II.* Joseph Henry Press, Washington, DC.

Remeš, V., Harmáčková, L., Matysioková, B., Rubáčová, L., and Remešová, E. (2022). Vegetation complexity and pool size predict species richness of forest birds. *Frontiers in Ecology and Evolution* **10**, 964180.

Richards, L. A., Dyer, L. A., Forister, M. L., Smilanich, A. M., Dodson, C. D., Leonard, M. D., and Jeffrey, C. S. (2015). Phytochemical diversity drives plant-insect community diversity. *Proceedings of the National Academy of Sciences* **112**, 10973–10978.

Ricklefs, R. E. (1990). *Ecology*, 3rd edition. W. H. Freeman, New York.

Righetti, D., Vogt, M., Gruber, N., Psomas, A., and Zimmermann, N. E. (2019). Global pattern of phytoplankton diversity driven by temperature and environmental variability. *Science Advances* **5**, eaau6253.

Ripley, S. D., and Beehler, B. M. (1990). Patterns of speciation in Indian birds. *Journal of Biogeography* **17**, 639–648.

Robbins, R. K., and Opler, P. A. (1997). Butterfly diversity and a preliminary comparison with bird and mammal diversity. In *Biodiversity II* (ed. M. L. Reaka-Kudla, D. E., Wilson, and E. O., Wilson), pp. 69–82. Joseph Henry Press, Washington, DC.

Romdal, T. S., Araújo, M. B., and Rahbek, C. (2013). Life on a tropical planet: niche conservatism and the global diversity gradient. *Global Ecology and Biogeography* **22**, 344–350.

Rosenberg, G. (2014). A new critical estimate of named species-level diversity of the recent Mollusca. *American Malacological Bulletin* **32**, 308–322.

Rosenzweig, M. L. (1995). *Species Diversity in Space and Time.* Cambridge University Press, New York.

Rosenzweig, M. L., and Sandlin, E. A. (1997). Species diversity and latitudes: listening to area's signal. *Oikos* **80**, 172–176.

Rosindell, J., Hubbell, S. P., and Etienne, R. S. (2011). The unified neutral theory of biodiversity and biogeography at age ten. *Trends in Ecology & Evolution* **26**, 340–348.

Roughgarden, J. (1974). Niche width: biogeographic patterns among *Anolis* lizard populations. *American Naturalist* **108**, 429–442.

Roughgarden, J. (1986). A comparison of food limited and space-limited animal competition communities. In *Community Ecology* (ed. J. Diamond and T. J. Case), pp. 492–516. Harper and Row, New York.

Sanders N. J., and Rahbek C. (2012). The patterns and causes of elevational diversity gradients. *Ecography* **35**, 1–3.

Santos, A. M., Field, R., and Ricklefs, R. E. (2016). New directions in island biogeography. *Global Ecology and Biogeography* **25**, 751–768.

Sato, H., Kelley, D. I., Mayor, S. J., Martin Calvo, M., Cowling, S. A., and Prentice, I. C. (2021). Dry corridors opened by fire and low CO2 in Amazonian rainforest during the Last Glacial Maximum. *Nature Geoscience* **14**, 578–585.

Saupe, E. E. (2023). Explanations for latitudinal diversity gradients must invoke rate variation. *Proceedings of the National Academy of Sciences* **120**, e2306220120.

Saupe, E. E., Myers, C. E., Townsend Peterson, A., Soberón, J., Singarayer, J., Valdes, P., and Qiao, H. (2019). Spatio-temporal climate change contributes to latitudinal diversity gradients. *Nature Ecology & Evolution* **3**, 1419–1429.

Scheffers, B. R., Joppa, L. N., Pimm, S. L., and Laurance, W. F. (2012). What we know and don't know about Earth's missing biodiversity. *Trends in Ecology & Evolution* **27**, 501–510.

Schemske, D. W., and Mittelbach, G. G. (2017). "Latitudinal gradients in species diversity": reflections on Pianka's 1966 article and a look forward. *American Naturalist* **189**, 599–603.

Schoener, A. (1988). Experimental island biogeography. In *Analytical Biogeography* (ed. A. A. Myers, and P. S. Giller), pp. 483–512. Chapman and Hall, New York.

Schuldt, A., Ebeling, A., Kunz, M., Staab, M., Guimarães-Steinicke, C., Bachmann, D., Buchmann, N., Durka, W., Fichtner, A., Fornoff, F., and Härdtle, W. (2019). Multiple plant diversity components drive consumer communities across ecosystems. *Nature Communications* **10**, 1460.

Segovia, R. A. (2023). Temperature predicts maximum tree-species richness and water availability and frost shape the residual variation. *Ecology* **104**, e4000.

Simberloff, D. S. (1974). Equilibrium theory of island biogeography and ecology. *Annual Review of Ecology and Systematics* **5**, 161–182.

Simberloff, D. S., and Wilson, E. O. (1969). Experimental zoogeography of islands: the colonization of empty islands. *Ecology* **50**, 278–296.

Simberloff, D. S., and Wilson, E. O. (1970). Experimental zoogeography of islands: a two-year record of colonization. *Ecology* **51**, 934–937.

Šímová, I., Storch, D., Keil, P., Boyle, B., Phillips, O. L., and Enquist, B. J. (2011). Global species-energy relationship in forest plots: role of abundance, temperature and species climatic tolerances. *Global Ecology and Biogeography* **20**, 842–856.

Simpson, G. G. (1964). Species density of North American recent mammals. *Systematic Zoology* **13**, 57–73.

Smith, T. B., Wayne, R. K., Girman, D. J., and Bruford, M. W. (1997). A role for ecotones in generating rainforest biodiversity. *Science* **276**, 1855–1857.

Stegen, J. C., Enquist, B. J., and Ferriere, R. (2009). Advancing the metabolic theory of biodiversity. *Ecology Letters* **12**, 1001–1015.

Stein, A., Gerstner, K., and Kreft, H. (2014). Environmental heterogeneity as a universal driver of species richness across taxa, biomes and spatial scales. *Ecology Letters* **17**, 866–880.

Stevens, G. C. (1989). The latitudinal gradient in geographical range: how so many species coexist in the tropics. *American Naturalist* **133**, 240–256.

Stevens, G. C. (1992). The elevational gradient in altitudinal range: an extension of Rapoport's latitudinal rule to altitude. *American Naturalist* **140**, 893–911.

Stevens, G. C., and Enquist, B. J. (1998). Macroecological limits to the abundance and distribution of *Pinus*. In *Ecology and Biogeography of the Genus Pinus* (ed. D. M. Richardson), pp. 183–190. Cambridge University Press, Cambridge.

Storch, D., Keil, P., and Jetz, W. (2012). Universal species-area and endemics-area relationships at continental scales. *Nature* **488**, 78–81.

Storch, D., and Okie, J. G. (2019). The carrying capacity for species richness. *Global Ecology and Biogeography* **28**, 1519–1532.

Stork, N. E. (2018). How many species of insects and other terrestrial arthropods are there on Earth? *Annual Review of Entomology* **63**, 31–45.

Svensson, J. R., Lindegarth, M., Jonsson, P. R., and Pavia, H. (2012). Disturbance-diversity models: what do they really predict and how are they tested? *Proceedings of the Royal Society B: Biological Sciences* **279**, 2163–2170.

Swann, G. E., Panizzo, V. N., Piccolroaz, S., Pashley, V., Horstwood, M. S., Roberts, S., Vologina, E., Piotrowska, N., Sturm, M., Zhdanov, A., and Granin, N. (2020). Changing nutrient cycling in Lake Baikal, the world's oldest lake. *Proceedings of the National Academy of Sciences* **117**, 27211–27217.

Tallis, J. H. (1991). *Plant Community History: Long-Term Changes in Plant Distribution and Diversity*. Chapman and Hall, New York.

Tenorio, E. A., Montoya, P., Norden, N., Rodríguez-Buriticá, S., Salgado-Negret, B., and Gonzalez, M. A. (2023). Mountains exhibit a stronger latitudinal diversity gradient than lowland regions. *Journal of Biogeography* **50**, 1026–1036.

Terborgh, J. (1986). The Lago Guri Islands. *Ecology* **78**, 1494–1501.

Terborgh, J. (2015). Toward a trophic theory of species diversity. *Proceedings of the National Academy of Sciences* **112**, 11415–11422.

Terbough, J., and Faaborg, J. (1973). Turnover and ecological release in the avifauna of Mona Island. *Auk* **90**, 759–779.

Thorne, R. F. (1996). Some guiding principles of biogeography. *Telopea* **6**, 845–850.

Thornton, I. (1996). *Krakatau: The Destruction and Reassembly of an Island Ecosystem*. Harvard University Press, Cambridge, MA.

Tietje, M., Antonelli, A., Baker, W. J., Govaerts, R., Smith, S. A., and Eiserhardt, W. L. (2022). Global variation in diversification rate and species richness are unlinked in plants. *Proceedings of the National Academy of Sciences, USA* **119**, e2120662119.

Tittensor, D. P., Mora, C., Jetz, W., Lotze, H. K., Ricard, D., Berghe, E. V., and Worm, B. (2010). Global patterns and predictors of marine biodiversity across taxa. *Nature* **466**, 1098–1101.

Tomašových, A. (2019). Biodiversity gradients emerge. *Nature Ecology and Evolution* **3**, 1376–1377.

Tu, H. M., Fan, M. W., and Ko, J. C. J. (2020). Different habitat types affect bird richness and evenness. *Scientific Reports* **10**, 1221.

Turner, G. F., Seehausen, O., Knight, M. E., Allender, C. J., and Robinson, R. L. (2001). How many species of cichlid fishes are there in African lakes? *Molecular Ecology* **10**, 793–806.

Vandermeer, J., Boucher, D., Perfecto, I., and Granzow de la Cerda, I. (1996). A theory of disturbance and species diversity: evidence from Nicaragua after Hurricane Joan. *Biotropica* **28**, 600–613.

VanDerWal, J., Shoo, L. P., and Williams, S. E. (2009). New approaches to understanding late Quaternary climate fluctuations and refugial dynamics in Australian wet tropical rain forests. *Journal of Biogeography* **36**, 291–301.

Wallace, A. R. (1878). *Tropical Nature and Other Essays*. Macmillan, New York.

Walter, H. S. (1998). Driving forces of island biodiversity: an appraisal of two theories. *Physical Geography* **19**, 351–377.

Walter, H. S. (2000). Stability and change in the bird communities of the Channel Islands. In *Proceedings of the 5th California Islands Symposium* (ed. D. R. Brown, K. L. Mitchell, and H. W. Chaney), pp. 307–314. U.S. Department of the Interior.

Wang, X., Edwards, R. L., Auler, A. S., Cheng, H., Kong, X., Wang, Y., Cruz, F. W., Dorale, J. A., and Chiang, H. W. (2017).

Hydroclimate changes across the Amazon lowlands over the past 45,000 years. *Nature* **541**, 204–207.

Wayne, R. K. (1996). Conservation genetics in the Canidae. In *Conservation Genetics: Case Histories from Nature* (ed. J. C. Avise and J. L. Hamrick), pp. 75–118. Chapman and Hall, New York.

Whittaker, R. H. (1997). Evolution of species diversity in land communities. *Evolutionary Biology* **10**, 1–67.

Whittaker, R. J. (1998). *Island Biogeography: Ecology, Evolution and Conservation*. Oxford University Press, Oxford.

Whittaker, R. J., and Bush, M. B. (1993). Dispersal and establishment of tropical forest assemblages, Krakatoa, Indonesia. In *Primary Succession on Land* (The British Ecological Society, special publication no. 12) (ed. J. Miles, and W. H. Walton), pp.147–160. Blackwell Scientific Publications, Oxford.

Whittaker, R. J., Fernández-Palacios, J. M., and Matthews, T. J. (2023). *Island Biogeography: Geo-environmental Dynamics, Ecology, Evolution, Human Impact, and Conservation*. Oxford University Press, Oxford.

Whittaker, R. J., Fernández-Palacios, J. M., Matthews, T. J., Borregaard, M. K., and Triantis, K. A. (2017). Island biogeography: taking the long view of nature's laboratories. *Science* **357**, eaam8326.

Whittaker, R. J., Triantis, K. A., and Ladle, R. J. (2008). A general dynamic theory of oceanic island biogeography. *Journal of Biogeography* **35**, 977–994.

Williams, P. H., Gaston, K. J., and Humphries, C. J. (1997). Mapping biodiversity value worldwide: combining higher-taxon richness from different groups. *Proceedings of the Royal Society of London B* **264**, 141–148.

Wilson, E. O. (1988). *Biodiversity*. National Academy Press, Washington, DC.

Wilson, E. O. (1988). The current state of biological diversity. In *Biodiversity* (ed. E. O. Wilson), pp. 3–18. National Academy Press, Washington, DC.

Wilson, E. O. (1992). *The Diversity of Life*. Belknap Press of Harvard University Press, Cambridge, MA.

World Conservation Monitoring Centre (1992). *Global Biodiversity: Status of the Earth's Living Resources*. Chapman and Hall, London.

Yasuhara, M., and Deutsch, C.A. (2023). Tropical biodiversity linked to polar climate. *Nature* **614**, 626–628.

Young, K. R. (1993). Biogeographical paradigms useful for the study of tropical montane forests and their biota. In *The Newotropical Montane Forest Biodiversity and Conservation Symposium* (ed. S. P. Churchill, H. Balslev, E. Forero, and J. L. Luteyn), pp. 79–87. New York Botanical Garden, New York.

Zamani, A., Dal Pos, D., Fric, Z. F., Orfinger, A. B., Scherz, M. D., Bartoňová, A. S., and Gante, H. F. (2022). The future of zoological taxonomy is integrative, not minimalist. *Systematics and Biodiversity* **20**, 1–14.

Zhang, Y., Song, Y. G., Zhang, C. Y., Wang, T. R., Su, T. H., Huang, P. H., Meng, H. H., and Li, J. (2022). Latitudinal diversity gradient in the changing world: retrospectives and perspectives. *Diversity* **14**(5), 334.

Zobel, M. (1997). The relative role of species pools in determining plant species richness: an alternative explanation of species coexistence? *Trends in Ecology and Evolution* **12**, 266–269.

CHAPTER 15

BIOGEOGRAPHY AND THE CONSERVATION CHALLENGES OF THE ANTHROPOCENE

In the eighteenth century, a small endemic butterfly called the bay checkerspot *(Euphydryas editha bayensis)* was probably relatively common in grassland habitat developed on serpentine soils adjacent to the San Francisco Bay. As the city of San Francisco and nearby communities grew and expanded during the nineteenth and twentieth centuries, the bay checkerspot became less and less common. By 1987 the butterfly had become so rare that it was listed as a threatened species by the Environmental Protection Agency. Things did not look completely bleak for the bay checkerspot, however. One population of the butterflies lived on Stanford University's Jasper Ridge Biological Preserve. The lands of the reserve were maintained as protected natural areas by the university and were under no direct threat from development. Yet, despite the best conservation efforts, by 1996 the butterfly had gone extinct at Jasper Ridge. The butterfly had also disappeared or suffered population declines in many of the other remaining small areas in which it was found. We have seen in earlier chapters that human activities, such as hunting, land use, and the introduction of alien species, have caused many extinctions. For most of these species, no attempts were made to conserve their habitat and save them from extinction. With regard to the bay checkerspot, however, an effort was made to preserve a relatively large population at Jasper Ridge, and that effort failed. The story of the bay checkerspot raises two important questions. First, why did the butterfly species go extinct at Jasper Ridge despite conservation efforts to save it? Second, what can be done to save the remaining bay checkerspot populations at other sites? The answers to these two questions are important not just for the preservation of the bay checkerspot; they are also important for the survival of literally tens of thousands of plant and animal species. Finding the answers to these questions to help conserve endangered species and habitat is the most important challenge facing conservation in the twenty-first century—a challenge that some refer to as the Anthropocene Crisis. As we shall see in this chapter, biogeography lies at the heart of the demise of the bay checkerspot at Jasper Ridge. Biogeography also lies at the heart of efforts to preserve this species elsewhere in the Bay Area and to preserve countless species around the globe.

Conservation biogeography finds itself closely aligned with the fields of conservation biology and **landscape ecology.** The field of conservation biogeography has been defined as "the application of biogeographic principles, theories, and analyses, being those concerned with the distributional dynamics of taxa individually and collectively, to problems concerning the conservation of biodiversity" (Whittaker et al., 2005, p. 3). The three fields overlap considerably. Conservation biology can be broadly defined as a synthetic discipline that applies principles from biology and other sciences to the maintenance of biodiversity. Conservation biogeography might be differentiated from conservation biology by the fact that most biogeographers concentrate specifically on how spatial distributions or temporal patterns and historical events affect the present abundance and potential conservation of biodiversity. Landscape ecology is concerned principally with spatial patterns of landscapes and the reciprocal effects of those patterns on ecology, including the distribution of species and habitat on local to regional scales. Landscape ecology seeks to understand how these patterns are developed and maintained by nature and human actions. In general, the spatial and temporal scales analyzed by biogeography are often somewhat greater than those analyzed for landscape ecology. In the end, however, it is difficult to draw firm lines between the three fields. Because the work of conservation is so important, biogeographers, conservation biologists, and landscape ecologists should put their effort into working together rather than trying to define and accentuate differences. All three fields described above are also being integrated with the continued revolution in conservation-oriented genetics. The growth of such conservation genetics is being fueled by the rapid growth of whole genome conservation genomics, targeted genetic barcoding, and the growing field of environmental DNA (eDNA) analysis for conservation (Technical Box 15.1).

In this chapter, we will consider some of the reasons why environmental conservation is so important, and then we will look at the challenges that face us as we try to preserve endangered species and their habitat around the world. We will investigate the role of biogeography in developing conservation strategies. We will then examine the challenges presented to conservation biogeography by the Anthropocene and associated global climate change.

TECHNICAL BOX 15.1

Environmental DNA (eDNA) and Conservation

Environmental DNA (eDNA) analysis is based on the collection and analysis of DNA from samples of water and sediments. It is developing to be an important tool for conservation. The DNA from plants and animals living in the vicinity of the sampled water or sediments is derived from cellular material that is shed by the organism, such as hair, feathers, scales, skin, blood, mucous, and waste products. The eDNA samples recovered from water or sediments can contain the DNA from dozens or more species that live near the collection site. During eDNA analysis, the DNA is amplified and separated out by species or higher taxa. Water samples or samples of lake sediment may contain the DNA of both aquatic and terrestrial species. The DNA of the terrestrial species can be washed into the waterbody or directly deposited in the water. In the case of water samples, the DNA may persist for only a few days or weeks. In the case of sediment samples, the DNA can bind to clays and silts and persist in analyzable form much longer. In some cases, eDNA recovered from sediments has revealed the past presence of extinct animals. DNA from extinct moa birds (Aves, Dinornithiformes) has been recovered from 3300-year-old cave sediments in New Zealand. Permafrost sediments from Siberia, ranging from 10,000 to 400,000 years in age, have yielded eDNA from Pleistocene plants and animals, including extinct mammoth *(Mammathus)*. Glacial ice cores can also contain eDNA. The analysis of basal silty ice from 2 kilometers below the present surface of the Greenland ice cap reveal that conifer forests and related insect assemblages lived on the island during a period of much warmer climate within the past 1 million years.

As eDNA analysis is rapid and cost-effective, it is being increasingly applied for biodiversity monitoring and conservation. The analysis of such DNA can help detect the presence of rare species, invasive species, and cryptospecies. In the scientific literature, the application of eDNA grew by an order of magnitude between 2008 and 2018. For biodiversity assays and monitoring, eDNA offers a technique that is noninvasive and nonlethal; it is also quicker and potentially more accurate than traditional survey methods such as spot identifications, camera and audio traps, and physically collecting living and dead organisms. These properties make eDNA particularly valuable for the monitoring and conservation of rare and endangered species.

An example of the use of eDNA to detect the presence of a critically endangered species is a study of the Alabama sturgeon *(Scaphirhynchus suttkusi)* and the Gulf sturgeon *(Acipenser oxyrinchus desotoi)* in the Mobile River Basin of Alabama. Over 85% of the world's sturgeon species are endangered or threatened. The Alabama sturgeon is considered the rarest and most endangered of the sturgeon group. Sturgeons are migratory species particularly sensitive to barriers such as dams. Traditional approaches to detect the presence and movements of the two sturgeon species in the Alabama River Basin had low success rates. eDNA was able to detect the presence or infer the absence of the two species at a number of sites in the basin and resulted in conservation recommendations such as the removal or mitigation of existing passage barriers created by dams to facilitate greater mobility.

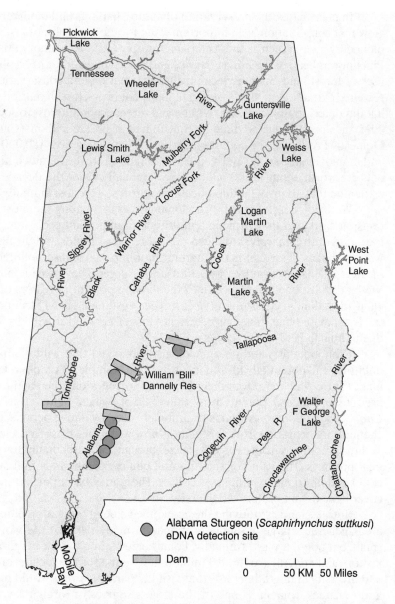

Data from Pfleger et al., 2016. Base map from GISGeography.com
https://gisgeography.com/alabama-lakes-rivers-map/

THE VALUE OF CONSERVATION

Why is the conservation of habitat and species important and worth so much effort? This is a legitimate question with many different potential answers. It would take an entire book to do justice to this issue. The term *ecosystem services* is often used to denote the benefits that humans derive from ecosystems. Those services may be active, quantifiable, and direct, such as the value of timber from a forest plot or the value of real estate that coastal marsh protects against storm surges. Ecosystem services can also be passive and less directly quantifiable. Examples of this are the relaxation derived from admiring a beautiful view of a peaceful forest or the contribution of the natural environment to spiritual values, cultural values, and identity. In addition to considering what ecosystems can do for people, some philosophers and environmentalists have argued that we must also consider our spiritual, moral, and ethical responsibility to other species that also inhabit the planet. Let's divide our summary into two broad categories: resources-economic and spiritual-ethical.

In many cases, the preservation of natural habitats and biodiversity has an economic value. The view that conservation is an important issue for the provision of resources such as timber was championed by people such as the American forester Gifford Pinchot at the start of the twentieth century. Pinchot was head of the Forest Service and an original author of its philosophy of promoting multiple uses of forest land. Such uses can include recreation and conservation, but also can include natural resource extraction such as logging. In many countries, loss of natural resources such as timber due to habitat deterioration can exact a huge and direct economic cost to society. In Canada, for example, in 2021 the forest industry directly contributed over \$25 billion (Canadian) to the Nominal Gross Domestic Product of the country and directly employed over 180,000 people. The ancillary economic and employment benefits derived from Canada's forests are much higher still. Globally, forest products are worth about \$1000 billion (U.S.) annually. Over the decade from 2010 to 2020, it has been estimated that about 4.7 million hectares of forest were lost annually on a global basis.

Wetlands, coastal marshes, and mangroves are among the most altered and threatened ecosystems on earth, yet they provide a multitude of important ecosystem services. These systems receive local and upstream waters and can mitigate against floods and drought. They help recharge groundwater supplies. They can act as natural biofilters to cleanse polluted waters. They produce coastal barriers against erosion by waves and storm surges. They provide habitat for valued animal species. Wetlands have been cited as one of the most important providers of ecosystem services. The global value of wetlands ecosystem services has been estimated at \$47 trillion per year. It has been conservatively estimated that there has been a net loss of 21% of the world's wetlands over the past 300 years due to human land-use practices.

Ethnobotanists and pharmacologists point out that wild plants may be able to provide many unknown and potentially life-saving drugs. For example, a compound from the bark of the Pacific yew plant *(Taxus brevifolia)* found in old growth forests along the Pacific Coast of North America can be used to produce a potent drug called taxol, which fights ovarian cancer. The Amazon plant *Chonodrodendron tomentosa* is traditionally used by native peoples to poison the tips of arrows. The chemical tubocurarine from this plant is now widely used as a muscle relaxant in surgical anesthesia. Today, compounds from over 120 species of plants are used in unaltered form to produce pharmaceutical products. The value of such natural clinical compounds that still remain unknown to science could add up to many billions of dollars. There may be important medicines that we will never discover if we do not protect biodiversity.

Social scientists studying the economic value of recreation point out that ecotourism in natural areas generates large amounts of money in many parts of the world. Consider one example of a small, but popular park. Annually, Point Pelee National Park in Ontario, Canada, sees over 400,000 visitors and generates over \$10 million in revenue for the local economy. A decline in the numbers and diversity of migrating birds that visit the park each year would produce a large decline in visitors and revenues. Tourism is one of the largest sources of foreign revenue for Kenya. Over 2 million people visit Kenya's parks each year. The interest of tourists in visiting African countries such as Kenya would be greatly diminished if animals such as elephants, giraffes, and lions were no longer found in the wild there. These large and attractive animals, called **charismatic megafauna,** are often the object of intense conservation measures for this reason. In 1989, a worldwide ban on the sale of new ivory was put into place to preserve Africa's native elephant populations. The ban was supported in part because of the importance of elephants *(Loxodonta africana)* as a tourist attraction for countries in Africa. There are also **charismatic flora** such as the giant redwood *(Sequoia sempervirens)* and sequoia *(Sequoiadendron giganteum)* trees of California, which attract tourists from around the world and are also the object of intense preservation efforts. Large areas of land with valuable timber have been set aside as parks and conservation areas in order to protect the redwood and sequoia trees of California. Some conservationists argue that the income generated by long-term tourism will greatly exceed the income that would be generated by harvesting the trees. In recent years, Redwood National Park and adjoining state parks in California have averaged over one million visitors each year, and the annual number of visitors to Sequoia National Park in the Sierra Nevada Mountains has also averaged over one million people. Ecotourism to these parks has added about \$130 million each year to the local economies.

An important ecosystem service provided by vegetation is the capture of carbon from the atmosphere via photosynthesis and its sequestration in biomass and underlying soils. The protection and promotion of the carbon capture potential of ecosystems is the fundamental goal of national and international programs to study and implement **natural climate solutions (NCSs)** to combat climate change. It has been estimated that natural terrestrial carbon sinks such as forests and wetlands have absorbed about 30% of the extra CO_2 added to the atmosphere by human activity since 1850. Marine systems have absorbed almost as much. Since 1960 terrestrial and marine ecosystems have absorbed a combined total of about 55% of human emissions. As human emissions rates have risen, so too have the natural rates of carbon uptake and sequestration by mechanisms such as vegetation growth, but not quickly enough to absorb all the additional CO_2. The capacity of ecosystems to absorb additional carbon is expected to decrease going forward in the twenty-first century. Protection of CO_2-absorbing terrestrial and marine ecosystems is an important component of national and international strategies to combat anthropogenic climate change. It has been suggested that natural climate solutions might be able to provide more than a third of the climate mitigation needed to keep twenty-first-century warming below 2° C.

Ecological economists have tried to place a monetary value on the ecosystem services provided by the natural world. In these estimates, they weigh factors such as the value of resources, the positive contribution natural areas make to world water resources and climate, and the recreational value of natural ecosystems. The American/Australian ecological economist Robert Constanza has been a leading pioneer in such efforts. In 2014, it was estimated that each year the earth's natural ecosystems produce $125 trillion to $145 trillion in economic value. That was nearly twice the gross domestic product of all the world's nations combined. It has further been estimated that depending upon how the natural environment is managed by humans, the annual value of global ecosystem services can either increase by $30 trillion or decline by $51 trillion by the year 2050. The economic value of ecosystem services differs by ecosystem. On a per-hectare basis, it has been estimated that wetlands in general, and tidal marshes and mangroves in particular, provide the greatest valuations in terms of ecosystem services (Table 15.1). It should be noted that there are wide variations in global estimates of the economic value of ecosystem services and this remains an area of important active research.

We can understand economic arguments for preserving resources or developing ecotourism attractions, but why preserve small, economically valueless, and seemingly insignificant plants or animals such as the bay checkerspot butterfly? In some cases, these organisms may serve as critically important species within economically valuable ecosystems, and their removal can cause large disruptions due to changes in competitive interactions and trophic structure. Early on, ecologists argued that biodiversity in and of itself helps to maintain ecosystems. They argued that the more species that are present, the more stable the system will be. Experimental manipulations of grasslands by G. David Tilman and his colleagues suggested that plots with high plant species diversity produce more biomass and are more resistant to disturbances such as droughts than are plots with less plant species diversity. The view that biodiversity itself imparts stability and the evidence used to support this hypothesis are contentious, however, and remain debated among ecologists and biogeographers. Recent experimental and modeling studies suggest that biodiversity can be a positive influence on ecosystem stability in systems that have generally low diversity to begin with, but ecosystems that exhibit and depend upon high biodiversity for functioning are more vulnerable to destabilization than lower diversity systems.

TABLE 15.1   Some Examples of Estimated per-Hectare Annual Economic Value of Ecosystems Services from Different Major Ecosystem Types

Ecosystem	2011 Estimated Unit Value per Hectare
Estuaries	$ 28,916
Inland swamps/floodplains	25,681
Tidal marshes/mangroves	193,843
Lakes/rivers	12,512
Forests	3,800
Grasslands	4,166

Source: Data from Mitsch and Gosselink, 2015.

In contrast to economic valuations, many people believe that the conservation of species and natural habitat is important because of the spiritual and aesthetic enrichment that nature provides. In North America, the concept that there is a spiritual value of conserving nature owes much to nineteenth-century writers such as the philosopher Henry David Thoreau and the naturalist John Muir. Although people like Thoreau and Muir put an emphasis on conservation that was divorced from monetary values, they still viewed conservation largely from the perspective of what nature could provide to people rather than in any sense that humans have an ethical responsibility to preserve nature. In the twentieth century, the wildlife ecologist Aldo Leopold espoused the concept that humans have an ethical responsibility to conserve and preserve the natural environment. He articulated his views in an influential book called *A Sand County Almanac* (1949). Leopold called this responsibility to nature the **land ethic.** He argued that human communities were part of a greater community that included the natural world. We should extend the same ethical treatment to nature that we extend within human communities. Leopold's views were fundamental to the development of the modern environmental movement.

Today, the basic philosophy that humans have an ethical responsibility to the natural world is being further developed by applied philosophers in the field of **enviroethics (ecological ethics and environmental ethics).** Such work focuses on the moral relationships between humans and the natural environment and the other species existing on the planet. It also considers how environmental ethics leads to sustainability of both human and natural systems. Thinkers such as Robert Cahn argue that progress within society depends on humans developing a broad environmental ethic with regard to nature. In this vein, the cofounder of the Society for Conservation Biology, Michael Soulé, articulated four postulates regarding the ethical value of nature: (1) diversity of organisms is good, (2) complexity in ecosystems is good, (3) natural evolutionary development is good, and (4) biological diversity should be valued and protected for itself regardless of utilitarian values. Modern thinkers have made connections between the goals of environmental ethics and the ethics embedded in feminism, diversity, equality and inclusion, combating poverty, and challenging traditional anthropomorphic western ethics with more ecocentric values. This latter approach is at the heart of the deep ecology movement and has been referred to as biospheric egalitarianism. We might summarize the spiritual-ethical view by simply saying that we should work to preserve other species and their natural habitat, not because of economic values, but because it is the right thing to do. In doing so, we become a better people and create a better world for humans and all other life.

ENDANGERED AND THREATENED SPECIES

In earlier chapters, we discussed some very specific definitions for the terms *local species extinction* and *global species extinction*. What exactly do we mean when we use terms such as *threatened species* or *endangered species*? The government of the United States tackled this question when an act of Congress was passed to protect species that were in danger of extinction. The Endangered Species Act of 1973 defines an **endangered species** as a species or subspecies that is at risk of extinction throughout all or a portion of its range. From a biogeographic perspective, we might define an endangered species as a species that is no longer sufficiently abundant and reproducing to ensure its survival. Species can be endangered at both the local and global level. A **threatened species** is a species that is likely to be endangered in the foreseeable future. It takes much research to demonstrate that a species is endangered or threatened. In general, conservation biologists consider the total population size, rate of mortality, reproductive rate, rate of overall population decline, and frequency of disturbances that produce significant population declines. Species that have too small a population size and too low a reproductive rate to offset normal mortality and the impact of population losses caused by disturbance are threatened or endangered.

The human activities that are endangering plant and animal species fall into several different broad categories. One such categorization is presented in Figure 15.1 and includes overexploitation, agricultural activity, human disturbance/modification, urban development, invasion and disease, pollution, climate change, transportation, and energy production. Of the species considered in this analysis, the largest proportion are birds and amphibians (Fig. 15.1). It is recognized that there are alternative ways of defining and categorizing these threats. It should also be remembered that there

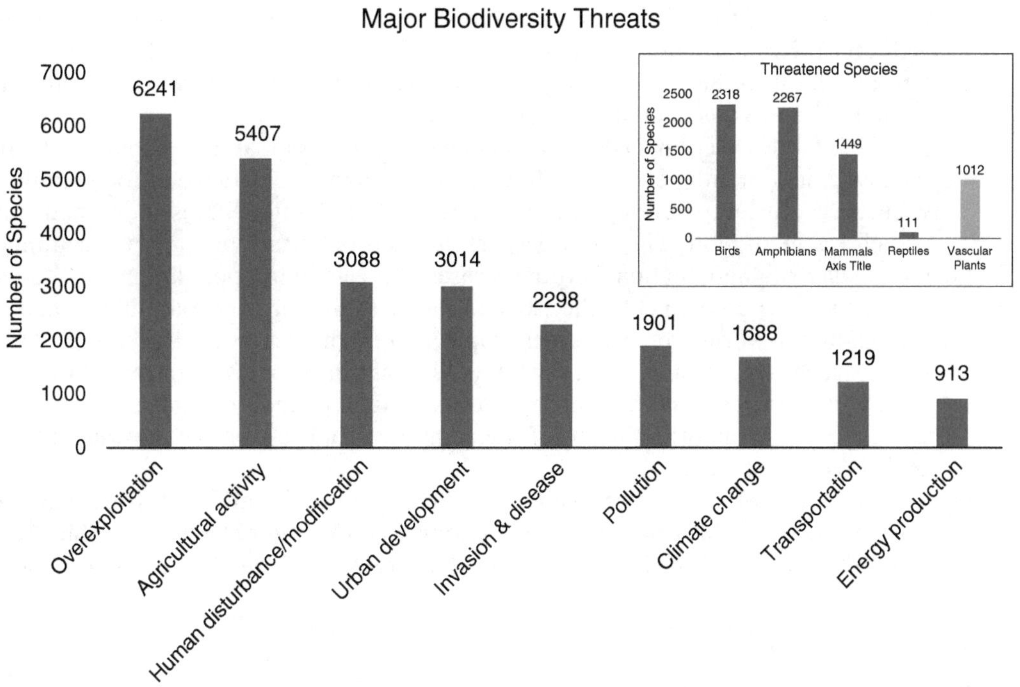

Major Biodiversity Threats

FIGURE 15.1 The major factors contributing to threat and endangerment to global biodiversity and the number of species under threat or endangerment according to the IUCN Redbook. The number of threatened or endangered bird, amphibian, mammal, reptile, and vascular plant species are presented in inset (data from Maxwell et al., 2016).

are many other species, particularly phyla such as fungi and insects, that are not included and are poorly known in terms of status at this time. If we group together all of the threats that are associated with human landscape use, disturbance, or modification (agricultural activity, human disturbance/ modification, urban development, transportation, and energy production) we can see that land use by humans is the greatest factor (53%) causing species to be threatened or endangered. This is consistent with studies that have suggested that habitat destruction is responsible for about 50% of bird, mammal, and plant species endangerment in the United States. Habit destruction from agriculture and logging is a particularly potent force causing species endangerments in tropical regions today.

Some of the greatest numbers of threatened and endangered species are found on the mainland and islands of southeastern Asia in tropical rainforest regions. Tropical rainforests have naturally high biodiversity and are being cleared throughout the world at a global average rate of over 2% per year, so it is not surprising that many endangered species are from tropical forest regions. About 48% of the world's threatened animal species are found in the tropics. Losses of habitat and species in the tropics often have repercussions beyond the equatorial regions. Tropical rainforests and deciduous forests are crucial for the wintering of many birds that breed in the high latitudes. For example, many species of flycatchers *(Empidonax)*, which are common during summer months in the United States and Canada, spend their winters in southern Mexico, Central America, and South America. The loss of habitat and species diversity in the tropics will undoubtedly cause the extinction of some of the bird species that breed in the temperate latitudes and overwinter in the tropics. Systematic bird surveys of the North American avifauna show a general decline over the past 30 years in the number of many migrant songbirds that winter in Central and South America.

As we have seen in previous chapters, many endemic species of birds may be found on islands such as Hawaii and the Galapagos. Because of the high number of endemic island bird species, the destruction of oceanic island habitat is just as significant a cause of bird endangerment as is the destruction of the rainforest. From 1500 to the present, the total number of animal species extinctions on islands was far higher than on continental areas. One analysis concluded that between the year 1500 and 2010 approximately 95% of recorded bird and mammal extinctions were on islands. Per unit of equal area, the rate of extinctions on islands is 187 times higher for birds and 177 times higher for mammals than for continental areas. However, as habitat destruction and the introduction of alien species have increased worldwide, the rate of extinctions on continental areas has been increasing.

Habitat destruction leads to both a decline in the total area of habitat and a fragmentation of remaining habitat into geographically separated patches. Clearance of forest in the nineteenth and twentieth centuries to develop farmland in the eastern United States and Canada is a good example of such fragmentation (Fig. 15.2). The fragmented patches separate once continuous populations into smaller metapopulations. In many cases, the remaining populations are too small to remain viable, and they go extinct in the isolated patches. Similar fragmentation of habitat has occurred throughout the world. In recent times, tropical forests have become increasingly fragmented, so that in areas such as Central America much of the rainforest and dry tropical forest exists as isolated patches of varying sizes. Even small corridors of human modified land, such as roads and powerline corridors, can be important forms of fragmentation that restrict the movement of many plants and small animal species.

A good example of a charismatic endangered species and the efforts made to preserve it is provided by the California condor *(Gymnogyps californianus),* one of the most endangered birds in North America (Fig. 15.3). It is also the largest bird on the continent, and it can obtain a

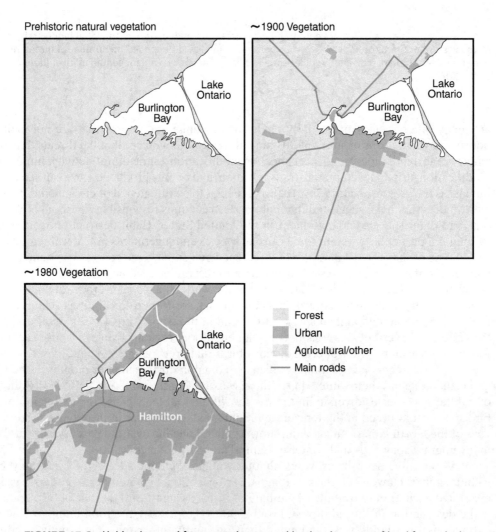

FIGURE 15.2   Habitat loss and fragmentation caused by the clearance of land for agriculture and urbanization in southern Ontario (*Source:* MacDonald 1987 / University of Toronto Press).

FIGURE 15.3 The largest and one of the most endangered bird species in North America, the California condor *(Gymnogyps californianus)*. This huge bird represents the type of charismatic megafauna that has been the object of an intensive combination of *ex situ* and *in situ* conservation programs.

wingspan of 3 meters. The condors require the carcasses of large animals to feed upon and the protection of rocky ledges for nest building. The females breed only every second year and produce one egg, a reproduction rate that is very low. The settlement of southern California by Europeans led to the loss of habitat for condor feeding and the introduction of damaging pollutants such as DDT. By 1985, only five condors remained alive in the wild. The threat of extinction to this endangered species was so great that the remaining wild condors were trapped in 1987 and placed in special breeding facilities at the Los Angeles and San Diego zoos. The condors were successfully bred in captivity, and by the 1990s their offspring were being released into a special preserve to try to develop new wild breeding populations. In 2022, there were over 340 California condors living in the wild.

The Vancouver Island marmot *(Marmota vancouverensis)* provides another good example of an endangered species and the intense efforts being made to preserve it. The Vancouver Island marmot is one of the most endangered mammal species in North America. Even prior to endangerment by recent human activity, the Vancouver Island marmot was not extremely abundant. It has a low reproductive rate, and it lives in small, scattered populations. The recent decline of the marmot can largely be attributed to the impact of logging on marmot habitat. Once regional populations began to decline and go extinct, even those located in protected areas such as parks could not escape local extinction because there were no nearby populations to replenish park populations during periods of natural population fluctuations. As a result, the marmot was found on only a few scattered sites on Vancouver Island (Fig. 15.4). The Canadian government declared it an endangered species in 1980 when there were still several hundred individuals in the wild. Despite protective measures, the Vancouver Island marmot population declined to fewer than 30 animals by 2003. To stave off potential extinction, some marmots were captured and are now being bred in captivity. The offspring are being released into the wild in an effort to save and support the non-captive populations. By the 2010s the population had rebounded to over 200 animals in the wild, but continues to experience fluctuations.

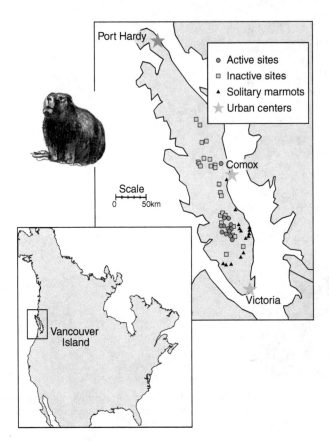

FIGURE 15.4   The past historical breeding sites of the Vancouver Island marmot *(Marmota vancouverensis)* and its modern distribution (from Marmot Recovery Foundation data). The population decline of the marmot has occurred in tandem with range collapse.

BIOGEOGRAPHY AND ENDANGERED SPECIES

Biogeographic studies of historical changes in species' ranges can provide evidence of species that are threatened or endangered. The history of the Vancouver Island marmot discussed earlier provides an example of the linkage between population decline and range decline that is typical of endangered species. In the case of the marmot, not only have population numbers fallen precipitously, but the geographic range of the species has declined. Documentary evidence shows that the marmots had once existed at many sites located hundreds of kilometers north of their present populations and at additional sites to the south (Fig. 15.4). The phenomenon of geographic range shrinkage has been documented for most endangered animal species. From our earlier considerations of what drives extinction in Chapter 9, we can appreciate that smaller geographic ranges lead to reduced carrying capacity, reduced population sizes, and higher probability of extinction.

In the case of the Vancouver Island marmot, range shrinkage has led to the last remaining populations surviving in the central area of the former natural range. Population size in the central portions of a species range is often higher than at peripheral sites, so we might assume that endangered species have a higher potential to survive near their range centers. This phenomenon is called **range collapse.** In selecting conservation strategies and refuge areas to save endangered species, we might be tempted to concentrate our efforts at the centers of the species' natural ranges. However, in many cases, range shrinkage has led to the species surviving only in small areas at the periphery of their original ranges. This phenomenon is referred to as **range eclipse.** Good examples of this type of range eclipse are provided by the giant panda *(Ailuropoda melanoleuca)* in China, the red wolf *(Canis rufus)* in the southeastern United States, and the western Quoll *(Dasyurus geoffroii)* in Australia (Fig. 15.5).

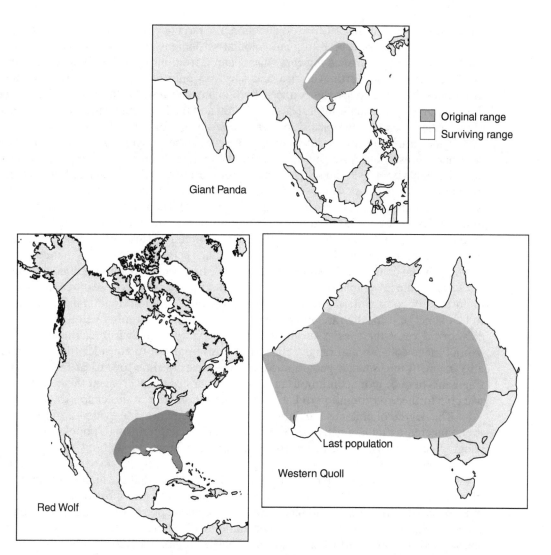

Original range
Surviving range

Giant Panda

Red Wolf

Last population

Western Quoll

FIGURE 15.5 Examples of geographic range eclipse with survival on the periphery of the former range; the giant panda *(Ailuropoda melanoleuca)* in China, the red wolf *(Canis rufus)* in the southeastern United States, and the western Quoll *(Dasyurus geoffroii)* in Australia (data from Lomolino and Channel, 1995; Brown and Lomolino, 1998).

This pattern of range eclipse to the periphery of a once larger geographic range is extremely common. The geographic pattern of range shrinkage caused by human activity was collected for over 250 different plant, invertebrate, and vertebrate species and analyzed by the biogeographers Mark Lomolino and Rob Channell. Only 2% of these species experienced range collapses that left the surviving populations at the center of the former geographic ranges. In total, 98% of all the species studied had some surviving populations at the periphery of their ranges. About 36% of the species studied survived only at the periphery of their former ranges.

Why do species that are declining in population size and geographic range tend to survive as geographically peripheral populations rather than as central populations? The likely answer is that human activity such as land clearance and hunting are often concentrated in certain areas and are less practiced in others. The peripheral sites where species survive human endangerment are often regions that are not heavily used by humans. For example, the remaining populations of red wolf in the United States survive in swampy and wooded areas along the Gulf of Mexico coastline, while the giant panda survives in the rugged mountains of southern China. Much of the former range of these two animals has been extensively cleared for farming or otherwise impacted by human land uses. Analysis of the history of range eclipse in North American and Australian mammals shows that many species went extinct in an east-to-west direction. The direction of extinction and range eclipse paralleled the westward expansion of European colonial settlement and land-use patterns. For example, the black-footed ferret *(Mustela nigripes)* disappeared from most of the intensively farmed portions of the Great Plains of North America and survived only along the western edge near the Rocky Mountains. Invasive species introduced by humans can also be factors causing range eclipse. The introduction of rats and monitor lizards to Gaum and the Mariana Islands has led to the range eclipse of the critically endangered lizard species *Emoia slevini,* for example.

There are some other interesting patterns in the geography of range shrinkage. In North America, Eurasia, and Australia, about 75% of all range shrinkages led to peripheral survivors, while in Africa only about 58% did so. The reasons for this difference are unclear but may relate to the lack of habitat refuges for savanna species at the periphery of the African savanna. Unlike continents, islands often display a pattern of central range collapse. The shrinkage of island species ranges tends to occur first along the coast and then proceeds inland. The remaining ranges of these species also tend to be found at higher altitudes in the island mountains. This pattern has been observed for many native species of the Hawaiian flora and fauna. Range eclipse occurs in this manner on islands because it is usually the coastal and lowland areas that are first occupied and modified by humans.

Taking the geographic patterns of range collapse and range eclipse into account can be extremely important for conservation efforts. A global study of 67 different mammal species that were thought to be extinct or critically rare and endangered but were then rediscovered in the wild found that in 60% of the cases the species survived at the periphery of their former range. We have focused on overall range collapse and range eclipse, but it is important to consider that geographic ranges can also fragment into smaller geographic units as species decline toward extinction. These geographic fragments of once more continuous ranges can occur in central areas, at the periphery, or in both. We will give further consideration to habitat fragmentation later in this chapter.

The degree of endemism exhibited by a species can provide evidence of the degree to which it is endangered by human activity. As shown in earlier chapters, small populations restricted to small geographic areas are more prone to extinction than large populations that are widely dispersed. Species that persist only in very small geographic areas are very prone to complete extinction because of factors such as disease or large disturbances. Thus, species that are highly endemic, as a result either of natural factors or range collapse/eclipse, are likely to be the most threatened or endangered by human activity. Analysis of endangered and threatened species of plants and animals in many geographic locations shows that a direct correlation exists between the present geographic range of a species and the degree to which it is threatened with global extinction. For example, an analysis of waterfowl found that all species with total geographic ranges of less than 500 square kilometers were threatened or endangered. In contrast, no waterfowl with ranges over 1000 square kilometers were considered endangered. A global meta-analysis of all taxonomic groups from over 170 studies found that small range size and narrow habitat breath within that range were important predictors for extinction vulnerability regardless of the type of organism.

Conservation ecologist Georgina Mace and evolutionary biologist Russell Lande have attempted to incorporate information on population size, rates of population decline, and geographic range to produce more analytic definitions for species endangerment (Table 15.2). The Mace-Lande criteria classify species as vulnerable, endangered, or critical based on present population size, observed and predicted rates of population decline, geographic range size, and degree of endemicity. Critical species have very small population sizes (less than 250 individuals); high rates of observed and projected population declines (80% in 10 years and 25% in 3 years); and single, small geographic ranges of less than 100 square kilometers. According to Mace and Lande, critical species have a 50% probability of going extinct within 10 years. A number of conservationists have accepted the Mace-Lande criteria and applied them to various species. One such analysis applied the Mace-Lande criteria to the study of primates and concluded that 12% of the world's primate species were in a critical state and at high risk for extinction.

Based on the premise that endemic species with relatively small ranges and small population sizes are more prone to be threatened or endangered, we can see some global patterns in the distribution of potentially endangered plant and animal species. The tropical regions and the southern hemisphere in general possess larger numbers of endemic animals and plants than the higher latitudes or the northern hemisphere. Brazil has about twice as many endemic amphibian, bird, and mammal

TABLE 15.2   A Summary of Mace and Lande (1991) Criteria for Assessing Threatened Species

	Degree of Threat		
Observations	Critical	Endangered	Vulnerable
Range	< 100 sq km	< 5000 sq km	< 20,000 sq km
	1 location	< 5 locations	< 10 locations
Population size	< 250	< 2500	< 10,000
	< 50 at each location	< 250 at each location	< 1000 at each location
Decline	80% decline	50% decline	20% decline
Population	Per decade or per 3 generations	Per decade or per 3 generations	Per decade or per 3 generations
Projected decline	> 25%	> 20%	> 20%
	Per 3 years or per 1 generation	Per 5 years or per 2 generations	Per 10 years or per 3 generations
Extinction probability	> 50%	> 20%	> 10%
	Per 10 years or per 3 generations	Per 20 years or per 5 generations	Per 100 years

Sources: Mace and Lande, 1991; Dobson, 1996.

species as the United States. Globally, a number of **biodiversity hot spots** have been identified that contain very large numbers of endemic plant and animal species that are prone to threat by human activity and require focused conservation efforts (Fig. 15.6). The hot spots represent areas where, due to the large numbers of endemic species, human activity has the potential for producing a large number of extinctions and significant declines in overall global biodiversity. Not surprisingly, most of these hot spots are in the tropics in regions such as equatorial South America, equatorial Africa, and southern Asia. Important extratropical hot spots are found in central Chile, southern Africa, and southern Australia. Three hot spots, the Caribbean, Hawaii, and California, are located in, or partially in, the United States.

One of the most important international organizations working to identify threatened and endangered species throughout the world today is the International Union for Conservation of Nature (IUCN). The IUCN is composed of over 1400 members including governmental and nongovernmental organizations working to identify and preserve endangered species around the world. Estimates of the total number of endangered species in the world are tabulated in the IUCN Red List (https://www.iucnredlist.org/). Tropical regions associated with the world biodiversity hot spots have the largest concentrations of vulnerable species. When viewed on a continental basis for the entire globe, the number of endangered plant and animal species as estimated by the IUCN is staggering (Table 15.3). The highest total number of known threatened or endangered species is found in Asia. On that continent, there are over 19,000 endangered species of plants, fish, amphibians, reptiles, birds, and mammals. Sub-Saharan Africa has over 18,000 endangered. The threatened or endangered species in North America includes over 2262 plant species, 65 mammal species, 113 bird species, 45 reptile species, 67 amphibian species, and 347 species of fish. Many islands also have very high numbers of endangered plant and animal species. For example, New Zealand has over 60 endangered bird species. This represents about 35% of the native New Zealand bird species. The global total of known species that are threatened or endangered on all continents and islands is more than 44,000. Species in the oceans are not immune to the impact of human activity, and so many marine species are similarly threatened or endangered. About 37% of all shark and ray species and 36% of coral species are threatened with extinction.

Although we have focused our attention on extinctions, many studies have revealed that in recent decades there have been alarming decreases in the population sizes of many taxa. Throughout the world, declines in the abundance of insects have been reported. In Germany, for example, a seasonal decline of 76% has been observed in the total biomass of flying insects in protected reserve areas. Significant declines in the abundance of a variety of insect taxa, ranging from bees to butterflies to beetles have been reported from the tropics to the Arctic regions. An analysis of population trends in European reptiles and amphibians found that 54% of the species studied had experienced declines over the period 1972–2016. Between 1970 and 2017, an abundance of over 300 North American bird

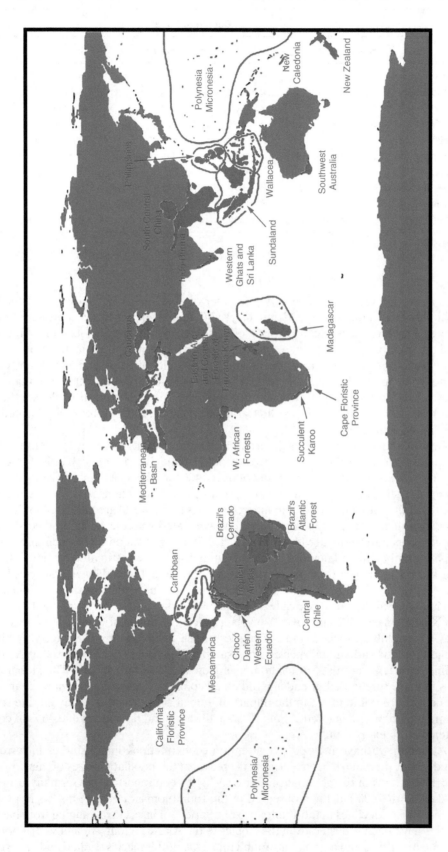

FIGURE 15.6 Global biodiversity hot spots have high numbers of vulnerable endemic species and are priority areas for focused conservation efforts (*Source:* Meyers et al., 2000 / Springer Nature).

TABLE 15.3 **Number of Threatened Species by Geographic Region**

Region	Mammals	Birds	Reptiles	Amphibians	Fish	Mollusks	Other Invertebrates	Plants	Fungi	Total
Sub-Saharan Africa	1,106	1,094	570	410	3,029	407	1,778	10,000	4	18,398
South and Southeast Asia	903	807	688	420	2,064	165	2,105	5,907	16	13,075
South America	430	633	400	1,196	1,440	83	302	6,098	137	10,720
Australia and Oceania Islands	212	398	308	79	1,055	503	2,184	2,807	33	7,580
Europe	248	463	74	83	1,348	1,117	1,804	1,149	856	7,143
Mesoamerica	178	177	266	552	973	20	361	3,420	19	5,968
Caribbean Islands	83	144	401	182	1,395	9	779	1,194	2	4,189
West and Central Asia	276	424	182	28	1,161	117	506	892	19	3,606
East Asia	156	285	108	225	755	85	402	969	34	3,019
North America	65	113	45	67	347	341	372	745	116	2,214
North Africa	98	85	46	5	411	65	147	123	10	990
Russian Region	56	99	11	1	101	21	88	101	59	537
Antarctica	2	5	0	0	1	0	0	0	0	8
TOTAL	3,813	4,727	3,099	3,248	14,080	2,933	10,828	33,405	1,305	77,447

Source: Data from the IUCN Red List 2023, https://www.iucnredlist.org/statistics.

species, both native and introduced, experienced population declines, representing an estimated net loss of almost 3 billion birds or 29% of overall bird abundance since 1970. It has further been estimated that worldwide 48% of bird species may be experiencing populations declines. A study of large mammal populations in protected areas of Africa found a 59% decline in net population numbers between 1970 and 2005. These widespread declines in population abundance are the result of the same factors driving extinctions and are of critical concern even in the case of currently abundant species. The declines of some populations produce declines in food and energy for higher trophic levels, increasing the potential for trophic cascades and extinctions of other species. Declines in the abundance of some insects decrease pollinator numbers and impacts this important ecological service for crops. Of course, smaller population sizes can push species into endangered status. There is also often a relationship between declining population size and declines in geographic range, which further exacerbates the threat of extinction.

Now, let's return briefly to the checkerspot butterfly. Given what we have learned, it is not surprising that the Jasper Ridge population went extinct. The overall population size of the butterfly had been in decline since the arrival of Europeans in the San Francisco Bay area. The habitat of the butterfly was becoming increasingly small and fragmented. During the time of its extinction at Jasper Ridge, a combination of a drought followed by very high rainfall served as a climatic disturbance that catastrophically reduced the butterfly population size. This pattern of habitat reduction, habitat fragmentation, population decline, and then extinction has been the local or global fate of thousands of species.

BIOGEOGRAPHY AND CONSERVATION PLANNING

We have seen that thousands of species are at risk of extinction throughout the world as a direct result of human activity. Many more such species will undoubtedly be identified in the future. These species are not yet extinct, however, and efforts can be made to preserve them. Many conservation efforts are aimed at preserving specific endangered species. In these cases, planning is tightly focused on the goal of preserving the identified species. We examined such targeted efforts in the case of the California condor and the Vancouver Island marmot. The advantage of such **species-based conservation** is that it provides simplified objectives and focuses resources on those clear objectives. An important question facing this approach is, how do you decide which species to conserve? There are not enough resources to support expensive efforts, such as we saw with the California condor, for all of the world's endangered species. One suggestion is to apply a rigorous multicriteria decision making (MCDM) approach that weighs quantitative variables for species under consideration for conservation efforts. Such an MCDM framework might consider (1) geographic rarity or endemism, (2) taxonomic rarity or uniqueness, (3) vulnerability to extinction, and (4) natural capital value. In this framework, natural capital value refers to the ecosystems service value, crop value, cultural value, and recreational value of the species.

In some cases, it is impossible to ensure the survival of a species in the wild. It may be that too much habitat has been destroyed or that the remaining population size is too low to sustain the species any longer. In these cases, species preservation must be done by maintaining the species in botanical gardens and zoos. In some cases, such as the California condor, the remaining wild individuals may even be captured and bred in captivity. These efforts, which rely on the preservation of the species in zoos and botanical gardens, are called *ex situ* **conservation.** Over 800 botanical gardens and a number of zoos are directly involved in *ex situ* conservation. In some cases, it is not the adult plant or animal that is preserved in *ex situ* facilities. Many botanical gardens have banks of preserved seeds for endangered species, while animal facilities may preserve frozen sperm, egg cells, and embryos. Despite the importance of *ex situ* conservation, it cannot be regarded as the complete answer to the conservation of biodiversity. For one thing, it is simply not possible from an economic standpoint to try to preserve huge numbers of endangered species in botanical gardens and zoos. Second, it is likely that the *ex situ* effort will not capture the full genetic diversity of the species. Third, *ex situ* conservation requires that scientists know that a species exists, recognize that it is endangered, and then collect it. Thousands of species, particularly invertebrates, remain unknown to science but are endangered. Fourth, in many cases it is difficult, if not impossible, to generate successful reproduction in captive populations. Passenger pigeons *(Ecopistes migratorius)* and the Carolina parakeet *(Conurpsis carolinensis)* both went extinct in North America despite efforts by zoos to maintain the species. Finally, it would be a rather dreary world if the only place we could see real biological diversity was in controlled environments such as botanical gardens and zoos. Despite these criticisms, *ex situ* conservation must be recognized as the most viable strategy of preserving some species.

Most conservation efforts are aimed at preserving species in their natural environment. This approach is called *in situ* **conservation.** Attempts to save the peninsular bighorn sheep *(Ovis canadensis cremnopbates)* in southern California is an example of an *in situ* conservation strategy. In 1998, only 280 individuals were believed to inhabit the desert region of southern California. Fences have been built to keep the sheep from crossing sections of highway where they have frequently been killed. Areas of sheep habitat have been protected from development and access has been restricted for hunting. By 2018, about 800 sheep inhabited the desert. U.S. attempts to save the California condor and Canadian efforts to save the Vancouver Island marmot are examples of hybrid programs that combine *ex situ* breeding efforts with *in situ* efforts to maintain a selected species in the wild. A similar effort of linking captive breeding with the release of offspring has been used to reintroduce black-footed ferret to the Great Plains and the Hawaiian goose *(Branta sandvicensis)* and ʻAlalā (Hawaiian crow—*Corvus hawaiiensis*) to the wild in Hawaii. The ferret and goose reintroductions have been moderately successful. In the case of the ʻAlalā, between 1993 and 2000, 27 were hatched in captivity and released. As of 2000, only 3 of the released birds had survived. By 2002, no ʻAlalā were observed in the wild. In the years following, *ex situ* breeding methods, forest management practices, and post-release support practices have improved. However, although over 100 ʻAlalā survive in captivity, release programs have had trouble reestablishing a viable wild population and work is ongoing. Individuals do not have to breed in captivity to be useful in proactive conservation. **Translocation** of individuals from one geographic site to an area with low abundance or where the species has been extirpated is a widely used strategy and has been important in the conservation of bighorn sheep in western North America, for example.

One of the biggest challenges facing all species conservation efforts is making sure that *ex situ* and *in situ* strategies preserve as much as possible the natural genetic variability of the endangered species. The field of conservation genetics studies genetic variations within endangered species and examines the importance of genetics in conservation practices. The importance of conservation genetics can be illustrated by the example of the Hawaiian silversword subspecies *Agroxiphium sandwicense sandwicense,* which is a narrow endemic on Mount Mauna Kea. In order to preserve the subspecies, plants were grown from seeds in botanical gardens and then outplanted to recolonize wildland sites. Unfortunately, the outplanted populations displayed very low levels of seed production. A subsequent genetic analysis showed that the majority of the individuals descended from the same female parent. In this species, the plants generally cannot fertilize themselves, and such close relatives cannot successfully reproduce with each other. Genetic screening prior to planting would have increased the success of the conservation effort. Because of its clear importance, it is not surprising that conservation genetics is a rapidly growing area of scientific interest.

Conservation genetics interfaces closely with biogeography in the field of **landscape genetics.** This emerging discipline combines the study of population and conservation genetics with landscape ecology to see how attributes of a landscape influence the genetic structure of populations residing there. Many species display high degrees of geographic variability in their genetic composition that reflect the influence of landscape on gene-flow or natural selection. The geographic variations in population genetics are particularly true of species that have large geographic distributions and occur as separated metapopulations. However, they also occur at much smaller geographic scales. If we protect or breed in captivity only a small sampling of the population of an endangered species, particularly if that sample is from only one of the environments or geographic sites in which the species lives, we may not preserve all of the genetic variability that allows the species to survive across its range and in different habitats. In captive breeding programs, genetic analysis is crucial to make sure that close relatives are not bred, for this promotes low genetic diversity and can also lead to congenital abnormalities and illness. Landscape genetics helps direct the geographic planning of conservation areas for *in situ* conservation as well as the selection of individuals or seeds for *ex situ* conservation.

An example of the application of landscape genetics for the development of conservation policy and practice is provided by work on the desert bighorn sheep *(Ovis canadensis nelson)* in southern California. In a landscape genetics study of 27 separated populations of desert bighorn sheep, the conservation scientist Clinton Epps and his colleagues found that fenced highways and fenced canals were producing significant barriers to gene-flow among the already fragmented sheep populations (Fig. 15.7). Genetic diversity was therefore declining rapidly, by about 15% over 40 years, and this presented a significant challenge to the long-term persistence of the species. It is ironic that fencing highways can protect individual animals from being killed by cars and trucks, but landscape genetics show that such alterations of the landscape can also have profound negative impacts on gene-flow, genetic diversity, and long-term viability of species. Recognition of this quandary is being addressed in many parts of the world through the building of wildlife overpasses to provide safe corridors for movement of individuals and facilitate gene-flow (Fig 15.7). Such wildlife overpasses can also reduce traffic accidents caused by vehicles hitting large animals. Landscape genetics has been important in identifying the importance of providing conservation corridors that link populations of endangered species and facilitate the movement of individuals and promote gene-flow between populations.

Narrowly focused species-based conservation approaches may provide little help to the other species and indeed may have a negative impact on some other species. Species-based conservation may not preserve important ecosystem services. When we apply species-based approaches, how do we choose which species to expend resources on and which species to ignore, or how do we know which species we may possibly harm in our efforts? An alternative to species-based conservation is **biodiversity conservation,** in which efforts are made to preserve the species richness of the entire ecosystem rather than focusing on one individual species. No matter how we approach biodiversity conservation, gauging the success of biodiversity programs is difficult without frequent censuses of species and genotypes within species. For many ecosystems, we may not even know the total biodiversity that is present at the start of the conservation efforts. In biodiversity management we must choose between preserving total biodiversity or just the biodiversity of native plants and animals. Almost all ecosystems now contain species introduced by humans. Nonnative species make up more than 50% of the total biodiversity of some regions and cities.

In many cases, these invasive alien species are deleterious to the native species we wish to preserve. In some cases, however, nonnative species provide important ecological services and resources used by native species. For example, in California, invasive mustard plants displace native plant species, but have also become a preferred food for some native birds and mammals. Urban areas typically have large numbers of nonnative species. It has been suggested that cities can serve as important refuges to preserve both nonnative and native plant and animal species that are under threat in their natural native ranges. For example, the red crowned parrot *(Amazona viridigenalis)* of Mexico is in

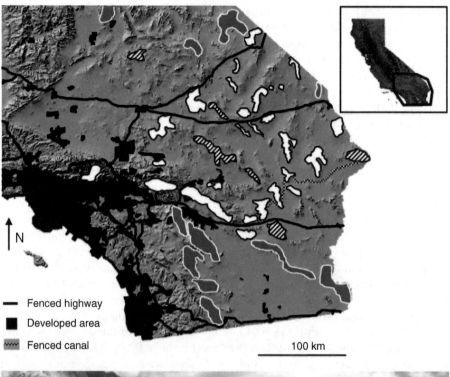

- Fenced highway
- Developed area
- Fenced canal

100 km

FIGURE 15.7   The fragmented populations of desert bighorn sheep *(Ovis canadensis nelson)* in southern California (white patches) studied by Epps et al. (2005) / John Wiley & Sons and the barriers to movement and gene-flow created by fenced highways and water canals. A wildlife overpass crossing a fenced section of the Trans-Canada Highway in Banff National Park, Alberta.

danger of extinction in its native range, but thriving in some metropolitan areas such as Los Angeles, California, where it was accidentally introduced. In a number of cases, there appears to be no negative impacts of such nonnative birds on the native bird populations of the city. A school of **biophilic urban design** has emerged with a focus on planning cities to foster biodiversity and potentially serve as **urban arks** to preserve nonnative and native species endangered in their natural habitat. Similarly, the field of **agroecology** seeks to develop tools and strategies for implementing biodiversity enhancing agricultural practices.

Today, many efforts are focused on **habitat-based conservation** in which the aim is to conserve the entire biological and physical environment in which endangered species live. Habitat-based conservation has the benefit of working to preserve not only those species that we can recognize as endangered, but all of the known and unknown species in the ecosystem as well. In most cases, habitat-based conservation is the only way to ensure the maintenance of important endangered species, biodiversity in general, and ecosystem services. In addition, it is often easier to monitor general habitat conditions using rapid ground observations and remote sensing than it is to conduct detailed census of all species when trying to understand the changes in the biodiversity of a region. In this approach, habitat conditions such as vegetation cover, vegetation greenness, locations of development, and roads can be linked to occurrences of endangered species or overall biodiversity using modeling. The models can then provide estimates of changing biodiversity based on the observed changes in habitat. One such study based on remote sensing of habitat in the Murray-Darling Basin of Australia concluded regions in the north Basin were experiencing decreases in biodiversity, while regions in the south Basin had likely experienced increases.

Habitat has sometimes been so profoundly disturbed by human activity that conservation efforts are no longer a matter of preservation but of restoration. The practice of **environmental restoration** requires rebuilding the biological and physical environment from damaged ecosystems and in some cases in areas where the natural habitat has been completely destroyed. Environmental restoration should be differentiated from **environmental reclamation** in which a severely damaged environment is modified to develop an ecosystem that is quite different from the one that existed prior to human modification. For example, stabilization of slopes and planting of lawn grasses and horticultural varieties of trees to cover up open mine tailings are examples of environmental reclamation, not restoration. Environmental restoration and the field of restoration ecology are extremely active areas of research and application of science and will be discussed further in this chapter.

A challenge of biodiversity and habitat-based conservation, and environmental restoration programs, involves determining what management strategies to implement and what specific measures to use for gauging the program's success. An alternative approach is to focus on the maintenance of key selected species that, because of their resource requirements and role in the ecosystem, provide evidence of strong biodiversity, habitat health, and robust ecological functioning. These indicator species are sometimes called **umbrella species** because it is thought that managing the ecosystem to preserve them provides an umbrella strategy that preserves the biodiversity and habitat integrity of the area in which they live. Some examples of proposed umbrella species include the California gnatcatcher *(Polioptila californica)* for California coastal sage scrubland biodiversity, the greater sage-grouse *(Centrocercus urophasianus)* for western North American sagebrush ecosystems, the white-backed woodpecker *(Dendrocopos leucotos)* for northern European forests, the Bale monkey *(Chlorocebus djamdjamensis)* for bamboo forests in Ethiopia, and the Asiatic black bear *(Ursus thibetanus)* in southern forests of Asia. The northern spotted owl *(Strix occidentalis caurina)* of the Pacific Northwest of North America is a well-studied example of one such umbrella species. By preserving the spotted owl through protection of the old growth forest stands in which it lives, the United States has protected a number of other, often less charismatic, animal and plant species with similar habitat requirements. The continued presence of the spotted owl in the protected stands is taken to be a relatively reliable indication that the ecosystem is functioning well and is providing proper habitat for the other species associated with it in mature closed forest habitat. However, recent research in California has shown that some bird species dependent upon more open forest cover and early successional vegetation receive no benefit from management strategies focused on spotted owls.

Two additional conservation concepts are often linked with umbrella species. In many cases, umbrella species can also be classified as **flagship species.** The term *flagship species* refers to charismatic species that are threatened with extinction and can be used to draw attention and resources to their conservation. This in turn aids in the conservation of habitat and other species. The preservation of the giant panda *(Ailuropoda melanoleuca)* of China has attracted enormous international attention and is an example of a high-profile flagship species that is also an umbrella species for mountain bamboo forest and other species in that habitat.

Keystone species are those species that through their impact on other species and/or on the physical and biological environment have a very large influence over the biodiversity and habitat of the areas in which they live. Keystone species may often be used as umbrella species and sometimes are flagship species. As an example, wolves *(Canis lupus)* are keystone species in the Rocky Mountains. Predation by wolves on elk *(Cervus canadensis)* reduces elk populations and browsing on aspen and shrubs. This increases the density of these plants, which in turn provides habitat and food for beavers *(Castor canadensis)* and other species such as riparian songbirds. The increased number of beavers leads to changes in surface hydrology due to dam building by beavers, increasing habitat for wetland and aquatic species. In the nineteenth and twentieth centuries, wolves were driven to extinction over much of their range in the Rocky Mountains. They have now become a flagship species for conservation and have seen reintroduction in U.S. and Canadian national parks. Research indicates that wolves can serve as important umbrella species for assessing conservation and biodiversity efforts in the parks.

Biogeography plays a particularly important role in developing *in situ* conservation strategies and specific plans to preserve both endangered species and endangered habitat. In particular, biogeographic analysis can help to determine the optimal size, spatial configuration, and geographic location of preserves and conservation areas. In addition, because biogeographers also work on issues of temporal changes in the environment, they can help provide the information required to manage conservation areas over long periods of time and to restore damaged habitat. In the next section, we will explore some of the ways in which biogeographic approaches can be applied to *in situ* species conservation, biodiversity conservation, and habitat restoration.

GEOGRAPHIC STRATEGIES FOR SPECIES CONSERVATION AND BIODIVERSITY CONSERVATION

To provide the best possible plan for the development of conservation areas, it would be ideal to know all of the physical resource requirements, demographic attributes, and biological interactions that are important to the species we wish to conserve. In many cases, however, such detailed autecological and synecological information is not available, and there may not be enough time and resources to gather this information. Most of the time we cannot preserve all of the habitat that was naturally occupied by the species we wish to protect. Indeed, much of the original habitat may already be occupied by urban areas or farmland. In these circumstances, biogeographic analysis can provide some guidelines in establishing protected areas to maximize the chances of success in our conservation efforts, despite our lack of detailed ecological information and the limitations we might face in setting aside large areas for preserves. Biogeography can help address two general geographic questions. What is the minimum area we should set aside to preserve a species or set of species, and what is the best spatial configuration of that reserve area?

Conservation biogeographers approach the question of the reserve area from the standpoint that the reserve must be large enough to support a **minimum viable population size (MVPS)** for the species they wish to preserve. Large reserve areas can support larger populations than small areas. Since large populations have a lower probability of suffering extinction, the larger the area, the more likely it is that it will support a minimum viable population size of a given species. Research on MVPS has focused on animals. The minimum viable population size concept was first outlined during a study of grizzly bears *(Ursus horribilis)* conducted in Yellowstone National Park by Mark Shaffer (1981). He defined the minimum viable population size for a species as the smallest isolated population having a 99% chance of remaining in existence over a 1000-year period. Many conservation scientists have embraced his concept. In one classic study focused on bighorn sheep *(Ovis canadensis)* in the southwestern United States, it was found that populations of 50 sheep or fewer were not able to persist more than 50 years. For biogeographers, the question is, what is the minimum area of habitat required to support the minimum viable population for a given species?

It takes many years of data collection on birth rates, death rates, and disturbance regimes to determine minimum viable population size and minimum habitat area for most species. The difficulties in calculating minimum viable population sizes are highlighted by the population sizes of grizzly bears and other mammals in the national parks of the United States. Some demographic models of

Yellowstone grizzlies suggest that the area of the current park is too small to support a minimum viable population and that the grizzlies should be in decline. Yet field censuses showed a slow increase in grizzly numbers over several decades. In other studies, empirical data show a decline in mammal species richness in the National Park system that is at odds with calculations of their potential mammal biodiversity based on park area. The main problem with some minimum viable population size estimates is that they are often using empirical data from one species, such as the bighorn sheep, and applying the data to other species that are quite different. In addition, models that predict minimum viable population size based on general relationships between birth rates, death rates, and extinction probabilities are sometimes applied without sufficient effort to verify that the results are applicable to the species we are attempting to preserve. If we base conservation efforts on an estimate of minimum viable population size that is too low, we can doom the species to extinction. The IUCN Red List suggested that species with populations of less than 250 individuals could be considered Critically Endangered, those with less than 2500 individuals could be considered Endangered, and those with less than 10,000 individuals could be considered Vulnerable.

In recent years, the techniques for estimating minimum viable population size have become increasingly sophisticated and explicitly geographic. The methodology of estimating minimum viable population size is known as **population viability analysis (PVA).** PVA specifies a survival probability for a species in a given area for a given number of years. In many cases, the target values used in PVA are 95% probability of survival for 100 years. Following the approach of Mace and Lande that we discussed earlier, the birth rates, death rates, disturbance rates, and geographic distributions of endangered species are examined by conservation planners. In addition, a target population size and geographic distribution of that target population required to meet the specified survival probability are determined. For example, we might look at a PVA focused on the red-cockaded woodpecker *(Picoides borealis).* Historical records show that the woodpecker had a natural range extending from central Texas and Oklahoma to Missouri and Kentucky, all the way to the tip of Florida. Due largely to land clearance, the range of the woodpecker has retreated southeastward, so it is now absent or rare in Missouri, Kentucky, Tennessee, and Maryland. Its distribution within its remaining range has become increasingly fragmented. The PVA study found that in order for the red-cockaded woodpecker to escape extinction in the next 100 years, the birds would have to reach a population size of 500 individuals in each of 15 different geographic regions in the southeastern United States. That would equal a total population size of 7500 individuals. The red-cockaded woodpecker is listed as an endangered species in the United States, and steps, including captive breeding and release programs, have been taken to meet the PVA goal. By the early 2000s, stabilization and some increases were observed in the extant woodpecker populations. Today, there are thought to be over 14,000 red-cockaded woodpeckers living in the wild.

Conservation genetics are an increasingly important component of modern PVA. To be viable for long into the future, species ideally must have enough genetic variability to avoid inbreeding depression, be adapted to persist in the different environments encompassed by their range, and be able to adapt to changing conditions through evolutionary processes. As early as 1980, conservation geneticists questioned the low estimates for MVPS such as 50 individuals and suggested that to allow species enough genetic variability to evolve and persist into perpetuity, population sizes of 500 or more individuals were required. From this arose the 50/500 Rule which posits that 50 individuals may allow a population to persist and possibly avoid inbreeding depression, but 500 are required to retain genetic evolutionary potential for long-term persistence. More recent genetically based PVA suggests that the 50/500 Rule may be overly optimistic, and effective population sizes to escape being considered Critically Endangered, Endangered, or Vulnerable should be twice as high as the IUCN Red List criteria cited above. Determining MVPS remains one of the most important and in many cases complex tasks facing conservation science.

Within conservation sciences the relationship between geographic range area and MVPS is encapsulated in the **minimum area requirement (MAR)** concept. MAR is defined as the amount of area of suitable habitat that is required for the long-term persistence of a species. The MAR for any species is influenced by a number of factors. These include demographic properties such as species rates of reproduction, mortality and dispersal properties, niche properties, and trophic level. The body size of the species has proven to be one of the best indicators of MAR as this variable often encapsulates many of the beforementioned properties. A global analysis of over 200 terrestrial animal species (mainly mammals and birds) reveals that body mass is the most significant variable predicting the MAR for the species in the analysis (Fig. 15.8).

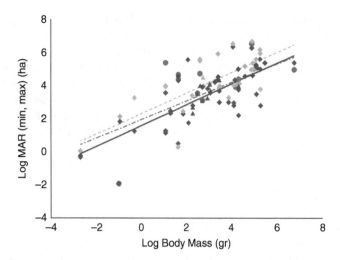

FIGURE 15.8   The relationship between animal body mass and the MAR associated with MVPS based on a global study that mainly incorporated mammals and birds (from Pe'er et al., 2014 / with permission of Elsevier).

The sobering message from these data is that very few species of large mammals or of birds appear to be sustainable with very small geographic ranges. Many parks and nature reserves may be smaller than the ranges that species appear to require for long-term survival. This is not to say that many species might not be sustainable in smaller geographic areas, particularly with intensive management efforts, but in nature few of these species are found to be restricted to such small areas. It is better to err on the side of caution and to make our reserves as large as possible. Unfortunately, the majority of national parks and wildlife refuges, even in the United States, are relatively small, often encompassing less than a few hundred square kilometers. Other semi-protected areas such as government forest lands may be larger in area but are often subject to habitat disturbance by logging, mining, or grazing.

Most natural preserves and conservation areas exist today as fragments surrounded by large areas of land that is used and modified by humans. Established national parks and wildlife areas are often like islands in a sea of developed and modified land. In 1975, Jared Diamond suggested that the theory of island biogeography may provide a general model for assessing the ability of conservation areas to preserve biodiversity. According to island biogeography theory, species diversity is highest on large islands close to the mainland and lowest on small isolated islands. The theory holds that rates of extinction are highest on small islands and isolated islands experience low rates of immigration. Together these factors produce low biodiversity on small isolated islands. If these simple rules from island biogeographic theory are applied to conservation areas, we would conclude that a large single conservation site would be preferable to a number of smaller separate and isolated conservation sites even if the separate sites together comprised the same total area as the large site (Fig. 15.9).

Diamond's conjectures about the application of island biogeography to conservation sparked a controversy in biogeography that continues today. The debate is between those who favor the establishment of single large conservation areas and those who maintain that several smaller areas are better for preserving biodiversity. This difference of opinion is known as the **SLOSS** (Single Large Or Several Small) **Debate.** Opponents of the single large reserve approach have posited several arguments. First, a single large area may not incorporate all of the habitat diversity that several small and strategically placed reserves might incorporate. Thus, a single large area might not provide the range of habitat and species diversity that several smaller reserves might. Second, the increase in species numbers that occurs with increasing island area is nonlinear, having the following form:

$$S = cA^z$$

where S is species richness, c is a constant based on the overall number of species per unit of area (geographic region) in which the site is located, A is the total area of the island, and z is a constant that represents the mathematical relationship between area and species richness. It can be thought of

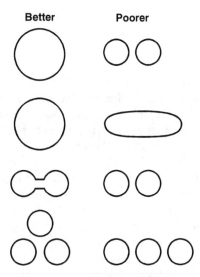

FIGURE 15.9   Different shapes and configurations of conservation areas compared (adapted from Wilson and Willis, 1975). All areas incorporate the same amount of area. Shapes that provide the maximum amount of continuous habitat with the lowest ratio of area to perimeter are optimal according to island biogeography theory. Close locations of similar habitat or corridors connecting reserves are highly beneficial. It is now questioned if a single large reserve is better than several smaller ones.

graphically as the slope of the line that describes the relationship between area and species richness. The value of z ranges from 0.15 to 0.40 (see Chapter 14). What this means is that an increase in island area of 10 times will generally lead to an increase in species richness of roughly only less than 2 to 4 times. Thus, since biodiversity increases in a less than one-to-one relationship with proportional increases in area, several smaller reserves could have the same or greater total biodiversity than one large reserve. Indeed, some empirical data have shown several instances in which small reserves have a combined total species richness that is greater than the species richness of single large reserves of approximately the same area as the total area of the smaller reserves. This result has been demonstrated for communities of arthropods, lizards, birds, and mammals. Third, a single conservation site might be more prone to the extinction of species owing to a large disturbance such as fire or disease than would be the case for several separate sites. This, however, assumes that any single disturbance is limited to one or only a few of the small conservation sites.

In response to the points in favor of several small reserves, some biogeographers have argued that the biota on most islands, particularly oceanic islands, provides a very poor model to use in trying to develop conservation strategies for fragments of continental habitat. They point out that much of the species richness on remote islands can be produced by endemic insular species that are adapted to survival on small and isolated areas of habitat. We have seen that many island species have special adaptations for limited dispersal or specialized feeding that aids their survival on small islands. Continental species, particularly birds and mammals, often do not have such adaptations and appear to rely on relatively large ranges and population sizes. In short, they are not evolutionarily adapted to survival on small isolated fragments of habitat.

Decades have passed since the start of the SLOSS Debate. Where do things currently stand? There remain scientific differences of opinion on the topic. However, it can be argued that the balance is in favor of multiple smaller reserves most frequently being superior to a single large reserve. The Canadian landscape ecologist Lenore Fahrig has put considerable thought into resolving the SLOSS Debate. In 2020, she published an analysis of 157 SLOSS comparisons from 58 different studies and found that in 72% of the comparison cases, several small reserves performed better than a single large area. Only 6% of the cases provided evidence that a single large area performed better than several small areas. Other comparisons were inconclusive. She concluded "conservation practitioners should understand that there is no ecological evidence supporting a general principle to preserve large, contiguous habitat areas rather than multiple small areas of the same total size"

(Fahrig, 2020, p. 615). However, as Fahrig and many others also have stated, the ecological reasons underlying the superiority of several small reserves over a single large reserve of equal area remain in need of further research. These reasons may vary from one geographic setting to the other, or from species to species.

Much effort in the study of geography and conservation has focused on the reduction in biodiversity that is seen when once continuous areas of habitat are fragmented into smaller units. Such geographic fragmentation of species habitats is a serious ongoing challenge today in many regions. We can recall from Chapter 14 that the decrease in biodiversity that occurs when habitat is fragmented into smaller and more isolated patches is called species relaxation. When a small habitat area is preserved, while the surrounding landscape is modified, the species richness of the newly isolated fragment of preserved habitat will likely decline over time. There are many reasons for this decline. For some species, the populations remaining in the small preserve will be below the minimum viable population size and will go extinct due to resource deprivation, chance factors, the deleterious impacts of inbreeding, and so on. For example, the fragment may not provide sufficient resources for larger animals that require large areas for hunting, browsing, or grazing. In very small fragments of forest, because of edge effects, the physical environment in terms of light penetration, wind, evaporation rates, and other factors may be so altered as to cause some forest-dwelling plants and animals to go extinct. The genetic bottleneck created by the reduced population may rob the species of the genetic variability needed to persist in variable habitat or during environmental change. When one species becomes extinct in the fragment, this might cause a trophic cascade and generate additional extinctions. In addition, if a keystone species becomes extinct, this may cause changes to competitive balances that lead to additional extinctions. Time and time again, it has been shown that the preservation of only small isolated fragments leads to a decline in the native species diversity of the fragments. This has been shown in many studies of tropical birds, primates, and insects in small fragments of Amazonian rainforest in Brazil that have been preserved experimentally following large-scale clearance of surrounding forest.

Aside from total area, other features of spatial configuration can be important to the success of conservation areas (Fig. 15.9). First, conservation area designs that decrease the ratio of perimeter to area are better than shapes that produce a large perimeter relative to the preserved habitat area. The edges of forest preserves, for example, are subject to higher wind speeds than the interiors of closed forests. The increased wind stress often causes windthrow and the destruction of old growth forest at the edges of the conservation areas. Windthrow due to this edge effect has been a significant problem in portions of northern California where very small stands of old growth redwoods (Sequoia sempervirens) have been set aside from logging. Large edges also allow greater penetration of light into forest floor ecosystems and also permit easier invasion of the conservation area by alien plant and animal species. The edge effect has been demonstrated to cause tree blowdowns, extra light and wind penetration, and other physical and biological changes for distances of between 100 and 500 meters in North American and Australian forests. The significant expansion of environmental changes inward to patches of forest from edges is one reason why the fragmentation of landscapes by relatively narrow obstructions such as roads can have significant impacts beyond the immediate vicinity of the road. In addition, roads often provide pathways that allow invasive alien plant and animal species to expand into landscapes. For many types of protected habitat, it is necessary to have buffer zones around the core habitat so that edge effects, including invasion by alien species, will not compromise our conservation efforts.

The degree of isolation of fragmented conservation areas is very important. The rescue effect (see Chapter 14) can be extremely beneficial in preventing extinction within a preserve. The more isolated a conservation area is, the more difficult it is for species to disperse to it from other areas of similar habitat. If a species goes extinct in an isolated reserve, it is less likely to be resurrected through immigration. Alternatively, conservation areas that are placed close to other similar conservation areas can help maintain species diversity through the rescue effect (Fig. 15.9). Reserves that are connected or close enough to allow for rescue effect events also help maintain genetic diversity within the reserves. Reserves that are connected by corridors of similar habitat are optimal for promoting the rescue effect. The proliferation of highway wildlife overpasses is an example of conservation planning recognizing the importance of corridors. Numerous recent conservation efforts around the world have focused on preserving or developing larger habitat corridors between conservation areas.

The recognition of the importance of conservation corridors has become so great that in 2019 the U.S. Congress passed the Wildlife Corridors Conservation Act. The act established a National

Wildlife Corridors System to designate national wildlife corridors on federal public lands and provide funds for corridors on state and tribal lands. However, the conservation corridor approach is not without criticisms. It can be argued that reliance on corridors connecting small fragments of habitat is not as effective as preserving larger areas of habitat. It has also been argued that corridors can provide routes by which invasive species or large numbers of predators can penetrate protected areas with deleterious effects. It has also been suggested that corridors present large lengths of perimeter area that facilitate the entry and spread of invasive species to conservation areas.

Geographic Tools for Species Conservation and Biodiversity Conservation

As we can see from above, the effective planning of protected areas for conservation is extremely important and requires spatial and environmental information. Biogeographers have long employed traditional field techniques of vegetation surveying, wildlife censusing, mapping, and so on. As described in Chapter 13, citizen scientists and biogeographic crowdsourcing can provide important on-the-ground information about species distributions, invasive species, and other aspects of conservation habitat. Biogeographic approaches to conservation have also come to be defined by the development and application of analytic tools based on geographic information systems and sciences (GISs) coupled with remote sensing. In particular, GIS is an immense help in collating and analyzing geospatial environmental information and making sure that conservation networks adequately capture the habitat diversity of a region. An important GIS-based approach to conservation planning is called **gap analysis.** In gap analysis, a number of variables relating to the physical and biological environment are placed in GIS databases. The variables commonly include information on topography, landforms, waterbodies, geology, hydrology, soils, climate, general vegetation cover, species distributions, and the locations of current and planned conservation lands. GIS allows maps to be made of these features, and we can overlay species maps to find areas of high biodiversity or to compare the distributions of endangered species to factors such as soil conditions or vegetation cover (Fig. 15.10).

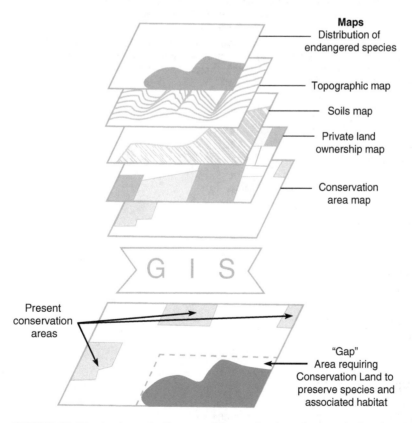

FIGURE 15.10   A schematic diagram of the application of gap analysis using a geographic information system to overlay species ranges, environmental conditions, and landownership information.

Areas of high biodiversity or the ranges of endangered species can be overlaid on maps of current or planned conservation areas to see if those areas are in the right location to preserve biodiversity or endangered species. Often this type of analysis identifies gaps in the geographic distributions of conservation sites where areas of high biodiversity are not protected by any existing conservation area. Gap analysis was pioneered by J. Michael Scott and colleagues in Hawaii to help preserve endangered bird species. The analysis showed that there were indeed major gaps in the conservation network in Hawaii and that this resulted in the establishment of the 6693-hectare Hakalau Forest National Wildlife Refuge. Since the publication of the early gap analysis work in Hawaii in the late 1980s, the remarkable increase in the availability and ease of use of GIS has led to hundreds of gap analyses being conducted at a wide variety of spatial scales for many different target species around the globe. In many cases, gap analyses have revealed significant deficits in the amount of suitable conservation area required to confidently protect the species being studied. A pioneering global gap analysis conducted in the early 2000s by the conservation biologist Ana Rodrigues and her colleagues found that 1400 terrestrial vertebrate species, representing 12% of the species analyzed, had no suitable protected areas anywhere in their current ranges.

In many instances, it is difficult to adequately monitor habitat conditions in large or remote conservation lands. This renders it difficult to impossible to gauge the ability of the conservation area to support overall biodiversity or the species that is being protected. Remote sensing of conservation area habitat, particularly from satellites, is an extremely important biogeographic tool for conservation. Over the twenty-first century, there has been a tremendous increase in the number and technological refinement of spaceborne remote sensing systems that can applied to monitoring conservation areas (Table 15.4). These include passive sensing multispectral systems that measure reflected and emitted light and other wavelengths of electromagnetic energy. Passive systems typically monitor vegetation cover, vegetation health, and vegetation phenology. Active sensing systems emit and then

TABLE 15.4   Various Spaceborne Remote Sensing Systems Used in the Mapping and Monitoring of Habitats and Their Properties

Satellite (sensor)	Pixel Size (m)	Number of Spectral Bands or Wavelength Region	Revisit Time (days)	Years of Operation
High spatial resolution				
QuickBird	0.65, 2.6	5	1–3.5	2001 to present
IKONOS	0.8, 3.2	5	3	1999 to present
GeoEye 1	0.46, 1.8	5	3	2008 to present
WorldView 3	0.3, 1.24, 3.7, 30	29	1	2014 to present
Moderate spatial resolution				
Landsat (ETM+)	15, 30, 60	8	16	1999 to present
Landsat 8	15, 30, 100	11	16	2013 to present
IRS (LISS III)	5, 23, 70	5	5	2003 to present
EOS (ASTER)	15, 30, 90	14	(on demand)	1999 to present
SPOT 5	2.5, 20, 1150	5	26	2002 to present
EOS (Hyperion)	30	220	(on demand)	2000 to present
Low spatial resolution				
NOAA (AVHRR)	1100	5	1	1979 to present
EOS (MODIS)	250, 500, 1000	36	1	1999 to present
Envisat (MERIS)	300	15	3	2002–2012
Active sensors				
QSCAT	2500	Ku	4	1999–2009
Radarsat 2	9–100	C	24	2007 to present
Envisat (ASAR)	30–150	C	35	2002–2012
Sentinel-1A	5–100	C	12	2014 to present
ICESat 1 (GLAS)	70	1064 nm, 532 nm	8	2003–2009

Source: Gillespie et al., 2015 / with permission of SAGE Publications.

capture the return signal from radar and laser beams. These systems can provide detailed information on vegetation canopy height and structure. The development and launch of these numerous space-borne systems have coincided with the availability of tremendous amounts of publicly available satellite data. The result has been a proliferation of publications and reports on habitat and conservation areas. Thousands of such remote sensing studies are published each year. The importance of such information for conservation is inestimable. Estimates of tropical forest habitat loss and fragmentation in critical areas such as the Amazon are dependent on data from satellite imagery. Sequences of images collected over different years provide quantitative data on rates of habitat change and loss throughout Amazonia and the tropics (Fig. 15.11). In addition, since overall biodiversity is often correlated with the diversity of vegetation structure, a property measurable from space, satellite imagery is used to estimate plant and animal diversity in a number of biomes. By using GIS, the

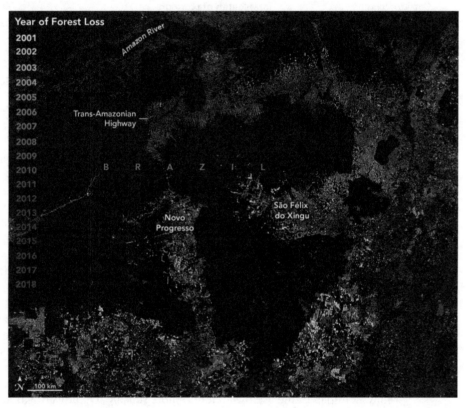

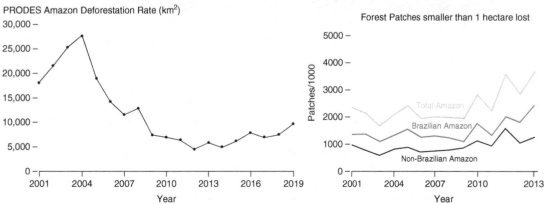

FIGURE 15.11 PRODES satellite image of Amazonian forest clearance (light areas) and resulting fragmentation between 2001 and 2018 in central Brazil. Rate of Amazonian forest loss from PRODES, and number of forest patches of less than 1 hectare in size cleared between 2001 and 2013 (*Source:* NASA / https://earthobservatory.nasa.gov/images/145988/tracking-amazon-deforestation-from-above / last accessed Feb 6, 2024 / Public Domain).

TABLE 15.5   Some Simple Common Landscape Metrics Derived from Remote Sensing Data and Applied in the Analysis of Habitat Fragmentation and Conservation Areas

Metric	Description	Unit of Measurement
Total patch area	Sum of the areas of all patches of the same habitat	Hectares/square kilometers
Individual patch area	Area of each patch of habitat	Hectares/square kilometers
Percentage of patch area	Percentage of total area of landscape covered by patches of the same habitat	Percentages
Patch core area	Core area of habitat protected from edge effects inside a patch buffer perimeter	Hectares
Number of patches	Number of patches in the landscape	Counts
Corridor area	Total area of corridors connecting patches	Hectares/square kilometers
Corridor length	Total length of corridors	Meters/kilometers
Total edge length	Sum of the lengths of all patch edges and/or corridor edges	Meters/kilometers
Perimeter-area ratio	Relationship between perimeter length and patch area	Ratios
Average distance to nearest neighbor	Proximity of similar patches, based on average distance between patches	Meters/kilometers

satellite-derived data on habitat distribution can be mapped and combined with ground-based measurements to aid in gap analysis.

The spatial distributions of habitat or conservation areas mapped via satellite remote sensing can be analyzed using an array of specialized landscape statistics that provide measures of the degree of clumping between patches of conservation habitat, the connectivity between patches, the ratio of edges to perimeter of habitat patches, and the homogeneity of habitat within patches. All of these features are important in conservation planning. There has been much research by biogeographers, landscape ecologists, and others in developing and refining conservation landscape metrics that can be applied to remotely sensed data (Table 15.5). There are, for example, over 30 landscape metrics applicable to quantifying the connectivity of habitat patches. Such metrics can also be applied to study the patch sizes and connectivity of cleared or disturbed patches on the landscape. In analyzing such landscape metrics, biogeography and landscape ecology are closely allied. Finally, by using GIS, time series of air photographs or satellite images can be linked with features such as topography or soils to trace the large-scale patterns and controls of the progress of invading plant species or other vegetation changes caused by humans.

Many examples of the use of satellite remote sensing to study critical habitat loss and fragmentation patterns come from work in the Amazon region (Fig. 15.11). Concerns about rates of deforestation became so acute in the late 1990s and early 2000s that two specialized satellite missions, PRODES and DETER, were instituted to study the problem from space. The data from these missions helped Brazil develop and institute measures to protect the forest. Between 2004 and 2012, there was an 80% decline in the rate of forest clearance and fragmentation. However, analysis of clearance patch sizes showed that there was also an increase in the number of small (less than 1 hectare) patches being cleared. This may have partially reflected an effort by those doing the clearance to avoid detection from space and on the ground. After 2013, the rate of deforestation and fragmentation began to increase, in part reflecting changed government priorities in Brazil.

The use of small drones (small unmanned aircraft system; sUAS) for the mapping and monitoring of smaller conservation areas is becoming an increasingly applied tool. Technological advances have allowed extremely sophisticated photographic and multispectral scanning systems to be carried on small drones. Not only are drones relatively inexpensive to purchase and operate, but they can sense features at much finer spatial scales than satellite and aircraft systems. Drones can, for instance, be used to monitor the locations of individual animals or the expansion of an invasive plant in a wetland. As of 2018, there were at least 256 published studies on the use of drones in conservation. Over

90 of these studies described the use of drone technologies in terrestrial and marine conservation areas. Drones will undoubtedly play an increasing role in conservation research and management, further becoming an important component of the biogeographer's toolkit.

Habitat Restoration and Conservation

As discussed previously, in many cases, the habitat in a targeted conservation area may be so compromised by previous human activity as to make it unsuitable for the species we are trying to protect. The environmental restoration of such disturbed habitat is often a very difficult process. It involves determining the type of ecosystem we wish to "restore" to, how to go about the restoration, and how to assess its success. In addition, although we may start on a restoration project by declaring that our goal is to restore the landscape to a "natural state," many so-called natural landscapes have been modified by human land use over thousands of years, and our idealized view of the natural state may actually be a human-modified system. For example, the sparse and shrubby vegetation typical of wildland areas in Mediterranean portions of southern Europe is the product of human land use, as is much of the lowland forest that Europeans encountered on Polynesian-occupied islands such as Hawaii. Should restoration efforts be aimed at restoring this earlier human modified landscape, or should the aim be to recreate a pre-human landscape that may not have existed for many thousands of years?

A rich scientific literature has developed on the theory and practice of habitat restoration and conservation. The term **restoration ecology** is applied to the theory and practice of restoring ecosystems. A simplified restoration approach can be summarized quite clearly in an elementary graph (Fig. 15.12). The first step is to determine the ideal restoration goal and how success in reaching that goal will be measured. In one instance, the goal may be to preserve an endangered species, such as the giant panda, by restoring the bamboo groves in which the pandas live and feed. In this case, the success of the project is determined by the size of panda populations in the restored habitat. If we can estimate the size of the panda population prior to recent human impact, we have an ideal target to shoot for and we can design our efforts accordingly. However, we may have to compromise on a lesser population size that will still be viable for long-term survival of the species. Restoration of the giant panda habitat is a project in which biogeographers have been very involved. In many cases, the restoration goal may be more complex, such as restoring the total biodiversity of native plant and animal species or restoring the primary productivity of an ecosystem. In addition, changing climate may make the restoration efforts of today ineffective in the future as the basic climatic conditions of a restoration site change over time.

The long temporal perspectives that many biogeographers develop in their research are particularly important for habitat restoration efforts. Without a clear idea of what the earlier environment

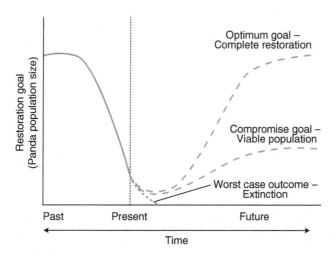

FIGURE 15.12   A graphical representation of restoration aims and tradeoffs (*Source:* National Research Council, 1992 / National Academy of Sciences). In this case, the goal is restoration of habitat to support giant pandas.

was like, as well as what natural variability, disturbance regimes, and human use it was subject to, it is impossible to know how to restore or conserve natural habitat. In addition, many environments are in natural states of change; unless we recognize this fact, our management policies may retard those changes and artificially alter the natural environment. In many cases, conservation managers must rely on short historical records to provide information regarding the original species composition, disturbance regimes, and population dynamics of the habitats they wish to restore. In North America, written accounts of landscapes and ecosystems often go back only 100 to 200 years. In many instances, these historical accounts commence after the landscape and environment have already been significantly impacted by the activities of European settlers. Information on natural conditions or native land-use patterns is often difficult to interpret from historical records.

In recent years, the importance of seeking knowledge of past environmental conditions and past and current land-use practices from the traditional knowledge of Indigenous peoples and local communities has been increasingly recognized as a critical part of conservation and restoration planning and practices. **Indigenous knowledge** and **traditional ecological knowledge** can be challenging to collect and integrate with conventional forms of environmental data and analysis. However, it can also be invaluable in environmental restoration and management efforts. For example, in an effort to restore the Dering-Dibru Saikhowa Elephant Corridor in India, researchers conducted interviews and discussions with local peoples to prioritize the tree species to be planted in the corridor region. Species selected after this process were used in experimental planting aimed to facilitate vegetation rehabilitation and enhance elephant movement. Some of the species were also identified as providing resources used by local peoples. This provided social, economic, and cultural benefits, as well as environmental restoration benefits. Efforts to integrate Indigenous knowledge and traditional ecological knowledge are also important in avoiding colonialist approaches to conservation and restoration which can alienate and prove injuries to local peoples.

During restoration attempts, the restoration goals must often be balanced against economic and cultural concerns. In many parts of the world, Indigenous peoples and their cultures depend on the use of land and resources from the same regions where we might wish to protect endangered species or general biodiversity. For example, to restore all of the pandas' original habitat in China would be impossible because it would mean displacing millions of people and disrupting societies and cultures. In most restorations, a compromise range of goals is identified. The compromise goal may not be to bring panda populations completely back to natural levels, but to bring the population up to a level higher than it is today that will ensure the survival of the species in the wild. In general, a time limit is set as to when the goal should be met, and different restoration strategies are then developed. The incorporation of the concerns and knowledge of local peoples in such strategies in indispensable.

In setting restoration goals and strategies, it is essential to know what the state of the ecosystem was in the past so that we can know the ideal state of the system we wish to recreate if that is our goal. We must also be able to monitor restorations and recognize if long-term variability and natural trends, such as increases or decreases in panda populations, are affecting the ecosystem. For example, there is very good evidence that the geographic range and population size of the California condor have been shrinking since the end of the Pleistocene. European settlement and land-use changes may have simply accelerated the movement of the species toward extinction. In this case, our conservation efforts would have to go beyond restoration of condor habitat if we wished to save the species. This raises an ethical issue regarding whether or not we should intervene to save species that are on a natural road to extinction.

Because biogeographers are concerned with changes that occur over relatively long periods of time, they have been very active in producing long environmental records that can cast light on the earlier natural state of ecosystems. Biogeographers have also been very active in developing records of the long-term variability and disturbance patterns of different ecosystems. The techniques that biogeographers use in such studies include analysis of fossil pollen, fossil diatoms, and other plant and animal remains from lake sediments, peatlands, and packrat middens; analysis of tree-rings and fire scars from forested sites; and study of historical records such as written material, old maps, paintings, and photographs. The application of these techniques to aid in restoration and conservation is sometimes called **applied historical ecology.** In most cases, applied ecology studies attempt to gather records from a number of different sources, ranging from historical documents to fossil evidence, and synthesize these into a coherent picture of what the past environment was like. This work can be quite difficult. Even when historical documents such as texts and maps are available, the task of determining what the past environment was like can be difficult. For example, in a study designed to aid in the restoration of coastal marshes along the San Francisco Bay, maps and documents ranging from eighteenth- and nineteenth-century Spanish land grant papers to nineteenth- and twentieth-century coastal surveys, topographic maps, and aerial photographs were available. In most cases, the maps, photos, and documents provided evidence of past conditions at different geographic locations in the marshes at different times and at different scales of resolution. These different sources required careful geographic reference, interpretation, and synthetization to provide a meaningful overview of past conditions. Often specific information regarding species presence or details on habitat conditions can only be inferred from such data sources.

Given the many aspects of research and planning that go into successful habitat restoration, it has been argued that biogeographers, with their broad training and perspectives, can play an extremely important role that extends beyond contributing to applied historical ecology. The American geographer Scott Markwith summed up the multiple roles for biogeographers in environmental restoration as follows: "The avenues for involvement include pre- and post-restoration monitoring, reference-site selection, data collection and analysis for sound project planning, development and testing of foundational theory, social perceptions and public involvement, and others" (2011, p. 531).

One example of the application of biogeographic research in helping to determine the natural state of an ecosystem prior to European settlement comes from research on ponderosa pine *(Pinus ponderosa)* forests in the southwestern United States. Extensive areas of ponderosa pine forest may be found today throughout Arizona and New Mexico. In many cases, these stands are characterized by large dominant trees that are several hundred years old and by dense understories of smaller trees and shrubs. Such stands are extremely prone to large fires that spread upward to the canopies and destroy both young and old trees. The region has been impacted by declines in the use of fire by Indigenous peoples, logging and the grazing of sheep and cattle, as well as 100 years of fire suppression, and it has been unclear if the present ponderosa pine stands are a natural vegetation type. Studies of old photographs and the writings of early travelers suggest that prior to extensive European settlement and forest management, many of these ponderosa pine stands were typified by widely scattered pine trees that were extremely large and generally old. The understory was dominated by grasses. Small trees and shrubs were rare. Analysis of twentieth-century aerial photographs further showed that there has been a progressive invasion of grassy meadows throughout much of the Southwest by coniferous trees, including ponderosa pines. To supplement the written evidence and photographic records, tree-rings and fire scars have been collected and analyzed. These records show that the many ponderosa pine forests were prone to frequent, but low-intensity, fires until European settlement and fire suppression began in the late nineteenth century. These frequent burns kept the understory of the pine stands clear of most shrubs and saplings and allowed grasses to dominate. In some areas the fires kept trees from establishing, and this produced grassy meadow. In other areas, the thick bark of the ponderosa pines protected the larger trees from ground fires that spread through the stands.

Because of logging, grazing, and fire suppression, the twentieth century experienced very low rates of burning, allowing the conifers to invade some meadows and dense understories of small trees and shrubs to become established under the large ponderosa pines. The current vegetation is now prone to very large, intense fires that destroy timber and structures. The U.S. Forest Service and the National Park Service used the information provided by analysis of historical records and tree-rings to institute a prescribed burn policy to attempt to restore the ponderosa pine stands and other southwestern ecosystems to their natural state. The prescribed burn program is extremely expensive and will take many decades to complete. Similar studies of natural and Indigenous cultural fire regimes and the impact of fire suppression are being conducted by biogeographers in large portions of North America, including many national parks in the United States and Canada. This is an economically important area of biogeographic research. In 1999, the U.S. Department of Agriculture and the U.S. Department of the Interior practiced prescribed burning and other forms of fuel reduction on over 2 million acres of federal land in the United States. The annual budget requested for such work grew to $380 million. In 2023, the president's proposed budget for 2024 called for more than $4.2 billion to support U.S. Department of Agriculture and U.S. Department of the Interior wildland fire and hazardous fuels management. Indigenous peoples in western North America have a long history of using fire to manage landscapes and associated resources. Greatly increased tribal involvement and responsibility is a notable advance in the restoration and management of some western forests. In recent years, a greater proportion of tribal lands have received forest management through prescribed burns than is the case for federally managed lands. Between 1998 and 2018, an average of about 7.5% of tribal lands received treatments each year.

It has been strongly argued that more needs to be done in terms of prescribed burning to restore western U.S. forests. However, prescribed burns and other manipulations of forest fuels, such as mechanical thinning and post-fire salvage logging, have also drawn criticism from some environmentalists. They cite that these practices are not appropriate for all settings, being themselves artificial and potentially causing additional harm to ecosystems. Indeed, research has shown that the fire regime in pre-European contact ponderosa pine stands experienced both low-severity fires and high-severity fires, leading to calls for accepting and allowing variability in fires within these ecosystems and other forests in western North America.

A classic study that sought to reconstruct a pre-European ecosystem and examine options to restore it was conducted at the Indiana Dunes National Lakeshore on Lake Michigan. The site is of special interest because it is where the ecologist Henry Cowles conducted his classic studies on plant succession during the nineteenth and early twentieth centuries (see Chapter 5). The site is also close to Chicago and affords a large urban population the chance to see a "natural" ecosystem. The current upland forest of the Indiana Dunes National Lakeshore is dominated by black oak *(Quercus velutina)* and white oak *(Quercus alba),* with open areas of wet grasslands and prairie. However, the site had been impacted by European land-use practices, and the Park Service wished to see what the natural vegetation was and if the park could be returned to that condition. In order to address this problem, Ken Cole of the U.S. Geological Survey and his colleagues collected data from a number of sources, including ground and aerial photographs, government documents, tree-rings, and fossil pollen records from small ponds and marshes. The ground photographs included a systematically collected time series of photos from survey points in the park. The systematic ground photos were taken regularly from 1978. Aerial photographs for the region span the period 1929 to the present.

The historical ecological evidence shows that the National Lakeshore has indeed been significantly changed by recent human activity. The evidence from the survey photographs shows that there has been a steady increase of fire-intolerant tree and shrub species such as black cherry *(Prunus serotina)* and witch hazel *(Hamamelis virginiana).* The increase in these species is directly attributable to fire suppression. The aerial photographic evidence shows that there has been a steady encroachment of oaks and other trees onto dry prairie areas. Again, it is likely that fire suppression has allowed this expansion of woody vegetation onto the prairie.

Documentary evidence of the vegetation of the park in the year 1834 is available from General Land Office (GLO) surveys conducted by the U.S. government. The GLO surveys provide data on the dominant tree cover encountered in 1834. The GLO records were compared with tree surveys conducted in 1985. Between 1835 and 1985, there has been a significant increase in the density of oak trees and a significant decrease in the density of pine trees. Today, there are small numbers of white pine *(Pinus strobus)* and jack pine *(Pinus banksiana)* in the area. The survey records indicate that in 1835 these pines were extremely abundant and were probably the dominant tree, or codominant with oak. Other historical documents show that the southern shores of Lake Michigan were heavily logged for pines in the mid-nineteenth century. In addition to logging, historical documents reveal that fires set by Europeans in the nineteenth century were often very extensive and likely contributed to the destruction of pine stands.

At this point, we might conclude that the shift from pine to oak was the result of logging, while subsequent oak expansion onto prairies and the increase of fire-intolerant plants in oak stands was the result of fire suppression. Would the reintroduction of fire lead to reestablishment of a pine-dominated vegetation? Should this be the restoration goal for the Indiana Dunes National Lakeshore? Here the story becomes more complex and requires the longer time perspective provided by fossil pollen and charcoal records. Such a record from a site called Howes Prairie Marsh shows that the onset of European land use in the nineteenth century produced a regional decrease in both oaks and pines. Some nearby areas that were dominated by these trees were transformed to agricultural fields. The pollen record shows increases in weedy field plants such as ragweed *(Ambrosia)* during this time. However, the pollen records also show something else that is very important in formulating a restoration strategy for the park. The decline in pines relative to oak and prairie grasses actually began almost 2000 years ago. It appears to coincide with increasing fires, as represented by increased amounts of charcoal in the pond sediments. The reason for the transition from pine to oaks and prairie likely reflects the fact that the southern shoreline of Lake Michigan is rising relative to the water level of the lake and the site is becoming progressively drier. The dry conditions favor oaks and grass over pines. In addition, the oaks and prairie grasses are well adapted to frequent burning, while the pines are not. Once the dry grass and oak vegetation are established, frequent fires serve to exclude pines. The decrease in pines in the National Lakeshore appears to have been a long-term trend that was simply accelerated by European logging and burning in the nineteenth century.

As might be anticipated, in formulating a restoration plan for the National Lakeshore, no serious thought has been given to restoring pine densities. Not only would it be somewhat unnatural, excessively difficult, and very expensive, but it would require intense fire suppression, which would in turn cause an artificial increase in fire-tolerant tree species in oak stands and the invasion of prairie by oaks. The course taken by the Park Service is to apply prescribed burns to limit the densities of fire-intolerant species and limit the invasion of prairies by oaks.

The example from the Indiana Dunes National Lakeshore shows how complex the issues associated with habitat restoration and conservation can be. The example also demonstrates that without adequate long records of past environmental conditions, disturbance regimes, and natural ecosystem trajectories, we cannot be confident that restoration and conservation strategies will truly serve to return sites to a desired previous condition. Finally, the Indiana Dunes region, like many ecosystems, was in a natural state of change prior to the European impact. These natural shifts will, and should, continue into the future. Attempting to restore a habitat should not be considered the equivalent of trying to freeze it in time and artificially preserving some idealized "natural" state.

It is important to recognize that in some cases, due to the extinction of important species, introduction of alien species, or alteration of the physical environment of the site, it is simply not possible to restore the pre-disturbance ecosystem; rather, decisions must be made to identify an optimal alternative. In addition, due to factors such as climate change or changing surrounding landcover, the restored ecosystem may not be sustainable into the future. Purdue University biologist Young Choi has suggested that a future-oriented paradigm is needed for restoration ecology and environmental restorations that "should (1) establish the ecosystems that are able to sustain in the future, not the past, environment; (2) have multiple alternative goals and trajectories for unpredictable endpoints; (3) focus on rehabilitation of ecosystem functions rather than recomposition of species or cosmetics of landscape surface; and (4) acknowledge its identity as a 'value-laden' applied science within economically and socially acceptable framework" (2007, p. 351).

The Anthropocene and Climate Change Challenges Ahead

As discussed previously, the profound changes that humans have wrought on the landcover of the earth, its atmosphere and climate, and the other species we share the planet with have led some scientists to propose that we have entered a new geologic Epoch—the Anthropocene. Dutch meteorologist Paul Crutzen and American limnologist Eugene Stoermer proposed the term over 20 years ago. However, the precise definition of the Anthropocene, the timing of its start, and if it qualifies as a formal geologic Epoch remain contentious. Designation as a new Epoch has thus far been rejected by the international governing body for geologic stratigraphy. Some say we should consider the Anthropocene to be a regionally variable event that commenced in some parts of the world as early as 40,000 years ago. Others contend its start should be pegged at 1950 and the start of the nuclear bomb era. The development of nuclear bombs and open-air testing of bombs released plutonium into the environment, and this serves as a chronological marker in sediment records. According to those proposing the designation of an Anthropocene Epoch commencing at 1950, this date is said to mark the start of the **Great Acceleration** in human society and the acceleration of global impacts of humans on the planet. The Great Acceleration, it is argued, is marked by quickening rates of human population growth (Fig 15.13), global economic growth, technological growth, energy use, resource use, and the impact of these and other factors on the environment. During the period after 1950, we can see that CO_2 emissions and global surface temperatures have shown marked increases (Fig. 15.13). We have also experienced increasing rates of extinctions.

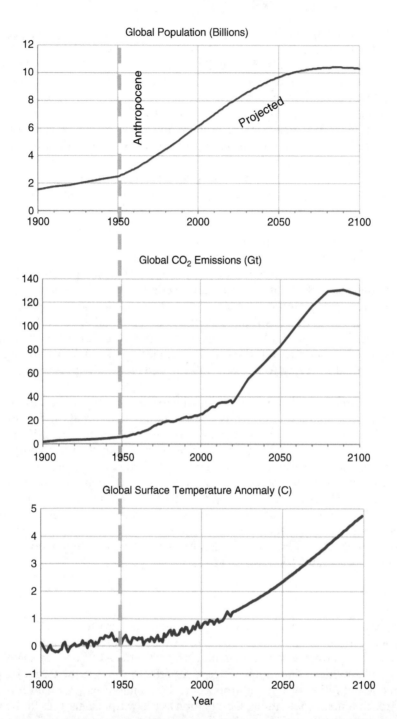

FIGURE 15.13 Observed changes in global populations, CO_2 emissions, and surface temperature anomalies from 1900 to 2022 and projected subsequent changes in these variables to 2100. Projected emissions and temperature are based on the IPCC scenario SSP5 8.5, which assumed no effective policy is implemented to reduce greenhouse gas emissions. The proposed start of the Anthropocene Epoch in 1950 is indicated by the dashed vertical line (data from U.S. Census Bureau, https://www.census.gov/data/tables/time-series/demo/international-programs/historical-est-worldpop.html; United Nations World Populations Prospects 2022 medium fertility variant, https://population.un.org/wpp/Download/Standard/MostUsed/; IPCC Sixth Assessment Report 2022 https://ipcc-browser.ipcc-data.org/).

Although it is clear that there has been increasing growth in human populations, the economy, and technology since 1950, and this has coincided with numerous environmental impacts, including increasing greenhouse gases and climate change, there are some valid arguments against the designation of 1950 as the start of a new geologic Epoch. After having read this book, you will recognize some of these. First, many of the attributes associated with the Great Acceleration and the Anthropocene began long before 1950. For example, human population growth has been experiencing a long-term upward trend since the development of agriculture over 10,000 years ago and a particularly strong increase after the Industrial Revolution about 300 years ago (see Chapter 11). The development of agriculture also produced massive changes in the landcover of the planet from the Early Holocene onward. Humans are associated with major extinction events of large mammals in the Late Pleistocene and Early Holocene (see Chapter 12). The current wave of extinctions began in the 1600s during the age of European expansion and colonization (see Chapter 12). Pacific island bird fauna experienced numerous extinction events during the spread of Polynesian agricultural societies commencing around 2000 years ago (see Chapter 12). The radioactive isotopes of plutonium have half-lives of between 87 and 24,065 years and will be undetectable in sediments after about 240,000 years if no other plutonium is released into the environment. If human population growth decreases as expected and more sustainable resource use patterns are developed, some of the environmental disruptions associated with the concept of the Anthropocene Epoch may also decline in the future. Remember that the shortest geologic Epoch prior to the Holocene is the Pleistocene, and that extended for over 2.5 million years. These and other arguments have been put forth by scientists such as Phil Gibbard at the University of Cambridge in opposition to the designation of a formal Anthropocene Epoch. They perceive the Anthropocene as a geologic event that is time-transgressive rather than a formal Epoch that can be defined as starting at a precise moment in time and evident from one geologic stratigraphic boundary. Debate will certainly continue on the nature and formal designation of the Anthropocene.

Regardless of if we accept the Anthropocene as a formal geologic Epoch or see it as a time-transgressive event, the Anthropocene concept has focused important attention on the startling growth in human populations, technology and resource use, and the profound changes that humans have wrought on the environment. The term *Anthropocene* is now embraced by a wide range of natural scientists, social scientists, humanities scholars, and others. As a result, there is a plethora of views on what the Anthropocene represents and what actions might be taken in response to it. The Anthropocene is often framed in terms of the problems and challenges presented to social and environmental sustainability. Going beyond the narrow geologic arguments of designating the Anthropocene as a formal Epoch, we might find it more useful to consider a broader definition. The political theorist Manuel Arias-Maldonado provides the following definition: "The Anthropocene is understood here as a concept that designates a state of the planet in which key natural systems are coupled with social systems at a global level, thus influencing [the] Earth system as a whole, thereby turning the human species into a global geophysical force that put[s] the future habitability of the Earth into question" (Arias-Maldonado, 2020, p. 97). As is clear from our discussion here, and from Figure 15.13, the changes to the planet and the environmental challenges posed by the Anthropocene not only extend back in time to well before 1950, but will also extend into the future and intensify as human population and resource requirements grow, greenhouse gases accumulate in the atmosphere, and climate continues to change. The central challenge of the Anthropocene is not whether it is a geologic Epoch; the central challenge is to provide for the sustainable well-being of humans and the wider environment. This has been called "one of the greatest challenges facing humanity" (Seddon et al., 2016, p. 1).

Earlier in this chapter, we looked at how factors such as overexploitation, agricultural activity, human disturbance/modification, urban development, invasion and disease, pollution, transportation, energy production, and climate change are contributing to extinctions, decreased biodiversity, and decreased population abundance for a number of taxa of plants and animals. These factors are all hallmarks of the Anthropocene and have the potential to greatly increase over the twenty-first century. It would take many books to address all of these factors in any depth. Let's focus on some unique overarching properties of the environmental challenge posed by the Anthropocene. These are the unprecedented geographic scope and global impacts of anthropogenic climate change, the rate at which climate change is occurring, and the development of novel new climates and ecological communities that have no analogs in the past.

The increasing levels of CO_2 and other greenhouse gases in the atmosphere caused by the burning of fossil fuels and other human activities are resulting in significant and rapid warming and changes in global climate that will accelerate over the next century (see Fig. 15.13 and Chapter 7). The impact of climate change on the biosphere is dramatic and is the subject of much study. Many of the areas of negative impact on people and the biosphere are in the tropics where we find numerous biodiversity hotspots (Fig. 15.6). Countries in these regions often have limited economic capacity to tackle climatic and environmental challenges. Arctic regions also face acute threat due to the amplified impacts of warming at high northern latitudes. This problem of climate change and the environment is of major interest to biogeographers. One approach to anticipating the biogeographic impacts of greenhouse warming is to examine how the geographic distribution of biomes might be affected. Biogeographers have long studied the link between climatic conditions and the present distribution of biomes. The primary productivity and vegetation structure of the biomes in turn provides energy and habitat required by higher trophic levels. By assuming that vegetation formations will move to adjust to geographic shifts in climate, we can project the future distributions of the biomes. There are two weaknesses to this approach. First, we do not know how long it will take species to shift their range boundaries. The rate of climatic warming in the next 100 years will likely be faster than any large-scale natural climatic changes since the end of the last ice age. Biomes may remain out of adjustment with new climatic conditions for some time, and some species may go extinct. Second, it is possible that combinations of temperature and precipitation that do not have a modern counterpart could develop and create plant and animal communities and biomes that have no exact modern counterpart, or any counterpart in the recent past. Of course, due to past and ongoing extinctions, coupled with invasive species, the community composition of many ecosystems is already becoming something without a past analog. In any case, a number of attempts have been made to predict the changes in the global distribution of biomes due to greenhouse climate change. Some studies have also incorporated the direct impact of increased CO_2 on rates of photosynthesis, water use, and plant growth. Although the results of such predictive modeling do not always agree with each other, many of these projections indicate that by 2100 large areas of the Arctic tundra will be replaced by boreal forest (Fig. 15.14). In turn, large areas presently occupied by boreal forest will be occupied by temperate deciduous forest of grassland/temperate savanna. Rainforest will expand in some areas and contract in others. The world of 2100 will likely look very different in terms of vegetation cover in many parts of the world.

As mentioned earlier, the shifts in species ranges and biomes that would accompany anticipated global warming will not occur instantly. Rather, some species will adjust their ranges quickly, while those that have lower dispersal rates may lag behind the climatic changes. Some biogeographers have looked at fossil records to see how quickly flora and fauna responded to the natural climatic changes that occurred over the Quaternary. Fossil pollen evidence shows that the rate at which tree species dispersed northward following the close of the last age ranged from less than 100 meters per year to more than 1000 meters per year, and this provides some evidence of the rate at which these species might adjust their ranges in response to future global warming. Unfortunately, these past dispersal events do not provide a perfect analog for future warming because most natural changes in global climate occurred at a much slower rate than anticipated greenhouse warming. Other biogeographers are looking for evidence of recent shifts in ecosystems and species range boundaries that indicate that organisms are already responding to warmer temperatures. It has already been found that southern species of coastal invertebrates are extending their ranges northward along the Pacific Coast of North America, bird species are extending their ranges northward in Europe and North America, and the size of tree populations is increasing at the northern edge of the boreal forest biome. Remote sensing and GIS are being combined to help detect large-scale shifts in ecotones. such as treeline, that are difficult to resolve from ground-based studies. Treeline location should be particularly sensitive to future climate warming.

Much work is being done to develop computer models of how organisms might adjust their ranges in the face of changing climatic conditions. The results from these models suggest that many species will face two difficult obstacles. First, the rates of climatic warming will require unusually rapid dispersal capabilities. Second, because many types of habitat have become extremely fragmented and smaller in size due to human activity, many species will have to cross large barriers, such as extensive farmlands or urban areas, as they migrate. The reliance on jump-dispersal events to cross such barriers will slow or even preclude the ability of some species to adjust to new climatic conditions. In these cases, *ex situ* conservation measures, such as maintaining breeding animal populations in zoos or holding plants and seed banks in botanical gardens, may be required to ensure the survival of some species during the period of rapid climatic change.

How should conservation biogeography engage with the Anthropocene? The global and explicitly geographical nature of biogeography, linked with its long temporal perspectives and inherent

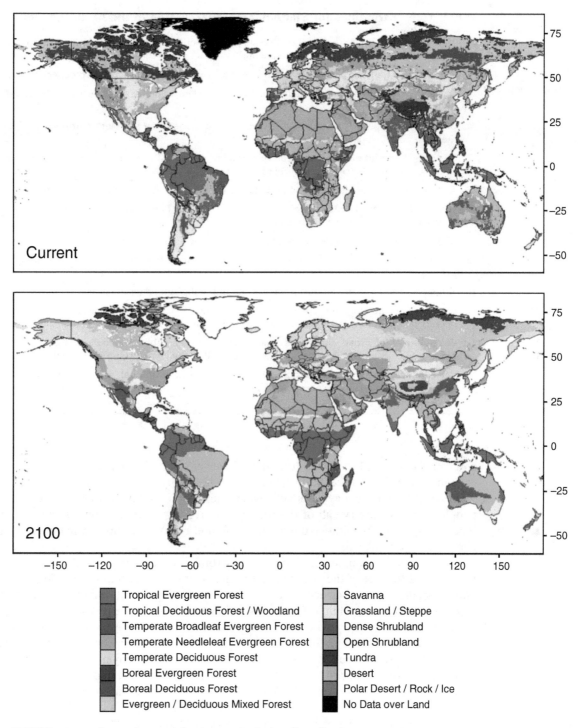

Tropical Evergreen Forest
Tropical Deciduous Forest / Woodland
Temperate Broadleaf Evergreen Forest
Temperate Needleleaf Evergreen Forest
Temperate Deciduous Forest
Boreal Evergreen Forest
Boreal Deciduous Forest
Evergreen / Deciduous Mixed Forest

Savanna
Grassland / Steppe
Dense Shrubland
Open Shrubland
Tundra
Desert
Polar Desert / Rock / Ice
No Data over Land

FIGURE 15.14 One estimate of the changes in the location of major terrestrial biomes that might result from global climate change (*Source:* Sato and Ise, 2022 / Copernicus Publications / CC BY 4.0.).

coupling of nature and humans, place it ideally to tackle the challenges posed by the Anthropocene. The geography of conservation strategies is of paramount importance. A global analysis of nearly 4000 terrestrial mammal species found that many of the current protected areas were too small and too poorly connected to provide reliable protection from extinction. It was estimated that over 1000 mammal species not under threat today will be threatened in the future. Perhaps 50% of all animal species are ultimately at risk of extinction. The loss in mammalian biodiversity will take millions of years to recover from, if ever. As noted earlier, gap analysis serves to identify such problems

TABLE 15.6   Strategies for Sustainable Conservation in the Anthropocene

1. Biodiversity and landscape conservation surveys to increase knowledge of how many species there are, their vulnerability, and their habitat requirements

2. Conservation genetics analysis to determine the scope of genotypes and cryptospecies for conservation

3. Gap analyses that are comprehensive and forward-looking to protect species in current ranges and anticipated future ranges

4. Connectivity analysis to make sure that protected areas have sufficient corridors, bridges, and so on to maintain populations and gene-flow

5. Systematic prioritization of species by vulnerability, potential for successful conservation, importance to other species, and importance to ecological services provision

6. Targeted ecological restorations to provide sustainable habitat now and for anticipated future conditions

7. Targeted translocation efforts to place endangered species in habitats and areas where they can be sustained now and under anticipated future conditions

8. Biophilic urban design planning so that cities can serve as biodiversity arks

9. Agroecologically informed agricultural practices to support biodiversity while providing sustainable food supplies

10. Targeted *ex situ* conservation efforts for important highly endangered species

11. Engagement with Indigenous peoples and local communities to inform conservation efforts and work in partnership on environmental sustainability and socioeconomic and cultural sustainability

12. Development of broad consensus among an informed public and policymakers on the importance of biodiversity, ecological services, and the sustainability challenges being faced now and in the future

13. Scaled-up regional, national, and global protected area development, corridor development, habitat restoration, translocation, and *ex situ* conservation efforts via governments, nongovernmental organizations, and the private sector

and can also be used to design solutions. However, such efforts must take into account the rapidly changing conditions of the Anthropocene and the reality that the environments and biological communities of the future may be novel new entities requiring new conservation strategies. While some people hope to conserve all of the present biodiversity of the planet, others feel that appreciable biodiversity loss is inevitable and attention should be focused only on habitat and species that can confidently be protected.

The rapidly changing climatic and environmental conditions and the interconnectedness of humans and the rest of the environment in the Anthropocene require solutions that incorporate the natural sciences, social sciences, and humanities to be successful. Importantly, these strategies must meet the test of sustainability. **Environmental sustainability** in the case of the Anthropocene requires the management of the environment and natural resources to meet current human and ecological needs while also providing for the ability to meet future needs. Climate change means that what is required to preserve a species, habitat, or ecological service today may not work in the future. Based on what we have learned in this book, we can make a list of some important biogeographic conservation strategies to meet the environmental sustainability of the Anthropocene in terms of biodiversity (Table 15.6). In reviewing the list, we can see that biogeographers have their work cut out for them, not only in terms of science, but also in terms of engagement with other stakeholders on our planet, including Indigenous peoples and local community members, policy makers, and the public at large.

REASONS FOR HOPE

It is clear that the Anthropocene poses unprecedented challenges for environmental sustainability. It is also clear that as of yet we have made insufficient progress in confronting greenhouse gas emissions and otherwise combating climate change. Nor have we succeeded in implementing adequate protected areas, conservation corridors, and other strategies to maintain the biodiversity and ecological services of the earth. More needs to be done and quickly. However, dwelling only on doom and gloom

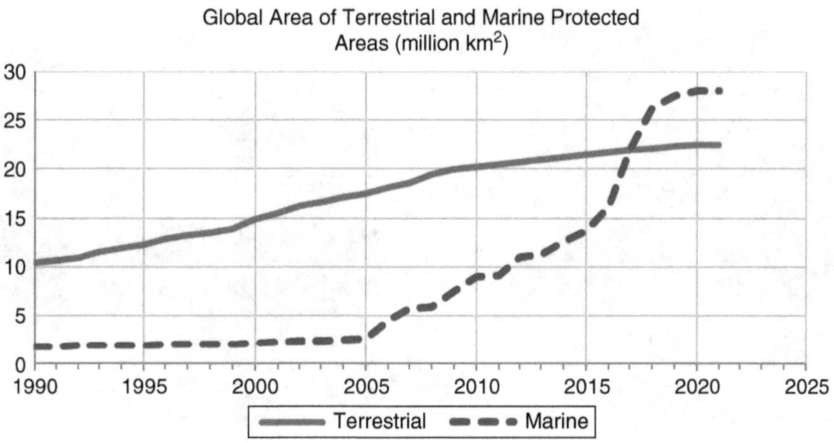

Global Area of Terrestrial and Marine Protected
Areas (million km^2)

FIGURE 15.15   Increase in the area of marine and terrestrial protected areas from 1990 to 2021 (data from UNEP-WCMC and IUCN Protected Planet Report 2020, https://livereport.protectedplanet.net/chapter-3).

and therefore giving up on our efforts is not productive. As we have seen in the earlier sections of this book, remarkable progress has been made in developing theory, analytic tools and important data sets in biogeography, macroecology, landscape ecology, conservation genetics, and conservation ecology. Technologies related to genetics, remote sensing, GIS, and modeling have evolved at an incredible pace. All of these can aid biogeographers and conservation scientists in tackling the environmental challenges of the Anthropocene.

Awareness by the public of the importance of conservation of the environment and biodiversity to humans has also increased remarkably. A survey conducted in the United States in 2023 found that 73% of Americans agreed that biodiversity is important and 67% believed protecting biodiversity should be a national priority. Prioritization of environmental sustainability and biodiversity conservation is not restricted to any one nation or continent. Between 1990 and 2021, 12 million square kilometers of terrestrial land area and 26 million square kilometers have been added to marine-protected areas globally (Fig. 15.15). During the 2022 United Nations COP15 conference, more than 190 nations agreed to 23 different targets to help halt the biodiversity crisis. One of the most important was the agreement to protect 30% of the earth's land and water area by 2030. At that time, about 16% of the land surface and 58% of marine waters were protected areas. The importance of Indigenous peoples to biodiversity conservation was also widely recognized at COP15. As an example, Canada committed $600 million to Indigenous conservation efforts and the creation of Indigenous Protected and Conserved Areas. Let's hope such progress continues and accelerates.

In closing this discussion, let's look beyond numbers and graphs and consider the conservation success story of two remarkable marine mammals as an example of hope. The sea otter *(Enhydra lutris)* is the smallest marine mammal in North America and lives close to shore in kelp and seagrass beds, making it a draw for ecotourism (Fig. 15.16). In precolonial times, the sea otter population along the northwestern coast of North America and northeastern coast of Asia has been estimated at between 150,000 to 300,000 individuals. Following European colonization of the west coast of North America in the mid-1700s, the fur of sea otters became a much sought after and overharvested commodity. By 1900, sea otters were almost completely extirpated across their entire range. The California sea otter *(Enhydra lutris nereis)* survived only as a virtually hidden population of about 30–50 individuals at a small remote spot on the coast of central California. This was a range collapse of hundreds of kilometers from the California otters' former northern and southern range limits. In 1911, an international treaty was enacted to prohibit hunting of sea otters, and populations began to recover. In 1972, the Marine Mammal Protection Act was enacted in the United States, and in 1977, the U.S. Fish and Wildlife Service listed California sea otters as threatened. A recovery plan was adopted for the otters in 1982. By 2016, the population of California sea otters had reached over 3000 individuals and their range had regrown to cover additional regions along the California coast where they had thrived prior to overhunting. Although they were on the razor edge of extinction, relatively modest efforts of protection and some translocations have produced a remarkable resurgence in the population and

FIGURE 15.16   California sea otters *(Enhydra lutris nereis)* in a seagrass bed at Morro Bay, California. Elephant seals *(Mirounga angustirostris)* warming themselves on a breeding beach near San Simeon, California. Both species have experienced large population growth and reexpansion of their geographic ranges following near extinction in the late nineteenth and early twentieth centuries.

geographic range of the California sea otters. Interestingly, genetic analysis indicates that all sea otter subspecies had relatively low genetic diversity originally and experienced declines in genetic diversity due to the bottleneck created by overhunting. Low genetic diversity has not seemed to impede recovery. California sea otters still do not yet occupy their entire former range, being absent from far northern California and Oregon. They also experience significant population variations. Maintenance and restoration of kelp habitat is seen as an important step in California sea otter conservation. At present, efforts are being made, including translocations, to reestablish sea otters in other areas of their former range in the Pacific Northwest and Asia.

In contrast to the sea otter, the northern elephant seal *(Mirounga angustirostris)* is the second largest seal in the world, with males reaching up to 2000 kilograms. Sea elephants spend most of their time at sea, but come ashore to breeding colonies along the coast of California and adjacent Mexico (Fig. 15.16). Easy to hunt when they are ashore, northern sea elephants were intensively harvested for their natural oil after the start of the colonial period. However, a small population of 20 to 100 individuals survived and continued to breed on the remote Isla Guadalupe off the coast of Baja California, Mexico. This was a case of extreme southward range eclipse from their former northern breeding sites. Appallingly, in 1892, a Smithsonian Museum expedition killed some of the few remaining sea elephants for their collections. The Smithsonian continued to visit and kill seals, justifying the actions under the premise that the species was doomed to extinction anyway. Thankfully, in 1922, the elephant seals were placed under protection by the government of Mexico. Much later, they came under the protection of the U.S. Marine Mammal Protection Act of 1972. Breeding beaches in California are also designated protected areas and disturbing the elephant seals is prohibited. The population of northern elephant seals has now grown to exceed 200,000 individuals and breeding sites can be found from portions of Baja California, Mexico, to as far north as Point Reyes in northern California. The elephant seals experienced a severe population bottleneck and decline in genetic diversity due to their near extinction. However, despite this, they have shown a strong recovery. More recently, there has been a decline in population size in sites in Mexico, possibly due to climate change and warming waters. More northerly populations have not experienced similar declines.

The population recovery and range reexpansion of the California sea otter and the northern elephant seal are examples of the ability of some species to rebound from near extinction—if we provide simple protection from exploitation and protection of their habitat. Every year, I take a field course to the coastline of central California and show students these remarkable marine mammals in their native habitat. How much better is this than having to show them empty waters and beaches and then lecture on the extinction of the two species. We can take hope that despite the clear challenges posed by the Anthropocene, our actions now can and will save other species from extinction. However, we cannot be complacent or assume that all species will survive and recover the Anthropocene Crisis. As biogeographers, we know we have much work to do over the coming years to preserve the wonderful biodiversity of our planet, but we also know we can make a positive difference.

FINAL REFLECTIONS

Throughout this book, we have examined and celebrated the great diversity of life that exists on our planet today. We have seen that geography and history have worked together to produce the incredibly complex and beautiful biosphere that we are a product and a part of. We have also learned that humans have played a lamentable role in the extinction of a great many species. In this final chapter, we have observed that we face many challenges in preserving the remaining biodiversity and natural habitats of the world. The recognition of the Anthropocene compels us to face these challenges with urgency. We have learned how biogeography and other sciences are tackling these challenges.

We have also seen how the study of biogeography can help us to be aware of, understand, and appreciate the spectacular diversity of life and habitats of the planet. Biogeography also helps us understand the long and fascinating story of our own species. Perhaps for some of you, this book will start you on the road to learning more about biogeography, ecology, paleontology, and the other life and earth sciences. You may actually take up the challenge of working as biogeographers to understand and protect the biosphere. If so, this book has been more than successful. For others, I will consider this book successful if a trip through a national park or even a simple walk through autumn leaves is just a little more enjoyable because you are now more aware of how you, and all the plants and animals you encounter, are part of the marvelous and continuing story of space, time, and life.

KEY WORDS AND TERMS

Agroecology
Applied historical ecology
Biodiversity conservation
Biodiversity hot spots
Biophilic urban design
Charismatic flora
Charismatic megafauna
Conservation biogeography
Endangered species
Enviroethics (ecological and
    environmental ethics)
Environmental reclamation
Environmental restoration
Environmental sustainability

Ex situ conservation
Flagship species
Gap analysis
Great Acceleration
Habitat-based conservation
Indigenous knowledge
In situ conservation
Land ethic
Landscape ecology
Landscape genetics
Minimum area requirement (MAR)
Minimum viable population
    size (MVPS)
Natural climate solutions (NCSs)

Population viability analysis (PVA)
Range collapse
Range eclipse
Restoration ecology
SLOSS debate
Species-based conservation
Threatened species
Traditional ecological knowledge
Translocation
Umbrella species
Urban arks

REFERENCES AND FURTHER READING

Abadía-Cardoso, A., Freimer, N. B., Deiner, K., and Garza, J. C. (2017). Molecular population genetics of the northern elephant seal *Mirounga angustirostris*. *Journal of Heredity* **108**, 618–627.

Ahmadi, M., Farhadinia, M. S., Cushman, S. A., Hemami, M. R., Nezami Balouchi, B., Jowkar, H., and Macdonald, D. W. (2020). Species and space: a combined gap analysis to guide management planning of conservation areas. *Landscape Ecology* **35**, 1505–1517.

Akçakaya, H. R., Bennett, E. L., Brooks, T. M., Grace, M. K., Heath, A., Hedges, S., Hilton-Taylor, C., Hoffmann, M., Keith, D. A., Long, B., and Mallon, D. P. (2018). Quantifying species recovery and conservation success to develop an IUCN Green List of Species. *Conservation Biology* **32**, 1128–1138.

Allendorf, F. W., Hohenlohe, P. A., and Luikart, G. (2010). Genomics and the future of conservation genetics. *Nature Reviews Genetics* **11**, 697–709.

Andrello, M., D'aloia, C., Dalongeville, A., Escalante, M. A., Guerrero, J., Perrier, C., Torres-Florez, J. P., Xuereb, A., and Manel, S. (2022) Evolving spatial conservation prioritization with intraspecific genetic data. *Trends in Ecology & Evolution* **37**, 553–564.

Arponen, A. (2012). Prioritizing species for conservation planning. *Biodiversity and Conservation* **21**, 875–893.

Avise, J. C., and Hamrick, J. L. (1996). *Conservation Genetics: Case Histories from Nature*. Chapman and Hall, New York.

Baker, B. C., and Hanson, C. T. (2022). Cumulative tree mortality from commercial thinning and a large wildfire in the Sierra Nevada, California. *Land* **11**, 995.

Baker, W. L. (1992). Effects of settlement and fire suppression on landscape structure. *Ecology* **73**, 1879–1887.

Baker, W. L. (1993). Spatially heterogeneous multi-scale response of landscapes to fire suppression. *Oikos* **66**, 66–71.

Baker, W. L. (1994). Restoration of landscape structure altered by fire suppression. *Conservation Biology* **8**, 763–769.

Baker, W. L., Honaker, J. J., and Weisberg, P. J. (1995). Using aerial photography and GIS to map the forest-tundra ecotone in Rocky Mountain National Park, Colorado, for global change research. *Photogrammetric Engineering and Remote Sensing* **61**, 313–320.

Baker, W. L., Veblen, T. T., and Sherriff, R. L. (2007). Fire, fuels and restoration of ponderosa pine–Douglas fir forests in the Rocky Mountains, USA. *Journal of Biogeography* **34**, 251–269.

Banerjee, P., Dey, G., Antognazza, C. M., Sharma, R. K., Maity, J. P., Chan, M. W., Huang, Y. H., Lin, P. Y., Chao, H. C., Lu, C. M., and Chen, C. Y. (2021). Reinforcement of environmental DNA based methods (sensu stricto) in biodiversity monitoring and conservation: a review. *Biology* **10**, 1223.

Barbier, E. B., Hacker, S. D., Kennedy, C., Koch, E. W., Stier, A. C., and Silliman, B. R. (2011). The value of estuarine and coastal ecosystem services. *Ecological Monographs* **81**, 169–193.

Barnes, M. A., and Turner, C. R. (2016). The ecology of environmental DNA and implications for conservation genetics. *Conservation Genetics* **17**, 1–17.

Bartlein, P. J., Whitlock, C., and Shafer, S. L. (1997). Future climate in the Yellowstone National Park region and its potential impact on vegetation. *Conservation Biology* **11**, 782–792.

Beatley, T. (2016). *Handbook of Biophilic City Planning and Design*. Island Press, Washington, DC.

Beichman, A. C., Kalhori, P., Kyriazis, C. C., DeVries, A. A., Nigenda-Morales, S., Heckel, G., Schramm, Y., Moreno-Estrada, A., Kennett, D. J., Hylkema, M., and Bodkin, J. (2023). Genomic analyses reveal range-wide devastation of sea otter populations. *Molecular Ecology* **32**, 281–298.

Bellard, C., Marino, C., and Courchamp, F. (2022). Ranking threats to biodiversity and why it doesn't matter. *Nature Communications* **13**, 2616.

Beng, K. C., and Corlett, R. T. (2020). Applications of environmental DNA (eDNA) in ecology and conservation: opportunities, challenges and prospects. *Biodiversity and Conservation* **29**, 2089–2121.

Berger, J. (1990). Persistence of different-sized populations: an empirical assessment of rapid extinctions in bighorn sheep. *Conservation Biology* **4**, 91–98.

Branton, M., and Richardson, J. S. (2011). Assessing the value of the umbrella-species concept for conservation planning with meta-analysis. *Conservation Biology* **25**, 9–20.

Braverman, I. (2014). Conservation without nature: the trouble with in situ versus ex situ conservation. *Geoforum* **51**, 47–57.

Brennan, A., and Lo, N. Y. S. (2022). Environmental ethics. In *The Stanford Encyclopedia of Philosophy* (ed. E. N. Zalta). Stanford Publishing, Stanford, CA.

Brooks, T. M. (2004). Coverage provided by the protected-area system: is it enough? *BioScience* **54**, 1081–1091.

Brown, J. H. (1995). *Macroecology*. University of Chicago Press, Chicago.

Brown, J. H., and Lomolino, M. V. (1998). *Biogeography*, 2nd edition. Sinauer Associates, Sunderland, MA.

Burdett, M. (2018). Case study: ecotourism in Kenya. GeographyCaseStudy.Com. https://geographycasestudy.com/case-study-ecotourism-in-kenya/.

Burnett, R. D., and Roberts, L. J. (2015). A quantitative evaluation of the conservation umbrella of spotted owl management areas in the Sierra Nevada. *PloS One* **10**, e0123778.

Bush, M. B. (1996). Amazonian conservation in a changing world. *Biological Conservation* **76**, 219–228.

Butler, D. R. (2020). The Anthropocene: a special issue. *Annals of the American Association of Geographers* **111**, 633–637.

Caughley, G., and Sinclair, A. R. E. (1994). *Wildlife Ecology and Management*. Blackwell Science, Cambridge, MA.

Channell, R., and Lomolino, M. V. (2000). Dynamic biogeography and conservation of endangered species. *Nature* **403**, 84–86.

Chetkiewicz, C. L. B., St. Clair, C. C., and Boyce, M. S. (2006). Corridors for conservation: integrating pattern and process. *Annual Review of Ecology and Evolutionary Systematics* **37**, 317–342.

Chichorro, F., Juslén, A., and Cardoso, P. (2019). A review of the relation between species traits and extinction risk. *Biological Conservation* **237**, 220–229.

Choi, Y. D. (2007). Restoration ecology to the future: a call for new paradigm. *Restoration Ecology* **15**, 351–353.

Cole, K. L. (2001). A multiple-scale history of past and ongoing vegetation change within the Indiana dunes. In *The Historical Ecology Handbook* (ed. D. Egan and E. A. Howell), pp. 391–412. Island Press, Washington, DC.

Corlatti, L., Hackländer, K., and Frey-Roos, F. R. E. D. Y. (2009). Ability of wildlife overpasses to provide connectivity and prevent genetic isolation. *Conservation Biology* **23**, 548–556.

Corlett, R. T. (2015). The Anthropocene concept in ecology and conservation. *Trends in Ecology & Evolution* **30**, 36–41.

COSEWIC (2008). *COSEWIC Assessment and Updated Status Report on the Vancouver Island marmot* (Marmota vancouverensis) *in Canada*. Committee on the Status of Endangered Wildlife in Canada, Ottawa. https://www.canada.ca/en/environment-climate-change/services/species-risk-public-registry/cosewic-assessments-status-reports/vancouver-island-marmot-2019.html.

Costanza, R. (1991). *Ecological Economics: The Science and Management of Sustainability*. Columbia University Press, New York.

Costanza, R., d'Arge, R., de Groot, R., Farber, S., Grasso, M., Hannon, B., Limburg, K., Naeem, S., O'Neil, R. V., Paruelo, J., Raskin, R. G., Sutton, P., and van den Belt, M. (1997). The value of the world's ecosystem services and natural capital. *Nature* **387**, 253–260.

Costanza, R., De Groot, R., Braat, L., Kubiszewski, I., Fioramonti, L., Sutton, P., Farber, S., and Grasso, M. (2017). Twenty years of ecosystem services: how far have we come and how far do we still need to go? *Ecosystem Services* **28**, 1–16.

Costanza, R., De Groot, R., Sutton, P., Van der Ploeg, S., Anderson, S. J., Kubiszewski, I., Farber, S., and Turner, R. K. (2014). Changes in the global value of ecosystem services. *Global Environmental Change* **26**, 152–158.

Covington, W. W., Fule, P. Z., Moore, M. M., Hart, S. C., Kolb, T. E., Mast, J. N., Sackett, S. S., and Wagner, M. R. (1997). Restoring ecosystem health in ponderosa pine forest of the southwest. *Journal of Forestry* **95**, 23–29.

Cowell, C. M. (1995). Presettlement Piedmont forests: patterns of composition and disturbance in central Georgia. *Annals of the Association of American Geographers* **85**, 65–83.

Cox, C. B., and Moore, P. D. (1993). *Biogeography: An Ecological and Evolutionary Approach*, 5th edition. Blackwell Scientific Publications, Oxford.

Craigie, I. D., Baillie, J. E., Balmford, A., Carbone, C., Collen, B., Green, R. E., and Hutton, J. M. (2010). Large mammal population declines in Africa's protected areas. *Biological Conservation* **143**, 2221–2228.

Craven, D., Eisenhauer, N., Pearse, W. D., Hautier, Y., Isbell, F., Roscher, C., Bahn, M., Beierkuhnlein, C., Bönisch, G., Buchmann, N., and Byun, C. (2018). Multiple facets of biodiversity drive the diversity-stability relationship. *Nature Ecology & Evolution* **2**, 1579–1587.

Crooks, K. R., Burdett, C. L., Theobald, D. M., King, S. R., Di Marco, M., Rondinini, C., and Boitani, L. (2017). Quantification of habitat fragmentation reveals extinction risk in terrestrial mammals. *Proceedings of the National Academy of Sciences* **114**, 7635–7640.

Cumming, G. S. (2011). Spatial resilience: integrating landscape ecology, resilience, and sustainability. *Landscape Ecology* **26**, 899–909.

Curry, P. (2011). *Ecological Ethics: An Introduction*. Polity Press, Cambridge, UK.

Dai, Y., Huang, H., Qing, Y., Li, J., and Li, D. (2023). Ecological response of an umbrella species to changing climate and land use: habitat conservation for Asiatic black bear in the Sichuan-Chongqing Region, Southwestern China. *Ecology and Evolution* **13**, e10222.

Daniel, T. C., Muhar, A., Arnberger, A., Aznar, O., Boyd, J. W., Chan, K. M., Costanza, R., Elmqvist, T., Flint, C. G., Gobster, P. H., and Grêt-Regamey, A. (2012). Contributions of cultural services to the ecosystem services agenda. *Proceedings of the National Academy of Sciences* **109**, 8812–8819.

Davidson, N. C., Van Dam, A. A., Finlayson, C. M., and McInnes, R. J. (2019). Worth of wetlands: revised global monetary values of coastal and inland wetland ecosystem services. *Marine and Freshwater Research* **70**, 1189–1194.

Davis, M., Faurby, S., and Svenning, J. C. (2018). Mammal diversity will take millions of years to recover from the current biodiversity crisis. *Proceedings of the National Academy of Sciences* **115**, 11262–11267.

Davis, M. B., and Zabinski, C. (1992). Changes in geographical range resulting from greenhouse warming effects on biodiversity in forests. In *Global Warming and Biological Diversity* (ed. R. L. Peters and T. E. Lovejoy), pp. 297–308. Yale University Press, New Haven, CT.

Davis, R. W., Bodkin, J. L., Coletti, H. A., Monson, D. H., Larson, S. E., Carswell, L. P., and Nichol, L. M. (2019). Future directions in sea otter research and management. *Frontiers in Marine Science* **5**, 510.

Davis, W. J. (1996). Focal species offer management tool. *Science* **271**, 1362–1363.

Deane, D. C., Nozohourmehrabad, P., Boyce, S. S., and He, F. (2020). Quantifying factors for understanding why several small patches host more species than a single large patch. *Biological Conservation* **249**, 108711.

Defenders of Wildlife (2023). *New Defenders Poll Shows American Public Overwhelmingly Supports the Endangered Species Act and Biodiversity Protection*. https://defenders.org/newsroom/new-defenders-poll-shows-american-public-overwhelmingly-supports-endangered-species-act.

Deiner, K., Yamanaka, H., and Bernatchez, L. (2021). The future of biodiversity monitoring and conservation utilizing environmental DNA. *Environmental DNA* **3**, 3–7.

DeMers, M. N., Simpson, J. W., Boerner, R. E., Silva, A., and Artigas, F. (1995). Fencerows, edges and implications of changing connectivity illustrated by two contiguous Ohio landscapes. *Conservation Biology* **9**, 1159–1168.

Dennhardt, A. J., Evans, M. E., Dechner, A., Hunt, L. E., and Maurer, B. A. (2016). Macroecology and the theory of island biogeography: abundant utility for applications in restoration ecology. In *Foundations of Restoration Ecology* (ed. M. A. Palmer, J. B. Zedler, and D. A. Falk), pp. 455–483. Island Press, Washington, DC.

Diamond, J. M. (1975). The island dilemma: lessons of modern biogeographic studies for the design of natural reserves. *Biological Conservation* **7**, 129–146.

Diamond, J. M., and May, R. M. (1976). Island biogeography and the design of natural preserves. In *Theoretical Ecology: Principles and Applications* (ed. R. M. May), pp. 163–186. W. B. Saunders, Philadelphia.

Dimson, M., Berio Fortini, L., Tingley, M. W., and Gillespie, T. W. (2023). Citizen science can complement professional invasive plant surveys and improve estimates of suitable habitat. *Diversity and Distributions* **29**, 1141–1156.

Dirzo, R., Young, H. S., Galetti, M., Ceballos, G., Isaac, N. J., and Collen, B. (2014). Defaunation in the Anthropocene. *Science* **345**, 401–406.

Dobson, A. P. (1996). *Conservation and Biodiversity*. Scientific American Library, New York.

Dobson, A. P., Rodriguez, J. P., Roberts, W. M., and Wilcove, D. S. (1997). Geographic distribution of endangered species in the United States. *Science* **275**, 550–553.

Dudley, N., and Alexander, S. (2017). Agriculture and biodiversity: a review. *Biodiversity* **18**, 45–49.

Dunstan, C. E., and Fox, B. J. (1996). The effects of fragmentation and disturbance of rainforest on ground-dwelling small mammals on the Robertson Plateau, New South Wales, Australia. *Journal of Biogeography* **23**, 187–201.

Duru, M., Therond, O., Martin, G., Martin-Clouaire, R., Magne, M. A., Justes, E., Journet, E. P., Aubertot, J. N., Savary, S., Bergez, J. E., and Sarthou, J. P. (2015). How to implement biodiversity-based agriculture to enhance ecosystem services: a review. *Agronomy for Sustainable Development* **35**, 1259–1281.

Dyer, J. M. (1994). Implications of habitat fragmentation on climate change-induced forest migration. *Professional Geographer* **46**, 449–459.

Dyer, J. M. (1995). Assessment of climatic warming using a model of forest species migration. *Ecological Modelling* **79**, 199–219.

Edgeworth, M., Ellis, E. C., Gibbard, P., Neal, C., and Ellis, M. (2019). The chronostratigraphic method is unsuitable for determining the start of the Anthropocene. *Progress in Physical Geography: Earth and Environment* **43**, 334–344.

Egan, D., and Howell, E. A. (2001). *The Historical Ecology Handbook*. Island Press, Washington, DC.

Ehrlich, P. R. (1961). Intrinsic barriers to dispersal in the checkerspot butterfly. *Science* **134**, 108–109.

Ehrlich, P. R. (1965). The population biology of the butterfly *Euphydryas editha*. II. The structure of the Jaspar Ridge colony. *Evolution* **19**, 327–336.

Ehrlich, P. R., and Murphy, D. D. (1987). Conservation lessons from long-term studies of checkerspot butterflies. *Conservation Biology* **1**, 122–131.

Eldredge, N. (1999). Cretaceous meteor showers, the human "niche" and the sixth extinction. In *Extinctions in Near Time* (ed. R. D. E. MacPhee), pp. 1–15. Kluwer Academic/Plenum, New York.

Ellis, R. D., McWhorter, T. J., and Maron, M. (2012). Integrating landscape ecology and conservation physiology. *Landscape Ecology* **27**, 1–12.

Epps, C. W., Palsboll, P. J., Wehausen, J. D., Roderick, G. K., Ramey, R. R., and McCullough, D. R. (2005). Highways block gene flow and cause a rapid decline in genetic diversity of desert bighorn sheep. *Ecology Letters* **8**, 1029–1038.

Epps, C. W., Wehausen, J. D., Bleich, V. C., Torres, S. G., and Brashares, J. S. (2007). Optimizing dispersal and corridor models using landscape genetics. *Journal of Applied Ecology* **44**, 714–724.

Fahrig, L. (2020). Why do several small patches hold more species than few large patches? *Global Ecology and Biogeography* **29**, 615–628.

Fahrig, L., Watling, J. I., Arnillas, C. A., Arroyo-Rodríguez, V., Jörger-Hickfang, T., Müller, J., Pereira, H. M., Riva, F., Rösch, V., Seibold, S., and Tscharntke, T. (2022). Resolving the SLOSS dilemma for biodiversity conservation: a research agenda. *Biological Reviews* **97**, 99–114.

Falaschi, M., Manenti, R., Thuiller, W., and Ficetola, G. F. (2019). Continental-scale determinants of population trends in European amphibians and reptiles. *Global Change Biology* **25**, 3504–3515.

Falcy, M. R., and Estades, C. F. (2007). Effectiveness of corridors relative to enlargement of habitat patches. *Conservation Biology* **21**, 1341–1346.

FAO (2020). *Global Forest Resources Assessment 2020: Key Findings*. United Nations, Rome.

Fiedler, P. L., and. Kareiva, P. M. (1998). *Conservation Biology for the Coming Decade*. Chapman and Hall, New York.

Fisher, B., Turner, R. K., and Morling, P. (2009). Defining and classifying ecosystem services for decision making. *Ecological Economics* **68**, 643–653.

Fisher, D. O. (2011). Trajectories from extinction: where are missing mammals rediscovered? *Global Ecology and Biogeography* **20**, 415–425.

Flanagan, A. M., Masuda, B., Komarczyk, L., Kuhar, A., Farabaugh, S., and Swaisgood, R. R. (2023). Adapting conservation breeding techniques using a data-driven approach to restore the 'Alalā (Hawaiian crow, *Corvus hawaiiensis*). *Zoo Biology* **14**, 834–839.

Flather, C. H. (1996). Fitting species-accumulation functions and assessing regional land use impacts on avian diversity. *Journal of Biogeography* **23**, 155–168.

Flather, C. H., Joyce, L. A., and Bloomgarden, C. A. (1994). *Species Endangerment Patterns in the United States*. U.S. Department of Agriculture, Forest Service, Washington, DC.

Fluet-Chouinard, E., Stocker, B. D., Zhang, Z., Malhotra, A., Melton, J. R., Poulter, B., Kaplan, J. O., Goldewijk, K. K., Siebert, S., Minayeva, T., and Hugelius, G. (2023). Extensive global wetland loss over the past three centuries. *Nature* **614**, 281–286.

Forister, M. L., Pelton, E. M., and Black, S. H. (2019). Declines in insect abundance and diversity: we know enough to act now. *Conservation Science and Practice* **1**, e80.

Forman, R. T. T. (1993). *Landscape and Regional Ecology*. Cambridge University Press, Cambridge.

Foster, D. R., Orwig, D. A., and McLachlan, J. S. (1996). Ecological and conservation insights from reconstructive studies of temperate old-growth forests. *Trends in Ecology and Evolution* **11**, 419–424.

Frankham, R., Bradshaw, C. J., and Brook, B. W. (2014). Genetics in conservation management: revised recommendations for the 50/500 rules, Red List criteria and population viability analyses. *Biological Conservation* **170**, 56–63.

Franklin, J. (2013). Species distribution models in conservation biogeography: developments and challenges. *Diversity and Distributions* **19**, 1217–1223.

Franklin, J., and Steadman, D. W. (1991). The potential for conservation of Polynesian birds through habitat mapping and species translocation. *Conservation Biology* **5**, 506–521.

Friedlingstein, P., Jones, M. W., O'Sullivan, M., Andrew, R. M., Bakker, D. C., Hauck, J., Le Quéré, C., Peters, G. P., Peters, W., Pongratz, J., and Sitch, S. (2022). Global carbon budget 2021. *Earth System Science Data* **14**, 1917–2005.

Gann, G. D., McDonald, T., Walder, B., Aronson, J., Nelson, C. R., Jonson, J., Hallett, J. G., Eisenberg, C., Guariguata, M. R., Liu, J., and Hua, F. (2019). International principles and standards for the practice of ecological restoration. *Restoration Ecology* **27**, S1–S46.

Gibbard, P., Walker, M., Bauer, A., Edgeworth, M., Edwards, L., Ellis, E., Finney, S., Gill, J. L., Maslin, M., Merritts, D., and Ruddiman, W. (2022). The Anthropocene as an event, not an epoch. *Journal of Quaternary Science* **37**, 395–399.

Gillespie, T. W. (2000). Rarity and conservation of forest birds in tropical dry forest regions of Central America. *Biological Conservation* **96**, 161–168.

Gillespie, T. W., Ostermann-Kelm, S., Dong, C., Willis, K. S., Okin, G. S., and MacDonald, G. M. (2018). Monitoring changes of NDVI in protected areas of southern California. *Ecological Indicators* **88**, 485–494.

Gillespie, T. W., Willis, K. S., and Ostermann-Kelm, S. (2015). Spaceborne remote sensing of the world's protected areas. *Progress in Physical Geography* **39**, 388–404.

Gillett, N. P., Malinina, E., Kaufman, D., and Neukom, R. (2023). Data for Figure SPM.1 (v20221116). *Summary for Policymakers of the Working Group I Contribution to the IPCC Sixth Assessment Report*. NERC EDS Centre for Environmental Data Analysis, 3 July 2023.

Goldman, M. (2009). Constructing connectivity: conservation corridors and conservation politics in East African range-lands. *Annals of the Association of American Geographers* **99**, 335–359.

Gorham, E. (1996). Lakes under a three-pronged attack. *Nature* **381**, 109–110.

Graumlich, L. J. (1991). Subalpine tree growth, climate, and increasing CO_2: an assessment of recent growth trends. *Ecology* **72**, 1–11.

Griscom, B. W., Adams, J., Ellis, P. W., Houghton, R. A., Lomax, G., Miteva, D. A., Schlesinger, W. H., Shoch, D., Siikamäki, J. V., Smith, P., and Woodbury, P. (2017). Natural climate solutions. *Proceedings of the National Academy of Sciences* **114**(44), 11645–11650.

Grossinger, R. (2005). Documenting local landscape change: The San Francisco Bay area historical project. In *The Historical Ecology Handbook* (ed. D. Egan and E. A. Howell), pp. 425–442. Island Press, Washington, DC.

Haddad, N. M., Brudvig, L. A., Damschen, E. I., Evans, D. M., Johnson, B. L., Levey, D. J., Orrock, J. L., Resasco, J., Sullivan, L. L., Tewksbury, J. J., and Wagner, S. A. (2014). Potential negative ecological effects of corridors. *Conservation Biology* **28**, 1178–1187.

Haig, S. M., and Avise, J. C. (1996). Avian conservation genetics. In *Conservation Genetics* (ed. J. C. Avise and J. C. Hamrick), pp. 160–189. Chapman and Hall, New York.

Haile, J., Holdaway, R., Oliver, K., Bunce, M., Gilbert, M. T. P., Nielsen, R., Munch, K., Ho, S. Y., Shapiro, B., and Willerslev, E. (2007). Ancient DNA chronology within sediment deposits: are paleobiological reconstructions possible and is DNA leaching a factor? *Molecular Biology and Evolution* **24**, 982–989.

Haines-Young, R., and Chopping, M. (1996). Quantifying landscape structure: a review of landscape indices and their application to forested landscapes. *Progress in Physical Geography* **20**, 418–445.

Hallmann, C. A., Sorg, M., Jongejans, E., Siepel, H., Hofland, N., Schwan, H., Stenmans, W., Müller, A., Sumser, H., Hörren, T., and Goulson, D. (2017). More than 75 percent decline over 27 years in total flying insect biomass in protected areas. *PloS One* **12**, e0185809.

Hamer, K. C., Hill, J. K., Lace, L. A., and Langan, A. M. (1997). Ecological and biogeographical effects of forest disturbance on tropical butterflies of Sumba, Indonesia. *Journal of Biogeography* **24**, 67–75.

Hannah, L., Carr, J. L., and Lankerani, A. (1995). Human disturbance and natural habitat: a biome level analysis of a global data set. *Biodiversity and Conservation* **4**, 128–155.

Hanson, C. T., Sherriff, R. L., Hutto, R. L., DellaSala, D. A., Veblen, T. T., and Baker, W. L. (2015). Setting the stage for mixed- and high-severity fire. In *The Ecological Importance of Mixed-Severity Fires* (ed. D. A. DellaSala and C. T. Hanson), pp. 3–22. Elsevier, New York.

Harfoot, M. B., Johnston, A., Balmford, A., Burgess, N. D., Butchart, S. H., Dias, M. P., Hazin, C., Hilton-Taylor, C.,

Hoffmann, M., Isaac, N. J., and Iversen, L. L. (2021). Using the IUCN Red List to map threats to terrestrial vertebrates at global scale. *Nature Ecology & Evolution* **5**, 1510–1519.

Harper, L. R., Handley, L. L., Carpenter, A. I., Ghazali, M., Di Muri, C., Macgregor, C. J., Logan, T. W., Law, A., Breithaupt, T., Read, D. S., and McDevitt, A. D. (2019). Environmental DNA (eDNA) metabarcoding of pond water as a tool to survey conservation and management priority mammals. *Biological Conservation* **238**, 108225.

Harrison, I. J., and Stiassny, M. L. J. (1999). The quiet crisis: a preliminary listing of the freshwater fishes of the world that are extinct or "missing in action." In *Extinctions in Near Time* (ed. R. D. E. MacPhee), pp. 271–331. Academic/Plenum, New York.

Head, M. J., Steffen, W., Fagerlind, D., Waters, C. N., Poirier, C., Syvitski, J., Zalasiewicz, J. A., Barnosky, A. D., Cearreta, A., Jeandel, C., and Leinfelder, R. (2022). The Great Acceleration is real and provides a quantitative basis for the proposed Anthropocene Series/Epoch. *Journal of International Geoscience* **45**, 359–376.

Head, M. J., Zalasiewicz, J. A., Waters, C. N., Turner, S. D., Williams, M., Barnosky, A. D., Steffen, W., Wagreich, M., Haff, P. K., Syvitski, J., and Leinfelder, R. (2022). The proposed Anthropocene Epoch/Series is underpinned by an extensive array of mid-20th century stratigraphic event signals. *Journal of Quaternary Science* **37**, 1181–1187.

Hellmann, J. J., Weiss, S. B., Mclaughlin, J. F., Boggs, C. L., Ehrlich, P. R., Launer, A. E., and Murphy, D. D. (2003). Do hypotheses from short-term studies hold in the long-term? An empirical test. *Ecological Entomology* **28**, 74–84.

Hemerik, L., Hengeveld, R., and Lippe, E. (2006). The eclipse of species ranges. *Acta Biotheoretica* **54**, 255–266.

Heywood, V. H. (1995). *Global Biodiversity, Assessment*. Cambridge University Press, Cambridge.

Higgs, E., Falk, D. A., Guerrini, A., Hall, M., Harris, J., Hobbs, R. J., Jackson, S. T., Rhemtulla, J. M., and Throop, W. (2014). The changing role of history in restoration ecology. *Frontiers in Ecology and the Environment* **12**, 499–506.

Hobbs, R. J., and Cramer, V. A. (2008). Restoration ecology: interventionist approaches for restoring and maintaining ecosystem function in the face of rapid environmental change. *Annual Review of Environment and Resources* **33**, 39–61.

Hoelle, J., and Kawa, N. C. (2020). Placing the anthropos in Anthropocene. *Annals of the American Association of Geographers* **111**, 655–662.

Houghton, J., Meira Filho, L. G., Callander, B. A., Harris, N., Kattenberg, A., and Maskell, K. (1996). *Climate Change 1995: The Science of Climate Change*, Cambridge University Press, Cambridge.

Houghton, J. T., Jenkins, G. J., and Ephraums, J. J. (1990). *Climate Change: The IPCC Scientific Assessment*. Cambridge University Press, Cambridge.

Huber, P. R., Greco, S. E., and Thorne, J. H. (2010). Boundaries make a difference: the effects of spatial and temporal parameters on conservation planning. *The Professional Geographer* **62**, 409–425.

IPCC (2022). Summary for policymakers. In Climate Change 2021: The Physical Science Basis. Contribution of Working Group I to the Sixth Assessment Report of the

Intergovernmental Panel on Climate Change (ed. V. Masson-Delmotte, P. Zhai, A. Pirani, S. L. Connors, C. Péan, S. Berger, N. Caud, Y. Chen, L. Goldfarb, M. I. Gomis, M. Huang, K. Leitzell, E. Lonnoy, J. B. R. Matthews, T. K. Maycock, T. Waterfield, O. Yelekçi, R. Yu, and B. Zhou), pp. 3–32. Cambridge University Press, Cambridge.

IUCN. (2023). *The IUCN Red List of Threatened Species.* Version 2023-1. https://www.iucnredlist.org.

Ivosevic, B., Han, Y. G., Cho, Y., and Kwon, O. (2015). The use of conservation drones in ecology and wildlife research. *Journal of Ecology and Environment* **38**, 113–118.

Jamieson, I. G., and Allendorf, F. W. (2012). How does the 50/500 rule apply to MVPs? *Trends in Ecology & Evolution* **27**, 578–584.

Janzen, D. H. (1988). Tropical ecological and biocultural restoration. *Science* **239**, 243–244.

Jeffries, M. J. (1997). *Biodiversity and Conservation.* Routledge, London.

Jiménez López, J., and Mulero-Pázmány, M. (2019). Drones for conservation in protected areas: present and future. *Drones* **3**, 10.

Kaufman, R. (2018). To save endangered species, should we bring them into our cities? *Smithsonian Magazine.* https://www.smithsonianmag.com/science-nature/save-endangered-species-should-we-bring-them-our-cities-180970611/.

Keeley, A. T., Beier, P., and Jenness, J. S. (2021). Connectivity metrics for conservation planning and monitoring. *Biological Conservation* **255**, 109008.

Keller, D., Holderegger, R., van Strien, M. J., and Bolliger, J. (2015). How to make landscape genetics beneficial for conservation management? *Conservation Genetics* **16**, 503–512.

Kellman, M., Tackaberry, R., and Rigg, L. (1998). Structure and function in two tropical gallery forest communities: implications for forest conservation in fragmented systems. *Journal of Applied Ecology* **35**, 368–370.

Kennel, C. F. (2021). The gathering Anthropocene crisis. *The Anthropocene Review* **8**, 83–95.

Kinghorn, A. D., Pan, L., Fletcher, J. N., and Chai, H. (2011). The relevance of higher plants in lead compound discovery programs. *Journal of Natural Products* **74**, 1539–1555.

Knapp, S., Aronson, M. F., Carpenter, E., Herrera-Montes, A., Jung, K., Kotze, D. J., La Sorte, F. A., Lepczyk, C. A., MacGregor-Fors, I., MacIvor, J. S., and Moretti, M. (2021). A research agenda for urban biodiversity in the global extinction crisis. *BioScience* **71**, 268–279.

Kolden, C. A. (2019). We're not doing enough prescribed fire in the Western United States to mitigate wildfire risk. *Fire* **2**, 30.

Kubiszewski, I., Costanza, R., Anderson, S., and Sutton, P. (2017). The future value of ecosystem services: global scenarios and national implications. *Ecosystem Services* **26**, 289–301.

Kupfer, J. A. (1995). Landscape ecology and biogeography. *Progress in Physical Geography* **19**, 18–34.

Kupfer, J. A. (2012). Landscape ecology and biogeography: rethinking landscape metrics in a post-FRAGSTATS landscape. *Progress in Physical Geography* **36**, 400–420.

Kupfer, J. A., and Cairns, D. M. (1996). The suitability of montane ecotones as indicators of global climatic change. *Progress in Physical Geography* **20**, 253–272.

Langhammer, P. F., Bakarr, M. I., Bennun, L. A., Brooks, T. M., Clay, R. P., Darwall, W., De Silva, N., Edgar, G. J., Eken, G., Fishpool, L. D. C., da Fonseca, G. A. B., Foster, M. N., Knox, D. H., Matiku, P., Radford, E. A., Rodrigues, A. S. L., Salaman, P., Sechrest, W., and Tordoff, A. W. (2007). *Identification and Gap Analysis of Key Biodiversity Areas: Targets for Comprehensive Protected Area Systems.* IUCN, Gland, Switzerland.

Laurance, W. F. (1998). A crisis in the making: response of Amazonian forests to land use and climate change. *Trends in Ecology and Evolution* **13**, 411–415.

Lees, A. C., Haskell, L., Allinson, T., Bezeng, S. B., Burfield, I. J., Renjifo, L. M., Rosenberg, K. V., Viswanathan, A., and Butchart, S. H. (2022). State of the world's birds. *Annual Review of Environment and Resources* **47**, 231–260.

Leopold, A. (1949). *A Sand County Almanac and Sketches Here and There.* Oxford University Press, New York.

Lewis, S. L., and Maslin, M. A. (2015). Defining the Anthropocene. *Nature* **519**, 171–180.

Lindenmayer, D. B., Barton, P. S., Lane, P. W., Westgate, M. J., McBurney, L., Blair, D., Gibbons, P., and Likens, G. E. (2014). An empirical assessment and comparison of species-based and habitat-based surrogates: a case study of forest vertebrates and large old trees. *PLoS One* **9**, e89807.

Liu, U., Kenney, S., Breman, E., and Cossu, T. A. (2019). A multicriteria decision making approach to prioritise vascular plants for species-based conservation. *Biological Conservation* **234**, 221–240.

Loehle, C., and Eschenbach, W. (2012). Historical bird and terrestrial mammal extinction rates and causes. *Diversity and Distributions* **18**, 84–91.

Lomolino, M. V., and Channel, R. (1995). Splendid isolation: patterns of range collapse in endangered mammals. *Journal of Mammalogy* **76**, 335–347.

Luque, S., Saura, S., and Fortin, M. J. (2012). Landscape connectivity analysis for conservation: insights from combining new methods with ecological and genetic data. *Landscape Ecology* **27**, 153–157.

Lurgi, M., López, B. C., and Montoya, J. M. (2012). Novel communities from climate change. *Philosophical Transactions of the Royal Society B: Biological Sciences* **367**, 2913–2922.

MacDonald, G. (2020). Climate, capital, conflict: geographies of success or failure in the twenty-first century. *Annals of the American Association of Geographers* **110**, 2011–2031.

MacDonald, G. M. (1987). Forests of the Hamilton region: past, present and future. In *Steel City: A Geography of Hamilton and Region* (ed. D. M. J. Dear, J. J. Reeds, and L. G. Reeds), pp. 65–84. University of Toronto Press, Toronto.

MacDonald, G. M., Bennett, K. D., Jackson, S. T., Parducci, L., Smith, F. A., Smol, J. P., and Willis, K. J. (2008). Impacts of climate change on species, populations and communities: palaeobiogeographical insights and frontiers. *Progress in Physical Geography* **32**, 139–172.

MacDonald, G. M., Larsen, C. P. S., Szeicz, J. M., and Moser, K. A. (1991). The reconstruction of boreal forest fire history from lake sediments: a comparison of charcoal, pollen, sedimentological, and geochemical indices. *Quaternary Science Reviews* **10**, 53–71.

MacDonald, G. M., Szeicz, J. M., Claricoates, J., and Dale, K. A. (1998). Response of the central Canadian treeline to recent climatic changes. *Annals of the Association of American Geographers* **88**, 183–208.

MacDonald, G., Wall, T., Enquist, C. A., LeRoy, S. R., Bradford, J. B., Breshears, D. D., Brown, T., Cayan, D., Dong, C., Falk, D. A., and Fleishman, E. (2023). Drivers of California's changing wildfires: a state-of-the-knowledge synthesis. *International Journal of Wildland Fire* **32**, 1039–1058.

Mace, G. M. (1993). An investigation into methods for categorizing the conservation status of species. In *Large-Scale Ecology and Conservation Biology* (ed. P. G. Edwards, R. M. May, and N. R. Webb), pp. 293–312. Blackwell Scientific Publications, Oxford.

Mace, G. M., and Lande, R. (1991). Assessing extinction threats: toward a reevaluation of IUCN threatened species categories. *Conservation Biology* **5**, 148–157.

Maizland, L. (2023). *The Push to Conserve 30 Percent of the Planet: What's at Stake?* Council on Foreign Relations. https://www.cfr.org/article/goal-conserve-30-percent-planet-2030-biodiversity-climate.

Malanson, G. P., and Cairns, D. M. (1997). Effects of dispersal, population delays and forest fragmentation on tree migration rates. *Plant Ecology* **134**, 67–79.

Manel, S., and Holderegger, R. (2013). Ten years of landscape genetics. *Trends in Ecology & Evolution* **28**, 614–621.

Marine Mammal Commission (2024). *Southern Sea Otter.* https://www.mmc.gov/priority-topics/species-of-concern/southern-sea-otter/.

Markwith, S. H. (2011). Biogeography and environmental restoration: an opportunity in applied research. *Geography Compass* **5**, 531–543.

Marshman, J., Blay-Palmer, A., and Landman, K. (2019). Anthropocene crisis: climate change, pollinators, and food security. *Environments* **6**(2), 22.

Mast, J. N., Veblen, T. T., and Hodgson, M. E. (1997). Tree invasion within a pine/grassland ecotone: an approach with historic aerial photography and GIS modeling. *Forest Ecology and Management* **93**, 181–194.

Maxwell, S. L., Fuller, R. A., Brooks, T. M., and Watson, J. E. (2016). Biodiversity: the ravages of guns, nets and bulldozers. *Nature* **536**, 143–145.

McCarthy, M. A., Thompson, C. J., Moore, A. L., and Possingham, H. P. (2011). Designing nature reserves in the face of uncertainty. *Ecology Letters* **14**, 470–475.

McGarrahan, E. (1997). Much-studied butterfly winks out on Stanford preserve. *Science* **275**, 479–480.

Meffe, G. K., Carroll, C. R., and Contributors. (1997). *Principles of Conservation Biology*, 2nd edition. Sinauer Associates, Sunderland, MA.

Mekonnen, A., Fashing, P. J., Chapman, C. A., Venkataraman, V. V., and Stenseth, N. C. (2022). The value of flagship and umbrella species for restoration and sustainable development: Bale monkeys and bamboo forest in Ethiopia. *Journal for Nature Conservation* **65**, 126117.

Mitsch, W. J., and Gosselink, J. G. (2015). *Wetlands*, 5th edition. Wiley, Hoboken, NJ.

Mokany, K., Ware, C., Harwood, T. D., Schmidt, R. K., and Ferrier, S. (2022). Habitat-based biodiversity assessment for ecosystem accounting in the Murray-Darling Basin. *Conservation Biology* **36**, e13915.

Montgomery, G. A., Dunn, R. R., Fox, R., Jongejans, E., Leather, S. R., Saunders, M. E., Shortall, C. R., Tingley, M. W., and Wagner, D. L. (2020). Is the insect apocalypse upon us? How to find out. *Biological Conservation* **241**, 108327.

Montibeller, B., Kmoch, A., Virro, H., Mander, Ü., and Uuemaa, E. (2020). Increasing fragmentation of forest cover in Brazil's Legal Amazon from 2001 to 2017. *Scientific Reports* **10**, 5803.

Mougi, A., and Kondoh, M. (2012). Diversity of interaction types and ecological community stability. *Science* **337**, 349–351.

Myers, N. (1988). Threatened biota: hotspots in tropical forests. *Environmentalist* **8**, 187–208.

Myers, N. (1990). The biodiversity challenge: expanded hot-spots analysis. *Environmentalist* **10**, 243–256.

Myers, N., Mittermeier, R. A., Mittermeier, C. G., Da Fonseca, G. A., and Kent, J. (2000). Biodiversity hotspots for conservation priorities. *Nature* **403**, 853–858.

Myneni, R. B., Keeling, C. D., Tucker, C. T., Asrar, G., and Nemani, R. R. (1997). Increased plant growth in the northern high latitudes from 1981 to 1991. *Nature* **386**, 698–702.

National Research Council (1992). *Restoration of Aquatic Ecosystems.* National Academy Press, Washington, DC.

Naveh, Z., and Lieberman, A. S. (1984). *Landscape Ecology: Theory and Application.* Springer-Verlag, New York.

Neilson, R. P., and Marks, D. (1994). A global perspective of regional vegetation and hydrologic sensitivities from climatic change. *Journal of Vegetation Science* **5**, 715–730.

Noever, D. A., Brittain, A., Matsos, H. C., Baskaran, S., and Obenhber, D. (1996). The effects of variable biome distribution on global climate. *BioSystems* **39**, 135–141.

O'Gara, K. A., Miles, J. E., and May, R. E. (2015). Feasibility of Red-cockaded woodpecker *(Picoides borealis)* reintroduction into the pine flatwoods communities of Jonathan Dickinson State Park, Florida. *Florida Field Naturalist* **43**, 47–61.

Ouborg, N. J., Pertoldi, C., Loeschcke, V., Bijlsma, R. K., and Hedrick, P. W. (2010). Conservation genetics in transition to conservation genomics. *Trends in Genetics* **26**, 177–187.

Overpeck, J. T., Hughen, K., Hardy, D., Bradley, R., Case, R., Douglas, M., Finney, B., Gajewski, K., Jacoby, G., Jennings, A., Lamoureux, S., Lasca, A., MacDonald, G. M., Moore, J., Retelle, M., Smith, S., Wolfe, A., and Zielinski, G. (1997). Arctic environmental change of the last four centuries. *Science* **278**, 1251–1256.

Parks Canada, 2010. Point Pelee National Park Management Plan. Parks Canada, Ottawa. https://parks.canada.ca/agence-agency/bib-lib/plans/~/media/BFFDE13D0B3F4965A305EE43F9BD4349.ashx.

Parmesan, C., and Singer, M. C. (2022). Mosaics of climatic stress across species' ranges: tradeoffs cause adaptive evolution to limits of climatic tolerance. *Philosophical Transactions of the Royal Society B* **377**, 20210003.

Parmesan, C., Singer, M. C., Wee, B., and Mikheyev, S. (2023). The case for prioritizing ecology/behavior and hybridization over genomics/taxonomy and species' integrity in conservation under climate change. *Biological Conservation* **281**, 109967.

Pedersen, M. W., Overballe-Petersen, S., Ermini, L., Sarkissian, C. D., Haile, J., Hellstrom, M., Spens, J., Thomsen, P. F., Bohmann, K., Cappellini, E., and Schnell, I. B. (2015). Ancient and modern environmental DNA. *Philosophical Transactions of the Royal Society B: Biological Sciences* **370**, 20130383.

Pe'er, G., Tsianou, M. A., Franz, K. W., Matsinos, Y. G., Mazaris, A. D., Storch, D., Kopsova, L., Verboom, J., Baguette, M., Stevens, V. M., and Henle, K. (2014). Toward better application of minimum area requirements in conservation planning. *Biological Conservation* **170**, 92–102.

Pennekamp, F., Pontarp, M., Tabi, A., Altermatt, F., Alther, R., Choffat, Y., Fronhofer, E. A., Ganesanandamoorthy, P., Garnier, A., Griffiths, J. I., and Greene, S. (2018). Biodiversity increases and decreases ecosystem stability. *Nature* **563**, 109–112.

Peterken, G. F., and Hughes, F. M. R. (1995). Restoration of floodplain forests in Britain. *Forestry* **68**, 187–202.

Pfleger, M. O., Rider, S. J., Johnston, C. E., and Janosik, A. M. (2016). Saving the doomed: using eDNA to aid in detection of rare sturgeon for conservation *(Acipenseridae)*. *Global Ecology and Conservation* **8**, 99–107.

Potvin, C., and Vasseur, L. (1997). Long-term CO2 enrichment of a pasture community: species richness, dominance, and succession. *Ecology* **78**, 666–677.

Prichard, S. J., Hessburg, P. F., Hagmann, R. K., Povak, N. A., Dobrowski, S. Z., Hurteau, M. D., Kane, V. R., Keane, R. E., Kobziar, L. N., Kolden, C. A., and North, M. (2021). Adapting western North American forests to climate change and wildfires: 10 common questions. *Ecological Applications* **31**, e02433.

Ramírez-Cruz, G. A., Solano-Zavaleta, I., Mendoza-Hernández, P. E., Méndez-Janovitz, M., Suárez-Rodríguez, M., and Zúñiga-Vega, J. J. (2019). This town ain't big enough for both of us . . . or is it? Spatial co-occurrence between exotic and native species in an urban reserve. *PloS One* **14**, e0211050.

Reaka-Kudla, M. L., Wilson, D. E., and Wilson, E. O. (1997). *Biodiversity II.* Joseph Henry Press, Washington, DC.

Reed, R. A., Johnson-Barnard, J., and Baker, W. L. (1996). Contribution of roads to forest fragmentation in the Rocky Mountains. *Conservation Biology* **10**, 1098–1106.

Reyes-García, V., Fernández-Llamazares, Á., McElwee, P., Molnár, Z., Öllerer, K., Wilson, S. J., and Brondizio, E. S. (2019). The contributions of Indigenous peoples and local communities to ecological restoration. *Restoration Ecology* **27**, 3–8.

Richmond, J. Q., Wostl, E., Reed, R. N., and Fisher, R. N. (2022). Range eclipse leads to tenuous survival of a rare lizard species on a barrier atoll. *Oryx* **56**, 63–72.

Roberge, J. M., Mikusiński, G., and Svensson, S. (2008). The white-backed woodpecker: umbrella species for forest conservation planning? *Biodiversity and Conservation* **17**, 2479–2494.

Robinson, J. M., Gellie, N., MacCarthy, D., Mills, J. G., O'Donnell, K., and Redvers, N. (2021). Traditional ecological knowledge in restoration ecology: a call to listen deeply, to engage with, and respect Indigenous voices. *Restoration Ecology* **29**, e13381.

Rodrigues, A. S., Akcakaya, H. R., Andelman, S. J., Bakarr, M. I., Boitani, L., Brooks, T. M., Chanson, J. S., Fishpool, L. D., Da Fonseca, G. A., Gaston, K. J., and Hoffmann, M. (2004). Global gap analysis: priority regions for expanding the global protected-area network. *BioScience* **54**, 1092–1100.

Rodrigues, A. S., Andelman, S. J., Bakarr, M. I., Boitani, L., Brooks, T. M., Cowling, R. M., Fishpool, L. D., Da Fonseca, G. A., Gaston, K. J., Hoffmann, M., and Long, J. S. (2004). Effectiveness of the global protected area network in representing species diversity. *Nature* **428**, 640–643.

Rodríguez-Echeverry, J., and Leiton, M. (2021). State of the landscape and dynamics of loss and fragmentation of forest critically endangered in the tropical Andes hotspot: implications for conservation planning. *Journal of Landscape Ecology* **14**, 73–91.

Rogelj, J., Smith, C., Plattner, G.-K., Meinshausen, M., Szopa, S., Milinski, S., and Marotzke, J. (2021). Data for Figure SPM.4 (v20210809). *Summary for Policymakers of the Working Group I Contribution to the IPCC Sixth Assessment Report.* NERC EDS Centre for Environmental Data Analysis. doi:10.5285/bd65331b1d344ccca44852e495d3a049.

Root, T. L., and Schneider, S. H. (1995). Ecology and climate: research strategies and implications. *Science* **269**, 334–341.

Rosenberg, K. V., Dokter, A. M., Blancher, P. J., Sauer, J. R., Smith, A. C., Smith, P. A., Stanton, J. C., Panjabi, A., Helft, L., Parr, M., and Marra, P. P. (2019). Decline of the North American avifauna. *Science* **366**, 120–124.

Rowland, M. M., Wisdom, M. J., Suring, L. H., and Meinke, C. W. (2006). Greater sage-grouse as an umbrella species for sagebrush-associated vertebrates. *Biological Conservation* **129**, 323–335.

Rubinoff, D. (2001). Evaluating the California gnatcatcher as an umbrella species for conservation of southern California coastal sage scrub. *Conservation Biology* **15**, 1374–1383.

Russell, J. C., and Kueffer, C. (2019). Island biodiversity in the Anthropocene. *Annual Review of Environment and Resources* **44**, 31–60.

Santini, L., Saura, S., and Rondinini, C. (2016). Connectivity of the global network of protected areas. *Diversity and Distributions* **22**, 199–211.

Sato, H., and Ise, T. (2022). Predicting global terrestrial biomes with the LeNet convolutional neural network. *Geoscientific Model Development* **15**, 3121–3132.

Savage, M. (1991). Structural dynamics of a southwestern pine forest under chronic human influence. *Annals of the Association of American Geographers* **81**, 271–289.

Savage, M., and Swetnam, T. W. (1990). Early 19th-century fire decline following sheep pasturing in a Navajo ponderosa pine forest. *Ecology* **71**, 2374–2378.

Schlaepfer, M. A. (2018). Do non-native species contribute to biodiversity? *PLoS Biology* **16**, e2005568.

Scott, J. M., Csuti, B., Jacobi, J. D., and Estes, J. E. (1987). Species richness. *BioScience* **37**, 782–788.

Scott, J. M., Davis, F., Csuti, B., Noss, R., Butterfield, B., Groves, C., Anderson, H., Caicco, S., D'Erchia, F., Edwards Jr., T. C., Ulliman, J., and Wright, R. G. (1993). Gap analysis: a geographic approach to protection of biological diversity. *Wildlife Monographs* **123**, 1–41.

Seager, J. (1995). *The New State of the Earth Atlas*, 2nd edition. Simon and Schuster, New York.

Seddon, N., Chausson, A., Berry, P., Girardin, C. A., Smith, A., and Turner, B. (2020). Understanding the value and limits of nature-based solutions to climate change and other global challenges. *Philosophical Transactions of the Royal Society B* **375**, 20190120.

Segelbacher, G., Cushman, S. A., Epperson, B. K., Fortin, M. J., Francois, O., Hardy, O. J., Holderegger, R., Taberlet, P., Waits, L. P., and Manel, S. (2010). Applications of landscape genetics in conservation biology: concepts and challenges. *Conservation Genetics* **11**, 375–385.

Seddon, N., Mace, G. M., Naeem, S., Tobias, J. A., Pigot, A. L., Cavanagh, R., Mouillot, D., Vause, J., and Walpole, M. (2016). Biodiversity in the Anthropocene: prospects and policy. *Proceedings of the Royal Society B: Biological Sciences* **283**, 20162094.

Serra-Diaz, J. M., and Franklin, J. (2019). What's hot in conservation biogeography in a changing climate? Going beyond species range dynamics. *Diversity and Distributions* **25**, 492–498.

Shaffer, M. L. (1981). Minimum population sizes for species conservation. *Bioscience* **31**, 131–134.

Shilling, F. (1997). Do habitat conservation plans protect endangered species? *Science* **276**, 1662–1663.

Shwartz, A. (2018). Designing nature in cities to safeguard meaningful experiences of biodiversity in an urbanizing world. *Urban Biodiversity* **200**(215), 200–215.

Simberloff, D. S., and Abele, L. G. (1982). Refuge design and island biogeographic theory: effects of fragmentation. *American Naturalist* **120**, 41–50.

Simpson, N. O., Stewart, K. M., Schroeder, C., Cox, M., Huebner, K., and Wasley, T. (2016). Overpasses and underpasses: effectiveness of crossing structures for migratory ungulates. *Journal of Wildlife Management* **80**, 1370–1378.

Sisk, T. D., Launer, A. E., Switky, K. R., and Ehrlich, P. R. (1994). Identifying extinction threats: global analyses of the distribution of biodiversity and the expansion of the human enterprise. *Bioscience* **44**, 592–604.

Smetzer, J. R., Greggor, A. L., Paxton, K. L., Masuda, B., and Paxton, E. H. (2021). Automated telemetry reveals post-reintroduction exploratory behavior and movement patterns of an endangered corvid, 'Alalā *(Corvus hawaiiensis)* in Hawai'i, USA. *Global Ecology and Conservation* **26**, e01522.

Soanes, K., and Lentini, P. E. (2019). When cities are the last chance for saving species. *Frontiers in Ecology and the Environment* **17**, 225–231.

Söderlund, J. (2019). Biophilic design: the stories of the pioneers of the social movement. In *The Emergence of Biophilic Design. Cities and Nature*, pp. 35–141. Springer, Cham, Switzerland.

Sommer, S., McDevitt, A. D., and Balkenhol, N. (2013). Landscape genetic approaches in conservation biology and management. *Conservation Genetics* **14**, 249–251.

Soulé, M. E. (1985). What is conservation biology? *BioScience* **35**, 727–738.

Soulé, M. E. (1987). *Viable Populations for Conservation*. Cambridge University Press, Cambridge.

Stan, A. B., Fulé, P. Z., and Hunter Jr., M. (2022). Reduced forest vulnerability due to management on the Hualapai Nation. *Trees, Forests and People* **10**, 100325.

Steenweg, R., Hebblewhite, M., Burton, C., Whittington, J., Heim, N., Fisher, J. T., Ladle, A., Lowe, W., Muhly, T., Paczkowski, J., and Musiani, M. (2023). Testing umbrella species and food-web properties of large carnivores in the Rocky Mountains. *Biological Conservation* **278**, 109888.

Steffen, W., Crutzen, P. J., and McNeill, J. R. (2007). The Anthropocene: are humans now overwhelming the great forces of nature? *Ambio-Journal of Human Environment Research and Management* **36**, 614–621.

Stewart, B. S., Yochem, P. K., Huber, H. R., DeLong, R. L., Jameson, R. J., Sydeman, W. J., Allen, S. G., and Le Boeuf, B. J. (1994). History and present status of the northern elephant seal population. *Elephant Seals: Population Ecology, Behavior, and Physiology*, pp. 29–48. University of California Press, Berkeley.

Strong, D. R., and Pemberton, R. W. (2000). Ecology: biological control of invading species—risk and reform. *Science* **288**, 1969–1970.

Swetnam, T. W., Allen, C. D., and Betancourt, J. L. (1999). Applied historical ecology: using the past to manage for the future. *Ecological Applications* **64**, 1189–1206.

Syvitski, J., Waters, C. N., Day, J., Milliman, J. D., Summerhayes, C., Steffen, W., Zalasiewicz, J., Cearreta, A., Gałuszka, A., Hajdas, I., and Head, M. J. (2020). Extraordinary human energy consumption and resultant geological impacts beginning around 1950 CE initiated the proposed Anthropocene Epoch. *Communications Earth & Environment* **1**, 32.

Thomsen, P. F., and Willerslev, E. (2015). Environmental DNA—emerging tool in conservation for monitoring past and present biodiversity. *Biological Conservation* **183**, 4–18.

Tilman, D. (1996). Biodiversity: population versus ecosystem stability. *Ecology* **77**, 350–363.

Tilman, D., Reich, P. B., and Knops, J. M. (2006). Biodiversity and ecosystem stability in a decade-long grassland experiment. *Nature* **441**, 629–632.

Turner, M. G. (2005). Landscape ecology: what is the state of the science? *Annual Review of Ecology, Evolution and Systematics* **36**, 319–344.

UNEP-WCMC, IUCN, and NGS (2020). *Protected Planet Report 2020*. UNEP-WCMC, IUCN, and NGS, Cambridge, UK; Gland, Switzerland; and Washington, DC. https://livereport.protectedplanet.net/.

USDA (2023). *President's 2024 Budget Advances Efforts to Address the Nation's Wildfire Crisis Through Workforce Reform and Other Investments in Wildland Fire Management Programs*. Press Release No. 0054.23. USDA, Washington, DC.

Van Blaricom, G. R., Gerber, L., and Brownell, R. L. (2013). Marine mammals, extinctions of. In *Encyclopedia of Biodiversity*, 2nd edition, pp. 64–93. Elsevier, New York.

Vimal, R., Rodrigues, A. S., Mathevet, R., and Thompson, J. D. (2011). The sensitivity of gap analysis to conservation targets. *Biodiversity and Conservation* **20**, 531–543.

Wagner, D. L. (2020). Insect declines in the Anthropocene. *Annual Review of Entomology* **65**, 457–480.

Washington, H., and Maloney, M. (2020). The need for ecological ethics in a new ecological economics. *Ecological Economics* **169**, 106478.

Waters, C. N., and Turner, S. D. (2022). Defining the onset of the Anthropocene. *Science* **378**, 706–708.

Watson, R., Baste, I., Larigauderie, A., Leadley, P., Pascual, U., Baptiste, B., Demissew, S., Dziba, L., Erpul, G., Fazel, A., and Fischer, M. (2019). *Summary for Policymakers of the Global Assessment Report on Biodiversity and Ecosystem Services of the Intergovernmental Science-Policy Platform on Biodiversity and Ecosystem Services*, pp. 22–47. IPBES Secretariat, Bonn, Germany.

Wessely, J., Hülber, K., Gattringer, A., Kuttner, M., Moser, D., Rabitsch, W., Schindler, S., Dullinger, S., and Essl, F. (2017). Habitat-based conservation strategies cannot compensate for climate-change-induced range loss. *Nature Climate Change* **7**, 823–827.

Whiting, J. C., Bleich, V. C., Bowyer, R. T., and Epps, C. W. (2023). Restoration of bighorn sheep: history, successes, and remaining conservation issues. *Frontiers in Ecology and Evolution* **11**, 1083350.

Whittaker, R. J., Araújo, M. B., Jepson, P., Ladle, R. J., Watson, J. E., and Willis, K. J. (2005). Conservation biogeography: assessment and prospect. *Diversity and Distributions* **11**, 3–23.

Wilson, E. D., and Willis, E. D. (1975). Applied biogeography. In *Ecology and Evolution of Communities* (ed. M. L. Cody and J. M. Diamond), pp. 552–534. Belknap Press, Cambridge, MA.

Wilson, M. C., Chen, X. Y., Corlett, R. T., Didham, R. K., Ding, P., Holt, R. D., Holyoak, M., Hu, G., Hughes, A. C., Jiang, L., and Laurance, W. F. (2016). Habitat fragmentation and biodiversity conservation: key findings and future challenges. *Landscape Ecology* **31**, 219–227.

Willerslev, E., Cappellini, E., Boomsma, W., Nielsen, R., Hebsgaard, M. B., Brand, T. B., Hofreiter, M., Bunce, M., Poinar, H. N., Dahl-Jensen, D., and Johnsen, S. (2007). Ancient biomolecules from deep ice cores reveal a forested southern Greenland. *Science* **317**, 111–114.

Willerslev, E., Hansen, A. J., Binladen, J., Brand, T. B., Gilbert, M. T. P., Shapiro, B., Bunce, M., Wiuf, C., Gilichinsky, D. A., and Cooper, A. (2003). Diverse plant and animal genetic records from Holocene and Pleistocene sediments. *Science* **300**, 791–795.

Williams, D. R., Rondinini, C., and Tilman, D. (2022). Global protected areas seem insufficient to safeguard half of the world's mammals from human-induced extinction. *Proceedings of the National Academy of Sciences* **119**, e2200118119.

Williams, J. W., and Jackson, S. T. (2007). Novel climates, no-analog communities, and ecological surprises. *Frontiers in Ecology and the Environment* **5**, 475–482.

Williams, J. W., Jackson, S. T., and Kutzbach, J. E. (2007). Projected distributions of novel and disappearing climates by 2100 AD. *Proceedings of the National Academy of Sciences* **104**, 5738–5742.

Willis, K. S. (2015). Remote sensing change detection for ecological monitoring in United States protected areas. *Biological Conservation* **182**, 233–242.

Wortley, L., Hero, J. M., and Howes, M. (2013). Evaluating ecological restoration success: a review of the literature. *Restoration Ecology* **21**, 537–543.

Wu, J. (2013). Key concepts and research topics in landscape ecology revisited: 30 years after the Allerton Park workshop. *Landscape Ecology* **28**, 1–11.

Zalasiewicz, J., Waters, C. N., Ellis, E. C., Head, M. J., Vidas, D., Steffen, W., Thomas, J. A., Horn, E., Summerhayes, C. P., Leinfelder, R., and McNeill, J. R. (2021). The Anthropocene: comparing its meaning in geology (chronostratigraphy) with conceptual approaches arising in other disciplines. *Earth's Future* **9**, e2020EF001896.

GLOSSARY OF KEY WORDS AND TERMS

Acheulean Tradition The stone tool–making tradition of *Homo erectus*.

Adaptive Radiation The development of many species from a single founding species. The new species evolve to occupy the different range of habitats and use the different resources that are present in the region.

Adventitious Roots Roots that develop from stems or other plant parts that are near the soil surface or get buried by soil due to floods or other depositional events.

Agricultural Revolution The initial development of agriculture and the associated evolution of human society representing a break from the earlier hunting-and-gathering economies of the Mesolithic.

Agriculture The deliberate cultivation of plants and animals for human use.

Agroecology Farming that acts in concert with nature, supports biodiversity, and is sustainable.

Alleles The gene forms that exist for a given locus.

Allelopathy Competition between plant species in which chemical compounds released by one species inhibits the germination or growth of competing species.

Allen's Rule The general observation that animals that live in cold environments have shorter extremities, such as limbs or ears, than related species that live in warm environments.

Allopatric Speciation The formation of new species by geographic isolation.

Alpha Diversity The degree of difference between species found at different sites of the same habitat or community type.

Amphioceanic Having a geographic distribution that is divided by an ocean.

Amphiregional Having a biogeographic distribution that is divided by a barrier into two distinct regions.

Amphitropical Having a biogeographic distribution that is divided by the tropics.

Analytical Biogeography The development of general rules that explain how geography affects the evolution and distribution of plants and animals and how past distributions and evolutionary history are reflected in modern distributions.

Anemochores Organisms dispersed by wind.

Anemohydrochores Organisms dispersed by wind and water.

Angiosperms Flowering plants. Angiosperms have seeds that are developed within an ovary. Some taxonomies refer to angiosperms as the *Magnoliophyta*.

Annuals Plants that germinate, grow, reproduce, and die all in one growing season.

Anthropocene Proposed geologic Epoch coinciding with growth in human populations and technologies and intensification of human impacts on the environment.

Anthropochores Organisms dispersed by humans.

Applied Historical Ecology The use of paleoecological techniques such as pollen analysis and dendrochronology to help in ecosystem restoration.

Areography The ecological analysis and interpretation of the large-scale geographic distributions of plant and animals.

Ark A geologic event that leads to the movement of a landmass and the geographic transport of flora and fauna.

Artificial Selection Conscious and unconscious genetic selection practiced by humans on species of plants and animals.

Assemblage A group of species that exist in a similar habitat, have similar characteristics, or are taxonomically related.

Association Groupings of plant species that are typically found growing together in similar habitats.

Autecology Ecological research that focuses on one species.

Autotrophs Photosynthetic plants that produce chemical energy using sunlight, water, and carbon dioxide.

Background Extinction Rate The long-term base-level rate of extinction.

Background Mortality The natural mortality rate for plants and animals caused by the death of individual organisms and not associated with large disturbance events.

Barrier Geographic features that block dispersal and colonization.

Batesian Mimicry When one species that is not poisonous or unpalatable has the same coloring or shape as a species that is poisonous or unpalatable. The palatable species benefits from coexisting with the unpalatable species because predators mistake it for the other species and avoid it. This can be considered a form of commensalism.

Bergmann's Rule Species found in colder climate regions are larger than related species found in warmer climate regions.

Beringia The region of eastern Siberia and Alaska that links North America with Asia during periods of lowered sea level.

Beta Diversity The degree of difference in the species composition between two different habitats or community types.

Big Five Mass Extinctions Past large-scale extinction events: (1) the Late Ordovician, 444 million years ago; (2) the Late Devonian, 375 million years ago; (3) the Late Permian, 250 million years ago; (4) the Late Triassic, 200 million years ago; and (5) the Late Cretaceous, 65 million years ago.

Biocontrol The use of predatory or pathogenic species from the native range of the invasive species to control an invading species.

Biodiversity Broadly defined as the number, variety, and variability of living organisms; often simply defined as the number of different species in a given geographic area. The greater the number of species present, the greater the biodiversity.

Biodiversity Conservation Efforts that are made to preserve the species richness of an entire ecosystem rather than focusing on one individual species.

Biodiversity Hot Spots Areas that contain high numbers of endemic species that are threatened by extinction due to human activities.

Biogeography: Introduction to Space, Time, and Life, Second Edition. Glen M. MacDonald.
© 2025 John Wiley & Sons, Inc. Published 2025 by John Wiley & Sons, Inc.
Companion website: www.wiley.com/go/macdonald/biogeography2e

Biogeographic Harmony When the flora and fauna of two or more areas are identical.

Biogeographic Kingdom A large biogeographic province.

Biogeographic Line A geographic boundary dividing two biogeographic regions.

Biogeographic Provinces Subcontinental areas that contain similar flora and fauna.

Biogeographic Provincialism The tendency for different regions to possess unique species, genera, or families.

Biogeographic Realms Supercontinental areas that contain similar flora and fauna.

Biogeographic Regions Continental and supercontinental areas that contain similar flora and fauna.

Biogeographic Relicts Taxa that once had larger distributions, but have become narrow endemics.

Biogeography The study of the past and present geographic distributions of plants and animals and other organisms, and the environmental and evolutionary forces that produce those distributions.

Biologging The use of highly miniaturized active tags, such as tiny recording devices or transmitters, to track animal movements.

Biomass The mass of living material in an ecosystem. Generally considered the dried weight of all living matter in a prescribed area.

Biome Very large areas of the earth's surface that have a similar climate and vegetation.

Biophilic Urban Design The planning of cities to foster biodiversity and ecological services, potentially serving as urban arks to preserve native species and nonnative endangered species.

Biosphere All life on earth.

Bipedal The ability to walk on two legs.

Bipolar Having a biogeographic distribution with populations located in both the north and south polar regions.

Blitzkrieg Theory of Pleistocene Overkill P. S. Martin's hypothesis of how a geographically expanding front of prehistoric human colonists led to the extinction of many large mammals in North and South America.

Bottleneck The decrease in genetic diversity that results from a significant decrease in population size such as might occur after a natural catastrophe.

C_3 Photosynthesis In C_3 plants, when carbon dioxide from the atmosphere is converted into a 3-carbon molecule called 3-phosphoglyceric acid.

C_4 Photosynthesis When C_4 plants convert carbon dioxide into two 4-carbon molecules, malic acid and aspartic acid.

Canopy Fire A fire that burns in the canopy cover of a forest.

Carrying Capacity The number of individuals of a given species that a given area of the environment can support.

Centers of Domestication Discrete areas of the world where a number of different plant and animal species were first domesticated.

Character Displacement When species display different behaviors or phenotypic attributes when living together with closely competing species than when each species is living in separate areas.

Charismatic Flora Showy, large, or otherwise interesting plants that are often the object of intense conservation measures.

Charismatic Megafauna Large, attractive, or otherwise interesting animals that are often the object of intense conservation measures.

Chemosynthesis The process by which primary producer organisms use chemicals, rather than sunlight and photosynthesis, as a source of energy.

Chloroplasts Photosynthetic bodies within plant cells.

Choropleth Maps Maps on which the geographic distributions of species (or other features and phenomena) are delineated by continuous shading or circumscribed by lines.

Chromosomes Threadlike structures within cells upon which genes are arranged.

Circumpolar Having a geographic distribution that circles the polar regions.

Citizen Scientist Nonspecialist member of the public who provides information of species presence on a volunteer basis, often using online tools and phone apps.

Clade The lineage of different related species that arise from a common ancestor.

Cladistics The reconstruction of evolutionary history and linkages between species by constructing branching diagrams *(cladograms)* based on the number of shared physical traits observed between different taxa.

Class A taxonomic category that includes groups of related orders.

Climax The idealized (and often unrealized or unrealistic) end point of succession.

Cline A geographic gradient in a genetically controlled trait.

Clovis Point An elongated stone point with a fluted base that is accepted by many archaeologists as representing the earliest tradition of tool making in North America. Most points of this type date from around 11,000 years ago.

Coefficient of Similarity Mathematical measures of how similar the flora and fauna of one geographic area are to the flora and fauna of another area.

Coevolution When two unrelated species evolve traits that are tied to their interactions.

Colonization The expansion of a species and establishment of a self-sustaining population in a new geographic region.

Columbian Exchange The unprecedented long-distance dispersal of plant and animal species between the world's continents by humans after 1492.

Commensalism A symbiotic relationship in which one species benefits and the other is not affected.

Community All the different populations of organisms that live and interact with each other within a prescribed area.

Community Type The collection of species that is generally found in a specific type of habitat.

Competitive Exclusion Principle The theory that species with identical niches cannot exist in the same geographic region because one would eventually drive the other to extinction.

Conifer Cone-bearing gymnosperms such as pines and spruces.

Conservation Biogeography The application of biogeographic data and principles for conservation planning and management.

Conservation Biology A synthetic discipline that applies principles from biology and other sciences to the maintenance of biodiversity.

Conservation Corridors Corridors of habitat that are created to link areas of similar habitat that would otherwise be isolated from each other.

Conservation Genetics The application of the principles and techniques of genetics to the conservation of biodiversity.

Continental Taxa Species that have evolved on continents.

Convergent Evolution The development of similar morphological or physiological traits in unrelated species living in geographically separated regions that have similar environments.

Cope's Rule A general trend toward larger size as the lineage evolves. This pattern occurs in the evolutionary history of many species.

Corridor Geographic features that promote dispersal and colonization.

Cosmopolitan Having a global geographic distribution.

Court Jester Hypothesis When species do not have the phenotypic plasticity and genetic diversity to adapt to long-term changes in their physical environment caused by an abiotic factor such as a climatic and environmental change.

Crassulacean Acid Metabolism (CAM) In CAM plants, CO_2 is absorbed at night and stored as malic acid. During the light of day, photosynthesis is conducted by the C_3 pathway. Cacti use CAM photosynthesis.

Crowdsourcing Obtaining information on species presence through large networks of many citizen scientists.

Crown Fire A fire that spreads to the upper layers and canopy of a forest.

Cryptic Species Morphologically identical, but genetically distinctive species.

Deciduous Perennial plants that lose their leaves during annual cold or dry seasons.

Density The number of individuals of a given species per unit of area.

Diffusion The expansion of the range of a species along a discrete front.

Dimorphic Leaves Different-sized or -shaped leaves on the same plant that may vary by season or by the height of the plant at which the leaf grows.

Diploid A cell or organism containing both sets of chromosome pairs.

Disjunct Distribution Having a biogeographic distribution in which two or more populations of the same taxon are separated by large geographic areas in which the taxon is absent. Disjunctions may be caused by dispersal, climate change, geologic changes or evolutionary processes.

Dispersal The movement of an organism away from its point of origin. Active dispersal involves locomotion by the organism. During passive dispersal, the organism is transported by gravity, wind, water, or other organisms.

Dispersal Kernel The mathematical function that describes the probability of dispersal of the propagule at different distances.

Disruptive Selection When individuals with extreme opposite values for genotypic traits are strongly selected for and individuals with intermediate values are selected against.

Disturbance A discrete and relatively rapid physical or biological event that disrupts ecosystem, community, or population structure and changes resources, substrate availability, or the physical environment.

DNA Deoxyribonucleic acid. Molecules made up of various combinations of sugars and phosphates that are joined together by nitrogenous compounds consisting of adenine, thymine, cytosine, and guanine bound by hydrogen. The DNA structure resembles a twisted ladder with the sugars and phosphates on the sides and the nitrogenous compounds forming the rungs. Genes are formed of DNA.

DNA Barcoding The use of genetic information from one or a few gene regions to identify a species or genotype.

Domain A proposed highest level of the taxonomic hierarchy. Divided into Archaea, Bacteria, and Eukarya.

Domestication The process of artificial selection by which plant and animal species come to depend upon humans for survival while in turn providing humans with practical or other benefits.

Dominance Type Plant associations that are defined and described on the basis of the one or several plant species that structurally dominate them.

Ecological Biogeography The biogeographic study of the modern relationships between organisms and the environment.

Ecological Equivalents Widely separated and unrelated species that are similar in morphology and behavior. Widely separated, but physiognomically and structurally similar vegetation formations.

Ecological Sustainability Ability of an ecosystem to maintain or restore its processes, composition, and structure.

Ecosystem The interacting system of organisms and the physical environment.

Ecosystem Services Physical and biological resources that ecosystems provide to humans.

Ecotone The geographic boundary between two different adjacent communities.

Ecotypes Genetically and phenotypically distinct members of the same species that occur in different environments.

Endangered Species A species or subspecies that is at risk of extinction throughout all or a portion of its range.

Endemic Restricted to one geographic area.

Enviroethics The area of ethics and philosophy that focuses on the moral and ethical relationships between humans and the environment/ecology.

Environmental Determinism The premise that human cultures and changes in culture are driven mainly by the natural environment and changes in environment.

Environmental DNA (eDNA) DNA that is recovered from environmental sources such as water samples or sediments.

Environmental Reclamation The replacement of a severely damaged or destroyed ecosystem with a landscape, such as a lawn, that does not attempt to replicate the original ecosystem.

Environmental Restoration The rebuilding of the original biological and physical environment in damaged ecosystems, and in some cases in areas where the natural habitat has been completely destroyed.

Environmental Sustainability Management of the environment and natural resources to meet current human and ecological needs while also providing for the ability to meet future needs.

Equilibrium Theories of Biodiversity Theories that assume that modern biodiversity is in equilibrium with modern environmental conditions.

Equilibrium Theory of Island Biogeography The theory advanced by R. H. MacArthur and E. O. Wilson that the biodiversity on islands is governed by rates of colonization and extinction, which are in turn controlled by island isolation and island size.

Euryhaline Organisms that can withstand a wide range of salinities.

Euryphagous Organisms that can consume a wide variety of different foods.

Eurythermic Organisms that can withstand a wide range of temperatures.

Eustatic Sea-Level Change Global changes in sea level caused by increases and decreases in the amount of water in the world's oceans. The storage of water in glacial ice is a major cause of eustatic sea-level declines.

Evapotranspiration Water that is added to the atmosphere by both evaporation from waterbodies and soils and by transpiration from plants.

Evolution Genetically controlled changes in physiology, anatomy, and behavior that occur to a clade over time. *Microevolution* refers to evolution at the scale of the species. *Macroevolution* refers to evolutionary changes viewed at higher taxonomic units such as genera and families.

Exponential Population Growth The geometric growth of a population in an environment where there are no limits to population size; $dN/dt = rN$.

Ex situ Conservation Attempts to save species from extinction by maintaining them outside of their natural habitat in areas such as zoos or botanical gardens.

Extinction The loss of all individuals of a species, genus, family, or order. Extinctions may be local or global.

Facilitation The process by which the establishment of one species changes the environment and allows the subsequent establishment of other species.

Family A taxonomic category that includes taxonomically similar and related genera.

Fertile Crescent A crescent of land that extends from the mouth of the Tigris and Euphrates rivers, north to eastern Turkey, and then south to the coastal regions of Lebanon and Israel. Some of the earliest evidence of plant and animal domestication comes from this region.

Filter Avenues of dispersal and colonization that are not equally favorable for all species.

Flagship Species Charismatic species that are threatened with extinction and can be used to draw attention and resources to their conservation and the conservation of associated habitat and species.

Food Web The complex flow of energy between different trophic levels.

Formation A classification of vegetation based upon structure.

Founder Principle The idea that populations founded by a very small number of individuals generally contain a small subset of the total genetic variability of the main population and are prone to allopatric speciation.

Fundamental Niche The niche space of a species if there was no competition with other species.

Gap Analysis The use of GIS to overlay species maps to find areas of high biodiversity or map endangered species and to compare the distributions of biodiversity or endangered species to factors such as soil conditions or vegetation cover. Areas of high biodiversity or the ranges of endangered species are then overlayed on maps of current or planned conservation areas to see if those areas are in the right location to preserve biodiversity or endangered species.

Gene Basic unit of heredity carried and transmitted by chromosomes. The genes of plants and animals consist of molecules of deoxyribonucleic acid (DNA).

General Circulation Models Computer-run mathematical models that incorporate the physical processes of the atmosphere of the earth to simulate past, present, and future climates.

General Dynamic Model (GDM) of Island Biogeography Model that integrates the impacts of changing island habitat diversity (and most likely size too) on carrying capacity, speciation rates, extinction rates, and species richness.

Generalist A species that has a wide range of environmental tolerances and a large niche.

Genetic Drift Stochastic changes in the genetic composition of a population that occur over time as new genes arise via mutation and other genes are lost through chance processes.

Genome The complete range of genes present in a species.

Genomics The branch of molecular biology concerned with the measurement, traits, evolution, and mapping of species genomes.

Genotypic Variations Differences in the genes between different species or members of the same species. Not all genotypic variations may be observable as phenotypic variations.

Genus A taxonomic category that includes groups of taxonomically similar and closely related species. The plural of genus is *genera*.

Geographic Range The entire geographic area in which a species can be found living.

Geologic Time Scale An internationally accepted system that serves as the standard means of subdividing the many millions of years of earth's history.

Glaciation A period of regional or global development and expansion of glaciers.

Global Extinction The worldwide extinction of a subspecies, species, or higher taxonomic level.

Gondwanaland A large continent that existed around 150 million years ago and included the modern continents of Africa, Australia, South America, and Antarctica.

Great Acceleration The post-1950 quickening in rates of human population growth, global economic growth, technological growth, energy use, and resource use, along with the impact of these and other factors on the environment.

Great American Biotic Exchange The movement of terrestrial fauna that occurred following the establishment of the Isthmus of Panama.

Great American Biotic Interchange The exchange of terrestrial and freshwater animal species that occurred between North American and South America facilitated by the development of the Isthmus of Panama in the Late Cenozoic.

Greenhouse Effect The phenomenon in which carbon dioxide, methane, and certain other gases in the atmosphere allow the downward transmission of incoming solar light energy and trap outgoing heat energy. Increases in carbon dioxide and other greenhouse gases due to the burning of fossil fuels and other human actions may lead to significant increases in temperature and global sea level over the next 100 years.

Greenhouse Gases Atmospheric gases that typically allow shortwave light energy to pass through the atmosphere, but absorb and reradiate longwave heat energy. Examples include carbon dioxide, methane, and nitrous oxide.

Growth Form A classification of plants on the basis of their morphology.

Guild Group of species within a community that have similar forms, habitat, and resource requirements.

Gymnosperm Seed-bearing plants in which the seeds are produced outside of a protective ovule.

Habitat The explicit spatial, physical, and biological environment in which species can be found.

Habitat-Based Conservation Efforts that attempt to conserve the entire biological and physical environment in which endangered species live.

Harmonization Two geographic areas are in harmony when they contain the same flora and fauna.

Heliophytes Plants that grow best in full sunlight.

Heterotrophs Nonphotosynthetic organisms that depend upon consuming other organisms to acquire energy.

Heterozygous Refers to a locus that has different alleles associated with it.

Hierarchy A system of organization in which components are ordered by rank.

Historical Biogeography The biogeographic study of the past distributions and evolution of life.

Historical Theories of Biodiversity Theories that assume that modern biodiversity is not in equilibrium with modern environmental conditions because it reflects past events of evolution and extinction.

Historic Extinctions The recent episode of extinctions (starting around 1600) caused by the combination of European colonization and socioeconomic expansion, and the rapid global increase in human population.

Hollow Curve Distribution A typical distribution of species ranges, consisting of many species with small ranges and few species with large ranges.

Homeotherm Animals that maintain a relatively steady body temperature through the metabolic generation of heat.

Hominid Moden human beings and their closest extinct relatives of the genera *Homo* and *Australopithecus*.

Human Appropriation of Net Primary Production (HANPP) The proportion of net primary productivity consumed by humans and their land use.

Hunters and Gatherers Human populations who rely upon naturally occurring vegetation, fruits, nuts, carrion, fish, and game for subsistence.

Hybridization Sexual reproduction between two different species.

Hydrochores Organisms dispersed by water.

Hydrophyte Plants that require high levels of soil moisture.

Hydrosphere All of the water in the atmosphere, on the surface, and below the surface of the earth.

Igneous Rocks Rocks formed by the cooling and solidification of molten material called magma. Intrusive igneous rocks are formed when the cooling occurs deep underground. Extrusive igneous rocks form when cooling occurs at the surface.

Individual A single organism.

Individualistic Community Concept Gleason's concept that species are distributed in the environment according to their individual resource requirements and not due to interactions with other species.

Industrial Revolution The rapid development of new technologies and the shift of the socioeconomic system away from an agrarian base toward a techno-industrial base that has occurred since about 1800.

Inhibition Model A model of succession that is based on the premise that the first species that arrive following disturbance come to dominate the site and competitively inhibit the establishment of later arriving species.

Initial Floristic Composition Model A successional model in which the species immediately at the start of primary or secondary succession have a strong impact on subsequent stages and nature of the final community composition.

***In situ* Conservation** Attempts to save species from extinction by maintaining them within their natural habitat.

Insular Taxa Species that have evolved on islands.

Intermediate Disturbance Hypothesis According to this hypothesis, if an ecosystem remains stable and free of disruption by disturbance, the stable homogeneous environmental conditions will favor some species but will lead to the extinction of species for which the stable habitat is not favorable or which are prone to competitive exclusion by species that are favored by the undisturbed environment. Disturbances cause physical and biological changes and spatial heterogeneity in the system that can favor species that would not survive in a stable undisturbed system. On the other hand, if disturbance occurs too frequently and is too severe, it will lead to the extinction of disturbance-sensitive species that have long generation times or occur in low numbers and are prone to change extinction.

Interspecific Competition Competition between different species.

Intraspecific Competition Competition between individuals of the same species.

Invasion The colonization of a new region by a species that has been introduced to that region by humans.

Invasive Species An introduced species that becomes established in a new environment, spreads geographically, and causes measurable environmental, economic, or human health impacts.

Irruption Episodic explosions in the population size and geographic ranges of insects and animals.

Isostatic Sea-Level Change Changes in regional sea level caused by the uplift or downward motion of land surfaces.

Jump Dispersal The dispersal of a species across a geographic area that is not occupied by the species.

Keystone Species A species that due to its presence or absence can greatly change the productivity, species composition, or species diversity of an ecosystem.

Kingdom The highest or penultimate division of the taxonomic hierarchy depending on the taxonomic system. Some systems divide the kingdoms into Monera, Protista, Fungi, Plantae, and Animalia.

Land Ethic The concept espoused by A. Leopold that humans have an ethical responsibility to conserve and preserve the natural environment.

Landscape Ecology The study of the spatial patterns of landscapes, including the distribution of species and habitat on local to regional scales, and how these patterns are developed and maintained by nature and human actions.

Landscape Genetics The study of population and conservation genetics combined with landscape ecology to understand how attributes of a landscape influence the genetic structure of species.

Latitudinal Diversity Gradient (LDG) The general decline in species richness from the tropics to the polar regions that is evident for many taxa.

Laurasia A large continent that existed around 150 million years ago and consisted of the modern continents of North America, Europe, and most of Asia.

Law of Superposition An assumption made in geology that overlying material is younger than underlying material.

Law of Uniformitarianism Am assumption made in geology that the natural processes of rock formation, weathering, erosion, and deposition that occur today also occurred in the past and can be applied to other historical studies in natural history. The law is often stated as "The present is the key to the past."

Life Form A classification of plants on the basis of physiognomy and life history.

Life Zone Elevational and latitudinal bands of similar climate and vegetation.

Liliopsida (Monocotyledon) A class of angiosperms in which the first leaf-like photosynthetic structures (cotyledons) develop as a single shoot. Monocotyledons also have leaves with parallel venation. Some newer taxonomies classify monocotyledons as the class Liliopsida.

Lithosphere The geological structure of the earth. It includes both solid rock and liquid forms such as magma.

Local Extinction Extinction of a subspecies, species, or higher taxonomic level in a specific region, but not in all areas of the earth.

Locus The point at which a gene is located on a chromosome.

Logistic Population Growth The *S*-shaped growth of a population in an environment in which there is a limitation on the ultimate size of the population that can be supported; $dN/dt = rN(K-N)$.

Lotka-Volterra Model A mathematical model that relates cyclic fluctuations in the size of predator and prey populations to the interaction between predators and prey; prey populations: $dN/dt = rN - a'PN$; predator populations: $dP/dt = fa'PN - qP$.

Lower Paleolithic Earliest part of the Early Stone Age. Typified by the earliest stone tools and spanning from about 3 million years ago to 300,000 years ago.

Macroecology The ecological analysis and interpretation of the large-scale geographic distributions of plants and animals.

Magnoliophyta (Angiosperm) Flowering plants. Angiosperms have seeds that are developed within an ovary. Some taxonomies refer to angiosperms as the Magnoliophyta.

Magnoliopsida (Dicotyledon) A class of angiosperms in which the earliest leaf-like photosynthetic structures (cotyledons) develop as a pair. Dicotyledons also have leaves with net-like venation. Some newer taxonomies classify dicotyledons as the class Magnoliopsida.

Megafauna Large birds and mammals that are greater than about 44 kilograms in weight.

Mesoamerica An archaeological term that refers to southern Mexico and northern central America.

Mesolithic (Middle Stone Age) Late hunting-and-gathering period prior to the Neolithic and development of agriculture.

Mesophyte Plants that require intermediate levels of soil moisture.

Metabolic Theory of Biodiversity (MTB) Theory that attempts to relate warmer temperatures to increased rates of metabolism and to increased evolutionary rates of speciation.

Metapopulations Members of the same species that are found in different and separated locations. Members of different metapopulations may not interact with other metapopulations on a regular or frequent basis.

Middle Paleolithic Middle part of the Early Stone Age. Typified by more refined stone tools and spanning from about 300,000 years ago to 30,000 years ago.

Milankovitch Orbital Theory of Glaciation The widely accepted theory that natural changes in the orbital geometry of the earth produced the cycles of glaciation and nonglaciation that typify the Late Quaternary.

Minimum Area Requirement (MAR) The minimum range area required to support a sustainable population of a species.

Minimum Viable Population Size (MVPS) The population size required to sustain a species and protect it from extinction within the foreseeable future.

Molecular Clocks Biomolecular approaches to reconstructing the timing of past divergences of populations, species, or higher taxonomic levels based on differences in DNA, RNA, or amino acid sequences.

Molecular Systematics The use of DNA to develop cladograms.

Monocotyledon (Liliopsida) A class of angiosperms in which the first leaf-like photosynthetic structures (cotyledons) develop as a single shoot. Monocotyledons also have leaves with parallel venation. Some newer taxonomies classify monocotyledons as the class Liliopsida.

Moustarian Tradition The stone tool–making tradition associated with the Neanderthals.

Müellerian Mimicry A type of mutualism that occurs when a species that is poisonous or unpalatable to predators possesses the same coloring or shape as another species that is also poisonous or unpalatable. Both species benefit since predators avoid either species.

Multiregional Model A model of human evolution that suggests that modern humans, and perhaps some other species of *Homo*, evolved independently through parallel evolution in different geographic regions.

Mutualism A symbiotic relationship in which both species benefit.

Natural Climate Solutions (NCS) The ecosystem service provided by vegetation to combat climate change by the capture of carbon from the atmosphere and sequestration in biomass and underlying soils and sediments.

Natural Selection The process by which the genes for genetically controlled traits become more common in a population over time because individuals with those traits are reproductively more successful than other individuals.

Neolithic (New Stone Age) Period of initial development of agriculture with reliance on stone tools.

Neolithic Revolution The initial development of agriculture and the associated evolution of human society representing a break from the earlier hunting-and-gathering economies of the Mesolithic.

Nested Pattern of Insular Communities The general tendency for the biota on islands to contain a subset of the flora and fauna found on nearby larger islands or continents.

Neutral Theory of Biodiversity and Biogeography. Posits that all trophically similar species in a community are ecologically identical and have little to no niche differentiation.

Niche The multidimensional resource space in which a species exists.

Niche Construction The way in which the interplay of ecology, genetics, and evolutionary changes and species themselves produce species niches.

Niche Divergence The divergence of the niches of closely competing species so as to decrease competition.

Oasis Theory The original theory by V. G. Childe that postulated that conditions in southwest Asia became drier and warmer at the close of the Pleistocene and that this led to the development of agriculture as people and domesticatable plants and animals became concentrated around oases.

Old Growth Forest Forest that has not experienced large disturbances or human land clearance for hundreds of years.

Oldowan Tradition Tool-making tradition, dating from 2.5 million to about 1.5 million years ago. It is associated with early *Homo* in Africa.

Oligophagous Species restricted to several specific food species.

Optimal Foraging Theory The theory that predators evolve to focus their foraging activities on prey species that provide the highest ratio of food energy relative to the energy required for foraging.

Order A taxonomic category that includes groups of related families.

Ordination Analysis The study of how species distributions relate to the environment, including how species ranges and densities are distributed along environmental gradients.

Orogeny The formation of mountains by geologic processes.

Out of Africa Model A model of human evolution that suggests most hominid species, including modern humans, first evolved in Africa and then dispersed from there to other parts of the world.

Out of Asia Model A model of human evolution that suggests that modern humans evolved in Asia and dispersed from there.

Overkill Extinction of a prey species caused by overhunting by predators, particularly by humans.

Panbiogeography A method of reconstructing evolutionary history based on modern geographic distributions and assumption that vicariance events were the most important way in which speciation occurred.

Pandemic Having a widespread distribution between and within biogeographic regions.

Pangaea A super continent that included most of the land masses of the modern continents and existed around 200 million years ago.

Pantropical Having a biogeographic distribution across the tropical regions.

Parallel Evolution Evolution that occurs when geographically isolated populations derived from the same ancestor evolve into morphologically and physiologically similar descendant species.

Parapatric Speciation Speciation caused by the evolutionary divergence of populations that occupy different habitats or niches in the same geographic area.

Parasites Organisms that are wholly dependent on other organisms for nutrients and, in many cases, microhabitat. Typically do not immediately kill their hosts.

Parasitism A symbiotic relationship in which one species benefits at the expense of another.

Pastoral Nomadicism A form of subsistence in which herders migrate and move along with their herds during seasonal changes in grazing areas.

Peninsula Effect The general observation that species richness on peninsulas generally declines toward the tip of the peninsula.

Perennial Plants that live for several growing seasons.

Peripatric Speciation Speciation that occurs when peripheral populations become geographically isolated from the main population and undergo genetic divergence and speciation.

PhAR Photosynthetically active radiation. This normally corresponds to visible light.

Phenology The seasonal timing of changes in the growth and reproduction of plants and animals.

Phenotypic Variation Observable differences in the physiology, anatomy, or behavior of different species or individuals of the same species.

Phloem Cells in vascular plants that are specialized for the flow of food.

Photosynthesis The process by which plants combine water and carbon dioxide in the presence of sunlight to capture visible light spectrum electromagnetic energy in the sunlight and transform it into chemical energy in the form of simple sugars.

Photosynthesis Light Response Curve The hyperbolic relationship between solar irradiance and photosynthesis.

Phyla A taxonomic category that contains groups of related classes.

Phyletic Gradualism A slow and gradual process of evolution during which new traits arise by mutations and traits that infer greater reproductive success are selected for and eventually become dominant over many generations.

Phylogenetic Biogeography The analysis of modern geographic distributions to infer evolutionary history.

Phylogeography The reconstruction of past biogeographic history on the basis of modern geographic distributions of genetic characteristics.

Phytogeography The study of the biogeography of plants.

Plate Tectonics The modern theory of how continental drift occurs due to movement of the mantle and outflows of magma that cause movement of the overlying crust.

Pluvial Wet period of time in the geologic record.

Pluvial Lakes Lakes that form during long periods of increased precipitation or decreased evaporation. The formation and evaporation of pluvial lakes are often linked to global glacial cycles.

Pneumatophores Elongated vertical root structures that rise above water-saturated soils and the level of standing water and allow roots to obtain oxygen.

Poikilotherm Organisms that assume the temperature of their environment.

Polymorphism Genetically controlled variation within a population or species.

Polyploid Organisms that have twice the chromosome number of either parent. This is a common mutation in plants.

Population All individuals of a given species in a prescribed area. Usually, members of the same population are assumed to be in close enough proximity to be able to interact and interbreed frequently.

Population Viability Analysis (PVA) The methods used for estimating minimum viable population size.

Predation A biological interaction in which one organism consumes another.

Prehistoric Extinctions Extinctions caused by the initial prehistoric geographic expansion of human beings from Africa and Eurasia.

Primary Forest Forest that has not experienced significant human modification.

Primary Productivity The energy fixed in an ecosystem by photosynthesis. Rates of biomass accumulation are sometimes used to determine primary productivity.

Primary Succession When a previously lifeless environment is first colonized by plants and animals.

Primates Members of the Primate Order that includes lemurs, tarsiers, monkeys, apes, and humans.

Propagule The stage in the life cycle (for example, a plant seed), part of an organism (for example, a piece of marram grass that can develop a new set of roots and grow into a full-sized grass plant), or group of organisms (for example, a male and female rabbit of mating age) that is required to establish a new reproducing population.

Punctuated Equilibria A model of evolution in which new genotypes and species arise as small isolated populations or populations at the edges of the main species population. These small populations increase and very rapidly become dominant when environmental changes cause them to be better adapted to new environmental conditions.

Pyramid The pattern of decreasing energy and biomass with increasing tropic level typical of terrestrial ecosystems.

Pyrogeography The study of the geography of fire as an environmental disturbance.

Random Model A model of succession that suggests that the patterns of species abundance that occur following a disturbance are highly variable and largely influenced by the random chance dispersal and establishment of post-disturbance colonizers.

Range The geographic area permanently occupied by a plant or animal taxon.

Range Collapse The linked phenomena of population decline and geographic range shrinkage from the periphery to the center of the range that is typical for some endangered species.

Range Eclipse The linked phenomena of population decline and geographic range decline toward the periphery of the range that is typical for some endangered species.

Rapoport's Rule The general pattern of decreasing range sizes of plant and animal taxa that live near the equator compared to those that live in higher latitudes.

Realized Niche The reduced niche space occupied by a species due to the impact of competition with other species.

Red Queen Hypothesis Theory that states that because all species competitors are continually evolving and becoming more competitive, if a species cannot evolve quickly enough to keep pace with the evolution of competing species, it will become extinct.

Releve A rapid, semi-quantitative way of conducting a census of a plant community.

Rescue Effect Effect that occurs when small populations of a species are rescued from the brink of extinction by the arrival of new immigrants of the same species.

Respiration The oxidative chemical reaction that breaks the high-energy bonds of carbohydrates to release energy for an organism's metabolism.

Restoration Ecology The application of ecological principles and techniques to restore ecosystems that have been damaged by human activity.

Resurrection Plant A plant that can survive extreme desiccation for years and begin to physiologically function again when moisture becomes available.

Rhizomes Horizontal root structures typical of many grasses and other plants such as cattails.

Riparian Forests Forests found along water bodies.

Sahul A landmass that existed during glacial periods and consisted of New Guinea, Australia, and Tasmania.

Samara Seeds Seeds with wing-like structures that aid in their dispersal by wind.

Scarification A physical or biological event that some seeds must undergo prior to germination.

Sciophytes Plants that grow best in shade.

Sclerophyllous Leaves Stiff leaves with hard and waxy cuticles and stomata that lie in small pits. These features decrease moisture loss.

Seasonal Migration The annual movements of organisms from one regularly occupied geographic region to another for purposes of avoiding harsh conditions or for feeding and mating.

Secondary Succession Occurs when an existing ecosystem recovers from a disturbance such as fire or flood.

Second Growth Forest The forest vegetation that initially develops following a large disturbance or human land clearance.

Sere An idealized or typical set of successional stages that occurs in an ecosystem following a disturbance.

Serotinous Cones The cones of certain species of conifers that remain closed and hold viable seeds. The seeds are released following disturbances such as fires that kill the parent trees and/or heat the cones.

Sixth Extinction The widespread episode of extinctions over the past 10,000 years or so caused by humans.

SLOSS (Single Large Or Several Small) **Debate** The scientific and political debate over whether a single large reserve is better than several small reserves for the conservation of biodiversity.

Soil The uppermost layer of mineral and organic matter found on the earth's surface.

Specialist A species with a narrow range of resource tolerances and a small niche.

Speciation The development of two or more genetically differentiable species from a single common ancestor species. Speciation results from evolutionary change, but not all evolutionary change leads to the development of two or more species from a common ancestor. Speciation is also referred to as *cladogenesis*.

Species There are several different definitions for species. *Biological species* can be thought of as a functional category that includes all individuals that can sexually reproduce and produce fertile offspring. *Phylogenetic species* are a group of sexually reproducing organisms that share at least one diagnostic character that is present in all members of the species, but absent in other organisms. *Evolutionary species* are organisms that have a direct ancestor-descendant relationship that is traceable in the fossil record.

Species Area Relationship (SAR) The relationship between the number of species and the size of an area of land.

Species-Based Conservation Conservation efforts that are aimed at the preservation of an individual species.

Species Distribution Models (SDMs) Computer models used to infer geographic distributions of species based on observed occurrences coupled with physiological, ecological, and environmental data.

Species Evenness The degree to which the number of individual organisms are evenly divided between the different species of the community. An estimate of species evenness alone is obtainable from the Shannon Index value (H') using the equation $E = H'/\ln S$.

Species Relaxation The decline in species richness that occurs when a region, such as a landbridge island, is cut off from a larger landmass or area of similar habitat. Species relaxation also occurs when land clearance produces small reserves in place of once extensive areas of natural habitat.

Species Richness The number of different species in a given area. In many large-scale biogeographic studies, biodiversity is synonymous with species richness.

Species Turnover Changes in the species represented at different sites or between different time periods at the same site.

Spermatophyta Seed-producing plants.

Stability-Time Hypothesis The hypothesis that long periods of environmental stability lead to high species richness.

Stand An individual plot of plants, particularly forest trees.

Stenophagous An organism with a very narrow range of foods.

Stenothermic Organisms that can withstand only a narrow range of temperature conditions.

Stepping Stones Chains of closely distributed islands or discrete areas of similar habitat, such as mountains surrounded by low-elevation deserts, that aid in the dispersal of some species.

Stomata Specialized cells on the surfaces of plants that open to allow the exchange of gases between the plant interior and the atmosphere.

Succession Changes in the physical and biological conditions of an ecosystem that follow disturbances.

Summit Metabolic Rate The maximum cold-induced metabolic rate of an animal.

Superorganism Community Concept Clements' concept that species in communities evolve together over long periods of time and are highly dependent upon each other, much like the individual organs in an animal depend upon the functioning of all the other organs.

Supertramp Species that are particularly well suited to long-distance dispersal and successful colonization.

Surface Fire A fire that is restricted to the ground level and lower vegetation and does not rise into the tree canopy if trees are present.

Sweepstakes Routes Geographic routes with low probability of successful dispersal, but which are occasionally crossed by species.

Symbiosis The close association between two species that generally develops through coevolution. In many cases, the symbiotic association is obligatory for the survival of one or both species.

Sympatric Speciation The development of new species within the same geographic area as the parent species.

Synecology Ecological research that focuses on the interactions between different species in communities.

Systematics Taxonomic studies that are directed at determining the evolutionary relationship between different groups of organisms.

Target Area Effect The observed positive relationship between island size and immigration rates due to the fact that larger islands provide a bigger target for dispersing organisms.

Taxon Any biological taxonomic category such as a family, genus, or species. The plural of taxon is *taxa*.

Taxonomy The subdiscipline of biology concerned with the classification and naming of organisms.

Temperate Having a geographic distribution in the temperate mid-latitude regions of the world.

Tephra Solid materials such as rock and ash ejected into the environment by a volcanic eruption.

Theory of Continental Drift The theory that latitudes, longitudes, and relative positions of the continental landmasses change over geologic time.

Threatened Species A species that is likely to be endangered in the foreseeable future.

Tolerance Model A model of succession that assumes that the species that will dominate the ecosystem following a disturbance are all established very quickly following the disturbance. All species can tolerate the physical and biological conditions that occur after the disturbance. As time progresses, some species disappear because they cannot tolerate the environment of later successional stages.

Tracheobionta Vascular plants.

Traditional Ecological Knowledge (TEK) and Indigenous Knowledge Knowledge on environmental conditions and management gained from Indigenous peoples and local communities.

Translocation The conservation strategy of moving individuals of a species from one geographic area to another.

Transpiration The release of water vapor to the air by plants.

Trophic Cascade What occurs when the loss of an important prey species causes further ecosystem disruptions and extinctions because of the loss of food for higher predators.

Trophic Levels The hierarchical levels of the food chain through which energy flows from primary producers to primary consumers, secondary consumers, and so on.

Tropical Niche Conservatism Hypothesis (TCH) Theory that niches for many species remained stable for long periods of time, producing extinctions as higher latitudes cooled, but allowing persistence in warmer tropical areas.

Umbrella Species Species that, because of their resource requirements and role in the ecosystem, provide evidence of biodiversity, habitat diversity, and healthy ecological functioning.

Upper Paleolithic The youngest subdivision of the Old Stone Age, dated broadly between 50,000 and 12,000 years ago, but this age range is geographically variable.

Urban Arks Cities used to foster biodiversity and to preserve native species and nonnative endangered species.

Vicariance Biogeography Reconstruction of evolutionary histories that assumes that splits in cladograms represent vicariance events that split the geographic ranges of taxa.

Vicariance Event Geologic events or environmental changes that divide the ranges of species into geographically isolated distributions.

Waif Dispersal Events Successful long-distance jump dispersal events.

Wallace's Line The boundary between the Australasian and Oriental biogeographic regions. First described by Alfred Russell Wallace.

Xerophyte Plants that require dry soils.

Xylem Cells in vascular plants that are specialized for the flow of water.

Younger Dryas A brief period of sharp cooling that affected the northern Atlantic region and many other parts of the world around 13,000 years ago.

Zoochores Organisms dispersed by other organisms.

Zoogeography The study of the biogeography of animals.

INDEX

Note: Letters following page references indicate the following: "f," illustration; "t," table; and "b," box.

acidity
  of precipitation, plant growth and, 57
  of water, plant growth and, 55
Abundance Classes, in Braun-Blanquet
    relevé approach, 125
Acheulean Tradition, 326, 328
active dispersal, 216, 217f
Adapids, as early primates, 315
adaptive radiation, in evolution, 271–274,
    272f, 300
adventitious roots, 54, 105
Aegyptopithecus, 316
areography, 371
Africanized bees, invasion of, 243
Agassiz, Louis, 192b
agriculture, 346–352
  advantages of, 348
  definition of, 342
  geographic spread of, 349–350
  plant domestication for, 341–342,
    342f, 344f
  reasons for developing, 346–352
  slash-and-burn, 137
  temperate deciduous forests cleared for,
    160
agroecology, 461
Agroxiphium sandwicense sandwicense,
    458–459
air
  composition of, 21
  in valley bottoms, 27
air pressure, *vs.* altitude, 21
air temperature
  vs. altitude, 21
  animal distribution and, 41, 46–51, 50f
  in coniferous boreal and montane forests,
    165–166
  continental *vs.* ocean, 23
  in deserts, 146–150
  in glaciation period, 194–195
  in grasslands, 154–157
  increase of, greenhouse gases and, 203,
    206–207, 207f, 208f, 208
  in last glacial maximum, 194
  vs. leaf temperature, 41
  in Mediterranean biome, 151–154

vs. photosynthesis, 59–61, 60f, 61f
plant distribution and, 41–46, 41f, 44f, 45f
precipitation and, 23–24, 25f
in savannas, 142–146
vs. season, 22–23, 24f
in temperate deciduous forests, 158, 159
in tropical rainforests, 133
in tropical seasonal forests, 140–141
in tundra, 169
alfisols, 28f, 30, 158
algae, light requirements of, 40
alleles, 252
allelopathy, 71–72
Allen's rule, 50
alligators, world distribution of, 2, 2f
allopatric speciation, 260–263, 263f, 392–393
alpha diversity, of community types, 126
alpine environments
  biodiversity in, 406–407, 407f
  coniferous forests in. *See* coniferous
    boreal and montane forests
  landslides in, 109
  precipitation in, 132f
  tundra in. *See* tundra
  wind damage in, 103
altithermal optimum, 202
altitude. *See* elevation
Amazona viridigenalis, 459–460
amictic lakes, 32
amphiatlantic taxa, 382
amphioceanic taxa, 382
amphipacific taxa, 382
amphiregional disjunct distributions, 382
amphitropical taxa, 382
analytical biogeography, 3
andisols, 28f, 30
anemochores, 217–221, 220f
anemohydrochores, 219
angiosperms, 13
  evolution of, 300–302, 302f, 303f
angle of insolation, *vs.* season, 21, 22f
animal(s). *See also specific animals*
  in biogeographic regions. *See*
    biogeographic regions; *specific*
    regions
  competition among, 71–75, 73f, 75f

dispersal by, 216–221, 220f
dispersal of. *See* dispersal
distribution of
  acidity and, 57
  vs. food sources, 66–71, 67f, 69f
  moisture and, 54–55
  oxygen and, 56
  vs. physical environment, 37–38, 38f
  salinity and, 56
diversity of, *vs.* elevation, 407, 407f
domestication of, 344, 352
evolution of. *See* evolution
flood effects on, 104
genetics of, 252–256, 254f
in geologic time scale, 180, 181, 182f
in glaciation period, 198–199, 198f
in grasslands, 155
Great American Exchange of, 235,
    236t, 304
invasions of, 239, 239f, 242–246, 243f
migration of. *See* migrations
in savannas, 145, 145f
species of, numbers of, 402–404, 402f
stenophagous, 66–67, 67f
symbiosis among, 75–78, 76f, 79
temperature and, 41, 46–51, 50f
in temperate rainforests, 165
in tropical rainforests, 137
in tropical seasonal forests, 140
in tundra, 171
Anna's hummingbird, range of, 380, 380f
annual plants
  in deserts, 147
  light conditions and, 40
anole lizards, parallel evolution of, 273–274,
    274t
anoxic conditions
  in floodplain soils, 105
  in oceans, 33, 56
anthrax epidemics, impact on vegetation,
    113
anthropocene, 4, 475
  challenges, 480–481
  strategies for sustainable conservation,
    480t
anthropochores, 219

Biogeography: Introduction to Space, Time, and Life, Second Edition. Glen M. MacDonald.
© 2025 John Wiley & Sons, Inc. Published 2025 by John Wiley & Sons, Inc.
Companion website: www.wiley.com/go/macdonald/biogeography2e

apes. *See* primates
aphelion, 21, 202
aphotic zone, of lakes, 31, 31f
apomorphs, 388
applied historical ecology, 472–473
aquatic plants
    invasions by, 242
    light requirements of, 40
Arctic fox, ecological hierarchy of, 14
Arctic regions
    fire frequency in, 94
    forests in. *See* coniferous boreal and
        montane forests
    precipitation in, 132f
area of, 132t
aridisols, 28f, 30, 147
arid regions. *See also* deserts
    area of, 146
Aristotle
    on stability of species, 256
    as taxonomy developer, 10
ark, as dispersal mechanism, 307
art forms, of *Homo sapiens sapiens*, 331
artificial selection, in domestication, 342
assemblage, 14
associations
    in communities, 122–143
    definition of, 122
    in temperate deciduous forests, 159–160,
        160f
asteroids, mass extinction due to, 279b
asthenosphere, 188
Atlantic fall-line hydrogeomorphic and
        pedological boundary, 107, 107f
atmosphere, 15
    general circulation of, precipitation and,
        23, 25f
    stratification of, 21, 21f
auk, extinction of, 360
Australian region, 293f, 294–295, 308
    extinctions in, 356
    mammals in, 296–300, 299f
    marsupials in. *See* marsupials
Australopithecus, 317–322, 317f, 320f, 321f,
        323f
    austral species, 382
autecology, 14
autotrophs, 17
autumn olive, 161
avalanches, 109, 109f, 110
avoidance, of freezing temperatures, 48

background mortality, of trees, in wind
        storms, 102–103
bald cypress, distribution of, flooding effects
        on, 105, 106f, 107f
bald eagle, range of, 380, 380f
barnacles, distribution of, 78–79, 79f
barriers
    amphiregional disjunct distributions and,
        382
    biogeographic regions and,
        294–295, 293f
    to dispersal, 233–237
    to mate selection, 262
    range size and, 377–378
    speciation and, 260

basalt formation, in plate tectonics, 186
Batesian mimicry, 78, 78f
bay checkerspot butterfly, conservation
        efforts for, 443, 457
beavers, 462
beech trees, distribution of, 301, 303f
benthos
    of lakes, 31
    of oceans, 33, 33f
Berberis thunbergii, 161
Bergmann's Rule, 50
Beringia, 195
beta diversity, of community types, 126
bighorn sheep, 459
biocontrol, 239–241
biodiversity, 399–435
    of community types, 126
    definition of, 399
    evolution and, 267
    factors affecting, 407–419
        competition, 415–416
        disturbance, 414–415
        elevation, 407, 407f
        environmental stability, 414
        equilibrium theories of, 410–417, 411f,
            413f, 419f
        evapotranspiration, 418, 419f
        habitat diversity, 411–413, 413f
        habitat fragmentation, 409, 466
        historical theories of, 408–410, 418
        large tropical areas, 413–414
        latitude, 404–407
        predation, 416
        productivity, 416–417
        synthesis of, 418–419, 419f
    genetic, 403–404
    intermediate disturbance hypothesis of,
        414–415
    vs. island size, 426f, 426, 430
    local differences in, 412
    peninsula effect in, 406
    rescue effect in, 430
    resource abundance differences in, 410, 411f
    resource gradients for, 410, 411f
    specialized niches in, 410, 411f
    species evenness comparisons in, 401–402
    species numbers and, 402–404, 402f
    species richness comparisons in, 399–402,
        401f
    species turnover in, 425, 427–429, 428t
    stability-time hypothesis of,
        408–409
    target area effect in, 430
biodiversity conservation
    definition of, 459
    strategies for, 462–480
biodiversity hot spots, 455, 456f
biogeographic (extra-range) dispersal, 216
biogeographic distributions. *See*
        distributions, biogeographic; *specific
        species*
biogeographic lines, 293f, 294–295
biogeographic maps, 372
biogeographic provinces, 293
biogeographic provincialism, 291
biogeographic realms, 293
biogeographic regions, 290, 291t, 294

Australian, 293f, 308
boundaries between, 293f, 294–295
endemic organisms in, 289
Ethiopian, 306–307
factors determining, 295–302, 299f, 302f, 303f
flowering plant evolution and, 300–302,
        302f, 303f
harmony in, 290
Holarctic, 303–305
mammal evolution and, 296–300, 299f
Nearctic, 303–305
Neotropical, 305–306
Oriental, 293f, 307
Palearctic, 303–305
plant-based, 293
provincialism in, 291
vs. realms, 293
Sclater-Wallace approach to, 292
similarity of, 290–291, 291t
subdivisions of, 293–294
biogeographic relicts, 385–386
biogeographic subdivisions, 289–309.
        See also biogeographic regions
    boundaries between, 293f, 294–295
    factors determining, 295–302, 299f, 302f, 303f
    types of, 289–294, 291t
biogeography
    analytical, 3
    conservation, 4, 443
    definition of, 1–2
    descriptive task of, 3
    ecological, 3
    explanatory task of, 3
    historical, 3
    multidisciplinary scope of, 9
    phylogenetic, 371
    subdisciplines of, 3–4
biologging, 222
biological diversity. *See* biodiversity
biological invasions. *See* invasions
biological species concept, 12
biology
    definition of, 9
    hierarchies in. *See* hierarchies of life
biomass
    in biomes, 132t
    in deserts, 149
    flammable, for fires, 93, 94f
    in grasslands, 157
    in savannas, 145
    in succession communities, 91
    in temperate deciduous forests, 132t, 160
    in tropical rainforests, 134
    in tropical seasonal forests, 140
    in tundra, 170–171
BIOME models, of plant functional
        types, 128
biomes, 128–171. *See also specific types*
    changes of, in glaciation period, 195,
        195b–196b, 198–200, 198f
    concept of, historical development of,
        128–133, 129f, 130f, 131f
    coniferous boreal and montane forests,
        132f, 132t, 165–168, 165f
    as corridors, 234
    definition of, 15, 121, 130
    deserts, 146–151

global warming predictions for, 477–478, 479f
Mediterranean, 151–154, 152f
savannas, 132f, 132t, 142–146, 144f, 145f
temperate deciduous forests, 158–162, 158f, 160f
temperate grasslands, 154–157, 155f, 156f. *See also* grasslands
temperate rainforests, 132f, 132t, 162–165, 164f
tropical rainforests, 133–140
tropical seasonal (dry) forests, 140–141, 141f
tundra, 168–171, 170f
Whittaker classification of, 131–133, 132f, 132t
biophilic urban design, 460
biosphere
definition of, 15
power for, 16, 17
bipedalism, in primates, 317–319
bipolar taxa, 382
birds
dispersal of, 228
distribution of, *vs.* temperature, 50, 50f
diversity of
vs. elevation, 407, 407f
vs. latitude, 405f, 408–410
endangered species of, 448–451, 449f–451f
extinction of, human-related, 357–362
flightless, evolution of, 269, 270f
hurricane effects on, 103
invasions by, 239, 244
migrations of
habitat destruction and, 449
seasonal, 222–233, 224f
ranges of, 377
species-area curve for, 400, 401f
birth rate/death rate, *vs.* extinction rate, 281, 283f
black ash trees, distribution of, fire and, 98
black cherry, 162, 474
black mustard, 154, 238
black oak, 474
Blitzkrieg Theory of Pleistocene Overkill, 356
bluebirds, geographic range of, 59, 59f
body temperature
animal distribution and, 47
lethal, 47
vs. metabolism, 47
regulation of, 47–50, 50f
body water, loss of, 55
Boiga irregularis, 244
boreal forests. *See* coniferous boreal and montane forests
boreal species, 382
Boreioglycaspis melaleucae, 240
botanical gardens, conservation in, 458
bottlenecks, in evolution, 268–269
boundaries, between biogeographic regions, 293f, 294–295
brachiation, in primates, 313
Brassica nigra, 154, 238
Braun-Blanquet releve approach, to communities, 125

brigalow, 142. *See also* savannas
Bromus tectorum, 240
Brown-Lomolino vegetation structural types, 128
Bruhnes Normal Chron, 187
buffalo, local extinction of, 274–275, 275f
Buffelgrass, 150b–151b
Burmese python, 245
burning bush, 161
bushveld, 142. *See also* savannas
buttresses, on tree trunks, 134, 163

cactus family
drought-resistance of, 149
freeze resistance of, 43–44, 44f
invasions by, 239
water stress avoidance in, 53
Caenorhabditis elegans, 253
calcium carbonate, of shells, in glaciation studies, 200
California condor, as endangered species, 450–451, 451f
California redwoods, 165
CAM (crassulacean acid metabolism), 16
in water stress, 53, 149
cambium, in tree rings, 90b
Cambrian Period, 181, 182f
campos, 142. *See also* savannas
Canada–United States Air Quality Agreement, 57
Canadian System of Soil Classification, 29
Canary Island pine, 238
Canis lupus, 462
cannibalism, 329
canopy
in coniferous boreal and montane forests, 166–167
fires in, 93
in temperate deciduous forests, 159, 160
in tropical rainforests, 133–140, 134f, 135f
in tropical seasonal forests, 140, 141f
captive breeding programs, 458–459
carbon dioxide
as greenhouse gas. *See* greenhouse gases
levels of, in glaciation period, 198
in photosynthesis, 16, 203
cardoon thistle, 154
Caribbean islands, biodiversity of, 421f, 422
range size of, 377
carnivores, 18, 66
carrying capacity
catastrophic (mass) extinction, 275f, 278, 279b, 282, 296
human-related, 352
mammal evolution after, 296–300
of populations, 226–228
of species, 60
Castor canadensis, 462
catena, 28
Catopithecus, 316
cattle egrets, dispersal of, 216, 217f, 230
Cenozoic Era, 181, 182f
centers of domestication, 349
centers of origin, evolutionary, 387–388
Centrocercus urophasianus, 240
Cervus canadensis, 462

chamaephytes, 127
chameleon, thermoregulation in, 48
Channel Islands, California, island biogeography of, 428
chaparral, 152, 152f, 153
charismatic flora, 446
charismatic megafauna, 446
cheatgrass, invasion by, 240, 241f
chemosynthesis, in oceanic thermal vents, 16b
chemotaxonomy, 10
Chicxulub Crater, Mexico, 279b
Childe Oasis Theory, 347
chilling stress, in plants, 42
Chinook winds, 25
chipmunks, interspecies competition among, 74, 75f
chloropleth maps, 371–376, 372f, 375f
Choristoneura fumiferana, 168
chromosomes, 252–253, 256
choropleth maps, 373
chromosomes, 252
chronosequence approach, to succession studies, 87
cichlid fish
adaptive radiation of, 271, 272f
extinction of, 360
circumpolar taxa, 381
Citizen scientist, 373
clade and cladogenesis, definition of, 251
cladistic biogeography, 388–390, 389f, 390f
class, taxonomic, 13
clay, in soils, 28
climate
definition of, 20
global, 20–27
domestication and, 347
extinctions related to, 278, 279b, 281
vs. fire frequency, 95
future, 203–209, 206f–207f, 209b
Köppen zones of, 26, 27f
lake classification and, 32
local, *vs.* biodiversity, 415
vs. microclimate, 27
precipitation in, 23, 25, 25f, 26
in Quaternary Period. *See* Quaternary Period, climate change in
soil formation and, 27
solar energy role in, 21–23
terminology of, 21
climatic disjunctions, 383–384, 384f
climatic optimum, 202
climatic relicts, 385
climax communities, 86–88, 91, 92f, 123
cline, 254, 254f
Clinton Merriam, on vegetation changes with elevation, 130, 130f
Clovis points, 356
clownfish, anemone commensalism with, 76, 76f
coal, burning of, carbon dioxide production from, 204
coastal floods, 103, 104
Coastal Plain–Appalachian Highlands fall line, 107
coconut palm, dispersal of, 218–219, 218f, 237

coefficients of similarity, 290–291, 291t
coevolution, 274–275
cold air drainage, 27
cold environments
  animals affected by, 47–51, 50f
  forests in. *See* coniferous boreal and
    montane forests
  plant tolerance of, 42–44
cold monomictic lakes, 32
collision zones, in plate tectonics, 188
colonization, 224, 226
  definition of, 216, 222
  by diffusion, 228–229, 229f, 231–233, 232f
  evolution after, 269
  extra-range dispersal for, 216
  factors favoring, 228
  impediments to, 228
  of islands, 236–237, 237f
    in theory of island biogeography, 420,
      424–429, 425f, 426f, 434f
  by jump dispersal, 228–233, 231f, 232f
  population growth after, 225, 225f,
    240–242, 227b, 228t
  by supertramps, 228
  of volcanic islands, 230
Columbanthus quitensis, 168
Columbian Exchange, 238
commensalism, 75, 76, 76f
communities, 121–126
  Braun-Blanquet releve approach to, 125
  Clements model of, 123
  climax, 86–88, 91, 92f, 123
  definition of, 121–122
  ecological, 14
  formations in, 121, 122f, 127–128
  Gleason individualistic community
    model of, 123
  hierarchical, 126
  models of, 121–126, 124f
  ordination analysis of, 123, 124f
  properties of, 122
  quantitative measures of, 125
  reticulate, 126
  superorganism model of, 123–125
community type, definition of, 121–122
competition, 71–75, 73f, 75f, 79, 80f, 81f
  biodiversity and, 416
  of colonizing species *vs.* resident species,
    228
  evolution and, 259–261
  in extinction, 280
  of marsupials with placental mammals,
    300
competitive exclusion principle, 80
complexity
  in evolution, 265
  extinction and, 280
Comprehensive Soil Classification System,
  28f, 29–30
condor
  as endangered species, 276, 451f
  extinction threat to, 276
cones, serotinous, 97
coniferous boreal and montane forests,
  165–168
  area of, 132t, 165

biomass of, 132t
fires in, 167
in glaciation period, 198, 199
human impact on, 168
location of, 132f, 165
plants in, 168–169
precipitation in, 132f, 165
primary production of, 132t
soils of, 166
temperature of, 165
trees in, 166–168
conifers, 13
  biogeographic similarity between, 290,
    291t
  cold adaptation by, 43
Connell, J.H., intermediate disturbance
  hypothesis of, 414–415
conservation, 443–483
  biodiversity, 459
  captive breeding with reintroduction for,
    458
  economic value of, 445–446
  edge effects in, 466
  of endangered species. *See* endangered
    species
  environmental reclamation in, 461
  environmental restoration in, 461
  ex situ, 458–459
  failure of, 446
  genetics in, 459
  global warming and, 477–480, 479f
  habitat-based, 461
  habitat restoration and,
    471–475, 471f
  in situ, 458–459
  island biogeography theory and, 464, 465f
  minimum viable population size for,
    462–463
  population viability analysis for, 463
  rescue effect in, 466
  reserve areas for, 462–467, 465f
  species-based, 459
  species diversity effects of, 466
  spiritual-ethical value of, 445–448
  strategies for, 462–480
  of threatened species. *See* threatened
    species
  value of, 445–448
conservation biogeography, 4, 443
conservation biology, 443
conservation genetics, 459
consumers
  of biomass, 19
  primary, 16
  secondary, 19
continental taxa, 433
continents
  endangered species statistics
    for, 457
  shifting of, 183–192
    biogeographic regions and, 295–302
    future, 189f
    in glaciation period, 199
    historical theories on, 183–186, 184f,
      185f, 187f
    modern theory of, 187–188, 189f

    during past 250 million years, 188, 189f
    splitting of, geographical isolation in,
      269–270, 270f, 271f
continuous extinction, 277
continuous zonal biogeographic
  distributions, 381–382
convective precipitation, 23
convergent evolution, 272, 273f
coordinated stasis, in evolution, 264
Cope's Rule, 266
corals
  mutualism of, 76
  temperature change effects on, 114–115
Cordilleran Orogeny, 191
corn, domestication of, 341–342, 342f, 349
corridors, for dispersal, 233, 234
Cortaderia jubata, 154
Cortaderia selloana, 154, 238
cosmopolitanism, 290, 291, 293, 381
cotton grass, 169
Court Jester Hypothesis, 280, 282
C_3 pathway, for photosynthesis, 16, 39
  plant distribution and, 39, 57–58
C_4 pathway, for photosynthesis, 16, 39–40
  plant distribution and, 39–40, 57–58, 58f
crassulacean acid metabolism (CAM), 16
  in water stress, 53, 149
creosote bush, 147f, 148
Cretaceous Period, 181, 182f
  continental movement during, 191,
    189f–190f
Croizat, Leon, panbiogeography method of,
  390–392, 391f
Cro-Magnons, 331
crowdsourcing, 373
crowned parrot, 459–460
crown fires, 94
cryptic species, 404
cryptophytes, 127
currents, ocean, 33, 34f, 35
  in glaciation period, 199
Cynara cardunculus, 154
cytotaxonomy, 10

dams, vegetation impact of, 108, 108t
Dansereau method, for plant growth form
  classification, 128
Dart, Raymond, *Australopithecus* discovery
  by, 319
Darwin, Charles
  on domestication, 342
  evolution theory of, 256, 257f, 258–261, 312
  on island life, 420
  on landmass breakup, 387
dawn redwood, as living fossil, 386
de Candolle classification, of plant
  functional types, 128
December solstice, insolation distribution
  at, 22, 22f
deciduous trees. *See also* temperate
  deciduous forests
  cold adaptation by, 43
decomposers, 18
dehydration, of plants, 51–54
dendrochronology, for disturbance
  detection, 89b

dendroclimatology, for global temperature changes, 209b
Dendroctonus ponderosae, 168
density, of species, 60
Deschampsia antarctica, 168
deserts
　animal adaptation to, 55
　area of, 132t, 146
　biomass of, 132t, 150
　definition of, 146
　fire frequency in, 93
　human impact on, 150
　location of, 132f, 146
　plants in, 52–54, 54f, 148–150, 147f
　precipitation in, 26, 146–147, 147f
　primary production of, 132t
　soils in, 147
　as structural type, 128
　temperature in, 146
desiccation, of plants, 52–54
DETER, 470
determinism, environmental, 347–348
deoxyribonucleic acid (DNA), 252
　in human evolution studies, 314–315, 314b
　in mammal evolution studies, 298
Devonian Period, 181, 182f
dicotyledons, 13
diffusion
　dispersal by, 228, 229f, 231–232, 231f, 232f
　of plant domestication, 349
dimictic lakes, 32
dimorphic leafing patterns, 46
dinosaurs, extinction of, 277, 279b, 282, 296–298
discontinuous distributions, 371–376, 375f
disjunctions
　amphiregional, 382
　climatic, 383–384
　dispersal, 382–383, 383f
　evolutionary, 385
　geologic, 384
dispersal, 216–221
　active, 216, 217f
　by ark, 307
　barriers to, 233–237
　colonization after. *See* colonization
　corridors for, 233, 234
　cosmopolitanism and, 290
　definition of, 216
　by diffusion into adjacent areas, 228, 229f, 232–233, 231f, 232f
　evolution after, 269
　vs. extinction rate, 281
　extra-range (biogeographic), 216
　filters for, 234–235
　in Great American Exchange, 235, 236t, 304
　to inhospitable location, 221
　intra-range (ecological), 216
　invasions in. *See* invasions
　irruption in, 224
　jump, 228–233, 231f, 232f, 382–383
　long-distance (waif dispersal events), 237
　migrations in. *See* migrations

modes for, 218–219, 218f, 220f
　passive, 216–218, 218f
　of seeds, 416
　speciation during, 387
　species adaptations for, 221
　stepping stones for, 236, 237f
　sweepstakes routes for, 237
　transport distance in, 219–220, 220f
dispersal disjunctions, 382–383, 383f
dispersalist biogeography, 395
disphotic zone, of lakes, 31, 31f
displacement, competitive, 74
distribution of physical environment, 37–38, 38f
distributions, biogeographic, 371–395. *See also* range(s)
　acidity and, 57
　amphiregional disjunct, 382
　biogeographic relicts, 385–386
　centers of origin, 387–388
　cladistic, 388–390, 389f, 390f, 393
　climatic disjunctions, 383–384, 384f
　competition and, 71–75, 73f, 75f, 80, 81f
　continuous zonal, 381–382
　cosmopolitan, 381
　discontinuous, 375, 375f
　dispersal disjunctions, 382–383, 383f
　dispersalist model, 387–388
　endemic, 381
　environmental gradients and, 58–61, 59f–61f, 79–82, 80f, 123–124, 124f
　evolutionary disjunctions, 385
　evolutionary history reconstruction and, 386–392
　vs. extinction rate, 381
　flooding and, 105, 107, 107f
　geologic disjunctions, 384
　interacting factors in, 57–58, 58f
　light and, 38–41, 39f
　mapping of, 371–376, 372f, 375f
　moisture and, 51–54
　multiple factors in, 78–79, 79f
　oxygen and, 56
　panbiogeography models, 390–392, 491f
　patterns of, 380–386
　phylogeographic revolution and, 392–395
　vs. physical environment, 37–38, 38f
　predation and, 66–71
　vs. range size and shape, 376–380, 376f, 378f, 379f, 380f
　salinity and, 56
　symbiosis and, 75–78, 76f, 79
　temperature and, 41–51
　vicariance models, 392–393, 394f
disturbance(s), 85–115. *See also* fires; flooding
　avalanches, 109, 109f, 110
　biodiversity and, 414–415
　definition of, 85
　extinction due to, 414
　landslides, 109–111
　marine, 114–115
　pathogen outbreaks, 112–114
　succession in, 86–88, 86f, 89b, 91–92, 91f, 92f
　terminology of, 86

in tropical rainforests, 415
　types of, 85
　volcanic, 111, 111f
　wind, 100–103, 101f, 102f
divisions, taxonomic, 13
DNA (deoxyribonucleic acid), 252
　in human evolution studies, 314–315, 314b
　in mammal evolution studies, 298
DNA barcoding, 372
dodo bird, extinction of, 360
dog, domestication of, 343
domains, taxonomic, 13
domestication, 341–346
　advantages of, 348
　artificial selection in, 342
　availability of plants and animals for, 352
　centers of, 349
　of corn, 341–342, 342f, 349
　definition of, 341
　of dog, 343
　geographic spread of, 349
　of herding animals, 343, 344f
　Oasis Theory of, 347
　of plants, 342, 342f, 344–345, 344f. *See also* agriculture
　reasons for, 346–352
　time required for, 348
　of wheat, 346
dominance types, of plant species, 122
dormancy
　in deserts, 148
　freezing temperatures and, 48
　of seeds, 52
　in water stress, 52
drift
　continental, theory of, 183–192, 185f
　genetic, 256
drip tips, on rainforest leaves, 136
drop stones, as glaciation evidence, 199
drought
　human dispersal by, 215
　in Mediterranean biome, 152
drugs, from wild plants, 446
dry climate. *See also* deserts
　in Köppen system, 26, 27f
Dubois, Eugence, hominid fossil discovery by, 324–325
dunes, succession on, 88, 90f

earth
　age of, 180
　composition of, 188
　orbit of
　　future changes in, 203–204
　　glaciation periods and, 201–202, 202f
earthquakes, at tectonic plate borders, 188
ecological biogeography, definition of, 3
ecological (intra-range) dispersal, 216
ecological equivalents, 129–130, 129f
ecological hierarchy, 14–15
ecological sustainability, 4
ecology
　applied historical, 472–473
　restoration, 471–475, 471f
economic value, of conservation, 446–447

ecosystem
  definition of, 14
  energy flow within, 18, 18f
  recovery of, after disturbance. *See* succession
ecosystem services, definition of, 1
ecotone
  definition of, 122
  grassland-forest, 155
ecotourism, conservation and, 446–447
ectotherms, 46
eDNA (environmental DNA), 372, 444b
einkorn wheat, domestication of, 346
Elaeis guineensis, 140
Eldredge, Neals, on speciation, 264–265
Eleagnus umbellata, 161
elevation
  vs. biodiversity, 406–407, 407f
  climate and, 21, 27
  precipitation and, 25
  vs. range size, 384, 379f
  vegetation zonation and, 129
elk, 462
El Niño, marine ecologic effects of, 114
eluviation, in soil formation, 29
emmer wheat, domestication of, 346
endangered species, 448–451
  activities impacting on, 448–451, 449f–450f
  biodiversity hot spots and, 455, 456f
  biogeography and, 452–457, 453f, 455t, 456f, 457t
  definition of, 448
  endemism and, 454
  habitat destruction and, 449–450, 449f, 450f
  organizations for, 455
  preservation of, 458, 451f, 452f. *See also* conservation
  range collapse and, 452–454, 453f, 455t
  statistics on, 457t
endemism, 290, 294, 390
  biogeographic relicts as, 385–386
  endangered species and, 454, 455
endotherms, 46
endo-zoochory, 219
energy
  capture of, in photosynthesis, 18
  efficient use of, *vs.* organism type, 18–19
  flow through ecosystems (trophic hierarchy), 15–20, 16b, 18f
English ivy, 161
entisols, 28f, 29, 147
environment
  human effects on, 3
  stability of, *vs.* biodiversity, 414
environmental determinism, 347–348
environmental DNA (eDNA), 372, 444b
environmental factors
  in extinction, 280–282
  in hominid geographic expansion, 326
  in speciation, 264–265

environmental gradients, species distribution and, 58–61, 59f–61f, 79–82, 80f
environmental reclamation, 461
environmental restorations, 461, 475
enviroethics, 448
environmental sustainability, 477, 480, 481
Eocene Epoch, 182f
  primate fossils in, 316
epicormic sprouting, after fire, 96
epilimnion, 31f, 32
epiphytes
  in tropical rainforests, 141, 135f
  in tropical seasonal forests, 140
equator, seasonal insolation at, 22, 22f, 23, 24f
equilibria, punctuated, 264
equilibrium theories
  environmental heterogeneity, 411–412
  niche-based species, 410
equilibrium theories, of biodiversity, 410–417, 411f, 413f, 419f
  for islands, 424–429, 425f, 426f, 428t
equinoxes, solar energy distribution at, 22, 22f
Equus ferus caballus, 240
ergs, 147
erosion
  of rocks, 180
  of thermokarst, 171
ethical issues, in conservation, 445–448, 472
Ethiopian region, 306
ethological isolation, 262
Eucalyptus allelopathy, 72
eucalyptus trees, 308
  allopatric speciation in, 260–263, 263f, 392
  invasions by, 239
Euonymus alatus, 161
euphotic zone, of lakes, 31, 31f
euryhaline species, 56
euryphagous species, 67
eurythermic species, 46
eustatic sea-level changes, 195
eutharian mammals, 298
eutrophic lakes, 32
evaporation
  from animals, 55
  from lakes, 32
  through stomata, 51
evapotranspiration, 51, 418, 419f
evenness, species, 122, 401
evolution, 251–283. *See also specific organisms*
  adaptive radiation in, 271–274, 272f, 300
  bottlenecks in, 268–269
  centers of origin of, 387–388
  cladograms in, 388–390, 389f
  coevolution, 274
  competition in, extinction and, 280
  complex life form development in, 265
  convergent, 272, 273f
  Cope's Rule in, 266
  cosmopolitanism and, 290
  definition of, 251
  direction in, 265–266
  dispersalist model of, 387–388
  extinction and, 282

founder principle in, 268–269
  genetics of, 252–256, 254f
  historical reconstruction of, biogeographic distributions in, 386–392
  on islands, 432
  in isolation, 260–263, 263f
  of mammals, 296–300, 299f
  of marsupials, 296–300, 299f
  natural selection in, 259–260, 266
  panbiogeography models for, 390–392, 391f
  parallel, 273–274, 274t
  perfection in, 266–267
  peripatric speciation in, 263f, 269
  phyletic gradualism in, 264
  punctuated equilibria in, 264
  quantum, 264–265
  simple life form development in, 265
  in small population, 269
  speciation resulting from. *See* speciation
  species diversity and, 267–268
  stasis in, 264
  temporal pattern of, 264–265
  terminology of, 251
  theory of, historical development of, 256–260, 257f, 258f
  transitional forms in, 264, 265
  types of, 251
  vicariance biogeography in, 392, 394f
  vicariance events in, 268–274, 271f
evolutionary disjunctions, 385
evolutionary relicts, 385–386
evolutionary species concept, 12
exclusion, competitive, 74
exo-zoochory, 219
exponential population growth, 225, 225f, 225
ex situ conservation, 458–459
extinction, 474–282
  caused by humans, 352–362
    historic, 358–362
    prehistoric, 353–358
    sixth, 353
  causes of, 281–282
  continuous, 277
  cosmopolitanism and, 290
  definition of, 274
  in disturbances, 414
  evolution and, 282
  global, 275–276, 352
  in habitat fragmentation, 466
  high rates of, 277, 281
  historic, 358–362, 359f
  interference with, ethical issues in, 472
  on islands, in theory of island biogeography, 420, 424–429, 426f, 434f
  living fossils escaping, 276–277
  local, 274–275, 275f
  mass (catastrophic), 278, 279b, 281,283
    human-related, 352
    mammal evolution after, 296–300
  of one species, 276
  phyletic, 276

pollution-related, 360
positive role of, 382
prehistoric, 353–358
rate of
    in fossil record, 276–277, 277f
    vs. latitude, 408
Red Queen Hypothesis of, 280–282
sixth, 352
threat of. *See* endangered species;
        threatened species
trophic cascades in, 276
true, 276
extrusive igneous rocks, 180

facilitation, of succession, 86–88
Fagus grandifolia, 261
family
    cosmopolitan, 290, 291, 293
    definition of, 13
    endemic, 290
farming, swidden, 137
fauna. *See* animal(s)
feral horses, 240
Fertile Crescent, plant domestication in,
        344–346, 344f, 349–362
filters, for dispersal, 234–235
fires, 92–100
    beneficial role of, 97, 98f, 473
    canopy, 93
    in coniferous boreal and montane forests,
        167–168
    crown, 94
    ecological effects of, *vs.* wind, 103
    vs. ecosystem type, 95
    evidence of, in tree rings, 90b
    flammable biomass available for, 93, 94f
    frequency of, 95, 96f
    in grasslands, 155, 240
    human practices affecting, 98–99, 99f
    ignition sources for, 94
    incidence of, 92
    intensity of, 94–95
    in invasive species stands, 240
    landslides due to, 110
    in Mediterranean biome, 152–154
    plant adaptation to, 95–96, 153, 168
    prescribed burns, 99, 473
    rate of spread of, 94–95
    regrowth after, 96, 97, 97f
    requirements for, 92–93
    in savannas, 143
    size of, 95
    species distribution and, 98
    succession after, 97, 98f
    surface, 94
    used by early hominids, 326–327
    of volcanic origin, 111
fish
    biodiversity of, *vs.* latitude, 406
    dispersal of, barriers to, 233
    extinction of, 358–360
    salinity and, 56
flagship species, 461
flash floods, 104
flooding, 103–109
    coastal, 104

definition of, 103
flash, 104
in floodplains, 104–105, 106f
frequency of, 104
human activity impact on, 107–109, 108t
plant species distribution and, 105, 106f,
    107f
positive role of, 104
seasonal, 103
severity of, 104
succession after, 105, 106f
terrestrial effects of, 104
types of, 103–104
floodplains, 104–105, 106f
flora. *See also* plant(s)
    charismatic, 446
foliar uptake, 162
food chain, 18
food webs, 17f, 18, 19
foraging, optimal, 67
foraminifera, 200
Forbes, Edward, 121
Forbes, Stephen, 123
forests. *See also* temperate deciduous
        forests; tree(s)
    boreal, 132f, 132t, 165–168, 165f
    coniferous. *See* coniferous boreal and
            montane forests
    fire in. *See* fires
    light requirements in, 40
    microclimates of, 27, 91, 92, 92f
    montane. *See* coniferous boreal and
            montane forests
    old growth, 161
    riparian, 105
    species analysis of, 121–128
    stands of, 122
    as structural type, 128
    temperate deciduous. *See* temperate
            deciduous forests
    tropical seasonal (dry), 140–141, 140f
    wind damage of, 100–103, 101f, 102f
    windthrow in, 102, 466
formations, 121, 122f, 127–128
fossils
    continental drift and, 184, 185f, 186
    Darwin studies of, 258
    evolutionary record in, 251, 264. *See also*
            humans, evolution of
    extinction record in, 276–277, 277f,
        279b–280b
    flowering plant evolution record in,
        300–303, 302f, 303f
    geologic time scale and, 181,
        182f, 183
    of megafauna, 353, 354f
    pollen
        in glaciation period, 194, 195b–196b,
            197
        for population growth studies, 226,
            227b, 228t
    in sedimentary rocks, 183
    shell, in glaciation studies, 200
    species diversity record in, 267
founder principle, in evolution, 268–269
fourfold model of glaciation, 193

fragmentation, habitat, 408–409, 409f, 450,
        450f, 465
Frederic Clements
    community view of, 123
    facilitative succession model of, 87
freezing
    animals adaptation to, 48
    deciduous trees and, 158–159, 158f
freezing stress, in plants, 42–44
frontal precipitation, 23
frost hardening
    in animals, 48
    in plants, 43
fuels
    for forest fires, 93, 94f
    fossil, burning of, carbon dioxide
            production from, 203
functional type classification, of plants, 128
fundamental niche, 80, 81f
fynbos (shrublands), 152

GABI (Great American Biotic
        Interchange), 235, 236t
Galapagos Islands
    biodiversity in, 422
    Darwin studies of, 258
gap analysis, for conservation planning, 467,
        468, 467f
gap, windthrow, 102
garrique (shrublands), 152
GDM (general dynamic model), 433
gelisols, 28f, 30
General Circulation Models, for global
        warming prediction, 206–207
general dynamic model (GDM) of island
        biogeography, 433
generalist, species, 61
genes, 252–253
    mutations of, 256, 261, 314b–315b
genetic biodiversity, 403–404
genetic drift, 256
genetics, 252–256
    chromosomes, 252–253, 256
    clines, 254, 254f
    conservation, 459
    genes, 252–253
    genetic drift, 256
    mutations, 256, 261, 314b–315b
    of primates, 313–315, 314b–315b
    variations in, 253–254, 254f
genome, 253
genomics, 253
genotypic variations, 253
genus (genera)
    abbreviation of, 13
    in binomial system, 12
    cosmopolitan, 290, 291, 293
    definition of, 10
    endemic, 289
    Latin use in, 10
    vs. species, 10–12, 11f
geographic information systems, for
        conservation planning, 467, 468, 467f
geographic isolation, 261–262, 269, 270f, 271f
geographic races, 254–255, 254f
geographic range, 59, 59f

geologic disjunctions, 384
geologic time scale, 179–183, 182f
giant pampas grass, 238
giant panda, range collapse of, 452, 453f
Gigantopithecus, 316
glacial relicts, 382
glacial tills, 180
glaciation(s)
    animal population in, 262, 263f
    biodiversity and, 408–410, 408f
    biome changes during, 195b–196b, 198f
    causes of, 201, 202f
    characteristics of, 194, 194f
    continental shifts during, 199
    extent of, 193
    fourfold model of, 193
    future, 203
    geographical isolation in, 269–270, 271f
    Milankovitch orbital theory of, 201–202,
        202f, 267
    numbers of, 200–201
    ocean currents in, 199
    ocean temperature in, 200
    orogenies in, 199
    in Quaternary Period. *See* Quaternary
        Period, climatic change in
    sea-level changes during, 194
    succession after, 196–198, 198f
    temperatures during, 200
    tree dispersal after, 228t, 229
    vegetation during, 196–198, 198f
glaciers
    ice core studies of, 193–194
    melting of
        with global warming, 207, 208f
    radiocarbon dating of, 193
    stones dropped from, 199
    succession communities after, 87
Glen Canyon Dam, vegetation impact of,
        108
global extinction, 275–276, 275f
global warming, 206–207, 207f, 208f, 209b
    conservation planning and, 475–480, 479f
glossy buckthorn, 161
goats, domestication of, 343, 344f
Gondwanaland, 184, 188–189
    breakup of, disjunction formation in, 394
    flowering plants in, 301–302, 302f
    mammals in, 306
gorillas, in climatic disjunctions, 384
Gould, Steven J., on speciation, 264
gradualism, phyletic, 263–265
grasslands, 154–157
    anthrax epidemic effects on, 113, 114
    area of, 132t, 154
    biomass of, 132t, 155
    competition in, 74
    evolution of, 305
    fires in, 240
        fuel for, 93, 94f
        prescribed burns and, 99, 99f
    human impact on, 157
    location of, 132f, 154
    in Mediterranean biome, 151
    plants in, 154–157, 155f, 156f
    precipitation in, 132f, 154, 157

primary production of, 132t
regional differences in, 156–157
roots of, 155
in savannas, 142–143, 144f, 145, 145f
soils in, 154
as structural type, 128
temperature in, 154
grazing
    in grasslands, 155, 157
    in tundra, 171
Great Acceleration, 475
Great American Biotic Interchange
    (GABI), 235, 236t
great auk, extinction of, 360
greenhouse effect, 21
greenhouse gases, 203–208
    activities producing, 204, 205f
    climatic impact of, 204–206, 207f, 208f
    conservation planning and, 475–480, 479f
    sea-level changes due too, 304
    types of, 216
Grisebach, August, 127
gross primary productivity, 19
growth forms, of plants, 128
guild, 14
guyots, 186
gymnosperms, 13, 300–301

habitat
    conservation of, value of, 445–448
    definition of, 61
    destruction of
        extinctions in, 358, 359, 361
        on islands, 449–450
        species endangerment in, 448–450,
            449f, 450f
    diversity, *vs.* biodiversity, 411–413, 413f
    fragmentation of, 450f, 450
        biodiversity and, 408–409, 409f, 466
    restoration of, 471–475, 471f
habitat-based conservation, 461
Habitat restoration, 471, 473
hamadas, 147
Hamamelis virginiana, 474
HANPP (human appropriation of net
    primary production), 20
harmonization, biogeographic, 234
harmony, in biogeographic regions, 290
haustoria, 153
Hawaiian Islands
    extinctions in, 357–358
    geologic history of, 191
hazel, 474
heat
    absorption of, by greenhouse gases, 203
    animal tolerance of, 47
    from fires, 95–96
    plant resistance and susceptibility to, 46
Hedera helix, 161
hekistotherms, 128
heliophytes, 39
heliotropism, in tundra, 169
hemicryptophytes, 127
hemiepiphytes, 136
Henry Gleason, individualistic community
    concept of, 123–125, 124f

herbivores
    food for, after fire, 98
    habitat supporting, 413
    as plant predators, 67
    plant protection against, 137
    range size of, 377
    stenophagous, 66–67, 67f
    toxins for, 68
heterotrophs, 18
heterozygosity, 252
heterozygous, 252
hibernation, freezing temperatures
    and, 48
hierarchical community types, 126
hierarchies of life, 9–20
    ecological, 14–15
    taxonomic, 9–14, 11f, 13t
    trophic, 15–20, 16b, 18f
historical biogeography, 3
historical theories of biodiversity, 408–410,
    409f, 418
historic extinctions, 358–362, 359f
histosols, 28f, 29–30, 169
Holarctic region, 303–305
Holdridge system, for vegetation
    temperature relationships, 130–131,
    131f
hollow curve distribution, of range size, 376,
    376f
Holocene Epoch, 181, 182f
homeotherms, 46, 47
hominids. *See also* Neanderthal man
    Australopithecus, 317–322, 317f, 320f, 321f,
        323f
    early *Homo,* 324–327
Homo, early species of, 324–327
Homo erectus, 325–328
    characteristics of, 325–327
    discovery of, 325
    geographic distribution of, 326
    tools of, 327
Homo ergaster, 327
Homo floresiensis, 330
Homo habilis, 324
Homo heidelbergensis, 327
Homo luzonensis, 330
Homo neanderthalensis, 327
Homo rhodesiensis, 327
Homo sapiens (archaic humans), 327
Homo sapiens neanderthalensis. See
    Neanderthal man
Homo sapiens sapiens (modern humans).
    See also humans
    arrival of, in America, 331–332
    characteristics of, 329
    evolution of, 324, 326
    geographic expansion of, 331–333, 330f
    vs. Neanderthal man, 328–330
    population of, 332–333
    tools of, 331
honeycreepers
    adaptive radiation of, 271, 272f
    evolution of, 270–272
    parasitism in, 77
honey mesquite tree, 157
honeysuckles, 161

Hooker, Joseph D.
  evolution studies of, 386
  on island life, 420
Horizontal Displacement of the Continents
    theory, 183
horizons, soil, 28–30, 28f
horse latitudes, 25
horseshoe crab, as living fossil, 277, 277f
hot spots, biodiversity, 455, 456f
house finch, spread of, 231, 231f
human appropriation of net primary
    production (HANPP), 20
humans
  adaptation of, to cold environments, 51
  biogeographic aspects of, 2–3
  conservation efforts of. *See* conservation
  dispersal of, by environmental
      change, 215
  environmental impact of. *See also*
      agriculture; domestication;
      endangered species; threatened
      species
    in coniferous boreal and montane
        forests, 165
    in deserts, 150
    dispersal of species, 219, 237–246, 239f,
        241f, 243f
    extinctions, 352–358, 354f, 359f
    fires, 94
    flood regime alterations,
        107–109, 108t
    forest management and fire control
        measures, 98–99, 99f
    in grasslands, 157
    greenhouse gas increases, 203–208,
        205f–207f, 209b
    in islands, 431
    in savannas, 145–146
    in temperate deciduous forests, 160
    in temperate rainforests, 164–165
    in tropical rainforests, 137, 139f
    in tropical seasonal forests, 141
    in tundra, 171
  evolution of, 312–333
    Australopithecus, 317–322, 317f, 320f,
        321f, 323f
    bipedalism in, 317–319
    Cro-Magnons and, 331
    early *Homo,* 324–327
    early primates, 315–316
    geographic expansion and, 331–333,
        330f
    Homo sapiens, 317–322
    Industrial Revolution and, 333
    molecular clock analysis of, 314,
        314b–315b
    Multiregional Model of, 328
    Neanderthal fossils and, 328, 329
    Out of Africa Model of, 327
    Out of Asia Model of, 328
    Pithecanthropus erectus, 325
    primate linkage in, 312–315, 314b
  extinctions caused by, 352–362
    historic, 358–362, 359f
    prehistoric, 353–358
    sixth, 352

  modern. *See Homo sapiens sapiens*
      (modern humans)
  species competition influenced by,
      74–75, 75f
humble roundworm, 253
hunters and gatherers
  advantages of, 347, 348
  definition of, 342
  disadvantages of, 348
  domestication by. *See* agriculture;
      domestication
  extinction caused by, 353, 354f, 356–362
  Homo habilis, 324
  Homo sapiens sapiens, 329, 352, 354f,
      356–362
  Neanderthal man, 328
  overkill by, 355–356
  vs. subsistence farmers, 347, 348
hurricanes, 101–104, 102f, 110
hybridization, 257
hydrilla, invasion by, 242
hydrochores, 218–219, 218f
hydrogeomorphic disturbance, 104
hydrophytes, 51, 127
hydroseral or hydrarch succession, 86
hydrosphere, 15
hydrothermal vents, in ocean, 15b, 16b
hypolimnion, 31–32, 31f
Hyracotherium, phyletic extinction of, 276

Ice Ages. *See* Quaternary Period, climatic
    change in
ice cores
  in glaciation studies, 193
  for greenhouse gas studies, 204, 205f
igneous rocks, 186
ignition sources, for fires, 94
illuviation, in soil formation, 29
immigration, to islands, in theory of island
    biogeography, 420, 424–435, 425f,
    426f, 434f
importance value, of plants, in community,
    125
inceptisols, 28f, 29
index of similarity, of species, 126
Indiana Dunes National Lakeshore,
    ecologic restoration of, 474–475
Indigenous knowledge, 472
individualistic community concept, 123–125,
    124f
Industrial Revolution, 333
inhibition model of succession, 87
initial floristic composition model, 87
insects
  dispersal of, by wind, 218
  extinction of, 359
  invasions of, 242
  species of, numbers of, 403–404, 402f
in situ conservation, 458, 459, 462
insolation
  angle of, *vs.* season, 21, 22f
  glaciation and, Milankovitch Theory of,
      201–202, 202f, 267
  vs. latitude, 22–23, 22f, 24f
  plants adaptation to, 38–41
  seasonal distribution of, 22–23, 22f, 24f

insular taxa, 433
interference competition, 71, 74
intermediate disturbance hypothesis,
    414–415
International Union for Conservation of
    Nature (IUCN), 455
interspecific competition, 71–72, 73f, 74
Inter-Tropical Convergence Zone, 25, 26
intra-range dispersal, 216
intraspecific competition, 71, 72, 75f
intrusive igneous rocks, 180
invasions, 237–246
  of amphibians, 244
  of birds, 240, 239f, 244
  definition of, 216
  extinctions related to, 360
  of insects, 242
  magnitude of, 238
  of mammals, 238, 244
  of microscopic organisms, 245
  of mollusks, 242–244, 243f
  of plants, 239–242, 241f
  of reptiles, 244
  success of, 246
  of water plants, 241
invasive species, 238
iridium measurement, in mass extinction
    studies, 279b–280b
irruptions, 224
islands, 420–435
  equilibrium theory of, 424–429, 425f,
      426f, 428t
  geographic patterns of, 420–423,
      421f–423f
  modern theory of, 429–435, 434f
  nested pattern of insular communities
      phenomenon in, 423
  related to nearby islands and mainland,
      423, 429
  biogeography, 433
  colonization of, 236–237, 237f
  concept of, in conservation biogeography,
      464, 465, 465f
  continental taxa on, 433
  early human migration to, 330f, 332
  evolution on, 269, 432
    parallel, 273–274, 274t
  extinctions on, 357
  habitat destruction on, 449–450
  historical interest in, 420
  insular taxa on, 433
  rescue effect on, 430
  size of, *vs.* biodiversity, 420–423, 421f,
      426f, 426–427, 430
  species richness on, 420–422, 421f, 423f,
      424–425, 426f, 430–435, 434f
  species turnover on, 425, 427–429, 428t
  as stepping stones for dispersal, 236–237,
      237f
  target area effect on, 430
isolation
  ethological, 262
  evolution in, 260–263, 263f
  geographic, 261–262, 268–270, 270f, 271f
  reproductive, 261–263
  speciation and, 260–263, 263f

510　INDEX

isostatic sea-level changes, 194
Isthmus of Panama, as dispersal barrier and corridor, 233–236
IUCN (International Union for Conservation of Nature), 455

Jaccard coefficient of similarity, 290
Japanese barberry, 161
Java man (*Pithecanthropus erectus*), 325
Johann Forester, on latitude and vegetation, 129
Johanson, Donald, *Australopithecus* discovery by, 319, 320f
jump dispersal, 228–233, 229f, 231f, 382–383, 383f
June solstice, insolation distribution at, 22, 22f
Jurassic Period, 181, 182f
　continental movement during, 183, 184f

kangaroo rat
　adaptation of, to arid environment, 55
　interspecies competition in, 73, 73f
keystone species, 69
kill front, 356
kingdoms, taxonomic, 13
Köppen climatic zones, 26, 27f, 28f, 29–30
Kosmos, 130
krummholtz, 166
KT Boundary, 279b
kudzu, 161

Lake Baikal, biodiversity of, 409
lakes
　acidification of, 57
　biodiversity in, 420–421
　classification of, 32
　depth zones in, 31, 31f
　flooding from, 103
　life forms in, 30–31
　organisms in, biodiversity of, *vs.* latitude, 406
　oxygenation of, *vs.* species, 56
　photosynthesis in, 31
　plants in, light requirements of, 40
　pluvial, 195
　salinity of, 56
　temperature zones of, 31–32
　zebra mussel invasion of, 243–244, 243f
Lamarck, Jean Baptiste, evolution theory of, 257
lampreys, predation by, 68, 69f
land ethic, 448
landmasses, as barriers to dispersal, 233
landscape ecology, 443–444
landscape genetics, 459
landslides, 109–111
last common ancestor (LCA), 314b
last glacial maximum, 193
　biodiversity in, 408–410, 409f
　characteristics of, 194, 294f
　summer insolation in, 201
　vegetation during, 195–198, 198f
latitude
　vs. biodiversity, 404–407, 405f, 407t

climate and, 22, 23, 25, 22f, 24f, 26, 27f
continuous biogeographic distributions following, 381–382
　vs. extinction rate, 408
　vs. ocean salinity, 33
　vs. range size, 376, 376f
latitudinal diversity gradient (LDG), 405
Laurasia, 184, 186
　flowering plants in, 301–302, 302f
　mammals in, 298
lava, disturbance by, 111, 111f
law of superposition, of strata, 180
law of uniformitarianism, in stratigraphy, 180
LCA (last common ancestor), 314b
LDG (latitudinal diversity gradient), 405
Leakey family
　Australopithecus discovery by, 319, 320f
　Homo erectus discovery by, 325
leaves
　in deserts, 149–150
　dimorphic, 46
　drip tips on, 136
　loss of, in water stress, 51, 53
　in Mediterranean biome, 152–153
　morphology of, *vs.* light intensity, 39–40
　in temperate deciduous forests, 158–159
　temperature of, *vs.* air temperature, 41
　in tropical seasonal forests, 140
　in tundra, 169
Leopold, Aldo, on conservation, 448
lianas, 134–136, 141
life forms, plant, Raunkiaer system of, 129
life zones, 130, 131f
Ligastrum spp., 161
light. *See also* photosynthetically active radiation (PhAR)
　intensity of, *vs.* photosynthesis rate, 39, 39f
　species distribution and, 39–41, 39f
　visible, in trophic hierarchy, 15, 17–19
lightning, fires due to, 94
lignotubers, growth of, after fire, 96
Liliopsida (monocotyledon), 13
lines, biogeographic, 293f, 294–295
Linnaeus, species naming system of, 12
lithification, of sediments, 180
lithosphere, 15, 188
littoral zone, of lakes, 31, 31f
llanos, 142. *See also* savannas
loam soils, 28
local extinction, 274–275, 275f
locus, of chromosomes, 252
locusts, irruptions of, 224, 224f
logging
　in coniferous boreal and montane forests, 168
　in temperate deciduous forests, 160
　in temperate rainforests, 164–165
　in tropical rainforests, 137–138
logistic population growth, 225f, 226, 227b, 231–232, 232f
Lonicera spp., 161
loose metapopulation, 14
Lotka-Volterra model, 70–71, 70f

Lower Paleolithic age, 322
Lucy (*Australopithecus*), 319, 320f

MacArthur, Robert, theory of island biogeography, 420, 424–431, 434f, 435
Mace-Lane criteria, for threatened species, 454, 455t
macro-areal species, 381
macroecology, 371
macroevolution, 251
magma, movement of, 188
magnetic field, polarity changes in, 187
magnetic studies, of continental drift, 187
magnolias
　cladogram for, 388, 389f
　in climatic disjunctions, 383–384, 384f
Magnoliophyta, 13
Magnoliopsida (dicotyledon), 13
malakophyllous xerophytes, 149
Malthus, Thomas, on population growth, 258–259
mammals
　diversity of
　　vs. elevation, 406–407, 407f
　　vs. latitude, 405f, 406–407, 403t
　eutharian, 298–299
　evolution of, 296–300, 299f
　placental, evolution of, 298–300, 299f
mammoth, extinction of, 275, 276, 353, 354f, 355–356
mangrove forests, 140
mantle, of earth, 191
maps, chloropleth, 371–372, 372f, 375f
maquis (shrublands), 152
March equinox, solar energy distribution at, 22, 22f
marine biogeographic region, 308–309, 309f
marine environment. *See* ocean(s)
marmots, Vancouver Island, as endangered species, 451–452, 452f
marsupials, 308
　evolution of, 272, 273f, 299–300, 299f
　extinction of, 356
　range of, 294
mass (catastrophic) extinction, 275f, 277, 279b–280b, 282, 296
　human-related, 352
　mammal evolution after, 296–297
mate selection, speciation and, 262–263
matorral (shrublands), 152
Matuyama Reversed Chron, 187
Mediterranean biome, 151–154
　area of, 132t, 150
　biomass of, 132t
　fires in, 152–153
　human impact on, 153–154
　location of, 132f, 151
　nutrients in, 152, 153
　plants of, 152–154, 152f
　precipitation in, 151, 152
　primary production of, 132t
　soils of, 151
　temperature in, 151

trees in, 152
megafauna
    charismatic, conservation of, 446
    extinction of, 353, 354f, 356
megatherms, 128
melaleuca trees, invasions by, 239, 240
meromictic lakes, 32
Mesoamerica, plant domestication in,
        351–352
meso-areal species, 381
mesophytes, 51
mesosphere, 188
mesotherms, 128
Mesozoic Era, 181, 182f
mesquite, grassland invasion by, 98, 99f
metabolic theory of biodiversity (MTB), 417
metabolism, *vs.* temperature, 47–49
metalimnion, 31, 31f
metamorphic rocks, 180
metapopulations
    definition of, 14
    mapping of, 375, 376, 375f
    ranges of, 377
methane, as greenhouse gas, 203
Mexican fan palm, 238
micro-areal species, 381
microblasts, of wind, 101, 102f
microclimates
    for body temperature regulation, 47, 48
    of forests, 91, 92, 92f
    of tundra, 169, 170f
microevolution, 251
microhabitats, post-fire, 98
microtherms, 128
midges, cladograms for, 390
migrations
    of birds, habitat destruction and, 449
    from cold environments, 50
    of *Homo sapiens sapiens* (modern
        humans), 331–333, 330f
    in pastoral nomadicism, 344, 351
    seasonal, 222–223, 224f
Milankovitch orbital theory of glaciation,
        201–202, 202f, 267
mild humid climate, in Köppen system,
        26, 27f
mimicry, 77–78
minimum area requirement (MAR), 463
minimum viable population size (MVPS),
        462–463
Miocene Epoch, 182f
    primate fossils in, 316
Mississippian Period, 181, 182f
mitochondrial DNA analysis, of human
        evolution, 314–315, 314b
moas, extinction of, 357
modern magnolias, 266
moisture. *See also* precipitation
    animal distribution and, 54–55
    forest fire frequency and, 95, 96f
    plant distribution and, 51–54, 53f, 54f
molecular clock analysis, of human
        evolution, 255, 314–315, 314b
moles, in climatic disjunctions, 384
mollisols, 28f, 30, 154

mollusks
    extinction of, 359
    invasions of, 242–244, 243f
monarch butterfly, distribution of, *vs.* food
        plants, 66, 67f
monocotyledons, 13
monsoon
    summer, 26
    in tropical seasonal forests, 140
montane coniferous forests. *See* coniferous
        boreal and montane forests
mountain(s)
    as barriers to dispersal, 233–234
    climate of, 26
    as corridors, 234
    formation of. *See also* orogenies
        from volcanic activity, 188
    microclimates of, 27
    precipitation induced by, 23, 25
    species ranges following, 379–380
    as stepping stones for dispersal, 236
most recent common ancestor (MRCA), 314b
Mousterian Tradition, 328
MRCA (most recent common ancestor),
        314b
MTB (metabolic theory of biodiversity), 417
Muir, John, on conservation, 448
mules, 261
Müllerian mimicry, 77
Multiregional Model, for human origin, 328
mutations, 256
    in human evolution, 314b
    reproductive isolation in, 261
mutualism, 75–78, 76f
Mycteria americana, 245

Nariokotome Boy (*Homo erectus*), 325
natural biofilters, 446
natural climate solutions (NCSs), 447
natural selection, 259–260, 266–267
    vs. artificial selection, 342
NCSs (natural climate solutions), 447
Neanderthal man, 327–329
    characteristics of, 328
    discovery of, 327
    evolution of, 328
    fate of, 329
    geographic distribution of, 328
    vs. Homo sapiens sapiens, 329
    tools of, 328–329
Nearctic region, 303–306
nekton, 31, 33
nene, extinction danger for, 269, 270f
neoendemic species, 381
Neogene Period, 181
Neotropical region, 304–306
nested pattern of insular communities
        phenomenon, 423
net primary productivity, 19
neutral theory, 411
niches, 58–61, 59f–61f, 79–82, 81f
    adaptive radiation and, 271–272, 272f
    competition and, 415
    cosmopolitanism and, 290
nitrous oxide, as greenhouse gas, 205

nomadicism, pastoral, 344, 351
nonequilibrium effect, in island
        biogeography
    biologic factors and, 430–431
    historical factors and, 432–435
nutrients
    for epiphytes, 134
    in Mediterranean biome, 151, 152
    in savannas, 142, 143
    in soil, 55
    in temperate rainforests, 162–163
    in tropical rainforests, 133, 134, 136

oaks, taxonomy of, 13
oases, 147
Oasis Theory, of domestication, 347
ocean(s)
    amphiregional disjunct distributions in,
        382
    as barrier to dispersal, 253
    currents in, 33, 34f, 35
    ecosystems of, disturbances in, 114–115
    flooding from, 103, 104
    floor of, mapping of, 186, 187f
    food web for, 17f
    glaciation evidence in, 199–202
    in glaciation period, 199
    hydrothermal vents in, 15b, 16b
    levels of, glaciation effects on, 194–195
    organisms in, biodiversity of, *vs.* latitude,
        404, 406
    oxygenation of, 33, 33f, 56
    plants in, light requirements of, 40
    pressures in, 35
    salinity of, 33
        vs. species, 56
    seed dispersal in, 217–218, 218f
    temperature of
        vs. continental temperature, 23
        in glaciation period, 200
        physical environment of, 32–35, 33f, 34f
        salinity of, 33, 33f, 34f
        seaweed viability and, 42
    volcanoes in, at midoceanic ridges,
        186, 187f
oil, burning of, carbon dioxide production
        from, 204
old growth forests, 161
Oldowan Tradition, 322
Oligocene Epoch, 182f
    primate fossils in, 316
oligomictic lakes, 32
oligotrophic lakes, 32
omnivores, 18
Omomyids, as early primates, 315
*On the Descent of Man, and Selection in
        Relation to Sex* (Darwin), 260, 312
optimal foraging theory, 67
orbit, of earth
    future changes in, 203
    glaciation periods and, 201–202, 202f
order, taxonomic, 13
ordination analysis
    of communities, 123, 124f
    of species, 60, 60f

Ordovician Period, 181, 182f
Oriental region, 293f, 294–296, 306–308
Origin of the Species by Means of Natural Selection, 260
orogenies, 188, 199
orographic precipitation, 23
Oryza sativa, 253
Out of Africa Model, for human origin, 327
Out of Asia Model, for human origin, 328
overkill, in hunting, 355–356
Ovis canadensis nelson, 459
oxisols, 28f, 30
oxygenation
    of lakes, 32
        vs. species, 56
    of oceans, 33, 33f
        vs. species, 56
oxygen isotopes, measurement of, in glaciation studies, 193, 194, 200, 201f
Oxyops vitiosa, 240

palatability, mimicry and, 78
Palearctic region, 303–305
Paleocene Epoch, 182f
    primate fossils in, 315
paleoendemic species, 381, 385–386
Paleogene Period, 181
Paleoindians, Clovis points of, 356
paleomagnetism, 186, 187f
Paleozoic Era, 181, 182f
palm, distribution of, *vs.* temperature, 44–45, 45f
palynology
    of glaciation period, 195, 195b–196b
    for population growth studies, 226, 227b, 228t
pampas, 154. *See also* grasslands
panbiogeography models, 390–392, 391f
panda, range collapse of, 452, 453f
pandemic species, 381
Pangaea, 184, 188, 191, 203
    flowering plants in, 301, 302f
    future existence of, 203
Panthalassa, 191
panting, for temperature regulation, 47
pantropical organisms, 382
parallel evolution, 273–274, 274t
parapatric speciation, 263–264, 263f
Parapithecus (primate), 316
parasites and parasitism, 75–77, 416
Pascal, as pressure unit, 52
passenger pigeons, extinction of, 360–362
passive dispersal, 216–219, 218f
pastoral nomadicism, 344, 351
pathogens
    invasions of, 245
    plant, 112–114, 160
peat moss, in forests, 166
Peking man. *See Homo erectus*
pelagic zone
    of lakes, 31
    of oceans, 32, 33f
peninsula effect, in biodiversity, 406
Pennsylvanian Period, 181, 182f, 183

perennial plants
    in deserts, 148, 149
    in grasslands, 155
    light conditions and, 40
    in tundra, 169
perfection, in evolution, 266–267
perihelion, 21, 201
peripatric speciation, 263f, 269
permafrost, 166, 167, 171
Permian Period, 182f, 183
    continental movement during, 189f, 191
phanerophytes, 127
Phanerozoic Eon, 181, 182f
PhAR (photosynthetically active radiation), 16, 31
phenology, 40
phenotypic variations, 253–254
phloem, 51
photic zone, of lakes, 31
photoinhibition, in shade tolerant plants, 39
photoperiodism, 40
photorespiration, 39
photosynthesis
    in aquatic plants, 40
    carbon dioxide for, 203
    in deserts, 149
    energy capture in, 18
    in lakes, 31
    vs. light intensity, 39–41, 39f
    vs. moisture availability, 51, 54
    multiple factors affecting, 57–58, 58f
    in oceans, 33
    primary productivity and, 19
    in stems, 149, 155
    in temperate deciduous forests, 158
    vs. temperature, 41, 41f, 46, 59–61, 60f, 61f
    in tropical rainforests, 133
Photosynthesis Light Response Curve, 39
photosynthetically active radiation (PhAR), 16, 31
phototrophs, 17
phyla, definition of, 13
phyletic extinction, 276, 282
phyletic gradualism, 264, 265
phylogenetic biogeography, 371
phylogenic species concept, 12
phylogeny, definition of, 9
phylogeography, 3, 392–395
physical geography, 20–35
    definition of, 20
    global climate, 21–27
    lakes, 31–32
    microclimate, 27
    oceans, 32–35, 33f, 34f
    soils, 27–30, 28f
physiognomy, of plants, 127–128
phytogeography, definition of, 3
phytoplankton, in lakes, 32
phytotoxins, in competition, 72
Pielou, E.C., biogeographic regions of, 292f, 294
Pinchot, Gifford, as conservationist, 446
pine beetle, 168
pine trees
    distribution of, 52, 53f, 302f, 303f

fires among, 473–474
flammability of, 93
interspecies competition among, 81
as glacial relicts, 385
regrowth of, after fire, 96, 97f
serotinous cones on, 97
taxonomy of, 10–11, 11f, 13, 13t
tree-ring analysis of, 90b
Pinophyta, 13
Pinus canariensis, 238
Pinus ponderosa, 240
pith, in tree rings, 89b
Pithecanthropus erectus, 325
placental mammals, evolution of, 298–300, 299f
plague, population effects of, 224
plank buttresses, on tree trunks, 134
plankton
    diversity of, *vs.* latitude, 406
    in lakes, 33
    in oceans, 33
Plantae, 13
plant(s). *See also specific plants*
    aquatic. *see* aquatic plants
    in biogeographic regions. *See* biogeographic regions; *specific regions*
    biomes of. *See* biomes
    cladograms for, 389, 389f
    classification of, 13, 13t
    communities of, 121–126, 124f
    competition among, 71–75, 80, 81f
    in coniferous boreal and montane forests, 168
    consumption of, species distribution and, 66–67, 67f
    in deserts, 147–150, 147f
    destruction of, carbon dioxide production from, 204
    dispersal of. *See* dispersal
    distribution of
        acidity and, 57
        moisture and, 51–54, 53f, 54f
        oxygen and, 56
        salinity and, 56
        temperature and, 41–46, 41f, 44f, 45f
    diversity of
        vs. elevation, 406–407, 407f
        vs. habitat diversity, 412, 413f
    domestication of, 341, 346, 342f, 344f. *See also* agriculture
    ecological equivalents of, 129–130, 129f
    evolution of. *See* evolution
    extinction of, 277, 358
    flood effects on, 104–108, 106f, 107f
    flowering
        evolution of, 300–303, 302f, 303f
        origin of, 388
    formations of, 121, 122f, 127–128
    functional type classification of, 127–128
    genetics of, 252–256, 254f
    in geologic time scale, 179, 180, 182f
    in grasslands, 154, 155, 155f, 156f, 157
    invasions of, 237–242, 241f

in lakes, 31
leaves of. *See* leaves
life forms of, 127
life zones of, 130, 130f
in Mediterranean biome, 152–154, 152f
parasitic, 77
pathogens of, disturbance by, 112–114
photosynthesis in. *See* photosynthesis
physiognomy of, 127–128
in savannas, 142–144, 144f
seasonal effects on, 40
seeds of. *See* seeds
shade-loving (sciophytes), 39–40
species of, numbers of, 402–403, 402f
succession of. *See* succession
symbiosis among, 75–78
in temperate deciduous forests, 158–162, 158f, 159f
in temperate rainforests, 162–163, 164f
in tundra, 168–170, 170f
vegetation structure in, 127–128
plate tectonics, 187–192, 189f, 190f
disjunctions created by, 384
Platyrrhinii primates, 316
playas, 147
Pleistocene Epoch, 182f, 183
megafauna extinction in, 353, 354f, 355–356
Plesiadapiformes, as early primates, 315
pleisiomorphs, 388
Pliocene Epoch, 182f
Pliopithecus, 316
pluvial lakes, 194, 195
pneumatophores, 105
poikilotherms
animals as, 46
plants as, 41
polar bear, adaptation of, to cold environment, 49
polar climate, in Köppen system, 26, 27f
polar regions
bipolar taxa in, 382
circumpolar taxa in, 381
in glaciation period, 199
precipitation in, 25f, 26
seasonal insolation at, 22, 22f, 23, 24f
pollen fossils
in glaciation period, 195, 195b–196b
for population growth studies, 226, 227b, 228t
pollination
control of, in domestication, 346
mutualism in, 75
seasons for, differences in, 262
in temperate deciduous forests, 159
in tropical rainforests, 136
in tropical seasonal forests, 140
pollution, extinction related to, 360
polymictic lakes, 32
polymorphism, 252
Polynesia, extinctions in, 332–333
polyploids, 256, 346
ponderosa pine, 240
ponds. *See* lakes

populations
carrying capacity of, 226, 228
carrying capacity overshoot of, 226, 228
colonization by. *See* colonization
competition among, 72–74, 73f
definition of, 14
dispersal of. *See* dispersal
of early humans, 332–333
geographic, 254–255, 255f
growth of
exponential, 225, 225f, 226
logistic, 225f, 226, 227b
irruptions of, 224, 224f
migration of. *See* migrations
minimum viable size, 462–463
predator-prey, Lotka-Volterra model of, 70–71, 70f
size of
endangered species and, 448, 452–454, 455t, 456f
vs. extinction rate, 281, 454, 455t
extinction risk and, 361
minimum viable, 463
population viability analysis (PVA), 463
post-fire, 97, 98f
post-landslide, 110
postzygotic isolation, 261
prairie, 154. *See also* grasslands
Precambrian Era, 182f
precipitation
in alpine environments, 132f
in arctic regions, 132f
in boreal forests, 132f
in cold temperate regions, 132f
in coniferous boreal and montane forests, 131f, 132t, 165
convective, 23
definition of, 23
in deserts, 146–148f
frontal, 23
geographic distribution of, 23–26, 25f
in grasslands, 132f, 154–155, 157
vs. Köppen zones, 26, 27f
in last glacial maximum, 194, 195
in Mediterranean biome, 151
orographic, 23
in savannas, 132f, 142
in scrublands, 132f
seasonal differences in, 26
in shrublands, 132f
in temperate deciduous forests, 132f, 157
in temperate rainforests, 132f, 162
in tropical deciduous forests, 132f
in tropical rainforests, 133–135, 132f
in tropical regions, 132f
in tropical seasonal forests, 140
in tundra, 132f, 169–170
in warm temperate regions, 132f
in woodlands, 132f
predation
biodiversity and, 416
definition of, 66
evolution and, 260
extinction and, 274–277, 281

for fighting exotic species, 239–240
of flightless birds, 269
generalist, 68, 69f
by *Homo habilis,* 322
by *Homo sapiens,* 329, 352, 354f, 355–362
in hunting and gathering society. *See* hunters and gatherers
by Neanderthal man, 328
nonselective, 353
overkill in, 355–356
populations of, *vs.* prey populations, 70–71, 70f
range size for, 376–377
seed, biodiversity and, 416
selectivity in, 66–67, 67f, 70
toxins preventing, 68
predatory reptiles, 244
prehensile extremities, in primates, 313
prehistoric extinctions, 353–358
Prentice BIOME models, 128
prey populations
extinction of, 353
vs. predator populations, 70–71, 70f
prezygotic isolation, 261
primary consumers, 16, 20
primary forests, 161
primary productivity, 19
biodiversity and, 416
of biomes, 132t
plants in, 16
primary succession, 86, 88, 91f
primates
DNA relationships among, 313–315, 314b
human evolution from, 312–315, 314b
Australopithecus, 317–322, 320f, 321f, 323f
early, 315–317
early *Homo,* 324–327
Homo sapiens, 327–331
physical characteristics of, 313–314
taxonomy of, 313
privet, 161
PRODES, 470
productivity, primary
biodiversity and, 416
of biomes, 132t
plants in, 16
profundal zone, of lakes, 31, 31f
progression rule, for evolution, 388
propagules
definition of, 215–216
dispersal of. *See* dispersal
Prosopis glandulosa, 157
provinces, biogeographic, 293
provincialism, biogeographic, 291
Prunus serotina, 162, 474
Pueraria montana, 161
punctuated equilibria, in evolution, 264
pyramid, 19
pyrogeography, 93
Python molurus bivittatus, 245

quadrat, for stand analysis, 125–126
quantum evolution, 264

Quaternary Period, 181, 182f
   climatic change in, 192–202
      biome changes with, 195, 195b–196b
      causes of, 199
      early studies of, 192–193
      evidence for, 199–200, 201f
      fourfold model of glaciation in, 193
      geochemical studies of, 200–201
      hydrological changes with, 195
      ice core studies of, 193–194
      Milankovitch orbital theory of, 201–202, 202f, 267
      vs. modern era, 194, 194f
      radiocarbon dating studies of, 193
      sea level changes with, 195
Quercus alba, 474
Quercus velutina, 474

races, geographic, 254–255, 254f
radiation
   adaptive, in evolution, 271–273, 272f, 300
   photosynthetically active, 16
radiocarbon dating
   of glaciers, 194
   of rocks, 180
rainfall. *See* precipitation
rainforests, temperate, 132f, 132t, 162–165, 164f
rainshadows, 26
random model of succession, 87
range(s), 376–380
   amphiregional, 382
   center of, survival near, 452, 453f, 454
   collapse of, species endangerment and, 452–454, 453f, 455t
   definition of, 371
   expansion of
      by diffusion, 228–229, 229f, 231–233, 231f, 232f
      by jump dispersal, 228–233, 231f, 232f
   limits of, 59, 59f
      competition and, 74
   of metapopulations, 377
   periphery of, survival at, 452–454, 453f
   shape of, 379–380
      vs. species viability, 463–468, 465f
   size of
      barriers and, 377–378
      cosmopolitanism and, 381
      vs. elevation, 379, 379f
      evolution effects on, 377–378
      for herbivores, 377
      hollow curve distribution of, 376, 376f
      vs. latitude, 379, 379f
      vs. organism size, 377, 378f
      vs. organism type, 377
      for predators, 377
      speciation events and, 377–378
      vs. species density, 377
      vs. species viability, 463–465, 465f
      statistics on, 376–377, 376f
range eclipse, 452–454, 482
Rapoport's Rule, 378
Raunkiaer system, of plant life forms, 127

Ray, John, species conception of, 12
realized niche, 80, 81f
realms, biogeographic, 293
reclamation, environmental, 461
recreation, conservation and, 446–447
Red Queen Hypothesis, of extinction, 280–281
red wolf, range collapse of, 452, 453f
redwood forests
   conservation of, 446
   as disjunct paleoendemic, 386
   as paleoendemic, 381
reforestation, of temperate deciduous forests, 161–162
regions, biogeographic. *See* biogeographic regions
regolith, 27
regs (deserts), 147
reindeer, carrying capacity overshoot of, 226–227
reintroduction, of captive populations, 458
relative density, of plants, in community, 125
relative dominance, of plants, in community, 125
relative frequency, of plants, in community, 125
relaxation, species, 466
releve approach, to plant communities, 125
relicts, biogeographic, 385–386
remote sensing, for conservation planning, 467–469, 467f
reproduction
   captive breeding programs, 458–459
   in cold environments, 51
   failure of, after colonization, 228
   fire role in, 153
   genetics of, 252–256, 254f
   isolation in, 260–263
   in plants
      light conditions and, 41
      temperature conditions and, 45
   in primates, 312–313
   between vicariant species, 270–271
reproductive isolation, 261
rescue effect
   in conservation, 466
   in island biodiversity, 430
resources
   vs. biodiversity, 410, 411f
   exploitation of, in competition, 71
respiration, 18
   vs. temperature, 47
   without energy production (photorespiration), 39
restoration, environmental, 461
restoration ecology, 471–475, 471f
resurrection plants, 54
reticulate community types, 126
Rhamnus cathartica, 161
rhizomes
   of floodplain plants, 105
   of grassland plants, 155–157
ribonucleic acid (RNA), 252
rice plant, 253
richness, species
   in biodiversity, 399–402, 401f

   vs. elevation, 406–407, 407f
   on islands, 420–422, 421f, 423f, 424–425, 426f, 430–435, 434f
Riddle, Brett R., on phylogeography, 393–394
ridges, midoceanic, 186–188, 187f
rift zones, 188
Ring of Fire, 186, 187f, 188
riparian forests, 105
rivers
   as corridors, 234
   flooding of. *See* flooding
   species ranges following, 378
RNA (ribonucleic acid), 252
rock(s)
   classification of, 180
   weathering of, soil formation in, 28f, 29–30
roots
   adventitious, 105
   in grasslands, 156
   in Mediterranean biome, 152
   in savannas, 142, 144f
   in tropical rainforests, 134, 137
   in water-logged soils, 54
   in water stress, 53
rule of deviation, for evolution, 388

sabertooth cat, extinction, of, 276, 354
sage brush, 240
sage grouse, 240
saguaro cactus, freeze resistance of, 43–44, 44f
saline lakes, 32, 56
salinity
   as barrier to fish dispersal, 233
   mangrove adaptation to, 136
   of oceans, 33, 33f, 34f
   species distribution and, 56
salt cedar, invasion by, 240, 241
salt pans, 147
samara seeds, 218
sand, in soils, 28
sand dunes, succession on, 88, 90f
satellite remote sensing, for conservation planning, 467–471, 467f
savannas, 142–146, 144f, 145f
   animals of, 143, 145, 145f
   area of, 132t
   Australopithecus living in, 317–319
   biomass of, 132t, 145
   human impact on, 145
   location of, 132f, 143
   plants in, 142–143, 144f
   precipitation in, 142
   primary production of, 132t
   regional names for, 142
   soils in, 142
   temperature in, 142
   trees in, 142–143, 144f
   tropical, 142–146, 144f, 145f
scarification, of seeds, 96
scavengers, extinction of, 276
Schema Avium Distributionis Geographicae, 291
sciophytes, 39

Sclater-Wallace approach, to biogeographic regions, 291–293, 292f
sclerophyllous leaves, 53, 153
sclerophyllous xerophytes, 149
scrub, definition of, 128
scrublands, 132f
sea anemone, clownfish commensalism with, 76, 76f
sea levels, glaciation effects on, 195
season(s)
  vs. biodiversity, 410
  earth's eccentric orbit and, 202
  lake stratification and, 31–32, 31f
  plant growth changes with, 40
  vs. solar radiation, 21–23, 22f, 24f
  vs. temperature, 22–23, 24f
  vs. wildfire frequency, 95
seasonal inundation, 104
seasonal migrations, 222–224, 224f
seaweed, chilling stress of, 42
secondary consumers, 20
secondary succession, 86, 86f
second growth forests, 162
sedimentary rocks, 180
seeds
  in deserts, 147–149
  desiccation of, 52
  dispersal of, 216, 416
    by animals, 219–221, 220f
    curves for, 220f, 221
    by flooding, 104
    by humans, 219
    by water, 218–219, 218f
    by wind, 218–219, 220f
  germination of, on volcanic landscape, 111, 111f
  in Mediterranean biome, 153
  predation of, 416
  samara, 218
  scarification of, 96
  size of, *vs.* shade mortality, 219, 220f
  in tropical rainforests, 136, 137
  in tropical seasonal forests, 140
selection, disruptive, 263, 263f
semicosmopolitan species, 381
September equinox, solar energy distribution at, 22, 22f
sere, in succession, 86, 87
serotinous cones, 97
shade, plants adapted to, 39, 40
Shannon index, of species diversity, 401–402
sheep, domestication of, 343–344, 344f
shrub(s)
  ancestor of, 304
  in coniferous forests, 166
  in deserts, 147–149, 147f
  invasion by, 241
  in Mediterranean biome, 151–153
  in savannas, 142–143, 144f
  in temperate rainforests, 164
  in tropical seasonal forests, 140
  in tundra, 168–171, 170f
shrublands
  location of, 132f
  precipitation in, 132f

as structural type, 128
silt, in soils, 28
Silurian Period, 181, 182f
similarity
  coefficients of, 290–291, 291t
  index of, 126
Simpson, George Gaylord, quantum evolution concept of, 264
Simpson coefficient of similarity, 290–291
Single Large Or Several Small (SLOSS) Debate, 464–466
sixth extinction, 352
sky islands, 234
slab pull, 188
slash-and-burn agriculture, in tropical rainforests, 137
SLOSS (Single Large Or Several Small) Debate, 464–466
slow penetration diffusion, of animals, 228–229
snails, extinction of, 360
snow, avalanches from, 109f, 110
snowy climate, in Köppen system, 26, 27f
Sociability Classes, in Braun-Blanquet releve approach, 125
soil(s)
  classification of, 29–30
  in coniferous boreal and montane forests, 166
  definition of, 27
  in deserts, 147
  formation of, 27
  in grasslands, 154
  horizons of, 28–30, 28f
  in Mediterranean biome, 151
  moisture content of, plant distribution and, 51–54, 53f, 54f
  nutrients in, 55
  profiles of, 28–29, 28f
  saline, 56
  in savannas, 142
  in temperate deciduous forests, 158–159
  in temperate rainforests, 162
  in tropical rainforests, 132, 136–137
  in tropical seasonal forests, 140
  in tundra, 169
  water-logged, 54
solar energy
  absorbed in lakes, 31
  absorbed in oceans, 33
  distribution of, in atmosphere, 21–22, 21f
  vs. latitude, 22, 22f
  in trophic hierarchy, 16–20
solstices, solar energy distribution at, 22, 22f
southern beeches, 393
specialist, species, 61
specialization, extinction and, 280
speciation, 260–263, 263f
  allopatric, 261–263, 263f, 392
  in binomial system, 12
  definition of, 10, 12–13, 251
  during dispersal, 387
  evolutionary disjunctions in, 385
  vs. extinction, 275f
  founder principle in, 268–269

vs. genus, 10–12, 11f
isolation and, 260–263, 263f
mate selection and, 262–263
parapatric, 263
peripatric, 263f, 269
phylogenetic, 12
rate of, *vs.* extinction, 282–283
reproduction timing differences and, 262
sympatric, 262, 263f
temporal pattern of, 262–265
vicariance events in, 269–271, 271f
species. See also speciation
  alien, invasions of. *See* invasions
  biological, 12
  carrying capacity of, 60
  colonization of. *See* colonization
  commensalism among, 88f, 75, 76, 76f
  competition between and among, 71–75, 73f, 75f, 80, 80f
  cosmopolitan, 290, 291, 293, 381
  density of, 60
  dispersal of. *See* dispersal
  distribution of. *See* distributions, biogeographic; *specific species*
  diversity of. *See* biodiversity
  in ecological hierarchy, 14–15
  endangered. *See* endangered species
  endemic, 290
  euryphagous, 67
  evolutionary, 12
  exotic, invasions by. *See* invasions
  extinction of. *See* extinction
  in formations, 121, 122f, 127–128
  genotypic variations of, 253
  interaction of
    of competition, 71–75, 73f, 75f, 80, 80f
    multiple factors in, 78–79, 79f
    predation, 66–71, 79
    symbiosis, 75–78, 76f
  keystone, 69
  Latin use in, 10
  macro-areal, 381
  meso-areal, 381
  micro-areal, 381
  mimicry in, 77–78
  minimum viable population size of, 462–463
  mutualism among, 75–78, 76f
  naming systems for, 12
  niches for, 58–61, 59f–61f, 81, 81f
  number of, 401–404, 402f, 403t
  ordination analysis of, 60, 60f
  paleoendemic, 381
  pandemic, 381
  parasitism among, 75, 77
  persistence of, 281–282, 281f
  phenotypic variations of, 253–256
  phylogenetic, 12
  population viability analysis of, 14, 463
  range limits for, 59, 59f
  vs. range size, 376–377
  semicosmopolitan, 381
  stenophagous, 66–67, 67f, 70
  threatened. *See* threatened species
  umbrella, 461
  undiscovered, 403

species-area curves, 400, 401f, 421f, 422–423
species-based conservation, 458–459
species evenness, 122, 401–402
species relaxation, 466
species richness
    in biodiversity, 399–402, 401f
    vs. elevation, 407, 407f
    on islands, 420–422, 421f, 423f, 424–425, 426f, 430–435, 434f
species turnover, 425, 427–429, 428t
spiritual-ethical value, of conservation, 448
spodosols, 28f, 30
    in coniferous forests, 166
    in temperate rainforests, 162
spore fossils, in glaciation period, 195, 195b–196b, 195
sporopollenin, 195b
spreading zones, in plate tectonics, 188
spruce bark beetle, 112, 113f
spruce budworm, 168
    outbreaks of, 113
spruce trees, geographical distribution of, 37, 38f
stability-time hypothesis, of biodiversity, 408, 409f
stands, of forests, 122, 125–126
stasis, in evolution, 264
stenohydric xerophytes, 149
    barriers to dispersal, 234
stenophagy, 66–67, 67f, 70
stenothermic species, 46
steppe, 154. See also grasslands
stepping stones, for dispersal, 236–237, 237f
stolons, in grassland plants, 155
stomata
    in glaciation period, 199
    vs. light environment, 39
storms, wind, 100–102, 100f–102f
strangler fig, 134–136, 135f
strata, law of superposition and, 180
stratification, thermal, in lakes, 31–32, 31f
stratigraphy, 180
stress, in plants
    chilling, 42–43
    freezing, 42–44
    water, 51–53
strike-slip fault zones, 188
structure, vegetation, analysis of, 127–128
subduction zones, 188
succession
    climax communities in, 86–88, 91, 92f, 123
    in coniferous boreal and montane forests, 166–167
    in coral reef systems, 114
    definition of, 86
    facilitative, 87–88
    hydroseral or hydrarch, 86
    multi-path, 88, 91f
    post-flood, 105, 106f
    post-glaciation, 195–199, 198f
    post-hurricane, 103
    post-slash-and-burn agriculture, 137
    post-volcanic eruption, 111, 111f, 226
    primary, 86, 88, 91f, 226
    secondary, 86, 86f
    stages of, 86, 86f

unidirectional, 88, 89b
Summer Monsoon, 26
summer range, migration to, 222, 224f
Summit Metabolic Rate, 49
sun. See also solar energy
    orbit of, climate and, 22, 22f
sun-loving (heliophytes), 39
supercooling
    in animals, 48
    in conifer needles, 43
superorganism model, of communities, 123–125
superposition, in stratigraphy, 180
supertramps, in colonization, 228
surface fires, 94
sweating, for temperature regulation, 47
sweepstakes routes, for dispersal, 237
swidden farming, 137
symbiosis, 75–78, 76f, 79
sympatric speciation, 262, 263f
synecology, 14
systematics (taxonomy), 9–10

taiga, 165. See also coniferous boreal and montane forests
Takhtajan, Armen, floral regions of, 293, 293f
Tansley, Arthur, 123
target area effect, in island biogeography, 430
taxa, 10
    amphioceanic, 382
    amphitropical, 382
    bipolar, 382
    circumpolar, 381
taxon, 9
taxonomic hierarchy, 9–14, 11f, 13t
taxonomic order, 13
taxonomy, definition of, 9–10
TCH (tropical niche conservatism hypothesis), 405
tectonics, plate, 187–192, 189f, 384
temperate deciduous forests, 158–162
    area of, 132t
    associations in, 159, 160f
    biomass of, 132t, 160
    canopy in, 159, 160
    in glaciation period, 199
    human impact on, 161
    leaves in, 158–159
    light transmission through, 159–160
    location of, 132f, 159
    old growth, 161
    pathogen attacks of, 162
    photosynthesis in, 158
    plants in, 158–161, 158f, 160f
    precipitation in, 132f, 158
    primary production of, 132t, 161
    recuperation of, 162
    second growth, 162
    soils of, 158
    structure of, 159
    temperature in, 158, 159
    trees in, 158–162, 158f, 160f
temperate grasslands. See grasslands
temperate plants and animals, 382

temperate rainforests
    animals in, 165
    area of, 132t
    biomass of, 132t
    human impact on, 162–165
    plants in, 162–164, 164f
    precipitation in, 132f, 162
    primary production of, 132t
    soils in, 162
    trees in, 162–163, 163f
temperature
    air. See air temperature
    body. See body temperature
    vs. metabolism, 47
    vs. respiration, 47
teosinte, domestication of, 341–342, 342f
tephra, 111
termites, in savannas, 142, 143
terra rosa soils, 151
Tertiary Period, 181, 182f
    continental movement during, 188–191, 189f–190f
    primate fossils in, 316
thalassochores, 218–219, 218f
theories
    of biodiversity
        equilibrium, 410–420, 411f, 413f, 419f
        historical, 408–410, 409f, 418
    of continental drift, 184–186, 185f
    of continental shifting
        historical, 183–187, 184f, 185f, 187f
        modern, 187–192, 189f–190f
    of evolution
        Darwinian, 257f, 257–260, 312
        historical development of, 256–260, 258f
    of island biodiversity
        equilibrium, 424–429, 426f, 428t
        modern, 429–435, 434f
    of island biogeography (MacArthur-Wilson), 420, 424–435, 425f, 426f, 434f
    Milankovitch orbital, 201–202, 202f, 267
    of optimal foraging, 67
thermal stratification, in lakes, 31–32, 31f
thermocline, of lakes, 31, 31f
thermokarst erosion, 171
therophytes, 127
Thoreau, Henry David, on conservation, 448
threatened species, 448–452
    activities impacting on, 448–452, 449f–452f
    biodiversity hot spots and, 454–455, 456f
    definition of, 448
    endemism and, 454
    habitat destruction and, 449–450, 449f, 450f
    Mace-Lane criteria for, 454–455, 455t
    organizations for, 455
    preservation of, 450–452, 451f, 452f. See also conservation
    range collapse and, 452–454, 453f, 455t
    statistics on, 455, 457t
thunderstorms, fires due to, 94
tight metapopulation, 14

tillers, 155–156
tillites, 180
time scale, geologic, 180–183, 182f
tinamou birds, 394
tolerance model of succession, 87
tools
    Clovis points, 356
    of *Homo erectus,* 325–327
    of *Homo habilis,* 324
    of *Homo sapiens sapiens,* 331
    of Neanderthal man, 328
tornadoes, 101, 102
total plant biomass, 160
topography, diversity of, diversity, 412
toxins
    in competition, 71–72
    mimicry and, 77
    plant, 68
tracks, in panbiogeographic analysis,
        390–392, 391f
traditional ecological knowledge, 472
Tracheobionta, 13
transect, for stand analysis, 126
transform zones, in plate tectonics, 188
translocation, 458
transpiration, 51
tree(s)
    background mortality of, in wind storms,
        102–103
    climax, 91
    cold adaptation by, 43–45
    competition among, 80
    in coniferous boreal and montane forests,
        166–167
    in deserts, 147
    destruction of
        by avalanches, 109, 109f
        by volcanoes, 111, 111f
    dispersal of by diffusion, 228–229, 229f
    distribution of
        moisture and, 51–54, 54f
        temperature and, 41, 41f, 43, 45f, 46
    diversity of
        vs. habitat diversity, 412, 413f
        vs. latitude, 406
    as evolutionary relicts, 385
    evolution of, 270, 300–303, 302f, 303f
    extinction of, in absence of disturbance, 414
    fires affecting. *See* fires
    flammability of, 93
    in glaciation period, 196–199
    for global temperature changes, 209b
    in grasslands, 154
    heat effects on, 95–96
    invasions by, 239–240
    jump, 231
    light requirements of, 40
    in Mediterranean biome, 152
    pathogens of, disturbance by, 112–114
    population doubling time for, after
        colonization, 226, 228t
    range size of, *vs.* elevation and latitude,
        377, 378f
    regrowth of, after fire, 97, 97f, 98
    in savannas, 142–143, 144f
    stratification of

    in tropical rainforests, 133–134, 134f
    in tropical seasonal forests, 140
    in succession, 86–88, 86f, 91–92, 91f, 92f
    in temperate deciduous forests, 158–162,
        158f, 160f
    in temperate rainforests, 162–163, 163f
    in tropical seasonal forests, 140, 141f
    in tundra, 168
    water stress in, 51–53
    wind damage of, 100–103, 101f, 102f
tree-ring analysis, for disturbance detection,
        90b
trenches, oceanic, 186, 188
Triassic Period, 181, 182f
trophic cascades, in extinction, 276
tropical deciduous forests area of, 132t
    biomass of, 132t
    location of, 132f
    primary production of, 132t
trophic hierarchy, 15–20, 15b, 16b, 17f
trophic level(s)
    biomass decrease with, 19
    definition of, 15
    energy transfer within, 17f, 18
    food webs within, 19–20
tropical niche conservatism hypothesis
        (TCH), 405
tropical rainforests, 133–140
    area of, 133, 132t
    biodiversity in, 137, 404, 406–407, 412
    biomass of, 132t, 134
    distribution of, 317f
    disturbances in, 414
    fragmentation of, 408–409, 409f, 450
    habitat destruction in, 449–450, 449f
    habitat diversity in, 412, 413f
    human impact on, 137–140, 137f
    large size of, *vs.* biodiversity, 413
    location of, 131, 132f
    plants of, 133–136, 135f
    pollination in, 136
    precipitation in, 133–134, 132f
    primary production of, 132t
    soils of, 133, 136–137
    temperature in, 133
    vs. tropical seasonal forests, 140
    vegetation stratification in,
        133–134, 134f
tropical rainy climate, in Köppen system,
        26, 27f
tropical regions
    biodiversity in, 404–407, 405f
    environment of, in glacial periods, 408,
        409f
    hurricanes in, 101–104, 102f, 110
    precipitation in, 132f
tropical savannas. *See* savannas
tropical seasonal (dry) forests,
        140–141, 141f
troposphere, 21, 21f
true extinction, 276
tuatara, as evolutionary relict, 385
tundra, 168–171
    area of, 132t, 162–169
    biomass of, 132t, 169–170
    in glaciation period, 196

    human impact on, 171
    location of, 132f
    plants in, 168–169, 170f
    precipitation in, 132f, 169
    primary production of, 132t
    soils in, 169
    temperature in, 169

ultisols, 28f, 30, 158
umbrella species, 461
Upper Paleolithic culture, 331
upwelling zones, in oceans,
        34f, 35, 114
urban arks, 460

Vancouver Island marmot, as endangered
        species, 451–452, 452f
vegetation. *See* plant(s)
vents, hydrothermal, in ocean, 15b, 16b
vertisols, 28f, 30
vicariance biogeography, 392
vicariance events, in evolution, 269–271, 271f
vines
    in tropical rainforests, 133–135
    in tropical seasonal forests, 140
vision, of primates, 313
Vitality Classes, in Braun-Blanquet releve
        approach, 125
volcanoes
    disturbance due to, 111, 111f
    eruption of, colonization after, 230, 427,
        431
    greenhouse gases from, 203
    of Hawaiian Islands, 191–192
    mountain formation from, 188
    in subduction zones, 188
    succession after, 226
    undersea
        ecosystems of, 16b
        at midoceanic ridges, 186, 187f
von Humboldt, Alexander, as father of
        biogeography, 129–130, 129f

wadis, 147
waif dispersal events, 237
Wallace, Alfred Russel
    biogeographic regions system of, 292–293,
        292f
    evolution theory of, 256, 258f, 259–260
    on island life, 420
    on landmass breakup, 387
Wallacea transition zone, 293f, 295
Wallace's Line, 294–295, 293f
warming, global, 206–208, 207f, 208f, 209b
    conservation planning and, 475–478, 479f
Warming, Johannes, 127
warm monomictic lakes, 32
Washingtonia robusta, 238
water. *See also* lakes; moisture; ocean(s);
        precipitation
    acidity of, plant growth and, 55
    as barrier to dispersal, 233–234
    dispersal by, 217–219, 220f
    distribution of, in glaciation period, 195
    flooding of. *See* flooding
    storage of, in plants, 53, 54, 54f, 149

water hyacinth, invasion by, 241
water pressure, in oceans, 35
water stress
    in glaciation period, 200
    in Mediterranean biome, 152
    in plants, 51–53
water temperature
    vs. animal distribution, 48
    of lakes, 31–32, 31f
    of oceans, 33, 33f
        changes of, 114–115
        seaweed viability and, 42
    reproduction and, 51
weather
    definition of, 20
    vs. fire frequency, 95
weathering, of rocks, 180
Wegener, Alfred, continental drift theory of, 183–186, 184f
western Quoll, range collapse of, 452, 453f
wheat, domestication of, 345–346
white oak, 474
white pine, taxonomy of, 10–11, 11f

Whittaker classification, of global vegetation formations, 131–133, 132f, 132t
wildfires. *See* fires
wildland-urban interface (WUI), 154
Wilson, Edward O., theory of island biogeography, 420, 424–435, 425f, 426f, 434f
wind(s)
    catastrophic, 100f–102f, 101–102
    dispersal by, 217–219, 220f
    fires propagated by, 95
    in horse latitudes, 25
windthrow, 164
    ecologic effects of, 466
    forest gaps in, 102
winter range, migration to, 222, 224f
winter, temperature in, plant distribution and, 45
wolf/wolves, 462
    domestication of, 343
    genetic biodiversity of, 403–404
woodlands
    location of, 132f

    in Mediterranean biome, 151–152
    as structural type, 128
wood stork, 245
WUI (wildland-urban interface), 154

xerophiles, 128
xerophytes, 51
    malakophyllous, 149
    sclerophyllous, 149
    stenohydric, 149
xylem, 51, 89b
xylopdia, 142

yarrow, genetic experiments with, 254–256, 254f
Y chromosome Adam, 314b
Younger Dryas cooling event, 347

zebra mussel, invasion by, 243–244, 243f
zoochores, 219–221, 220f
zoogeography, definition of, 3
zooplankton, in lakes, 30
zoos, conservation in, 458